Multivalued Fields
in Condensed Matter, Electromagnetism, and Gravitation

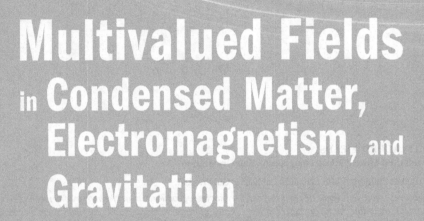

Multivalued Fields
in Condensed Matter, Electromagnetism, and Gravitation

Hagen Kleinert

Freie Universität Berlin, Germany,
ICRANet, Pescara, Italy and Nice, France

World Scientific

NEW JERSEY · LONDON · SINGAPORE · BEIJING · SHANGHAI · HONG KONG · TAIPEI · CHENNAI

Published by

World Scientific Publishing Co. Pte. Ltd.

5 Toh Tuck Link, Singapore 596224

USA office: 27 Warren Street, Suite 401-402, Hackensack, NJ 07601

UK office: 57 Shelton Street, Covent Garden, London WC2H 9HE

British Library Cataloguing-in-Publication Data
A catalogue record for this book is available from the British Library.

MULTIVALUED FIELDS
in Condensed Matter, Electromagnetism, and Gravitation

ISBN-13 978-981-279-170-2
ISBN-10 981-279-170-1
ISBN-13 978-981-279-171-9 (pbk)
ISBN-10 981-279-171-X (pbk)

Printed in Singapore.

To Annemarie and Hagen II

Preface

The theory presented in this book has four roots. The first lies in Dirac's seminal paper of 1931 [1] in which he pointed out that Maxwell's equations can accommodate magnetic monopoles, in spite of the vanishing divergence of the magnetic field, thanks to quantum mechanics It is always possible to create a magnetic field emerging from a point by importing the field from far distance to the point through an infinitely thin magnetic flux tube. But it is only due to quantum mechanics, that such a flux tube can be made physically undetectable. This is true provided the famous *Dirac charge quantization condition* is fulfilled which states that all electric charges are integer multiples of $2\pi\hbar c/g$, where g is the total magnetic flux through the tube. The undetectable flux tube is called the *Dirac string*. From the endpoint of the string, magnetic field lines emerge radially outwards in the same way as electric field lines emerge from an electric point charge, so that the endpoint acts as a magnetic monopole. The shape of the undetectable string is completely irrelevant. It is a mathematical artifact. For this stunning observation, Pauli gave Dirac the nickname *Monopoleon*. The Dirac quantization condition was subsequently sharpened by Schwinger [2] who showed that the double-valuedness of the spin-$\frac{1}{2}$ wave functions of electron restricted the integer multiples to even multiples. Experimentally, no magnetic monopole was found in spite of intensive search, and the Dirac theory was put ad acta for a long time. It resurfaced, however, in the last 35 years, in the attempt to explain the phenomenon of quark confinement.

The second root in this book lies in the theory of the superfluid phase transition. Here the crucial papers were written by Berezinski [3] and by Kosterlitz and Thouless [4]. They showed that the phase transition in a film of superfluid helium can be understood by the statistical mechanics of vortices of superflow. Their description attaches to each point a phase angle of the condensate wave function which lies in the interval $(0, 2\pi)$. When encircling a vortex, this angle must jump somewhere by 2π. A jumping line connects a vortex with an antivortex and forms an analog of a "Dirac string", whose precise shape is irrelevant. If these ideas are carried over to bulk superfluid helium in three dimensions, as done in my textbook [5], one is led to the statistical mechanics of vortex loops. These interact with the same long-range forces as electric current loops.

The third root of the theory in this book comes from a completely different direction — the theory of plastic deformations, which is the basis of our understanding of work hardening of metals and material fatigue. This theory was developed after the discovery of dislocations in crystals in 1934 [6]. With the help of field-theoretic techniques, this theory was extended to a statistical mechanics of line-like defects in my textbook [7], where I explained the important melting transitions by the condensation of line-like defects.

The fourth root lies in the work of Bilby, Bullough, Smith [8], Kondo [9], and Kröner [10], who showed that line-like defects can also be described in geometric terms. Elastic distortions of crystals do not change the defect geometry, thus playing a similar role as Einstein's coordinate transformations. Crystals with defects form a special version of a *Riemann-Cartan space*. The theory of such spaces was set up in 1922 by Cartan who extended the curved Riemannian space by another geometric property: *torsion* [11]. Cartan's work instigated Einstein to develop a theory of gravitation in a Riemann-Cartan spacetime with teleparallelism [12].

Twenty years later, Schrödinger attempted to relate torsion to electromagnetism [13]. He noticed that the presence of torsion in the universe would make photons massive and limit the range of magnetic fields emerging from planets and stars. From the observed ranges of his time he deduced upper bounds on the photon mass which were, even then, extremely small. Further twenty years passed before Utiyama, Sciama, and Kibble [14, 15, 16] clarified the intimate relationship between torsion and the spin density of the gravitational field. A detailed review of the theory was given in my textbook [7]. The recent status of the subject is summarized by Hammond [17].

I ran into the subject in the eighties after having developed a disorder field theory of line-like objects in my textbook [5]. My first applications dealt with vortex lines in superfluids and superconductors, where the disorder formulation helped me to solve the long-standing problem of theoretically predicting where the second-order phase transition of a superconductor becomes first-order [18].

After this I turned to the application of the disorder field theory to line-like defects in crystals. The original description of such defects was based on functions which are discontinuous on surfaces, whose boundaries are the defect lines. The shape of these surfaces is arbitrary, as long as the boundaries are fixed. I realized that the deformations of the surfaces can be formulated as gauge transformatios of a new type of gauge fields which I named *defect gauge fields*.

By a so-called *duality transformation* it was possible to reformulate the theory of defects and their interactions as a more conventional type of gauge theory. This brought about another freedom in the description which I named *stress gauge invariance*. The dual formulation can be viewed as a linearized form of yet another geometric Einstein-Cartan space in which the gauge transformations are a combination of Einstein's local translations and a local generalization of *Lorentz invariance*.

The relation between the dual and the original description of defects in terms of jump surfaces is completely analogous to the well-known relation between Maxwell's theory of magnetism formulated in terms of a gauge field, the vector potential, and an alternative formulation in which the magnetic field is the gradient of a multivalued scalar field.

While the above developments were in progress, field theorists were searching for a simple explanation of the phenomenon of quark confinement by color-electric field lines. Here the physics of superconductors became an important source of inspiration. Since London's theory of superconductivity [19] it was known that superconductors would confine magnetic charges if they exist. The reason is the *Meissner effect*, which tries to expel magnetic flux lines from a superconductor. As a consequence, flux lines emerging from a magnetic monopole are compressed into flux tubes of a fixed thickness. The energy of such tubes is proportional to their length implying that opposite magnetic charges are held together forever. From the BCS theory of superconductivity [20] we know that this effect is caused by a condensate of electric charges, the famous *Cooper pairs* of electrons.

This phenomenon suggested the presently accepted viewpoint on quark confinement. The vacuum state of the world is imagined to contain a condensate of color-magnetic monopoles. This condensate acts upon color-electric fields in the same way as the Cooper pairs in a superconductor act upon the magnetic field, causing a Meissner effect and confinement of color-electric charges. Models utilizing this confinement mechanism were developed by Nambu [21], Mandelstam [22], 't Hooft [23], and Polyakov [24], and on a lattice by Wilson [25].

In studying this phenomenon I observed the close mathematical analogies between Dirac's magnetic monopoles and the above defect structures. Dirac used a vector potential with a jump surface to construct an infinitely thin magnetic flux tube with a magnetic point source at its end. Thus the world line of a monopole in spacetime could be viewed as a kind of "vortex line" in a Maxwell field. Knowing how to construct a disorder theory of vortex lines it was easy to set up a disorder field theory of monopole worldlines, which presntly serves as the simplest model of quark confinement [26].

When extending the statistical mechanics of vortex lines to defect lines in the second volume of the textbook [7], I used the dual description of defect lines, and expressed it as a linear approximation to a geometric description in Riemann-Cartan space. This suggested to me that it would be instructive to reverse the development in the theory of defects from multivalued fields to geometry and reformulate the theory of gravity, which is conventionally treated as a geometric theory, in an alternative way with the help of jumping surfaces of translation and rotation fields. In the theory of plasticity, such singular transformations are used to carry an ideal crystal into crystals with translational and rotational defects. Their geometric analogs carry a flat spacetime into a spacetime with curvature and torsion. The mathematical basis expressing the new geometry are *multivalued* tetrad fields $e^\alpha{}_\mu(x)$.

In the traditional literature on gravity with spinning particles, a special role is played by single-valued vierbein fields $h^\alpha{}_\mu(x)$. They define local nonholonomic coordinate differentials dx^α. These are reached from the physical coordinate differentials dx^μ by a transformation $dx^\alpha = h^\alpha{}_\mu(x)dx^\mu$. Only infinitesimal vectors dx^α are defined, and the transformation cannot be extended over finite domains. For the description of spinning particles, such an extension is not needed since the infinitesimal nonholonomic coordinates dx^α are completely sufficient to specify the transformation properties of spin in Riemannian spacetime.

The theory in terms of multivalued tetrad fields to be presented here goes an important step further, leading to a drastic simplification of the description of non-Riemannian geometry. The key is the efficient use of a set of completely new nonholonomic coordinates dx^a which are more nonholonomic than the traditional dx^α. To emphasize this one might call them *hyper-nonholonomic coordinates*. They are related to dx^α by a *multivalued* Lorentz transformation $dx^a = \Lambda^a{}_\alpha(x)dx^\alpha$, and to the physical dx^μ by the above *multivalued* tetrad fields as $dx^a = e^a{}_\mu(x)dx^\mu \equiv \Lambda^a{}_\alpha(x)h^\alpha{}_\mu(x)dx^\mu$. The gradients $\partial_\mu e^a{}_\nu(x)$ determine directly the full affine connection, and their antisymmetric combination $\partial_\mu e^a{}_\nu(x) - \partial_\nu e^a{}_\mu(x)$ determines the torsion. This is in contrast to the curl of the usual vierbein fields $h^\alpha{}_\mu(x)$ which determines the object of anholonomy, a quantity existing also in purely Riemannian spacetime, i.e., in curved spacetime without torsion.

One of the purposes of this book is to make students and colleagues working in electromagnetism and gravitational physics appreciate the many advantages brought about by the use of the multivalued tetrad fields $e^a{}_\mu(x)$. Apart from a simple intuitive reformulation of Riemann-Cartan geometry, it suggests a new principle in physics [27], which I have named *multivalued mapping principle* or *nonholonomic mapping principle*, to be explained in detail in this book. Multivalued coordinate transformations enable us to transform the physical laws governing the behavior of fundamental particles from flat spacetime to spacetimes with curvature and torsion. It is therefore natural to postulate that the images of these laws describe correctly the physics in such general affine spacetimes. As a result I am able to make *predictions* which cannot be made with Einstein's construction method based merely on the postulate of covariance under ordinary coordinate transformations, since those are unable to connect different geometries.

It should be emphasized that it is not the purpose of this book to propose repeating all geometric calculations of gravitation with the help of multivalued coordinate transformations. In fact, I shall restrict much of the discussion to almost flat auxiliary spacetimes. This will be enough to derive the general form of the physical laws in the presence of curvature and torsion. At the end I shall always return to the usual geometric description. The intermediate auxiliary spacetime with defects will be referred to as *world crystal*.

The reader will be pleased to see in Subsection 4.5 that the standard minimal coupling of electromagnetism is a simple consequence of the multivalued mapping principle. The similar minimal coupling to gravity will be derived from this principle in Chapter 17.

At the end I shall argue that torsion fields in gravity, if they exist, would lead quite a hidden life, unless they are of a special form. They would not be observable for many generations to come since they could exist only in an extremely small neighborhood of material point particles, limited to distances of the order of the Planck length 10^{-33} cm, which no presently conceivable experiment can probe.

The detailed development in this book in gravity with torsion is thus at present a purely theoretical endeavor. Its main merit lies in exposing the multivalued approach to Riemann-Cartan geometry, which has turned out to be quite useful in teaching the geometrical basis of gravitational physics to beginning students, and to explain what is omitted in Einstein's theory by assuming the absence of torsion.

The definitions of parallel displacements and covariant derivatives appear naturally as nonholonomic images of truly parallel displacements and ordinary derivatives in flat spacetime. So do the rules of minimal coupling.

Valuable insights are gained by realizing the universality of the multivalued defect description in various fields of physics. The predictions based on the multivalued mapping principle remain to be tested experimentally.

Thanks go to my secretary S. Endrias for her help in preparing the manuscript in LaTeX. Most importantly, I am grateful to my wife Dr. Annemarie Kleinert for her sacrifices, inexhaustible patience, constant encouragement, and a critical reading of the manuscript.

H. Kleinert
Berlin, November 2007

Notes and References

[1] P.A.M. Dirac, *Quantized Singularities in the Electromagnetic Field*, Proceedings of the Royal Society, A **133**, 60 (1931). Can be read on the web under the URL `kl/files`, where `kl` is short for the URL `www.physik.fu-berlin.de/~kleinert`.

[2] J. Schwinger, Phys. Rev. **144**, 1087 (1966).

[3] V.L. Berezinski, Zh. Eksp. Teor. Fiz. **59**, 907 (1970) [Sov. Phys. JETP **32**, 493 (1971)].

[4] J.M. Kosterlitz and D.J. Thouless, J. Phys. C **5**, L124 (1972); J. Phys. C **6**, 1181 (1973);
J.M. Kosterlitz, J. Phys. C **7**, 1046 (1974).

[5] H. Kleinert, *Gauge Fields in Condensed Matter*, Vol. I, *Superflow and Vortex Lines*, World Scientific, Singapore, 1989, pp. 1–742 (`kl/re.html#b1`).

[6] E. Orowan, Z. Phys. **89**, 605, 634 (1934);
M. Polany, Z. Phys. **89**, 660 (1934);
G.I. Taylor, Proc. Roy. Soc. A **145**, 362 (1934).

[7] H. Kleinert, *Gauge Fields in Condensed Matter*, Vol. II, *Stresses and Defects*, World Scientific, Singapore, 1989, pp. 743-1456 (`kl/re.html#b2`).

[8] B.A. Bilby, R. Bullough, and E. Smith, *Continuous distributions of disloca-tions: A New Application of the Methods of Non-Riemannian Geometry*, Proc. Roy. Soc. London, A **231**, 263-273 (1955).

[9] K. Kondo, in *Proceedings of the II Japan National Congress on Applied Me-chanics*, Tokyo, 1952, publ. in *RAAG Memoirs of the Unified Study of Basic Problems in Engeneering and Science by Means of Geometry*, Vol. 3, 148, ed. K. Kondo, Gakujutsu Bunken Fukyu-Kai, 1962.

[10] E. Kröner, in *The Physics of Defects*, eds. R. Balian et al., North-Holland, Amsterdam, 1981, p. 264.

[11] E. Cartan, Comt. Rend. Acad. Science **174**, 593 (1922); Ann. Ec. Norm. Sup. **40**, 325 (1922); **42**, 17 (1922).

[12] E. Cartan and A. Einstein, *Letters of Absolute Parallelism*, Princeton Univer-sity Press, Princeton, NJ.
See Chapter 21 for more details.

[13] E. Schrödinger, Proc. R. Ir. Acad. A **49**, 135 (1943); **52**, 1 (1948); **54**, 79 (1951).

[14] R. Utiyama, Phys. Rev. **101**, 1597 (1956).

[15] D.W. Sciama, Rev. Mod. Phys. **36**, 463 (1964).

[16] T.W.B. Kibble, J. Math. Phys. **2**, 212 (1961).

[17] R.T. Hammond, Rep. Prog. Phys. **65**, 599 (2002).

[18] H. Kleinert, *Disorder Version of the Abelian Higgs Model and the Order of the Superconductive Phase Transition*, Lett. Nuovo Cimento **35**, 405 (1982) (`kl/97`).

[19] F. London and H. London, Proc. R. Soc. London, A **149**, 71 (1935); Physica A **2**, 341 (1935);
H. London, Proc. R. Soc. A **155**, 102 (1936);
F. London, *Superfluids*, Dover, New York, 1961.

[20] J. Bardeen, L.N. Cooper, J.R. Schrieffer, Phys. Rev. **108**, 1175 (1957);
M. Tinkham, *Introduction to Superconductivity*, McGraw-Hill, New York, 1975.

[21] Y. Nambu, Phys. Rev. D **10**, 4262 (1974).

[22] S. Mandelstam, Phys. Rep. C **23**, 245 (1976); Phys. Rev. D **19**, 2391 (1979).

[23] G. 't Hooft, Nucl. Phys. B **79**, 276 (1974); and in *High Energy Physics*, ed. by A. Zichichi, Editrice Compositori, Bologna, 1976.

[24] A.M. Polyakov, JEPT Lett. **20**, 894 (1974).

[25] K.G. Wilson, *Confinement of Quarks*, Phys. Rev. D **10**, 2445 (1974).

[26] H. Kleinert, *The Extra Gauge Symmetry of String Deformations in Electromagnetism with Charges and Dirac Monopoles*, Int. J. Mod. Phys. A **7**, 4693 (1992) (kl/203); *Double-Gauge Invariance and Local Quantum Field Theory of Charges and Dirac Magnetic Monopoles*, Phys. Lett. B **246**, 127 (1990) (kl/205); *Abelian Double-Gauge Invariant Continuous Quantum Field Theory of Electric Charge Confinement*, Phys. Lett. B **293**, 168 (1992) (kl/211).

[27] H. Kleinert, *Quantum Equivalence Principle for Path Integrals in Spaces with Curvature and Torsion*, in Proceedings of the XXV International Symposium Ahrenshoop on *Theory of Elementary Particles* in Gosen/Germany 1991, ed. by H.J. Kaiser (quant-ph/9511020); *Quantum Equivalence Principle*, Lectures presented at the 1996 Cargèse Summer School on *Functional Integration: Basics and Applications* (quant-ph/9612040).

Contents

List of Figures

1

Basics

A book on multivalued fields must necessarily review some basic concepts of classical mechanics and the theory of single-valued fields. This will be done in the first three chapters. Readers familiar with these subjects may move directly Chapter 4.

In his fundamental work on theoretical mechanics entitled *Principia*, Newton (1642–1727) assumed the existence of an absolute spacetime. Space is parametrized by vectors $\mathbf{x} = (x^1, x^2, x^3)$, and the movement of point particles is described by trajectories $\mathbf{x}(t)$ whose components $q^i(t)$ $(i = 1, 2, 3)$ specify the coordinates $x^i = q^i(t)$ along which the particles move as a function of time t. In Newton's absolute spacetime, a single free particle moves without acceleration. Mathematically, this is expressed by the differential equation

$$\ddot{\mathbf{x}}(t) \equiv \frac{d^2}{dt^2}\mathbf{x}(t) = 0. \tag{1.1}$$

The dots denote derivatives with respect to the argument.

A set of N point particles $\mathbf{x}_n(t)$ $(n = 1, \ldots, N)$ with masses m_n is subject to gravitational forces which change the free equations of motion to

$$m_n\ddot{\mathbf{x}}_n(t) = G_{\mathrm{N}} \sum_{m \neq n} m_n m_m \frac{\mathbf{x}_m(t) - \mathbf{x}_n(t)}{|\mathbf{x}_m(t) - \mathbf{x}_n(t)|^3}, \tag{1.2}$$

where G_{N} is Newton's gravitational constant

$$G_{\mathrm{N}} \approx 6.67259(85) \times 10^{-8}\mathrm{cm}^3/\mathrm{g\ sec}^2. \tag{1.3}$$

1.1 Galilean Invariance of Newtonian Mechanics

The parametrization of absolute spacetime in which the above equations of motion hold is not unique. There is substantial freedom in choosing the coordinates.

1

1.1.1 Translations

The coordinates $\mathbf{x}$ may always be changed by translated coordinates

$$\mathbf{x}' = \mathbf{x} - \mathbf{x}_0. \tag{1.4}$$

It is obvious that the translated trajectories $\mathbf{x}'_n(t) = \mathbf{x}_n(t) - \mathbf{x}_0$ will again satisfy the equations of motion (1.2). The equations remain also true for a translated time

$$t' = t - t_0, \tag{1.5}$$

i.e., the trajectories

$$\mathbf{x}'(t) \equiv \mathbf{x}(t + t_0) \tag{1.6}$$

satisfy (1.2). This property of Newton's equations (1.2) is referred to as *translational symmetry* in spacetime.

An alternative way of formulating this invariance is by keeping the coordinate frame fixed and displacing the physical system in spacetime, moving all particles to new coordinates $\mathbf{x}' = \mathbf{x} + \mathbf{x}_0$ at a new time $t' = t + t_0$. The equations of motion are again invariant. The first procedure of reparametrizing the same physical system is called *passive symmetry transformation*, the second *active symmetry transformation*. One may use either procedure to discuss symmetries. In this book we shall use active or passive transformations, depending on the circumstance.

1.1.2 Rotations

The equations of motion are invariant under more transformations which mix different coordinates linearly with each other, for instance the *rotations*:

$$x'^i = R^i{}_j x^j, \tag{1.7}$$

where $R^i{}_j$ is the rotation matrix

$$R^i{}_j = \cos\theta \, \delta_{ij} + (1 - \cos\theta) \, \hat{\theta}_i \hat{\theta}_j + \sin\theta \, \epsilon_{ijk} \hat{\theta}_k, \tag{1.8}$$

in which $\hat{\theta}_i$ denotes the *directional* unit vector of the rotation axis. The matrices satisfy the *orthogonality relation*

$$R^T R = 1. \tag{1.9}$$

In Eq. (1.7) a sum from 1 to 3 is implied over the repeated spatial index j. This is called the *Einstein summation convention*, which will be followed throughout this text. As for the translations, the rotations can be applied in the passive or active sense.

The active rotations are obtained from the above passive ones by changing the sign of θ. For example, the active rotations around the z-axis with a rotation vector $\hat{\varphi} = (0, 0, 1)$ are given by the orthogonal matrices

$$R_3(\varphi) = \begin{pmatrix} \cos\varphi & -\sin\varphi & 0 \\ \sin\varphi & \cos\varphi & 0 \\ 0 & 0 & 1 \end{pmatrix}. \tag{1.10}$$

1.1.3 Galilei Boosts

A further set of transformations mixes space and time coordinates:

$$x'^i = x^i - v^i t, \qquad (1.11)$$
$$t' = t. \qquad (1.12)$$

These are called *pure Galilei transformations* of *Galilei boosts*. The coordinates x'^i, t' are positions and time of a particle observed in a frame of reference that moves uniformly through absolute spacetime with velocity $\mathbf{v} \equiv (v^1, v^2, v^3)$. In the active description, the transformation $x'^i = x^i + v^i t$ specifies the coordinates of a physical system moving past the observer with uniform velocity $\mathbf{v}$.

1.1.4 Galilei Group

The combined set of all transformations

$$x'^i = R^i{}_j x^j - v^i t - x_0^i, \qquad (1.13)$$
$$t' = t - t_0, \qquad (1.14)$$

forms a group. Group multiplication is defined by performing the transformations successively. This multiplication law is obviously associative, and each element has an inverse. The set of transformations (1.13) and (1.14) is referred to as the *Galilei group*.

Newton called all coordinate frames in which the equations of motion have the simple form (1.2) *inertial frames*.

1.2 Lorentz Invariance of Maxwell Equations

Problems with Newton's theory arose when J. C. Maxwell (1831–1879) formulated in 1864 his theory of electromagnetism. His equations for the *electric field* $\mathbf{E}(x)$ and the *magnetic flux density* or *magnetic induction* $\mathbf{B}(x)$ in empty space

$$\boldsymbol{\nabla} \cdot \mathbf{E} = 0 \qquad \text{(Coulomb's law)}, \qquad (1.15)$$

$$\boldsymbol{\nabla} \times \mathbf{B} - \frac{1}{c}\frac{\partial \mathbf{E}}{\partial t} = 0 \qquad \text{(Ampère's law)}, \qquad (1.16)$$

$$\boldsymbol{\nabla} \cdot \mathbf{B} = 0 \qquad \text{(absence of magnetic monopoles)}, \qquad (1.17)$$

$$\boldsymbol{\nabla} \times \mathbf{E} + \frac{1}{c}\frac{\partial \mathbf{B}}{\partial t} = 0 \qquad \text{(Faraday's law)}, \qquad (1.18)$$

can be combined to obtain the second-order differential equations

$$\left(\frac{1}{c^2}\partial_t^2 - \boldsymbol{\nabla}^2\right)\mathbf{E}\,(\mathbf{x}, t) = 0, \qquad (1.19)$$

$$\left(\frac{1}{c^2}\partial_t^2 - \boldsymbol{\nabla}^2\right)\mathbf{B}\,(\mathbf{x}, t) = 0. \qquad (1.20)$$

The equations contain explicitly the light velocity

$$c \equiv 299\,792\,458\,\frac{\text{m}}{\text{sec}}, \tag{1.21}$$

and are not invariant under the Galilei group (1.14). Indeed, they contradict Newton's postulate of the existence of an absolute spacetime. If light propagates with the velocity c in absolute spacetime, it could not do so in other inertial frames which have a nonzero velocity with respect to the absolute frame. A precise measurement of the light velocity could therefore single out the absolute spacetime. However, experimental attempts to do this did not succeed. The experiment of Michelson (1852–1931) and Morley (1838–1923) in 1887 showed that light travels parallel and orthogonal to the earth's orbital motion with the same velocity up to ± 5 km/sec [1, 2]. This led Fitzgerald (1851–1901) [3], Lorentz (1855–1928) [4], Poincaré (1854–1912) [5], and Einstein (1879–1955) [6] to suggest that Newton's postulate of the existence of an absolute spacetime was unphysical [7].

1.2.1   Lorentz Boosts

The conflict was resolved by modifying the Galilei transformations (1.11) and (1.12) in such a way that Maxwell's equations remain invariant. This is achieved by the coordinate transformations

$$x'^i = x^i + (\gamma - 1)\frac{v^i v^j}{v^2}x^j - \gamma v^i t, \tag{1.22}$$

$$t' = \gamma t - \frac{1}{c^2}\gamma v^i x^i, \tag{1.23}$$

where γ is the velocity-dependent parameter

$$\gamma = \frac{1}{\sqrt{1 - v^2/c^2}}. \tag{1.24}$$

The transformations (1.22) and (1.23) are referred to as *pure Lorentz transformations* or *Lorentz boosts*. The parameter γ has the effect that in different moving frames of reference, time elapses differently. This is necessary to make the light velocity the same in all frames.

Pure Lorentz transformations are conveniently written in a four-dimensional vector notation. Introducing the *four-vectors* x^a labeled by indices $a, b, c, \ldots$ running through the values $0, 1, 2, 3$,

$$x^a = \begin{pmatrix} ct \\ x^1 \\ x^2 \\ x^3 \end{pmatrix}, \tag{1.25}$$

we rewrite (1.22) and (1.23) as

$$x'^a = \Lambda^a{}_b x^b, \tag{1.26}$$

where $\Lambda^a{}_b$ are the 4×4-matrices

$$\Lambda^a{}_b \equiv \left(\begin{array}{c|c} \gamma & -\gamma v^i/c \\ \hline -\gamma v^i/c & \delta_{ij} + (\gamma - 1)v_i v_j/v^2 \end{array} \right). \tag{1.27}$$

Note that we adopt Einstein's summation convention also for repeated labels $a, b, c, \ldots = 0, \ldots, 3$. The matrices $\Lambda^a{}_b$ satisfy the *pseudo-orthogonality relation* [compare (1.9)]:

$$\Lambda^T{}_a{}^c \, g_{cd} \, \Lambda^d{}_b = g_{ab}, \tag{1.28}$$

where g_{ab} is the *Minkowski metric* with the matrix elements

$$g_{ab} = \left(\begin{array}{cccc} 1 & & & \\ & -1 & & \\ & & -1 & \\ & & & -1 \end{array} \right). \tag{1.29}$$

Equation (1.28) has the consequence that for any two four-vectors x^a and y^a, the scalar product formed with the help of the Minkowski metric

$$xy \equiv x^a g_{ab} y^b \tag{1.30}$$

is invariant under Lorentz transformation.

In order to verify the relation (1.28) it is convenient to introduce a dimensionless vector $\boldsymbol{\zeta}$ called *rapidity*, which points in the direction of the velocity $\mathbf{v}$ and has a length $\zeta \equiv |\boldsymbol{\zeta}|$ given by

$$\cosh \zeta = \gamma, \qquad \sinh \zeta = \gamma v/c. \tag{1.31}$$

We also define the unit vectors in three-space

$$\hat{\boldsymbol{\zeta}} \equiv \boldsymbol{\zeta}/\zeta = \hat{\mathbf{v}} \equiv \mathbf{v}/v, \tag{1.32}$$

so that

$$\boldsymbol{\zeta} = \zeta \hat{\boldsymbol{\zeta}} = \operatorname{atanh}\frac{v}{c}\, \hat{\mathbf{v}}. \tag{1.33}$$

Then the matrices $\Lambda^a{}_b$ of the pure Lorentz transformations (1.27) take the form

$$\Lambda^a{}_b = B^a{}_b(\boldsymbol{\zeta}) \equiv \left(\begin{array}{c|c} \cosh \zeta & -\sinh \zeta \, \hat{\zeta}_1 \;\; -\sinh \zeta \, \hat{\zeta}_2 \;\; -\sinh \zeta \, \hat{\zeta}_3 \\ \hline \begin{array}{c} -\sinh \zeta \, \hat{\zeta}_1 \\ -\sinh \zeta \, \hat{\zeta}_2 \\ -\sinh \zeta \, \hat{\zeta}_3 \end{array} & \delta_{ij} + (\cosh \zeta - 1)\, \hat{\zeta}_i \hat{\zeta}_j \end{array} \right). \tag{1.34}$$

The notation $B^a{}_b(\boldsymbol{\zeta})$ emphasizes that the transformations are boosts. The pseudo-orthogonality property (1.28) follows directly from the identities $\hat{\zeta}^2 = 1$, $\cosh^2 \zeta - \sinh^2 \zeta = 1$.

For active transformations of a physical system, the above transformations have to be inverted. For instance, the active boosts with a rapidity $\zeta = \zeta(0,0,1)$ pointing in the z-direction, have the pseudo-orthogonal matrix

$$\Lambda^a{}_b = B_3(\zeta) = \left(\begin{array}{c|ccc} \cosh\zeta & 0 & 0 & \sinh\zeta \\ \hline 0 & 1 & 0 & 0 \\ 0 & 0 & 1 & 0 \\ \sinh\zeta & 0 & 0 & \cosh\zeta \end{array} \right). \tag{1.35}$$

1.2.2   Lorentz Group

The set of Lorentz boosts (1.34) can be extended by rotations to form the *Lorentz group*. In 4×4 -matrix notation, the rotation matrices (1.8) have the block form

$$\Lambda^a{}_b(R) = R^a{}_b \equiv \left(\begin{array}{c|ccc} 1 & 0 & 0 & 0 \\ \hline 0 & & & \\ 0 & & R^i{}_j & \\ 0 & & & \end{array} \right). \tag{1.36}$$

It is easy to verify that these satisfy the relation (1.28), which becomes here an orthogonality relation (1.9).

The four-dimensional versions of the active rotations (1.10) around the z-axis with a rotation vector $\hat{\varphi} = (0,0,1)$ are given by the orthogonal matrices

$$\Lambda^b{}_a = R_3(\varphi) = \left(\begin{array}{c|ccc} 1 & 0 & 0 & 0 \\ \hline 0 & \cos\varphi & -\sin\varphi & 0 \\ 0 & \sin\varphi & \cos\varphi & 0 \\ 0 & 0 & 0 & 1 \end{array} \right). \tag{1.37}$$

The rotation matrix (1.37) differs from the boost matrix (1.35) mainly in the presence of trigonometric functions instead of hyperbolic functions. In addition, there is a sign change under transposition accounting for the opposite sign in the time- and space-like parts of the metric (1.29).

When combining all possible Lorentz boosts and rotations in succession, the resulting set of transformations forms a group called the *Lorentz group*.

1.3   Infinitesimal Lorentz Transformations

The transformation laws of continuous groups such as rotation and Lorentz group are conveniently expressed in an infinitesimal form. By combining successively many infinitesimal transformations it is always possible to reconstruct from these the finite transformation laws. This is a consequence of the fact that the exponential function e^x can always be obtained by a product of many small-x approximations $e^{\epsilon x} \approx 1 + \epsilon x$:

$$e^x = \lim_{\epsilon \to 0} (1 + \epsilon x)^{1/\epsilon}. \tag{1.38}$$

1.3.1 Generators of Group Transformations

Let us illustrate this procedure for the active rotations (1.37). These can be written in the exponential form

$$
R_3(\varphi) = \exp\left\{ \begin{pmatrix} 0 & 0 & 0 & 0 \\ 0 & 0 & -1 & 0 \\ 0 & 1 & 0 & 0 \\ 0 & 0 & 0 & 0 \end{pmatrix} \varphi \right\} \equiv e^{-iL_3\varphi}. \tag{1.39}
$$

The matrix

$$
L_3 = -i \begin{pmatrix} 0 & 0 & 0 & 0 \\ 0 & 0 & 1 & 0 \\ 0 & -1 & 0 & 0 \\ 0 & 0 & 0 & 0 \end{pmatrix} \tag{1.40}
$$

is called the *generator* of this rotation in the Lorentz group. There are similar generators for rotations around x- and y-directions

$$
L_1 = -i \begin{pmatrix} 0 & 0 & 0 & 0 \\ 0 & 0 & 0 & 0 \\ 0 & 0 & 0 & 1 \\ 0 & 0 & -1 & 0 \end{pmatrix}, \tag{1.41}
$$

$$
L_2 = -i \begin{pmatrix} 0 & 0 & 0 & 0 \\ 0 & 0 & 0 & -1 \\ 0 & 0 & 0 & 0 \\ 0 & 1 & 0 & 0 \end{pmatrix}. \tag{1.42}
$$

The three generators may compactly be written as

$$
L_i \equiv -i \left(\begin{array}{c|c} 0 & 0 \\ \hline 0 & \epsilon_{ijk} \end{array} \right), \tag{1.43}
$$

where ϵ_{ijk} is the completely antisymmetric *Levi-Civita tensor* with $\epsilon_{123} = 1$.

Introducing a vector notation for the three generators, $\mathbf{L} \equiv (L_1, L_2, L_2)$, the general pure rotation matrix (1.36) is given by the exponential

$$
\Lambda(R(\boldsymbol{\varphi})) = e^{-i\boldsymbol{\varphi}\cdot\mathbf{L}}. \tag{1.44}
$$

This follows from the fact that all orthogonal 3×3-matrices in the spatial block of (1.36) can be written as an exponential of i times all antisymmetric 3×3-matrices, and that these can all be reached by the linear combinations $\boldsymbol{\varphi} \cdot \mathbf{L}$.

Let us now find the generators of the active boosts, first in the z-direction. From Eq. (1.35) we see that the boost matrix can be written as an exponential

$$
\begin{aligned}
B_3(\zeta) &= \exp\left\{ \begin{pmatrix} 0 & 0 & 0 & 1 \\ 0 & 0 & 0 & 0 \\ 0 & 0 & 0 & 0 \\ 1 & 0 & 0 & 0 \end{pmatrix} \zeta \right\} \\
&= e^{-iM_3\zeta},
\end{aligned} \tag{1.45}
$$

with the generator

$$M_3 = i \begin{pmatrix} 0 & 0 & 0 & 1 \\ 0 & 0 & 0 & 0 \\ 0 & 0 & 0 & 0 \\ 1 & 0 & 0 & 0 \end{pmatrix}. \qquad (1.46)$$

Similarly we find the generators for the x- and y-directions:

$$M_1 = i \begin{pmatrix} 0 & 1 & 0 & 0 \\ 1 & 0 & 0 & 0 \\ 0 & 0 & 0 & 0 \\ 0 & 0 & 0 & 0 \end{pmatrix}, \qquad (1.47)$$

$$M_2 = i \begin{pmatrix} 0 & 0 & 1 & 0 \\ 0 & 0 & 0 & 0 \\ 1 & 0 & 0 & 0 \\ 0 & 0 & 0 & 0 \end{pmatrix}. \qquad (1.48)$$

Introducing a vector notation for the three boost generators, $\mathbf{M} \equiv (M_1, M_2, M_2)$, the general Lorentz transformation matrix (1.34) is given by the exponential

$$\Lambda(B(\boldsymbol{\zeta})) = e^{-i\boldsymbol{\zeta}\cdot\mathbf{M}}. \qquad (1.49)$$

The proof is analogous to the proof of the exponential form (1.44).

The Lorentz group is therefore generated by the six matrices L_i, M_i, to be collectively denoted by $G_a(a = 1, \ldots, 6)$. Every element of the group can be written as

$$\Lambda = e^{-i(\boldsymbol{\varphi}\cdot\mathbf{L}+\boldsymbol{\zeta}\cdot\mathbf{M})} \equiv e^{-i\alpha_a G_a}. \qquad (1.50)$$

There exists a Lorentz-covariant way of specifying the generators of the Lorentz group. We introduce the 4×4-matrices

$$(L^{ab})^{cd} = i(g^{ac}g^{bd} - g^{ad}g^{bc}), \qquad (1.51)$$

labeled by the antisymmetric pair of indices ab, i.e.,

$$L^{ab} = -L^{ba}. \qquad (1.52)$$

There are six independent matrices which coincide with the generators of rotations and boosts as follows:

$$L_i = \frac{1}{2}\epsilon_{ijk}L^{jk}, \qquad M_i = L^{0i}. \qquad (1.53)$$

With the help of the generators (1.51), we can write every element (1.50) of the Lorentz group as follows

$$\Lambda = e^{-i\frac{1}{2}\omega_{ab}L^{ab}}, \qquad (1.54)$$

where the antisymmetric angular matrix $\omega_{ab} = -\omega_{ba}$ collects both, rotation angles and rapidities:

$$\omega_{ij} = \epsilon_{ijk}\varphi^k, \tag{1.55}$$

$$\omega_{0i} = \zeta^i. \tag{1.56}$$

Summarizing the notation we have set up an exponential representation of all Lorentz transformations

$$\Lambda = e^{-i(\boldsymbol{\varphi}\cdot\mathbf{L}+\boldsymbol{\zeta}\cdot\mathbf{M})} = e^{-i(\frac{1}{2}\varphi^i\epsilon_{ijk}L^{jk}+\zeta^i L^{0i})} = e^{-i(\frac{1}{2}\omega_{ij}L^{ij}+\omega_{0i}L^{0i})} = e^{-i\frac{1}{2}\omega_{ab}L^{ab}}. \tag{1.57}$$

Note that for a Euclidean metric

$$g_{ab} = \begin{pmatrix} 1 & & & \\ & 1 & & \\ & & 1 & \\ & & & 1 \end{pmatrix}, \tag{1.58}$$

the above representation are familiar from basic matrix theory. Then Eq. (1.28) implies that Λ comprises all real orthogonal matrices in four dimensions, which can be written as an exponential of all real antisymmetric 4×4-matrices. For the pseudo-orthogonal matrices satisfying (1.28) with the Minkowski metric (1.29), only the iL_i are antisymmetric while iM_i are symmetric.

1.3.2 Group Multiplication and Lie Algebra

The reason for expressing the group elements as exponentials of the six generators is that, in this way, the *multiplication rules* of infinitely many group elements can be completely reduced to the knowledge of the finite number of *commutation rules* among the six generators L_i, M_i. This is a consequence of the *Baker-Campbell-Hausdorff formula* [8]:

$$e^A e^B = e^{A+B+\frac{1}{2}[A,B]+\frac{1}{12}[A-B,[A,B]]-\frac{1}{24}[A,[B,[A,B]]]+\cdots}. \tag{1.59}$$

According to this formula, the product of exponentials can be written as an exponential of commutators. Adapting the general notation $G_r = (L_i, M_i)$ for the six generators in Eqs. (1.53) and (1.57), the product of two group elements is

$$\begin{aligned}
\Lambda_1\Lambda_2 &= e^{-i\alpha_r^1 G_r} e^{-i\alpha_s^2 G_s} \\
&= \exp\Big\{ -i\alpha_r^1 G_r - i\alpha_s^2 G_s + \frac{1}{2}[-i\alpha_r^1 G_r, -i\alpha_s^2 G_s] \\
&\qquad + \frac{1}{12}[-i(\alpha_t^1 - \alpha_t^2)G_t, [-i\alpha_r^1 G_r, -i\alpha_s^2 G_s]] + \ldots \Big\}.
\end{aligned} \tag{1.60}$$

The exponent involves only commutators among G_r's. For the Lorentz group these can be calculated from the explicit 4×4 -matrices (1.40)–(1.42) and (1.46)–(1.48). The result is

$$[L_i, L_j] = i\epsilon_{ijk}L_k, \tag{1.61}$$
$$[L_i, M_j] = i\epsilon_{ijk}M_k, \tag{1.62}$$
$$[M_i, M_j] = -i\epsilon_{ijk}L_k. \tag{1.63}$$

This algebra of generators is called the *Lie algebra* of the group. In the general notation with generators G_r, the algebra reads

$$[G_r, G_s] = if_{rst}G_t. \tag{1.64}$$

The number of linearly independent matrices G_r (here 6) is called the *rank* of the Lie algebra.

In any Lie algebra, the commutator of two generators is a linear combination of generators. The coefficients f_{abc} are called *structure constants*. They are completely antisymmetric in a, b, c, and satisfy the relation

$$f_{rsu}f_{utv} + f_{stu}f_{urv} + f_{tru}f_{usv} = 0. \tag{1.65}$$

This guarantees that the generators obey the *Jacobi identity*

$$[[G_r, G_s], G_t] + [[G_s, G_t], G_r] + [[G_t, G_r], G_s] = 0, \tag{1.66}$$

which ensures that multiplication of three exponentials $\Lambda_j = e^{-i\alpha_r^j G_r}$ $(i = 1, 2, 3)$ obeys the law of associativity $(\Lambda_1\Lambda_2)\Lambda_3 = \Lambda_1(\Lambda_2\Lambda_3)$ when evaluating the products via the expansion Eq. (1.60).

The relation (1.65) can easily be verified explicitly for the structure constants (1.61)–(1.63) of the Lorentz group using the identity for the ϵ-tensor

$$\epsilon_{ijl}\epsilon_{lkm} + \epsilon_{jkl}\epsilon_{lim} + \epsilon_{kil}\epsilon_{ljm} = 0. \tag{1.67}$$

The Jacobi identity implies that the r matrices with $r \times r$ elements

$$(F_r)_{st} \equiv -if_{rst} \tag{1.68}$$

satisfy the commutation rules (1.64). They are the generators of the so-called *adjoint representation* of the Lie algebra. The matrix in the spatial block of Eq. (1.43) for L_i is precisely of this type.

In terms of the matrices F_r of the adjoint representation, the commutation rules can also be written as

$$[G_r, G_s] = -(F_t)_{rs}G_t. \tag{1.69}$$

Inserting for G_r the generators (1.68), we reobtain the relation (1.65).

Continuing the expansion in terms of commutators in the exponent of (1.60), all commutators can be evaluated successively and one remains at the end with an expression

$$\Lambda_{12} = e^{-i\alpha_r^{12}(\alpha^1, \alpha^2)G_r}, \tag{1.70}$$

in which the parameters of the product α_r^{12} are completely determined from those of the factor, α_r^1, α_r^2. The result depends only on the structure constants f_{abc}, not on the representation.

If we employ the tensor notation L^{ab} for L_i and M_i of Eqs. (1.53), (1.53), and perform multiplication covariantly, so that products $L^{ab}L^{cd}$ have the matrix elements $(L^{ab})_{\sigma\tau}(L^{cd})^\tau{}_\delta$, the commutators (1.61)–(1.63) can be written as

$$[L^{ab}, L^{cd}] = -i(g^{ac}L^{bd} - g^{ad}L^{bc} + g^{bd}L^{ac} - g^{bc}L^{ad}). \tag{1.71}$$

Due to the antisymmetry in $a \leftrightarrow b$ and $c \leftrightarrow d$ it is sufficient to specify only the simpler commutators

$$[L^{ab}, L^{ac}] = -ig^{aa}L^{bc}, \quad \text{no sum over } a. \tag{1.72}$$

This list of commutators omits only commutation rules of (1.71) which vanish since none of the indices ab is equal to one of the indices cd.

For infinitesimal transformations, the matrices (1.54) have the general form

$$\Lambda \equiv 1 - i\frac{1}{2}\omega_{ab}L^{ab}. \tag{1.73}$$

Inserting the e 4×4-generators (1.51), their matrix elements are

$$\Lambda^a{}_b = \delta^a{}_b + \omega^a{}_b, \quad \left(\Lambda^{-1}\right)^a{}_b = \delta^a{}_b - \omega^a{}_b, \tag{1.74}$$

where $\omega^a{}_b$ and $\omega_a{}^b$ are related to the antisymmetric angular matrix ω_{ab} by

$$\omega^a{}_b = g^{aa'}\omega_{a'b}, \quad \omega_a{}^b = g^{bb'}\omega_{ab'}. \tag{1.75}$$

1.4 Vector-, Tensor-, and Scalar Fields

We shall frequently consider four-component physical quantities v^a which, under Lorentz transformation, change in the same way as the coordinates x^a:

$$v'^a = \Lambda^a{}_b v^b. \tag{1.76}$$

This transformation property defines a *Lorentz vector*, or *four-vector*. In addition to such vectors, there are quantities with more indices $t^{ab}, t^{abc}, \ldots$ which transform like products of vectors:

$$t'^{ab} = \Lambda^a{}_c\Lambda^b{}_d t^{cd}, \quad t'^{abc} = \Lambda^a{}_d\Lambda^b{}_e\Lambda^c{}_f t^{def}, \ldots . \tag{1.77}$$

These are the transformation properties of *Lorentz tensors* of rank two, three, ...

Given any two four-vectors u^a and v^a, we define their scalar product in the same way as in (1.30) for two coordinate vectors x^a and y^a:

$$uv = u^a g_{ab} v^b. \tag{1.78}$$

Scalar products are, of course, invariant under Lorentz transformations due to their pseudo-orthogonality (1.28).

If $v^a, t^{ab}, t^{abc}, \ldots$ are functions of x, they are called *vector and tensor fields*. Derivatives with respect to x of such a field obey vector and tensor transformation laws. Indeed, since

$$x'^a = \Lambda^a{}_b x^b, \tag{1.79}$$

we see that the derivative $\partial/\partial x^b$ satisfies

$$\frac{\partial}{\partial x'^a} = \left(\Lambda^{T-1}\right)_a{}^b \frac{\partial}{\partial x^b}, \tag{1.80}$$

i.e., it transforms with the inverse of the transposed Lorentz matrix $\Lambda^a{}_b$. Using the pseudo-orthogonality relation (1.28),

$$\frac{\partial}{\partial x'^a} = \left(g\Lambda g^{-1}\right)_a{}^b \frac{\partial}{\partial x^b}. \tag{1.81}$$

It will be useful to define the matrix elements

$$\Lambda_a{}^b \equiv \left(g\Lambda g^{-1}\right)_a{}^b = g_{ac}\,\Lambda^c{}_d\,g_{db}. \tag{1.82}$$

The we can rewrite (1.81) as

$$\partial'_a = \Lambda_a{}^{b'} \partial_b. \tag{1.83}$$

In general, any four-component quantity v_a which transforms like the derivatives

$$v'_a = \Lambda_a{}^b v_b \tag{1.84}$$

is called a *covariant* four-vector or Lorentz vector, as opposed to the vector v^a transforming like the coordinates x^a, which is called *contravariant* vector.

A covariant vector v_a can be produced from a contravariant one v^b by multiplication with the metric tensor:

$$v_a = g_{ab} v^b. \tag{1.85}$$

This operation is called *lowering the index*. The operation can be inverted to what is called *raising the index*:

$$v^a = g^{ab} v_b, \tag{1.86}$$

where g^{ab} are the matrix elements of the *inverse metric*

$$g^{ab} \equiv \left(g^{-1}\right)_{ab}. \tag{1.87}$$

With Einstein's summation convention, the inverse metric $g^{ab} \equiv (g^{-1})^{ab}$ satisfies the equation

$$g^{ab} g_{bc} = \delta^a{}_c. \tag{1.88}$$

The sum over a common upper and lower index is called *contraction*.

Note that the notation (1.82) is perfectly compatible with the rules for raising and lowering indices.

In Minkowski spacetime, the matrices g and g^{-1} happen to be the same and so are the matrix elements g_{ab} and g^{ab}, both being equal to (1.29). This is no longer true in the general geometries of gravitational physics. For this reason it will be useful to keep separate symbols for the metric g and its inverse g^{-1}, and for their matrix elements g_{ab} and g^{ab}.

The contraction of a covariant vector with a contravariant vector is a scalar product, as is obvious if we rewrite the scalar product (1.78) as

$$uv = u^a g_{ab} v^b = u^a v_a = u_a v^a. \tag{1.89}$$

Of course, we can form also the scalar product of two covariant vectors with the help of the inverse metric g^{-1}:

$$uv = u_a g^{ab} v_b. \tag{1.90}$$

The invariance under Lorentz transformations (1.84) is easily verified using the pseudo-orthogonality property (1.28):

$$u'_a g^{ab} v'_b = u'^T g^{-1} v' = u^T g^{-1} \Lambda^T g\, g^{-1} g \Lambda g^{-1} v = u^T g^{-1} v = u_a g^{ab} v_b. \tag{1.91}$$

Since $\partial/\partial x^a$ transforms like a covariant vector, it is useful to emphasize this behavior by the notation

$$\partial_a \equiv \frac{\partial}{\partial x^a}. \tag{1.92}$$

Extending the definition of covariant vectors, one defines covariant tensors of rank two t_{ab}, three t_{abc}, etc. as quantities transforming like

$$t'_{ab} = \Lambda_a{}^c \Lambda_b{}^d t_{cd}, \quad t'_{abc} = \Lambda_a{}^c \Lambda_b{}^f \Lambda_c{}^g t_{efg}, \dots. \tag{1.93}$$

Co- and contravariant vectors and tensors can always be multiplied with each other to form new co- and contravariant quantities if the indices to be contracted are raised and lowered appropriately. If no uncontracted indices are left, one obtains an invariant, a *Lorentz scalar*.

It is useful to introduce a contravariant version of the covariant derivative vector

$$\partial^a \equiv g^{ab} \partial_b, \tag{1.94}$$

and covariant versions of the contravariant coordinate vector

$$x_a \equiv g_{ab} x^b. \tag{1.95}$$

The invariance of Maxwell's equations (1.20) is a direct consequence of these contraction rules since the differential operator on the left-hand side can be written covariantly as

$$\frac{1}{c^2} \partial_t^2 - \nabla^2 = \frac{\partial}{\partial x^a} g^{ab} \frac{\partial}{\partial x^b} = \partial_a g^{ab} \partial_b = \partial^a \partial_a = \partial^2. \tag{1.96}$$

The right-hand side is obviously a Lorentz scalar.

1.4.1 Discrete Lorentz Transformations

The Lorentz group can be extended to include space reflections in any of the four spacetime directions

$$x^a \rightarrow -x^a, \tag{1.97}$$

without destroying the defining property (1.28). The determinant of Λ, however, is then negative. If only x^0 is reversed, the reflection is also called _time reversal_ and denoted by

$$T = \begin{pmatrix} -1 & & & \\ & 1 & & \\ & & 1 & \\ & & & 1 \end{pmatrix}. \tag{1.98}$$

The simultaneous reflection of the three spatial coordinates is called _parity transformation_ and denoted by the 4×4 -matrix P:

$$P = \begin{pmatrix} 1 & & & \\ & -1 & & \\ & & -1 & \\ & & & -1 \end{pmatrix}. \tag{1.99}$$

After this extension, the entire Lorentz group can no longer be obtained from the neighborhood of the identity by a product of infinitesimal transformations, i.e., by an exponential of the Lie algebra in Eq. (1.57). It consists of four topologically disjoint pieces which can be obtained by a product of infinitesimal transformations multiplied with 1, P, T, and PT. The four pieces of the group are

$$e^{-i\frac{1}{2}\omega_{ab}L^{ab}}, \quad e^{-i\frac{1}{2}\omega_{ab}L^{ab}}P, \quad e^{-i\frac{1}{2}\omega_{ab}L^{ab}}T, \quad e^{-i\frac{1}{2}\omega_{ab}L^{ab}}PT. \tag{1.100}$$

The Lorentz transformations Λ of the pieces associated with P and T have a negative determinant. This leads to the definition of _pseudotensors_ which transform like a tensor, but with an additional determinantal factor $\det \Lambda$. A vector with this property is also called _axial vector_. In three dimensions, the angular momentum $\mathbf{L} = \mathbf{x} \times \mathbf{p}$ is an axial vector since it does not change sign under space reflections, as the vector $\mathbf{x}$ does, but remains invariant.

1.4.2 Poincaré group

Just as the Galilei transformations, the Lorentz transformations can be extended by the group of spacetime translations

$$x^a = x^a - a^a \tag{1.101}$$

to form the _inhomogeneous Lorentz group_ or _Poincaré group_.

Inertial frames may be defined as all those frames in which Maxwell's equations are valid. They differ from each other by Poincaré transformations.

$$x'^a = \Lambda^a{}_b x^a - a^a. \tag{1.102}$$

1.5 Differential Operators for Lorentz Transformations

The physical laws in four-dimensional spacetime are formulated in terms of Lorentz-invariant field theories. The fields depend on the spacetime coordinates x^a. In order to perform transformations of the Lorentz group we need differential operators for the generators of this group.

For Lorentz transformations Λ with small rotation angles and rapidities, we can approximate the exponential in (1.57) as

$$\Lambda \equiv 1 - i\frac{1}{2}\omega_{ab}L^{ab}. \tag{1.103}$$

The Lorentz transformation of the coordinates

$$x \xrightarrow{\Lambda} x' = \Lambda x \tag{1.104}$$

is conveniently characterized by the infinitesimal change

$$\delta_\Lambda x = x' - x = -i\frac{1}{2}\omega_{ab}L^{ab}x. \tag{1.105}$$

Inserting the 4×4 -matrix generators (1.51), this becomes more explicitly [compare (1.74)]

$$\delta_\Lambda x^a = \omega^a{}_b x^b. \tag{1.106}$$

We now observe that (1.105) can be expressed in terms of the differential operators

$$\hat{L}^{ab} \equiv i(x^a\partial^b - x^b\partial^a) = -\hat{L}^{ba} \tag{1.107}$$

as a commutator

$$\delta_\Lambda x = i\frac{1}{2}\omega_{ab}[\hat{L}^{ab}, x]. \tag{1.108}$$

The differential operators (1.107) satisfy the same commutation relations (1.71), (1.72) as the 4×4 -generators L^{ab} of the Lorentz group. They form a representation of the Lie algebra (1.71), (1.72). By exponentiation we can thus form the operator representation of finite Lorentz transformations

$$\hat{D}(\Lambda) \equiv e^{-i\frac{1}{2}\omega_{ab}\hat{L}^{ab}}, \tag{1.109}$$

which satisfy the same group multiplication rules as the 4×4-matrices Λ.

The relation between the finite Lorentz transformations (1.104) and the operator version (1.109) is

$$x' = \Lambda x = e^{-i\frac{1}{2}\omega_{ab}L^{ab}}x = e^{i\frac{1}{2}\omega_{ab}\hat{L}^{ab}}x\,e^{-i\frac{1}{2}\omega_{ab}\hat{L}^{ab}} = \hat{D}^{-1}(\Lambda)\,x\,\hat{D}(\Lambda). \tag{1.110}$$

This is proved by expanding, on the left-hand side, $e^{-i\frac{1}{2}\omega_{ab}L^{ab}}x$ in powers of ω_{ab}, and doing the same on the right-hand expression $e^{i\frac{1}{2}\omega_{ab}\hat{L}^{ab}}x\,e^{-i\frac{1}{2}\omega_{ab}\hat{L}^{ab}}$ with the help of *Lie's expansion formula*

$$e^{-i\hat{A}}\,\hat{B}\,e^{i\hat{A}} = 1 - i[\hat{A},\hat{B}] + \frac{i^2}{2!}[\hat{A},[\hat{A},\hat{B}]] + \dots . \tag{1.111}$$

This operator representation (1.109) can be used to generate Lorentz transformations on the spacetime argument of any function of x:

$$f'(x) \equiv f(\Lambda^{-1}x) = f\left(\hat{D}(\Lambda)x\,\hat{D}^{-1}(\Lambda)\right) = \hat{D}(\Lambda)f(x)\,\hat{D}^{-1}(\Lambda). \qquad (1.112)$$

The latter step follows from a power series expansion of $f(x)$. Take for example an expansion term $f_{a,b}x^a x^b$ of $f(x)$. In the transformed function $f'(x)$, this becomes

$$f_{a,b}\hat{D}(\Lambda)x^a\,\hat{D}^{-1}(\Lambda)\hat{D}(\Lambda)x^b\hat{D}^{-1}(\Lambda) = \hat{D}(\Lambda)\left(f_{a,b}x^a x^b\right)\hat{D}^{-1}(\Lambda). \qquad (1.113)$$

1.6   Vector and Tensor Operators

In working out the commutation rules among the differential operators $\hat{L}^{ab}$ one conveniently uses the commutation rules between $\hat{L}^{ab}$ and x^c, $\hat{p}^c$:

$$[\hat{L}^{ab}, x^c] = -i(g^{ac}x^b - g^{bc}x^a) = -(L^{ab})^c{}_d x^d, \qquad (1.114)$$

$$[\hat{L}^{ab}, \hat{p}^c] = -i(g^{ac}\hat{p}^b - g^{bc}\hat{p}^a) = -(L^{ab})^c{}_d \hat{p}^d. \qquad (1.115)$$

These commutation rules identify x^c and $\hat{p}^c$ as *vector operators*

In general, an operator $\hat{t}^{c_1,\dots,c_n}$ is said to be a *tensor operator* of rank n if each of its tensor indices is transformed under commutation with L^{ab} like the index of x^a or $\hat{p}^a$ in (1.114) and (1.115):

$$[\hat{L}^{ab}, \hat{t}^{c_1,\dots,c_n}] = -i[(g^{ac_1}\hat{t}^{b,\dots,c_n} - g^{bc_1}\hat{t}^{a,\dots,c_n}) + \dots + (g^{ac_n}\hat{t}^{c_1,\dots,b} - g^{bc_n}\hat{t}^{c_1,\dots,a})]$$
$$= -(L^{ab})^{c_1}{}_d \hat{t}^{dc_2,\dots,c_n} - (L^{ab})^{c_2}{}_d \hat{t}^{c_1 d,\dots,c_n} - \dots - (L^{ab})^{c_n}{}_d \hat{t}^{c_1 c_2,\dots,d}. \qquad (1.116)$$

The commutators (1.71) between the generators imply that these are themselves tensor operators.

The simplest examples for such tensor operators are the direct products of vectors such as $\hat{t}^{c_1,\dots,c_n} = x^{c_1}\cdots x^{c_n}$ or $\hat{t}^{c_1,\dots,c_n} = \hat{p}^{c_1}\cdots\hat{p}^{c_n}$. In fact, the right-hand side can be found for such direct products using the commutation rules between products of operators

$$[\hat{a}, \hat{b}\hat{c}] = [\hat{a}, \hat{b}]\hat{c} + \hat{b}[\hat{a}, \hat{c}], \quad [\hat{a}\hat{b}, \hat{c}] = \hat{a}[\hat{b}, \hat{c}] + [\hat{a}, \hat{c}]\hat{b}. \qquad (1.117)$$

These are the analogs of the *Leibnitz chain rule* for derivatives

$$\partial(fg) = (\partial f)g + f(\partial g). \qquad (1.118)$$

1.7   Behavior of Vectors and Tensors under Finite Lorentz Transformations

Let us apply such a finite operator representation (1.109) to the vector x^c to form

$$\hat{D}(\Lambda)x^c\hat{D}^{-1}(\Lambda). \qquad (1.119)$$

We shall do this separately for rotations and Lorentz transformations.

1.7.1 Rotations

An arbitrary three-vector (x^1, x^2, x^3) is rotated around the 3-axis by the operator $\hat{D}(R_3(\varphi)) = e^{-i\varphi \hat{L}_3}$ with $\hat{L}_3 = -i(x^1 \partial_2 - x^2 \partial_1)$ by the operation

$$\hat{D}(R_3(\varphi)) x^i \hat{D}^{-1}(R_3(\varphi)) = e^{-i\varphi \hat{L}_3} x^i e^{i\varphi \hat{L}_3}. \tag{1.120}$$

Since $\hat{L}_3$ commutes with x^3, this component is invariant under the operation (1.120):

$$\hat{D}(R_3(\varphi)) x^3 \hat{D}^{-1}(R_3(\varphi)) = e^{-i\varphi \hat{L}_3} x^3 e^{i\varphi \hat{L}_3} = x^3. \tag{1.121}$$

For x^1 and x^2, the Lie expansion of (1.119) contains the commutators

$$-i[L_3, x^1] = x^2, \qquad -i[L_3, x^2] = -x^1. \tag{1.122}$$

Thus, the first-order expansion term on the right-hand side of (1.120) transforms the two-dimensional vector (x^1, x^2) into $(x^2, -x^1)$. The second-order term is obtained by commuting the operator $-i\hat{L}_3$ with $(x^2, -x^1)$, yielding $-(x^1, x^2)$. To third-order, this is again transformed into $-(x^2, -x^1)$, and so on. Obviously, all even orders reproduce the initial two-dimensional vector (x^1, x^2) with an alternating sign, while all odd powers are proportional to $(x^2, -x^1)$. Thus we obtain the expansion

$$
\begin{aligned}
e^{-i\varphi \hat{L}_3}(x^1, x^2) e^{i\varphi \hat{L}_3} &= \left(1 - \frac{1}{2!}\varphi^2 + \frac{1}{4!}\varphi^4 + \dots\right)(x^1, x^2) \\
&+ \left(\varphi - \frac{1}{3!}\varphi^3 + \frac{1}{5!}\varphi^5 + \dots\right)(x^2, -x^1).
\end{aligned} \tag{1.123}
$$

The even and odd powers can be summed up to a cosine and a sine, respectively, resulting in

$$e^{-i\varphi \hat{L}_3}(x^1, x^2) e^{i\varphi \hat{L}_3} = \cos\varphi \, (x^1, x^2) + \sin\varphi \, (x^2, -x^1). \tag{1.124}$$

Together with the invariant x^3 in (1.121), the right-hand side forms a vector arising from x^i by an *inverse* rotation (1.37). Thus

$$\hat{D}(R_3(\varphi)) x^i \hat{D}^{-1}(R_3(\varphi)) = e^{-i\varphi \hat{L}_3} x^i e^{i\varphi \hat{L}_3} = \left(e^{i\varphi L_3}\right)^i{}_j x^j = R_3^{-1}(\varphi)^i{}_j x^j. \tag{1.125}$$

By performing successive rotations around the three axes we can generate in this way any inverse rotation:

$$\hat{D}(R(\boldsymbol{\varphi})) x^i \hat{D}^{-1}(R(\boldsymbol{\varphi})) = e^{-i\boldsymbol{\varphi}\cdot\hat{\mathbf{L}}} x^i e^{i\boldsymbol{\varphi}\cdot\hat{\mathbf{L}}} = \left(e^{i\boldsymbol{\varphi}\cdot\hat{\mathbf{L}}}\right)^i{}_j x^j = R^{-1}(\boldsymbol{\varphi})^i{}_j x^j. \tag{1.126}$$

This is the finite transformation law associated with the commutation relation

$$[\hat{L}_i, x_k] = x_j (L_i)_{jk}, \tag{1.127}$$

which characterizes the vector operator nature of x^i [compare (1.114)]. Thus also (1.126) holds for finite rotations of any vector operator $\hat{v}^i$.

The time component x^0 is obviously unchanged by rotations since $\hat{L}_3$ commutes with x^0. Hence we can extend (1.126) trivially to a four-vector, replacing $\hat{D}(R(\boldsymbol{\varphi}))$ by $\hat{D}(\Lambda(R(\boldsymbol{\varphi})))$ [recall (1.44)].

1.7.2   Lorentz Boosts

A similar calculation may be done for Lorentz boosts. Here we first consider a boost in the 3-direction $B_3(\zeta) = e^{-i\zeta \hat{M}_3}$ generated by $\hat{M}_3 = \hat{L}^{03} = -i(x^0 \partial_3 + x^3 \partial_0)$ [recall (1.57), (1.53), and (1.107)]. Note the positive relative sign of the two terms in the generator $\hat{L}^{03}$ is caused by the fact that $\partial_i = -\partial^i$, in contrast to $\partial_0 = \partial^0$. Thus we form

$$\hat{D}(B_3(\zeta)) x^i \hat{D}^{-1}(B_3(\zeta)) = e^{-i\zeta \hat{M}_3} x^i e^{i\zeta \hat{M}_3}. \tag{1.128}$$

The Lie expansion of the right-hand side involves the commutators

$$- i[M_3, x^0] = -x^3, \quad -i[M_3, x^3] = -x^0, \quad -i[M_3, x^1] = 0, \quad -i[M_3, x^2] = 0. \tag{1.129}$$

Here the two-vector (x^1, x^2) is unchanged, while the two-vector (x^0, x^3) is transformed into $-(x^3, x^0)$. In the second expansion term, the latter becomes (x^0, x^3), and so on, yielding

$$e^{-i\zeta \hat{M}_3}(x^0, x^3) e^{i\zeta \hat{M}_3} = \left(1 + \frac{1}{2!}\zeta^2 + \frac{1}{4!}\zeta^4 + \ldots\right)(x^0, x^3)$$
$$- \left(\zeta + \frac{1}{3!}\zeta^3 + \frac{1}{5!}\zeta^5 + \ldots\right)(x^3, x^0). \tag{1.130}$$

The right-hand sides can be summed up to hyperbolic cosines and sines:

$$e^{-i\zeta \hat{M}_3}(x^0, x^3) e^{i\zeta \hat{M}_3} = \cosh \zeta \; (x^0, x^3) - \sinh \zeta \; (x^3, x^0). \tag{1.131}$$

Together with the invariance of (x^1, x^2), this corresponds precisely to the inverse of the boost transformation (1.35):

$$\hat{D}(B_3(\zeta)) x^a \hat{D}^{-1}(B_3(\zeta)) = e^{-i\zeta \hat{M}_3} x^a e^{i\zeta \hat{M}_3} = \left(e^{i\zeta M_3}\right)^a{}_b x^b = B_3^{-1}(\zeta)^a{}_b x^b. \tag{1.132}$$

1.7.3   Lorentz Group

By performing successive rotations and boosts in all directions we find all Lorentz transformations

$$\hat{D}(\Lambda) x^c \hat{D}^{-1}(\Lambda) = e^{-i\frac{1}{2}\omega_{ab}\hat{L}^{ab}} x^c e^{i\frac{1}{2}\omega_{ab}\hat{L}^{ab}} = (e^{i\frac{1}{2}\omega_{ab}L^{ab}})^c{}_{c'} x^{c'} = (\Lambda^{-1})^c{}_{c'} x^{c'}, \tag{1.133}$$

where ω_{ab} are the parameters (1.55) and (1.56). In the last term on the right-hand side we have expressed the 4×4 -matrix Λ as an exponential of its generators, to emphasize the one-to-one correspondence between the generators L^{ab} and their differential-operator representation $\hat{L}^{ab}$.

At first it may seem surprising that the group transformations appearing as a left-hand factor of the two sides of these equations are *inverse* to each other. However, we may easily convince ourselves that this is necessary to guarantee the correct group multiplication law. Indeed, if we perform two successive transformations they appear in opposite order on the right- and left-hand sides:

$$\hat{D}(\Lambda_2 \Lambda_1) x^c \hat{D}^{-1}(\Lambda_2 \Lambda_1) = \hat{D}(\Lambda_2) \hat{D}(\Lambda_1) x^c \hat{D}^{-1}(\Lambda_1) \hat{D}^{-1}(\Lambda_2)$$
$$= (\Lambda_1^{-1})^c{}_{c'} \hat{D}(\Lambda_2) x^{c'} \hat{D}^{-1}(\Lambda_2) = (\Lambda_1^{-1})^c{}_{c'} (\Lambda_2^{-1})^{c'}{}_{c''} x^{c''} = [(\Lambda_2 \Lambda_1)^{-1}]^c{}_{c'} x^{c'}. \tag{1.134}$$

If the right-hand side of (1.133) would contain Λ instead of Λ^{-1}, the order of the factors in $\Lambda_2\Lambda_1$ on the right-hand side of (1.134) would be opposite to the order in $\hat{D}(\Lambda_2\Lambda_1)$ on the left-hand side.

A straightforward extension of the operation (1.133) yields the transformation law for a tensor $\hat{t}^{c_1,\dots,c_n} = x^{c_1}\cdots x^{c_n}$:

$$
\begin{aligned}
\hat{D}(\Lambda)\hat{t}^{c_1,\dots,c_n}\hat{D}^{-1}(\Lambda) &= e^{-i\frac{1}{2}\omega_{ab}\hat{L}^{ab}}\,\hat{t}^{c_1,\dots,c_n}\,e^{i\frac{1}{2}\omega_{ab}\hat{L}^{ab}} \\
&= (\Lambda^{-1})^{c_1}{}_{c_1'}\cdots(\Lambda^{-1})^{c_n}{}_{c_n'}\,\hat{t}^{c_1',\dots,c_n'} \\
&= (e^{i\frac{1}{2}\omega_{ab}L^{ab}})^{c_1}{}_{c_1'}\cdots(e^{i\frac{1}{2}\omega_{ab}L^{ab}})^{c_n}{}_{c_n'}\,\hat{t}^{c_1',\dots,c_n'}.
\end{aligned}
\tag{1.135}
$$

This follows directly by inserting an auxiliary unit factor $1 = \hat{D}(\Lambda)\hat{D}^{-1}(\Lambda) = e^{-i\frac{1}{2}\omega_{ab}\hat{L}^{ab}}e^{i\frac{1}{2}\omega_{ab}\hat{L}^{ab}}$ into the product $x^{c_1}\cdots x^{c_n}$ between neighboring factors x^{c_i}, and performing the operation (1.135) on each of them. The last term in (1.135) can also be written as

$$
\left[e^{i\frac{1}{2}\omega_{ab}(L^{ab}\times 1\times 1\cdots\times 1\,+\,\cdots\,+\,1\times L^{ab}\times 1\cdots\times 1)}\right]^{c_1\dots c_n}{}_{c_1'\dots c_n'}\,t^{c_1'\dots c_n'}.
\tag{1.136}
$$

Since the commutation relations (1.116) determine the result completely, the transformation formula (1.135) is true for any tensor operator $\hat{t}^{c_1,\dots,c_n}$, not only for those composed from a product of vectors x^{c_i}.

The result can easily be extended to an exponential function e^{-ipx}, and further to any function $f(x)$ which possesses a Fourier representation

$$
\hat{D}(\Lambda)f(x)\hat{D}^{-1}(\Lambda) = f(\Lambda^{-1}x) = e^{-i\frac{1}{2}\omega_{ab}\hat{L}^{ab}}\,f(x)\,e^{i\frac{1}{2}\omega_{ab}\hat{L}^{ab}}.
\tag{1.137}
$$

Since the last differential operator has nothing to act on, it can also be omitted and we can also write

$$
\hat{D}(\Lambda)f(x)\hat{D}^{-1}(\Lambda) = f(\Lambda^{-1}x) = e^{-i\frac{1}{2}\omega_{ab}\hat{L}^{ab}}\,f(x).
\tag{1.138}
$$

1.8 Relativistic Point Mechanics

The Lorentz invariance of the Maxwell equations explains the observed invariance of the light velocity in different inertial frames. It is, however, incompatible with Newton's mechanics. There exists a modification of Newton's laws which makes them Lorentz-invariant as well, while differing very little from Newton's equations in their description of slow macroscopic bodies, for which Newton's equations were originally designed. Let us introduce the Poincaré-invariant distance measure in spacetime

$$
ds \equiv \sqrt{dx^2} = \left(g_{ab}dx^a dx^b\right)^{1/2}.
\tag{1.139}
$$

At a fixed coordinate point of an inertial frame, ds is equal to c times the elapsed time:

$$
ds = \sqrt{g_{00}dx^0 dx^0} = dx^0 = cdt.
\tag{1.140}
$$

Einstein called the quantity

$$\tau \equiv s/c \tag{1.141}$$

the *proper time*.

When going from one inertial frame to another, two simultaneous events at different points in the first frame will take place at *different* times in the other frame. Their invariant distance, however, remains the same, due to the pseudo-orthogonality relation (1.28) which ensures that

$$ds' = \left(g_{ab}dx'^a dx'^b\right)^{1/2} = \left(g_{ab}dx^a dx^b\right)^{1/2} = ds. \tag{1.142}$$

A particle moving with a constant velocity along a trajectory $\mathbf{x}(t)$ in one Minkowski frame remains at rest in another frame moving with velocity $\mathbf{v} = \dot{\mathbf{x}}(t)$ relative to the first. Its proper time is then related to the coordinate time in the first frame by the Lorentz transformation

$$cd\tau = ds = \sqrt{c^2 dt^2 - d\mathbf{x}^2} = cdt\sqrt{1 - \frac{1}{c^2}\left(\frac{d\mathbf{x}}{dt}\right)^2} = cdt\sqrt{1 - \frac{\mathbf{v}^2}{c^2}} = \frac{cdt}{\gamma}. \tag{1.143}$$

This is the famous Einstein relation implying that a moving particle lives longer by a factor γ. There exists direct experimental evidence for this phenomenon. For example, the meson π^+ lives on the average $\tau_a = 2.60 \times 10^{-8}$ sec, after which it decays into a muon and a neutrino. If the pion is observed in a bubble chamber with a velocity equal to 10% of the light velocity $c \equiv 299\,792\,458$ m/sec, it leaves trace of an average length $l \approx \tau_a \times c \times 0.1/\sqrt{1 - 0.1^2} \approx 0.78$ cm. A very fast muon moving with 90% of the light velocity, however, leaves a trace which is longer by a factor $(0.9/0.1) \times \sqrt{1 - 0.10^2}/\sqrt{1 - 0.9^2} \equiv 20.6$. Massless particles move with light velocity and have $d\tau = 0$, i.e., the proper time stands still along their paths. This implies that massless particles can never decay — they are necessarily stable particles.

Another way to see the time dilation is by observing the spectral lines of a moving atom, say a hydrogen atom. If the atom is at rest, the frequency of the line is given by

$$\nu = -\mathrm{Ry}\left(\frac{1}{n^2} - \frac{1}{m^2}\right) \tag{1.144}$$

where $\mathrm{Ry} = m_e c^2 \alpha^2/2 \approx 13.6$ eV, is the *Rydberg constant*,

$$\alpha \equiv \frac{e^2}{4\pi\hbar c} \approx 1/137.035\,989 \tag{1.145}$$

is the *fine-structure constant*, and n and m are the principal quantum numbers of initial and final electron orbits. If the atom emits a light quantum while moving with velocity **v** through the laboratory orthogonal to the direction of observation, this frequency is *lowered* by a factor $1/\gamma$:

$$\frac{\nu_{\text{obs}}}{\nu} = \frac{1}{\gamma} = \sqrt{1 - \frac{v^2}{c^2}}. \tag{1.146}$$

If the atom runs away from the observer or towards him, the frequency is further changed by the *Doppler shift*. Due to the growing or decreasing distance, the wave trains arrive with a smaller or higher frequency given by

$$\frac{\nu_{\text{obs}}}{\nu} = \left(1 \pm \frac{v}{c}\right)^{-1} \frac{1}{\gamma} = \sqrt{\frac{1 \mp v/c}{1 \pm v/c}}. \tag{1.147}$$

In the first case the observer sees an additional *red shift*, in the second a *violet shift* of the spectral lines.

Without external forces, the trajectories of free particles are straight lines in four-dimensional spacetime. If the particle positions are parametrized by the proper time τ, they satisfy the equation of motion

$$\frac{d^2}{d\tau^2} x^a(\tau) = \frac{d}{d\tau} p^a(\tau) = 0. \tag{1.148}$$

The first derivative of $x^a(\tau)$ is the *relativistic four-vector of momentum* $p^a(\tau)$, briefly called *four-momentum*:

$$p^a(\tau) \equiv m \frac{d}{d\tau} x^a(\tau) \equiv m u^a(\tau). \tag{1.149}$$

On the right-hand side we have introduced the *relativistic four-vector of velocity* $u^a(\tau)$, or *four-velocity*. Inserting (1.143) into (1.149) we identify the components of $u^a(\tau)$ as

$$u^a = \begin{pmatrix} \gamma c \\ \gamma v^a \end{pmatrix}, \tag{1.150}$$

and see that $u^a(\tau)$ is normalized to the light-velocity:

$$u^a(\tau) u_a(\tau) = c^2. \tag{1.151}$$

The time and space components of (1.149) are

$$p^0 = m\gamma c = mu^0, \qquad p^i = m\gamma v^i = mu^i. \tag{1.152}$$

This shows that the time dilation factor γ is equal to p^0/mc, and the same factor increases the spatial momentum with respect to the nonrelativistic momentum mv^i. This correction becomes important for particles moving near the velocity of light, which are called *relativistic*. The light particle has $m = 0$ and $v = c$. It is ultra-relativistic.

Note that by Eq. (1.152), the hyperbolic functions of the rapidity in Eq. (1.31) are related to the four velocity and to energy and momentum by

$$\cosh \zeta = u^0/c = p^0/mc, \quad \sinh \zeta = |\mathbf{u}|/c = |\mathbf{p}|/mc. \tag{1.153}$$

Under a Lorentz transformation of space and time, the four-momenta p^a transform in exactly the same way as the coordinate four-vectors x^a. This is, of course, due to the Lorentz invariance of the proper time τ in Eq. (1.149). Indeed, from Eq. (1.152) we derive the important relation

$$p^{0^2} - \mathbf{p}^2 = m^2 c^2, \tag{1.154}$$

which shows that the square of the four-momentum taken with the Minkowski metric is an invariant:

$$p^2 \equiv p^a g_{ab} p^b = m^2 c^2. \tag{1.155}$$

Since both x^a and p^a are Lorentz vectors, the scalar product of them,

$$xp \equiv g_{ab} x^a p^b, \tag{1.156}$$

is an invariant. In the canonical formalism, the momentum p^i is the conjugate variable to the space coordinate x^i. Equation (1.156) suggests that the quantity cp^0 is conjugate to $x^0/c = t$. As such it must be the energy of the particle:

$$E = cp^0. \tag{1.157}$$

From relation (1.154), we calculate the energy as a function of the momentum of a relativistic particle:

$$E = c\sqrt{\mathbf{p}^2 + m^2 c^2}. \tag{1.158}$$

For small velocities, this can be expanded as

$$E = mc^2 + \frac{m}{2}\mathbf{v}^2 + \dots . \tag{1.159}$$

The first term gives a nonvanishing *rest energy* which is unobservable in nonrelativistic physics. The second term is Newton's kinetic energy.

The first term has dramatic observable effects. Particles can be produced and disappear in collision processes. In the latter case, their rest energy mc^2 can be transformed into kinetic energy of other particles. The large factor c makes unstable particles a source of immense energy, with disastrous consequences for Hiroshima and Nagasaki in 1945.

1.9 Quantum Mechanics

In quantum mechanics, free spinless particles of momentum p are described by plane waves of the form

$$\phi_p(x) = \mathcal{N}\, e^{-ipx/\hbar}, \tag{1.160}$$

where $\mathcal{N}$ is some normalization factor. The momentum components are the eigen-value of the differential operators

$$\hat{p}_a = i\hbar\frac{\partial}{\partial x^a}, \tag{1.161}$$

which satisfy with x^b the commutation rules

$$[\hat{p}_a, x^b] = i\hbar\delta_a{}^b. \tag{1.162}$$

In terms of these, the generators (1.107) can be rewritten as

$$\hat{L}^{ab} \equiv \frac{1}{\hbar}(x^a\hat{p}^b - x^b\hat{p}^a). \tag{1.163}$$

Apart from the factor $1/\hbar$, this is the tensor version of the four-dimensional angular momentum.

It is worth observing that the differential operators (1.163) can also be expressed as a sandwich of the 4×4 -matrix generators (1.51) between x^c and $\hat{p}^d$:

$$\hat{L}^{ab} = -\frac{i}{\hbar}(L^{ab})_{cd}x^c\hat{p}^d = -\frac{i}{\hbar}x^T L^{ab}\hat{p} = i\hat{p}^T L^{ab}x. \tag{1.164}$$

This way of forming operator representations of the 4×4 -Lie algebra (1.71) is a special application of a general construction technique of higher representations of a defining matrix representations. In fact, the procedure of second quantization is based on this construction, which extends the single-particle Schrödinger operators to the Fock space of many-particle states.

In general, one may always introduce vectors of creation and annihilation operators $\hat{a}^\dagger_c$ and $\hat{a}^d$ with the commutation rules

$$[\hat{a}^c, \hat{a}^d] = [\hat{a}^\dagger_c, \hat{a}^\dagger_d] = 0; \quad [\hat{a}^c, \hat{a}^\dagger_d] = \delta^c{}_d, \tag{1.165}$$

and form sandwich operators

$$\hat{L}^{ab} = \hat{a}^\dagger_c(L^{ab})^c{}_d\hat{a}^d. \tag{1.166}$$

These satisfy the same commutation rules as the sandwiched matrices due to the Leibnitz chain rule (1.117). Since $-i\hat{p}_a/\hbar$ and x^a commute in the same way as $\hat{a}$ and $\hat{a}^\dagger$, the commutation rules of the matrices go directly over to the sandwich operators (1.164). The higher representations generated by them lie in the Hilbert space of square-integrable functions.

Under a Lorentz transformation, the momentum of the particle described by the wave function (1.160) goes over into $p' = \Lambda p$, so that the wave function transforms as follows:

$$\phi_p(x) \xrightarrow{\Lambda} \phi'_p(x) \equiv \phi_{p'}(x) = \mathcal{N}e^{-i(\Lambda p)x} = \mathcal{N}e^{-ip\Lambda^{-1}x} = \phi_p(\Lambda^{-1}x). \tag{1.167}$$

This can also be written as $\phi'_{p'}(x') = \phi_p(x)$. An arbitrary superposition of such waves transforms like

$$\phi(x) \xrightarrow{\Lambda} \phi'(x) = \phi(\Lambda^{-1}x), \tag{1.168}$$

which is the defining relation for a *scalar field*.

The transformation (1.168) may be generated by the differential-operator representation of the Lorentz group (1.138) as follows:

$$\phi(x) \xrightarrow{\Lambda} \phi'(x) = \hat{D}(\Lambda)\phi(x). \tag{1.169}$$

1.10   Relativistic Particles in Electromagnetic Field

Lorentz and Einstein formulated a theory of relativistic massive particles with electromagnetic interactions referred to as *Maxwell-Lorentz theory*. It is invariant under the Poincaré group and describes the dynamical properties of charged particles such as electrons moving with nonrelativistic and relativistic speeds.

The motion for a particle of charge e and mass m in an electromagnetic field is governed by the *Lorentz equations*

$$\frac{dp^a(\tau)}{d\tau} = m\frac{d^2x^a(\tau)}{d\tau^2} = f^a(\tau), \tag{1.170}$$

where f^a is the four-vector associated with the *Lorentz force*

$$f^a = \frac{e}{c}F^a{}_b\frac{dx^b}{d\tau} = \frac{e}{mc}F^a{}_b(x(\tau))\,p^b(\tau), \tag{1.171}$$

and $F^a{}_b(x)$ is a 4×4-combination of electric and magnetic fields with the components

$$F^i{}_j = \epsilon^{ijk}B^k, \quad F^0{}_i = E^i. \tag{1.172}$$

By raising the second index of $F^a{}_b$ one obtains the tensor

$$F^{ac} = g^{cb}F^a{}_b \tag{1.173}$$

associated with the antisymmetric matrix of the six electromagnetic fields

$$F^{ab} = \begin{pmatrix} 0 & -E^1 & -E^2 & -E^3 \\ \hline E^1 & 0 & B^3 & -B^2 \\ E^2 & -B^3 & 0 & B^1 \\ E^3 & B^2 & -B^1 & 0 \end{pmatrix}. \tag{1.174}$$

This tensor notation is useful since F^{ab} transforms under the Lorentz group in the same way as the direct product $x^a x^b$, which goes over into $x'^a x'^b = \Lambda^a{}_c\Lambda^b{}_d\,x^c x^d$. In

$F^{ab}(x)$, also the arguments must be transformed as in the scalar field in Eq. (1.168), so that we find the generic transformation behavior of a *tensor field*:

$$F^{ab}(x) \xrightarrow{\ \Lambda\ } F'^{ab}(x) = \Lambda^a{}_c \Lambda^b{}_d \, F^{cd}(\Lambda^{-1}x). \tag{1.175}$$

Recalling the exponential representation (1.136) of the direct product of the Lorentz transformations and the differential operator generation (1.138) of the transformation of the argument x, this can also be written as

$$F^{ab}(x) \xrightarrow{\ \Lambda\ } F'^{ab}(x) = [e^{-i\frac{1}{2}\omega_{ab}\hat{J}^{ab}} F]^{ab}(\Lambda^{-1}x), \tag{1.176}$$

where

$$\hat{J}^{cd} \equiv L^{cd} \times 1 + 1 \times \hat{L}^{cd} \tag{1.177}$$

are the generators of the total four-dimensional angular momentum of the tensor field. The factors in the direct products apply successively to the representation spaces associated with the two Lorentz indices and the spacetime coordinates. The generators $\hat{J}^{ab}$ obey the same commutation rules (1.71) and (1.72) as L_{ab} and $\hat{L}_{ab}$.

In order to verify the transformation law (1.175), we recall the basic result of electromagnetism that, under a change to a coordinate frame $x \to x' = \Lambda x$ moving with a velocity $\mathbf{v}$, the electric and magnetic fields change as follows

$$\mathbf{E}'_\parallel(x') \ = \ \mathbf{E}_\parallel(x), \quad \mathbf{E}'_\perp(x') \ = \ \gamma\left[\mathbf{E}_\perp(x) + \frac{1}{c}\mathbf{v} \times \mathbf{B}(x)\right], \tag{1.178}$$

$$\mathbf{B}'_\parallel(x') \ = \ \mathbf{B}_\parallel(x), \quad \mathbf{B}'_\perp(x') \ = \ \gamma\left[\mathbf{B}_\perp(x) - \frac{1}{c}\mathbf{v} \times \mathbf{E}(x)\right], \tag{1.179}$$

where the subscripts $\parallel$ and $\perp$ denote the components parallel and orthogonal to $\mathbf{v}$. Recalling the matrices (1.27) we see that (1.178) and (1.179) correspond precisely to the transformation law (1.175) of a tensor field.

The field tensor in the electromagnetic force of the equation of motion (1.170) transforms accordingly:

$$F^a{}_b(x(\tau)) \xrightarrow{\ \Lambda\ } F'^a{}_b(x(\tau)) = \Lambda^a{}_c \Lambda^T{}_b{}^d F'^c{}_d(\Lambda^{-1}x(\tau)). \tag{1.180}$$

This can be verified by rewriting $F^a{}_b(x(\tau))$ as

$$F^a{}_b(x(\tau)) = \int d^4x \, F^a{}_b(x) \, \delta^{(4)}(x - x(\tau)), \tag{1.181}$$

and applying the transformation (1.175).

Separating time and space components of the four-vector of the Lorentz force (1.171) we find

$$\frac{d}{d\tau}p^0 \ = \ f^0 = \frac{e}{Mc}\mathbf{E} \cdot \mathbf{p}, \tag{1.182}$$

$$\frac{d}{d\tau}\mathbf{p} \ = \ \mathbf{f} \ = \ \frac{e}{Mc}\left(\mathbf{E}\,p^0 + \mathbf{p} \times \mathbf{B}\right). \tag{1.183}$$

The Lorentz force can also be stated in terms of velocity as

$$f^a = \frac{e}{c} F^a{}_b \frac{dx^b}{d\tau} = \gamma \begin{pmatrix} \dfrac{e}{c} \mathbf{v} \cdot \mathbf{E} \\ e\mathbf{E}^i + \dfrac{1}{c}(\mathbf{v} \times \mathbf{B})^i \end{pmatrix}. \tag{1.184}$$

It should be noted that if we do not use the proper time τ to describe the particle orbits but the coordinate time $dt = \gamma d\tau$, the equation of motion reads

$$\frac{dp^a}{dt} = \frac{1}{\gamma} f^a, \tag{1.185}$$

so that the acceleration is governed by the three-vector of the Lorentz force

$$\mathbf{f}^{\mathrm{em}} = e \left[\mathbf{E}(x) + \frac{\mathbf{v}}{c} \mathbf{B}(x) \right]. \tag{1.186}$$

The above equations rule the movement of charged point particles in a given external field. The moving particles will, however, also give rise to additional electromagnetic fields. These are calculated by solving the *Maxwell equations* in the presence of charge and current densities ρ and $\mathbf{j}$, respectively:

$$\nabla \cdot \mathbf{E} = \rho \qquad \text{(Coulomb's law)}, \tag{1.187}$$

$$\nabla \times \mathbf{B} - \frac{1}{c}\frac{\partial \mathbf{E}}{\partial t} = \frac{1}{c}\mathbf{j} \qquad \text{(Ampère's law)}, \tag{1.188}$$

$$\nabla \cdot \mathbf{B} = 0 \qquad \text{(absence of magnetic monopoles)}, \tag{1.189}$$

$$\nabla \times \mathbf{E} + \frac{1}{c}\frac{\partial \mathbf{B}}{\partial t} = 0 \qquad \text{(Faraday's law)}. \tag{1.190}$$

In a dielectric and paramagnetic medium with dielectric constant ϵ and magnetic permeability μ one defines the displacement field $\mathbf{D}(x)$ and the magnetic field $\mathbf{H}(x)$ by the relations

$$\mathbf{D}(x) = \epsilon \mathbf{E}(x), \quad \mathbf{B}(x) = \epsilon \mathbf{B}(x), \tag{1.191}$$

and the Maxwell equations become

$$\nabla \cdot \mathbf{D} = \rho \qquad \text{(Coulomb's law)}, \tag{1.192}$$

$$\nabla \times \mathbf{H} - \frac{1}{c}\frac{\partial \mathbf{D}}{\partial t} = \frac{1}{c}\mathbf{j} \qquad \text{(Ampère's law)}, \tag{1.193}$$

$$\nabla \cdot \mathbf{B} = 0 \qquad \text{(absence of magnetic monopoles)}, \tag{1.194}$$

$$\nabla \times \mathbf{E} + \frac{1}{c}\frac{\partial \mathbf{B}}{\partial t} = 0 \qquad \text{(Faraday's law)}. \tag{1.195}$$

On the right-hand sides of (1.187), (1.188) and (1.192), (1.193) we have omitted factors 4π, for convenience. This makes the charge of the electron equal to $-e = -\sqrt{4\pi\alpha\hbar c}$.

In the vacuum, the two inhomogeneous Maxwell equations (1.187) and (1.188) can be combined to a single equation

$$\partial_b F^{ab} = -\frac{1}{c} j^a,$$
(1.196)

where j^a is the *four-vector of current density*

$$j^a(x) = \begin{pmatrix} c\rho(\mathbf{x}, t) \\ \mathbf{j}(\mathbf{x}, t) \end{pmatrix}.$$
(1.197)

Indeed, the zeroth component of (1.196) is equal to (1.187):

$$\partial_i F^{0i} = -\boldsymbol{\nabla} \cdot \mathbf{E} = -\rho,$$
(1.198)

whereas the spatial components with $a = i$ reduce to Eq. (1.188):

$$\partial_0 F^{i0} + \partial_j F^{ij} = \partial_j \epsilon^{ijk} B^k + \frac{1}{c}\frac{\partial}{\partial t} E^i = -(\boldsymbol{\nabla} \times \mathbf{B})^i + \frac{1}{c}\frac{\partial}{\partial t} E^i = -\frac{1}{c} j^i.$$
(1.199)

The remaining homogeneous Maxwell equations (1.189) and (1.190) can also be rephrased in tensor form as

$$\partial_b \tilde{F}^{ab} = 0.$$
(1.200)

Here $\tilde{F}^{ab}$ is the so-called *dual field tensor* defined by

$$\tilde{F}^{ab} = \frac{1}{2} \epsilon^{abcd} F_{cd},$$
(1.201)

where ϵ^{abcd} is the totally antisymmetric unit tensor with $\epsilon^{0123} = 1$. Its properties are summarized in Appendix A.

The antisymmetry of F^{ab} in (1.196) implies the vanishing of the four-divergence of the current density:

$$\partial_a j^a(x) = 0.$$
(1.202)

This is the four-dimensional way of expressing the *local conservation law* of charges. Written out in space and time components it reads

$$\partial_t \rho(\mathbf{x}, t) + \boldsymbol{\nabla} \cdot \mathbf{j}(\mathbf{x}, t) = 0.$$
(1.203)

Integrating this over a finite volume gives

$$\partial_t \left[\int d^3x \, \rho(\mathbf{x}, t) \right] = -\int d^3x \, \boldsymbol{\nabla} \cdot \mathbf{j}(\mathbf{x}, t) = 0.$$
(1.204)

The right-hand side vanishes by the Gauss divergence theorem, according to which the volume integral over the divergence of a current density is equal to the surface integral over the flux through the boundary of the volume. This vanishes if currents do not leave a finite spatial volume, which is usually true for an infinite system. Thus we find that, as a consequence of local conservation law (1.202), the charge of the system

$$Q(t) \equiv \int d^3 \, \rho(\mathbf{x}, t) \equiv \frac{1}{c} \int d^3 x \, j^0(x) \tag{1.205}$$

satisfies the *global conservation law* according to which charge is time-independent

$$Q(t) \equiv Q. \tag{1.206}$$

For a set of point particles of charges e_n, the charge and current densities are

$$\rho(\mathbf{x}, t) = \sum_n e_n \delta^{(3)} \left(\mathbf{x} - \mathbf{x}_n(t) \right), \tag{1.207}$$

$$\mathbf{j}(\mathbf{x}, t) = \sum_n e_n \dot{\mathbf{x}}_n(t) \delta^{(3)} \left(\mathbf{x} - \mathbf{x}_n(t) \right). \tag{1.208}$$

Combining these expressions to a four-component current density (1.197), we can easily verify that $j^a(x)$ transforms like a *vector field* [compare with the behaviors (1.168) of scalar field and (1.175) of tensor fields]:

$$j^a(x) \xrightarrow{\Lambda} j'^a(x) = \Lambda^a{}_b \, j^b(\Lambda^{-1} x). \tag{1.209}$$

To verify this we note that $\delta^{(3)} \left(\mathbf{x} - \mathbf{x}(t) \right)$ can also be written as an integral along the path of the particle parametrized with the help of the proper time τ. This is done with the help of the identity

$$\int_{-\infty}^{\infty} d\tau \, \delta^{(4)}(x - x(\tau)) = \int_{-\infty}^{\infty} d\tau \, \delta(x^0 - x^0(\tau)) \delta^{(3)} \left(\mathbf{x} - \mathbf{x}(\tau) \right)$$

$$= \frac{d\tau}{dx^0} \delta^{(3)} \left(\mathbf{x} - \mathbf{x}(t) \right) = \frac{1}{c\gamma} \delta^{(3)} \left(\mathbf{x} - \mathbf{x}(t) \right). \tag{1.210}$$

This allows us to rewrite (1.207) and (1.208) as

$$c\rho(\mathbf{x}, t) = c \sum_n \int_{-\infty}^{\infty} d\tau_n e_n \gamma_n c \, \delta^{(4)} \left(x - x_n(\tau) \right), \tag{1.211}$$

$$\mathbf{j}(\mathbf{x}, t) = c \sum_n \int_{-\infty}^{\infty} d\tau_n e_n \gamma_n \mathbf{v}_n \delta^{(4)} \left(x - x_n(\tau) \right). \tag{1.212}$$

These equations can be combined in a single four-vector equation

$$j^a(x) = c \sum_n \int_{-\infty}^{\infty} d\tau_n e_n \dot{x}_n^a(\tau) \delta^{(4)} \left(x - x_n(\tau) \right), \tag{1.213}$$

which makes the transformation behavior (1.209) an obvious consequence of the vector nature of $\dot{x}_n^a(\tau)$.

In terms of the four-dimensional current density, the inhomogeneous Maxwell equation (1.196) becomes the *Maxwell-Lorentz equation*

$$\partial_b F^{ab} = -\frac{1}{c} j^a = -\sum_n \int_{-\infty}^{\infty} d\tau_n e_n \dot{x}_n^a(\tau) \delta^{(4)} \left(x - x_n(\tau) \right). \tag{1.214}$$

It is instructive to verify the conservation law (1.202) for the current density (1.213). Applying the derivative ∂_a to the δ-function gives $\partial_a \delta^{(4)}(x - x_n(\tau)) = -\partial_{x_n^a} \delta^{(4)}(x - x_n(\tau))$, and therefore

$$\partial_a j^a(x) = -c \sum_n \int_{-\infty}^{\infty} d\tau_n e_n \frac{dx_n^a(\tau)}{d\tau} \frac{\partial}{\partial x_n^a} \delta^{(4)}(x - x_n(\tau))$$

$$= -c \sum_n \int_{-\infty}^{\infty} d\tau_n e_n \partial_\tau \delta^{(4)}(x - x_n(\tau)). \tag{1.215}$$

If the particle orbits $x(\tau)$ are stable, they are either closed in spacetime, or they come from negative infinite x_0 and run to positive infinite x_0. Then the right-hand side vanishes in any finite volume so that the current density is indeed conserved.

We end this section by remarking that the vector transformation law (1.209) can also be written by analogy with the tensor law (1.175) as

$$j^a(x) \xrightarrow{\Lambda} j'^a(x) = [e^{-i\frac{1}{2}\omega_{ab}\hat{J}^{ab}} j]^a(\Lambda^{-1}x), \tag{1.216}$$

where

$$\hat{J}^{cd} \equiv L^{cd} \times \hat{1} + 1 \times \hat{L}^{cd} \tag{1.217}$$

are the generators of the total four-dimensional angular momentum of the vector field. As in (1.177), the factors in the direct products apply separately to the representation spaces associated with the Lorentz index and the spacetime coordinates, and the generators $\hat{J}^{ab}$ obey the same commutation rules (1.71) and (1.72) as L_{ab} and $\hat{L}_{ab}$.

1.11 Dirac Particles and Fields

The observable matter of the universe consists mainly of electrons and nucleons, the latter being predominantly bound states of three quarks. Electrons and quarks are spin-1/2 particles which may be described by four-component Dirac fields $\psi(x)$. These obey the Dirac equation

$$(i\gamma^a \partial_a - m)\psi(x) = 0, \tag{1.218}$$

where γ^a are the 4×4 -*Dirac matrices*

$$\gamma^a = \begin{pmatrix} 0 & \sigma^a \\ \tilde{\sigma}^a & 0 \end{pmatrix}, \tag{1.219}$$

in which the 2×2 -submatrices σ^a and $\tilde{\sigma}^a$ with $a - 0, \ldots, 3$ form the *four-vectors of Pauli matrices*

$$\sigma^a \equiv (\sigma^0, \sigma^i), \quad \tilde{\sigma}^a \equiv (\sigma^0, -\sigma^i). \tag{1.220}$$

The spatial components σ^i are the ordinary *Pauli matrices*

$$\sigma^1 = \begin{pmatrix} 0 & 1 \\ 1 & 0 \end{pmatrix}, \quad \sigma^2 = \begin{pmatrix} 0 & -i \\ i & 0 \end{pmatrix}, \quad \sigma^3 = \begin{pmatrix} 1 & 0 \\ 0 & -1 \end{pmatrix}, \tag{1.221}$$

while the zeroth component σ^0 is defined as the 2×2 -unit matrix:

$$\sigma^0 \equiv \begin{pmatrix} 1 & 0 \\ 0 & 1 \end{pmatrix}. \tag{1.222}$$

From the algebraic properties of these matrices

$$(\sigma^a)^2 = \sigma^0 = 1, \quad \sigma_i \sigma_j = \delta_{ij} + i\epsilon_{ijk}\sigma_k, \quad \sigma^a \tilde{\sigma}^b + \sigma^b \tilde{\sigma}^a = 2g^{ab}, \tag{1.223}$$

we deduce that the Dirac matrices γ^a satisfy the anticommutation rules

$$\left\{ \gamma^a, \gamma^b \right\} = 2g^{ab}. \tag{1.224}$$

Under Lorentz transformations, the Dirac field transforms according to the spinor representation of the Lorentz group

$$\psi_A(x) \xrightarrow{\Lambda} \psi'_A(x) = D_A{}^B(\Lambda)\psi_B(\Lambda^{-1}x), \tag{1.225}$$

by analogy with the transformation law (1.209) of a vector field. The 4×4 -matrices Λ of the defining representation of the Lorentz group in (1.209) are replaced by the 4×4 -matrices $D(\Lambda)$ representing the Lorentz group in spinor space.

It is easy to find these matrices. If we denote the spinor representation of the Lie algebra (1.72) by 4×4 -matrices Σ^{ab}, these have to satisfy the commutation rules

$$[\Sigma^{ab}, \Sigma^{ac}] = -ig^{aa}\Sigma^{bc}, \quad \text{no sum over } a. \tag{1.226}$$

These can be solved by the matrices

$$\Sigma^{ab} \equiv \frac{1}{2}\sigma^{ab}, \tag{1.227}$$

where σ^{ab} is the antisymmetric tensor of matrices

$$\sigma^{ab} \equiv \frac{i}{2}[\gamma^a, \gamma^b]. \tag{1.228}$$

The representation matrices of finite Lorentz transformations may now be expressed as exponentials of the form (1.54):

$$D(\Lambda) = e^{-i\frac{1}{2}\omega_{ab}\Sigma^{ab}}, \tag{1.229}$$

where ω_{ab} is the same antisymmetric matrix as in (1.54), containing the rotation and boost parameters as specified in (1.55) and (1.56). Comparison with (1.57) shows that pure rotations and pure Lorentz transformations are generated by the spinor representations of L^{ab} in (1.57):

$$\Sigma^{ij} = \epsilon_{ijk}\frac{1}{2}\begin{pmatrix} \sigma^k & 0 \\ 0 & \sigma^k \end{pmatrix}, \quad \Sigma^{0i} = \frac{i}{2}\begin{pmatrix} -\sigma^i & 0 \\ 0 & \sigma^i \end{pmatrix}. \tag{1.230}$$

The generators of the rotation group $\Sigma^i = \frac{1}{2}\epsilon_{ijk}\Sigma^{jk}$ corresponding to L_i in (1.53) consist of a direct sum of two Pauli matrices, the 4×4 -spin matrix:

$$\Sigma \equiv \frac{1}{2}\begin{pmatrix} \boldsymbol{\sigma} & 0 \\ 0 & \boldsymbol{\sigma} \end{pmatrix}. \tag{1.231}$$

The generators Σ^{0i} of the pure Lorentz transformations corresponding to M_i in (1.53) can also be expressed as $\Sigma^{0i} = i\alpha^i/2$ with the vector of 4×4 -matrices

$$\boldsymbol{\alpha} = \begin{pmatrix} -\boldsymbol{\sigma} & 0 \\ 0 & \boldsymbol{\sigma} \end{pmatrix}. \tag{1.232}$$

In terms of Σ and α, the representation matrices (1.229) for pure rotations and pure Lorentz transformations are seen to have the explicit form

$$D(R) = e^{-i\boldsymbol{\varphi}\cdot\boldsymbol{\Sigma}} = \begin{pmatrix} e^{-i\boldsymbol{\varphi}\cdot\boldsymbol{\sigma}/2} & 0 \\ 0 & e^{-i\boldsymbol{\varphi}\cdot\boldsymbol{\sigma}/2} \end{pmatrix}, \quad D(B) = e^{\boldsymbol{\zeta}\cdot\boldsymbol{\alpha}} = \begin{pmatrix} e^{-\boldsymbol{\zeta}\cdot\boldsymbol{\sigma}/2} & 0 \\ 0 & e^{\boldsymbol{\zeta}\cdot\boldsymbol{\sigma}/2} \end{pmatrix}. \tag{1.233}$$

The commutation relations (1.226) are a direct consequence of the commutation relations of the generators Σ^{ab} with the gamma matrices:

$$[\Sigma^{ab}, \gamma^c] = -(L^{ab})^c{}_d\,\gamma^d = -i(g^{ac}\gamma^b - g^{bc}\gamma^a). \tag{1.234}$$

Comparison with (1.114) and (1.115) shows that the matrices γ^a transform like x^a, i.e., they form a vector operator. The commutation rules (1.226) follow directly from (1.234) upon using the Leibnitz chain rule (1.117).

For global transformations, the vector property (1.234) implies that γ^a behaves like the vector x^a in Eq. (1.133):

$$D(\Lambda)\gamma^c D^{-1}(\Lambda) = e^{-i\frac{1}{2}\omega_{ab}\Sigma^{ab}}\gamma^c e^{i\frac{1}{2}\omega_{ab}\Sigma^{ab}} = (e^{i\frac{1}{2}\omega_{ab}L^{ab}})^c{}_{c'}\,\gamma^{c'} = (\Lambda^{-1})^c{}_{c'}\,\gamma^{c'}. \tag{1.235}$$

In terms of the generators Σ^{ab}, we can write the field transformation law (1.225) more explicitly as

$$\psi(x) \xrightarrow{\Lambda} \psi'_\Lambda(x) = D(\Lambda)\psi(\Lambda^{-1}x) = e^{-i\frac{1}{2}\omega_{ab}\Sigma^{ab}}\psi(\Lambda^{-1}x), \tag{1.236}$$

in perfect analogy with the transformation laws of scalar, tensor, and vector fields in Eqs. (1.168), (1.175), and (1.209).

It is useful to re-express the transformation of the spacetime argument on the right-hand side in terms of the differential operator of four-dimensional angular momentum and rewrite (1.236) as in (1.177) and (1.217) as

$$\psi(x) \xrightarrow{\Lambda} \psi'_\Lambda(x) = \hat{D}(\Lambda) \times D(\Lambda)\psi(x) = e^{-i\frac{1}{2}\omega_{ab}\hat{J}^{ab}}\psi(x), \tag{1.237}$$

where

$$\hat{J}^{cd} \equiv \Sigma^{cd} \times \hat{1} + 1 \times \hat{L}^{cd} \tag{1.238}$$

are the generators of the total four-dimensional angular momentum of the Dirac field.

1.12   Energy-Momentum Tensor

The four-dimensional current density $j^a(x)$ contains all information on the electric properties of relativistic particle orbits. It is possible to collect also the mechanical properties in a tensor, the *energy-momentum tensor*.

1.12.1   Point Particles

The *energy density* of the particles can be written as

$$\overset{\text{part}}{\mathcal{E}}(\mathbf{x}, t) = \sum_n m_n \gamma c^2 \delta^{(3)}(\mathbf{x} - \mathbf{x}_n(t)). \tag{1.239}$$

We have previously seen that the energy transforms like a zeroth component of a four-vector [recall (1.152)]. The energy density measures the energy per spatial volume element. An infinitesimal four-volume d^4x is invariant under Lorentz transformations, due to the unit determinant $|\Lambda^a{}_b| = 1$ implied by the pseudo-orthogonality relation (1.28), so that indeed

$$d^4x' = \left| \frac{\partial x'^a}{\partial x^b} \right| d^4x = |\Lambda^a{}_b| d^4x = d^4x. \tag{1.240}$$

This shows that $\delta^{(3)}(\mathbf{x})$ which transforms like an inverse spatial volume

$$\frac{1}{d^3x} = \frac{dx^0}{d^4x} \tag{1.241}$$

behaves like the zeroth component of a four-vector. The energy density (1.239) can therefore be viewed as a 00-component of a Lorentz tensor called the symmetric energy-momentum tensor. By convention, this is chosen to have the dimension of momentum density, so that we must identify the energy density with $c \overset{\text{part}}{T}{}^{ab}$. In fact, using the identity (1.210), we may rewrite (1.239) as

$$\overset{\text{part}}{\mathcal{E}}(\mathbf{x}, t) = c \sum_n \int_{-\infty}^{\infty} d\tau_n \frac{1}{m_n} p_n^0(\tau) p_n^0(\tau) \delta^{(4)}(x - x(\tau)), \tag{1.242}$$

which is equal to c times the 00-component of the energy-momentum tensor

$$\overset{\text{part}}{T}{}^{ab}(\mathbf{x}, t) = \sum_n \int_{-\infty}^{\infty} d\tau_n \frac{1}{m_n} p_n^a(\tau) p_n^b(\tau) \delta^{(4)}(x - x(\tau)). \tag{1.243}$$

The spatial momenta of the particles

$$\overset{\text{part}}{\mathcal{P}}{}^i(\mathbf{x}, t) = \sum_n m_n \gamma_n \dot{x}_n^i(\tau) \delta^{(3)}(\mathbf{x} - \mathbf{x}(\tau)) \tag{1.244}$$

are three-vectors. Their densities transform therefore like $0i$-components of a Lorentz tensor. Indeed, using once more the identity (1.210), we may rewrite (1.244) as

$$\overset{\text{part}}{\mathcal{P}}{}^i(\mathbf{x}, t) = \overset{\text{part}}{T}{}^{0i}(\mathbf{x}, t) = \sum_n \int_{-\infty}^{\infty} d\tau_n \frac{1}{m_n} p_n^0(\tau) p_n^i(\tau) \delta^{(4)}(x - x(\tau)), \tag{1.245}$$

which shows precisely the tensor character. The four-vector of the total energy-momentum of the many-particle system is given by the integrals over the $0a$-components

$$\overset{\text{part}}{P}{}^{a}(t) \equiv \int d^3x \, \overset{\text{part}}{T}{}^{0a}(\mathbf{x}, t). \tag{1.246}$$

Inserting here (1.242) and (1.245), we obtain the sum over all four-momenta

$$\overset{\text{part}}{P}{}^{a}(t) = \sum_n p_n^a(\tau). \tag{1.247}$$

By analogy with the four-dimensional current density $j^a(\mathbf{x})$, let us calculate the four-divergence $\partial_b \overset{\text{part}}{T}{}^{ab}$. A partial integration yields

$$\sum_n \int_{-\infty}^{\infty} d\tau_n \, p_n^a(\tau) \dot{x}_n^b(\tau) \partial_b \delta^{(4)}(x - x(\tau)) = -\sum_n \int_{-\infty}^{\infty} d\tau_n \, p_n^a(\tau) \partial_\tau \delta^{(4)}(x - x(\tau))$$

$$= -\sum_n \int_{-\infty}^{\infty} d\tau_n \partial_\tau \left[p_n^a(\tau) \delta^{(4)}(x - x(\tau)) \right] + \sum_n \int_{-\infty}^{\infty} d\tau_n \dot{p}_n^a(\tau) \delta^{(4)}(x - x(\tau)). \tag{1.248}$$

The first term on the right-hand side disappears if the particles are stable, i.e., if the orbits are closed or come from negative infinite x^0 and disappear into positive infinite x^0. The derivative $\dot{p}_n^a(\tau)$ in the second term can be made more explicit if only electromagnetic forces act on the particles. Then it is equal to the Lorentz force, i.e., the four-vector $f^a(\tau)$ of Eq. (1.184), and we obtain

$$\partial_b \overset{\text{part}}{T}{}^{ab} = \sum_n \int_{-\infty}^{\infty} d\tau_n f_n(\tau) \delta^{(4)}(x - x(\tau)) \tag{1.249}$$

$$= \frac{1}{c} \sum_n \int_{-\infty}^{\infty} d\tau_n e_n F^a{}_b(x_n(\tau)) \dot{x}_n^b(\tau) \delta^{(4)}(x - x(\tau)).$$

Expressed in terms of the current four-vector (1.213), this reads

$$\partial_b \overset{\text{part}}{T}{}^{ab}(x) = \frac{1}{c^2} F^a{}_b(x) j^b(x). \tag{1.250}$$

In the absence of electromagnetic fields, the energy-momentum tensor of the particles is conserved.

Integrating (1.246) over the spatial coordinates gives the time change of the total four-momentum

$$\partial_t \overset{\text{part}}{P}{}^{a}(t) = c \partial_0 \left[\int d^3x \, \overset{\text{part}}{T}{}^{a0} \right] = c \int d^3x \, \partial_b \overset{\text{part}}{T}{}^{ab} - c \int d^3x \, \partial_i \overset{\text{part}}{T}{}^{0i}$$

$$= \frac{e}{c} \sum_n F^a{}_b(x_n(\tau)) \dot{x}_n^b(\tau) \gamma_n(\tau). \tag{1.251}$$

This agrees, of course, with the Lorentz equations (1.170) since by (1.247)

$$\partial_t \overset{\text{part}}{P}{}^{a}(t) = \partial_t \sum_n p_n^a(\tau) = \sum_n \dot{p}_n^a(\tau) \gamma_n. \tag{1.252}$$

If there are no electromagnetic forces, then $\overset{\text{part}}{P}{}^{a}$ is time-independent.

1.12.2   Perfect Fluid

A perfect fluid is defined as an idealized uniform material medium moving with velocity $\mathbf{v}(\mathbf{x}, t)$. The uniformity is an acceptable approximation as long as the microscopic mean free paths are short with respect to the length scale recognizable by the observer. Consider such a fluid at rest. Then the energy-momentum tensor has no momentum density:

$$\overset{\text{fluid}}{T}{}^{0i} = 0. \tag{1.253}$$

The energy density is given by

$$c \, \overset{\text{fluid}}{T}{}^{00} = c^2 \rho, \tag{1.254}$$

where ρ is the mass density.

Due to the isotropy, the *purely spatial* part of the energy-momentum tensor must be diagonal:

$$\overset{\text{fluid}}{T}{}^{ij} = \frac{p}{c} \delta_{ij}, \tag{1.255}$$

where p is the *pressure* of the fluid. We can now calculate the energy-momentum tensor of a moving perfect fluid by performing a Lorentz transformation on the energy-momentum tensor at rest:

$$\overset{\text{fluid}}{T}{}^{ab} \rightarrow \Lambda^a{}_c \Lambda^b{}_d \, \overset{\text{fluid}}{T}{}^{cd}. \tag{1.256}$$

Applying to this the Lorentz boosts from rest to momentum $\mathbf{p}$ of Eq. (1.34), and expressing the hyperbolic functions in terms of energy and momentum according to Eq. (1.153), we obtain

$$\overset{\text{fluid}}{T}{}_{ab} = \frac{1}{c} \left[\left(\frac{p}{c^2} + \rho \right) u^a u^b - p g^{ab} \right], \tag{1.257}$$

where u^a is the four-velocity (1.150) of the fluid with $u^a u_a = c^2$.

1.12.3   Electromagnetic Field

The energy density of an electromagnetic field is well-known:

$$\mathcal{E}(x) = \frac{1}{2} \left[\mathbf{E}^2(x) + \mathbf{B}^2(x) \right]. \tag{1.258}$$

The associated energy current density is given by the *Poynting vector*:

$$\mathbf{S}(x) = c \, \mathbf{E}(x) \times \mathbf{B}(x). \tag{1.259}$$

From these we find four components of the energy-momentum tensor:

$$\overset{\text{em}}{T}{}^{00}(x) \equiv \frac{1}{c} \mathcal{E}(x), \qquad \overset{\text{em}}{T}{}^{0i} = \overset{\text{em}}{T}{}^{i0} \equiv \frac{1}{c^2} \mathbf{S}^i(x). \tag{1.260}$$

The remaining components are determined by the tensor

$$\overset{em}{T}{}^{ab}(x) = \frac{1}{c}\left[-F^a{}_cF^{bc} + \frac{1}{4}g^{ab}F^{cd}F_{cd}\right].$$

(1.261)

The four-divergence of this is

$$\partial_b\overset{em}{T}{}^{ab} = \frac{1}{c}\left[-F^a{}_c\partial_bF^{ab} - (\partial_bF^a{}_c)F^{bc} + \frac{1}{4}\partial^a\left(F^{cd}F_{cd}\right)\right].$$

(1.262)

The second and third terms cancel each other, due to the homogeneous Maxwell equations (1.189) and (1.190). In order to see this, take the trivial identity $\partial_b\epsilon^{abcd}F_{cd} = 2\epsilon^{abcd}\partial_b\partial_cA_d = 0$, and multiply this by $\epsilon_{aefg}F_{fg}$. Using the identity (1A.23):

$$\epsilon^{abcd}\epsilon_{aefg} = -\left(\delta^b_e\delta^c_f\delta^d_g + \delta^c_e\delta^d_f\delta^b_g + \delta^d_e\delta^b_f\delta^c_g - \delta^b_e\delta^d_f\delta^c_g - \delta^d_e\delta^c_f\delta^b_g - \delta^c_e\delta^b_f\delta^d_g\right),$$

(1.263)

we find

$$-F^{cd}\partial_eF_{cd} - F^{db}\partial_bF_{ed} - F^{bc}\partial_bF_{ce} + F^{dc}\partial_eF_{cd} + F^{cb}\partial_bF_{ce} + F^{bd}\partial_bF_{ed} = 0.$$

(1.264)

Due to the antisymmetry of F_{ab}, this gives

$$-\partial_e\left(F^{cd}F_{cd}\right) + 4F^{bd}\partial_bF_{bd} = 0,$$

(1.265)

so that we obtain the conservation law

$$\partial_b\overset{em}{T}{}^{ab}(x) = -\frac{1}{c}\left[F^a{}_c(x)\partial_bF^{bc}(x)\right] = 0.$$

(1.266)

In the last step we have used Maxwell's equation Eq. (1.196) with zero currents.

The timelike component of the conservation law (1.266) reads

$$\partial_t\overset{em}{T}{}^{00}(x) + c\partial_i\overset{em}{T}{}^{0i}(x) = 0,$$

(1.267)

which can be rewritten with (1.258) and (1.260) as the well-known *Poynting law* of energy flow:

$$\partial_t\mathcal{E}(x) + \boldsymbol{\nabla}\cdot\mathbf{S}(x) = 0.$$

(1.268)

If currents are present, the Maxwell equation (1.196) changes the conservation law (1.266) to

$$c\partial_b\overset{em}{T}{}^{ab}(x) = -\frac{1}{c}F^a{}_c(x)j^c(x) = 0,$$

(1.269)

which modifies (1.268) to

$$\partial_t\mathcal{E}(x) + \boldsymbol{\nabla}\cdot\mathbf{S}(x) = -\mathbf{j}(x)\cdot\mathbf{E}(x).$$

(1.270)

A current parallel to the electric field reduces the field energy.

In a medium, the energy density and Poynting vector become

$$\mathcal{E}(x) \equiv \frac{1}{2}\left[\mathbf{E}(x) \cdot \mathbf{D}(x) + \mathbf{B}(x) \cdot \mathbf{H}(x)\right], \qquad \mathbf{S}(x) \equiv c\mathbf{E}(x) \times \mathbf{H}(x), \quad (1.271)$$

and the conservation law can easily be verified using the Maxwell equations (1.193) and (1.195):

$$\boldsymbol{\nabla} \cdot \mathbf{S}(x) = c\boldsymbol{\nabla} \cdot [\mathbf{E}(x) \times \mathbf{H}(x)] = c[\boldsymbol{\nabla} \times \mathbf{E}(x)] \cdot \mathbf{H}(x) - c\mathbf{E}(x) \cdot [\boldsymbol{\nabla} \times \mathbf{B}(x)]$$
$$= \{\partial_t \mathbf{B}(x) \cdot \mathbf{H}(x) + \mathbf{E}(x) \cdot [\partial_t \mathbf{D}(x) + \mathbf{j}(x)]\} = \partial_t \mathcal{E}(x) + \mathbf{j}(x) \cdot \mathbf{E}(x). \,(1.272)$$

We now observe that the force on the right-hand side of (1.269) is precisely the opposite of the right-hand side of (1.250), as required by Newton's third axiom of actio = reactio. Thus, the total energy-momentum tensor of the combined system of particles and electromagnetic fields

$$T^{ab}(x) = \overset{\text{part}}{T}{}^{ab}(x) + \overset{\text{em}}{T}{}^{ab}(x) \tag{1.273}$$

has a vanishing four-divergence,

$$\partial_b T^{ab}(x) = 0 \tag{1.274}$$

implying that the total four-momentum $P^a \equiv \int d^3x\, T^{0a}$ is a conserved quantity

$$\partial_t P^a(t) = 0. \tag{1.275}$$

1.13   Angular Momentum and Spin

Similar considerations apply to the total angular momentum of particles and fields. Since $T^{i0}(x)$ is a momentum density, we may calculate the spatial tensor of total angular momentum from the integral

$$J^{ij}(t) = \int d^3x \left[x^i T^{j0}(x) - x^j T^{i0}(x)\right]. \tag{1.276}$$

In three space dimensions one describes the angular momentum by a vector $J^i = \frac{1}{2}\epsilon^{ijk} J^{jk}$. The angular momentum (1.276) may be viewed as the integral

$$J^{ij}(t) = \int d^3x\, J^{ij,0}(x) \tag{1.277}$$

over the $i, j, 0$-component of the Lorentz tensor

$$J^{ab,c}(x) = x^a T^{bc}(x) - x^b T^{ac}(x). \tag{1.278}$$

It is easy to see that due to (1.274) and the symmetry of the energy-momentum tensor, the Lorentz tensor $J^{ab,c}(x)$ is divergenceless in the index c

$$\partial_c J^{ab,c}(x) = 0. \tag{1.279}$$

As a consequence, the spatial integral

$$J^{ab}(t) = \int d^3x \, J^{ab,0}(x)$$

(1.280)

is a conserved quantity. This is the four-dimensional extension of the conserved total angular momentum. The conservation of the components J^{0i} is the center-of-mass theorem.

A set of point particles with the energy-momentum tensor (1.243) possesses a four-dimensional angular momentum

$$\overset{\text{part}}{J}{}^{ab}(\tau) = \sum_n \left[x_n^a(\tau) p_n^b(\tau) - x_n^b(\tau) p_n^a(\tau) \right].$$

(1.281)

In the absence of electromagnetic fields, this is conserved, otherwise the τ-dependence is important.

The spin of a particle is defined by its total angular momentum in its rest frame. It is the *intrinsic angular momentum* of the particle. Electrons, protons, neutrons, and neutrinos have spin $1/2$. For nuclei and atoms, the spin can take much larger values.

There exists a four-vector $S^a(\tau)$ along the orbit of a particle whose spatial part reduces to the angular momentum in the rest frame. It is defined by a combination of the angular momentum (1.281) and the four-velocity $u^d(\tau)$ [recall (1.150)]

$$S^a(\tau) \equiv \frac{1}{2c} \epsilon^{abcd} \overset{\text{part}}{J}{}_{bc}(\tau) u_d(\tau).$$

(1.282)

In the rest frame where

$$u_R^a = (c, 0, 0, 0),$$

(1.283)

this reduces indeed to the three-vector of total angular momentum

$$S_R^a(\tau) = (0, \overset{\text{part}}{J}{}_{23}(\tau), \overset{\text{part}}{J}{}_{31}(\tau), \overset{\text{part}}{J}{}_{12}(\tau)) = (0, \overset{\text{part}}{\mathbf{J}}(\tau)).$$

(1.284)

For a free particle we find, due to conservation of momentum and total angular momentum

$$\frac{d}{d\tau} u_d(\tau) = 0, \qquad \frac{d}{d\tau} \overset{\text{part}}{J}{}_{bc}(\tau) = 0,$$

(1.285)

that also the spin vector $S^a(\tau)$ is conserved:

$$\frac{d}{d\tau} S^a(\tau) = 0.$$

(1.286)

The spin four-vector is useful to understand an important phenomenon in atomic physics called the *Thomas precession* of the electron spin in an atom. It explains why the observed fine structure of atomic physics determines the gyromagnetic ratio g_e of the electron to be close to 2.

The relation between spin (1.284) and its four-vector is exhibited clearly by applying the pure Lorentz transformation matrix (1.27) to (1.284) yielding

$$S^i = S_R^i + \frac{\gamma^2}{\gamma+1} \frac{v^i v^j}{c^2} S_R{}^j, \quad S^0 = \gamma \frac{v^i}{c} S_R^i. \tag{1.287}$$

Note that S^0 and S^i satisfy $S^0 = v^i S^i / c$, which can be rewritten covariantly as

$$u^a S_a = 0. \tag{1.288}$$

The inverse of the transformation (1.287) is found with the help of the identity $v^2/c^2 = (\gamma^2 - 1)/\gamma^2$ as follows:

$$S_R^i = S^i - \frac{\gamma}{\gamma+1} \frac{v^i v^j}{c^2} S^j = S^i - \frac{\gamma-1}{\gamma} \frac{v^i v^j}{v^2} S^j. \tag{1.289}$$

If external forces act on the system, the spin vector starts moving. This movement is called *precession*. If the point particle moves in an orbit under the influence of a *central force* (for example, an electron around a nucleus in an atom), there is no torque on the particle so that the total angular momentum in its rest frame is conserved. Hence $dS_R^i(\tau)/d\tau = 0$, which is expressed covariantly as $dS^a(\tau)/d\tau \propto u^a(\tau)$. In the rest frame of the atom, however, the spin shows precession. Let us calculate its rate. From the definition (1.282) we have

$$\frac{dS_a}{d\tau} = \frac{1}{2} \epsilon_{abcd} \overset{\text{part}}{J}{}^{bc} \frac{du^d}{d\tau}. \tag{1.290}$$

There is no contribution from

$$\frac{d}{d\tau} \overset{\text{part}}{J}{}^{bc} = x^a(\tau)\ddot{p}^b(\tau) - x^b(\tau)\ddot{p}^a(\tau), \tag{1.291}$$

since $\dot{p} = m\dot{u}$, and the ϵ-tensor is antisymmetric.

The right-hand side of (1.290) can be simplified by multiplying it with the trivial expression

$$g_{st} u^s u^t = c^2, \tag{1.292}$$

and using the identity for the ϵ-tensor

$$\epsilon^{abcd} g^{st} = \epsilon^{abcs} g^{dt} + \epsilon^{absd} g^{ct} + \epsilon^{ascd} g^{bt} + \epsilon^{sbcd} g^{at}. \tag{1.293}$$

This can easily be proved by taking advantage of the antisymmetry of ϵ^{abcd} and choosing a, b, c, d to be equal to $0, 1, 2, 3$, respectively. After this, the right-hand side of (1.290) becomes a sum of the four terms

$$\frac{1}{2} \left(\epsilon_{abcs} \overset{\text{part}}{J}{}^{bc} u^s u^d u'^d + \epsilon_{absd} \overset{\text{part}}{J}{}^{bc} u_c u^s u'^d + \epsilon_{ascd} \overset{\text{part}}{J}{}^{bc} u_b u^a u'^d + \epsilon_{sbcd} \overset{\text{part}}{J}{}^{bc} u^s u^a \dot{u}^d \right).$$

The first term vanishes, since $u^d \dot{u}_d = (1/2)du^2/d\tau = (1/2)dc^2/d\tau = 0$. The last term is equal to $-S_d \dot{u}^d u_a/c^2$. Inserting the identity (1.293) into the second and third terms, we obtain twice the left-hand side of (1.290). Taking this to the left-hand side, we find the equation of motion

$$\frac{dS_a}{d\tau} = \frac{1}{c^2} S_c \frac{du^c}{d\tau} u_a. \tag{1.294}$$

Note that on account of this equation, the time derivative $dS_a/d\tau$ points in the direction of u^a, in accordance with the initial assumption of a torque-free force.

We are now prepared to calculate the rate of the Thomas precession. Denoting in the final part of this section the derivatives with respect to the physical time $t = \gamma\tau$ by a dot, we can rewrite (1.294) as

$$\dot{\mathbf{S}} \equiv \frac{d\mathbf{S}}{dt} = \frac{1}{\gamma}\frac{d\mathbf{S}}{d\tau} = -\frac{1}{c^2}\left(S^0\dot{u}^0 + \mathbf{S}\cdot\dot{\mathbf{u}}\right)\mathbf{u} = \frac{\gamma^2}{c^2}\left(\mathbf{S}\cdot\dot{\mathbf{v}}\right)\mathbf{v}, \tag{1.295}$$

$$\dot{S}_0 \equiv \frac{dS_0}{dt} = \frac{1}{c}\frac{d}{dt}\left(\mathbf{S}\cdot\mathbf{v}\right) = \frac{\gamma^2}{c^2}\left(\mathbf{S}\cdot\dot{\mathbf{v}}\right). \tag{1.296}$$

We now differentiate Eq. (1.289) with respect to the time using the relation $\dot{\gamma} = \gamma^3 \dot{\mathbf{v}}\mathbf{v}/c^2$, and find

$$\dot{\mathbf{S}}_R = \dot{\mathbf{S}} - \frac{\gamma}{\gamma+1}\frac{1}{c^2}\dot{S}^0\mathbf{v} - \frac{\gamma}{\gamma+1}\frac{1}{c^2}S^0\dot{\mathbf{v}} - \frac{\gamma^3}{(\gamma+1)^2}\frac{1}{c^4}(\mathbf{v}\cdot\dot{\mathbf{v}})S^0\mathbf{v}. \tag{1.297}$$

Inserting here Eqs. (1.295) and (1.296), we obtain

$$\dot{\mathbf{S}}_R = \frac{\gamma^2}{\gamma+1}\frac{1}{c^2}(\mathbf{S}\cdot\dot{\mathbf{v}})\mathbf{v} - \frac{\gamma}{\gamma+1}\frac{1}{c^2}S^0\dot{\mathbf{v}} - \frac{\gamma^3}{(\gamma+1)^2}(\mathbf{v}\cdot\dot{\mathbf{v}})S^0\mathbf{v}. \tag{1.298}$$

On the right-hand side we return to the spin vector $\mathbf{S}_R$ using Eqs. (1.287), and find

$$\dot{\mathbf{S}}_R = \frac{\gamma^2}{\gamma+1}\frac{1}{c^2}\left[(\mathbf{S}_R\cdot\dot{\mathbf{v}})\mathbf{v} - (\mathbf{S}_R\cdot\mathbf{v})\dot{\mathbf{v}}\right] = \mathbf{\Omega}_T \times \mathbf{S}_R, \tag{1.299}$$

with the Thomas precession frequency

$$\mathbf{\Omega}_T = -\frac{\gamma^2}{(\gamma+1)}\frac{1}{c^2}\mathbf{v}\times\dot{\mathbf{v}}. \tag{1.300}$$

This is a purely kinematic effect. If an electromagnetic field is present, there will be an additional dynamic precession. For slow particles, it is given by

$$\dot{\mathbf{S}} \equiv -\mathbf{S}\times\mathbf{\Omega}_{em} \approx \mathbf{\mu}\times\left(\mathbf{B} - \frac{\mathbf{v}}{c}\times\mathbf{E}\right), \tag{1.301}$$

where $\mathbf{\mu}$ is the magnetic moment

$$\mathbf{\mu} = g\mu_B\frac{\mathbf{S}}{\hbar} = \frac{eg}{2Mc}\mathbf{S}, \tag{1.302}$$

and g the dimensionless *gyromagnetic ratio*, also called *Landé factor*. Recall the value of the *Bohr magneton*

$$\mu_B \equiv \frac{e\hbar}{2Mc} \approx 3.094 \times 10^{-30} \, \text{C cm} \approx 0.927 \times 10^{-20} \, \frac{\text{erg}}{\text{gauss}} \approx 5.788 \times 10^{-8} \, \frac{\text{eV}}{\text{gauss}}. \tag{1.303}$$

If the electron moves fast, we transform the electromagnetic field to the electron rest frame by a Lorentz transformation (1.178), (1.179), and obtain an equation of motion for the spin

$$\dot{\mathbf{S}}_R = \boldsymbol{\mu} \times \mathbf{B}' = \boldsymbol{\mu} \times \left[\gamma \left(\mathbf{B} - \frac{\mathbf{v}}{c} \times \mathbf{E} \right) - \frac{\gamma^2}{\gamma + 1} \frac{\mathbf{v}}{c} \left(\frac{\mathbf{v}}{c} \cdot \mathbf{B} \right) \right]. \tag{1.304}$$

Expressing $\boldsymbol{\mu}$ via Eq. (1.302), this becomes

$$\dot{\mathbf{S}}_R \equiv -\mathbf{S}_R \times \boldsymbol{\Omega}_{\text{em}} = \frac{eg}{2mc} \mathbf{S}_R \times \left[\left(\mathbf{B} - \frac{\mathbf{v}}{c} \times \mathbf{E} \right) - \frac{\gamma}{\gamma + 1} \frac{\mathbf{v}}{c} \left(\frac{\mathbf{v}}{c} \cdot \mathbf{B} \right) \right], \tag{1.305}$$

which is the relativistic generalization of Eq. (1.301). It is easy to see that the associated fully covariant equation is

$$S^{a\prime} = \frac{g}{2mc} \left[eF^{ab}S_b + \frac{1}{mc} p^a S_c \frac{d}{d\tau} p^c \right] = \frac{eg}{2mc} \left[F^{ab}S_b + \frac{1}{m^2c^2} p^a S_c F^{c\kappa} p_\kappa \right]. \tag{1.306}$$

On the right-hand side we have inserted the relativistic equation of motion (1.170) of a point particle in an external electromagnetic field.

If we add to this the torque-free Thomas precession rate (1.294), we obtain the covariant *Bargmann-Michel-Telegdi equation* [9]

$$S^{a\prime} = \frac{1}{2mc} \left[egF^{ab}S_b + \frac{g-2}{mc} p^a S_c \frac{d}{d\tau} p^c \right] = \frac{e}{2mc} \left[gF^{ab}S_b + \frac{g-2}{m^2c^2} p^a S_c F^{c\kappa} p_\kappa \right]. \tag{1.307}$$

For the spin vector $\mathbf{S}_R$ in the electron rest frame this implies a change in the electromagnetic precession rate in Eq. (1.305) to [10]

$$\frac{d\mathbf{S}}{dt} = \boldsymbol{\Omega}_{\text{em T}} \times \mathbf{S} \equiv (\boldsymbol{\Omega}_{\text{em}} + \boldsymbol{\Omega}_{\text{T}}) \times \mathbf{S} \tag{1.308}$$

with a frequency given by the *Thomas equation*

$$\boldsymbol{\Omega}_{\text{em T}} = -\frac{e}{mc} \left[\left(\frac{g}{2} - 1 + \frac{1}{\gamma} \right) \mathbf{B} - \left(\frac{g}{2} - 1 \right) \frac{\gamma}{\gamma+1} \left(\frac{\mathbf{v}}{c} \cdot \mathbf{B} \right) \frac{\mathbf{v}}{c} - \left(\frac{g}{2} - \frac{\gamma}{\gamma+1} \right) \frac{\mathbf{v}}{c} \times \mathbf{E} \right]. \tag{1.309}$$

The contribution of the Thomas precession is the part of the right-hand side without the gyromagnetic factor g:

$$\boldsymbol{\Omega}_{\text{T}} = -\frac{e}{mc} \left[-\left(1 - \frac{1}{\gamma} \right) \mathbf{B} + \frac{\gamma}{\gamma+1} \frac{1}{c^2} (\mathbf{v} \cdot \mathbf{B}) \mathbf{v} + \frac{\gamma}{\gamma+1} \frac{1}{c} \mathbf{v} \times \mathbf{E} \right]. \tag{1.310}$$

This agrees with the Thomas frequency in Eq. (1.300) if we insert the acceleration

$$\dot{\mathbf{v}}(t) = c\frac{d}{dt}\frac{\mathbf{p}}{p^0} = \frac{e}{\gamma m}\left[\mathbf{E} + \frac{\mathbf{v}}{c}\times\mathbf{B} - \frac{\mathbf{v}}{c}\left(\frac{\mathbf{v}}{c}\cdot\mathbf{E}\right)\right], \tag{1.311}$$

which follows directly from (1.182) and (1.183).

The Thomas equation (1.309) can be used to calculate the time dependence of the helicity $h \equiv \mathbf{S}_R \cdot \hat{\mathbf{v}}$ of an electron, i.e., its component of the spin in the direction of motion. Using the chain rule of differentiation,

$$\frac{d}{dt}(\mathbf{S}_R\cdot\hat{\mathbf{v}}) = \dot{\mathbf{S}}_R\cdot\hat{\mathbf{v}} + \frac{1}{v}[\mathbf{S}_R - (\hat{\mathbf{v}}\cdot\mathbf{S}_R)\hat{\mathbf{v}}]\frac{d}{dt}\mathbf{v} \tag{1.312}$$

and inserting (1.308) as well as the equation for the acceleration (1.311), we obtain

$$\frac{dh}{dt} = -\frac{e}{mc}\mathbf{S}_{R\perp}\cdot\left[\left(\frac{g}{2}-1\right)\hat{\mathbf{v}}\times\mathbf{B} + \left(\frac{gv}{2c}-\frac{c}{v}\right)\mathbf{E}\right], \tag{1.313}$$

where $\mathbf{S}_{R\perp}$ is the component of the spin vector orthogonal to $\mathbf{v}$. This equation shows that for a Dirac electron which has $g = 2$ the helicity remains constant in a purely magnetic field. Moreover, if the electron moves ultra-relativistically ($v \approx c$), the value $g = 2$ makes the last term extremely small, $\approx (e/mc)\gamma^{-2}\mathbf{S}_{R\perp}\cdot\mathbf{E}$, so that the helicity is almost unaffected by an electric field. The anomalous magnetic moment of the electron $a \equiv (g-2)/2$, however, changes this to a finite value $\approx -(e/mc)a\mathbf{S}_{R\perp}\cdot\mathbf{E}$. This drastic effect was used to measure the experimental values of a for electrons, positrons, and muons:

$$\begin{aligned}a(e^-) &= (115\,965.77\pm0.35)\times10^{-8}, &(1.314)\\ a(e^+) &= (116\,030\pm120)\times10^{-8}, &(1.315)\\ a(\mu^\pm) &= (116\,616\pm31)\times10^{-8}. &(1.316)\end{aligned}$$

1.14 Spacetime-Dependent Lorentz Transformations

The theory of gravitation to be developed in this book will not only be Lorentz-invariant, but also invariant under local Lorentz transformations

$$x'^a = \Lambda^a{}_b(x)x^b. \tag{1.317}$$

As a preparation for dealing with such theories let us derive a group-theoretic formula which is useful for many purposes.

1.14.1 Angular Velocities

Consider a time-dependent 3×3-rotation matrix $R(\boldsymbol{\varphi}(t)) = e^{-i\boldsymbol{\varphi}(t)\cdot\mathbf{L}}$ with the generators $(L_i)_{jk} = -i\epsilon_{ijk}$ [compare (1.43)]. As time proceeds, the rotation angles change with an *angular velocity* $\boldsymbol{\omega}(t)$ defined by the relation

$$R^{-1}(\boldsymbol{\varphi}(t))\,\dot{R}(\boldsymbol{\varphi}(t)) = -i\boldsymbol{\omega}(t)\cdot\mathbf{L}. \tag{1.318}$$

The components of $\boldsymbol{\omega}(t)$ can be specified more explicitly by parametrizing the rotations in terms of *Euler angles* α, β, γ:

$$R(\alpha, \beta, \gamma) = R_3(\alpha) R_2(\beta) R_3(\gamma), \tag{1.319}$$

where $R_3(\alpha)$, $R_3(\gamma)$ are rotations around the z-axis by angles α, γ, respectively, and $R_2(\beta)$ is a rotation around the y-axis by β, i.e.,

$$R(\alpha, \beta, \gamma) \equiv e^{-i\alpha \hat{L}_3} e^{-i\beta \hat{L}_2} e^{-i\gamma \hat{L}_3}. \tag{1.320}$$

The relations between the vector $\boldsymbol{\varphi}$ of rotation angles in (1.57) and the Euler angles α, β, γ can be found by purely geometric considerations. Most easily, we equate the 2×2-representation of the rotations $R(\boldsymbol{\varphi})$,

$$R(\boldsymbol{\varphi}) = \cos \frac{\varphi}{2} - i \boldsymbol{\sigma} \cdot \hat{\boldsymbol{\varphi}} \sin \frac{\varphi}{2}, \tag{1.321}$$

with the 2×2-representation of the Euler decomposition (1.320):

$$R(\alpha, \beta, \gamma) = \left(\cos \frac{\alpha}{2} - i\sigma_3 \sin \frac{\alpha}{2} \right) \left(\cos \frac{\beta}{2} - i\sigma_2 \sin \frac{\beta}{2} \right) \left(\cos \frac{\gamma}{2} - i\sigma_3 \sin \frac{\gamma}{2} \right). \tag{1.322}$$

The desired relations follow directly from the multiplication rules for the Pauli matrices (1.223).

In the Euler decomposition, we may calculate the derivatives:

$$\begin{aligned}
i\hbar \partial_\alpha R &= R \left[\cos \beta \, L_3 - \sin \beta (\cos \gamma \, L_1 - \sin \gamma \, L_2) \right], & (1.323) \\
i\hbar \partial_\beta R &= R \left(\cos \gamma \, L_2 + \sin \gamma \, L_1 \right), & (1.324) \\
i\hbar \partial_\gamma R &= R \, L_3. & (1.325)
\end{aligned}$$

The third equation is trivial, the second follows from the rotation of the generator

$$e^{i\gamma L_3/\hbar} L_2 e^{-i\gamma L_3/\hbar} = \cos \alpha \, L_2 + \sin \gamma \, L_1, \tag{1.326}$$

which is a consequence of *Lie's expansion formula*

$$e^{iA} B e^{-iA} = 1 + i[A, B] + \frac{i^2}{2!}[A, [A, B]] + \dots, \tag{1.327}$$

and the commutation rules (1.61) of the 3×3-matrices L_i. The derivation of the first equation (1.323) requires, in addition, the rotation

$$e^{i\beta L_2/\hbar} L_3 e^{-i\beta L_2/\hbar} = \cos \beta L_3 - \sin \beta L_1. \tag{1.328}$$

We may now calculate the time derivative of $R(\alpha, \beta, \gamma)$ using Eqs. (1.323)–(1.325) and the chain rule of differentiation, and find the right-hand side of (1.318) with the angular velocities

$$\begin{aligned}
\omega_1 &= \dot{\beta} \sin \gamma - \dot{\alpha} \sin \beta \cos \gamma, & (1.329) \\
\omega_2 &= \dot{\beta} \cos \gamma + \dot{\alpha} \sin \beta \sin \gamma, & (1.330) \\
\omega_3 &= \dot{\alpha} \cos \beta + \dot{\gamma}. & (1.331)
\end{aligned}$$

Only commutation relations have been used to derive (1.323)–(1.325), so that the formulas (1.329)–(1.331) hold for *all* representations of the rotation group.

1.14.2   Angular Gradients

The concept of angular velocities can be generalized to spacetime-dependent Euler angles $\alpha(x)$, $\beta(x)$, $\gamma(x)$, replacing (1.318) by *angular gradients*

$$R^{-1}(\boldsymbol{\varphi}(x))\,\partial_a R(\boldsymbol{\varphi}(x)) = -i\boldsymbol{\omega}_a(x)\cdot\mathbf{L}, \tag{1.332}$$

with the generalization of the vector of angular velocity

$$\omega_{a;1} = \partial_a\beta\sin\gamma - \partial_a\alpha\sin\beta\cos\gamma, \tag{1.333}$$

$$\omega_{a;2} = \partial_a\beta\cos\gamma + \partial_a\alpha\sin\beta\sin\gamma, \tag{1.334}$$

$$\omega_{a;3} = \partial_a\alpha\cos\beta + \partial_a\gamma. \tag{1.335}$$

The derivatives ∂_a act only upon the functions right after it. These equations are again valid if $R(\boldsymbol{\varphi}(x))$ and $\mathbf{L}$ in (1.332) are replaced by any representation of the rotation group and its generators.

A relation of type (1.332) exists also for the Lorentz group where $\Lambda(\omega_{ab}(x)) = e^{-i\frac{1}{2}\omega_{ab}(x)L^{ab}}$ [recall (1.57)], and the generalized angular velocities are defined by

$$\Lambda^{-1}(\omega_{ab}(x))\,\partial_c\Lambda(\omega_{ab}(x)) = -i\,\frac{1}{2}\omega_{c;ab}(x)L^{ab}. \tag{1.336}$$

Inserting the explicit 4×4 -generators (1.51) on the right-hand side, we find for the matrix elements the relation

$$[\Lambda^{-1}(\omega_{ab}(x))\,\partial_c\Lambda(\omega_{ab}(x))]_{ef} = \omega_{c;ef}(x). \tag{1.337}$$

As before, the matrices $\Lambda(\omega_{ab}(x))$ and L^{ab} in (1.336) can be replaced by any representations of the Lorentz group and its generators, in particular in the spinor representation (1.229) where

$$D^{-1}(\Lambda(\omega_{ab}(x)))\,\partial_c D(\Lambda(\omega_{ab}(x))) = -i\,\frac{1}{2}\omega_{c;ab}(x)\Sigma^{ab}. \tag{1.338}$$

Appendix 1A   Tensor Identities

In the tensor calculus of Euclidean as well as Minkowski space in d spacetime dimensions, a special role is played by the contravariant *Levi-Civita tensor*

$$\epsilon^{a_1 a_2\cdots a_d},\quad a_i = 0, 1, \ldots, d-1. \tag{1A.1}$$

This is a totally antisymmetric unit tensor with the normalization

$$\epsilon^{012\cdots(d-1)} = 1. \tag{1A.2}$$

It vanishes if any two indices coincide, and is equal to ± 1 if they differ from the natural ordering $0, 1, \ldots, (d-1)$ by an even or odd perturbation. The Levi-Civita tensor serves to calculate a determinant of a tensor t_{ab} as follows

$$\det(t_{ab}) = \frac{1}{d!}\epsilon^{a_1, a_2\cdots a_d}\epsilon^{b_1 b_2\cdots b_d}\,t_{a_1 b_1}\cdots t_{a_d b_d}. \tag{1A.3}$$

In order to see this it is useful to introduce also the covariant version of $\epsilon^{a_1\cdots a_d}$ defined by

$$\epsilon_{a_1 a_2 \ldots a_d} \equiv g_{a_1 b_1} g_{a_2 b_2} \cdots g_{a_d b_d}\, \epsilon^{b_1 b_2 \ldots b_d}. \tag{1A.4}$$

This is again a totally antisymmetric unit tensor with

$$\epsilon_{012\ldots(d-1)} = (-1)^{d-1}. \tag{1A.5}$$

The contraction of the two is easily seen to be

$$\epsilon_{a_1\ldots a_d}\epsilon^{a_1\ldots a_d} = -d!. \tag{1A.6}$$

Now, by definition, a determinant is a totally antisymmetric sum

$$\det(t_{ab}) = \epsilon^{a_1\ldots a_d} t_{a_1 0}\cdots t_{a_d(d-1)}. \tag{1A.7}$$

We may also write

$$\det(t_{ab})\,\epsilon_{b_1\ldots b_d} = -\epsilon^{a_1\ldots a_d} t_{a_1 b_1}\cdots t_{a_d b_d}. \tag{1A.8}$$

By contracting with $\epsilon^{b_1\ldots b_d}$ and using (1A.6) we find

$$\det(t_{ab}) = -\frac{1}{d!}\epsilon^{a_1\ldots a_d}\epsilon^{b_1\ldots b_d} t_{a_1 b_1}\cdots t_{a_d b_d}, \tag{1A.9}$$

which agrees with (1A.7).

In the same way we can derive the formula

$$\det\!\left(t_a{}^b\right) = -\frac{1}{d!}\epsilon^{a_1\ldots a_d}\epsilon_{b_1\ldots b_d} t_{a_1}{}^{b_1}\cdots t_{a_d}{}^{b_d}. \tag{1A.10}$$

Under mirror reflection, the Levi-Civita tensor behaves like a pseudotensor. Indeed, if we subject it to a Lorentz transformation $\Lambda^a{}_b$, we obtain

$$\epsilon'{}^{a_1\ldots a_d} = \Lambda^{a_1}{}_{b_1}\cdots\Lambda^{a_d}{}_{b_d}\,\epsilon^{b_1\ldots b_d} = \det(\Lambda)\,\epsilon^{a_1\ldots a_d}. \tag{1A.11}$$

As long as $\det\Lambda = 1$, the tensor $\epsilon^{a_1\ldots a_d}$ is covariant under Lorentz transformations. If space or time inversion are included, then $\det\Lambda = -1$, and (1A.11) exhibits the pseudotensor nature of $\epsilon^{a_1\ldots a_d}$.

We now collect a set of useful identities of the Levi-Civita tensor which will be needed in this text.

1A.1 Product Formulas

a) $d = 2$ Euclidean space with $g_{ij} = \delta_{ij}$.

The antisymmetric Levi-Civita tensor ϵ_{ij} with the normalization $\epsilon_{12} = 1$ satisfies the identities

$$\epsilon_{ij}\epsilon_{kl} = \delta_{ik}\delta_{il} - \delta_{il}\delta_{jk}, \tag{1A.12}$$

$$\epsilon_{ij}\epsilon_{ik} = \delta_{jk}, \tag{1A.13}$$

$$\epsilon_{ij}\epsilon_{ij} = 2, \tag{1A.14}$$

$$\epsilon_{ij}\delta_{kl} = \epsilon_{ik}\delta_{jl} + \epsilon_{kj}\delta_{il}. \tag{1A.15}$$

b) $d = 3$ Euclidean space with $g_{ij} = \delta_{ij}$.

The antisymmetric Levi-Civita tensor ϵ_{ijk} with the normalization $\epsilon_{123} = 1$ satisfies the identities

$$\epsilon_{ijk}\epsilon_{lmn} = \delta_{il}\delta_{jm}\delta_{kn} + \delta_{im}\delta_{jn}\delta_{kl} + \delta_{in}\delta_{jl}\delta_{km},$$
$$- \delta_{il}\delta_{jn}\delta_{km} - \delta_{in}\delta_{jm}\delta_{kl} - \delta_{im}\delta_{jl}\delta_{kn}, \tag{1A.16}$$

$$\epsilon_{ijk}\epsilon_{imn} = \delta_{jm}\delta_{kn} - \delta_{in}\delta_{km}, \tag{1A.17}$$

$$\epsilon_{ijk}\epsilon_{ijn} = 2\delta_{kn}, \tag{1A.18}$$

$$\epsilon_{ijk}\epsilon_{ijk} = 6, \tag{1A.19}$$

$$\epsilon_{ijk}\delta_{lm} = \epsilon_{ijl}\delta_{km} + \epsilon_{ilk}\delta_{jm} + \epsilon_{ljk}\delta_{im}, \tag{1A.20}$$

c) $d = 4$ Minkowski space with metric

$$g_{ab} = \begin{pmatrix} 1 & & & \\ & -1 & & \\ & & -1 & \\ & & & -1 \end{pmatrix}. \tag{1A.21}$$

The antisymmetric Levi-Civita tensor with the normalization $\epsilon^{0123} = -\epsilon_{0123} = 1$ satisfies the product identities

$$\begin{aligned}
\epsilon_{abcd}\epsilon^{efgh} = & -\left(\delta_a^e\delta_b^f\delta_c^g\delta_d^h + \delta_a^f\delta_b^g\delta_c^h\delta_d^c + \delta_a^g\delta_b^k\delta_c^e\delta_d^f + \delta_a^h\delta_b^e\delta_c^f\delta_d^g \right. \\
& + \delta_a^f\delta_b^e\delta_c^h\delta_d^g + \delta_a^e\delta_b^h\delta_c^g\delta_d^f + \delta_a^h\delta_b^g\delta_c^f\delta_d^e + \delta_a^g\delta_b^f\delta_c^e\delta_d^h \\
& + \delta_a^h\delta_b^g\delta_c^f\delta_d^e + \delta_a^g\delta_b^f\delta_c^e\delta_d^h + \delta_a^f\delta_b^e\delta_c^h\delta_d^g + \delta_a^c\delta_b^h\delta_c^g\delta_d^f \\
& - \delta_a^e\delta_b^f\delta_c^h\delta_d^g - \delta_a^f\delta_b^h\delta_c^g\delta_d^e - \delta_a^h\delta_b^g\delta_c^e\delta_d^f - \delta_a^g\delta_b^c\delta_c^f\delta_d^h \\
& - \delta_a^f\delta_b^e\delta_c^g\delta_d^h - \delta_a^e\delta_b^g\delta_c^h\delta_d^f - \delta_a^g\delta_b^h\delta_c^f\delta_d^e - \delta_a^h\delta_b^f\delta_c^e\delta_d^g \\
& \left. - \delta_a^g\delta_b^h\delta_c^f\delta_d^e - \delta_a^h\delta_b^f\delta_c^e\delta_d^g - \delta_a^f\delta_b^e\delta_c^g\delta_d^h - \delta_a^e\delta_b^g\delta_c^h\delta_d^f \right), \tag{1A.22}
\end{aligned}$$

$$\epsilon_{abcd}\epsilon^{afgh} = -\left(\delta_b^f\delta_c^g\delta_d^h + \delta_b^g\delta_c^h\delta_d^f + \delta_b^h\delta_c^f\delta_d^g - \delta_b^f\delta_c^h\delta_d^g - \delta_b^h\delta_c^g\delta_d^f - \delta_b^g\delta_c^f\delta_d^h \right), \tag{1A.23}$$

$$\epsilon_{abcd}\epsilon^{abgh} = -2\left(\delta_c^g\delta_d^h - \delta_c^h\delta_d^g \right), \tag{1A.24}$$

$$\epsilon_{abcd}\epsilon^{abch} = -6\delta_d^h, \tag{1A.25}$$

$$\epsilon_{abcd}\epsilon^{abcd} = -24, \tag{1A.26}$$

$$\epsilon_{abcd}g_{ef} = \epsilon_{abce}g_{dt} + \epsilon_{abcd}g_{cf} + \epsilon_{aecd}g_{bf} + \epsilon_{ebcd}g_{af}. \tag{1A.27}$$

1A.2   Determinants

a) $d = 2$ Euclidean:

$$g \;=\; \det(g_{ij}) = \frac{1}{2!}\epsilon_{ik}\epsilon_{il}g_{ij}g_{kl} \equiv \frac{1}{2}g_{ij}C^{ij}, \tag{1A.28}$$

$$C^{ij} \;=\; \epsilon_{ik}\epsilon_{jl}g_{kl} = \text{cofactor},$$

$$g^{ij} \;=\; \frac{1}{g}C_{ij} = \text{inverse of } g_{ij}.$$

b) $d = 3$ Euclidean:

$$g \;=\; \det(g_{ij}) = \frac{1}{3!}\epsilon_{ikl}\epsilon_{jmn}g_{ij}g_{km}g_{ln} = g_{ij}C^{ij}, \tag{1A.29}$$

$$C^{ij} \;=\; \frac{1}{2!}\epsilon_{ikl}\epsilon_{jmn}g_{km}g_{ln} = \text{cofactor},$$

$$g^{ij} \;=\; \frac{1}{g}C^{ij} = \text{inverse of } g_{ij}.$$

c) $d = 4$ Minkowski:

$$g = \det(g_{ab}) \;=\; -\frac{1}{4!}\epsilon^{abcd}\epsilon^{efgh}g_{ac}g_{bf}g_{cg}g_{dh} = \frac{1}{4}g^{ae}C^{ae}, \tag{1A.30}$$

$$C^{ae} \;=\; -\frac{1}{3!}\epsilon^{abcd}\epsilon^{efgh}g_{bf}g_{cg}g_{dh} = \text{cofactor},$$

$$g^{ab} \;=\; \frac{1}{g}C^{ab} = \text{inverse of } g_{ab}.$$

1A.3   Expansion of Determinants

From Formulas (1A.28)–(1A.30) together with (1A.12), (1A.16), (1A.22), we find

$$d=2 : \; \det(g_{ij}) = \frac{1}{2!}\left[(\text{tr}g)^2 - \text{tr}(g^2)\right],$$

$$d=3 : \; \det(g_{ij}) = \frac{1}{3!}\left[(\text{tr}g)^3 + 2\,\text{tr}(g^3) - 3\,\text{tr}g\,\text{tr}(g^2)\right], \tag{1A.31}$$

$$d=4 : \; \det(g_{ab}) = \frac{1}{24}\left[(\text{tr}g)^4 - 6(\text{tr}g)^2\,\text{tr}(g^2) + 3[\text{tr}(g^2)]^2 + 8\,\text{tr}(g)\,\text{tr}(g^3) - 6\,\text{tr}(g^4)\right].$$

Notes and References

[1] A.A. Michelson, E.W. Morley, Am. J. Sci. **34**, 333 (1887), reprinted in *Relativity Theory: Its Origins and Impact on Modern Thought* ed. by L.P. Williams, J. Wiley and Sons, N.Y. (1968).

[2] The newer limit is 1 km/sec. See T.S. Jaseja, A. Jaxan, J. Murray, C.H. Townes, Phys. Rev. **133**, 1221 (1964).

[3] G.F. Fitzgerald, as told by O. Lodge, Nature **46**, 165 (1982).

[4] H.A. Lorentz, Zittingsverslag van de Akademie van Wetenschappen **1**, 74 (1892), Proc. Acad. Sci. Amsterdam **6**, 809 (1904).

[5] J.H. Poincaré, Rapports présentés au Congrès International de Physique réuni á Paris (Gauthier-Villiers, Paris, 1900).

[6] A. Einstein, Ann. Phys. **17**, 891 (1905), **18**, 639 (1905).

[7] For a list of shortcomings of Newton's mechanics see the web page
`www.physics.gmu.edu/classinfo/astr228/CourseNotes/ln_ch14.htm`.

[8] For a derivation see
J.E. Campbell, Proc. London Math. Soc. **28**, 381 (1897); **29**, 14 (1898);
H.F. Baker, ibid., **34**, 347 (1902); **3**, 24 (1905);
F. Hausdorff, Berichte Verhandl. Sächs. Akad. Wiss. Leipzig, Math. Naturw. Kl. **58**, 19 (1906);
J.A. Oteo, J. Math. Phys. **32**, 419 (1991),
or Chapter 2 in the textbook
H. Kleinert, *Path Integrals in Quantum Mechanics, Statistics, Polymer Physics, and Financial Markets*, World Scientific Publishing Co., Singapore 2004, 3rd extended edition, pp. 1–1460 (`kl/b5`), where `kl` is short for the URL: `www.physik.fu-berlin.de/~kleinert`.

[9] V. Bargmann, L. Michel, and V.L. Telegdi, Phys. Rev. Lett. **2**, 435 (1959).

[10] L.T. Thomas, Phil. Mag. **3**, 1 (1927).

2

Action Approach

The most efficient way of describing the physical properties of a system is based on its *action* $\mathcal{A}$. The extrema of $\mathcal{A}$ yield the equations of motion, and the sum over all histories of the phase factors $e^{i\mathcal{A}/\hbar}$ renders the quantum-mechanical time evolution amplitude [1, 2]. The sum over all histories is performed with the help of a *path integral*. Historically, the action approach was introduced in classical mechanics to economize Newton's procedure of setting up equations of motion, and to extend its applicability to a larger variety of mechanical problems with generalized coordinates. In quantum mechanics, the sum over all paths with phase factors involving the action $e^{i\mathcal{A}/\hbar}$ replaces and generalizes the Schrödinger theory. The path integral runs over all position and momentum variables at each time and specifies what are called *quantum fluctuations*. Their size is controlled by Planck's quantum $\hbar$, and there is great similarity with thermal fluctuations whose size is controlled by the temperature T. In the limit $\hbar \to 0$, paths with highest amplitude run along the extrema of the action, thus explaining the emergence of classical mechanics from quantum mechanics.

The pleasant property of the action approach is that it can be generalized directly to field theory. Classical fields were discovered by Maxwell as a useful concept to describe the phenomena of electromagnetism. In particular, his equations allow us to study the propagation of free electromagnetic waves without considering the sources. In the last century, Einstein constructed his theory of gravity by assuming the metric of spacetime to become a spacetime-dependent field, which can propagate in the form of gravitational waves. In condensed matter physics, fields were introduced to describe excitations in many systems, and Landau made them a universal tool for understanding phase transitions [3]. Such fields are called *order fields*. A more recently discovered domain of applications of fields is in the statistical mechanics of grand-canonical ensembles of line-like excitations, such as vortex lines in superfluids and superconductors [4], or defect lines in crystals [5]. Such excitations disturb the order of the system, and the associated fields are known as *disorder fields* [4].

2.1 General Particle Dynamics

Given an arbitrary classical system with generalized coordinates $q_n(t)$ and velocities $\dot{q}_n(t)$, the typical action has the form

$$A[q_k] = \int_{t_a}^{t_b} dt\, L\left(q_k(t), \dot{q}_k(t), t\right), \tag{2.1}$$

where $L\left(q_k(t), \dot{q}_k(t), t\right)$ is called the *Lagrangian* of the system, which is usually at most quadratic in the velocities $\dot{q}_k(t)$. A Lagrangian with this property is called *local* in time. If a theory is governed by a local Lagrangian, the action and the entire theory are also called local. The quadratic dependence on $\dot{q}(t)$ may emerge only after an integration by parts in the action. For example, $-\int dt\, q(t)\ddot{q}(t)$ is a local term in the Lagrangian since it turns into $\int dt\, \dot{q}^2(t)$ after a partial integration in the action (2.1).

The physical trajectories of the system are found from the *extremal principle*. One compares the action for one orbit $q_k(t)$ connecting the endpoints

$$q_k(t_a) = q_{k,a}, \qquad q_k(t_b) = q_{k,b}, \tag{2.2}$$

with that of an infinitesimally different orbit $q'_k(t) \equiv q_k(t) + \delta q_k(t)$ connecting the same endpoints, where $\delta q_k(t)$ is called the *variation* of the orbit. Since the endpoints of $q_k(t) + \delta q_k(t)$ are the same as those of $q_k(t)$, the variations of the orbit vanish at the endpoints:

$$\delta q(t_a) = 0, \qquad \delta q(t_b) = 0. \tag{2.3}$$

The associated variation of the action is

$$\delta A \equiv A[q_k + \delta q_k] - A[q_k] = \int_{t_a}^{t_b} dt \sum_k \left(\frac{\partial L}{\partial q_k(t)} \delta q_k(t) + \frac{\partial L}{\partial \dot{q}_k(t)} \delta \dot{q}_k(t) \right). \tag{2.4}$$

After an integration by parts, this becomes

$$\delta A = \int_{t_a}^{t_b} dt \sum_k \left(\frac{\partial L}{\partial q_k} - \frac{d}{dt}\frac{\partial L}{\partial \dot{q}_k} \right) \delta q_n(t) + \sum_k \frac{\partial L}{\partial \dot{q}_k} \delta q_k(t) \Big|_{t_a}^{t_b}. \tag{2.5}$$

In going from (2.4) to (2.5) one uses the fact that by definition of $\delta q_k(t)$ the variation of the time derivative is equal to the time derivative of the variation:

$$\delta \dot{q}_k(t) = \dot{q}'_k(t) - \dot{q}_k(t) = \frac{d}{dt}[q_k(t) + \delta q_k(t)] - \dot{q}_k(t) = \frac{d}{dt}\delta q_k(t). \tag{2.6}$$

Expressed more formally, the time derivative commutes with the variations of the orbit:

$$\delta \frac{d}{dt} q_k(t) \equiv \frac{d}{dt} \delta q_k(t). \tag{2.7}$$

Using the property (2.3), the boundary term on the right-hand side of (2.5) vanishes. Since the action is extremal for a classical orbit, $\delta\mathcal{A}$ must vanish for all variations $\delta q_k(t)$, implying that $q_k(t)$ satisfies the *Euler-Lagrange equation*

$$\frac{\partial L}{\partial q_k(t)} - \frac{d}{dt}\frac{\partial L}{\partial \dot{q}_k(t)} = 0, \tag{2.8}$$

which is the *equation of motion* of the system. For a local Lagrangian $L(q_k(t), \dot{q}_k(t))$, which contains $\dot{q}_k$ at most quadratically, in $L(q_k(t), \dot{q}_k(t))$, the Euler-Lagrange equation is a second-order differential equation for the orbit $q_k(t)$.

The local Lagrangian of a set of gravitating mass points is

$$L(\mathbf{x}(t), \dot{\mathbf{x}}(t)) = \sum_k \frac{m_k}{2}\dot{\mathbf{x}}_k^2(t) + G_N \sum_{k\neq k'} \frac{m_k m_{k'}}{|\mathbf{x}_k(t) - \mathbf{x}_{k'}(t)|}. \tag{2.9}$$

If we identify the $3N$ coordinates x_n^i ($n = 1, \ldots, N$) with the $3N$ generalized coordinates q_k ($k = 1, \ldots, 3N$), the Euler-Lagrange equations (2.8) reduce precisely to Newton's equations (1.2).

The energy of a general Lagrangian system is found from the Lagrangian by forming the so-called *Hamiltonian*. It is defined by the *Legendre transform*

$$H = \sum_k p_k \dot{q}_k - L, \tag{2.10}$$

where

$$p_k \equiv \frac{\partial L}{\partial \dot{q}_k} \tag{2.11}$$

are called *canonical momenta*. The energy (2.10) forms the basis of the Hamiltonian formalism. If expressed in terms of p_k, q_k, the equations of motion are

$$\dot{q}_k = \frac{\partial H}{\partial p_k}, \quad \dot{p}_k = -\frac{\partial H}{\partial q_k}. \tag{2.12}$$

For the Lagrangian (2.9), the generalized momenta are equal to the physical momenta $p_n = m_n \dot{x}_n$, and the Hamiltonian is given by Newton's expression

$$H = T + V \equiv \sum_k \frac{m_k}{2}\dot{\mathbf{x}}_k^2 - G_N \sum_{k\neq k'} \frac{m_k m_{k'}}{|\mathbf{x}_k - \mathbf{x}_{k'}|}. \tag{2.13}$$

The first term is the kinetic energy, the second the potential energy of the system.

2.2   Single Relativistic Particle

For a single relativistic massive point particle, the mechanical action reads

$$\overset{m}{\mathcal{A}} = \int_{t_a}^{t_b} dt\,\overset{m}{L} = -mc^2 \int_{t_a}^{t_b} dt \sqrt{1 - \frac{\dot{\mathbf{x}}^2(t)}{c^2}}. \tag{2.14}$$

The canonical momenta (2.11) yield directly the spatial momenta of the particle:

$$\mathbf{p}(t) = \frac{\partial \overset{m}{L}}{\partial \dot{\mathbf{x}}(t)}. \tag{2.15}$$

In a relativistic notation, the derivative with respect to the contravariant vectors $\partial \overset{m}{L}/\partial \dot{x}^i$ is a covariant vector with a lower index i. To ensure the nonrelativistic identification (2.15) and maintain the relativistic notation we must therefore identify

$$p_i \equiv -\frac{\partial \overset{m}{L}}{\partial \dot{x}^i} = m\gamma \dot{x}_i. \tag{2.16}$$

The energy is obtained from the Legendre transform

$$\overset{m}{H} = \mathbf{p}\dot{\mathbf{x}} - \overset{m}{L} = -p_i \dot{x}^i - \overset{m}{L} = m\gamma \mathbf{v}^2 + mc^2\sqrt{1 - \frac{\mathbf{v}^2}{c^2}} = m\gamma \mathbf{v}^2 + mc^2 \frac{1}{\gamma}$$
$$= m\gamma c^2, \tag{2.17}$$

in agreement with the energy in Eq. (1.157) [recalling (1.152)].

The action (2.14) can be written in a more covariant form by observing that

$$\int_{t_a}^{t_b} dt \sqrt{1 - \frac{\dot{\mathbf{x}}^2}{c^2}} = \frac{1}{c} \int_{t_a}^{t_b} dt \sqrt{\left(\frac{dx^0}{dt}\right)^2 - \left(\frac{d\mathbf{x}}{dt}\right)^2}. \tag{2.18}$$

In this expression, the infinitesimal time element dt can be replaced by an arbitrary time-like parameter $t \to \sigma = f(t)$, so that the action takes the more general form

$$\overset{m}{\mathcal{A}} = \int_{\sigma_a}^{\sigma_b} d\sigma \overset{m}{L} = -mc \int_{\sigma_a}^{\sigma_b} d\sigma \sqrt{g_{ab}\dot{x}^a(\sigma)\dot{x}^b(\sigma)}. \tag{2.19}$$

For this action we may define generalized four-momentum with respect to the parameter σ by forming the derivatives

$$p_a(\sigma) \equiv -\frac{\partial \overset{m}{L}}{\partial \dot{x}^a(\sigma)} = \frac{mc}{\sqrt{g_{ab}\dot{x}^a(\sigma)\dot{x}^b(\sigma)}} g_{ab}\dot{x}^b(\sigma), \tag{2.20}$$

where the dots denote the derivatives with respect to the argument. Note the minus sign in the definition of the canonical momentum with respect to the nonrelativistic case. This is introduced to make the canonical formalism compatible with the negative signs in the spatial part of the Minkowski metric (1.29). The derivatives with respect to $\dot{x}^a$ transforms like the covariant components of a vector with a subscript a, whereas the physical momenta are given by the contravariant components p^a.

If σ is chosen to be the proper time τ, then the square root in (2.87) becomes τ-independent

$$\sqrt{g_{ab}\dot{x}^a(\tau)\dot{x}^b(\tau)} = c, \tag{2.21}$$

so that

$$p_a(\tau) = m \, g_{ab}\dot{x}^b(\tau) = m \, \dot{x}_a(\tau) = mu_a(\tau), \tag{2.22}$$

in agreement with the previously defined four-momenta in Eq. (1.152).

In terms of the proper time, the Euler-Lagrange equation reads

$$\frac{d}{d\tau}p_a(\tau) = m \, \frac{d}{d\tau}g_{ab}\dot{x}^b(\tau) = m \, \ddot{x}_a(\tau) = 0, \tag{2.23}$$

implying that free particles run along straight lines in Minkowski space.

Note that the Legendre transform with respect to the momentum $p_{\sigma,0}$ has nothing to do with the physical energy. In fact, it vanishes identically:

$$\overset{m}{H}_\sigma = -p_a(\sigma)\,\dot{x}^a(\sigma) - \overset{m}{L} = -\frac{mc}{\sqrt{\dot{x}_a(\sigma)\dot{x}^a(\sigma)}}\dot{x}_a(\sigma)\dot{x}^a(\sigma) + mc\,\sqrt{\dot{x}_a(\sigma)\dot{x}^a(\sigma)} \equiv 0. \tag{2.24}$$

The reason for this is the invariance of the action (2.19) under arbitrary reparametrizations of the time $\sigma \to \sigma' = f(\sigma)$. We shall understand this better in Chapter 3 when discussing generators of continuous symmetry transformations in general (see in particular Subsection 3.5.3).

The role of the physical energy is played by c times the zeroth component $p_0(\tau) = mc\gamma$ in Eq. (2.22), which is equal to the energy H in (2.17).

2.3   Scalar Fields

The free classical point particles of the last section are quanta of a relativistic local scalar free-field theory.

2.3.1   Locality

Generalizing the concept of temporal locality described in the paragraph after Eq. (2.1), locality in field theory implies that the action is a spacetime integral over the *Lagrangian density*:

$$\mathcal{A} = \int_{t_a}^{t_b} dt \int d^3x \, \mathcal{L}(x) = \frac{1}{c} \int d^4x \, \mathcal{L}(x). \tag{2.25}$$

According to the concept of temporal locality in Section 2.1, there should only be a quadratic dependence on the time derivatives of the fields. Due to the equal footing of space and time in relativistic theories, the same restriction applies to the space derivatives. A local Lagrangian density is at most quadratic in the first spacetime derivatives of the fields at the same point. Physically this implies that a field at a point x interacts at most with the field at the infinitesimally close neighbor point $x + dx$, just like the mass points on a linear chain with nearest-neighbor spring interactions. If the derivative terms are not of this form, they must at least be equivalent to it by a partial integration in the action integral (2.25). If the Lagrangian density is local we also call the action and the entire theory local.

A local free-field Lagrangian density is quadratic in both the fields and their derivatives at the same point, so that it reads for a scalar field,

$$\mathcal{L}(x) = \frac{1}{2} \left[\hbar^2 \partial_a \phi(x) \partial^a \phi(x) - m^2 c^2 \phi(x) \phi(x) \right]. \tag{2.26}$$

If the particles are charged, the fields are complex, and the Lagrangian density becomes

$$\mathcal{L}(x) = \hbar^2 \partial_a \varphi^*(x) \partial^a \varphi(x) - m^2 c^2 \varphi^*(x) \varphi(x). \tag{2.27}$$

2.3.2 Lorenz Invariance

In addition to being local, any relativistic Lagrangian density $\mathcal{L}(x)$ must be a scalar, i.e., transform under Lorentz transformations in the same way as the scalar field $\phi(x)$ in Eq. (1.168):

$$\mathcal{L}(x) \xrightarrow{\Lambda} \mathcal{L}'(x) = \mathcal{L}(\Lambda^{-1}x). \tag{2.28}$$

We verify this for the Lagrangian density (2.26) by showing that $\mathcal{L}'(x') = \mathcal{L}(x)$. By definition, $\mathcal{L}'(x')$ is equal to

$$\mathcal{L}'(x') = \hbar^2 \partial'_a \phi'(x') \partial'^a \phi(x') - m^2 c^2 \phi'(x') \phi'(x'). \tag{2.29}$$

Using the transformation behavior (1.168) of the scalar field, we obtain

$$\mathcal{L}'(x') = \hbar^2 \partial'_a \phi(x) \partial'^a \phi(x) - m^2 c^2 \phi(x) \phi(x). \tag{2.30}$$

Inserting here

$$\partial'_a = \Lambda_a{}^b \partial_b, \quad \partial'^a = \Lambda^a{}_b \partial^b \tag{2.31}$$

with

$$\Lambda_a{}^b \equiv g_{ac} g^{bd} \Lambda^c{}_d, \tag{2.32}$$

we see that ∂^2 is Lorentz-invariant,

$$\partial'^2 = \partial^2, \tag{2.33}$$

so that the transformed Lagrangian density (2.29) coincides indeed with the original one in (2.27).

As a spacetime integral over a scalar Lagrangian density, the action (2.25) is Lorentz invariant. This follows directly from the Lorentz invariance of the spacetime volume element

$$dx'^0 d^3x' = d^4x' = d^4x, \tag{2.34}$$

proved in Eq. (1.240). This is verified by direct calculation:

$$\mathcal{A}' = \int d^4x \, \mathcal{L}'(x) = \int d^4x' \, \mathcal{L}'(x') = \int d^4x' \, \mathcal{L}(x) = \int d^4x \, \mathcal{L}(x) = \mathcal{A}. \tag{2.35}$$

2.3.3   Field Equations

The equation of motion for the scalar field is derived by varying the action (2.25) with respect to the fields. Consider the case of complex fields $\varphi(x), \varphi^*(x)$ which must be varied independently. The independence of the field variables is expressed by the functional differentiation rules

$$\frac{\delta\varphi(x)}{\delta\varphi(x')} = \delta^{(4)}(x-x'), \qquad \frac{\delta\varphi^*(x)}{\delta\varphi^*(x')} = \delta^{(4)}(x-x'),$$

$$\frac{\delta\varphi(x)}{\delta\varphi^*(x')} = 0, \qquad \frac{\delta\varphi^*(x)}{\delta\varphi(x')} = 0. \tag{2.36}$$

With the help of these rules and the Leibnitz chain rule (1.118), we calculate the functional derivatives of the action (2.25) as follows:

$$\frac{\delta\mathcal{A}}{\delta\varphi^*(x)} = \int d^4x' \left[\hbar^2 \partial_a' \delta^{(4)}(x'-x)\partial'^a\varphi(x') - m^2c^2\delta^{(4)}(x'-x)\varphi(x') \right]$$

$$= (-\hbar^2\partial^2 - m^2c^2)\varphi(x) = 0. \tag{2.37}$$

$$\frac{\delta\mathcal{A}}{\delta\varphi(x)} = \int d^4x' \left[\hbar^2 \partial_a'\varphi^*(x')\partial'^a\delta^{(4)}(x'-x) - m^2c^2\varphi^*(x')\delta^{(4)}(x'-x) \right]$$

$$= \varphi^*(x)(-\hbar^2\overleftarrow{\partial}^2 + m^2c^2) = 0, \tag{2.38}$$

where the arrow pointing to the left on top of the last derivative indicates that it acts on the field to the left. The second equation is just the complex conjugate of the previous one.

The field equations can also be derived directly from the Lagrangian density (2.27) by forming ordinary partial derivatives of $\mathcal{L}$ with respect to all fields and their derivatives. Indeed, a functional derivative of a local action can be expanded in terms of derivatives of the Lagrangian density according to the general rule

$$\frac{\delta\mathcal{A}}{\delta\varphi(x)} = \frac{\partial\mathcal{L}(x)}{\partial\varphi(x)} - \partial_a\frac{\partial\mathcal{L}(x)}{\partial\left[\partial_a\varphi(x)\right]} + \partial_a\partial_b\frac{\partial\mathcal{L}(x)}{\partial\left[\partial_a\partial_b\varphi(x)\right]} + \dots, \tag{2.39}$$

and a complex-conjugate expansion for $\varphi^*(x)$. These expansions follow directly from the defining relations (2.36). At the extremum of the action, the field satisfies the Euler-Lagrange equation

$$\frac{\partial\mathcal{L}(x)}{\partial\varphi(x)} - \partial_a\frac{\partial\mathcal{L}(x)}{\partial\partial_a\varphi(x)} + \partial_a\partial_b\frac{\partial\mathcal{L}(x)}{\partial\partial_a\partial_b\varphi(x)} + \dots = 0. \tag{2.40}$$

Inserting the Lagrangian density (2.27), we obtain the field equation for $\varphi(x)$:

$$\frac{\delta\mathcal{A}}{\delta\varphi^*(x)} = \frac{\partial\mathcal{L}(x)}{\partial\varphi^*(x)} - \partial_a\frac{\partial\mathcal{L}(x)}{\partial\left[\partial_a\varphi^*(x)\right]} = (-\hbar^2\partial^2 + m^2c^2)\varphi(x) = 0, \tag{2.41}$$

and its complex conjugate for $\varphi^*(x)$.

The Euler-Lagrange equations are invariant under partial integrations in the action integral (2.25). Take for example a Lagrangian density which is equivalent to (2.27) by a partial integration:

$$\mathcal{L} = -\hbar^2 \varphi^*(x)\partial^2 \varphi(x) - m^2 c^2 \varphi^*(x)\varphi(x). \tag{2.42}$$

Inserted into (2.40), the field equation for $\varphi(x)$ becomes particularly simple:

$$\frac{\delta \mathcal{A}}{\delta \varphi^*(x)} = \frac{\partial \mathcal{L}(x)}{\partial \varphi^*(x)} = (-\hbar^2 \partial^2 + m^2 c^2)\varphi(x) = 0. \tag{2.43}$$

The derivation of the equation for $\varphi^*(x)$, on the other hand, becomes now more complicated. Evaluating all nonzero derivatives in Eq. (2.39), we simply find the complex-conjugate of Eq. (2.43):

$$\frac{\delta \mathcal{A}}{\delta \varphi(x)} = \frac{\partial \mathcal{L}(x)}{\partial \varphi(x)} - \partial_a \frac{\partial \mathcal{L}(x)}{\partial \left[\partial_a \varphi(x)\right]} + \partial_a \partial_b \frac{\partial \mathcal{L}(x)}{\partial \left[\partial_a \partial_b \varphi(x)\right]} = (-\hbar^2 \partial^2 + m^2 c^2)\varphi^*(x) = 0. \tag{2.44}$$

2.3.4 Plane Waves

The field equations (2.43) and (2.44) are solved by the quantum mechanical plane waves

$$f_{\mathbf{p}}(x) = \mathcal{N}_{p_0}\, e^{-ipx/\hbar}, \qquad f_{\mathbf{p}}^*(x) = \mathcal{N}_{p_0}\, e^{ipx/\hbar}, \tag{2.45}$$

where $\mathcal{N}_{p_0}$ is some normalization factor which may depend on the energy, and the four-momenta satisfy the so-called *mass shell condition*

$$p^a p_a - m^2 c^2 = 0. \tag{2.46}$$

It is important to realize that the two sets of solutions (2.45) are independent of each other. Physically, the main difference between them is the sign of energy

$$i\partial_0 f_{\mathbf{p}}(x) = p^0 f_{\mathbf{p}}(x), \qquad i\partial_0 f_{\mathbf{p}}^*(x) = -p^0 f_{\mathbf{p}}^*(x). \tag{2.47}$$

For this reason they will be referred to as positive- and negative-energy wave functions, respectively. The physical significance of the latter can only be understood after quantizing the field. Then they turn out to be associated with antiparticles. Field quantization, however, lies outside the scope of this text. Only at the end, in Subsection 22.2, will its effects on gravity be discussed.

2.3.5 Schrödinger Quantum Mechanics as Nonrelativistic Limit

The nonrelativistic limit of the action (2.25) for a scalar field with Lagrangian density (2.26) is obtained by removing from the positive frequency part of the field $\phi(x)$ a rapidly oscillating factor corresponding to the rest energy mc^2, replacing

$$\phi(x) \rightarrow e^{-imc^2 t/\hbar} \frac{1}{\sqrt{2M}} \psi(\mathbf{x}, t). \tag{2.48}$$

For a plane wave $f_p(x)$ in (2.45), the field $\psi(x)$ becomes $\mathcal{N}\sqrt{2M}e^{-i(p^0c-mc^2)t/\hbar}e^{i\mathbf{px}/\hbar}$. In the limit of large c, the first exponential becomes $e^{-ip^2t/2M}$ [recall (1.159)]. The result is a plane-wave solution to the Schrödinger equation

$$\left[i\hbar\partial_t + \frac{\hbar^2}{2M}\partial_{\mathbf{x}}^2\right]\psi(\mathbf{x}, t) = 0. \tag{2.49}$$

This is the Euler-Lagrange equation extremizing the nonrelativistic action

$$\mathcal{A} = \int dt d^3x\, \psi^*(\mathbf{x}, t)\left[i\hbar\partial_t + \frac{\hbar^2}{2M}\partial_{\mathbf{x}}^2\right]\psi(\mathbf{x}, t). \tag{2.50}$$

Note that the plane wave $f_p^*(x)$ in (2.45) with negative frequency does not possess a nonrelativistic limit since it turns into $\mathcal{N}\sqrt{2M}e^{i(p^0c+mc^2)t/\hbar}e^{i\mathbf{px}/\hbar}$ which has a temporal prefactor $e^{2imc^2t/\hbar}$. This oscillates infinitely fast for $c \to \infty$, and is therefore equivalent to zero by the *Riemann-Lebesgue Lemma*. This statement holds in the sense of distribution theory where a zero distribution means that all integrals over its products with arbitrary smooth test functions vanishes. According to the Riemann-Lebesgue lemma, this is indeed the case for all integral containing $e^{2imc^2t/\hbar}$ for large c multiplied by a smooth function.

2.3.6   Natural Units

The appearance of the constants $\hbar$ and c in all future formulas can be avoided by working with fundamental units l_0, m_0, t_0, E_0 different from the ordinary physical SMI or cgs units. They are chosen to give $\hbar$ and c the value 1. Expressed in terms of the conventional length, time, mass, and energy, these *natural units* are

$$l_0 = \frac{\hbar}{m_0 c}, \qquad t_0 = \frac{\hbar}{m_0 c^2}, \qquad m_0 = M, \qquad E_0 = m_0 c^2, \tag{2.51}$$

where M is some special mass. If we are dealing, for example, with a proton, we choose $M = m_p$, and the fundamental units are

$$
\begin{aligned}
l_0 &= 2.103138 \times 10^{-11}\text{cm} \\
&= \text{Compton wavelength of proton,}
\end{aligned}
\tag{2.52}
$$

$$
\begin{aligned}
t_0 &= l_0/c = 7.0153141 \times 10^{-22}\text{sec} \\
&= \text{time it takes light to cross the Compton wavelength,}
\end{aligned}
\tag{2.53}
$$

$$m_0 = m_p = 1.6726141 \times 10^{-24}\text{g}, \tag{2.54}$$

$$E_0 = 938.2592\,\text{MeV}. \tag{2.55}$$

For any other mass, they can easily be rescaled.

With these natural units we can drop c and $\hbar$ in all formulas and write the action simply as

$$\mathcal{A} = \int d^4x\, \varphi^*(x)(-\partial^2 - m^2)\varphi(x). \tag{2.56}$$

Actually, since we are dealing with relativistic particles, there is no fundamental reason to assume $\varphi(x)$ to be a complex field. In nonrelativistic field theory, this was necessary in order to find the time derivative term

$$\int dt d^3x \, \psi^*(\mathbf{x}, t) i\hbar \partial_t \psi(\mathbf{x}, t) \tag{2.57}$$

in the action (2.50). For a real field, this would be a pure surface term and thus have no influence upon the dynamics of the system. The second-order time derivatives of a relativistic field in (2.56), however, does have an influence and leads to the correct field equation for a real field. As we shall understand better in the next chapter, the complex scalar field describes charged spinless particles, the real field neutral particles.

Thus we may also consider a real scalar field with an action

$$\mathcal{A} = \frac{1}{2} \int d^4x \, \phi(x)(-\partial^2 - m^2)\phi(x). \tag{2.58}$$

In this case it is customary to use a prefactor $1/2$ to normalize the field.

For either Lagrangian (2.56) or (2.58), the Euler-Lagrange equation (2.40) renders *Klein-Gordon equations*

$$(-\partial^2 - m^2)\phi(x) = 0, \quad (-\partial^2 - m^2)\varphi(x) = 0, \quad (-\partial^2 - m^2)\varphi^*(x) = 0. \tag{2.59}$$

2.3.7 Hamiltonian Formalism

It is possible to set up a Hamiltonian formalism for the scalar fields. For this we introduce an appropriate generalization of the canonical momentum (2.11). The labels k in that equation are now replaced by the continuous spatial labels $\mathbf{x}$, and we define a density of *field momentum*:

$$\pi(x) \equiv \frac{\partial \mathcal{L}}{\partial \partial^0 \phi(x)} = \partial_0 \phi^*(x), \quad \pi^*(x) \equiv \frac{\partial \mathcal{L}}{\partial \partial^0 \phi^*(x)} = \partial_0 \phi(x), \tag{2.60}$$

and a *Hamiltonian density*:

$$\begin{aligned} \mathcal{H}(x) &= \pi(x) \, \partial^0 \phi(x) + \partial^0 \phi(x) \, \pi^*(x) - \mathcal{L}(x) \\ &= \pi^*(x) \, \pi(x) + \boldsymbol{\nabla} \phi^*(x) \, \boldsymbol{\nabla} \phi(x) + m^2 \phi^*(x) \, \phi(x). \end{aligned} \tag{2.61}$$

For a real field, we simply drop the complex conjugation symbols. The spatial integral over $\mathcal{H}(x)$ is the field Hamiltonian

$$H = \int d^3x \, \mathcal{H}(x). \tag{2.62}$$

2.3.8 Conserved Current

For the solutions $\psi(\mathbf{x}, t)$ of the Schrödinger equation (2.49), the probability density is

$$\rho(\mathbf{x}, t) \equiv \psi^*(\mathbf{x}, t)\psi(\mathbf{x}, t) \qquad (2.63)$$

and there exists an associated particle current density

$$\mathbf{j}(\mathbf{x}, t) \equiv \frac{\hbar}{2m\,i}\psi^*(\mathbf{x}, t)(\vec{\nabla} - \overleftarrow{\nabla})\psi(\mathbf{x}, t) \equiv \frac{\hbar}{2m\,i}\psi^*(\mathbf{x}, t)\overleftrightarrow{\nabla}\psi(\mathbf{x}, t), \qquad (2.64)$$

which satisfies the conservation law

$$\mathbf{\nabla} \cdot \mathbf{j}(\mathbf{x}, t) = -\partial_t\rho(\mathbf{x}, t). \qquad (2.65)$$

This can be verified with the help of the Schrödinger equation (2.49). It is this property which permits normalizing the Schrödinger field $\psi(\mathbf{x}, t)$ to unity at all times, since

$$\partial_t \int d^3x\, \psi^*(\mathbf{x}, t)\psi(\mathbf{x}, t) = \int d^3x\, \partial_t\rho(\mathbf{x}, t) = -\int d^3x\, \mathbf{\nabla} \cdot \mathbf{j}(\mathbf{x}, t) = 0. \qquad (2.66)$$

For a complex relativistic field $\varphi(x)$, there exists a similar local conservation law. We define the four-vector of probability current density (now in natural units with $\hbar = c = 1$)

$$j_a(x) = -i\varphi^* \overleftrightarrow{\partial}_a \varphi, \qquad (2.67)$$

which describes the probability flow of the charged scalar particle. The double arrow on top of the derivative is defined as in (2.64) by $\overleftrightarrow{\partial}_a \equiv \vec{\partial}_a - \overleftarrow{\partial}_a$, i.e.,

$$\varphi^* \overleftrightarrow{\partial}_a \varphi \equiv \varphi^*\partial_a\varphi - (\partial_a\varphi^*)\varphi. \qquad (2.68)$$

It is easy to verify that, on account of the Klein-Gordon equations in (2.59), the current density has no four-divergence:

$$\partial_a j^a(x) = 0. \qquad (2.69)$$

This four-dimensional *current conservation law* permits us to couple electromagnetism to the field and identify $ej^a(x)$ as the electromagnetic current of the charged scalar particles.

The deeper reason for the existence of a conserved current will be understood in Chapter 3, where we shall see that it is intimately connected with an invariance of the action (2.56) under arbitrary changes of the phase of the field

$$\phi(x) \to e^{-i\alpha}\phi(x). \qquad (2.70)$$

The zeroth component of $j^a(x)$,

$$\rho(x) = j^0(x), \qquad (2.71)$$

describes the charge density. The spatial integral over $\rho(x)$ measures the total probability. It measures the total charge in natural units:

$$Q(t) = \int d^3x \, j^0(x). \tag{2.72}$$

Due to the local conservation law (2.69), the total charge does not depend on time. This is seen by rewriting

$$\dot{Q}(t) = \int d^3x \, \partial_0 j^0(x) = \int d^3x \, \partial_a j^a(x) - \int d^3x \, \partial_i j^i(x) = - \int d^3x \, \partial_i j^i(x). \tag{2.73}$$

The right-hand side vanishes because of the Gauss divergence theorem, assuming the currents to vanish at spatial infinity [compare (1.204)].

2.4 Maxwell's Equation from Extremum of Field Action

The above action approach is easily generalized, and applied to electromagnetic fields. By setting up an appropriate action, Maxwell's field equations can be derived by extremization. The relevant fields are the Coulomb potential $\phi(\mathbf{x}, t)$ and the vector potential $\mathbf{A}(\mathbf{x}, t)$. Recall that electric and magnetic fields $\mathbf{E}(x)$ and $\mathbf{B}(x)$ can be written as derivatives of the Coulomb potential $A^0(\mathbf{x}, t)$ and the vector potential $\mathbf{A}(\mathbf{x}, t)$ as

$$\mathbf{E}(x) = -\boldsymbol{\nabla}\phi(x) - \frac{1}{c}\dot{\mathbf{A}}(x), \tag{2.74}$$

$$\mathbf{B}(x) = \boldsymbol{\nabla} \times \mathbf{A}(x), \tag{2.75}$$

with the components

$$E^i(x) = -\partial_i\phi(x) - \frac{1}{c}\partial_t A^i(x), \tag{2.76}$$

$$B^i(x) = \epsilon^{ijk}\partial_j A^k(x). \tag{2.77}$$

Recalling the identifications (1.172) of electric and magnetic fields with the components F^{i0} and $-F^{jk}$ of the covariant field tensor F^{ab}, we can also write

$$F^{i0}(x) = \partial^i\phi(x) - \frac{1}{c}\partial_t A^i(x), \tag{2.78}$$

$$F^{jk}(x) = \partial^j A^k(x) - \partial^k A^j(x), \tag{2.79}$$

where $\partial^i = -\partial_i$. This suggests combining the Coulomb potential and the vector potential into a four-component vector potential

$$A^a(x) = \begin{pmatrix} \phi(\mathbf{x}, t) \\ A^i(\mathbf{x}, t) \end{pmatrix}, \tag{2.80}$$

in terms of which the field tensor is simply the four-dimensional curl:

$$F^{ab}(x) = \partial^a A^b(x) - \partial^b A^a(x). \tag{2.81}$$

The field $A^a(x)$ transforms in the same way as the vector field $j^a(x)$ in Eq. (1.209):

$$A^a(x) \xrightarrow{\Lambda} A'^a(x) = \Lambda^a{}_b \, A^b(\Lambda^{-1}x). \tag{2.82}$$

2.4.1   Electromagnetic Field Action

Maxwell's equations can be derived from the electromagnetic field action

$$\overset{\text{em}}{\mathcal{A}} = \frac{1}{c} \int d^4x \, \overset{\text{em}}{\mathcal{L}}(x), \tag{2.83}$$

where the temporal integral runs from t_a to t_b, as in (2.1) and (2.25), and the Lagrangian density reads

$$\overset{\text{em}}{\mathcal{L}}(x) \equiv \overset{\text{em}}{\mathcal{L}}\left(A^a(x), \partial^b A^a(x)\right) = -\frac{1}{4}F^{ab}(x)F_{ab}(x) - \frac{1}{c}j^a(x)A_a(x). \tag{2.84}$$

It depends quadratically on the fields $A^a(x)$ and its derivatives, thus defining a local field theory [recall Subsection 2.3.1]. All Lorentz indices are fully contracted.

If (2.84) is decomposed into electric and magnetic parts using Eqs. (2.74) and (2.75), it reads

$$\overset{\text{em}}{\mathcal{L}}(x) = \frac{1}{2}\left[\mathbf{E}^2(x) - \mathbf{B}^2(x)\right] - \rho(x)A^0(x) + \frac{1}{c}\mathbf{j}(x)\mathbf{A}(x). \tag{2.85}$$

From the transformation laws (1.175), (1.209), and (2.82) it follows that (2.84) behaves under Lorentz transformations like a scalar field, as in (2.28). Together with (2.34) this implies that the action is Lorentz-invariant.

The field equations are obtained from the Euler-Lagrange equation (2.40) with the field $A^0(x)$ replaced by the four-vector potential $A^a(x)$, so that it reads

$$\frac{\partial \mathcal{L}}{\partial A^a} - \partial_b \frac{\partial \mathcal{L}}{\partial_b \partial A^a} + \partial_b \partial_c \frac{\partial \mathcal{L}(x)}{\partial \left[\partial_b \partial_c A^a(x)\right]} = 0. \tag{2.86}$$

Inserting the Lagrangian density (2.84), we obtain

$$\partial_b F^{ab} = -\frac{1}{c}j^a, \tag{2.87}$$

which is precisely the inhomogeneous Maxwell equation (1.196). Note that the homogeneous Maxwell equation (1.200),

$$\partial_b \tilde{F}^{ab} = 0, \tag{2.88}$$

is automatically fulfilled by the antisymmetric combination of derivatives in the *four-curl* (2.81). This is true as long as the four-component vector potential is smooth and single-valued, so that it satisfies the integrability condition

$$(\partial_a \partial_b - \partial_b \partial_a)A^c(x) = 0. \tag{2.89}$$

In this book we shall call any identity which follows from the single-valuedness of a field and the associated Schwarz integrability condition a *Bianchi identity*. The name emphasizes the close analogy with the identity discovered by Bianchi in Riemannian geometry as a consequence of the single-valuedness of the Christoffel symbols. For the derivation see Section 12.5, where the Schwarz integrability condition (12.106) leads to Bianchi's identity (12.115).

In this sense, the homogeneous Maxwell equation (2.88) is a Bianchi identity, since it follows directly from the commuting derivatives of A^c in Eq. (2.89).

2.4.2   Alternative Action for Electromagnetic Field

There exists an alternative form of the electromagnetic Lagrangian density (2.84) due to Schwinger which contains directly the field tensor as independent variables and uses the vector potential only as Lagrange multipliers to enforce the inhomogeneous Maxwell equations (2.87):

$$\overset{em}{\mathcal{L}}(x) = \overset{em}{\mathcal{L}}\left(A^a(x), F_{ab}(x)\right) = -\frac{1}{4}F^{ab}(x)F_{ab}(x) - \frac{1}{c}\left[j^a(x) + \partial_b F^{ab}(x)\right]A_a(x). \quad (2.90)$$

Extremizing this with respect to F_{ab} show that F_{ab} is a four-curl of the vector potential, as in Eq. (2.81). As a consequence, F_{ab} satisfies the Bianchi identity (1.201).

If (2.90) is decomposed into electric and magnetic parts, it reads

$$\overset{em}{\mathcal{L}}(x) = \overset{em}{\mathcal{L}}\left(A^0(x), \mathbf{A}(x), \mathbf{E}(x), \mathbf{B}(x)\right) = \frac{1}{4}\left[\mathbf{E}^2(x) - \mathbf{B}^2(x)\right]$$
$$+ \left[\boldsymbol{\nabla}\cdot\mathbf{E}(x) - \rho(x)\right]A^0(x) - \left[\boldsymbol{\nabla}\times\mathbf{B}(x) - \frac{1}{c}\partial_t\mathbf{E}(x) - \frac{1}{c}\mathbf{j}(x)\right]\cdot\mathbf{A}(x), \quad (2.91)$$

where the Lagrange multipliers $A^0(x)$ and $\mathbf{A}(x)$ enforce directly the Coulomb law (1.187) and the Ampère law (1.188).

The above equations hold only in the vacuum. In homogeneous materials with nonzero dielectric constant ε and magnetic permeability μ determining the displacement fields $\mathbf{D} = \varepsilon\mathbf{E}$ and the magnetic fields $H = \mathbf{B}/\mu$, the Lagrangian density (2.90) reads

$$\overset{em}{\mathcal{L}}(x) = \frac{1}{4}\left[\mathbf{E}(x)\cdot\mathbf{D}(x) - \mathbf{B}(x)\cdot\mathbf{H}(x)\right]$$
$$+ \left[\boldsymbol{\nabla}\cdot\mathbf{D}(x) - \rho(x)\right]A^0(x) - \left[\boldsymbol{\nabla}\times\mathbf{H}(x) - \frac{1}{c}\partial_t\mathbf{D}(x) - \frac{1}{c}\mathbf{j}(x)\right]\cdot\mathbf{A}(x). \quad (2.92)$$

Now variation with respect to the Lagrange multipliers $A^0(x)$ and $\mathbf{A}(x)$ yields the Coulomb and Ampère laws in a medium Eqs. (1.192) and (1.193):

$$\boldsymbol{\nabla}\cdot\mathbf{D}(x) = \rho(x), \qquad \boldsymbol{\nabla}\times\mathbf{H}(x) - \frac{1}{c}\partial_t\mathbf{D}(x) = \frac{1}{c}\mathbf{j}(x). \quad (2.93)$$

Variation with respect to $\mathbf{D}(x)$ and $\mathbf{H}(x)$ yields the same curl equations (2.74) and (2.75) as in the vacuum, so that the homogeneous Maxwell equations (1.189) and (1.190), i.e., the Bianchi identities (1.201), are unaffected by the medium.

2.4.3   Hamiltonian of Electromagnetic Fields

As in Eqs. (2.60)–(2.62), we can find a Hamiltonian for the electromagnetic fields, by defining a density of field momentum:

$$\pi_a(x) = \frac{\partial \overset{em}{\mathcal{L}}}{\partial \partial^0 A^a(x)} = -F_{0a}(x), \quad (2.94)$$

and a *Hamiltonian density*

$$\overset{\text{em}}{\mathcal{H}}(x) = \pi_a(x)\,\partial^0 A^a(x) - \overset{\text{em}}{\mathcal{L}}(x). \tag{2.95}$$

It is important to realize that $\partial\overset{\text{em}}{\mathcal{L}}/\partial\dot{A}^0$ vanishes, so that A^0 possesses no conjugate field momentum. Hence A^0 is not a proper dynamical variable. Indeed, by inserting (2.84) and (2.94) into (2.95) we find

$$
\begin{aligned}
\overset{\text{em}}{\mathcal{H}} &= -F_{0a}\partial^0 A^a - \overset{\text{em}}{\mathcal{L}} = -\frac{1}{c}F_{0a}F^{0a} - \overset{\text{em}}{\mathcal{L}} - F_{0a}\partial^a A^0 \\
&= \frac{1}{2}\left(\mathbf{E}^2 + \mathbf{B}^2\right) + \mathbf{E}\cdot\boldsymbol{\nabla}A^0 + \frac{1}{c}j^a A_a.
\end{aligned}
\tag{2.96}
$$

Integrating this over all space gives

$$\overset{\text{em}}{H} = c\int d^3x\,\overset{\text{em}}{\mathcal{H}} = \int d^3x\left[\frac{1}{2}\left(\mathbf{E}^2 + \mathbf{B}^2\right) - \frac{1}{c}\mathbf{j}\cdot\mathbf{A}\right]. \tag{2.97}$$

The result is the well-known energy of the electromagnetic field in the presence of external currents [6]. To obtain this expression from (2.96), an integration by part is necessary, in which the surface terms at spatial infinity is neglected, where the charge density $\rho(x)$ is always assumed to be zero. After this, Coulomb's law (1.187) leads directly to (2.97).

At first sight, one may wonder why the electrostatic energy does not show up explicitly in (2.97). The answer is that it is contained in the $\mathbf{E}^2$-term which, by Coulomb's law (1.187), satisfies

$$\boldsymbol{\nabla}\cdot\mathbf{E} = -\boldsymbol{\nabla}^2 A^0 - \frac{1}{c}\partial_t\boldsymbol{\nabla}\cdot\mathbf{A} = \rho. \tag{2.98}$$

Splitting $\mathbf{E}$ into transverse and longitudinal parts

$$\mathbf{E} = \mathbf{E}_t + \mathbf{E}_l, \tag{2.99}$$

which satisfy $\boldsymbol{\nabla}\cdot\mathbf{E}_t = 0$ and $\boldsymbol{\nabla}\times\mathbf{E}_l = 0$, respectively, we see that (2.98) implies

$$\boldsymbol{\nabla}\cdot\mathbf{E}_l = \rho. \tag{2.100}$$

The longitudinal part can be written as a derivative of some scalar potential ϕ',

$$\mathbf{E}_l = \boldsymbol{\nabla}\phi', \tag{2.101}$$

which can be calculated due to (2.100) from the equation

$$\phi'(x) = \frac{1}{\boldsymbol{\nabla}^2}\rho(x) = -\int d^3x'\frac{1}{4\pi|\mathbf{x}-\mathbf{x}'|}\rho(\mathbf{x}', t). \tag{2.102}$$

Using this we see that

$$
\begin{aligned}
\frac{1}{2}\int d^3x\,\mathbf{E}^2 &= \frac{1}{2}\int d^3x\left(\mathbf{E}_t^2 + \mathbf{E}_l^2\right) = \frac{1}{2}\int d^3x\left[\mathbf{E}_t^2 + \left(\partial_i\frac{1}{\boldsymbol{\nabla}^2}\phi'\right)\left(\partial_i\frac{1}{\boldsymbol{\nabla}^2}\phi'\right)\right] \\
&= \frac{1}{2}\int d^3x\,\mathbf{E}_t^2 + \frac{1}{2}\int d^3x\,d^3x'\rho(\mathbf{x}, t)\frac{1}{4\pi|\mathbf{x}-\mathbf{x}'|}\rho(\mathbf{x}', t).
\end{aligned}
\tag{2.103}
$$

The last term is the Coulomb energy associated with the charge density $\rho(\mathbf{x}, t)$.

2.4.4   Gauge Invariance of Maxwell's Theory

The four-dimensional curl (2.81) is manifestly invariant under the gauge transformations

$$A_a(x) \longrightarrow A'_a(x) = A_a(x) + \partial_a \Lambda(x), \tag{2.104}$$

where $\Lambda(x)$ is any smooth field which satisfies the integrability condition

$$(\partial_a \partial_b - \partial_b \partial_a)\Lambda(x) = 0. \tag{2.105}$$

Gauge invariance implies that a scalar field degree of freedom contained in $A^a(x)$ does not contribute to the physically observable electromagnetic fields $\mathbf{E}(x)$ and $\mathbf{B}(x)$. This degree of freedom can be removed by *fixing a gauge*. One way of doing this is to require the vector potential to satisfy the *Lorentz gauge condition*

$$\partial_a A^a(x) = 0. \tag{2.106}$$

For such a vector field, the field equations (2.87) decouple and each of the four components of the vector potential $A^a(x)$ satisfies the massless Klein-Gordon equation:

$$- \partial^2 A_b(x) = 0. \tag{2.107}$$

If a vector potential $A^a(x)$ does not satisfy the Lorentz gauge condition (2.106), one may always perform a gauge transformation (2.104) to a new field $A'^a(x)$ that has no four-divergence. We merely have to choose a gauge function $\Lambda(x)$ in (2.104) which solves the inhomogeneous differential equation

$$- \partial^2 \Lambda(x) = \partial_a A^a(x). \tag{2.108}$$

Then $A'^a(x)$ will satisfy $\partial_a A'^a(x) = 0$.

There are infinitely many solutions to equation (2.108). Given one solution $\Lambda(x)$ which leads to the Lorentz gauge, one can add any solution of the homogenous Klein-Gordon equation without changing the four-divergence of $A^a(x)$. The associated gauge transformations

$$\Lambda_a(x) \longrightarrow \Lambda_a(x) + \partial_a \Lambda'(x), \qquad \partial^2 \Lambda'(x) = 0, \tag{2.109}$$

are called *restricted gauge transformations* or *gauge transformation of the second kind*. If a vector potential $A^a(x)$ in the Lorentz gauge solves the field equations (2.87), the gauge transformations of the second kind can be used to remove its spatial divergence $\nabla \cdot \mathbf{A}(\mathbf{x}, t)$. Under (2.109), the components $A^0(\mathbf{x}, t)$ and $\mathbf{A}(\mathbf{x}, t)$ go over into

$$\begin{aligned} A^0(x) &\rightarrow A'^0(\mathbf{x}, t) = A^0(\mathbf{x}, t) + \partial_0 \Lambda'(\mathbf{x}, t), \\ \mathbf{A}(x) &\rightarrow \mathbf{A}'(\mathbf{x}, t) = \mathbf{A}(\mathbf{x}, t) - \nabla \Lambda'(\mathbf{x}, t). \end{aligned} \tag{2.110}$$

Thus, if we choose the gauge function

$$\Lambda'(\mathbf{x}, t) = - \int d^3 x' \frac{1}{4\pi |\mathbf{x} - \mathbf{x}'|} \nabla \cdot \mathbf{A}(\mathbf{x}', t), \tag{2.111}$$

then

$$\nabla^2 \Lambda'(\mathbf{x}, t) = \nabla \cdot \mathbf{A}(\mathbf{x}, t) \tag{2.112}$$

makes the gauge-transformed field $\mathbf{A}'(\mathbf{x}, t)$ divergence-free:

$$\nabla \cdot \mathbf{A}'(\mathbf{x}, t) = \nabla \cdot [\mathbf{A}(\mathbf{x}, t) - \nabla \Lambda(\mathbf{x}, t)] = 0. \tag{2.113}$$

The condition

$$\nabla \cdot \mathbf{A}'(\mathbf{x}, t) = 0 \tag{2.114}$$

is known as the *Coulomb gauge* or *radiation gauge*.

The solution (2.111) to the differential equation (2.112) is still undetermined up to an arbitrary solution $\Lambda''(x)$ of the homogeneous Poisson equation

$$\nabla^2 \Lambda''(\mathbf{x}, t) = 0. \tag{2.115}$$

Together with the property $\partial^2 \Lambda''(\mathbf{x}, t) = 0$ implied by (2.109), one also has

$$\partial_t^2 \Lambda''(\mathbf{x}, t) = 0. \tag{2.116}$$

This leaves only trivial linear functions $\Lambda''(\mathbf{x}, t)$ of $\mathbf{x}$ and t which contribute constants to (2.110). These, in turn, are zero since the fields $A^a(x)$ are always assumed to vanish at infinity before and after the gauge transformation.

Another possible gauge is obtained by removing, in the field equation (2.87), the zeroth component of the vector potential $A^a(x)$. This is achieved by performing the gauge transformation (2.104) with a gauge function

$$\Lambda(\mathbf{x}, t) = -\int^t dt'\, A_0(\mathbf{x}, t'), \tag{2.117}$$

instead of (2.111). The new field $A'^a(x)$ has no zeroth component:

$$A'^0(x) = 0. \tag{2.118}$$

It is said to be in the *axial gauge*. The solutions of Eqs. (2.117) are determined up to a trivial constant, and no further gauge freedom is left.

For free fields, the Coulomb gauge and the axial gauge coincide. This is a consequence of the charge-free Coulomb law $\nabla \cdot \mathbf{E} = 0$ in Eq. (2.98). By expressing $\mathbf{E}(x)$ explicitly in terms of the space- and time-like components of the vector potential as

$$\mathbf{E}(x) = -\partial_0 \mathbf{A}(x) - \nabla A^0(x), \tag{2.119}$$

Coulomb's law reads

$$\nabla^2 A^0(\mathbf{x}, t) = -\nabla \cdot \dot{\mathbf{A}}(\mathbf{x}, t). \tag{2.120}$$

This shows that if $\boldsymbol{\nabla} \cdot \mathbf{A}(x) = 0$, also $A^0(x) = 0$ (assuming zero boundary values at infinity), and vice versa.

The differential equation (2.120) can be integrated to

$$A^0(\mathbf{x}, t) = \frac{1}{4\pi} \int d^3x' \frac{1}{|\mathbf{x}' - \mathbf{x}|} \boldsymbol{\nabla} \cdot \dot{\mathbf{A}}(\mathbf{x}', t). \tag{2.121}$$

In an infinite volume with asymptotically vanishing fields there is no freedom of adding to the left-hand side a nontrivial solution of the homogenous Poisson equation

$$\boldsymbol{\nabla}^2 A^0(\mathbf{x}, t) = 0. \tag{2.122}$$

In the presence of charges, Coulomb's law has a source term [see Eq. (2.98)]:

$$\boldsymbol{\nabla} \cdot \mathbf{E}(\mathbf{x}, t) = \rho(\mathbf{x}, t), \tag{2.123}$$

where $\rho(\mathbf{x}, t)$ is the electric charge density. In this case the vanishing of $\boldsymbol{\nabla} \cdot \mathbf{A}(\mathbf{x}, t)$ no longer implies $A^0(\mathbf{x}, t) \equiv 0$. Then one has the possibility of choosing $\Lambda(\mathbf{x}, t)$ either to satisfy the Coulomb gauge

$$\boldsymbol{\nabla} \cdot \mathbf{A}(\mathbf{x}, t) \equiv 0, \tag{2.124}$$

or the axial gauge

$$A^0(\mathbf{x}, t) \equiv 0. \tag{2.125}$$

Only for free fields the two gauges coincide.

In a fixed gauge, the vector potential $A^a(x)$ does not, in general, transform as a four-vector field under Lorentz transformations, which according to (1.209) and (1.216) would imply

$$A^a(x) \xrightarrow{\Lambda} A'^a_\Lambda(x) = \Lambda^a{}_b A^b(\Lambda^{-1}x) = [e^{-i\frac{1}{2}\omega_{ab}\hat{J}^{ab}} A]^a(\Lambda^{-1}x). \tag{2.126}$$

This is only true if the gauge is fixed in a Lorentz-invariant way, for instance by the Lorentz gauge condition (2.106). In the Coulomb gauge, the right-hand side of (2.126) will be modified by an additional gauge transformation depending on Λ which ensures the Coulomb gauge for the transformed vector potential.

2.5 Maxwell-Lorentz Action for Charged Point Particles

Consider now charged relativistic massive particles interacting with electromagnetic fields and derive the Maxwell-Lorentz equations of Section 1.10 from the action approach. A single particle of charge e carries a current

$$j^a(x) = ec \int_{-\infty}^{\infty} d\tau \, \dot{q}^a(\tau) \delta^{(4)}(x - q(\tau)), \tag{2.127}$$

and the total action in an external field is given by the sum of (2.83) and (2.19):

$$\mathcal{A} = \overset{\text{em}}{\mathcal{A}} + \overset{\text{m}}{\mathcal{A}} = -\frac{1}{4} \int d^4x \, F^{ab}(x) F_{ab}(x) - mc^2 \int_{\tau_a}^{\tau_b} d\tau - \frac{1}{c} \int d^4x \, j^a(x) A_a(x). \tag{2.128}$$

In terms of the physical time t, the last two terms can be separated into spatial and time-like components as follows:

$$-mc^2 \int_{t_a}^{t_b} dt \sqrt{1 - \frac{\dot{\mathbf{q}}^2}{c^2}} - e \int_{t_a}^{t_b} dt \, A^0(\mathbf{q}(t), t) + \frac{e}{c} \int_{t_a}^{t_b} dt \, \mathbf{v} \cdot \mathbf{A}(\mathbf{q}(t), t). \quad (2.129)$$

The equations of motion are obtained by writing the free-particle action in the form (2.19), and extremizing (2.128) with respect to variations $\delta q^a(\tau)$. This yields the *Maxwell-Lorentz equations* (1.170):

$$
\begin{aligned}
m \frac{d^2 q^a}{d\tau^2} &= \frac{e}{c} \left[-\frac{\partial}{\partial \tau} A^a + \frac{dq^b}{d\tau} \partial^a A_b \right] = \frac{e}{c} \left[-\frac{dq^b}{d\tau} \partial_b A^a + \frac{dq^b}{d\tau} \partial^a A_b \right] \\
&= \frac{e}{c} F_{ab} \frac{dq^b}{d\tau}.
\end{aligned}
\quad (2.130)
$$

On the right-hand side we recognize the Lorentz force (1.184).

Note that in the presence of electromagnetic fields, the canonical momenta (2.11) are no longer equal to the physical momenta as in (2.15), but receive a contribution from the vector potential:

$$P_i = -\frac{\partial L}{\partial \dot{q}^i} = -(m\gamma \dot{q}^i + \frac{e}{c} A^i) = p_i + \frac{e}{c} A_i. \quad (2.131)$$

Including the zeroth component, the canonical four-momentum is

$$P_a = p_a + \frac{e}{c} A_a. \quad (2.132)$$

The zeroth component of P_a coincides with $1/c$ times the energy defined by the Legendre transform [recall (2.97)]:

$$P_0 = \frac{1}{c}(H + eA^0) = -\frac{1}{c}(P_i \dot{q}^i - L). \quad (2.133)$$

2.6   Scalar Field with Electromagnetic Interaction

The spacetime derivatives of a plane wave such as (1.160) yields the energy-momentum of the particle whose probability amplitude is described by the wave:

$$i\hbar \partial_a \phi_p(x) = p_a \phi_p(x). \quad (2.134)$$

In the presence of electromagnetism, the role of the momentum four-vector is taken over by the momenta (2.132). In the Lagrangian density (2.27) of the scalar field, this is accounted for by the so-called *minimal replacement*, in which ordinary derivatives are replaced by the *covariant derivatives*:

$$\partial_a \phi(x) \to D_a \phi(x) \equiv \left[\partial_a + i \frac{e}{c\hbar} A_a(x) \right] \phi(x). \quad (2.135)$$

The Lagrangian density of a scalar field with electromagnetic interactions is therefore

$$\mathcal{L}(x) = \hbar^2[D_a\phi(x)]^*D^a\phi(x) - m^2c^2\phi^*(x)\phi(x) - \frac{1}{4}\int d^4x\, F^{ab}(x)F_{ab}(x). \qquad (2.136)$$

It governs systems of charged spinless particles with the laws of *scalar electrodynamics*.

This expression is invariant under local gauge transformations (2.104) of the electromagnetic field, if we simultaneously multiply the scalar field by an x-dependent phase factor:

$$\varphi(x) \rightarrow e^{ie\Lambda(x)/c}\varphi(x). \qquad (2.137)$$

By extremization of the action in natural units $\mathcal{A} = \int d^4x\,\mathcal{L}(x)$ we find the Euler-Lagrange equation and its conjugate

$$\frac{\delta\mathcal{A}}{\delta\varphi^*(x)} = (-D^2 - m^2)\varphi(x), \qquad \frac{\delta\mathcal{A}}{\delta\varphi(x)} = (-D^{*2} - m^2)\varphi^*(x). \qquad (2.138)$$

In the presence of the electromagnetic field, the particle current density (2.67) turns into the charge current density

$$j_a(x) = e\frac{i}{2}\phi^*D_a\phi + \text{c.c.} = e\frac{i}{2}\phi^*\overleftrightarrow{\partial_a}\phi - \frac{e^2}{c}A_a(x)\phi^*\phi. \qquad (2.139)$$

This satisfies the same conservation law (2.69) as the current density of the free scalar field, as we can verify by a short calculation:

$$\partial_a j^a = \partial_a\left[\frac{i}{2}\phi^*D^a\phi\right] + \text{c.c.} = \frac{i}{2}\partial_a\phi^*D^a\phi + \frac{i}{2}\phi^*\partial_aD^a\phi + \text{c.c.} \qquad (2.140)$$

$$= \frac{i}{2}\partial_a\phi^*D^a\phi + \frac{i}{2}\phi^*D^2\phi - \frac{i}{2}\frac{e}{c}A^a\phi^*D_a\phi + \text{c.c.} = \frac{i}{2}D_a^*\phi^*D^a\phi - m^2\frac{i}{2}\phi^*\phi + \text{c.c.} = 0.$$

2.7   Dirac Fields

An action whose extremum yields the Dirac equation (1.218) is, in natural units,

$$\overset{D}{\mathcal{A}} = \int d^4x\, \overset{D}{\mathcal{L}}(x) \equiv \int d^4x\, \bar{\psi}(x)\,(i\gamma^a\partial_a - m)\,\psi(x) \qquad (2.141)$$

where

$$\bar{\psi}(x) \equiv \psi^\dagger(x)\gamma^0, \qquad (2.142)$$

and the matrices γ^a satisfy the anticommutation rules (1.224). The Dirac equation and its conjugate are obtained from the extremal principle

$$\frac{\delta\overset{D}{\mathcal{A}}}{\delta\bar{\psi}(x)} = (i\gamma^a\partial_a - m)\psi(x) = 0, \qquad \frac{\delta\overset{D}{\mathcal{A}}}{\delta\psi(x)} = \bar{\psi}(x)(-i\gamma^a\overleftarrow{\partial_a} - m)\psi(x) = 0. \qquad (2.143)$$

The action (2.141) is invariant under the Lorentz transformations of spinors (1.236). The invariance of the mass term follows from the fact that

$$D^\dagger(\Lambda)\gamma^0 D(\Lambda) = \gamma^0. \tag{2.144}$$

This equation is easily verified by inserting the explicit matrices (1.219) and (1.233). If we define

$$\bar{D} \equiv \gamma^0 D^\dagger \gamma^0, \tag{2.145}$$

this implies

$$\bar{D}(\Lambda)D(\Lambda) = 1, \tag{2.146}$$

so that the mass term in the Lagrangian density transforms like a scalar field in (1.168):

$$\bar{\psi}(x)\psi(x) \xrightarrow{\Lambda} \bar{\psi}'_\Lambda(x)\psi'_\Lambda(x) = \bar{\psi}(\Lambda^{-1}x)\psi(\Lambda^{-1}x). \tag{2.147}$$

Consider now the gradient term in the action (2.141). Its invariance is a consequence of the vector property of the Dirac matrices with respect to the spinor representation of the Lorentz group derived in Eq. (1.235). From (2.146) we deduce that $D^{-1}(\Lambda) = \bar{D}(\Lambda)$, which allows us to write the vector transformation law (2.146) as

$$D(\Lambda)\gamma^a \bar{D}(\Lambda) = (\Lambda^{-1})^a{}_b\gamma^b, \qquad \bar{D}(\Lambda)\gamma^a D(\Lambda) = D^{-1}(\Lambda)\gamma^a D(\Lambda) = \Lambda^a{}_b\gamma^b. \tag{2.148}$$

From this we derive at once that

$$\bar{\psi}(x)\gamma^a\psi(x) \xrightarrow{\Lambda} \bar{\psi}'(x)\gamma^a\psi'(x) = \Lambda^a{}_b\bar{\psi}(\Lambda^{-1}x)\gamma^b\psi(\Lambda^{-1}x), \tag{2.149}$$

and

$$\bar{\psi}(x)\gamma^a\partial_a\psi(x) \xrightarrow{\Lambda} \bar{\psi}'(x)\gamma^a\partial_a\psi'(x) = [\bar{\psi}\gamma^b\partial_a\psi](\Lambda^{-1}x). \tag{2.150}$$

Thus also the gradient term in the Dirac Lagrangian density transforms like a scalar field, and so does the full Lagrangian density as in (2.28), which makes the action (2.141) invariant under Lorentz transformations, due to (2.34).

From the discussion in Section 2.6 we know how to couple the Dirac field to electromagnetism. We simply have to replace the derivative in the Lagrangian density by the covariant derivative (2.135), and obtain the gauge-invariant Lagrangian density of the electrodynamics

$$\mathcal{L}(x) = \bar{\psi}(x)\left(i\gamma^a D_a - m\right)\psi(x) - \frac{1}{4}\int d^4x\, F^{ab}(x)F_{ab}(x). \tag{2.151}$$

This equation is invariant under local gauge transformations (2.104), if we simultaneously multiply the Dirac field by an x-dependent phase factor

$$\psi(x) \rightarrow e^{ie\Lambda(x)/c}\psi(x). \tag{2.152}$$

The interaction term in this Lagrangian density comes entirely from the covariant derivative and reads, more explicity,

$$\mathcal{L}^{\text{int}}(x) = -\frac{1}{c}\int d^4x\, A_a(x)j^a(x), \tag{2.153}$$

where

$$j^a(x) \equiv e\,\bar{\psi}(x)\gamma^a\psi(x) \tag{2.154}$$

is the current density of the electrons.

By extremizing the action $\overset{D}{\mathcal{A}} = \int d^4x\, \overset{D}{\mathcal{L}}(x)$ we now find the Euler-Lagrange equation and its conjugate

$$\frac{\delta \overset{D}{\mathcal{A}}}{\delta \bar{\psi}(x)} = (i\gamma^a D_a - m)\psi(x) = 0, \qquad \frac{\delta \overset{D}{\mathcal{A}}}{\delta \psi(x)} = \bar{\psi}(x)(-i\gamma^a \overset{\leftarrow}{D}{}^*_a - m)\psi(x) = 0. \tag{2.155}$$

For classical fields obeying these equations, the current density (2.154) satisfies the same local conservation law as the scalar field in Eq. (2.140):

$$\partial_a j^a(x) = 0. \tag{2.156}$$

This can be verified by a much simpler calculation than that in Eq. (2.140):

$$\partial_a j^a = e\partial_a(\bar{\psi}\gamma^a\psi) = e\bar{\psi}\gamma^a\overset{\leftarrow}{\partial}_a\psi + e\bar{\psi}\gamma^a\partial_a\psi = e\bar{\psi}\gamma^a\overset{\leftarrow}{D}{}^*_a\psi + e\bar{\psi}\gamma^a D_a\psi = 0. \tag{2.157}$$

2.8 Quantization

Given the action of a field, there exist two ways of quantizing the theory. One is based on the good-old operator approach. This will not be followed here. We shall proceed on the basis of Feynman's theory of path integrals. Feynman observed that the physical amplitude for an event to happen is found by a sum over all classical histories leading to this event. Each history carries a probability amplitude $e^{i\mathcal{A}/\hbar}$, where $\mathcal{A}$ is the action of the system.

For a field theory the sum over histories runs over all possible time-dependent field configurations

$$\text{Amplitude} = \sum_{\text{field configurations}} e^{i\mathcal{A}/\hbar}. \tag{2.158}$$

An important advantage of this formulation of quantum theory is that the time t may be continued analytically to an imaginary time $\tau = -it$ leading to a corresponding statistical theory. This is obvious from the similarity of the time evolution operator of quantum mechanics $e^{it\hat{H}/\hbar}$ and the Boltzmann factor $e^{-\beta\hat{H}}$ of statistical physics, where β is related to the inverse temperature by the Boltzmann constants:

$$\beta = 1/k^B T \tag{2.159}$$

The analytic continuation can be performed directly on the action $\mathcal{A}$ in which case one obtains

$$\mathcal{A} = \int dt \int d^3x \mathcal{L}(\mathbf{x}, t) = i\int d\tau \int d^3x \mathcal{L}(\mathbf{x}, -i\tau) \equiv i\int d\tau \int d^3x \mathcal{L}^E(\mathbf{x}, \tau) \equiv i\mathcal{A}^E. \tag{2.160}$$

The resulting $\mathcal{A}^E$ is called the *Euclidean action*, and $\mathcal{L}^E$ the associated *Euclidean Lagrangian density*. The name alludes to the fact that under the continuation to $t = -i\tau$, the Minkowski scalar products $xx' = c^2tt' - \mathbf{xx'}$ go over into the Euclidean ones $xx' = -(\tau\tau' + \mathbf{xx'}) = -x^E x'^E$. Here we have introduced the Euclidean vector $x^E \equiv (c\tau, \mathbf{x})$. Denoting the volume element Euclidean spacetime $d\tau d^3x$ by d^4x^E, we may write the Euclidean action as

$$\mathcal{A}^E = \int d^4x^E \, \mathcal{L}^E(\tau, \mathbf{x}). \tag{2.161}$$

The quantum mechanical amplitude goes over into

$$Z = \sum_{\text{field configurations}} e^{-\mathcal{A}^E/\hbar}. \tag{2.162}$$

This is the quantum statistical partition function of the same system. A finite temperature T can be imposed by letting all fields be periodic in the interval $(0, \hbar\beta)$. Fermi fields have to be antiperiodic to account for the anticommuting statistics.

Notes and References

[1] R.P. Feynman and A.R. Hibbs, *Quantum Mechanics and Path Integrals*, McGraw-Hill, New York, 1965.

[2] H. Kleinert, *Path Integrals in Quantum Mechanics, Statistics, Polymer Physics, and Financial Markets*, World Scientific Publishing Co., Singapore 2004, 4th extended edition, pp. 1–1547 (kl/b5), where kl is short for the www address http://www.physik.fu-berlin.de/~kleinert.

[3] L.D. Landau and E.M. Lifshitz, *The Classical Theory of Fields*, Pergamon Press, Oxford, 1975.

[4] H. Kleinert, *Gauge Fields in Condensed Matter*, Vol. I: *Superflow and Vortex Lines, Disorder Fields, Phase Transitions*, World Scientific, Singapore, 1989 (kl/b1).

[5] H. Kleinert, *Gauge Fields in Condensed Matter*, Vol. II: *Stresses and Defects, Differential Geometry, Crystal Defects*, World Scientific, Singapore, 1989 (kl/b2).

[6] J.D. Bjorken and S.D. Drell, *Relativistic Quantum Fields*, McGraw-Hill, New York, 1956, Sect. 15.2.

3

Continuous Symmetries and Conservation Laws. Noether's Theorem

In many physical systems, the action is invariant under some continuous set of transformations. If this is the case, there exist local and global *conservation laws* analogous to current and charge conservation in electrodynamics. With the help of Poisson brackets, the analogs of the charges can be used to generate the symmetry transformation from which they were derived. After field quantization, the Poisson brackets become commutators of operators associated with these charges.

3.1 Continuous Symmetries and Conservation Laws

Consider first a simple mechanical system with a generic action

$$\mathcal{A} = \int_{t_a}^{t_b} dt\, L(q(t), \dot{q}(t)), \tag{3.1}$$

and subject it to a continuous set of local transformations of the dynamical variables:

$$q(t) \rightarrow q'(t) = f(q(t), \dot{q}(t)), \tag{3.2}$$

where $f(q(t), \dot{q}(t))$ is some function of $q(t)$ and $\dot{q}(t)$. In general, $q(t)$ will carry various labels as in (2.1) which are suppressed, for brevity. If the transformed action

$$\mathcal{A}' \equiv \int_{t_a}^{t_b} dt\, L(q'(t), \dot{q}'(t)) \tag{3.3}$$

is the same as $\mathcal{A}$, up to boundary terms, then (3.2) is called a symmetry transformation.

3.1.1 Group Structure of Symmetry Transformations

For any two *symmetry transformations*, we may define a product by performing the transformations successively. The result is certainly again a symmetry transformation. All such transformations can, of course, be undone, i.e., they possess an inverse. Thus, symmetry transformations form a group called the *symmetry group* of the system. When testing the equality of the actions $\mathcal{A}'$ and $\mathcal{A}$, up to boundary terms, it is important *not* to use the equations of motion.

3.1.2   Substantial Variations

For infinitesimal symmetry transformations (3.2), the difference

$$\delta_s q(t) \equiv q'(t) - q(t) \tag{3.4}$$

is called a *symmetry variation*. It has the general form

$$\delta_s q(t) = \epsilon \Delta(q(t), \dot{q}(t)), \tag{3.5}$$

where ϵ is a small parameter. Symmetry variations must not be confused with the variations $\delta q(t)$ which were used in Section 2.1 to derive the Euler-Lagrange equations (2.8). Those variations always vanish at the ends, $\delta q(t_b) = \delta q(t_a) = 0$ [recall (1.4)]. For symmetry variation $\delta_s q(t)$, this is in general not true.

Another name for the symmetry variation (3.5) is *substantial variation*. It is defined for any function of spacetime $f(x)$ as the difference between $f(x)$ and a transformed function $f'(x)$:

$$\delta_s f(x) \equiv f(x) - f'(x). \tag{3.6}$$

Note that the functions $f(x)$ and $f'(x)$ are evaluated at the *same values of the coordinates* x which may correspond to two *different points* in space.

3.1.3   Conservation Laws

Let us calculate the change of the action under a substantial variation (3.5). Using the chain rule of differentiation and a partial integration we obtain

$$\delta_s \mathcal{A} = \int_{t_a}^{t_b} dt \left[\frac{\partial L}{\partial q(t)} - \partial_t \frac{\partial L}{\partial \dot{q}(t)} \right] \delta_s q(t) + \frac{\partial L}{\partial \dot{q}(t)} \delta_s q(t) \Big|_{t_a}^{t_b}. \tag{3.7}$$

For solutions of the Euler-Lagrange equations (2.8), the first term vanishes, and only the second term survives. We shall denote such solutions by $q_{cl}(t)$ and call them *classical orbits*. For a classical orbit, the action changes by

$$\delta_s \mathcal{A} = \frac{\partial L}{\partial \dot{q}(t)} \delta_s q(t) \Big|_{t_a}^{t_b} = \epsilon \frac{\partial L}{\partial \dot{q}} \Delta(q, \dot{q}) \Big|_{t_b}^{t_a}, \quad \text{for} \quad q(t) = q_{cl}(t). \tag{3.8}$$

By assumption, $\delta_s q(t)$ is a symmetry transformation of $\mathcal{A}$ implying that $\delta_s \mathcal{A}$ vanishes or is equal to a boundary term for *all* paths $q(t)$. In the first case, the quantity

$$Q(t) \equiv \frac{\partial L}{\partial \dot{q}} \Delta(q, \dot{q}), \quad \text{for} \quad q(t) = q_{cl}(t) \tag{3.9}$$

is the same at times $t = t_a$ and $t = t_b$. Since t_b is arbitrary, $Q(t)$ is *independent* of the time t:

$$Q(t) \equiv Q. \tag{3.10}$$

Thus $Q(t)$ is a *constant of motion* along the orbit, it is a *conserved quantity*, called *Noether charge*.

In the second case where $\delta_s \mathcal{A}$ is equal to a boundary term,

$$\delta_s \mathcal{A} = \epsilon \Lambda(q, \dot{q})\Big|_{t_a}^{t_b}, \qquad (3.11)$$

the conserved Noether charge becomes

$$Q(t) = \frac{\partial L}{\partial \dot{q}} \Delta(q, \dot{q}) - \Lambda(q, \dot{q}), \quad \text{for} \ \ q(t) = q_{\text{cl}}(t). \qquad (3.12)$$

It is possible to derive the constant of motion (3.12) without invoking the action, starting from the Lagrangian $L(q, \dot{q})$. We expand its substantial variation of $L(q, \dot{q})$ as follows:

$$\delta_s L \equiv L\left(q + \delta_s q, \dot{q} + \delta_s \dot{q}\right) - L(q, \dot{q}) = \left[\frac{\partial L}{\partial q(t)} - \partial_t \frac{\partial L}{\partial \dot{q}(t)}\right] \delta_s q(t) + \frac{d}{dt}\left[\frac{\partial L}{\partial \dot{q}(t)} \delta_s q(t)\right].$$
$$(3.13)$$

On account of the Euler-Lagrange equations (2.8), the first term on the right-hand side vanishes as before, and only the last term survives. The assumption of invariance of the action up to a possible surface term in Eq. (3.11) is equivalent to assuming that the substantial variation of the Lagrangian is at most a *total time derivative* of some function $\Lambda(q, \dot{q})$:

$$\delta_s L(q, \dot{q}, t) = \epsilon \frac{d}{dt} \Lambda(q, \dot{q}). \qquad (3.14)$$

Inserting this into the left-hand side of (3.13), we find the equation

$$\epsilon \frac{d}{dt}\left[\frac{\partial L}{\partial \dot{q}} \Delta(q, \dot{q}) - \Lambda(q, \dot{q})\right] = 0, \quad \text{for} \ \ q(t) = q_{\text{cl}}(t), \qquad (3.15)$$

thus recovering again the conserved Noether charge (3.12).

3.1.4 Alternative Derivation of Conservation Laws

Let us subject the action (3.1) to an arbitrary variation $\delta q(t)$, which may be nonzero at the boundaries. Along a classical orbit $q_{\text{cl}}(t)$, the first term in (3.7) vanishes, and the action changes at most by the boundary term:

$$\delta \mathcal{A} = \frac{\partial L}{\partial \dot{q}} \delta q \Big|_{t_a}^{t_b}, \quad \text{for} \ \ q(t) = q_{\text{cl}}(t). \qquad (3.16)$$

This observation leads to another derivation of Noether's theorem. Suppose we subject $q(t)$ to a so-called *local symmetry transformations*, which generalizes the previous substantial variations (3.5) to a *time-dependent* parameter $\epsilon(t)$:

$$\delta_s^t q(t) = \epsilon(t) \Delta(q(t), \dot{q}(t)). \qquad (3.17)$$

The superscript t on $\delta_{\rm s}^t$ emphasized the extra time dependence in $\epsilon(t)$. If the variations (3.17) are performed on a classical orbit $q_{\rm cl}(t)$, the action changes by the boundary term (3.16).

This will now be expressed in a more convenient way. For this purpose we introduce the infinitesimally transformed orbit

$$q^{\epsilon(t)}(t) \equiv q(t) + \delta_{\rm s}^t q(t) = q(t) + \epsilon(t)\Delta(q(t), \dot{q}(t)), \tag{3.18}$$

and the transformed Lagrangian

$$L^{\epsilon(t)} \equiv L(q^{\epsilon(t)}(t), \dot{q}^{\epsilon(t)}(t)). \tag{3.19}$$

Then the local substantial variation of the action with respect to the time-dependent parameter $\epsilon(t)$ is

$$\delta_{\rm s}^t \mathcal{A} = \int_{t_a}^{t_b} dt \left[\frac{\partial L^{\epsilon(t)}}{\partial \epsilon(t)} - \frac{d}{dt} \frac{\partial L^{\epsilon(t)}}{\partial \dot{\epsilon}(t)} \right] \epsilon(t) + \frac{d}{dt} \left[\frac{\partial L^{\epsilon(t)}}{\partial \dot{\epsilon}} \right] \epsilon(t) \Bigg|_{t_a}^{t_b}. \tag{3.20}$$

Along a classical orbit, the action is extremal. Hence the infinitesimally transformed action

$$\mathcal{A}^\epsilon \equiv \int_{t_a}^{t_b} dt \, L(q^{\epsilon(t)}(t), \dot{q}^{\epsilon(t)}(t)) \tag{3.21}$$

must satisfy the equation

$$\frac{\delta \mathcal{A}^\epsilon}{\delta \epsilon(t)} = 0. \tag{3.22}$$

This holds for an arbitrary time dependence of $\epsilon(t)$, in particular for $\epsilon(t)$ which vanishes at the ends. In this case, (3.22) leads to an Euler-Lagrange type of equation

$$\frac{\partial L^{\epsilon(t)}}{\partial \epsilon(t)} - \frac{d}{dt} \frac{\partial L^{\epsilon(t)}}{\partial \dot{\epsilon}(t)} = 0, \quad \text{for} \quad q(t) = q_{\rm cl}(t). \tag{3.23}$$

This can also be checked explicitly by differentiating (3.19) according to the chain rule of differentiation:

$$\frac{\partial L^{\epsilon(t)}}{\partial \epsilon(t)} = \frac{\partial L}{\partial q(t)} \Delta(q, \dot{q}) + \frac{\partial L}{\partial \dot{q}(t)} \dot{\Delta}(q, \dot{q}), \tag{3.24}$$

$$\frac{\partial L^{\epsilon(t)}}{\partial \dot{\epsilon}(t)} = \frac{\partial L}{\partial \dot{q}(t)} \Delta(q, \dot{q}), \tag{3.25}$$

and inserting on the right-hand side the ordinary Euler-Lagrange equations (1.5). Note that (3.25) can also be written as

$$\frac{\partial L^{\epsilon(t)}}{\partial \dot{\epsilon}(t)} = \frac{\partial L}{\partial \dot{q}(t)} \frac{\delta_{\rm s} q(t)}{\epsilon(t)}. \tag{3.26}$$

We now invoke the symmetry assumption that the action is a pure surface term under the time-independent transformations (3.17). This implies that

$$\frac{\partial L^\epsilon}{\partial \epsilon} = \frac{\partial L^{\epsilon(t)}}{\partial \epsilon(t)} = \frac{d}{dt}\Lambda. \tag{3.27}$$

Combining this with (3.23), we derive a conservation law for the charge:

$$Q = \frac{\partial L^{\epsilon(t)}}{\partial \dot{\epsilon}(t)} - \Lambda, \quad \text{for} \ \ q(t) = q_{\mathrm{cl}}(t). \tag{3.28}$$

Inserting here Eq. (3.25) we find that this is the same charge as in the previous Eq. (3.12).

3.2 Time Translation Invariance and Energy Conservation

As a simple but physically important example consider the case where the Lagrangian does not depend explicitly on time, i.e., $L(q(t), \dot{q}(t), t) \equiv L(q(t), \dot{q}(t))$. Let us perform a time translation on the system, shifting events at time t to the new time

$$t' = t - \epsilon. \tag{3.29}$$

The time-translated orbit has the time dependence

$$q'(t') = q(t), \tag{3.30}$$

i.e., the translated orbit $q'(t)$ has at the time t' the same value as the orbit $q(t)$ at the original time t. For the Lagrangian, this implies that

$$L'(t') \equiv L(q'(t'), \dot{q}'(t')) = L(q(t), \dot{q}(t)) \equiv L(t). \tag{3.31}$$

This makes the action (3.3) equal to (3.1), up to boundary terms. Thus time-independent Lagrangians possess time translation symmetry.

The associated substantial variations of the form (3.5) are

$$\begin{aligned}
\delta_s q(t) &= q'(t) - q(t) = q(t' + \epsilon) - q(t) \\
&= q(t') + \epsilon \dot{q}(t') - q(t) = \epsilon \dot{q}(t).
\end{aligned} \tag{3.32}$$

Under these, the Lagrangian changes by

$$\delta_s L = L(q'(t), \dot{q}'(t)) - L(q(t), \dot{q}(t)) = \frac{\partial L}{\partial q}\delta_s q(t) + \frac{\partial L}{\partial \dot{q}}\delta_s \dot{q}(t). \tag{3.33}$$

Inserting $\delta_s q(t)$ from (3.32) we find, without using the Euler-Lagrange equation,

$$\delta_s L = \epsilon \left(\frac{\partial L}{\partial \dot{q}}\dot{q} + \frac{\partial L}{\partial \dot{q}}\ddot{q} \right) = \epsilon \frac{d}{dt}L. \tag{3.34}$$

This has precisely the derivative form (3.14) with $\Lambda = L$, thus confirming that time translations are symmetry transformations.

According to Eq. (3.12), we find the Noether charge

$$Q = H \equiv \frac{\partial L}{\partial \dot{q}} \dot{q} - L(q, \dot{q}), \quad \text{for} \quad q(t) = q_{\text{cl}}(t) \tag{3.35}$$

to be a constant of motion. This is recognized as the *Legendre transform* of the Lagrangian, which is the *Hamiltonian* (2.10) of the system.

Let us briefly check how this Noether charge is obtained from the alternative formula (3.12). The time-dependent substantial variation (3.17) is here

$$\delta_s^t q(t) = \epsilon(t) \dot{q}(t) \tag{3.36}$$

under which the Lagrangian is changed by

$$\delta_s^t L = \frac{\partial L}{\partial q} \epsilon \dot{q} + \frac{\partial L}{\partial \dot{q}} (\dot{\epsilon} \dot{q} + \epsilon \ddot{q}) = \frac{\partial L^\epsilon}{\partial \dot{\epsilon}} \epsilon + \frac{\partial L^\epsilon}{\partial \dot{\epsilon}} \dot{\epsilon}, \tag{3.37}$$

with

$$\frac{\partial L^\epsilon}{\partial \dot{\epsilon}} = \frac{\partial L}{\partial \dot{q}} \dot{q} \tag{3.38}$$

and

$$\frac{\partial L^\epsilon}{\partial \epsilon} = \frac{\partial L}{\partial q} \dot{q} + \frac{\partial L}{\partial \dot{q}} \epsilon \ddot{q} = \frac{d}{dt} L. \tag{3.39}$$

The last equation confirms that time translations fulfill the symmetry condition (3.27), and from (3.38) we see that the Noether charge (3.28) coincides with the Hamiltonian found in Eq. (3.12).

3.3   Momentum and Angular Momentum

While the conservation law of energy follow from the symmetry of the action under time translations, conservation laws of momentum and angular momentum are found if the action is invariant under translations and rotations, respectively.

Consider a Lagrangian of a point particle in a Euclidean space

$$L = L(x^i(t), \dot{x}^i(t)). \tag{3.40}$$

In contrast to the previous discussion of time translation invariance, which was applicable to systems with arbitrary Lagrange coordinates $q^i(t)$, we denote the coordinates here by x^i, with the superscripts i emphasizing the fact that we are dealing here with Cartesian coordinates. If the Lagrangian depends only on the velocities $\dot{x}^i$ and not on the coordinates x^i themselves, the system is *translationally invariant*. If it depends, in addition, only on $\dot{\mathbf{x}}^2 = \dot{x}^i \dot{x}^i$, it is also rotationally invariant.

The simplest example is the Lagrangian of a point particle of mass m in Euclidean space:

$$L = \frac{m}{2} \dot{\mathbf{x}}^2. \tag{3.41}$$

It exhibits both invariances, leading to conserved Noether charges of momentum and angular momentum, as we shall now demonstrate.

3.3.1 Translational Invariance in Space

Under a spatial translation, the coordinates x^i of the particle change to

$$x'^i = x^i + \epsilon^i, \tag{3.42}$$

where ϵ^i are small numbers. The infinitesimal translations of a particle path are [compare (3.5)]

$$\delta_s x^i(t) = \epsilon^i. \tag{3.43}$$

Under these, the Lagrangian changes by

$$
\begin{aligned}
\delta_s L &= L(x'^i(t), \dot{x}'^i(t)) - L(x^i(t), \dot{x}^i(t)) \\
&= \frac{\partial L}{\partial x^i} \delta_s x^i = \frac{\partial L}{\partial x^i} \epsilon^i = 0.
\end{aligned}
\tag{3.44}
$$

By assumption, the Lagrangian is independent of x^i, so that the right-hand side vanishes for any path $q(t)$. This zero is to be equated with the substantial variation of the Lagrangian around a classical orbit calculated with the help of the Euler-Lagrange equation:

$$\delta_s L = \left(\frac{\partial L}{\partial x^i} - \frac{d}{dt} \frac{\partial L}{\partial \dot{x}^i} \right) \delta_s x^i + \frac{d}{dt} \left[\frac{\partial L}{\partial \dot{x}^i} \delta_s x^i \right] = \frac{d}{dt} \left[\frac{\partial L}{\partial \dot{x}^i} \right] \epsilon^i. \tag{3.45}$$

The result has the form (3.8), from which we extract a conserved Noether charge (3.9) for each coordinate x^i, to be called p^i:

$$p^i = \frac{\partial L}{\partial \dot{x}^i}. \tag{3.46}$$

Thus the Noether charges associated with translational invariance are simply the canonical momenta of the point particle.

3.3.2 Rotational Invariance

Under rotations, the coordinates x^i of the particle change to

$$x'^i = R^i{}_j x^j, \tag{3.47}$$

where $R^i{}_j$ are the orthogonal 3×3 -matrices (1.8). Infinitesimally, these can be written as

$$R^i{}_j = \delta^i{}_j - \varphi_k \epsilon_{kij}, \tag{3.48}$$

where φ is the infinitesimal rotation vector in Eq. (1.57). The corresponding rotation of a particle path is

$$\delta_s x^i(t) = x'^i(t) - x^i(t) = -\varphi^k \epsilon_{kij} x^j(\tau). \tag{3.49}$$

In the antisymmetric tensor notation (1.55) with $\omega_{ij} \equiv \varphi_k \epsilon_{kij}$, we write

$$\delta_s x^i = -\omega_{ij} x^j. \tag{3.50}$$

Under this, the substantial variation of the Lagrangian (3.41)

$$
\begin{aligned}
\delta_s L &= L(x'^i(t), \dot{x}'^i(t)) - L(x^i(t), \dot{x}^i(t)) \\
&= \frac{\partial L}{\partial x^i} \delta_s x^i + \frac{\partial L}{\partial \dot{x}^i} \delta_s \dot{x}^i
\end{aligned} \tag{3.51}
$$

becomes

$$\delta_s L = -\left(\frac{\partial L}{\partial x^i} x^j + \frac{\partial L}{\partial \dot{x}^i} \dot{x}^j \right) \omega_{ij} = 0. \tag{3.52}$$

For any Lagrangian depending only on the rotational invariants $\mathbf{x}^2, \dot{\mathbf{x}}^2, \mathbf{x} \cdot \dot{\mathbf{x}}$, and powers thereof, the right-hand side vanishes on account of the antisymmetry of ω_{ij}. This ensures the rotational symmetry for the Lagrangian (3.41).

We now calculate the substantial variation of the Lagrangian once more using the Euler-Lagrange equations:

$$
\begin{aligned}
\delta_s L &= \left(\frac{\partial L}{\partial x^i} - \frac{d}{dt} \frac{\partial L}{\partial \dot{x}^i} \right) \delta_s x^i + \frac{d}{dt} \left[\frac{\partial L}{\partial \dot{x}^i} \delta_s x^i \right] \\
&= -\frac{d}{dt} \left[\frac{\partial L}{\partial \dot{x}^i} x^j \right] \omega_{ij} = \frac{1}{2} \frac{d}{dt} \left[x_i \frac{\partial L}{\partial \dot{x}^j} - (i \leftrightarrow j) \right] \omega_{ij}.
\end{aligned} \tag{3.53}
$$

The right-hand side yields the conserved Noether charges of the type (3.9), one for each antisymmetric pair i, j:

$$L^{ij} = x^i \frac{\partial L}{\partial \dot{x}^j} - x^j \frac{\partial L}{\partial \dot{x}^i} \equiv x^i p^j - x^j p^i. \tag{3.54}$$

These are the conserved components of angular momentum for a Cartesian system in any dimension.

In three dimensions, we may prefer working with the original rotation angles φ^k, in which case we find the angular momentum in the standard form

$$L_k = \frac{1}{2} \epsilon_{kij} L^{ij} = (\mathbf{x} \times \mathbf{p})^k. \tag{3.55}$$

3.3.3 Center-of-Mass Theorem

Let us now study symmetry transformations corresponding to a uniform motion of the coordinate system described by Galilei transformations (1.11), (1.12). Consider a set of free massive point particles in Euclidean space described by the Lagrangian

$$L(\dot{q}_n^i) = \sum_n \frac{m_n}{2} \dot{q}_n^{i\,2}. \tag{3.56}$$

The infinitesimal substantial variation associated with the Galilei transformations are

$$\delta_s x_n^i(t) = \dot{x}_n^i(t) - x_n^i(t) = -v^i t, \tag{3.57}$$

where v^i is a small relative velocity along the ith axis. This changes the Lagrangian by

$$\delta_s L = L(x_n^i - v^i t, \dot{x}_n^i - v^i) - L(x_n^i, \dot{x}_n^i). \tag{3.58}$$

Inserting here (3.56), we find

$$\delta_s L = \sum_n \frac{m_n}{2} \left[(\dot{x}_n^i - v^i)^2 - (\dot{x}_n{}^i)^2 \right], \tag{3.59}$$

which can be written as a total time derivative

$$\delta_s L = \frac{d}{dt} \Lambda = \frac{d}{dt} \sum_n m_n \left[-\dot{x}_n^i v^i + \frac{v^2}{2} t \right], \tag{3.60}$$

proving that Galilei transformations are symmetry transformations in the Noether sense. Note that terms quadratic in v^i are omitted in the last expression since the velocities v^i in (3.57) are infinitesimal, by assumption.

By calculating $\delta_s L$ once more via the chain rule, inserting the Euler-Lagrange equations. Equating the result with (3.60), as in the derivation of (3.12), we find the conserved Noether charge

$$
\begin{aligned}
Q &= \sum_n \frac{\partial L}{\partial \dot{x}_n^i} \delta_s x_n^i - \Lambda \\
&= \left(-\sum_n m_n \dot{x}_n^i t + \sum_n m_n x_n^i \right) v^i.
\end{aligned} \tag{3.61}
$$

Since the direction of the velocities v^i is arbitrary, each component is a separate constant of motion:

$$N^i = -\sum_n m_n \dot{x}^i t + \sum_n m_n x_n{}^i = \text{const.} \tag{3.62}$$

This is the well-known *center-of-mass theorem* [1]. Indeed, introducing the center-of-mass coordinates

$$x_{\mathrm{CM}}^i \equiv \frac{\sum_n m_n x_n{}^i}{\sum_n m_n}, \tag{3.63}$$

and the velocities

$$v_{\mathrm{CM}}^i = \frac{\sum_n m_n \dot{x}_n{}^i}{\sum_n m_n}, \tag{3.64}$$

the conserved charge (3.62) can be written as

$$N^i = \sum_n m_n(-v_{\rm CM}^i\, t + x_{\rm CM}^i).$$

(3.65)

The time-independence of N^i implies that the center-of-mass moves with uniform velocity according to the law

$$x_{\rm CM}^i(t) = x_{\rm CM,0}^i + v_{\rm CM}^i t,$$

(3.66)

where

$$x_{\rm CM,0}^i = \frac{N^i}{\sum_n m_n}$$

(3.67)

is the position of the center of mass at $t = 0$.

Note that in nonrelativistic physics, the center of mass theorem is a consequence of momentum conservation, since momentum $\equiv$ mass $\times$ velocity. In relativistic physics, this is no longer true.

3.3.4   Conservation Laws from Lorentz Invariance

In relativistic physics, particle orbits are described by functions in Minkowski spacetime $x^a(\sigma)$, where σ is a Lorentz-invariant length parameter. The action is an integral over some Lagrangian:

$$\overset{\rm m}{\mathcal{A}} = \int_{\sigma_a}^{\sigma_b} d\sigma\, \overset{\rm m}{L}\left(x^a(\sigma), \dot{x}^a(\sigma)\right),$$

(3.68)

where the dot denotes the derivative with respect to the parameter σ. If the Lagrangian depends only on invariant scalar products $x^a x_a, x^a \dot{x}_a, \dot{x}^a \dot{x}_a$, then it is invariant under Lorentz transformations

$$x^a \rightarrow x'^a = \Lambda^a{}_b\, x^b,$$

(3.69)

where $\Lambda^a{}_b$ are the pseudo-orthogonal 4×4 -matrices (1.28).

A free massive point particle in spacetime has the Lagrangian [see (2.19)]

$$\overset{\rm m}{L}\left(\dot{x}(\sigma)\right) = -mc\sqrt{g_{ab}\dot{x}^a\dot{x}^b},$$

(3.70)

so that the action (3.68) is invariant under arbitrary reparametrizations $\sigma \rightarrow f(\sigma)$. Since the Lagrangian depends only on $\dot{x}(\sigma)$, it is invariant under arbitrary translations of the coordinates:

$$\delta_{\rm s} x^a(\sigma) = x^a(\sigma) - \epsilon^a(\sigma),$$

(3.71)

for which $\delta_{\rm s} L = 0$. Calculating this variation once more with the help of the Euler-Lagrange equations, we find

$$\delta_{\rm s} \overset{\rm m}{L} = \int_{\sigma_a}^{\sigma_b} d\sigma \left(\frac{\partial \overset{\rm m}{L}}{\partial x^a}\delta_{\rm s}x^a + \frac{\partial \overset{\rm m}{L}}{\partial \dot{x}^a}\delta_{\rm s}\dot{x}^a\right) = -\epsilon^a \int_{\sigma_a}^{\sigma_b} d\sigma \frac{d}{d\sigma}\left(\frac{\partial \overset{\rm m}{L}}{\partial \dot{x}^a}\right).$$

(3.72)

From this we obtain the conserved Noether charges

$$p_a \equiv -\frac{\partial \overset{m}{L}}{\partial \dot{x}^a} = m\frac{\dot{x}_a}{\sqrt{g_{ab}\dot{x}^a\dot{x}^b/c^2}} = mu^a, \tag{3.73}$$

which satisfy the conservation law

$$\frac{d}{d\sigma}p_a(\sigma) = 0. \tag{3.74}$$

The Noether charges $p_a(\sigma)$ are the conserved four-momenta (1.149) of the free relativistic particle, derived in Eq. (2.20) from the canonical formalism. The four-vector

$$u^a \equiv \frac{\dot{x}^a}{\sqrt{g_{ab}\dot{x}^a\dot{x}^b/c^2}} \tag{3.75}$$

is the relativistic four-velocity of the particle. It is the reparametrization-invariant expression for the four-velocity $\dot{q}_a(\tau) = u_a(\tau)$ in Eqs. (2.22) and (1.149). A sign change is made in Eq. (3.73) to agree with the canonical definition of the covariant momentum components in (2.20). By choosing for σ the physical time $t = x^0/c$, we can express u^a in terms of the physical velocities $v^i = dx^i/dt$, as in (1.150):

$$u^a = \gamma(1, v^i/c), \quad \text{with} \quad \gamma \equiv \sqrt{1 - v^2/c^2}. \tag{3.76}$$

For small Lorentz transformations near the identity we write

$$\Lambda^a{}_b = \delta^a{}_b + \omega^a{}_b \tag{3.77}$$

where

$$\omega^a{}_b = g^{ac}\omega_{cb} \tag{3.78}$$

is an arbitrary infinitesimal antisymmetric matrix. An infinitesimal Lorentz transformation of the particle path is

$$\begin{aligned}
\delta_s x^a(\sigma) &= \dot{x}^a(\sigma) - x^a(\sigma) \\
&= \omega^a{}_b x^b(\sigma).
\end{aligned} \tag{3.79}$$

Under it, the substantial variation of any Lorentz-invariant Lagrangian vanishes:

$$\delta_s L = \left(\frac{\partial L}{\partial x^a}x^b + \frac{\partial L}{\partial \dot{x}^a}\dot{x}^b\right)\omega^a{}_b = 0. \tag{3.80}$$

This is to be compared with the substantial variation of the Lagrangian calculated via the chain rule with the help of the Euler-Lagrange equation

$$\begin{aligned}
\delta_s L &= \left(\frac{\partial L}{\partial x^a} - \frac{d}{d\sigma}\frac{\partial L}{\partial \dot{x}^a}\right)\delta_s x^a + \frac{d}{d\sigma}\left[\frac{\partial L}{\partial \dot{x}^a}\delta_s x^a\right] \\
&= \frac{d}{d\sigma}\left[\frac{\partial L}{\partial \dot{x}^a}\dot{x}^b\right]\omega^a{}_b \\
&= \frac{1}{2}\omega_a{}^b\frac{d}{d\sigma}\left(x^a\frac{\partial L}{\partial \dot{x}_b} - x^b\frac{\partial L}{\partial \dot{x}_a}\right).
\end{aligned} \tag{3.81}$$

By equating this with (3.80) we obtain the conserved rotational Noether charges

$$L^{ab} = -x^a \frac{\partial L}{\partial \dot{x}_b} + x^b \frac{\partial L}{\partial \dot{x}_a} = x^a p^b - x^b p^a. \tag{3.82}$$

They are the four-dimensional generalizations of the angular momenta (3.54).

The Noether charges L^{ij} coincide with the components (3.54) of angular momentum. The conserved components

$$L^{0i} = x^0 p^i - x^i p^0 \equiv M_i \tag{3.83}$$

yield the relativistic generalization of the center-of-mass theorem (3.62):

$$M_i = \text{const.} \tag{3.84}$$

3.4   Generating the Symmetries

As mentioned in the introduction to this chapter, there is a second important relation between invariances and conservation laws. The charges associated with continuous symmetry transformations can be used to *generate* the symmetry transformation from which it was derived. In the classical theory, this is done with the help of Poisson brackets:

$$\delta_s \hat{x} = \epsilon \{\hat{Q}, \hat{x}(t)\}. \tag{3.85}$$

After canonical quantization, the Poisson brackets turn into $-i$ times commutators, and the charges become operators, generating the symmetry transformation by the operation

$$\delta_s \hat{x} = -i\epsilon [\hat{Q}, \hat{x}(t)]. \tag{3.86}$$

The most important example for this quantum-mechanical generation of symmetry transformations is the effect of the Noether charge (3.35) derived in Section 3.2 from the invariance of the system under time displacement. That Noether charge Q was the *Hamiltonian H*, whose operator version generates the infinitesimal time displacements (3.32) by the *Heisenberg equation of motion*

$$\delta_s x(t) = \epsilon \dot{\hat{x}}(t) = -i\epsilon [\hat{H}, \hat{x}(t)], \tag{3.87}$$

which is thus a special case of the general Noether relation (3.86).

The canonical quantization is straightforward if the Lagrangian has the standard form

$$L(x, \dot{x}) = \frac{m}{2} \dot{x}^2 - V(x). \tag{3.88}$$

Then the operator version of the canonical momentum $p \equiv \dot{x}$ satisfies the equal-time commutation rules

$$[\hat{p}(t), \hat{x}(t)] = -i, \quad [\hat{p}(t), \hat{p}(t)] = 0, \quad [\hat{x}(t), \hat{x}(t)] = -i. \tag{3.89}$$

The Hamiltonian

$$H = \frac{p^2}{2m} + V(\hat{x}) \tag{3.90}$$

turns directly into the *Hamiltonian operator*

$$\hat{H} = \frac{\hat{p}^2}{2m} + V(\hat{x}). \tag{3.91}$$

If the Lagrangian does not have the standard form (3.88), quantization is a nontrivial problem [3].

Another important example is provided by the charges (3.46) derived in Section 3.3.1 from translational symmetry. After quantization, the commutator (3.86) generating the transformation (3.43) becomes

$$\epsilon^j = i\epsilon^i [\hat{p}^i(t), \hat{x}^j(t)]. \tag{3.92}$$

This coincides with one of the canonical commutation relations (3.89) in three dimensions.

The relativistic charges (3.73) of spacetime generate translations via

$$\delta_s \hat{x}^a = \epsilon^a = -i\epsilon^b [\hat{p}_b(t), \hat{x}^a(\tau)], \tag{3.93}$$

implying the relativistic commutation rules

$$[\hat{p}_b(t), \hat{x}^a(\tau)] = i\delta_b{}^a, \tag{3.94}$$

in agreement with the relativistic canonical commutation rules (1.162) (in natural units with $\hbar = 1$).

Note that all commutation rules derived from the Noether charge according to the rule (3.86) hold for the operators in the Heisenberg picture, where they are time-dependent. The commutation rules in the purely algebraic discussion in Chapter 3, on the other hand, apply to the time-independent Schrödinger picture of the operators.

Similarly we find that the quantized versions of the conserved charges L_i in Eq. (3.55) generate infinitesimal rotations:

$$\delta_s \hat{x}^j = -\omega^i \epsilon_{ijk} \hat{x}^k(t) = i\omega^i [\hat{L}_i, \hat{x}^j(t)], \tag{3.95}$$

whereas the quantized conserved charges N^i of Eq. (3.62) generate infinitesimal Galilei transformations. The charges M_i of Eq. (3.83) generate pure Lorentz transformations [compare (1.129)]:

$$\delta_s \hat{x}^j = \epsilon_i \hat{x}^0 = i\epsilon_i [M_i, \hat{x}^j], \qquad \delta_s \hat{x}^0 = \epsilon_i \hat{x}^i = i\epsilon_i [M_i, \hat{x}^0]. \tag{3.96}$$

Since the quantized charges generate the symmetry transformations, they form a *representation* of the generators of the Lorentz group. As such they must have the same commutation rules between each other as the generators of the symmetry group in Eq. (1.71) or their short version (1.72). This is indeed true, since the operator versions of the Noether charges (3.82) correspond to the operators (1.163) (in natural units).

3.5   Field Theory

A similar relation between continuous symmetries and constants of motion holds in field theories, where the role of the Lagrange coordinates is played by fields $q_{\mathbf{x}}(t) = \varphi(\mathbf{x}, t)$.

3.5.1   Continuous Symmetry and Conserved Currents

Let $\mathcal{A}$ be the local action of an arbitrary field $\varphi(x) \to \varphi(\mathbf{x}, t)$,

$$\mathcal{A} = \int d^4x \, \mathcal{L}(\varphi, \partial\varphi, x), \tag{3.97}$$

and suppose that a transformation of the field

$$\delta_s\varphi(x) = \epsilon\Delta(\varphi, \partial\varphi, x) \tag{3.98}$$

changes the Lagrangian density $\mathcal{L}$ merely by a total derivative

$$\delta_s\mathcal{L} = \epsilon\partial_a\Lambda^a, \tag{3.99}$$

which makes the change of the action $\mathcal{A}$ a surface integral, by Gauss's divergence theorem:

$$\delta_s\mathcal{A} = \epsilon\int d^4x \, \partial_a\Lambda^a = \epsilon\int_S ds_a \, \Lambda^a, \tag{3.100}$$

where S is the surface of the total spacetime volume. Then $\delta_s\varphi$ is called a *symmetry transformation*.

Under the assumption of symmetry we derive a local conservation law in the same way as for the mechanical action (3.1). We calculate the variation of $\mathcal{L}$ under infinitesimal symmetry transformations (3.98) in a similar way as in Eq. (3.13), and find

$$\begin{aligned}
\delta_s\mathcal{L} &= \left(\frac{\partial\mathcal{L}}{\partial\varphi} - \partial_a\frac{\partial\mathcal{L}}{\partial\partial_a\varphi}\right)\delta_s\varphi + \partial_a\left(\frac{\partial\mathcal{L}}{\partial\partial_a\varphi}\delta_s\varphi\right) \\
&= \epsilon\left(\frac{\partial\mathcal{L}}{\partial\varphi} - \partial_a\frac{\partial\mathcal{L}}{\partial\partial_a\varphi}\right)\Delta + \epsilon\partial_a\left(\frac{\partial\mathcal{L}}{\partial\partial_a\varphi}\Delta\right).
\end{aligned} \tag{3.101}$$

The Euler-Lagrange equation removes the first term, and equating the second term with (3.99) we see that the four-dimensional current density

$$j^a = \frac{\partial\mathcal{L}}{\partial\partial_a\varphi}\Delta - \Lambda^a \tag{3.102}$$

has no divergence

$$\partial_a j^a(x) = 0. \tag{3.103}$$

This is *Noether's theorem* for field theory [4]. The expression (3.102) is called a *Noether current density* and (3.103) is a *local conservation law*, just as in the electromagnetic equation (1.202).

We have seen in Eq. (1.204) that a local conservation law (3.103) always implies a global conservation law of the type (3.9) for the charge, which is now the Noether charge $Q(t)$ defined as in (1.205) by the spatial integral over the zeroth component (here in natural units with $c = 1$)

$$Q(t) = \int d^3x \, j^0(\mathbf{x}, t). \tag{3.104}$$

3.5.2 Alternative Derivation

There is again an alternative derivative of the conserved current analogous to Eqs. (3.17)–(3.28). It is based on a variation of the fields under symmetry transformations whose parameter ϵ is made artificially spacetime-dependent $\epsilon(x)$, thus extending (3.17) to

$$\delta_s^x \varphi(x) = \epsilon(x) \Delta(\varphi(x), \partial_a \varphi(x)). \tag{3.105}$$

As before in Eq. (3.19), let us calculate the Lagrangian density for a slightly transformed field

$$\varphi^{\epsilon(x)}(x) \equiv \varphi(x) + \delta_s^x \varphi(x), \tag{3.106}$$

calling it

$$\mathcal{L}^{\epsilon(x)} \equiv \mathcal{L}(\varphi^{\epsilon(x)}, \partial \varphi^{\epsilon(x)}). \tag{3.107}$$

The associated action differs from the original one by

$$\delta_s^x \mathcal{A} = \int dx \left\{ \left[\frac{\partial \mathcal{L}^{\epsilon(x)}}{\partial \epsilon(x)} - \partial_a \frac{\partial \mathcal{L}^{\epsilon(x)}}{\partial \partial_a \epsilon(x)} \right] \delta \epsilon(x) + \partial_a \left[\frac{\partial \mathcal{L}^{\epsilon(x)}}{\partial \partial_a \epsilon(x)} \delta \epsilon(x) \right] \right\}. \tag{3.108}$$

For classical fields $\varphi(x) = \varphi_{\mathrm{cl}}(x)$ satisfying the Euler-Lagrange equation (2.40), the extremality of the action implies the vanishing of the first term, and thus the Euler-Lagrange-like equation

$$\frac{\partial \mathcal{L}^{\epsilon(x)}}{\partial \epsilon(x)} - \partial_a \frac{\partial \mathcal{L}^{\epsilon(x)}}{\partial \partial_a \epsilon(x)} = 0. \tag{3.109}$$

By assumption, the action changes by a pure surface term under the x-independent transformation (3.105), implying that

$$\frac{\partial \mathcal{L}^\epsilon}{\partial \epsilon} = \partial_a \Lambda^a. \tag{3.110}$$

Inserting this into (3.109) we find that

$$j^a = \frac{\partial \mathcal{L}^{\epsilon(x)}}{\partial \partial_a \epsilon(x)} - \Lambda^a \tag{3.111}$$

has no four-divergence. This coincides with the previous Noether current density (3.102), as can be seen by differentiating (3.107) with respect to $\partial_a \epsilon(x)$:

$$\frac{\partial \mathcal{L}^{\epsilon(x)}}{\partial \partial_a \epsilon(x)} = \frac{\partial \mathcal{L}}{\partial \partial_a \varphi} \Delta(\varphi, \partial \varphi). \tag{3.112}$$

3.5.3   Local Symmetries

In Chapter 2 we observed that charged particles and fields coupled to electromagnetism possess a more general symmetry. They are invariant under local gauge transformations (2.104). The scalar Lagrangian (2.136), for example, is invariant under the gauge transformations (2.104) and (2.137), and the Dirac Lagrange density (2.151) under (2.104) and (2.152). These are all of the generic form (3.98), but with a parameter ϵ depending now on spacetime. Thus the action is invariant under local substantial variations of the type (3.105), which were introduced in the last section only as an auxiliary tool for an alternative derivation of the Noether current density (2.136).

For a locally gauge-invariant Lagrangian density, the Noether expression (3.111) reads

$$j_a = \frac{\delta \mathcal{L}}{\partial \partial_a \Lambda},$$ (3.113)

and vanishes identically. This does not mean, however, that the system has no conserved current, as we have seen in Eqs. (2.140) and (2.156). Only Noether's derivation breaks down. Let us study this phenomenon in more detail for the Lagrangian density (2.151).

If we restrict the gauge transformations (2.152) to spacetime-independent gauge transformations

$$\psi(x) \to e^{ie\Lambda/c}\psi(x),$$ (3.114)

we can easily derive a conserved Noether current density of the type (3.102) for the Dirac field. The result is the Dirac current density (2.154). It is the source of the electromagnetic field coupled in a minimal way. The minimal coupling makes the globally gauge-invariant theory locally invariant. Nature has used this gauge principle in many other circumstances. Many global internal symmetries are really local due to the existence of minimally coupled gauge fields, some of which being nonabelian versions of electromagnetism. The most important examples are the nonabelian gauge symmetries associated with strong and weak interactions.

Let us see what happens to Noether's derivation of conservation laws in such theories. Since the expression (3.111) for the current density vanishes identically, due to local gauge invariance, but renders a conserved current density (2.154) for the Dirac field in the absence of a gauge field, we may calculate $j^a(x)$ from the functional derivative

$$j_a \equiv \left.\frac{\partial \mathcal{L}}{\partial \partial_a \Lambda}\right|_{A^a}$$ (3.115)

at *fixed* gauge fields.

Alternatively, we can use the fact that the complete change under local gauge transformations $\delta_s^x \mathcal{L}$ vanishes identically, and vary *only* the gauge fields keeping the particle orbit fixed. This yields the same current density from the derivative

$$j_a = -\left.\frac{\partial \mathcal{L}}{\partial \partial_a \Lambda}\right|_{\psi}.$$ (3.116)

Yet another way of calculating this expression is by forming the functional derivative with respect to the gauge field A_a thereby omitting the contribution of $\overset{em}{\mathcal{L}}$, i.e., by applying it only to the Lagrangian of the charge particles $\overset{e}{\mathcal{L}} \equiv \mathcal{L} - \overset{em}{\mathcal{L}}$:

$$j^a = -\frac{\partial \overset{e}{\mathcal{L}}}{\partial \partial_a \Lambda} = -\frac{\partial \overset{e}{\mathcal{L}}}{\partial A_a}. \tag{3.117}$$

As a check we use the rule (3.117) to calculate the conserved current densities of Dirac and complex Klein-Gordon fields with the Lagrangian densities in Eqs. (2.141) and (2.27), and re-obtain the expressions (2.154) and (2.139) (the extra factor c is a convention). For the Schrödinger Lagrangian density in (2.50) we derive the conserved current density of electric charge

$$\mathbf{j}(\mathbf{x}, t) \equiv e \frac{i}{2m} \psi^*(\mathbf{x}, t) \overset{\leftrightarrow}{\nabla} \psi(\mathbf{x}, t) - \frac{e^2}{c} \mathbf{A}\, \psi^*(\mathbf{x}, t) \psi(\mathbf{x}, t), \tag{3.118}$$

differing from the particle current density (2.64) by a factor e. The current conservation law (2.65) holds now with the charge density $\rho(\mathbf{x}, t) \equiv e\psi^*(\mathbf{x}, t)\psi(\mathbf{x}, t)$ on the right-hand side.

An important consequence of local gauge invariance can be found for the gauge field itself. If we form the variation of the pure gauge field action

$$\delta_s \overset{em}{\mathcal{A}} = \int d^4x \; \text{tr}\left(\delta_s^x A_a \frac{\delta \overset{em}{\mathcal{A}}}{\delta A_a}\right), \tag{3.119}$$

and insert for $\delta_s^x A$ an infinitesimal pure gauge field configuration

$$\delta_s^x A_a = -\partial_a \Lambda(x), \tag{3.120}$$

the right-hand side must vanish for all $\Lambda(x)$. After a partial integration this implies the local conservation law $\partial_a j^a(x) = 0$ for the Noether current

$$\overset{em}{j}{}^a(x) = -\frac{\delta \overset{em}{\mathcal{A}}}{\delta A_a}. \tag{3.121}$$

Recalling the explicit form of the action in Eqs. (2.83) and (2.84), we find

$$\overset{em}{j}{}^a(x) = -\partial_b F^{ab}. \tag{3.122}$$

The Maxwell equation (2.87) can therefore be written as

$$\overset{em}{j}{}^a(x) = -\overset{e}{j}{}^a(x). \tag{3.123}$$

The superscript e emphasizes the fact that the current density $j^a(x)$ contains only the fields of the charged particles. In the form (3.123), the Maxwell equation implies the vanishing of the total current density consisting of the sum of the conserved current (3.116) of the charges and the Noether current (3.121) of the electromagnetic field:

$$\overset{\text{tot}}{j}{}^{a}(x) = \overset{\text{e}}{j}{}^{a}(x) + \overset{\text{em}}{j}{}^{a}(x) = 0. \tag{3.124}$$

This unconventional way of phrasing the Maxwell equation (2.87) will be useful for understanding later the Einstein field equation (17.157) by analogy.

At this place we make an important observation. In contrast to the conservation laws derived for matter fields, which are valid only if the matter fields obey the Euler-Lagrange equations, the current conservation law for the Noether current (3.122) of the gauge fields

$$\partial_a \overset{\text{em}}{j}{}^{a}(x) = -\partial_a \partial_b F^{ab} = 0 \tag{3.125}$$

is valid for *all* field configurations. The right-hand side vanishes *identically* since the vector potential A^a is an observable field in any fixed gauge and thus satisfies the *Schwarz integrability condition* (2.89).

3.6   Canonical Energy-Momentum Tensor

As an important example for the field-theoretic version of the Noether theorem consider a Lagrangian density that does not depend explicitly on the spacetime coordinates x:

$$\mathcal{L}(x) = \mathcal{L}(\varphi(x), \partial\varphi(x)). \tag{3.126}$$

We then perform a translation of the coordinates along an arbitrary direction $b = 0, 1, 2, 3$ of spacetime

$$x'^{a} = x^{a} - \epsilon^{a}, \tag{3.127}$$

under which field $\varphi(x)$ transforms as

$$\varphi'(x') = \varphi(x), \tag{3.128}$$

so that

$$\mathcal{L}'(x') = \mathcal{L}(x). \tag{3.129}$$

If ϵ^a is infinitesimally small, the field changes by

$$\delta_s\varphi(x) = \varphi'(x) - \varphi(x) = \epsilon^b \partial_b \varphi(x), \tag{3.130}$$

and the Lagrangian density by

$$\begin{aligned}
\delta_s\mathcal{L} &\equiv \mathcal{L}(\varphi'(x), \partial\varphi'(x)) - \mathcal{L}(\varphi(x), \partial\varphi(x)) \\
&= \frac{\partial\mathcal{L}}{\partial\varphi(x)}\delta_s\varphi(x) + \frac{\partial\mathcal{L}}{\partial\partial_a\varphi}\partial_a\delta_s\varphi(x),
\end{aligned} \tag{3.131}$$

which is a pure divergence term

$$\delta_s \mathcal{L}(x) = \epsilon^b \partial_b \mathcal{L}(x).$$ (3.132)

Hence the requirement (3.99) is satisfied and $\delta_s \varphi(x)$ is a symmetry transformation, with a function Λ which happens to coincide with the Lagrangian density

$$\Lambda = \mathcal{L}.$$ (3.133)

We can now define four four-vectors of current densities $j_b{}^a$, one for each component of ϵ^b. For the spacetime translation symmetry, they are denoted by $\Theta_b{}^a$:

$$\Theta_b{}^a = \frac{\partial \mathcal{L}}{\partial \partial_a \varphi} \partial_b \varphi - \delta_b{}^a \mathcal{L}.$$ (3.134)

Since ϵ^b is a vector, this 4×4-object is a tensor field, the so-called *energy-momentum tensor* of the scalar field $\varphi(x)$. According to Noether's theorem, this has no divergence in the index a [compare (3.103)]:

$$\partial_a \Theta_b{}^a(x) = 0.$$ (3.135)

The four conserved charges Q_b associated with these current densities [see the definition (3.104)]

$$P_b = \int d^3x \, \Theta_b{}^0(x),$$ (3.136)

are the components of the *total four-momentum* of the system.

The alternative derivation of this conservation law follows Subsection 3.1.4 by introducing the local variations

$$\delta_s^x \varphi(x) = \epsilon^b(x) \partial_b \varphi(x)$$ (3.137)

under which the Lagrangian density changes by

$$\delta_s^x \mathcal{L}(x) = \epsilon^b(x) \partial_b \mathcal{L}(x).$$ (3.138)

Applying the chain rule of differentiation we obtain, on the other hand,

$$\delta_s^x \mathcal{L} = \frac{\partial \mathcal{L}}{\partial \varphi(x)} \epsilon^b(x) \partial_b \varphi(x) + \frac{\partial \mathcal{L}}{\partial \partial_a \varphi(x)} \left\{ [\partial_a \epsilon^b(x)] \partial_b \varphi + \epsilon^b \partial_a \partial_b \varphi(x) \right\},$$ (3.139)

which shows that

$$\frac{\partial \mathcal{L}^\epsilon}{\partial \partial_a \epsilon^b(x)} = \frac{\partial \mathcal{L}}{\partial \partial_a \varphi} \partial_b \varphi.$$ (3.140)

By forming, for each b, the combination (3.102), we obtain again the conserved energy-momentum tensor (3.134).

Note that, by analogy with (3.26), we can write (3.140) as

$$\frac{\partial \mathcal{L}^\epsilon}{\partial \partial_a \epsilon^b(x)} = \frac{\partial \mathcal{L}}{\partial \partial_a \varphi} \frac{\partial \delta_s^x \varphi}{\partial \epsilon^b(x)}. \tag{3.141}$$

Note further that the component $\Theta_0{}^0$ of the *canonical energy momentum tensor*

$$\Theta_0{}^0 = \frac{\partial \mathcal{L}}{\partial \partial_0 \varphi} \partial_0 \varphi - \mathcal{L} \tag{3.142}$$

coincides with the Hamiltonian density (2.61) derived in the canonical formalism by a Legendre transformation of the Lagrangian density.

3.6.1 Electromagnetism

As an important physical application of the field-theoretic Noether theorem, consider the free electromagnetic field with the action

$$\mathcal{L} = -\frac{1}{4} F_{cd} F^{cd}, \tag{3.143}$$

where F_d are the field strength $F_{cd} \equiv \partial_c A_d - \partial_d A_c$. Under a translation of the spacetime coordinates from x^a to $x^a - \epsilon^a$, the vector potential undergoes a similar change as the scalar field in (3.128):

$$A'^a(x') = A^a(x). \tag{3.144}$$

For infinitesimal translations, this can be written as

$$\begin{aligned} \delta_s A^c(x) &\equiv A'^c(x) - A^c(x) \\ &= A'^c(x' + \epsilon) - A^c(x) \\ &= \epsilon^b \partial_b A^c(x), \end{aligned} \tag{3.145}$$

and the field tensor changes by

$$\delta_s F^{cd} = \epsilon^b \partial_b F^{cd}. \tag{3.146}$$

Inserting this into (3.143) we see the Lagrangian density changes by a total four divergence:

$$\delta_s \mathcal{L} = -\epsilon^b \frac{1}{2} \left(\partial_b F_{cd} F^{cd} + F_{cd} \partial_b F^{cd} \right) = \epsilon^b \partial_b \mathcal{L}. \tag{3.147}$$

Hence the spacetime translations (3.145) are symmetry transformations, and Eq. (3.102) yields the four Noether current densities, one for each ϵ^b:

$$\Theta_b{}^a = \frac{1}{c} \left[\frac{\partial \mathcal{L}}{\partial \partial_a A^c} \partial_b A^c - \delta_b{}^a \mathcal{L} \right]. \tag{3.148}$$

These form the *canonical energy-momentum tensor* of the electromagnetic field, which satisfies the local conservation laws

$$\partial_a \Theta_b{}^a(x) = 0. \tag{3.149}$$

The factor $1/c$ is introduced to give the Noether current densities the dimension of the previously introduced energy-momentum tensors in Eq. (1.261), which are momentum densities. Inserting the derivatives $\partial \mathcal{L}/\partial \partial_a A^c = -F^a{}_c$ into (3.148), we obtain

$$\Theta_b{}^a = \frac{1}{c}\left[-F^a{}_c \partial_b A^c + \frac{1}{4}\delta_b{}^a F^{cd}F_{cd} \right]. \tag{3.150}$$

3.6.2 Dirac Field

We now turn to the Dirac field whose transformation law under spacetime translations

$$x'^a = x^a - \epsilon^a \tag{3.151}$$

is

$$\psi'(x') = \psi(x). \tag{3.152}$$

Since the Lagrangian density in (2.141) does not depend explicitly on x we calculate, as in (3.129):

$$\overset{D}{\mathcal{L}}{}'(x') = \overset{D}{\mathcal{L}}(x). \tag{3.153}$$

The infinitesimal variations

$$\delta_s \psi(x) = \epsilon^a \partial_a \psi(x) \tag{3.154}$$

produce the pure derivative term

$$\delta_s \overset{D}{\mathcal{L}}(x) = \epsilon^a \partial_a \overset{D}{\mathcal{L}}(x), \tag{3.155}$$

and the combination (3.102) yields the Noether current densities

$$\Theta_b{}^a = \frac{\partial \overset{D}{\mathcal{L}}}{\partial \partial_a \psi^c}\partial_b \psi^c + \text{cc} - \delta_b{}^a \overset{D}{\mathcal{L}}, \tag{3.156}$$

which satisfy the local conservation laws

$$\partial_a \Theta_b{}^a(x) = 0. \tag{3.157}$$

From (2.141) we see that

$$\frac{\partial \overset{D}{\mathcal{L}}}{\partial \partial_a \psi^c} = \frac{1}{2}\bar{\psi}\gamma^a, \tag{3.158}$$

and obtain the *canonical energy-momentum tensor* of the Dirac field:

$$\Theta_b{}^a = \frac{1}{2}\bar{\psi}\gamma^a \partial_b \psi^c + \text{cc} - \delta_b{}^a \overset{D}{\mathcal{L}}. \tag{3.159}$$

3.7   Angular Momentum

Let us now turn to angular momentum in field theory. Consider first the case of a scalar field $\varphi(x)$. Under a rotation of the coordinates

$$x'^i = R^i{}_j x^j, \tag{3.160}$$

the field does not change if considered at the same point in space with different coordinates x^i and x'^i:

$$\varphi'(x'^i) = \varphi(x^i). \tag{3.161}$$

The infinitesimal substantial variation is:

$$\delta_s \varphi(x) = \varphi'(x) - \varphi(x). \tag{3.162}$$

For infinitesimal rotations (3.48),

$$\delta_s x^i = -\varphi_k \epsilon_{kij} x^j = -\omega_{ij} x^j, \tag{3.163}$$

we see that

$$\begin{aligned}
\delta_s \varphi(x) &= \varphi'(x^0, x'^i - \delta x^i) - \varphi(x) \\
&= \partial_i \varphi(x) x^j \omega_{ij}.
\end{aligned} \tag{3.164}$$

For a rotationally Lorentz-invariant Lagrangian density which has no explicit x-dependence:

$$\mathcal{L}(x) = \mathcal{L}(\varphi(x), \partial \varphi(x)), \tag{3.165}$$

the substantial variation is

$$\begin{aligned}
\delta_s \mathcal{L}(x) &= \mathcal{L}(\varphi'(x), \partial \varphi'(x)) - \varphi(\varphi(x), \partial \varphi(x)) \\
&= \frac{\partial \mathcal{L}}{\partial \varphi(x)} \delta_s \varphi(x) + \frac{\partial \mathcal{L}}{\partial \partial_a \varphi(x)} \partial_a \delta_s \varphi(x).
\end{aligned} \tag{3.166}$$

Inserting (3.164), this becomes

$$\begin{aligned}
\delta_s \mathcal{L} &= \left[\frac{\partial \mathcal{L}}{\partial \varphi} \partial_i \varphi x^j + \frac{\partial \mathcal{L}}{\partial_a \varphi} \partial_a (\partial_i \varphi x^j) \right] \omega_{ij} \\
&= \left[(\partial_i \mathcal{L}) x^j + \frac{\partial \mathcal{L}}{\partial \partial_j \varphi} \partial_i \varphi \right] \omega_{ij}.
\end{aligned} \tag{3.167}$$

Since we are dealing with a rotation-invariant local Lagrangian density $\mathcal{L}(x)$, by assumption, the derivative $\partial \mathcal{L} / \partial \partial_a \varphi$ is a vector proportional to $\partial_a \varphi$. Hence the second term in the brackets is symmetric and vanishes upon contraction with the antisymmetric ω_{ij}. This allows us to express $\delta_s \mathcal{L}$ as a pure derivative term

$$\delta_s \mathcal{L} = \partial_i \left(\mathcal{L} x^j \omega_{ij} \right). \tag{3.168}$$

Calculating $\delta_s\mathcal{L}$ once more using the chain rule and inserting the Euler-Lagrange equations yields

$$
\begin{aligned}
\delta_s\mathcal{L} &= \frac{\partial\mathcal{L}}{\partial\mathcal{L}}\delta_s\varphi + \frac{\partial\mathcal{L}}{\partial\partial_a\varphi}\partial_a\delta_s\varphi && (3.169)\\
&= \left(\frac{\partial\mathcal{L}}{\partial\varphi} - \partial_a\frac{\partial\mathcal{L}}{\partial\partial_a\varphi}\right)\delta_s\varphi + \partial_a\left(\frac{\partial\mathcal{L}}{\partial\partial_a\varphi}\delta_s\varphi\right)\\
&= \partial_a\left(\frac{\partial\mathcal{L}}{\partial\partial_a\varphi}\partial_i\varphi\, x^j\right)\omega_{ij}.
\end{aligned}
$$

Thus we find the Noether current densities (3.102):

$$
L^{ij,a} = \left(\frac{\partial\mathcal{L}}{\partial\partial_a\varphi}\partial_i\varphi\, x^j - \delta_i{}^a\mathcal{L}\, x^j\right) - (i \leftrightarrow j), \tag{3.170}
$$

which have no four-divergence

$$
\partial_a L^{ij,a} = 0. \tag{3.171}
$$

The current densities can be expressed in terms of the canonical energy-momentum tensor (3.134) as

$$
L^{ij,a} = x^i\Theta^{ja} - x^j\Theta^{ia}. \tag{3.172}
$$

The associated Noether charges

$$
L^{ij} = \int d^3x\, L^{ij,a} \tag{3.173}
$$

are the time-independent components of the *total angular momentum* of the field system.

3.8 Four-Dimensional Angular Momentum

Consider now pure Lorentz transformations (1.27). An infinitesimal boost to a rapidity ζ^i is described by a coordinate change [recall (1.34)]

$$
x'^a = \Lambda^a{}_b x^b = x^a - \delta^a{}_0\zeta^i x^i - \delta^a{}_i\zeta^i x^0. \tag{3.174}
$$

This can be written as

$$
\delta x^a = \omega^a{}_b x^b, \tag{3.175}
$$

where for passive boosts

$$
\omega_{ij} = 0, \quad \omega_{0i} = -\omega_{i0} = \zeta^i. \tag{3.176}
$$

With the help of the tensor $\omega^a{}_b$, the boosts can be treated on the same footing as the passive rotations (1.36), for which (3.175) holds with

$$\omega_{ij} = \omega_{ij} = \epsilon_{ijk}\varphi^k, \quad \omega_{0i} = \omega_{i0} = 0. \tag{3.177}$$

For both types of transformations, the substantial variations of the field are

$$\begin{aligned} \delta_s\varphi(x) &= \varphi'(x'^a - \delta x^a) - \varphi(x) \\ &= -\partial_a\varphi(x)x^b\omega^a{}_b. \end{aligned} \tag{3.178}$$

For a Lorentz-invariant Lagrangian density, the substantial variation can be shown, as in (3.168), to be a total derivative:

$$\delta_s\varphi = -\partial_a(\mathcal{L}x^b)\omega^a{}_b, \tag{3.179}$$

and we obtain the Noether current densities

$$L^{ab,c} = -\left(\frac{\partial\mathcal{L}}{\partial\partial_c\varphi}\partial^c\varphi\, x^b - \delta^{ac}\mathcal{L}\,x^b\right) - (a \leftrightarrow b). \tag{3.180}$$

The right-hand side can be expressed in terms of the canonical energy-momentum tensor (3.134), yielding

$$L^{ab,c} = x^a\Theta^{bc} - x^b\Theta^{ac}. \tag{3.181}$$

According to Noether's theorem (3.103), these current densities have no four-divergence:

$$\partial_c L^{ab,c} = 0. \tag{3.182}$$

The charges associated with these current densities

$$L^{ab} \equiv \int d^3x\, L^{ab,0} \tag{3.183}$$

are independent of time. For the particular form (3.176) of ω_{ab}, we recover the time independent components L^{ij} of angular momentum.

The time-independence of L^{i0} is the relativistic version of the *center-of-mass theorem* (3.66). Indeed, since

$$L^{i0} = \int d^3x\, (x^i\Theta^{00} - x^0\Theta^{i0}), \tag{3.184}$$

we can define the relativistic center of mass

$$x^i_{\text{CM}} = \frac{\int d^3x\, \Theta^{00}x^i}{\int d^3x\, \Theta^{00}} \tag{3.185}$$

and the average velocity

$$v^i_{\text{CM}} = c\frac{d^3x\Theta^{i0}}{\int d^3x\, \Theta^{00}} = c\frac{P^i}{P^0}. \tag{3.186}$$

Since $\int d^3x\, \Theta^{i0} = P^i$ is the constant momentum of the system, also v_{CM}^i is a constant. Thus, the constancy of L^{0i} implies the center of mass to move with the constant velocity

$$x_{CM}^i(t) = x_{CM,0}^i + v_{CM,0}^i t, \tag{3.187}$$

with $x_{CM,0}^i = L^{0i}/P^0$.

The Noether charges L^{ab} are the four-dimensional angular momenta of the system.

It is important to point out that the vanishing divergence of $L^{ab,c}$ makes Θ^{ba} symmetric:

$$
\begin{aligned}
\partial_c L^{ab,c} &= \partial_c(x^a \Theta^{bc} - x^b \Theta^{ac}) \\
&= \Theta^{ba} - \Theta^{ba} = 0.
\end{aligned} \tag{3.188}
$$

Thus, field theories which are invariant under spacetime translations and Lorentz transformations must have a symmetric canonical energy-momentum tensor

$$\Theta^{ab} = \Theta^{ba}. \tag{3.189}$$

3.9 Spin Current

If the field $\varphi(x)$ is no longer a scalar but has several spatial components, then the derivation of the four-dimensional angular momentum becomes slightly more involved.

3.9.1 Electromagnetic Fields

Consider first the case of electromagnetism where the relevant field is the four-vector potential $A^a(x)$. When going to a new coordinate frame

$$x'^a = \Lambda^a{}_b x^b \tag{3.190}$$

the vector field at the same point remains unchanged in absolute spacetime. But since the components A^a refer to two different basic vectors in the different frames, they must be transformed simultaneously with x^a. Since $A^a(x)$ is a vector, it transforms as follows:

$$A'^a(x') = \Lambda^a{}_b A^b(x). \tag{3.191}$$

For an infinitesimal transformation

$$\delta_s x^a = \omega^a{}_b x^b \tag{3.192}$$

this implies the substantial variation

$$
\begin{aligned}
\delta_s A^a(x) &= A'^a(x) - A^a(x) = A'^a(x - \delta x) - A^a(x) \\
&= \omega^a{}_b A^b(x) - \omega^c{}_b x^b \partial_c A^a.
\end{aligned} \tag{3.193}
$$

The first term is a *spin transformation*, the other an *orbital transformation*. The orbital transformation can also be written in terms of the generators $\hat{L}_{ab}$ of the Lorentz group defined in (3.82) as

$$\delta_{\mathrm{s}}^{\mathrm{orb}} A^a(x) \;=\; -i\omega^{bc}\hat{L}_{bc} A^a(x).$$ (3.194)

The spin transformation of the vector field is conveniently rewritten with the help of the 4×4 -generators L_{ab} in Eq. (1.51). Adding the two together, we form the operator of total four-dimensional angular momentum

$$\hat{J}_{ab} \equiv 1 \times \hat{L}_{ab} + L_{ab} \times 1,$$ (3.195)

and can write the transformation (3.193) as

$$\delta_{\mathrm{s}}^{\mathrm{orb}} A^a(x) \;=\; -i\omega^{ab}\hat{J}_{ab} A(x).$$ (3.196)

If the Lagrangian density involves only scalar combinations of four-vectors A^a and if it has no explicit x-dependence, it changes under Lorentz transformations like a scalar field:

$$\mathcal{L}'(x') \equiv \mathcal{L}(A'(x'), \partial' A'(x')) = \mathcal{L}(A(x), \partial A(x)) \equiv \mathcal{L}(x).$$ (3.197)

Infinitesimally, this amounts to

$$\delta_{\mathrm{s}}\mathcal{L} = -(\partial_a \mathcal{L}\, x^b)\omega^a{}_b.$$ (3.198)

With the Lorentz transformations being symmetry transformations, we calculate as in (3.170) the current density of total four-dimensional angular momentum:

$$J^{ab,c} = \frac{1}{c}\left[\frac{\partial \mathcal{L}}{\partial \partial_c A_a} A^b - \left(\frac{\partial \mathcal{L}}{\partial \partial_c A_d}\partial^a A^d x^b - \delta^{ac}\mathcal{L}\, x^b\right) - (a \leftrightarrow b)\right].$$ (3.199)

The prefactor $1/c$ is chosen to give these Noether currents of the electromagnetic field the conventional physical dimension. In fact, the last two terms have the same form as the current density $L^{ab,c}$ of the four-dimensional angular momentum of the scalar field. Here they are the corresponding quantities for the vector potential $A^a(x)$:

$$L^{ab,c} = -\frac{1}{c}\left(\frac{\partial \mathcal{L}}{\partial \partial_c A_d}\partial^a A^d x^b - \delta^{ac}\mathcal{L}\, x^b\right) + (a \leftrightarrow b).$$ (3.200)

This can also be written as

$$L^{ab,c} = \frac{1}{c}\left\{-i\frac{\partial \mathcal{L}}{\partial \partial_c A_d}\hat{L}^{ab} A^d + \left[\delta^{ac}\mathcal{L}\, x^b - (a \leftrightarrow b)\right]\right\},$$ (3.201)

where $\hat{L}^{ab}$ are the differential operators of four-dimensional angular momentum (1.107) satisfying the commutation rules (1.71) and (1.72).

The current densities (3.200) can be expressed in terms of the canonical energy-momentum tensor as

$$L^{ab,c} = x^a \Theta^{bc} - x^b \Theta^{ac}, \tag{3.202}$$

just as the scalar case (3.181). The first term in (3.199),

$$\Sigma^{ab,c} = \frac{1}{c} \left[\frac{\partial \mathcal{L}}{\partial \partial_c A_b} A^b - (a \leftrightarrow b) \right], \tag{3.203}$$

is referred to as the *spin current density*. It can be written in terms of the 4×4-generators (1.51) of the Lorentz group as

$$\Sigma^{ab,c} = -\frac{i}{c} \frac{\partial \mathcal{L}}{\partial \partial_c A^d} (L^{ab})_{d\sigma} A^\sigma. \tag{3.204}$$

The two current densities together

$$J^{ab,c}(x) \equiv L^{ab,c}(x) + \Sigma^{ab,c}(x) \tag{3.205}$$

have zero divergence:

$$\partial_c J^{ab,c}(x) = 0. \tag{3.206}$$

Hence the total angular momentum given by the charge

$$J^{ab} = \int d^3x \, J^{ab,0}(x) \tag{3.207}$$

is a constant of motion.

Individually, angular momentum and spin are not conserved. Using the conservation law of the energy-momentum tensor we find, just as in (3.188), that the current density of orbital angular momentum satisfies

$$\partial_c L^{ab,c}(x) = - \left[\Theta^{ab}(x) - \Theta^{ba}(x) \right]. \tag{3.208}$$

From this we find the divergence of the spin current

$$\partial_c \Sigma^{ab,c}(x) = \left[\Theta^{ab}(x) - \Theta^{ba}(x) \right]. \tag{3.209}$$

For the charges associated with orbital and spin currents

$$L^{ab}(t) \equiv \int d^3x L^{ab,0}(x), \quad \Sigma^{ab}(t) \equiv \int d^3x \Sigma^{ab,0}(x), \tag{3.210}$$

this implies the following time dependence:

$$\begin{aligned} \dot{L}^{ab}(t) &= -\int d^3x \left[\Theta^{ab}(x) - \Theta^{ba}(x) \right], \\ \dot{\Sigma}^{ab}(t) &= \int d^3x \left[\Theta^{ab}(x) - \Theta^{ba}(x) \right]. \end{aligned} \tag{3.211}$$

Fields with nonzero spin always have a nonsymmetric energy momentum tensor.

Then the current $J^{ab,c}$ becomes, now back in natural units,

$$J^{ab,c} = \left(\frac{\partial \delta_s^x \mathcal{L}}{\partial \partial_c \omega_{ab}(x)} - \delta^{ac} \mathcal{L} x^b \right) - (a \leftrightarrow b). \tag{3.212}$$

By the chain rule of differentiation, the derivative with respect to $\partial, \omega_{ab}(x)$ can come only from field derivatives. For a scalar field we obtain

$$\frac{\partial \delta_s^x \mathcal{L}}{\partial \partial_c \omega_{ab}(x)} = \frac{\partial \mathcal{L}}{\partial \partial_c \varphi} \frac{\partial \delta_s^x \varphi}{\partial \omega_{ab}(x)}, \tag{3.213}$$

and for a vector field

$$\frac{\partial \delta_s^x \mathcal{L}}{\partial \partial_c \omega_{ab}(x)} = \frac{\partial \mathcal{L}}{\partial \partial_c A^d} \frac{\partial \delta_s^x A^d}{\partial \omega_{ab}}. \tag{3.214}$$

The alternative rule of calculating angular momenta is to introduce spacetime-dependent transformations

$$\delta^x x = \omega^a{}_b(x) x^b \tag{3.215}$$

under which the scalar fields transform as

$$\delta_s \varphi = -\partial_c \varphi \omega^c{}_b(x) x^b \tag{3.216}$$

and the Lagrangian density as

$$\delta_s^x \varphi = -\partial_c \mathcal{L} \, \omega^c{}_b(x) x^b = -\partial_c (x^b \mathcal{L}) \omega^c{}_b(x). \tag{3.217}$$

By separating spin and orbital transformations of $\delta_s^x A^d$ we find the two contributions $\sigma^{ab,c}$ and $L^{ab,c}$ to the current $J^{ab,c}$ of the total angular momentum, the latter receiving a contribution from the second term in (3.212).

3.9.2 Dirac Field

We now turn to the Dirac field. Under a Lorentz transformation (3.190), this transforms according to the law

$$\psi(x') \xrightarrow{\Lambda} \psi'_\Lambda(x) = D(\Lambda) \psi(x), \tag{3.218}$$

where $D(\Lambda)$ are the 4×4-spinor representation matrices of the Lorentz group. Their matrix elements can most easily be specified for infinitesimal transformations. For an infinitesimal Lorentz transformation

$$\Lambda_a{}^b = \delta_a{}^b + \omega_a{}^b, \tag{3.219}$$

under which the coordinates are changed by

$$\delta_s x^a = \omega^a{}_b x^b \tag{3.220}$$

the spin transforms under the representation matrix

$$D(\delta_a{}^b + \omega_a{}^b) = \left(1 - i\frac{1}{2}\omega_{ab}\sigma^{ab}\right)\psi(x), \tag{3.221}$$

where σ_{ab} are the 4×4 -matrices acting on the spinor space defined in Eq. (1.228). We have shown in (1.226) that the spin matrices $\Sigma_{ab} \equiv \sigma_{ab}/2$ satisfy the same commutation rules (1.71) and (1.72) as the previous orbital and spin-1 -generators $\hat{L}_{aba}$ and L_{ab} of Lorentz transformations.

The field has the substantial variation [compare (3.193)]:

$$
\begin{aligned}
\delta_s\psi(x) &= \psi'(x) - \psi(x) = D(\delta_a{}^b + \omega_a{}^b)\psi(x - \delta x) - \psi(x) \\
&= -i\frac{1}{2}\omega_{ab}\sigma^{ab}\psi(x) - \omega^c{}_b x^b \partial_c\psi(x) \\
&= -i\frac{1}{2}\omega_{ab}\left[S^{ab} + \hat{L}^{ab}\right]\psi(x) \equiv -i\frac{1}{2}\omega_{ab}\hat{J}^{ab}\psi(x),
\end{aligned}
\tag{3.222}
$$

the last line showing the separation into spin and orbital transformation for a Dirac particle.

Since the Dirac Lagrangian is Lorentz-invariant, it changes under Lorentz transformations like a scalar field:

$$\mathcal{L}'(x') = \mathcal{L}(x). \tag{3.223}$$

Infinitesimally, this amounts to

$$\delta_s\mathcal{L} = -(\partial_a\mathcal{L}x^b)\omega^a{}_b. \tag{3.224}$$

With the Lorentz transformations being symmetry transformations in the Noether sense, we calculate the *current density of total four-dimensional angular momentum*, extending the formulas (3.180) and (3.205) for scalar and vector fields. The result is

$$J^{ab,c} = \left(-i\frac{\partial\mathcal{L}}{\partial\partial_c\psi}\Sigma^{ab}\psi - i\frac{\partial\mathcal{L}}{\partial\partial_c\psi}\hat{L}^{ab}\psi + \mathrm{cc}\right) + \left[\delta^{ac}\mathcal{L}x^b - (a \leftrightarrow b)\right]. \tag{3.225}$$

As in (3.181) and (3.202), the orbital part of (3.225) can be expressed in terms of the canonical energy-momentum tensor as

$$L^{ab,c} = x^a\Theta^{bc} - x^b\Theta^{ac}. \tag{3.226}$$

The first term in (3.225) is the *spin current density*

$$\Sigma^{ab,c} = -i\frac{\partial\mathcal{L}}{\partial\partial_c\psi}\Sigma^{ab}\psi + \mathrm{cc} . \tag{3.227}$$

Inserting (3.158), this becomes explicitly

$$\Sigma^{ab,c} = -i\bar{\psi}\gamma^c\Sigma^{ab}\psi = -\frac{i}{2}\epsilon^{abcd}\bar{\psi}\gamma^d\gamma_5\psi, \tag{3.228}$$

with $\gamma_5 \equiv i\gamma^0\gamma^1\gamma^2\gamma^3$. The spin current density is completely antisymmetric in its three indices, an important property for the construction of a consistent quantum mechanics in a space with torsion (see Ref. [3]).

The conservation properties of these current densities are completely identical to those in Eqs. (3.206), (3.208), and (3.209).

Due to the presence of spin, the energy-momentum tensor is not symmetric.

3.10   Symmetric Energy-Momentum Tensor

In Eq. (3.209) we have seen that the presence of spin is the cause for the asymmetry of the canonical energy-momentum tensor. It is therefore suggestive to use the spin current density for the construction of a new symmetric energy-momentum tensor

$$T^{ab} = \Theta^{ab} + \Delta\Theta^{ba}. \tag{3.229}$$

In order to deserve its name, this should, of course, still possess the fundamental property of Θ^{ab}, that the integral

$$P^a = \int d^3x\, T^{a0} \tag{3.230}$$

yields the total energy-momentum vector of the system. This is certainly the case if $\Delta\Theta^{a0}$ is a three-divergence of a spatial vector. The appropriate construction was found by Belinfante in 1939. He introduced the tensor [5]

$$T^{ab} = \Theta^{ab} - \frac{1}{2}\partial_c(\Sigma^{ab,c} - \Sigma^{bc,a} + \Sigma^{ca,b}), \tag{3.231}$$

whose symmetry is manifest, due to (3.209) and the symmetry of the last two terms in ab. Moreover, the components

$$T^{a0} = \Theta^{a0} - \frac{1}{2}\partial_c(\Sigma^{a0,c} - \Sigma^{0c,a} + \Sigma^{ca,0}) \tag{3.232}$$

differ from Θ^{a0} by a pure three-divergence, as required for the property (3.230).

Another important property of the symmetric energy-momentum tensor (3.231) that if we form the current density of total angular momentum

$$J^{ab,c} \equiv x^a T^{bc} - x^b T^{ac}, \tag{3.233}$$

the spatial integral over the zeroth component

$$J^{ab} = \int d^3x\, J^{ab,0}. \tag{3.234}$$

leads to the same total angular momentum as the canonical expression (3.205). Indeed, the zeroth component of (3.233) is

$$x^a\Theta^{b0} - x^b\Theta^{a0} - \frac{1}{2}\left[\partial_k(\Sigma^{a0,k} - \Sigma^{0k,a} + \Sigma^{ka,0})x^b - (a \leftrightarrow b)\right], \tag{3.235}$$

and if we integrate the term in brackets over d^3x, a partial integration yields for $a = 0, b = i$:

$$-\frac{1}{2} \int d^3x \left[x^0 \partial_k (\Sigma^{i0,k} - \Sigma^{0k,i} + \Sigma^{ki,0}) - x^i \partial_k (\Sigma^{00,k} - \Sigma^{0k,0} + \Sigma^{k0,0}) \right] = \int d^3x \, \Sigma^{0i,0}. \tag{3.236}$$

For $a = i, b = j$, it yields

$$-\frac{1}{2} \int d^3x \left[x^i \partial_k (\Sigma^{j0,k} - \Sigma^{0k,j} + \Sigma^{kj,0}) - (i \leftrightarrow j) \right] = \int d^3x \, \Sigma^{ij,0}. \tag{3.237}$$

The right-hand sides of (3.236) and (3.237) are the contributions of the spin to the total angular momentum.

For the electromagnetic field, the spin current density (3.203) reads, explicitly

$$\Sigma^{ab,c} = -\frac{1}{c} \left[F^{ca} A^b - (a \leftrightarrow b) \right]. \tag{3.238}$$

From this we calculate the Belinfante correction

$$\begin{aligned}
\Delta \Theta^{ab} &= \frac{1}{2c} [\partial_c (F^{ca} A^b - F^{cb} A^a) - \partial_c (F^{ab} A^c - F^{ac} A^b) + \partial_c (F^{bc} A^a - F^{ba} A^c)] \\
&= \frac{1}{c} \partial_c (F^{bc} A^a). \tag{3.239}
\end{aligned}$$

Adding this to the canonical energy-momentum tensor (3.150)

$$\Theta^{ab} = \frac{1}{c} \left[-F^b{}_c \partial^a A^c + \frac{1}{4} g^{ab} F^{cd} F_{cd} \right], \tag{3.240}$$

we find the symmetric energy-momentum tensor

$$T^{ab} = \frac{1}{c} \left[-F^b{}_c F^{ac} + \frac{1}{4} g^{ab} F^{cd} F_{cd} + (\partial_c F^{bc}) A^a \right]. \tag{3.241}$$

The last term vanishes for a free Maxwell field which satisfies $\partial_c F^{ab} = 0$ [recall (2.87)], and can be dropped, so that T^{ab} agrees with the previously constructed symmetric energy-momentum tensor (1.261) of the electromagnetic field. The symmetry of T^{ab} can easily be verified using once more the Maxwell equation $\partial_c F^{ab} = 0$.

Recall that according to Eq. (1.258) the component $cT^{00}(x)$ is the known expression for the energy density of the electromagnetic field $\mathcal{E}(x) = (\mathbf{E}^2 + \mathbf{B}^2)/2$. The components $c^2 T^{0i}(x)$ coincide with the *Poynting vector* of energy current density $\mathbf{S}(x) = c \mathbf{E} \times \mathbf{B}$ of Eq. (1.259), and the conservation law $c^2 \partial_a T^{0a}(0)$ reproduces Poynting's law $\partial_t \mathcal{E}(x) + \mathbf{\nabla} \cdot \mathbf{S}(x) = 0$ of Eq. (1.268).

In the presence of an external current, where the Lagrangian density is (2.84), the canonical energy-momentum tensor becomes

$$\Theta^{ab} = \frac{1}{c} \left[-F^b{}_c \partial^a A^c + \frac{1}{4} g^{ab} F^{cd} F_{cd} + \frac{1}{c} g^{ab} j^c A_c \right], \tag{3.242}$$

generalizing (3.240).

The spin current is again given by Eq. (3.238), leading to the Belinfante energy-momentum tensor

$$
\begin{aligned}
T^{ab} &= \Theta^{ab} + \frac{1}{c}\partial_c(F^{bc}A^a) \\
&= \frac{1}{c}\left[-F^b{}_c F^{ac} + \frac{1}{4}g^{ab}F^{cd}F_{cd} + \frac{1}{c}g^{ab}j^c A_c - \frac{1}{c}j^b A^a\right].
\end{aligned}
\tag{3.243}
$$

The last term prevents T^{ab} from being symmetric, unless the current vanishes. Due to the external current, the conservation law $\partial_b T^{ab} = 0$ is modified to

$$
\partial_b T^{ab} = \frac{1}{c^2}A_c(x)\partial^a j^c(x).
\tag{3.244}
$$

3.11 Internal Symmetries

In quantum field theory, an important role in classifying various actions is played by *internal symmetries*. They do not involve any change in the spacetime coordinate of the fields. For an N-component real field $\phi(x)$, they have the form

$$
\phi'(x) = e^{-i\alpha_r G_r}\phi(x)
\tag{3.245}
$$

where G_r are the generators of some Lie group, and α_r the associated transformation parameters. The generators G_r are $N \times N$-matrices, satisfying commutation rules with structure constants f_{rst} [recall (1.65)]:

$$
[G_r, G_s] = if_{rst}G_t, \qquad (r, s, t = 1, \ldots, \text{rank}),
\tag{3.246}
$$

where f_{rst} are the structure constants of the Lie algebra.

The infinitesimal symmetry transformations are substantial variations of the form

$$
\delta_s\phi = -i\alpha_r G_r\phi.
\tag{3.247}
$$

The associated conserved current densities read

$$
j_r^a = -i\frac{\partial\mathcal{L}}{\partial\partial_a\phi}G_r\phi.
\tag{3.248}
$$

They can also be written as

$$
j_r^a = -i\pi\, G_r\phi,
\tag{3.249}
$$

where $\pi(x) \equiv \partial\mathcal{L}(x)/\partial\partial_a\phi(x)$ is the canonical momentum of the field $\phi(x)$ [compare (2.60)].

The most important example is that of a complex field ϕ and a generator $G = 1$, where the symmetry transformation (3.245) is simply a multiplication by a constant phase factor. One also speaks of U(1)-symmetry. Other important examples are those of a triplet or an octet of fields ϕ with G_r being the generators of an SU(2) or SU(3) representation. The U(1)-symmetry leads to charge conservation in electromagnetic interactions, the other two are responsible for isospin SU(2) and SU(3) invariance in strong interactions. The latter symmetries are, however, not exact.

3.11.1   U(1)-Symmetry and Charge Conservation

Consider the Lagrangian density of a complex scalar field

$$\mathcal{L}(x) = \mathcal{L}(\varphi(x), \varphi^*, \partial\varphi(x), \partial\varphi^*(x), x). \tag{3.250}$$

It is invariant under U(1)-transformations

$$\delta_s\varphi(x) = -i\alpha\varphi(x), \quad \delta_s\phi^*(x) = i\alpha\phi^*(x), \tag{3.251}$$

i.e., $\delta_s\mathcal{L} = 0$. By the chain rule of differentiation we obtain, using the Euler-Lagrange equation (2.40),

$$\delta_s\mathcal{L} = \left(\frac{\partial\mathcal{L}}{\partial\varphi} - \partial_a\frac{\partial\mathcal{L}}{\partial_a\varphi}\right)\delta_s\varphi + \partial_a\left[\frac{\partial\mathcal{L}}{\partial\partial_a\varphi}\delta_s\varphi\right] + cc = \partial_a\left[\frac{\partial\mathcal{L}}{\partial\partial_a\varphi}\delta_s\varphi\right] + cc. \tag{3.252}$$

Equating this with zero according to the symmetry assumption, and inserting (3.251), we find that

$$j_r^a = -i\frac{\partial\mathcal{L}}{\partial\partial_a\varphi}\varphi + cc \tag{3.253}$$

is a conserved current.

For the free Lagrangian density (2.27) in natural units

$$\mathcal{L}(x) = \partial_\mu\varphi^*\partial_\mu\varphi - m^2\varphi^*\varphi \tag{3.254}$$

we obtain the conserved current density of Eq. (2.67):

$$j_\mu = -i\varphi^*\overleftrightarrow{\partial}_\mu\varphi, \tag{3.255}$$

where the symbol $\varphi^*\overleftrightarrow{\partial}_\mu\varphi$ denotes the right-minus-left derivative (2.68).

For a free Dirac field, the current density (3.253) takes the form

$$j^\mu(x) = \bar{\psi}(x)\gamma^\mu\psi(x). \tag{3.256}$$

3.11.2   Broken Internal Symmetries

The physically important symmetries SU(2) of isospin and SU(3) are not exact. The symmetry variation of the Lagrange density is not strictly zero. In this case we make use of the alternative derivation of the conservation law based on Eq. (3.109). We introduce the spacetime-dependent parameters $\alpha(x)$ and conclude from the extremality property of the action that

$$\partial_a\frac{\partial\mathcal{L}^\epsilon}{\partial\partial_a\alpha_r(x)} = \frac{\partial\mathcal{L}^\epsilon}{\partial\alpha_r(x)}. \tag{3.257}$$

This implies the divergence law for the above currents

$$\partial_a j_r^a(x) = \frac{\partial\mathcal{L}^\epsilon}{\partial\alpha_r}. \tag{3.258}$$

3.12   Generating the Symmetry Transformations for Quantum Fields

As in quantum mechanical systems, the charges associated with the conserved currents obtained in the previous section can be used to generate the transformations of the fields from which they were derived. One merely has to invoke the canonical field commutation rules.

For the currents (3.248) of an N-component real field $\phi(x)$, the charges are

$$Q_r = -i \int d^3x \frac{\partial \mathcal{L}}{\partial \partial_a \phi} G_r \phi \qquad (3.259)$$

and can be written as

$$Q_r = -i \int d^3x\, \pi\, G_r \phi, \qquad (3.260)$$

where $\pi(x) \equiv \partial \mathcal{L}(x)/\partial \partial_a \phi(x)$ is the canonical momentum of the field $\phi(x)$. After quantization, these fields satisfy the canonical commutation rules:

$$\begin{aligned}
[\hat{\pi}(\mathbf{x},t), \hat{\phi}(\mathbf{x}',t)] &= -i\delta^{(3)}(\mathbf{x} - \mathbf{x}'), \\
[\hat{\phi}(\mathbf{x},t), \hat{\phi}(\mathbf{x}',t)] &= 0, \\
[\hat{\pi}(\mathbf{x},t), \hat{\pi}(\mathbf{x}',t)] &= 0.
\end{aligned} \qquad (3.261)$$

From this we derive directly the commutation rule between the quantized version of the charges (3.260) and the field operator $\hat{\phi}(x)$:

$$[\hat{Q}_r, \hat{\phi}(x)] = -\alpha_r G_r \phi(x). \qquad (3.262)$$

We also find that the commutation rules among the quantized charges $\hat{Q}_r$ are the same as those of the generators G_r in (3.246):

$$[\hat{Q}_r, \hat{Q}_s] = f_{rst}\hat{Q}_t, \quad (r,s,t = 1, \ldots, \text{rank}). \qquad (3.263)$$

Hence the operators $\hat{Q}_r$ form a representation of the generators of symmetry group in the many-particle Hilbert space generated by the quantized fields $\hat{\phi}(x)$ (*Fock space*).

As an example, we may derive in this way the commutation rules of the conserved charges associated with the Lorentz generators (3.205):

$$J^{ab} \equiv \int d^3x J^{ab,0}(x). \qquad (3.264)$$

They are obviously the same as those of the 4×4-matrices (1.51), and those of the quantum mechanical generators (1.107):

$$[\hat{J}^{ab}, \hat{J}^{ac}] = -ig^{aa}\hat{J}^{bc}. \qquad (3.265)$$

The generators $J^{ab} \equiv \int d^3x J^{ab,0}(x)$, are sums $J^{ab} = L^{ab}(t) + \Sigma^{ab}(t)$ of charges (3.210) associated with orbital and spin rotations. According to (3.211), these individual charges are time dependent, only their sum being conserved. Nevertheless, they both generate Lorentz transformations: $L^{ab}(t)$ on the spacetime argument of the fields, and $\Sigma^{ab}(t)$ on the spin indices. As a consequence, they both satisfy the commutation relations (3.265):

$$[\hat{L}^{ab}, \hat{L}^{ac}] = -ig^{aa}\hat{L}^{bc}, \qquad [\hat{\Sigma}^{ab}, \hat{\Sigma}^{ac}] = -ig^{aa}\hat{\Sigma}^{bc}. \qquad (3.266)$$

It is important to realize that the commutation relations (3.262) and (3.263) remain also valid in the presence of symmetry-breaking terms as long as these do not contribute to the canonical momentum of the theory. Such terms are called *soft symmetry-breaking terms*. The charges are no longer conserved, so that we must attach a time argument to the commutation relations (3.262) and (3.263). All times in these relations must be the same, in order to invoke the equal-time canonical commutation rules.

The commutators (3.263) have played an important role in developing a theory of strong interactions, where they first appeared in the form of a *charge algebra* of the broken symmetry SU(3) × SU(3) of weak and electromagnetic charges. This symmetry will be discussed in more detail in Chapter 10.

3.13 Energy-Momentum Tensor of Relativistic Massive Point Particle

If we want to study energy and momentum of charged relativistic point particles in an electromagnetic field it is useful to consider the action (3.68) with (3.70) as an integral over a Lagrangian density:

$$\mathcal{A} = \int d^4x\, \mathcal{L}(x), \quad \text{with} \quad \mathcal{L}(x) = \int_{\tau_a}^{\tau_b} d\tau\, \overset{m}{L}\left(\dot{x}^a(\tau)\right)\delta^{(4)}(x - x(\tau)). \qquad (3.267)$$

This allows us to derive local conservation laws for point particles in the same way as for fields. Instead of doing this, however, we shall take advantage of the previously derived global conservation laws and convert them into local ones by inserting appropriate δ-functions with the help of the trivial identity

$$\int d^4x\, \delta^{(4)}(x - x(\tau)) = 1. \qquad (3.268)$$

Consider for example the conservation law (3.72) for the momentum (3.73). With the help of (3.268) this becomes

$$0 = -\int d^4x \int_{-\infty}^{\infty} d\tau \left[\frac{d}{d\tau}p_c(\tau)\right]\delta^{(4)}(x - x(\tau)). \qquad (3.269)$$

Note that in this expression the boundaries of the four-volume contain the information on initial and final times. We then perform a partial integration in τ, and rewrite (3.269) as

$$0 = -\int d^4x \int_{-\infty}^{\infty} d\tau \frac{d}{d\tau} \left[p_c(\tau)\delta^{(4)}(x - x(\tau)) \right] + \int d^4x \int_{-\infty}^{\infty} d\tau p_c(\tau)\partial_\tau \delta^{(4)}(x - x(\tau)).$$
(3.270)

The first term vanishes if the orbits come from and disappear into infinity. The second term can be rewritten as

$$0 = -\int d^4x\, \partial_b \left[\int_{-\infty}^{\infty} d\tau p_c(\tau)\dot{x}^b(\tau)\delta^{(4)}(x - x(\tau)) \right].$$
(3.271)

This shows that

$$\overset{m}{\Theta}{}^{cb}(x) \equiv m\int_{-\infty}^{\infty} d\tau\, \dot{x}^c(\tau)\dot{x}^b(\tau)\delta^{(4)}(x - x(\tau))$$
(3.272)

satisfies the local conservation law

$$\partial_b \overset{m}{\Theta}{}^{cb}(x) = 0,$$
(3.273)

which is the conservation law for the energy-momentum tensor of a massive point particle.

The total momenta are obtained from the spatial integrals over Θ^{c0}:

$$P^a(t) \equiv \int d^3x\, \Theta^{c0}(x).$$
(3.274)

For point particles, they coincide with the canonical momenta $p^a(t)$. If the Lagrangian depends only on the velocity $\dot{x}^a$ and not on the position $x^a(t)$, the momenta $p^a(t)$ are constants of motion: $p^a(t) \equiv p^a$.

The Lorentz invariant quantity

$$M^2 = P^2 = g_{ab}P^a P^b$$
(3.275)

is called the *total mass* of the system. For a single particle it coincides with the mass of the particle.

Subjecting the orbits $x^a(\tau)$ to Lorentz transformations according to the rules of the last section we find the currents of total angular momentum

$$L^{ab,c} \equiv x^a \Theta^{bc} - x^b \Theta^{ac}$$
(3.276)

to satisfy the conservation law:

$$\partial_c L^{ab,c} = 0.$$
(3.277)

A spatial integral over the zeroth component of the current $L^{ab,c}$ yields the conserved charges:

$$L^{ab}(t) \equiv \int d^3x\, L^{ab,0}(x) = x^a p^b(t) - x^b p^a(t).$$
(3.278)

3.14 Energy-Momentum Tensor of Massive Charged Particle in Electromagnetic Field

Let us also consider an important combination of a charged point particle and an electromagnetic field Lagrangian

$$\mathcal{A} = -mc \int_{\tau_a}^{\tau_b} d\tau \sqrt{g_{ab}\dot{x}^a(\tau)\dot{x}^b(\tau)} - \frac{1}{4}\int d^4x F_{ab}F^{ab} - \frac{e}{c}\int_{\tau_a}^{\tau_b} d\tau \dot{x}^a(\tau)A_a(x(\tau)). \quad (3.279)$$

By varying the action in the particle orbits, we obtain the Lorentz equation of motion

$$\frac{dp^a}{d\tau} = \frac{e}{c}F^a{}_b\dot{x}^b(\tau). \quad (3.280)$$

By varying the action in the vector potential, we find the Maxwell-Lorentz equation

$$-\partial_b F^{ab} = \frac{e}{c}\dot{x}^b(\tau). \quad (3.281)$$

The action (3.279) is invariant under translations of the particle orbits and the electromagnetic fields. The first term is obviously invariant, since it depends only on the derivatives of the orbital variables $x^a(\tau)$. The second term changes under translations by a pure divergence [recall (3.132)]. Also the interaction term changes by a pure divergence. Indeed, under infinitesimal spacetime translations $x^b(\tau) \rightarrow x^b(\tau) - \epsilon^b$, the velocities $\dot{x}^a(\tau)$ are invariant:

$$\dot{x}^a(\tau) \quad \rightarrow \quad \dot{x}^a(\tau), \quad (3.282)$$

and $A_a(x^b)$ changes as follows:

$$A_a(x^b) \rightarrow A'_a(x^b) = A_a(x^b + \epsilon^b) = A_a(x^b) + \epsilon^b\partial_a A_a(x^b). \quad (3.283)$$

For the Lagrangian density of the action (3.279), this implies that the substantial variation is a pure derivative term:

$$\delta_s \mathcal{L} = \epsilon^b \partial_b \mathcal{L}. \quad (3.284)$$

We now we calculate the same variation once more invoking the Euler-Lagrange equations. This gives

$$\delta_s \mathcal{A} = \int d\tau \frac{d}{d\tau}\frac{\partial \overset{m}{L}}{\partial x'^a}\delta_s x^a + \int d^4x \frac{\partial \overset{em}{\mathcal{L}}}{\partial \partial_c A^a}\delta_s A^a. \quad (3.285)$$

The first term can be treated as in (3.270) and (3.271), after which it acquires the form

$$-\int_{\tau_a}^{\tau_b} d\tau \frac{d}{d\tau}\left(p_a + \frac{e}{c}A_a\right) = -\int d^4x \int_{-\infty}^{\infty} d\tau \frac{d}{d\tau}\left[\left(p_a + \frac{e}{c}A_a\right)\delta^{(4)}(x - x(\tau)\right] \quad (3.286)$$

$$+ \int d^4x \int_{-\infty}^{\infty} d\tau \left(p_a + \frac{e}{c}A_a\right)\frac{d}{d\tau}\delta^{(4)}(x - x(\tau))$$

and becomes, after dropping boundary terms,

$$-\int_{\tau_a}^{\tau_b} d\tau \frac{d}{d\tau}\left(p_a + \frac{e}{c}A_a\right) = \partial_c \int d^4x \int_{-\infty}^{\infty} d\tau \left(p_a + \frac{e}{c}A_a\right)\frac{dx^c}{d\tau}\delta^{(4)}(x - x(\tau)). \quad (3.287)$$

The electromagnetic part of (3.279) is the same as before in Subsection 3.6.1, since the interaction contains no derivative of the gauge field. In this way we find the canonical energy-momentum tensor

$$\Theta^{ab}(x) = \int d\tau \left(p^a + \frac{e}{c}A^a\right)\dot{x}^b(\tau)\delta^{(4)}(x - x(\tau)) - F^b{}_c\partial^a A^c + \frac{1}{4}g^{ab}F^{cd}F_{cd}. \quad (3.288)$$

Let us check its conservation by calculating the divergence:

$$\partial_b\Theta^{ab}(x) = \int d\tau \left(p + \frac{e}{c}A_a\right)\dot{x}^b(\tau)\partial_b\delta^{(4)}(x - x(\tau))$$

$$-\partial_b F^b{}_c\partial^a A^c - F^b{}_c\partial_b\partial^a A^c + \frac{1}{4}\partial^a(F^{cd}F_{cd}). \quad (3.289)$$

The first term is, up to a boundary term, equal to

$$-\int d\tau \left(p^a + \frac{e}{\tau}A^a\right)\frac{d}{d\tau}\delta^{(4)}(x - x(\tau)) = \int d\tau \left[\frac{d}{d\tau}\left(p^a + \frac{e}{c}A^a\right)\right]\delta^{(4)}(x - x(\tau)). \quad (3.290)$$

Using the Lorentz equation of motion (3.280), this becomes

$$\frac{e}{c}\int_{-\infty}^{\infty} d\tau \left(F^a{}_b\dot{x}^b(\tau) + \frac{d}{d\tau}A^a\right)\delta^{(4)}(x - x(\tau)). \quad (3.291)$$

Inserting the Maxwell equation

$$\partial_b F^{ab} = -e\int d\tau (dx^a/d\tau)\delta^{(4)}(x - x(\tau)), \quad (3.292)$$

the second term in Eq. (3.289) can be rewritten as

$$-\frac{e}{c}\int_{-\infty}^{\infty} d\tau \frac{dx_c}{d\tau}\partial^a A^c\delta^{(4)}(x - x(\tau)), \quad (3.293)$$

which is the same as

$$-\frac{e}{c}\int d\tau \left(\frac{dx_a}{d\tau}F^{ac} + \frac{dx_c}{d\tau}\partial^c A^a\right)\delta^{(4)}(x - x(\tau)), \quad (3.294)$$

thus canceling (3.291). The third term in (3.289) is, finally, equal to

$$-F^b{}_c\partial^a F_b{}^c + \frac{1}{4}\partial^a(F^{cd}F_{cd}), \quad (3.295)$$

due to the antisymmetry of F^{bc}. By rewriting the homogeneous Maxwell equation, the Bianchi identity (2.89), in the form

$$\partial_c F_{ab} + \partial_a F_{bc} + \partial_b F_{ca} = 0, \quad (3.296)$$

and contracting it with F^{ab}, we see that the term (3.295) vanishes identically.

It is easy to construct from (3.288) Belinfante's symmetric energy momentum tensor. We merely observe that the spin density is entirely due to the vector potential, and hence the same as before [see (3.238)]

$$\Sigma^{ab,c} = -\left[F^{ca} A^b - (a \leftrightarrow b) \right]. \tag{3.297}$$

Thus the additional piece to be added to the canonical energy momentum tensor is again [see (3.239)]

$$\Delta\Theta^{ab} = \partial_c(F^{ab} A^a) = \frac{1}{2}(\partial_c F^{bc} A^a + F^{bc} \partial_c A^a). \tag{3.298}$$

The last term in this expression serves to symmetrize the electromagnetic part of the canonical energy-momentum tensor, bringing it to the Belinfante form (3.241):

$$\overset{\text{em}}{T}{}^{ab} = -F^b{}_c F^{ac} + \frac{1}{4} g^{ab} F^{cd} F_{cd}. \tag{3.299}$$

The term containing $\partial_c F^{bc}$ in (3.298), which vanishes in the absence of charges, is needed to symmetrize the matter part of Θ^{ab}. Indeed, using once more Maxwell's equation, it becomes

$$-\frac{e}{c} \int d\tau \, \dot{x}^b(\tau) A^a \delta^{(4)}(x - x(\tau)), \tag{3.300}$$

thus canceling the corresponding term in (3.288). In this way we find that the total energy-momentum tensor of charged particles plus electromagnetic fields is simply the sum of the two symmetric energy-momentum tensors:

$$T^{ab} = \overset{\text{m}}{T}{}^{ab} + \overset{\text{em}}{T}{}^{ab} \tag{3.301}$$

$$= m \int_{-\infty}^{\infty} d\tau \, \dot{x}^a \dot{x}^b \delta^{(4)}(x - x(\tau)) - F^b{}_c F^{ac} + \frac{1}{4} g^{ab} F^{cd} F_{cd}.$$

For completeness, let us check its conservation. Forming the divergence $\partial_b T^{ab}$, the first term gives

$$\frac{e}{c} \int d\tau \, \dot{x}^b(\tau) F^a{}_b(x(\tau)), \tag{3.302}$$

in contrast to (3.291), which is canceled by the divergence in the second term

$$-\partial_b F^b{}_c F^{ac} = -\frac{e}{c} \int d\tau \, \dot{x}_c(\tau) F^{ac}(x(\tau)), \tag{3.303}$$

in contrast to (3.294).

Notes and References

For more details on classical electromagnetic fields see
L.D. Landau, E.M. Lifshitz, *The Classical Theory of Fields*, Addison-Wesley, Reading, MA, 1951;
A.O. Barut, *Electrodynamics and Classical Theory of Fields and Particles*, MacMillan, New York, N.Y. 1964;
J.D. Jackson, *Classical Electrodynamics*, John Wiley & Sons, New York, N.Y., 1975.

The individual citations refer to:

[1] S. Coleman and J.H. Van Vleck, Phys. Rev. **171**, 1370 (1968).

[2] H. Kleinert, *Path Integrals in Quantum Mechanics, Statistics, Polymer Physics, and Financial Markets*, World Scientific, Singapore 2006, 4th extended edition, pp. 1–1546 (kl/b5), where kl is short for the www address http://www.physik.fu-berlin.de/~kleinert.

[3] This problem is discussed in my textbook
H. Kleinert, *Path Integrals in Quantum Mechanics, Statistics, Polymer Physics, and Financial Markets* World Scientific, Singapore 2006, 4th extended edition, pp. 1–1546 (kl/b5), where kl is short for the www address http://www.physik.fu-berlin.de/~kleinert.

[4] E. Noether, Nachr. d. vgl. Ges. d. Wiss. Göttingen, Math-Phys. Klasse, **2**, 235 (1918);
See also
E. Bessel-Hagen, Math. Ann. **84**, 258 (1926);
L. Rosenfeld, Me. Acad. Roy. Belg. **18**, 2 (1938);
F.J. Belinfante, Physica **6**, 887 (1939).

[5] The Belinfante energy-momentum tensor is discussed in detail in
H. Kleinert, *Gauge Fields in Condensed Matter*, Vol. II *Stresses and Defects*, World Scientific Publishing, Singapore 1989, pp. 744-1443 (kl/b2).

4

Multivalued Gauge Transformations in Magnetostatics

For the development of a theory of gravitation it is crucial to realize that physical laws in Euclidean space can be transformed directly into spaces with curvature and torsion. As mentioned in the Introduction, this will be possible by a geometric generalization of a field-theoretic technique invented by Dirac to introduce magnetic monopoles into electrodynamics. So far, no magnetic monopoles have been discovered in nature, but the mathematics used by Dirac will suggest us how to proceed in the geometric situation.

4.1 Vector Potential of Current Distribution

Let us begin by recalling the standard description of magnetism in terms of vector potentials. Since there are no magnetic monopoles in nature, a magnetic field $\mathbf{B}(\mathbf{x})$ satisfies the identity $\boldsymbol{\nabla} \cdot \mathbf{B}(\mathbf{x}) = 0$, implying that only two of the three field components of $\mathbf{B}(\mathbf{x})$ are independent. To account for this, one usually expresses a magnetic field $\mathbf{B}(\mathbf{x})$ in terms of a vector potential $\mathbf{A}(\mathbf{x})$, setting $\mathbf{B}(\mathbf{x}) = \boldsymbol{\nabla} \times \mathbf{A}(\mathbf{x})$. Then Ampère's law, which relates the magnetic field to the electric current density $\mathbf{j}(\mathbf{x})$ by $\boldsymbol{\nabla} \times \mathbf{B} = \mathbf{j}(\mathbf{x})$, becomes a second-order differential equation for the vector potential $\mathbf{A}(\mathbf{x})$ in terms of an electric current

$$\boldsymbol{\nabla} \times [\boldsymbol{\nabla} \times \mathbf{A}](\mathbf{x}) = \mathbf{j}(\mathbf{x}). \tag{4.1}$$

In this chapter we are using natural units with $c = 1$ to save recurring factors of c.

The vector potential $\mathbf{A}(\mathbf{x})$ is a *gauge field*. Given $\mathbf{A}(\mathbf{x})$, any locally gauge-transformed field

$$\mathbf{A}(\mathbf{x}) \to \mathbf{A}'(\mathbf{x}) = \mathbf{A}(\mathbf{x}) + \boldsymbol{\nabla}\Lambda(\mathbf{x}) \tag{4.2}$$

yields the same magnetic field $\mathbf{B}(\mathbf{x})$. This reduces the number of physical degrees of freedom in the gauge field $\mathbf{A}(\mathbf{x})$ to two, just as those in $\mathbf{B}(\mathbf{x})$. In order for this to hold, the transformation function must be single-valued, i.e., it must have commuting derivatives

$$(\partial_i\partial_j - \partial_j\partial_i)\Lambda(\mathbf{x}) = 0. \tag{4.3}$$

The equation expressing the absence of magnetic monopoles $\nabla \cdot \mathbf{B} = 0$ is ensured if the vector potential has commuting derivatives

$$(\partial_i \partial_j - \partial_j \partial_i)\mathbf{A}(\mathbf{x}) = 0. \tag{4.4}$$

This integrability property makes $\nabla \cdot \mathbf{B} = 0$ a *Bianchi identity* in this gauge field representation of the magnetic field [recall the generic definition after Eq. (2.88)].

In order to solve (4.1), we remove the gauge ambiguity by choosing a particular gauge, for instance the *transverse gauge* $\nabla \cdot \mathbf{A}(\mathbf{x}) = 0$ in which $\nabla \times [\nabla \times \mathbf{A}(\mathbf{x})] = -\nabla^2 \mathbf{A}(\mathbf{x})$, and obtain

$$\mathbf{A}(\mathbf{x}) = \frac{1}{4\pi} \int d^3 x' \frac{\mathbf{j}(\mathbf{x}')}{|\mathbf{x} - \mathbf{x}'|}. \tag{4.5}$$

The associated magnetic field is

$$\mathbf{B}(\mathbf{x}) = \frac{1}{4\pi} \int d^3 x' \frac{\mathbf{j}(\mathbf{x}') \times \mathbf{R}'}{R'^3}, \qquad \mathbf{R}' \equiv \mathbf{x}' - \mathbf{x}. \tag{4.6}$$

This standard representation of magnetic fields is not the only possible one. There exists another one in terms of a scalar potential $\Lambda(\mathbf{x})$, which must, however, be multivalued to account for the two physical degrees of freedom in the magnetic field.

4.2   Multivalued Gradient Representation of Magnetic Field

Consider an infinitesimally thin closed wire carrying an electric current I along the line L. It corresponds to a current density

$$\mathbf{j}(\mathbf{x}) = I\, \boldsymbol{\delta}(\mathbf{x}; L), \tag{4.7}$$

where $\boldsymbol{\delta}(\mathbf{x}; L)$ is the δ-function on the closed line L:

$$\boldsymbol{\delta}(\mathbf{x}; L) = \int_L d\mathbf{x}'\, \delta^{(3)}(\mathbf{x} - \mathbf{x}'). \tag{4.8}$$

For a closed line L, this function has zero divergence:

$$\nabla \cdot \boldsymbol{\delta}(\mathbf{x}; L) = 0. \tag{4.9}$$

This follows from the property of the δ-function on an arbitrary open line $L_{\mathbf{x}_1}^{\mathbf{x}_2}$ connecting the points $\mathbf{x}_1$ and $\mathbf{x}_2$ defined by

$$\boldsymbol{\delta}(\mathbf{x}; L_{\mathbf{x}_1}^{\mathbf{x}_2}) = \int_{\mathbf{x}_1}^{\mathbf{x}_2} d\mathbf{x}'\, \delta^{(3)}(\mathbf{x} - \mathbf{x}'), \tag{4.10}$$

which satisfies

$$\nabla \cdot \boldsymbol{\delta}(\mathbf{x}; L_{\mathbf{x}_1}^{\mathbf{x}_2}) = \delta(\mathbf{x}_1) - \delta(\mathbf{x}_2). \tag{4.11}$$

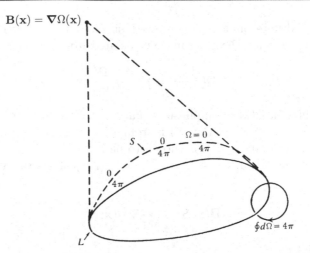

FIGURE 4.1 Infinitesimally thin closed current loop L. The magnetic field $\mathbf{B}(\mathbf{x})$ at the point $\mathbf{x}$ is proportional to the solid angle $\Omega(\mathbf{x})$ under which the loop is seen from $\mathbf{x}$. In any single-valued definition of $\Omega(\mathbf{x})$, there is some surface S across which $\Omega(\mathbf{x})$ jumps by 4π. In the multivalued definition, this surface is absent.

For closed lines L, the right-hand side of (4.11) vanishes:

$$\nabla \cdot \boldsymbol{\delta}(\mathbf{x}; L) = 0. \tag{4.12}$$

As an example, take a line $L_{\mathbf{x}_1}^{\mathbf{x}_2}$ which runs along the positive z-axis from z_1 to z_2, so that

$$\boldsymbol{\delta}(\mathbf{x}; L_{\mathbf{x}_1}^{\mathbf{x}_2}) = \int_{z_1}^{z_2} dz'\, \delta(x)\delta(y)\delta(z - z') = \delta(x)\delta(y)[\Theta(z - z_1) - \Theta(z - z_2)], \tag{4.13}$$

and

$$\nabla \cdot \boldsymbol{\delta}(\mathbf{x}; L_{\mathbf{x}_1}^{\mathbf{x}_2}) = \delta(x)\delta(y)\left[\delta(z - z_1) - \delta(z - z_2)\right] = \delta(\mathbf{x}_1) - \delta(\mathbf{x}_2). \tag{4.14}$$

From Eq. (4.5) we obtain the associated vector potential

$$\mathbf{A}(\mathbf{x}) = \frac{I}{4\pi} \int_L d\mathbf{x}' \frac{1}{|\mathbf{x} - \mathbf{x}'|}, \tag{4.15}$$

yielding the magnetic field

$$\mathbf{B}(\mathbf{x}) = \frac{I}{4\pi} \int_L \frac{d\mathbf{x}' \times \mathbf{R}'}{R'^3}, \qquad \mathbf{R}' \equiv \mathbf{x}' - \mathbf{x}. \tag{4.16}$$

The same result will now be derived from a multivalued scalar field. Let $\Omega(\mathbf{x}; S)$ be the solid angle under which the current loop L is seen from the point $\mathbf{x}$ (see

Fig. 4.1). If S denotes an arbitrary smooth surface enclosed by the loop L, and $d\mathbf{S}'$ a surface element, then $\Omega(\mathbf{x}; S)$ can be calculated from the surface integral

$$\Omega(\mathbf{x}; S) = \int_S \frac{d\mathbf{S}' \cdot \mathbf{R}'}{R'^3}. \tag{4.17}$$

The argument S in $\Omega(\mathbf{x}; S)$ emphasizes that the definition depends on the choice of the surface S. The range of $\Omega(\mathbf{x}; S)$ is from -2π to 2π, as can be seen most easily if L lies in the xy-plane and S is chosen to lie in the same place. Then we find for $\Omega(\mathbf{x}; S)$ the value 2π for $\mathbf{x}$ just below S, and -2π just above. We form the vector field

$$\mathbf{B}(\mathbf{x}; S) = \frac{I}{4\pi} \boldsymbol{\nabla} \Omega(\mathbf{x}; S), \tag{4.18}$$

which is equal to

$$\mathbf{B}(\mathbf{x}; S) = \frac{I}{4\pi} \int_S dS'_k \boldsymbol{\nabla} \frac{R'_k}{R'^3} = -\frac{I}{4\pi} \int_S dS'_k \boldsymbol{\nabla}' \frac{R'_k}{R'^3}. \tag{4.19}$$

This can be rearranged to

$$B_i(\mathbf{x}; S) = -\frac{I}{4\pi} \left[\int_S \left(dS'_k \partial'_i \frac{R'_k}{R'^3} - dS'_i \partial'_k \frac{R'_k}{R'^3} \right) + \int_S dS'_i \partial'_k \frac{R'_k}{R'^3} \right]. \tag{4.20}$$

With the help of Stokes' theorem

$$\int_S (dS_k \partial_i - dS_i \partial_k) f(\mathbf{x}) = \epsilon_{kil} \int_L dx_l f(\mathbf{x}), \tag{4.21}$$

and the relation $\partial'_k(R'_k/R'^3) = 4\pi \delta^{(3)}(\mathbf{x} - \mathbf{x}')$, this becomes

$$\mathbf{B}(\mathbf{x}; S) = -I \left[\frac{1}{4\pi} \int_L \frac{d\mathbf{x}' \times \mathbf{R}'}{R'^3} + \int_S d\mathbf{S}' \delta^{(3)}(\mathbf{x} - \mathbf{x}') \right]. \tag{4.22}$$

The first term is recognized to be precisely the magnetic field (4.16) of the current I. The second term is the singular magnetic field of an infinitely thin magnetic dipole layer lying on the arbitrarily chosen surface S enclosed by L.

The second term is a consequence of the fact that the solid angle $\Omega(\mathbf{x}; S)$ was defined by the surface integral (4.17). If $\mathbf{x}$ crosses the surface S, the solid angle jumps by 4π.

It is useful to re-express Eq. (4.19) in a slightly different way. By analogy with (4.23) we define a δ-function on a surface as

$$\boldsymbol{\delta}(\mathbf{x}; S) = \int_S d\mathbf{S}' \delta^{(3)}(\mathbf{x} - \mathbf{x}'), \tag{4.23}$$

and observe that Stokes' theorem (4.21) can be written as an identity for δ-functions:

$$\boldsymbol{\nabla} \times \boldsymbol{\delta}(\mathbf{x}; S) = \boldsymbol{\delta}(\mathbf{x}; L), \tag{4.24}$$

where L is the boundary of the surface S. This equation proves once more the zero divergence (4.9).

Using the δ-function on a surface S, we can rewrite (4.17) as

$$\Omega(\mathbf{x}; S) = \int d^3x' \, \delta(\mathbf{x}'; S) \cdot \frac{\mathbf{R}'}{R'^3}, \tag{4.25}$$

and (4.19) as

$$\mathbf{B}(\mathbf{x}; S) = -\frac{I}{4\pi} \int d^3x' \, \delta_k(\mathbf{x}'; S) \boldsymbol{\nabla}' \frac{R'_k}{R'^3}, \tag{4.26}$$

and (4.20), after an integration by parts, as

$$B_i(\mathbf{x}; S) = I \left\{ \frac{1}{4\pi} \int d^3x' \, [\partial'_i \delta_k(\mathbf{x}'; S) - \partial'_k \delta_i(\mathbf{x}'; S)] \frac{R'_k}{R'^3} - \int d^3x' \delta_i(\mathbf{x}'; S) \boldsymbol{\nabla}' \cdot \frac{\mathbf{R}'}{R'^3} \right\}. \tag{4.27}$$

The divergence at the end yields a $\delta^{(3)}$-function, and we obtain

$$B_i(\mathbf{x}; S) = -I \left[\frac{1}{4\pi} \int d^3x' [\nabla \times \boldsymbol{\delta}(\mathbf{x}; S)] \times \frac{\mathbf{R}'}{R'^3} + \int d^3x' \, \boldsymbol{\delta}(\mathbf{x}'; S) \, \delta^{(3)}(\mathbf{x} - \mathbf{x}') \right]. \tag{4.28}$$

Using (4.24) and (4.23), this is once more equal to (4.22).

Stokes' theorem written in the form (4.24) displays an important property. If we move the surface S to S' with the same boundary, the δ-function $\delta(\mathbf{x}; S)$ changes by the gradient of a scalar function

$$\boldsymbol{\delta}(\mathbf{x}; S) \to \boldsymbol{\delta}(\mathbf{x}; S') = \boldsymbol{\delta}(\mathbf{x}; S) + \boldsymbol{\nabla}\delta(\mathbf{x}; V), \tag{4.29}$$

where

$$\delta(\mathbf{x}; V) \equiv \int d^3x' \, \delta^{(3)}(\mathbf{x} - \mathbf{x}'), \tag{4.30}$$

is the δ-function on the volume V over which the surface has swept. Under this transformation, the curl on the left-hand side of (4.24) is invariant. Comparing (4.29) with (4.2) we identify (4.29) as a novel type of gauge transformation [1, 2]. Under this, the magnetic field in the first term of (4.28) is invariant, the second is not. It is then obvious how to find a gauge-invariant magnetic field: we simply subtract the singular S-dependent term and form

$$\mathbf{B}(\mathbf{x}) = \frac{I}{4\pi} \left[\boldsymbol{\nabla}\Omega(\mathbf{x}; S) + 4\pi\boldsymbol{\delta}(\mathbf{x}; S) \right]. \tag{4.31}$$

This field is independent of the choice of S and coincides with the magnetic field (4.16) derived in the usual gauge theory. To verify this explicitly we calculate the change of the solid angle (4.17) under a change of S. For this we rewrite (4.25) as

$$\Omega(\mathbf{x}; S) = -\int d^3x' \, \boldsymbol{\nabla}' \frac{1}{R'} \cdot \boldsymbol{\delta}(\mathbf{x}'; S) = -\frac{4\pi}{\boldsymbol{\nabla}^2} \boldsymbol{\nabla} \cdot \boldsymbol{\delta}(\mathbf{x}; S). \tag{4.32}$$

Performing the gauge transformation (4.29), the solid angle changes as follows:

$$\Omega(\mathbf{x}; S) \rightarrow \Omega(\mathbf{x}; S') = \Omega(\mathbf{x}; S) - \frac{4\pi}{\nabla^2} \boldsymbol{\nabla} \cdot \boldsymbol{\nabla}\delta(\mathbf{x}; V) = \Omega(\mathbf{x}; S) - 4\pi\delta(\mathbf{x}; V), \quad (4.33)$$

so that the magnetic field (4.31) is indeed invariant. Hence the description of the magnetic field as a gradient of field $\Omega(\mathbf{x}; S)$ combined with the gauge field $4\pi\boldsymbol{\delta}(\mathbf{x}; S)$ is completely equivalent to the usual gauge field description in terms of the vector potential $\mathbf{A}(\mathbf{x})$. Both are gauge theories, but of a completely different type.

The gauge freedom (4.29) can be used to move the surface S into a standard configuration. One possibility is to choose S in a way that the third component of $\boldsymbol{\delta}(\mathbf{x}; S)$ vanishes. This is called the *axial gauge*. If $\boldsymbol{\delta}(\mathbf{x}; S)$ does not have this property, we can always shift S by a volume V determined by the equation

$$\delta(V) = -\int_{-\infty}^{z} \delta_z(\mathbf{x}; S). \quad (4.34)$$

Then the transformation (4.29) will produce a $\boldsymbol{\delta}(\mathbf{x}; S)$ in the axial gauge $\delta_3(\mathbf{x}; S) = 0$.

A general differential relation between δ-functions on volumes and surfaces related to (4.34) is

$$\boldsymbol{\nabla}\delta(\mathbf{x}; V) = -\boldsymbol{\delta}(\mathbf{x}; S). \quad (4.35)$$

There exists another possibility of defining a solid angle $\Omega(\mathbf{x}; L)$ which is independent of the shape of the surface S and depends only on the boundary line L of S. This is done by analytic continuation of $\Omega(\mathbf{x}; S)$ through the surface S. The continuation removes the jump and produces a *multivalued function* $\Omega(\mathbf{x}; L)$ ranging from $-\infty$ to ∞. At each point in space, there are infinitely many Riemann sheets whose branch line is L. The values of $\Omega(\mathbf{x}; L)$ on the sheets differ by integer multiples of 4π. From this multivalued function, the magnetic field (4.16) can be obtained as a simple gradient:

$$\mathbf{B}(\mathbf{x}) = \frac{I}{4\pi}\boldsymbol{\nabla}\Omega(\mathbf{x}; L). \quad (4.36)$$

Ampère's law (4.1) implies that the multivalued solid angle $\Omega(\mathbf{x}; L)$ satisfies the equation

$$(\partial_i\partial_j - \partial_j\partial_i)\Omega(\mathbf{x}; L) = 4\pi\epsilon_{ijk}\delta_k(\mathbf{x}; L). \quad (4.37)$$

Thus, as a consequence of its multivaluedness, $\Omega(\mathbf{x}; L)$ violates the *Schwarz integrability condition*. This makes it an unusual mathematical object to deal with. It is, however, perfectly suited to describe the magnetic field of an electric current along L.

Let us see explicitly how Eq. (4.37) is fulfilled by $\Omega(\mathbf{x}; L)$. For simplicity, we consider the two-dimensional situation where the loop corresponds to two points (in which the loop intersects a plane). In addition, we move one point to infinity, and place the other at the coordinate origin. The role of the solid angle $\Omega(\mathbf{x}; L)$ is now played by the azimuthal angle $\varphi(\mathbf{x})$ of the point $\mathbf{x}$ with respect to the origin:

$$\varphi(\mathbf{x}) = \arctan \frac{x^2}{x^1}. \tag{4.38}$$

The function $\arctan(x^2/x^1)$ is usually made unique by cutting the $\mathbf{x}$-plane from the origin along some line C to infinity, preferably along a straight line to $\mathbf{x} = (-\infty, 0)$, and assuming $\varphi(\mathbf{x})$ to jump from π to $-\pi$ when crossing the cut, as shown in Fig. 4.2a. The cut corresponds to the magnetic dipole surface S in the integral (4.17).

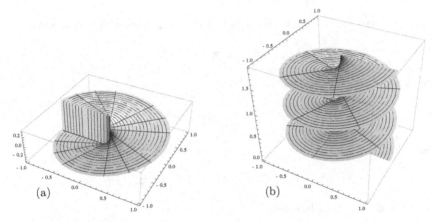

(a)   (b)

FIGURE 4.2 Single- and multi-valued definitions of $\arctan \varphi$.

In contrast to this, we shall take $\varphi(\mathbf{x})$ to be the *multivalued* analytic continuation of this function. Then the derivative ∂_i yields

$$\partial_i \varphi(\mathbf{x}) = -\epsilon_{ij} \frac{x_j}{(x^1)^2 + (x^2)^2}. \tag{4.39}$$

This is in contrast to the derivative $\partial_i \varphi(\mathbf{x})$ of the single-valued definition of $\partial_i \varphi(\mathbf{x})$ which would contain an extra δ-function $\epsilon_{ij} \delta_j(\mathbf{x}; C)$ across the cut C, corresponding to the second term in (4.22). When integrating the curl of the derivative (4.39) across the surface s of a small circle c around the origin, we obtain by Stokes' theorem

$$\int_s d^2x (\partial_i \partial_j - \partial_j \partial_i) \varphi(\mathbf{x}) = \int_c dx_i \partial_i \varphi(\mathbf{x}), \tag{4.40}$$

which is equal to 2π for the multivalued definition of $\varphi(\mathbf{x})$ shown in Fig. 4.2b and the cover of this book. This result implies the violation of the integrability condition like (4.49):

$$(\partial_1 \partial_2 - \partial_2 \partial_1) \varphi(\mathbf{x}) = 2\pi \delta^{(2)}(\mathbf{x}), \tag{4.41}$$

whose three-dimensional generalization is Eq. (4.37). In the single-valued definition of $\varphi(\mathbf{x})$ with the jump by 2π across the cut C, the right-hand side of (4.40) would vanish, since the contribution from the jump would cancel the integral along c. Thus the single-valued $\varphi(\mathbf{x})$ would satisfy Eq. (4.41) with zero on the right-hand side, thus being an integrable function.

On the basis of Eq. (4.41) we may construct a Green function for solving the corresponding differential equation with an arbitrary source, which is a superposition of infinitesimally thin line-like currents piercing the two-dimensional space at the points $\mathbf{x}_n$:

$$j(\mathbf{x}) = \sum_n I_n \delta^{(2)}(\mathbf{x} - \mathbf{x}_n), \qquad (4.42)$$

where I_n are currents. We may then easily solve the differential equation

$$(\partial_1 \partial_2 - \partial_2 \partial_1) f(\mathbf{x}) = j(\mathbf{x}), \qquad (4.43)$$

with the help of the Green function

$$G(\mathbf{x}, \mathbf{x}') = \frac{1}{2\pi} \varphi(\mathbf{x} - \mathbf{x}') \qquad (4.44)$$

which satisfies

$$(\partial_1 \partial_2 - \partial_2 \partial_1) G(\mathbf{x} - \mathbf{x}') = \delta^{(2)}(\mathbf{x} - \mathbf{x}'). \qquad (4.45)$$

The solution of (4.43) is obviously

$$f(\mathbf{x}) = \int d^2 x' \, G(\mathbf{x}, \mathbf{x}') j(\mathbf{x}). \qquad (4.46)$$

The gradient of $f(\mathbf{x})$ yields the magnetic field of an arbitrary set of line-like currents vertical to the plane under consideration.

It is interesting to realize that the Green function (4.44) is the imaginary part of the complex function $(1/2\pi) \log(z - z')$ with $z = x^1 + ix^2$, whose real part $(1/2\pi) \log|z - z'|$ is the Green function $G_\Delta(\mathbf{x} - \mathbf{x}')$ of the two dimensional Poisson equation:

$$(\partial_1^2 + \partial_2^2) G_\Delta(\mathbf{x} - \mathbf{x}') = \delta^{(2)}(\mathbf{x} - \mathbf{x}'). \qquad (4.47)$$

It is important to point out that the superposition of line-like currents cannot be smeared out into a continuous distribution. The integral (4.46) yields the superposition of multivalued functions

$$f(\mathbf{x}) = \frac{1}{2\pi} \sum_n I_n \arctan \frac{x^2 - x_n^2}{x^1 - x_n^1}, \qquad (4.48)$$

which is properly defined only if one can clearly continue it analytically into all Riemann sheets branching off from the endpoints of the cuts at $\mathbf{x}_n$. If we were to replace the sum by an integral, this possibility would be lost. Thus it is, strictly speaking, impossible to represent arbitrary continuous magnetic fields as gradients of superpositions of scalar potentials $\Omega(\mathbf{x}; L)$. This, however, is not a severe disadvantage of this representation since arbitrary currents can be approximated by a superposition of line-like currents with any desired accuracy. The same will be true for the associated magnetic fields.

The arbitrariness of the shape of the jumping surface is the origin of a further interesting gauge structure which has important physical consequences to be discussed in Subsection 4.6.

4.3 Generating Magnetic Fields by Multivalued Gauge Transformations

After this first exercise in multivalued functions, we turn to another example in magnetism which will lead directly to our intended geometric application. We observed before that the local gauge transformation (4.2) produces the same magnetic field $\mathbf{B}(\mathbf{x}) = \boldsymbol{\nabla} \times \mathbf{A}(\mathbf{x})$ only as long as the function $\Lambda(\mathbf{x})$ satisfies the Schwarz integrability criterion (4.37):

$$(\partial_i \partial_j - \partial_j \partial_i)\Lambda(\mathbf{x}) = 0. \tag{4.49}$$

Any function $\Lambda(\mathbf{x})$ violating this condition would change the magnetic field by

$$\Delta B_k(\mathbf{x}) = \epsilon_{kij}(\partial_i \partial_j - \partial_j \partial_i)\Lambda(\mathbf{x}), \tag{4.50}$$

thus being no proper gauge function. The gradient of $\Lambda(\mathbf{x})$

$$\mathbf{A}(\mathbf{x}) = \boldsymbol{\nabla}\Lambda(\mathbf{x}) \tag{4.51}$$

would be a *nontrivial* vector potential.

By analogy with the multivalued coordinate transformations violating the integrability conditions of Schwarz as in (4.37), the function $\Lambda(\mathbf{x})$ will be called *nonholonomic gauge function*.

Having just learned how to deal with multivalued functions we may change our attitude towards gauge transformations and decide to generate *all* magnetic fields approximately in a field-free space by such improper gauge transformations $\Lambda(\mathbf{x})$. By choosing for instance

$$\Lambda(\mathbf{x}) = \frac{\Phi}{4\pi}\Omega(\mathbf{x}), \tag{4.52}$$

we see from (4.37) that this generates a field

$$B_k(\mathbf{x}) = \epsilon_{kij}(\partial_i \partial_j - \partial_j \partial_i)\Lambda(\mathbf{x}) = \Phi\delta_k(\mathbf{x}; L). \tag{4.53}$$

This is a magnetic field of total flux Φ inside an infinitesimal tube. By a superposition of such infinitesimally thin flux tubes analogous to (4.46) we can obviously generate a discrete approximation to any desired magnetic field in a field-free space.

4.4   Magnetic Monopoles

Multivalued fields have also been used to describe magnetic monopoles [4, 5, 6]. A monopole charge density $\rho_m(\mathbf{x})$ is the source of a magnetic field $\mathbf{B}(\mathbf{x})$ as defined by the equation

$$\mathbf{\nabla} \cdot \mathbf{B}(\mathbf{x}) = \rho_m(\mathbf{x}). \tag{4.54}$$

If $\mathbf{B}(\mathbf{x})$ is expressed in terms of a vector potential $\mathbf{A}(\mathbf{x})$ as $\mathbf{B}(\mathbf{x}) = \mathbf{\nabla} \times \mathbf{A}(\mathbf{x})$, equation (4.54) implies the noncommutativity of derivatives in front of the vector potential $\mathbf{A}(\mathbf{x})$:

$$\frac{1}{2}\epsilon_{ijk}(\partial_i\partial_j - \partial_j\partial_i)A_k(\mathbf{x}) = \rho_m(\mathbf{x}). \tag{4.55}$$

Thus $\mathbf{A}(\mathbf{x})$ must be multivalued.

In his famous theory of monopoles [7, 8, 9], Dirac made the field single-valued by attaching to the world line of the particle a jumping world surface, whose intersection with a coordinate plane at a fixed time forms the *Dirac string*. Inside the string, the magnetic field of the monopole is imported from infinity. This world surface can be made physically irrelevant by quantizing it appropriately with respect to the charge. Its shape in space is just as irrelevant as that of the jumping surface S in Fig. 4.1. The invariance under shape deformations constitute once more a second gauge structure of the type mentioned earlier and discussed in Refs. [2, 4, 10, 11, 12].

Once we admit the use of multivalued fields, we may easily go one step further and express also $\mathbf{A}(\mathbf{x})$ as a gradient of a scalar field as in (4.51). Then the condition becomes

$$\epsilon_{ijk}\partial_i\partial_j\partial_k\Lambda(\mathbf{x}) = \rho_m(\mathbf{x}). \tag{4.56}$$

Let us explicitly construct the field of a magnetic monopole of charge g at a point $\mathbf{x}_0$, which satisfies (4.54) with $\rho_m(\mathbf{x}) = g\,\delta^{(3)}(\mathbf{x} - \mathbf{x}_0)$. We set up an infinitely thin solenoid along an arbitrary line $L^{\mathbf{x}_0}$ to import the flux from some point at infinity to the point $\mathbf{x}_0$, where the flux emerges. This is the physical version of Dirac string. The superscript $\mathbf{x}_0$ indicates that the line ends at $\mathbf{x}_0$. Inside the solenoid, the magnetic field is infinite, and equal to

$$\mathbf{B}_{\text{inside}}(\mathbf{x}; L^{\mathbf{x}_0}) = g\,\boldsymbol{\delta}(\mathbf{x}; L^{\mathbf{x}_0}), \tag{4.57}$$

where $\boldsymbol{\delta}(\mathbf{x}; L^{\mathbf{x}_0})$ is a modification of (4.10) in which the integral runs along the line $L^{\mathbf{x}_0}$ to $\mathbf{x}_0$:

$$\boldsymbol{\delta}(\mathbf{x}; L^{\mathbf{x}_0}) = \int^{\mathbf{x}_0} d^3x' \delta^{(3)}(\mathbf{x} - \mathbf{x}'). \tag{4.58}$$

The divergence of this function is concentrated at the endpoint $\mathbf{x}_0$ of the solenoid:

$$\mathbf{\nabla} \cdot \boldsymbol{\delta}(\mathbf{x}; L^{\mathbf{x}_0}) = -\delta^{(3)}(\mathbf{x} - \mathbf{x}_0). \tag{4.59}$$

Similarly we may define a δ-function along a line $L_{\mathbf{x_0}}$ which starts at $\mathbf{x_0}$ and runs to some point at infinity:

$$\delta(\mathbf{x}; L_{\mathbf{x_0}}) = \int_{\mathbf{x_0}} dx' \delta^{(3)}(\mathbf{x} - \mathbf{x'}), \qquad (4.60)$$

which satisfies

$$\nabla \cdot \delta(\mathbf{x}; L_{\mathbf{x_0}}) = \delta^{(3)}(\mathbf{x} - \mathbf{x_0}). \qquad (4.61)$$

Here the magnetic flux is exported from $\mathbf{x_0}$ to infinity, corresponding to an anti-monopole at $\mathbf{x_0}$.

As an example, take a Dirac string L^0 which imports the flux along the z-axis from positive infinity to the origin. If $\hat{\mathbf{z}}$ denotes the unit vector along the z-axis, then

$$\delta(\mathbf{x}; L^0) = \hat{\mathbf{z}} \int_{\infty}^{0} dz' \, \delta(x)\delta(y)\delta(z - z') = -\hat{\mathbf{z}} \, \delta(x)\delta(y)\Theta(z), \qquad (4.62)$$

so that $\nabla \cdot \delta(\mathbf{x}; L^0) = -\delta(x)\delta(y)\delta(z) = -\delta^{(3)}(\mathbf{x})$. This agrees with (4.11) if we move the initial point $\mathbf{x_1}$ to $(0,0,\infty)$. If the flux is imported from negative infinity to the origin, the δ-function is

$$\delta(\mathbf{x}; L^0) = \hat{\mathbf{z}} \int_{-\infty}^{0} dz' \, \delta(x)\delta(y)\delta(z - z') = \hat{\mathbf{z}} \, \delta(x)\delta(y)\Theta(-z), \qquad (4.63)$$

which has the same divergence $\nabla \cdot \delta(\mathbf{x}; L^0) = -\delta^{(3)}(\mathbf{x})$.

By analogy with the curl relation (4.24) we observe a further gauge invariance. If we deform the line $L^{\mathbf{x_0}}$ with fixed endpoint $\mathbf{x_0}$, the δ-function (4.58) changes by what we call a *monopole gauge transformation*:

$$\delta(\mathbf{x}; L^{\mathbf{x_0}}) \to \delta(\mathbf{x}; L'^{\mathbf{x_0}}) = \delta(\mathbf{x}; L^{\mathbf{x_0}}) + \nabla \times \delta(\mathbf{x}; S), \qquad (4.64)$$

where S is the surface over which $L^{\mathbf{x_0}}$ has swept on its way to $L'^{\mathbf{x_0}}$. Under this gauge transformation, the equation (4.59) is obviously invariant. The magnetic field (4.57) is therefore a *monopole gauge field*, whose divergence gives the magnetic charge density of a monopole in a gauge-invariant way.

Note that with respect to the previous gauge transformations (4.29) which shifted the surface S, the gradient is exchanged by a curl, with a corresponding exchange of the gauge-invariant quantities. In the previous case of the surface S, the invariant was the boundary line L found from a curl in Eq. (4.24), here the invariant is the starting point $\mathbf{x_0}$ of the boundary line $L^{\mathbf{x_0}}$, found from the divergence in Eq. (4.59).

It is straightforward to construct the associated ordinary gauge field $\mathbf{A}(\mathbf{x})$ of the monopole. Consider first the $L^{\mathbf{x_0}}$-dependent field

$$\mathbf{A}(\mathbf{x}; L^{\mathbf{x_0}}) = \frac{g}{4\pi} \int d^3x' \frac{\nabla' \times \delta(\mathbf{x'}; L^{\mathbf{x_0}})}{R'} = -\frac{g}{4\pi} \int d^3x' \, \delta(\mathbf{x'}; L^{\mathbf{x_0}}) \times \frac{\mathbf{R'}}{R'^3}. \qquad (4.65)$$

The curl of the first expression is

$$\boldsymbol{\nabla} \times \mathbf{A}(\mathbf{x}; L^{\mathbf{x}_0}) = \frac{g}{4\pi} \int d^3x' \frac{\boldsymbol{\nabla}' \times [\boldsymbol{\nabla}' \times \boldsymbol{\delta}(\mathbf{x}'; L^{\mathbf{x}_0})]}{R'}, \tag{4.66}$$

and consists of two terms

$$\frac{g}{4\pi} \int d^3x' \frac{\boldsymbol{\nabla}'[\boldsymbol{\nabla}' \cdot \boldsymbol{\delta}(\mathbf{x}'; L^{\mathbf{x}_0})]}{R'} - \frac{g}{4\pi} \int d^3x' \frac{\boldsymbol{\nabla}'^2 \boldsymbol{\delta}(\mathbf{x}'; L^{\mathbf{x}_0})}{R'}. \tag{4.67}$$

After an integration by parts, and using (4.59), the first term is $L^{\mathbf{x}_0}$-independent:

$$\frac{g}{4\pi} \int d^3x' \delta^{(3)}(\mathbf{x} - \mathbf{x}_0) \boldsymbol{\nabla}' \frac{1}{R'} = \frac{g}{4\pi} \frac{\mathbf{x} - \mathbf{x}_0}{|\mathbf{x} - \mathbf{x}_0|^3}. \tag{4.68}$$

The second term becomes, after two integration by parts,

$$g\,\boldsymbol{\delta}(\mathbf{x}'; L^{\mathbf{x}_0}). \tag{4.69}$$

The first term is the desired magnetic field of the monopole. Its divergence is $\delta^{(3)}(\mathbf{x} - \mathbf{x}_0)$, which we wanted to achieve. The second term is the magnetic field inside the solenoid, the monopole gauge field (4.57). The total divergence of the field (4.66) is, of course, equal to zero.

By analogy with (4.31) we now subtract the latter term and find the $L^{\mathbf{x}_0}$-independent magnetic field of the monopole

$$\mathbf{B}(\mathbf{x}) = \boldsymbol{\nabla} \times \mathbf{A}(\mathbf{x}; L^{\mathbf{x}_0}) - g\,\boldsymbol{\delta}(\mathbf{x}; L^{\mathbf{x}_0}), \tag{4.70}$$

which depends only on $\mathbf{x}_0$ and satisfies $\boldsymbol{\nabla} \cdot \mathbf{B}(\mathbf{x}) = g\,\delta^{(3)}(\mathbf{x} - \mathbf{x}_0)$.

Let us calculate the vector potential explicitly for the monopole where the Dirac string imports the flux along the z-axis from positive infinity to the origin along L^0. Inserting (4.62) into the right-hand side of (4.65), we obtain

$$\mathbf{A}^{(g)}(\mathbf{x}; L^0) = \frac{g}{4\pi} \int_{\infty}^{0} dz' \frac{\hat{\mathbf{z}} \times \mathbf{x}}{\sqrt{x^2 + y^2 + (z' - z)^2}^{3/2}}$$

$$= -\frac{g}{4\pi} \frac{\hat{\mathbf{z}} \times \mathbf{x}}{R(R - z)} = \frac{g}{4\pi} \frac{(y, -x, 0)}{R(R - z)}. \tag{4.71}$$

For the alternative string (4.63) we obtain

$$\mathbf{A}^{(g)}(\mathbf{x}; L^0) = \frac{g}{4\pi} \int_{-\infty}^{0} dz' \frac{\hat{\mathbf{z}} \times \mathbf{x}}{\sqrt{x^2 + y^2 + (z' - z)^2}^{3/2}}$$

$$= \frac{g}{4\pi} \frac{\hat{\mathbf{z}} \times \mathbf{x}}{R(R + z)} = -\frac{g}{4\pi} \frac{(y, -x, 0)}{R(R + z)}. \tag{4.72}$$

The vector potential has only azimuthal components. If we parametrize (x, y, z) in terms of spherical coordinates as $r(\sin\theta\cos\varphi, \sin\theta\sin\varphi, \cos\theta)$, these are

$$A_\varphi^{(g)}(\mathbf{x}; L^0) = \frac{g\sin\theta}{4\pi R(1 + \cos\theta)} \quad \text{or} \quad A_\varphi^{(g)}(\mathbf{x}; L^0) = -\frac{g\sin\theta}{4\pi R(1 - \cos\theta)}, \tag{4.73}$$

respectively.

In general, the shape of the line $L^{\mathbf{x}_0}$ (or $L_{\mathbf{x}_0}$) can be brought to a standard form, which corresponds to fixing a gauge of the gauge field $\boldsymbol{\delta}(\mathbf{x}; L^{\mathbf{x}_0})$ or $\boldsymbol{\delta}(\mathbf{x}; L_{\mathbf{x}_0})$. For example, we may always choose $L^{\mathbf{x}_0}$ to run along the positive z-axis.

An interesting observation is the following: If the gauge function $\Lambda(\mathbf{x})$ is considered as a nonholonomic displacement in some fictitious crystal dimension, then the magnetic field arising from a gauge transformation $\Lambda(\mathbf{x})$ with noncommuting derivatives along a line as in Eq. (4.53) gives a thin magnetic flux tube along the line. This will turn out in Section (9.8) to be the analog of a translational defect called dislocation line, corresponding to torsion in the geometry [see (14.3)]. A magnetic monopole, on the other hand, arises from noncommuting derivatives $(\partial_i\partial_j - \partial_j\partial_i)\partial_k\Lambda(\mathbf{x}) \neq 0$ in Eq. (4.56). We shall see that it is the analog of a rotational defect called disclination and corresponds in the crystal geometry to a curvature concentrated at the point of the monopole [see Eq. (14.4)].

4.5 Minimal Magnetic Coupling of Particles from Multivalued Gauge Transformations

Multivalued gauge transformations are the perfect tool to minimally couple electromagnetism to any type of matter. Consider for instance a free nonrelativistic point particle with a Lagrangian

$$L = \frac{M}{2}\dot{\mathbf{x}}^2. \tag{4.74}$$

The equations of motion are invariant under a gauge transformation

$$L \to L' = L + \boldsymbol{\nabla}\Lambda(\mathbf{x})\,\dot{\mathbf{x}}, \tag{4.75}$$

since this changes the action $\mathcal{A} = \int_{t_a}^{t_b} dt\, L$ merely by a surface term:

$$\mathcal{A}' \to \mathcal{A} = \mathcal{A} + \Lambda(\mathbf{x}_b) - \Lambda(\mathbf{x}_a). \tag{4.76}$$

The invariance is absent if we take $\Lambda(\mathbf{x})$ to be a multivalued gauge function. In this case, a nontrivial vector potential $\mathbf{A}(\mathbf{x}) = \boldsymbol{\nabla}\Lambda(\mathbf{x})$ (working in natural units with $e = 1$) is created in the field-free space, and the nonholonomically gauge-transformed Lagrangian corresponding to (4.75),

$$L' = \frac{M}{2}\dot{\mathbf{x}}^2 + \mathbf{A}(\mathbf{x})\,\dot{\mathbf{x}}, \tag{4.77}$$

describes correctly the dynamics of a free particle in an external magnetic field.

The coupling derived by multivalued gauge transformations is automatically invariant under additional ordinary single-valued gauge transformations of the vector potential

$$\mathbf{A}(\mathbf{x}) \to \mathbf{A}'(\mathbf{x}) = \mathbf{A}(\mathbf{x}) + \boldsymbol{\nabla}\Lambda(\mathbf{x}), \tag{4.78}$$

since these add to the Lagrangian (4.77) once more the same pure derivative term which changes the action by an irrelevant surface term as in (4.76).

The same procedure leads in quantum mechanics to the minimal coupling of the Schrödinger field $\psi(\mathbf{x})$. The action is $\mathcal{A} = \int dt d^3x\, L$ with a Lagrangian density (in natural units with $\hbar = 1$)

$$L = \psi^*(\mathbf{x}) \left(i\partial_t + \frac{1}{2M}\boldsymbol{\nabla}^2 \right) \psi(\mathbf{x}). \tag{4.79}$$

The physics described by a Schrödinger wave function $\psi(\mathbf{x})$ is invariant under arbitrary local phase changes

$$\psi(\mathbf{x},t) \rightarrow \psi'(\mathbf{x}) = e^{i\Lambda(\mathbf{x})}\psi(\mathbf{x},t), \tag{4.80}$$

called local U(1) transformations. This implies that the Lagrangian density (4.79) may equally well be replaced by the gauge-transformed one

$$L = \psi^*(\mathbf{x},t) \left(i\partial_t + \frac{1}{2M}\mathbf{D}^2 \right) \psi(\mathbf{x},t), \tag{4.81}$$

where $-i\mathbf{D} \equiv -i\boldsymbol{\nabla} - \boldsymbol{\nabla}\Lambda(\mathbf{x})$ is the operator of physical momentum.

We may now go over to nonzero magnetic fields by admitting gauge transformations with multivalued $\Lambda(\mathbf{x})$ whose gradient is a nontrivial vector potential $\mathbf{A}(\mathbf{x})$ as in (4.51). Then $-i\mathbf{D}$ turns into the covariant momentum operator

$$\hat{\mathbf{P}} = -i\mathbf{D} = -i\boldsymbol{\nabla} - \mathbf{A}(\mathbf{x}), \tag{4.82}$$

and the Lagrangian density (4.81) describes correctly the magnetic coupling in quantum mechanics.

As in the classical case, the coupling derived by multivalued gauge transformations is automatically invariant under ordinary single-valued gauge transformations under which the vector potential $\mathbf{A}(\mathbf{x})$ changes as in (4.78), whereas the Schrödinger wave function undergoes a local U(1)-transformation (4.80). This invariance is a direct consequence of the simple transformation behavior of $\mathbf{D}\psi(\mathbf{x},t)$ under gauge transformations (4.78) and (4.80) which is

$$\mathbf{D}\psi(\mathbf{x},t) \rightarrow \mathbf{D}\psi'(\mathbf{x},t) = e^{i\Lambda(\mathbf{x})}\mathbf{D}\psi(\mathbf{x},t). \tag{4.83}$$

Thus $\mathbf{D}\psi(\mathbf{x},t)$ transforms just like $\psi(\mathbf{x},t)$ itself, and for this reason, $\mathbf{D}$ is called *covariant derivative*. The generation of magnetic fields by a multivalued gauge transformation is the simplest example for the power of the nonholonomic mapping principle.

After this discussion it is quite suggestive to introduce the same mathematics into differential geometry, where the role of gauge transformations is played by reparametrizations of the space coordinates.

4.6 Equivalence of Multivalued Scalar and Singlevalued Vector Fields

In the previous sections we have given examples for the use of multivalued fields in describing magnetic phenomena. The multivalued gauge transformations by which we created line-like nonzero field configurations were shown to be the natural origin of the minimal couplings to the classical actions as well as to the Schrödinger equation. It is interesting to establish the complete equivalence of the multivalued scalar theory with the usual vector potential theory of magnetism. This is done by a proper treatment of the degrees of freedom of the jumping surfaces S. For this purpose we set up an action formalism for calculating the magnetic energy of a current loop in the gradient representation of the magnetic field. In Euclidean field theory, the action is provided by the field energy

$$H = \frac{1}{2} \int d^3x \, \mathbf{B}^2(\mathbf{x}). \tag{4.84}$$

Inserting the gradient representation (4.36) of the magnetic field, we can write this as

$$H = \frac{I^2}{2(4\pi)^2} \int d^3x \, [\boldsymbol{\nabla}\Omega(\mathbf{x})]^2. \tag{4.85}$$

This holds for the multivalued solid angle $\Omega(\mathbf{x})$ which is independent of S. In order to perform field theoretic calculations, we must go over to the single-valued representation (4.31) of the magnetic field for which the energy is

$$H = \frac{I^2}{2(4\pi)^2} \int d^3x \, [\boldsymbol{\nabla}\Omega(\mathbf{x}; S) + 4\pi\boldsymbol{\delta}(\mathbf{x}; S)]^2. \tag{4.86}$$

The δ-function removes the unphysical field energy on the artificial magnetic dipole layer on S.

The Hamiltonian is extremized by the scalar field (4.25). Moreover, due to infinite field strength on the surface, all field configurations $\Omega(\mathbf{x}; S')$ with a jumping surface S' different from S will have an infinite energy. Thus we may omit the argument S in $\Omega(\mathbf{x}; S)$ and admit an arbitrary field $\Omega(\mathbf{x})$ to the Hamiltonian (4.86). Only the field (4.25) will give a finite contribution.

Let us calculate the magnetic field energy of the current loop from the energy (4.86). For this we rewrite the energy (4.86) in terms of an *independent auxiliary vector field* $\mathbf{B}(\mathbf{x})$ as

$$H = \int d^3x \left\{ -\frac{1}{2}\mathbf{B}^2(\mathbf{x}) + \frac{I}{4\pi}\mathbf{B}(\mathbf{x}) \cdot [\boldsymbol{\nabla}\Omega(\mathbf{x}) + 4\pi\boldsymbol{\delta}(\mathbf{x}; S)] \right\}. \tag{4.87}$$

A partial integration brings the second term to

$$\int d^3x \, \frac{1}{4\pi} \boldsymbol{\nabla} \cdot \mathbf{B}(\mathbf{x}) \, \Omega(\mathbf{x}).$$

Extremizing this in $\Omega(\mathbf{x})$ yields the equation

$$\boldsymbol{\nabla} \cdot \mathbf{B}(\mathbf{x}) = 0, \tag{4.88}$$

implying that the field lines of $\mathbf{B}(\mathbf{x})$ form closed loops. This equation may be enforced identically (as a Bianchi identity) by expressing $\mathbf{B}(\mathbf{x})$ as a curl of an auxiliary vector potential $\mathbf{A}(\mathbf{x})$, setting

$$\mathbf{B}(\mathbf{x}) \equiv \boldsymbol{\nabla} \times \mathbf{A}(\mathbf{x}). \tag{4.89}$$

This ansatz brings the energy (4.87) to the form

$$H = \int d^3x \left\{ -\frac{1}{2}[\boldsymbol{\nabla} \times \mathbf{A}(\mathbf{x})]^2 - I\,[\boldsymbol{\nabla} \times \mathbf{A}(\mathbf{x})] \cdot \boldsymbol{\delta}(\mathbf{x}; S) \right\}. \tag{4.90}$$

A partial integration of the second term leads to

$$H = \int d^3x \left\{ -\frac{1}{2}[\boldsymbol{\nabla} \times \mathbf{A}(\mathbf{x})]^2 - I\,\mathbf{A}(\mathbf{x}) \cdot [\boldsymbol{\nabla} \times \boldsymbol{\delta}(\mathbf{x}; S)] \right\}. \tag{4.91}$$

The factor of $\mathbf{A}(\mathbf{x})$ in the linear term is identified as an *auxiliary current*

$$\mathbf{j}(\mathbf{x}) \equiv I\,\boldsymbol{\nabla} \times \boldsymbol{\delta}(\mathbf{x}; S)^{\cdot} = I\,\boldsymbol{\delta}(\mathbf{x}; L). \tag{4.92}$$

In the last step we have used Stokes' law (4.24). According to Eq. (4.9), this current is conserved for loops L.

The representation (4.91) of the energy is called the *dually transformed version* of the original energy (4.86).

By extremizing the energy (4.90), we obtain Ampère's law (4.1). Thus the auxiliary quantities $\mathbf{B}(\mathbf{x})$, $\mathbf{A}(\mathbf{x})$, and $\mathbf{j}(\mathbf{x})$ coincide with the usual magnetic quantities of the same name. If we insert the explicit solution (4.5) of Ampère's law into the energy, we obtain the *Biot-Savart* or *Ampère* energy for an arbitrary current distribution

$$H = \frac{1}{8\pi} \int d^3x\, d^3x'\, \mathbf{j}(\mathbf{x}) \frac{1}{|\mathbf{x} - \mathbf{x}'|} \mathbf{j}(\mathbf{x}'). \tag{4.93}$$

Inserting here two current filaments running parallel in thin wires, the energy (4.93) decreases with increasing distance, suggesting for a moment that the force between them is repulsive. The experimental force, however, is attractive. The sign change is due to the fact that when increasing the distance of the wires we must perform work against the inductive forces in order to maintain the constant currents. This work is not calculated above. It turns out to be exactly twice the energy gain implied by (4.93).

The energy responsible for discussing the forces of given external current distributions is the *free magnetic energy*

$$F = \frac{1}{2} \int d^3x\, (\boldsymbol{\nabla} \times \mathbf{A})^2(\mathbf{x}) - \int d^3x\, \mathbf{j}(\mathbf{x}) \cdot \mathbf{A}(\mathbf{x}). \tag{4.94}$$

Extremizing this in $\mathbf{A}(\mathbf{x})$ yields the vector potential (4.5), and reinserting this potential into (4.94) we find that the *free Biot-Savart energy* is, indeed, the opposite of (4.93):

$$F|_{\text{ext}} = -\frac{1}{8\pi} \int d^3x \, d^3x' \, \mathbf{j}(\mathbf{x}) \frac{1}{|\mathbf{x} - \mathbf{x}'|} \mathbf{j}(\mathbf{x}'). \tag{4.95}$$

As a consequence, parallel wires with fixed currents attract rather than repel each other.

Note that the energy (4.90) is invariant under two mutually dual gauge transformations: the usual magnetic one in (4.2), by which the vector potential receives a gradient of an arbitrary scalar field, and the gauge transformation (4.29), by which the irrelevant surface S is moved to another configuration S'.

Thus we have proved the complete equivalence of the gradient representation of the magnetic field to the usual gauge field representation. In the gradient representation, there exists a new type of gauge invariance which expresses the physical irrelevance of the jumping surface appearing when using single-valued solid angles.

The energy (4.91) describes magnetism in terms of a *double gauge theory* [13], in which both the gauge of $\mathbf{A}(\mathbf{x})$ and the shape of S can be changed arbitrarily. By setting up a grand-canonical partition function of many fluctuating surfaces it is possible to describe a large family of phase transitions mediated by the proliferation of line-like defects. Examples are vortex lines in the superfluid-normal transition in helium, to be discussed in the next chapter, and dislocation and disclination lines in the melting transition of crystals, to be discussed later [2, 4, 10, 11, 12].

4.7 Multivalued Field Theory of Magnetic Monopoles and Electric Currents

Let us now go through the analogous discussion for a gas of monopoles at $\mathbf{x}_n$ with strings $L^{\mathbf{x}_n}$ importing their fluxes from infinity, and electric currents along closed $L_{n'}$. The free energy of fixed currents is given by the energy of the magnetic field (4.70) coupled to the currents as in the action Eq. (4.94):

$$F = \int d^3x \left\{ \frac{1}{2} \left[\boldsymbol{\nabla} \times \mathbf{A} - g \sum_n \boldsymbol{\delta}(\mathbf{x}; L^{\mathbf{x}_n}) \right]^2 - I \mathbf{A}(\mathbf{x}) \cdot \sum_{n'} \boldsymbol{\delta}(\mathbf{x}, L_{n'}) \right\}. \tag{4.96}$$

Extremizing this in $\mathbf{A}(\mathbf{x})$ we obtain

$$\mathbf{A}(\mathbf{x}) = -\frac{1}{\boldsymbol{\nabla}^2} \left[g \sum_n \boldsymbol{\nabla} \times \boldsymbol{\delta}(\mathbf{x}; L^{\mathbf{x}_n}) + I \sum_{n'} \boldsymbol{\delta}(\mathbf{x}, L_{n'}) \right]. \tag{4.97}$$

Reinserting this $\mathbf{A}(\mathbf{x})$ into (4.96) yields three terms. First, there is an interaction between the current lines

$$H_{II} = -\frac{I^2}{2} \int d^3x \sum_{n,n'} \boldsymbol{\delta}(\mathbf{x}; L_n) \frac{1}{-\boldsymbol{\nabla}^2} \boldsymbol{\delta}(\mathbf{x}; L_{n'}) = -\frac{I^2}{2} \sum_{n,n'} \int_{L_n} d\mathbf{x}_n \int_{L_{n'}} d\mathbf{x}_{n'} \frac{1}{|\mathbf{x}_n - \mathbf{x}_{n'}|}, \tag{4.98}$$

which corresponds to (4.95). Second, there is an interaction between monopole strings

$$\frac{g^2}{2}\int d^3x \left\{ \left[\sum_n \boldsymbol{\delta}(\mathbf{x}; L^{\mathbf{x}_n})\right]^2 + \left[\sum_n \boldsymbol{\nabla} \times \boldsymbol{\delta}(\mathbf{x}; L^{\mathbf{x}_n})\right] \frac{1}{\boldsymbol{\nabla}^2} \left[\sum_n \boldsymbol{\nabla} \times \boldsymbol{\delta}(\mathbf{x}; L^{\mathbf{x}_n})\right] \right\}, (4.99)$$

which can be brought to the form

$$
\begin{aligned}
H_{gg} &= \frac{g^2}{2}\int d^3x \left[\sum_n \boldsymbol{\nabla} \cdot \boldsymbol{\delta}(\mathbf{x}; L^{\mathbf{x}_n})\right]^2 = \frac{g^2}{2}\int d^3x \left[\sum_n \delta(\mathbf{x} - \mathbf{x}_n)\right]^2 \\
&= \frac{g^2}{8\pi}\sum_{n,n'} \frac{1}{|\mathbf{x}_n - \mathbf{x}_{n'}|}.
\end{aligned}
\tag{4.100}
$$

Finally, there is an interaction between the monopoles and the currents

$$H_{gI} = -gI\int d^3x \sum_{n,n'} \boldsymbol{\nabla} \times \boldsymbol{\delta}(\mathbf{x}; L^{\mathbf{x}_n})\frac{1}{\boldsymbol{\nabla}^2}\boldsymbol{\delta}(\mathbf{x}; L_{n'}). \tag{4.101}$$

An integration by parts brings this to the form

$$
\begin{aligned}
H_{gI} &= -gI\int d^3x \sum_{n,n'} \boldsymbol{\delta}(\mathbf{x}; L^{\mathbf{x}_n})\frac{1}{\boldsymbol{\nabla}^2}\boldsymbol{\nabla} \times \boldsymbol{\delta}(\mathbf{x}; L_{n'}) \\
&= -gI\int d^3x \sum_{n,n'} \boldsymbol{\delta}(\mathbf{x}; L^{\mathbf{x}_n})\frac{1}{\boldsymbol{\nabla}^2}\boldsymbol{\nabla} \times [\boldsymbol{\nabla} \times \boldsymbol{\delta}(\mathbf{x}; S_{n'})],
\end{aligned}
\tag{4.102}
$$

which is equal to

$$H_{gI} = H'_{gI} + \Delta H_{gI}, \tag{4.103}$$

with

$$H'_{gI} = -gI\int d^3x \sum_{n,n'} \boldsymbol{\delta}(\mathbf{x}; L^{\mathbf{x}_n})\boldsymbol{\nabla}\frac{1}{\boldsymbol{\nabla}^2}[\boldsymbol{\nabla} \cdot \boldsymbol{\delta}(\mathbf{x}; S_{n'})], \tag{4.104}$$

and

$$\Delta H_{gI} = gI\int d^3x \sum_{n,n'} \boldsymbol{\delta}(\mathbf{x}; L^{\mathbf{x}_n})\boldsymbol{\delta}(\mathbf{x}; S_{n'}). \tag{4.105}$$

Each integral in the sum yields an integer number which counts how often the lines L_n pierce the surface $S_{n'}$, so that

$$\Delta H_{gI} = gIk, \quad k = \text{integer}. \tag{4.106}$$

Recalling (4.32), the interaction (4.104) can be rewritten as

$$H_{gI} = -\frac{gI}{4\pi}\int d^3x \sum_{n,n'} \boldsymbol{\delta}(\mathbf{x}; L^{\mathbf{x}_n})\boldsymbol{\nabla}\Omega(\mathbf{x}; S_{n'}). \tag{4.107}$$

An integration by parts and the relation (4.59) brings this to the form

$$H_{gI} \;=\; \frac{gI}{4\pi} \sum_{n,n'} \Omega(\mathbf{x}_n; S_{n'}). \qquad (4.108)$$

It is proportional to the sum of the solid angles $\Omega(\mathbf{x}_n; S_{n'})$ under which the current loops $L_{n'}$ are seen from the monopoles at $\mathbf{x}_n$. The result does not depend on the surfaces S_n, but only on the boundary lines L_n along which the currents flow.

The total interaction is obviously invariant under shape deformations of S, except for the term (4.106). This term, however, is physically irrelevant provided we subject the charges Q in the currents to the quantization rule

$$Qg = 2\pi k, \quad k = \text{integer}. \qquad (4.109)$$

This rule was first found by Dirac [7] and will be rederived and in Section 8.2.

Notes and References

[1] This gauge freedom is *independent* of the electromagnetic one. See
H. Kleinert, Phys. Lett. B **246**, 127 (1990) (kl/205); Int. J. Mod. Phys. A **7**, 4693 (1992) (kl/203); Phys. Lett. B **293**, 168 (1992) (kl/211), where kl is short for the www address http://www.physik.fu-berlin.de/~kleinert.

[2] H. Kleinert, *Gauge Fields in Condensed Matter*, Vol. I, *Superflow and Vortex Lines*, World Scientific, Singapore, 1989 (kl/b1).

[3] The theory of multivalued fields developed first in detail in textbook [2] is so unfamiliar to field theorists that Physical Review Letters found it of broad interest to publish a Comment on Eq. (4.41) by
C. Hagen, Phys. Rev. Lett. **66**, 2681 (1991),
together with the prompt illuminating reply by
R. Jackiw and S.-Y. Pi, Phys. Rev. Lett. **66**, 2682 (1991).

[4] H. Kleinert, Int. J. Mod. Phys. **A 7**, 4693 (1992) (kl/203).

[5] H. Kleinert, Phys. Lett. **B 246**, 127 (1990) (kl/205).

[6] H. Kleinert, Phys. Lett. **B 293**, 168 (1992) (kl/211).

[7] P.A.M. Dirac, Proc. Roy. Soc. A **133**, 60 (1931); Phys. Rev. **74**, 817 (1948), Phys. Rev. **74**, 817 (1948).

[8] J. Schwinger, Phys. Rev. **144**, 1087 (1966).

[9] M.N. Saha, Ind. J. Phys. **10**, 145 (1936);
J. Schwinger, *Particles, Sources and Fields*, Vols. 1 and 2, Addison Wesley, Reading, Mass., 1970 and 1973;
G. Wentzel, Progr. Theor. Phys. Suppl. **37**, 163 (1966);

E. Amaldi, in *Old and New Problems in Elementary Particles*, ed. by G. Puppi, Academic Press, New York (1968);
D. Villaroel, Phys. Rev. D **14**, 3350 (1972);
Yu.D. Usachev, Sov. J. Particles & Nuclei **4**, 92 (1973);
A.O. Barut, J. Phys. A **11**, 2037 (1978);
J.D. Jackson, *Classical Electrodynamics*, John Wiley and Sons, New York, 1975, Sects. 6.12-6.13.

[10] H. Kleinert, *Gauge Fields in Condensed Matter*, Vol. II, *Stresses and Defects*, World Scientific, Singapore, 1989 (kl/b2).

[11] H. Kleinert, *Nonholonomic Mapping Principle for Classical and Quantum Mechanics in Spaces with Curvature and Torsion*, Gen. Rel. Grav. **32**, 769 (2000) (kl/258); Act. Phys. Pol. B **29**, 1033 (1998) (gr-qc/9801003).

[12] H. Kleinert, *Theory of Fluctuating Nonholonomic Fields and Applications: Statistical Mechanics of Vortices and Defects and New Physical Laws in Spaces with Curvature and Torsion*, in: Proceedings of NATO Advanced Study Institute on Formation and Interaction of Topological Defects, Plenum Press, New York, 1995, pp. 201–232 (kl/227).

[13] H. Kleinert, *Double Gauge Theory of Stresses and Defects*, Phys. Lett. A **97**, 51 (1983) (kl/107).

5

Multivalued Fields in Superfluids and Superconductors

Multivalued fields play an important role in understanding a great variety of phase transitions. In this chapter we shall discuss two simple but important examples.

5.1 Superfluid Transition

The simplest phase transitions which can be explained by multivalued field theory is the so-called λ-transition of superfluid helium. The name indicates the shape of the peak in the specific heat observed at a critical temperature $T_c \approx 2.18\,\text{K}$ shown in Fig. 5.1.

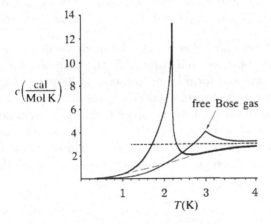

FIGURE 5.1 Specific heat of superfluid ^{4}He. For very small T, it shows the typical power behavior $\propto T^3$ characteristic for massless excitations in three dimensions in the Debye theory of specific heat. Here these excitations are phonons of the second sound. The peak is caused by the proliferation of vortex loops at the superfluid-normal transition.

For temperatures T below T_c, the fluid shows no friction and possesses only massless excitations. These are the quanta of the *second sound*, called phonons. They cause the typical temperature behavior of the specific heat which in D dimensions starts out like

$$C \sim T^D. \tag{5.1}$$

This was first explained in 1912 by Debye in his theory of specific heat [1], in which he generalized Planck's theory of black-body radiation to solid bodies.

As the temperature rises, another type of excitations appears in the superfluid. These are the famous *rotons* whose existence was deduced in 1947 by Landau from the thermodynamic properties of the superfluid [2, 3]. Rotons are the excitations of wavenumber near $2/\text{Å}$ where the phonon dispersion curve has a minimum. The full shape of this curve can be measured by neutron scattering and is displayed in Fig. 5.2.

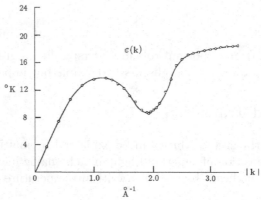

FIGURE 5.2 Energies of the elementary excitations in superfluid ^{4}He measured by neutron scattering showing the roton minimum near $k \approx 2/\text{Å}$ (data are taken from Ref. [4]).

As long as T stays sufficiently far below T_c, the thermodynamic properties of the superfluid are dominated by phonons and rotons. If the temperature approaches T_c, the rotons join side by side and form large surfaces, as shown in Fig. 5.3. The adjacent boundaries of the rotons cancel each other, so that the memory of the surfaces is lost, their shape becomes irrelevant, and only the boundaries of the surfaces are physical objects, observable as *vortex loops*. At T_c, the vortex loops become infinitely long and proliferate. The large activation energies for creating single rotons are overcome by the high configurational entropy of the long vortex loops.

FIGURE 5.3 Near T_c, more and more rotons join side by side to form surfaces whose boundary appears as a large vortex loop. The adjacent roton boundaries cancel each other.

The inside of a vortex line consists of normal fluid since the large rotation velocity destroys superfluidity. For this reason, the proliferation of vortex loops fills the system with normal fluid, and the fluid looses its superfluid properties. The existence of such a mechanism for a phase transition was realized more than fifty years ago by Onsager in 1949 [5]. It was re-emphasized by Feynman in 1955 [6], and turned into a proper disorder field theory in the 1980's by the author [7]. The same idea was advanced in 1952 by Shockley [8] who proposed a proliferation of defect lines in solids to be responsible for the melting transition. His work instigated the author to develop a detailed theory of melting in textbook [9].

The disorder field theory of superfluids and superconductors will be derived in Subsection 5.1.10, the melting theory in Chapter 10.

5.1.1 Configuration Entropy

There exists a simple estimate for the temperature of a phase transition based on the proliferation of line-like excitations. A long line of length l with an energy per length ϵ is suppressed strongly by a Boltzmann factor $e^{-\epsilon l/T}$. This suppression is, however, counteracted by configurational entropy. Suppose the line can bend easily on a length scale ξ which is of the order of the coherence length of the system. Hence it can occur in approximately $(2D)^{l/\xi}$ possible configurations, where D is the space dimension [10]. A rough approximation to the partition function of a single loop of arbitrary length is given by the integral

$$Z_1 \approx \oint \frac{dl}{l}\, (2D)^{l/\xi} e^{-\epsilon l/T}. \tag{5.2}$$

The factor $1/l$ in the integrand accounts for the cyclic invariance of the loop. By exponentiating this one-loop expression we obtain the partition function of a grand-canonical ensemble of loops of arbitrary length l, whose free energy is therefore $F = -Z_1/\beta$.

The integral (5.2) converges only below a critical temperature

$$T < T_c = \epsilon\xi/\log(2D). \tag{5.3}$$

Above T_c, the integral diverges and the ensemble undergoes a phase transition in which the loops proliferate and become infinitely long. This process will be called *condensation* of loops. A Monte-Carlo simulation of this process is shown in Fig. 5.4.

From Eq. (5.3) we can immediately deduce a relation between the critical temperature and the roton energy in superfluids. The size of a roton will be roughly $\pi\xi$. Its energy is therefore $E_{\text{roton}} \approx \pi\xi\epsilon$. Inserting this into Eq. (5.3) we estimate the critical temperature of a line ensemble as

$$T_c = c_{\text{lines}} E_{\text{rot}}. \tag{5.4}$$

It is proportional to the roton energy with a proportionality constant in $D = 3$ dimensions

$$c_{\text{lines}} \approx 1/\pi \log 6 \approx 1/5.6. \tag{5.5}$$

This prediction was recently verified experimentally [11].

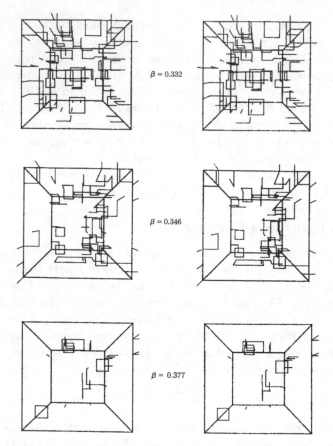

FIGURE 5.4 Vortex loops in XY-model with periodic boundary condition for different values of $\beta = 1/k_B T$. Close to $T_c \equiv 3$, the loops proliferate, with some becoming infinitely long (from Ref. [12]). The plots show the views of left and right eye. To perceive the loops three-dimensionally, place a sheet of paper vertically between the pictures and point the eyes parallel until you see only one picture.

5.1.2 Origin of Massless Excitations

The massless excitations in superfluid helium are a consequence of spontaneous breakdown of a continuous symmetry of the Hamiltonian. Such massless excitations are called *Nambu-Goldstone modes*. These arise as follows. Superfluid ^{4}He is described by a complex *order field* $\phi(\mathbf{x})$ which is the wave function of the condensate. Near the transition and for smooth spatial variations, the energy density is given by the Hamiltonian of Landau, Ginzburg, and Pitaevskii [13],

$$H[\phi] = \frac{1}{2} \int d^3x \left\{ |\boldsymbol{\nabla}\phi|^2 + \tau|\phi|^2 + \frac{\lambda}{2}|\phi|^4 \right\}. \tag{5.6}$$

The parameter τ is proportional to the relative temperature distance from the critical temperature

$$\tau \equiv \frac{1}{\xi_0^2}\left(\frac{T}{T_c} - 1\right). \tag{5.7}$$

The parameter ξ_0 is a length parameter determining the coherence length below in Eq. (5.13). Below the critical temperature where $\tau < 0$, the ground state lies at

$$\phi(\mathbf{x}) = \phi_0 = \sqrt{\frac{-\tau}{\lambda}}e^{i\alpha}. \tag{5.8}$$

This field value is called the *order parameter* of the superfluid.

The ground state is not unique but infinitely degenerate. Only its absolute value of $|\phi_0|$ is fixed, the phase α is arbitrary. For this reason, the entropy does not go to zero at zero temperature. The degeneracy in α is due to the fact that the Hamiltonian density (5.6) is invariant under constant U(1) phase transformations

$$\phi(\mathbf{x}) \rightarrow e^{i\alpha}\phi(\mathbf{x}). \tag{5.9}$$

The Nambu-Goldstone theorem states that such a degenerate ground state possesses massless excitations, unless there is another massless excitation which prevents this by mixing with the Goldstone excitation. In the field theory with Hamiltonian (5.6), the appearance of massless excitations is easily understood by decomposing the order field $\phi(\mathbf{x})$ into size and phase variables,

$$\phi(\mathbf{x}) = \rho(\mathbf{x})e^{i\theta(\mathbf{x})}, \tag{5.10}$$

and rewriting (5.6) as

$$H[\rho, \theta] = \frac{1}{2}\int d^3x \left[(\boldsymbol{\nabla}\rho)^2 + \rho^2(\boldsymbol{\nabla}\theta)^2 + \tau\rho^2 + \frac{\lambda}{2}\rho^4\right]. \tag{5.11}$$

If τ is negative, the size of the order field fluctuations stay close to the minimum (5.8) where $\rho_0 = \sqrt{-\tau/\lambda}$.

By expanding the Hamiltonian (5.11) in powers of the fluctuations $\delta\rho \equiv \rho - \rho_0$, we find that the ρ-fluctuations have a quadratic energy

$$H_{\delta\rho} = \frac{1}{2}\int d^3x \left[(\boldsymbol{\nabla}\delta\rho)^2 - 2\tau(\delta\rho)^2\right], \tag{5.12}$$

implying that these have a finite *coherence length*

$$\xi = \frac{1}{\sqrt{-2\tau}} = \frac{\xi_0}{\sqrt{2(1 - T/T_c)}}. \tag{5.13}$$

Ignoring the fluctuations, the Hamiltonian (5.6) can be approximated by its so-called *hydrodynamic limit*, also called *London limit* (more in Section 7.2):

$$H^{\mathrm{hy}}[\theta] = \rho_0^2 \int d^3x \, (\boldsymbol{\nabla}\theta)^2. \tag{5.14}$$

We have omitted a constant *condensation energy*

$$E_c = -\int d^3x \frac{\tau^2}{4\lambda}.$$

(5.15)

The Hamiltonian density (5.14) shows that the energy of a plane-wave excitation $\theta(\mathbf{x}) \propto e^{i\mathbf{kx}}$ of the phase grows with the square of the wave vector $\mathbf{k}$, and goes to zero for $\mathbf{k} \to 0$. These are the massless Nambu-Goldstone modes of the spontaneously broken U(1)-symmetry. By rewriting (5.14) as

$$H^{\mathrm{hy}}[\theta] = \frac{\rho_s \hbar^2}{2M} \int d^3x \, (\boldsymbol{\nabla}\theta)^2 \,,$$

(5.16)

we obtain the usual hydrodynamic kinetic energy, and identify

$$\rho_s = M\rho_0^2/\hbar^2$$

(5.17)

as the *superfluid density*, and

$$\mathbf{v}(\mathbf{x}) \equiv \frac{\hbar}{M} \boldsymbol{\nabla}\theta(\mathbf{x})$$

(5.18)

as the *superfluid velocity*.

This expression is found by inserting the field decomposition (5.10) on the current density of the Hamiltonian (5.6) [compare (2.64)]:

$$\mathbf{j}(\mathbf{x}) = \frac{1}{2\hbar i}\phi^*(\mathbf{x}) \overleftrightarrow{\boldsymbol{\nabla}} \phi(\mathbf{x}),$$

(5.19)

and going to the London limit, where

$$\mathbf{j}(\mathbf{x}) = \rho_0^2 \boldsymbol{\nabla}\theta(\mathbf{x}).$$

(5.20)

Apart from the constant field $\phi(\mathbf{x}) = \phi_0$, there exist nontrivial field configurations which extremize the Hamiltonian (5.6). They represent vortex lines which play a crucial role for many phenomena encountered in superfluid helium. Some relevant properties of these lines are discussed in Appendix 5A. Here we only note that at the center of each line, the size $\rho(\mathbf{x})$ of order field vanishes. The question arises as to what happens to these solutions in the hydrodynamic limit where $\rho(\mathbf{x})$ is constant everywhere?

The alert reader may have noticed that in going from (5.6) to (5.11) we have made an important error which for the discussion of the Nambu-Goldstone mechanism was irrelevant but becomes important for the understanding of the λ-transition. We have used the chain rule of differentiation to express

$$\boldsymbol{\nabla}\phi(\mathbf{x}) = \{i[\boldsymbol{\nabla}\theta(\mathbf{x})]\rho + \boldsymbol{\nabla}\rho(\mathbf{x})\}e^{i\theta(\mathbf{x})}.$$

(5.21)

However, this rule cannot be applied here. Since $\theta(\mathbf{x})$ is the phase of the complex order field $\phi(\mathbf{x})$, it is a *multivalued field*. At every point $\mathbf{x}$ it is possible to add an arbitrary integer-multiple of 2π without changing $e^{i\theta(\mathbf{x})}$.

The correct chain rule is [14]

$$\nabla\phi(\mathbf{x}) = \{i\left[\nabla\theta(\mathbf{x}) - 2\pi\boldsymbol{\delta}(\mathbf{x};S)\right]\rho(\mathbf{x}) + \nabla\rho(\mathbf{x})\}\,e^{i\theta(\mathbf{x})} \tag{5.22}$$

where $\boldsymbol{\delta}(\mathbf{x};S)$ is the δ-function on the surface S across which $\theta(\mathbf{x})$ jumps by 2π [recall the definition Eq. (4.23)]. This brings the gradient term in the Hamiltonian (5.6) to the form

$$|\nabla\phi(\mathbf{x})|^2 = [\nabla\rho(\mathbf{x})]^2 + \rho^2\left[\nabla\theta(\mathbf{x}) - \boldsymbol{\theta}^{\mathrm{v}}(\mathbf{x})\right]^2, \tag{5.23}$$

where we have introduced the field

$$\boldsymbol{\theta}^{\mathrm{v}}(\mathbf{x}) \equiv 2\pi\boldsymbol{\delta}(\mathbf{x},S). \tag{5.24}$$

The correct version of the Hamiltonian (5.11) reads therefore

$$H[\rho,\theta] = \frac{1}{2}\int d^3x\left[(\nabla\rho)^2 + \rho^2\left(\nabla\theta - \boldsymbol{\theta}^{\mathrm{v}}\right)^2 + \tau\rho^2 + \frac{\lambda}{2}\rho^4\right]. \tag{5.25}$$

The London limit of this Hamiltonian is now

$$H_{\mathrm{v}}^{\mathrm{hy}}[\theta] = \frac{1}{2}\int d^3x\left(\nabla\theta - \boldsymbol{\theta}^{\mathrm{v}}\right)^2, \tag{5.26}$$

and the correct form of the superfluid velocity (5.18) is

$$\mathbf{v}(\mathbf{x}) \equiv \nabla\theta(\mathbf{x}) - \boldsymbol{\theta}^{\mathrm{v}}(\mathbf{x}), \tag{5.27}$$

if we use natural units with $\rho_s/M = 1$. This Hamiltonian describes now *all* important excitations of the superfluid, phonons and rotons.

An important property of the superfluid velocity (5.27), the Hamiltonian (5.25), and its London limit (5.26) is its invariance under the so-called *vortex gauge transformations*. These are deformations of the surface $S \to S'$ accompanied by a change in the field $\theta(\mathbf{x})$:

$$\boldsymbol{\theta}^{\mathrm{v}}(\mathbf{x}) \to \boldsymbol{\theta}^{\mathrm{v}}(\mathbf{x}) + \nabla\Lambda_{\delta}^{\mathrm{v}}(\mathbf{x}), \quad \theta(\mathbf{x}) \to \theta(\mathbf{x}) + \Lambda_{\delta}^{\mathrm{v}}(\mathbf{x}), \tag{5.28}$$

where $\Lambda_{\delta}^{\mathrm{v}}(\mathbf{x})$ is the gauge functions

$$\Lambda_{\delta}^{\mathrm{v}}(\mathbf{x}) = 2\pi\delta(\mathbf{x};V). \tag{5.29}$$

Thus we encounter again the gauge transformations (4.29) and (4.33) of the gradient representation of magnetic fields (4.31). The field $\boldsymbol{\theta}^{\mathrm{v}}(\mathbf{x})$ is the *vortex gauge field* of the superfluid.

In the sequel we shall see that all essential physical properties of the full complex field theory with Hamiltonian (5.6) can be found in the theory of the multivalued field $\theta(\mathbf{x})$ with the hydrodynamic vortex gauge-invariant Hamiltonian (5.26).

5.1.3   Vortex Density

As in the magnetic discussion in Section 4.2, the physical content of the vortex gauge field $\boldsymbol{\theta}^{\mathrm{v}}(\mathbf{x})$ appears when forming its curl. By Stokes' theorem (4.24) we find the *vortex density*

$$\boldsymbol{\nabla} \times \boldsymbol{\theta}^{\mathrm{v}}(\mathbf{x}) \equiv \mathbf{j}^{\mathrm{v}}(\mathbf{x}) = 2\pi\boldsymbol{\delta}(\mathbf{x}; L). \tag{5.30}$$

As a consequence of Eq. (4.9), the vortex density satisfies the conservation law

$$\boldsymbol{\nabla} \cdot \mathbf{j}^{\mathrm{v}}(\mathbf{x}) = 0, \tag{5.31}$$

implying that vortex lines are closed.

The conservation law is a trivial consequence of $\mathbf{j}^{\mathrm{v}}$ being the curl of $\boldsymbol{\theta}^{\mathrm{v}}$. It is therefore a Bianchi identity associated with the vortex gauge field structure.

The expression (5.26) is in general not the complete energy of a vortex configuration. It is possible to add a gradient energy in the vortex gauge field, which introduces an extra *core energy* to the vortex line. The extended Hamiltonian of the hydrodynamic limit of the Ginzburg-Landau Hamiltonian containing phonons and vortex lines with an extra core energy reads

$$H_{\mathrm{vc}}^{\mathrm{hy}} = \int d^3x \left[\frac{1}{2} (\boldsymbol{\nabla}\theta - \boldsymbol{\theta}^{\mathrm{v}})^2 + \frac{\epsilon_c}{2} (\boldsymbol{\nabla} \times \boldsymbol{\theta}^{\mathrm{v}})^2 \right]. \tag{5.32}$$

The extra core energy does not destroy the invariance under vortex gauge transformations (5.28).

The core energy term is proportional to the square of a δ-function which is highly singular. The singularity is a consequence of the hydrodynamic limit in which the field $\rho(\mathbf{x})$ in (5.11) is completely frozen at the minimum of (5.25). Moreover, the coherence length of the ρ-field is zero, which is the origin of the above δ-functions. With this in mind we may regularize the δ-functions in the core energy physically by smearing them out over the actual small coherence length ξ of the superfluid, which is of the order of a few Å. Whatever the size of ξ, the regularized last term yields an energy proportional to the total length of the vortex lines.

5.1.4   Partition Function

The partition function of the Nambu-Goldstone modes and all fluctuating vortex lines may be written as a functional integral

$$Z_{\mathrm{vc}}^{\mathrm{hy}} = \sum_{\{S\}} \int_{-\pi}^{\pi} \mathcal{D}\theta \, e^{-\beta H_{\mathrm{vc}}^{\mathrm{hy}}}, \tag{5.33}$$

where β is the inverse temperature $\beta \equiv 1/T$ in natural units where the Boltzmann constant k_B is equal to unity. The measure $\int_{-\pi}^{\pi} \mathcal{D}\theta$ is defined by discretizing space to a fine-grained simple cubic lattice of spacing a and integrating $\theta(\mathbf{x})$ at each lattice point $\mathbf{x}$ over all $\theta(\mathbf{x}) \in (-\pi, \pi)$. The sum over all surface configurations $\sum_{\{S\}}$ is defined on the lattice by setting at each lattice point $\mathbf{x}$

$$\theta_i^{\mathrm{v}}(\mathbf{x}; S) \equiv 2\pi n_i(\mathbf{x}), \tag{5.34}$$

where $n_i(\mathbf{x})$ is an integer-valued version of the vortex gauge field $\theta_i^{\mathrm{v}}(\mathbf{x}; S)$, and by summing over all integer numbers $n_i(\mathbf{x})$:

$$\sum_{\{S\}} \equiv \sum_{\{n_i(\mathbf{x})\}}. \tag{5.35}$$

The partition function (5.33) is the continuum limit of the lattice partition function

$$Z_V = \sum_{\{n_i(\mathbf{x})\}} \left[\prod_{\mathbf{x}} \int_{-\pi}^{\pi} d\theta(\mathbf{x}) \right] e^{-\beta H_V}, \tag{5.36}$$

where H_V is the lattice version of the Hamiltonian (5.32):

$$H_V = \frac{1}{2} \sum_{\mathbf{x}} [\boldsymbol{\nabla}\theta(\mathbf{x}) - 2\pi\mathbf{n}(\mathbf{x})]^2 + \frac{1}{2}\epsilon_c[\boldsymbol{\nabla} \times \mathbf{n}(\mathbf{x})]^2. \tag{5.37}$$

Here the symbol $\boldsymbol{\nabla}$ denotes the *lattice derivative* whose components ∇_i act on an arbitrary function $f(\mathbf{x})$ as

$$\nabla_i f(\mathbf{x}) \equiv a^{-1}[f(\mathbf{x} + a\mathbf{e}_i) - f(\mathbf{x})], \tag{5.38}$$

where $\mathbf{e}_i$ are the unit vectors to the nearest neighbors in the plane, and a is the lattice spacing. There exists also a conjugate lattice derivative

$$\overline{\nabla}_i f(\mathbf{x}) \equiv a^{-1}[f(\mathbf{x}) - f(\mathbf{x} - a\mathbf{e}_i)]. \tag{5.39}$$

It arises naturally in the lattice version of partial integration

$$\sum_{\mathbf{x}} f(\mathbf{x}) \nabla_i g(\mathbf{x}) = -\sum_{\mathbf{x}} [\overline{\nabla}_i f(\mathbf{x})] g(\mathbf{x}), \tag{5.40}$$

which holds for functions $f(\mathbf{x})$, $g(\mathbf{x})$ vanishing on the surface of the system, or satisfying periodic boundary conditions. This will always be assumed in the remainder of this discussion. In Fourier space, the eigenvalues of ∇_i, $\overline{\nabla}_i$ are $K_i = (e^{ik_i a} - 1)/a$, $\overline{K}_i = (1 - e^{-ik_i a})/a$, respectively, where k_i are the wave numbers in the i-direction.

The lattice version of the Laplace operator is the *lattice Laplacian* $\overline{\nabla}\nabla$. Its eigenvalues in D dimensions are [15]

$$\overline{\mathbf{K}}\mathbf{K} = 2\sum_{i=1}^{D}(1 - \cos k_i a), \tag{5.41}$$

where $k_i \in (-\pi/a, \pi/a)$ are the wave numbers in the Brillouin zone of the lattice [15]. In the continuum limit $a \to 0$, both lattice derivatives reduce to the ordinary derivative ∂_i, and $\overline{\mathbf{K}}\mathbf{K}$ goes over into $\mathbf{k}^2$. In the Hamiltonian (5.37), the lattice spacing a has been set equal to unity, for simplicity.

In the partition function (5.36), the integer-valued vortex gauge fields $n_i(\mathbf{x})$ are summed without restriction. Alternatively, we may fix a gauge of $n_i(\mathbf{x})$ by some functional $\Phi[\mathbf{n}]$, and obtain: [7]

$$Z_V = \sum_{\{n_i(\mathbf{x})\}} \Phi[\mathbf{n}] \left[\prod_{\mathbf{x}} \int_{-\infty}^{\infty} d\theta(\mathbf{x}) \right] e^{-\beta H_V}. \tag{5.42}$$

On the lattice we can always enforce the axial gauge [16]:

$$n_3(\mathbf{x}) = 0. \tag{5.43}$$

Note that in contrast to continuum gauge fields it is impossible to choose the Lorentz gauge $\boldsymbol{\nabla} \cdot \mathbf{n}(\mathbf{x}) = 0$.

In the formulation (5.42), the gauge freedom has been absorbed into the $\theta(\mathbf{x})$-field which now runs, for each $\mathbf{x}$, from $\theta = -\infty$ to ∞ rather than from $-\pi$ to π in (5.36). This has the advantage that the integrals over all $\theta(\mathbf{x})$ can be done, yielding

$$Z_V = \mathrm{Det}^{1/2}[-\overline{\boldsymbol{\nabla}}\boldsymbol{\nabla}^{-1}] \sum_{\{n_i(\mathbf{x})\}} \Phi[\mathbf{n}] e^{-\beta H'_V}, \tag{5.44}$$

with

$$\beta H'_V = \sum_{\mathbf{x}} \left[\frac{4\pi^2}{2} \left\{ \mathbf{n}^2(\mathbf{x}) - [\boldsymbol{\nabla} \cdot \mathbf{n}(\mathbf{x})] \frac{1}{-\overline{\boldsymbol{\nabla}}\boldsymbol{\nabla}} [\boldsymbol{\nabla} \cdot \mathbf{n}(\mathbf{x})] \right\} + \frac{1}{2}\epsilon_c [\boldsymbol{\nabla} \times \mathbf{n}(\mathbf{x})]^2 \right]. \tag{5.45}$$

In calculating partition functions we shall always ignore trivial overall factors. If we introduce lattice curls of the integer-valued jump fields (5.34):

$$\mathbf{l}(\mathbf{x}) \equiv \boldsymbol{\nabla} \times \mathbf{n}(\mathbf{x}), \tag{5.46}$$

we can rewrite the Hamiltonian (5.45) as

$$\beta H'_V = \sum_{\mathbf{x}} \left[\frac{4\pi^2}{2} \mathbf{l}(\mathbf{x}) \frac{1}{-\overline{\boldsymbol{\nabla}}\boldsymbol{\nabla}} \mathbf{l}(\mathbf{x}) + \frac{\epsilon_c}{2} \mathbf{l}^2(\mathbf{x}) \right]. \tag{5.47}$$

Being lattice curls, the fields $\mathbf{l}(\mathbf{x})$ satisfy $\boldsymbol{\nabla} \cdot \mathbf{l}(\mathbf{x}) = 0$. They are, of course, integer-valued versions of the vortex density $\mathbf{j}^v(\mathbf{x})/2\pi$ defined in Eq. (5.30). The energy (5.47) is the interaction energy between the vortex loops.

The inverse lattice Laplacian $-\overline{\boldsymbol{\nabla}}\boldsymbol{\nabla}^{-1}$ in (5.44) and (5.47) is the lattice version of the inverse Laplacian $-\boldsymbol{\nabla}^{-2}$. Its local matrix elements $\langle \mathbf{x}_2 | -\boldsymbol{\nabla}^{-2} | \mathbf{x}_1 \rangle$ yield the Coulomb potential of the coordinate difference $\mathbf{x} = \mathbf{x}_2 - \mathbf{x}_1$:

$$V_0(\mathbf{x}) \equiv \int \frac{d^3 k}{(2\pi)^3} \frac{e^{i\mathbf{k}\mathbf{x}}}{\mathbf{k}^2} = \frac{1}{4\pi r}, \qquad r \equiv |\mathbf{x}|. \tag{5.48}$$

The corresponding matrix elements on the lattice $\langle \mathbf{x}_2 | -\overline{\boldsymbol{\nabla}}\boldsymbol{\nabla}^{-1} | \mathbf{x}_1 \rangle$ are given by

$$v_0(\mathbf{x}) = \int_{\mathrm{BZ}} \frac{d^3 k}{(2\pi)^3} \frac{e^{i\mathbf{k}\mathbf{x}}}{\overline{\mathbf{K}}\mathbf{K}} = \frac{1}{a} \left[\prod_{i=1}^{3} \int_{-\pi}^{\pi} \frac{d^3(ak_i)}{(2\pi)^3} \right] \frac{e^{i\sum_{i=1}^{3} k_i x_i}}{2\sum_{i=1}^{3}(1 - \cos ak_i)}, \tag{5.49}$$

where the subscript BZ of the momentum integral indicates the restriction to the Brillouin zone.

The lattice Coulomb potential (5.49) is the zero-mass limit of the *lattice Yukawa potential*

$$v_m(\mathbf{x}) = \frac{1}{a} \left[\prod_{i=1}^{3} \int_{-\pi}^{\pi} \frac{d^3(ak_i)e^{ik_ix_i}}{(2\pi)^3} \right] \frac{1}{2\sum_{i=1}^{3}(1 - \cos ak_i) + m^2a^2}, \tag{5.50}$$

whose continuum limit is the ordinary Yukawa potential

$$V_m(r) \equiv \int \frac{d^3k}{(2\pi)^3} e^{i\mathbf{k}\mathbf{x}} \frac{1}{\mathbf{k}^2 + m^2} = \frac{e^{-mr}}{4\pi r}, \qquad r \equiv |\mathbf{x}|. \tag{5.51}$$

In terms of the lattice Coulomb potential, we can write the partition function (5.44) for zero extra core energy as

$$Z_V = \mathrm{Det}^{1/2}[\hat{v}_0] \sum_{\mathbf{l}, \boldsymbol{\nabla}\cdot\mathbf{l}=0} e^{-(4\pi^2\beta a/2)\Sigma_{\mathbf{x},\mathbf{x}'}\,\mathbf{l}(\mathbf{x})v_0(\mathbf{x}-\mathbf{x}')\,\mathbf{l}(\mathbf{x}')}, \tag{5.52}$$

where $\hat{v}_0$ abbreviates the operator $-\boldsymbol{\nabla}\boldsymbol{\nabla}^{-1}$.

The momentum integrals over the different lattice directions can be done separately by applying the Schwinger trick to express the denominator as an integral over an exponential

$$\frac{1}{a} = \int_0^\infty ds\, e^{-sa}, \tag{5.53}$$

so that

$$\frac{e^{i\Sigma_{i=1}^{3}k_ix_i}}{2\sum_{i=1}^{3}(1 - \cos ak_i) + m^2a^2} = \int_0^\infty ds\, e^{-s(6+m^2a^2)} \prod_{i=1}^{3} e^{i(k_ia)(x_i/a)+s\cos k_ia}. \tag{5.54}$$

The ratios x_i/a are integer numbers, so that we may use the integral representation of the modified Bessel functions of the first kind

$$I_n(z) = \frac{1}{2\pi} \int_{-\pi}^{\pi} d\theta\, e^{in\theta + z\cos\theta}, \qquad n = \text{integer}, \tag{5.55}$$

we find

$$v_m(\mathbf{x}) = \frac{1}{a} \int_0^\infty ds\, e^{-(6+m^2a^2)s} I_{x_1/a}(2s)I_{x_2/a}(2s)I_{x_3/a}(2s). \tag{5.56}$$

In contrast to the continuum version (5.51), the lattice potential $v_m(\mathbf{x})$ is finite at the origin. The values of $v_m(\mathbf{0})$ as a function of m^2a^2 are plotted in Fig. 5.5. The Coulomb potential has the value $v_0(\mathbf{0}) \approx 0.2527/a$ [17].

The functional determinant of the lattice Laplacian appearing in (5.44) and (5.52) as a prefactor can be calculated easily from the Yukawa potential (5.56) for $m = 0$. We simply use the relation

$$\mathrm{Det}^{-1/2}(-\boldsymbol{\nabla}\boldsymbol{\nabla}+m^2) = \mathrm{Det}^{1/2}(\hat{v}_m) = e^{-\frac{1}{2}\mathrm{Tr}\log(-\boldsymbol{\nabla}\boldsymbol{\nabla}+m^2)} = e^{-\frac{a^3}{2}\Sigma_{\mathbf{x}}\langle\mathbf{x}|\log(-\boldsymbol{\nabla}\boldsymbol{\nabla}+m^2)|\mathbf{x}\rangle}, \tag{5.57}$$

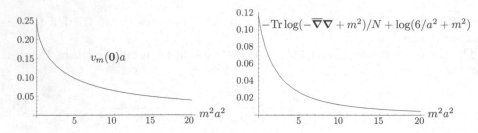

FIGURE 5.5 Lattice Yukawa potential at the origin and the associated Tracelog. The plot shows the subtracted expression $\mathrm{Tr}\log(-\overline{\nabla}\nabla + m^2)/N$, where N is the number of sites on the lattice.

and calculate

$$\mathrm{Tr}\log(-\overline{\nabla}\nabla + m^2) = a^3 \langle \mathbf{x}| \int dm^2 \sum_{\mathbf{x}} (-\overline{\nabla}\nabla + m^2)^{-1}|\mathbf{x}\rangle$$

$$= Na^3 \int dm^2 \langle \mathbf{0}|(-\overline{\nabla}\nabla + m^2)^{-1}|\mathbf{0}\rangle = Na^3 \int dm^2\, v_m(\mathbf{0}), \quad (5.58)$$

where N is the number of lattice sites. We have omitted a constant of integration which is fixed by the leading small-m-behavior in D-dimensions

$$\mathrm{Tr}\log(-\overline{\nabla}\nabla + m^2) \underset{\text{small } m}{\approx} \int \frac{d^D k}{(2\pi)^D} \log(\mathbf{k}^2 + m^2) \underset{\text{small } m}{\approx} -\frac{\Gamma(-D/2)}{(4\pi)^{D/2}} m^D, \quad (5.59)$$

$$v_m(\mathbf{0}) \underset{\text{small } m}{\approx} \int \frac{d^D k}{(2\pi)^D} \frac{1}{\mathbf{k}^2 + m^2} \underset{\text{small } m}{\approx} \frac{\Gamma(1 - D/2)m^{D-2}}{(4\pi)^{D/2}}. \quad (5.60)$$

Performing the integral over m^2 in (5.56), we obtain

$$a^3 \int dm^2\, v_m(\mathbf{0}) = -\int_0^\infty \frac{ds}{s} e^{-(6+m^2 a^2)s} I_0^2(2s). \quad (5.61)$$

The divergence at $s = 0$ can be removed by subtracting a similar integral representation

$$a^2 \int dm^2\, (6 + m^2 a^2)^{-1} = -\int_0^\infty \frac{ds}{s} e^{-(6+m^2 a^2)s}. \quad (5.62)$$

This leads to the finite result

$$\frac{1}{N}\mathrm{Tr}\log(-\overline{\nabla}\nabla + m^2) - \log(6/a^2 + m^2) = -\int_0^\infty \frac{ds}{s} e^{-(6+m^2 a^2)s} \left[I_0^3(2s) - 1 \right]. \,(5.63)$$

The m^2-behavior of this expression is displayed in Fig. 5.5.

The lattice expression (5.52) makes it easy to perform a graphical expansion of the partition function. It becomes a sum of terms associated with longer and longer loops, each term carrying a Boltzmann factor $e^{-\beta\text{const}/2}$. This expansion converges fast for low temperatures. As the temperature is raised, fluctuations create more and longer loops. If there is no extra core energy ϵ_c, the loops become infinitely long and dense at a critical value $T_c = 1/\beta_c \approx 1/0.33 \approx 3$, where the sum in (5.44) diverges since the configurational entropy of long lines overwhelms the Boltzmann suppression factor in a similar way as in the simple model integral (5.2). At that point the system is filled with vortex loops. Since the inside of each vortex loop consists of normal fluid, this condensation of vortex loops makes the entire fluid normal. In Fig. 5.4 this condensation process is visualized. The associated specific heat is plotted in Fig. 5.6.

Without the extra core energy, Z_V defines the famous *Villain model* [18], a discrete Gaussian approximation to the so-called *XY-model* whose Hamiltonian is

$$H_{XY} = \sum_{\mathbf{x}} \sum_{i=1}^{3} \cos[\nabla_i \theta(\mathbf{x})]. \qquad (5.64)$$

Both the XY-model and the Villain model with Hamiltonian (5.37) can be simulated on a computer using Monte Carlo techniques. Both display a second-order phase transition. In the XY-model, this occurs at an inverse temperature $\beta_c \approx 0.45$ (in natural units), in the Villain model with zero extra core energy ϵ_c at $\beta_c \approx 0.33$ [7]. The critical exponents of the two models coincide. The specific heat of the Villain model is shown in Fig. 5.6. It has the typical λ-shape observed in ^{4}He in Fig. 5.1.

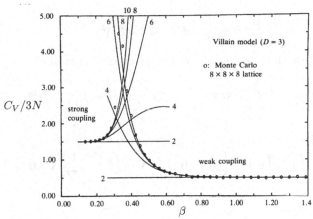

FIGURE 5.6 Specific heat of Villain model in three dimensions plotted against $\beta = 1/T$ in natural units. The λ-transition is seen as a sharp peak, with properties near T_c similar to the experimental curve in Fig. 5.1. The solid curves stem from analytic expansions once in powers of $T \equiv 1/\beta$ (low-temperature or weak-coupling expansion) and once in powers of $T^{-1} = \beta$ (high-temperature or strong-coupling expansion) to increasing order in these variables (see Ref. [19]).

By analogy with the lattice formulation, we fix the gauge in the continuum partition function (5.33) with the energy (5.26) or (5.32), by inserting a gauge-fixing functional $\Phi[\boldsymbol{\theta}^v]$. The axial gauge is fixed by the δ-functional

$$\Phi[\boldsymbol{\theta}^v] = \delta[\theta_3^v]. \qquad (5.65)$$

Note that due to the presence of the sum over all vortex gauge fields $\boldsymbol{\theta}^v$, the partition function (5.33) describes superfluid ^{4}He not only near zero temperature, where the Nambu-Goldstone modes are dominant, but also at all not too large temperatures. In particular, the most interesting temperature regime around the superfluid phase transition is included.

The vortex gauge field extends the partition function of fluctuating Nambu-Goldstone modes in the same way as the size of the order field ψ does in a Landau description of the phase transition. In fact, it is easy to show that near the transition, the partition function (5.33) can be transformed into a $|\psi|^4$-theory of the Landau type [7].

5.1.5 Continuum Derivation of Interaction Energy

Let us calculate the interaction energy (5.47) between vortex loops once more in the continuum formulation. Omitting the core energy, for simplicity, the partition function with a fixed vortex gauge is given by

$$Z_v^{hy} = \sum_{\{S\}} \Phi[\boldsymbol{\theta}^v] \int_{-\infty}^{\infty} \mathcal{D}\theta \, e^{-\beta H_v^{hy}}, \tag{5.66}$$

where

$$H_v^{hy} = \frac{1}{2} \int d^3x \, (\boldsymbol{\nabla}\theta - \boldsymbol{\theta}^v)^2 \tag{5.67}$$

is the energy (5.32) without core energy. Let us expand this into two parts

$$H_v^{hy} = \frac{1}{2} \int d^3x \left[(\boldsymbol{\nabla}\theta)^2 + 2\theta\boldsymbol{\nabla}\boldsymbol{\theta}^v + \boldsymbol{\theta}^{v2} \right] = H_{v1}^{hy} + H_{v2}^{hy}, \tag{5.68}$$

where

$$H_{v1}^{hy} = \frac{1}{2} \int d^3x \left(\theta + \frac{1}{-\boldsymbol{\nabla}^2}\boldsymbol{\nabla}\cdot\boldsymbol{\theta}^v \right)(-\boldsymbol{\nabla}^2)\left(\theta + \frac{1}{-\boldsymbol{\nabla}^2}\boldsymbol{\nabla}\cdot\boldsymbol{\theta}^v \right) \tag{5.69}$$

and

$$H_{v2}^{hy} = \frac{1}{2} \int d^3x \left(\boldsymbol{\theta}^{v2} - \boldsymbol{\nabla}\cdot\boldsymbol{\theta}^v \frac{1}{-\boldsymbol{\nabla}^2}\boldsymbol{\nabla}\cdot\boldsymbol{\theta}^v \right). \tag{5.70}$$

Inserting this into (5.66), we can perform the Gaussian integrals over $\theta(\mathbf{x})$ at each $\mathbf{x}$ using the generalization of the Gaussian formula

$$\int_{-\infty}^{\infty} \frac{d\theta}{2\pi} e^{-a(\theta-c)^2/2} = a^{-1/2} \tag{5.71}$$

to fields $\theta(\mathbf{x})$ and differential operators $\hat{\mathcal{O}}$ in $\mathbf{x}$-space

$$\int_{-\infty}^{\infty} \mathcal{D}\theta \, e^{-\int d^3x \, [\theta(\mathbf{x})-c(\mathbf{x})]\, \hat{\mathcal{O}}\, [\theta(\mathbf{x})-c(\mathbf{x})]/2} = [\text{Det}\,\hat{\mathcal{O}}]^{-1/2}. \tag{5.72}$$

Applying this formula to (5.66), we obtain

$$Z_{\mathrm{v}}^{\mathrm{hy}} = \left[\mathrm{Det}(-\boldsymbol{\nabla}^2)\right]^{-1/2} \sum_{\{S\}} \Phi[\boldsymbol{\theta}^{\mathrm{v}}]\, e^{-\beta H_{\mathrm{v}}}, \tag{5.73}$$

where H_{v} is the interaction energy of the vortex loops corresponding to (5.47):

$$\begin{aligned} H_{\mathrm{v}} &= \frac{1}{2}\int d^3x (\boldsymbol{\nabla}\times\boldsymbol{\theta}^{\mathrm{v}})\frac{1}{-\boldsymbol{\nabla}^2}(\boldsymbol{\nabla}\times\boldsymbol{\theta}^{\mathrm{v}}) = \frac{1}{2}\int d^3x\, \mathbf{j}^{\mathrm{v}}\frac{1}{-\boldsymbol{\nabla}^2}\mathbf{j}^{\mathrm{v}} \\ &= \frac{1}{8\pi}\int d^3x d^3x'\, \mathbf{j}^{\mathrm{v}}(\mathbf{x})\frac{1}{|\mathbf{x}-\mathbf{x}'|}\mathbf{j}^{\mathrm{v}}(\mathbf{x}'). \end{aligned} \tag{5.74}$$

This has the same form as the magnetic Biot-Savart energy (4.93) for current loops, implying that parallel vortex lines repel each other [as currents would do if no work were required to keep them constant against the inductive forces, which reverses the sign. Recall the discussion of the free magnetic energy (4.95)]. On a lattice, the partition function (5.73) takes once more the form (5.44).

The process of removing some variables from a partition function by integration will occur frequently in the sequel and will be referred to as *integrating out*. It will be used also in discussions of Hamiltonians without writing always down the associated partition function in which the integrals are actually performed.

5.1.6 Physical Jumping Surfaces

The invariance of the energy (5.32) under vortex gauge transformations guarantees the physical irrelevance of the jumping surfaces S. If we destroy this invariance, the surfaces become physical objects. This may be done by destroying the original U(1)-symmetry explicitly. This will give a mass to the Nambu-Goldstone modes. To lowest approximation, it adds a mass term $m^2\theta(\mathbf{x})^2$ to the energy (5.32) without core energy:

$$H_{\mathrm{v}m}^{\mathrm{hy}} = \frac{1}{2}\int d^3x\left\{[\boldsymbol{\nabla}\theta(\mathbf{x})-\boldsymbol{\theta}^{\mathrm{v}}(\mathbf{x})]^2 + m^2\theta(\mathbf{x})^2\right\}. \tag{5.75}$$

The mass term gives an energy to the surfaces S. To see this we write the energy as

$$H_{\mathrm{v}m}^{\mathrm{hy}} = \frac{1}{2}\int d^3x\left\{[(\boldsymbol{\nabla}\theta)^2 + m^2\theta^2] + 2\theta\,\boldsymbol{\nabla}\cdot\boldsymbol{\theta}^{\mathrm{v}} + \boldsymbol{\theta}^{\mathrm{v}2}]\right\}, \tag{5.76}$$

and decompose this into two parts as in (5.68):

$$H_{\mathrm{v}m1}^{\mathrm{hy}} = \frac{1}{2}\int d^3x\left(\theta + \frac{1}{-\boldsymbol{\nabla}^2+m^2}\boldsymbol{\nabla}\cdot\boldsymbol{\theta}^{\mathrm{v}}\right)(-\boldsymbol{\nabla}^2+m^2)\left(\theta + \frac{1}{-\boldsymbol{\nabla}^2+m^2}\boldsymbol{\nabla}\cdot\boldsymbol{\theta}^{\mathrm{v}}\right) \tag{5.77}$$

and

$$H_{\mathrm{v}m2}^{\mathrm{hy}} = \frac{1}{2}\int d^3x\left(\boldsymbol{\theta}^{\mathrm{v}2} - \boldsymbol{\nabla}\cdot\boldsymbol{\theta}^{\mathrm{v}}\frac{1}{-\boldsymbol{\nabla}^2+m^2}\boldsymbol{\nabla}\cdot\boldsymbol{\theta}^{\mathrm{v}}\right). \tag{5.78}$$

The Gaussian integrals over $\theta(\mathbf{x})$ can be done as before, and the partition function (5.73) becomes

$$Z_{\mathrm{v}m}^{\mathrm{hy}} = \mathrm{Det}^{-1/2}[-\boldsymbol{\nabla}^2 + m^2] \sum_{\{S\}} \Phi[\boldsymbol{\theta}^{\mathrm{v}}]\, e^{-\beta H_{\mathrm{v}m2}^{\mathrm{hy}}}. \tag{5.79}$$

The energy $H_{\mathrm{v}m2}^{\mathrm{hy}}$ in the exponent can be rewritten as

$$H_{\mathrm{v}m2}^{\mathrm{hy}} = \frac{1}{2} \int d^3x \left[(\boldsymbol{\nabla} \times \boldsymbol{\theta}^{\mathrm{v}}) \frac{1}{-\boldsymbol{\nabla}^2 + m^2} (\boldsymbol{\nabla} \times \boldsymbol{\theta}^{\mathrm{v}}) + m^2 \boldsymbol{\theta}^{\mathrm{v}} \frac{1}{-\boldsymbol{\nabla}^2 + m^2} \boldsymbol{\theta}^{\mathrm{v}} \right]. \tag{5.80}$$

The first term is a modification of the Biot-Savart-type of energy (5.74):

$$\begin{aligned}
H_{\mathrm{v}m}^{\mathrm{hy}} &= \frac{1}{2} \int d^3x\, (\boldsymbol{\nabla} \times \boldsymbol{\theta}^{\mathrm{v}}) \frac{1}{-\boldsymbol{\nabla}^2 + m^2} (\boldsymbol{\nabla} \times \boldsymbol{\theta}^{\mathrm{v}}) = \frac{1}{2} \int d^3x\, \mathbf{j}^{\mathrm{v}} \frac{1}{-\boldsymbol{\nabla}^2 + m^2} \mathbf{j}^{\mathrm{v}} \\
&= \frac{1}{8\pi} \int d^3x\, d^3x'\, \mathbf{j}^{\mathrm{v}}(\mathbf{x}) \frac{e^{-m|\mathbf{x}-\mathbf{x}'|}}{|\mathbf{x}-\mathbf{x}'|} \mathbf{j}^{\mathrm{v}}(\mathbf{x}').
\end{aligned} \tag{5.81}$$

The presence of the mass m changes the long-range Coulomb-like interaction $1/R$ in Eq. (5.74) into a finite-range Yukawa-like interaction e^{-mR}/R.

The second term in (5.80),

$$H_{Sm} = \frac{m^2}{2} \int d^3x\, \boldsymbol{\theta}^{\mathrm{v}} \frac{1}{-\boldsymbol{\nabla}^2 + m^2} \boldsymbol{\theta}^{\mathrm{v}} = \frac{m^2}{8\pi} \int d^3x\, d^3x'\, \boldsymbol{\theta}^{\mathrm{v}}(\mathbf{x}) \frac{e^{-m|\mathbf{x}-\mathbf{x}'|}}{|\mathbf{x}-\mathbf{x}'|} \boldsymbol{\theta}^{\mathrm{v}}(\mathbf{x}'), \tag{5.82}$$

is of a completely new type. It describes a Yukawa-like interaction between the normal vectors of the surface elements, and gives rise to a field energy within a layer of thickness $1/m$ around the surfaces S. As a consequence, the surfaces acquire *tension*. Their shape is no longer irrelevant, but for a given set of vortex loops L at their boundaries, the surface will span minimal surfaces. For $m = 0$, the tension disappears and the shape of the surface becomes again irrelevant, thus restoring vortex gauge invariance.

This mechanism of generating surface tension will be used in Chapter 8 to construct a simple model of quark confinement.

5.1.7   Canonical Representation of Superfluid

We can set up an alternative representation of the partition function of the superfluid in which the vortex loops are more directly described by their physical vortex density, instead of their jumping surfaces S. This is possible by eliminating the Nambu-Goldstone modes in favor of a new gauge field. It is canonically conjugate to the angular field θ and called generically *stress gauge field* [7]. In the particular case of the superfluid under discussion it is a *gauge field of superflow*.

In order to understand how this new gauge field arises recall that the canonically conjugate momentum variable $p(t)$ in an ordinary path integral [20]

$$\int \mathcal{D}\mathbf{x} \exp\left(-\frac{M}{2} \int_{t_a}^{t_b} dt\, \dot{\mathbf{x}}^2 \right) \tag{5.83}$$

is introduced by a quadratic completion, rewriting (5.83) as

$$\int \mathcal{D}\mathbf{x}\mathcal{D}\mathbf{p} \exp\left[\int_{t_a}^{t_b} dt \left(i\mathbf{p}\dot{\mathbf{x}} - \frac{\mathbf{p}^2}{2M}\right)\right]. \tag{5.84}$$

By analogy, we introduce a canonically conjugate vector field $\mathbf{b}(\mathbf{x})$ to rewrite the partition function (5.66) as

$$Z_{\mathrm{v}}^{\mathrm{hy}} = \int_{-\infty}^{\infty} \mathcal{D}\mathbf{b} \sum_{\{S\}} \Phi[\boldsymbol{\theta}^{\mathrm{v}}] \int_{-\infty}^{\infty} \mathcal{D}\theta\, e^{-\beta\bar{H}_{\mathrm{v}}^{\mathrm{hy}}} \tag{5.85}$$

where [21]

$$\beta\bar{H}_{\mathrm{v}}^{\mathrm{hy}} = \int d^3x \left\{\frac{1}{2\beta}\mathbf{b}^2(\mathbf{x}) - i\mathbf{b}(\mathbf{x})\left[\boldsymbol{\nabla}\theta(\mathbf{x}) - \boldsymbol{\theta}^{\mathrm{v}}(\mathbf{x})\right]\right\}. \tag{5.86}$$

In principle, the gradient energy could contain higher derivatives and higher powers of $\partial_i\theta$. Then the canonical representation (5.86) would contain more complicated functions of $b_i(\mathbf{x})$.

It is useful to observe at this point that if we go over to a Minkowski space formulation in which $x^0 = -ix^3$ plays the role of time, the integral

$$\int_{-\infty}^{\infty} \mathcal{D}b_0\, e^{-ib_0(\mathbf{x})\partial_0\theta(\mathbf{x})} \tag{5.87}$$

creates, on a discretized time axis, a product of δ-functions

$$\langle\theta_{n+1}|\theta_n\rangle\langle\theta_n|\theta_{n-1}\rangle\langle\theta_{n-1}|\theta_{n-2}\rangle \tag{5.88}$$

with

$$\langle\theta_n|\theta_{n-1}\rangle = \delta_n(\theta_n - \theta_{n-1}). \tag{5.89}$$

These can be interpreted as Dirac scalar products in the Hilbert space of the system. On this Hilbert space, there exists an operator $\hat{b}_i(\mathbf{x})$ whose zeroth component is given by

$$\hat{b}_0 = -i\partial_\theta \tag{5.90}$$

which satisfies the equal-time canonical communication rule

$$\left[\hat{b}_0(\mathbf{x}_\perp, x_0), \theta(\mathbf{x}'_\perp, x_0)\right] = -i\delta^{(2)}(\mathbf{x}_\perp - \mathbf{x}'_\perp), \tag{5.91}$$

where $\mathbf{x}_\perp = (x^1, x^2)$ denotes the spatial components of the vector (x^0, x^1, x^2). The charge associated with $\hat{b}_0(\mathbf{x})$,

$$\hat{Q}(x_0) = \int d^2x\, \hat{b}_0(\mathbf{x}_\perp, x_0), \tag{5.92}$$

generates a constant shift in θ:

$$e^{-i\alpha\hat{Q}(x_0)}\theta(\mathbf{x}_\perp, x_0)e^{i\alpha\hat{Q}(x_0)} - \theta(\mathbf{x}_\perp, x_0) + \alpha. \tag{5.93}$$

Thus it multiplies the original field $e^{i\theta(\mathbf{x})}$ by a phase factor $e^{i\alpha}$. The charge $\hat{Q}(\mathbf{x}_0)$ is the generator of the U(1)-symmetry transformation whose spontaneous breakdown is responsible for the Nambu-Goldstone nature of the fluctuations of $\theta(\mathbf{x})$. Since the original theory is invariant under the transformations $\phi \rightarrow e^{i\alpha}\phi$, the energy (5.86) does not depend on θ itself, but only on $\partial_i\theta$.

In the partition function (5.85) we may use the formula

$$\int_{-\infty}^{\infty} \mathcal{D}\theta\, e^{i\int d^3x\, f(\mathbf{x})\theta(\mathbf{x})} = \delta[f(\mathbf{x})], \tag{5.94}$$

to obtain from $f(\mathbf{x}) = \boldsymbol{\nabla} \cdot \mathbf{b}(\mathbf{x})$ the conservation law

$$\boldsymbol{\nabla} \cdot \mathbf{b}(\mathbf{x}) = 0. \tag{5.95}$$

This implies that $\hat{Q}(\mathbf{x}_0)$ is a time-independent charge and $e^{i\alpha\hat{Q}}$ is a symmetry transformation.

If the energy (5.86) would depend on θ itself, then the charge $\hat{Q}(\mathbf{x}_0)$ would no longer be time-independent. However, it would still generate the above U(1)-transformation.

In general, the conjugate variable to the phase angle of a complex field is the particle number (recall Subsection 3.5.3). This role is played here by $\hat{Q}(\mathbf{x}_0)$ which counts the number of particles in the superfluid. Thus we may identify the vector field $\mathbf{b}(\mathbf{x})$ as the particle current density of the superfluid condensate:

$$\mathbf{j}_s(\mathbf{x}) \equiv \mathbf{b}(\mathbf{x}), \tag{5.96}$$

also called the *supercurrent density* of the superfluid.

After integrating out the θ-fields in the partition function (5.85), we can also perform the sum over all surface configurations of the vortex gauge field $\boldsymbol{\theta}^{\mathrm{v}}(\mathbf{x})$. For this we employ the following useful formula applicable to any function $\mathbf{b}(\mathbf{x})$ with $\boldsymbol{\nabla} \cdot \mathbf{b}(\mathbf{x}) = 0$:

$$\sum_{\{S\}} e^{2\pi i \int d^3x\, \boldsymbol{\delta}(\mathbf{x};S)\mathbf{b}(\mathbf{x})} = \sum_{\{L\}} \delta\left[\mathbf{b}(\mathbf{x}) - \boldsymbol{\delta}(\mathbf{x};L)\right]. \tag{5.97}$$

This can easily be proved by going on a lattice where this formula reads [recall (5.35)]

$$\sum_{\{n_i\}} e^{2\pi i \Sigma_\mathbf{x} n_i(\mathbf{x}) f_i(\mathbf{x})} = \sum_{\{m_i\}} \prod_{\mathbf{x},i} \delta\left(f_i(\mathbf{x}) - m_i(\mathbf{x})\right), \tag{5.98}$$

and using for each $\mathbf{x}, i$ the Poisson formula [20]

$$\sum_n e^{2\pi i n f} = \sum_m \delta(f - m). \tag{5.99}$$

Then we obtain for (5.85) the following alternative representation

$$Z_{\mathrm{v}}^{\mathrm{hy}} = \sum_{\{L\}} e^{-\int d^3x\, \mathbf{b}^2/2\beta}, \tag{5.100}$$

where $\mathbf{b} = \boldsymbol{\delta}(\mathbf{x}; L)$. On the lattice, this partition function becomes

$$Z_v^{\mathrm{hy}} = \sum_{\mathbf{b}, \boldsymbol{\nabla} \cdot \mathbf{b} = 0} e^{-\sum_{\mathbf{x}} \mathbf{b}^2 / 2\beta}, \tag{5.101}$$

where $\mathbf{b}(\mathbf{x})$ is now an integer-valued divergenceless field representing the closed lines of superflow.

The partition function (5.101) can be evaluated graphically adding terms of longer and longer loops each term carrying a Boltzmann factor $e^{-\mathrm{const}/2\beta}$. This expansion converges fast for high temperature. The specific heat following from the lowest approximations obtained in this way is plotted in Fig. 5.6. For very high temperature, there is no loop of superflow. As the temperature is lowered, fluctuations create more and longer loops. At the critical value

$$T_c \equiv 1/\beta_c \approx 1/0.33 \approx 3, \tag{5.102}$$

the loops grow infinitely long, and the sum in (5.101) diverges. The system becomes a superfluid.

Note that the superflow partition function (5.101) looks quite similar to the vortex loop partition function (5.52). Both contain the same type of sum over non-selfbacktracking loops. The main difference is the long-range Coulomb interaction between the loop elements. Suppose we forget for a moment the nonlocal parts of the Coulomb interaction and approximate the vortex loop partition function (5.52) keeping only the self-energy part of the Coulomb interaction:

$$Z_{V\,app} = \mathrm{Det}^{1/2}[\hat{v}_0] \sum_{\mathbf{l}, \boldsymbol{\nabla} \cdot \mathbf{l} = 0} e^{-(4\pi^2 \beta a/2) v_0(\mathbf{0}) \sum_{\mathbf{x}} \mathbf{l}^2(\mathbf{x})}. \tag{5.103}$$

Apart from a constant overall factor, this approximation coincides with the superflow partition function (5.101). By comparing the prefactors of the energy with that in the partition function (5.101) whose critical point is determined by (5.102), we see that (5.103) has a second-order phase transition at

$$4\pi^2 a \beta v_0(\mathbf{0}) \approx T_c \approx 3, \tag{5.104}$$

implying that $\beta \approx 3/4\pi^2 a v_0(\mathbf{0}) \approx 0.30$, corresponding to an approximate critical temperature

$$T_c^{\mathrm{appr}} \approx \frac{4\pi^2 a v_0(\mathbf{0})}{3} \approx 3.3. \tag{5.105}$$

This is only 10% larger than the precise value $T_c = 1/\beta_c \approx 3$. Hence we conclude that the nonlocal parts of the Coulomb interaction in (5.52) have little effect upon the transition temperature.

5.1.8   Yukawa Loop Gas

The above observation allows us to estimate the transition temperature in the partition function closely related to (5.52)

$$Z_V^Y = \text{Det}^{1/2}[\hat{v}_m] \sum_{\mathbf{l},\nabla\cdot\mathbf{l}=0} e^{-(4\pi^2\beta a/2)\Sigma_{\mathbf{x},\mathbf{x}'}\,\mathbf{l}(\mathbf{x})v_m(\mathbf{x}-\mathbf{x}')\mathbf{l}(\mathbf{x}')}, \tag{5.106}$$

where $v_m(\mathbf{x})$ is the lattice version of the Yukawa potential (5.51), and $\hat{v}_m$ the associated operator $(-\overline{\nabla}\nabla + m^2)^{-1}$.

Performing also here the local approximation of the type (5.103),

$$Z_{V\,app}^Y = \text{Det}^{1/2}[\hat{v}_m] \sum_{\mathbf{l},\nabla\cdot\mathbf{l}=0} e^{-(4\pi^2\beta a/2)v_m(\mathbf{0})\Sigma_{\mathbf{x}}\,\mathbf{l}^2(\mathbf{x})}, \tag{5.107}$$

we estimate the critical value $\beta_{m,c}$ of the Yukawa loop gas by the equation corresponding to (5.104):

$$4\pi^2 a \beta_{m,c} v_m(\mathbf{0}) \approx T_c \approx 3. \tag{5.108}$$

Since the Yukawa potential becomes more and more local for increasing m, the local approximation (5.107) becomes exact. Thus we conclude that the error in the estimate of the critical temperature $T_{m,c} = 1/\beta_{m,c}$ from Eq. (5.108) drops from 10% at $m = 0$ to zero as m goes to infinity. We have plotted the resulting critical values of $T_{m,c} = 1/\beta_{m,c}$ in Fig. 5.7.

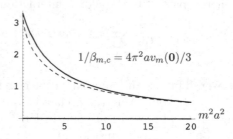

FIGURE 5.7   Critical temperature $1/\beta_{m,c}$ of a loop gas with Yukawa interactions between line elements, estimated by Eq. (5.108). The error is with 10% the largest at $m = 0$, and decreases to zero for increasing m. The dashed curve is the analytic approximation (5.112).

We conclude that the Yukawa loop gas (5.106) has a second-order phase transition as the Villain- and the XY-models. The critical exponents of the Yukawa loop gas are all of the same as those of the Villain-model, and thus also of the XY-model. In the terminology of the theory of critical phenomena, the Yukawa loop gases lie, for all m, in the same universality class as the XY-model.

It is possible to find a simple analytic approximation for the critical temperature plotted in Fig. 5.7. For this we use the so-called *hopping expansion* [7] of the lattice

Yukawa potential (5.56). It is found by expanding the modified Bessel function $I_{x_i/a}(2s)$ in Eq. (5.56) in powers of s using the series representation

$$I_n(2s) = \sum_{k=0}^{\infty} \frac{s^{2k}}{k!\Gamma(n+k+1)}. \tag{5.109}$$

At the origin $\mathbf{x} = 0$, the integral over s in (5.56) yields the expansion

$$v_m(\mathbf{0}) = \frac{1}{a} \sum_{n=0,2,4} \frac{H_n}{(m^2a^2+6)^{n+1}}, \quad H_0 = 1, H_2 = 6,\ldots. \tag{5.110}$$

To lowest order, this implies the approximate ratio $v_m(\mathbf{0})/v_0(\mathbf{0}) \equiv 1/(m^2a^2/6+1)$. A somewhat more accurate fit to the ratio is

$$\frac{v_m(\mathbf{0})}{v_0(\mathbf{0})} \approx \frac{1}{\sigma\, m^2a^2/6+1}, \quad \text{with} \quad \sigma \approx 1.6. \tag{5.111}$$

Together with (5.105) this leads to the analytic approximation

$$T_{m,c} = \frac{1}{\beta_{m,c}} \approx \frac{4\pi^2 a v_m(\mathbf{0})}{3} \approx \frac{4\pi^2 a v_0(\mathbf{0})}{3} \frac{1}{\sigma\, m^2a^2/6+1}. \tag{5.112}$$

A comparison with the numerical evaluation of (5.108) is shown in Fig. 5.7. The fit has only a 10% error for $m = 0$ and becomes accurate for large m.

5.1.9   Gauge Field of Superflow

The current conservation law $\boldsymbol{\nabla}\cdot\mathbf{b}(\mathbf{x}) = 0$ can be ensured automatically as a Bianchi identity, if we represent $\mathbf{b}(\mathbf{x})$ as a curl of a gauge field of superflow

$$\mathbf{b}(\mathbf{x}) = \boldsymbol{\nabla} \times \mathbf{a}(\mathbf{x}). \tag{5.113}$$

The energy (5.86), with the core energy reinserted, goes over into what is called the *dual representation*:

$$\beta H_{\text{avc}} = \int d^3x \left[\frac{1}{2\beta}(\boldsymbol{\nabla}\times\mathbf{a})^2 + i\mathbf{a}\cdot(\boldsymbol{\nabla}\times\boldsymbol{\theta}^{\text{v}}) + \frac{\beta\epsilon_c}{2}(\boldsymbol{\nabla}\times\boldsymbol{\theta}^{\text{v}})^2\right]. \tag{5.114}$$

The second term is obtained after a partial integration of $\int d^3x\, i(\boldsymbol{\nabla}\times\mathbf{a})\cdot\boldsymbol{\theta}^{\text{v}}$.

This form of the energy is now double-gauge invariant. Apart from the invariance under the vortex gauge transformation (5.28), there is now the additional invariance under the gauge transformations of superflow

$$\mathbf{a}(\mathbf{x}) \to \mathbf{a}(\mathbf{x}) + \boldsymbol{\nabla}\Lambda(\mathbf{x}), \tag{5.115}$$

with arbitrary functions $\Lambda(\mathbf{x})$.

The energy (5.114) can be expressed in terms of the vortex density of Eq. (5.30) as

$$\beta H'_{\text{avc}} = \int d^3x \left[\frac{1}{2\beta} (\boldsymbol{\nabla} \times \mathbf{a})^2 + i\mathbf{a} \cdot \mathbf{j}^{\text{v}} + \frac{\beta \epsilon_c}{2} \mathbf{j}^{\text{v}2} \right]. \tag{5.116}$$

In this expression, the freely deformable jumping surfaces have disappeared and the energy depends only on the vortex lines. For a fixed set of vortex lines along L, the energy (5.116) has a similar form as the free magnetic energy of a given current distribution in Eq. (4.94). The only difference is a factor i. Around a vortex line, the field $\mathbf{b}(\mathbf{x}) = \boldsymbol{\nabla} \times \mathbf{a}(\mathbf{x})$ looks precisely like a magnetic field $\mathbf{B}(\mathbf{x}) = \boldsymbol{\nabla} \times \mathbf{A}(\mathbf{x})$ around a current line, except for the factor i. Extremizing the energy in $\mathbf{a}$ and reinserting the extremum yields once more the Biot-Savart interaction energy of the form Eq. (5.74) [which is of the form (4.93), not (4.95) due to the factor i].

If we want to express the partition function (5.85) in terms of the gauge field of superflow $\mathbf{a}(\mathbf{x})$, we must fix its gauge. Here we may choose the transverse gauge:

$$\Phi_T[\mathbf{a}] = \delta[\boldsymbol{\nabla} \cdot \mathbf{a}], \tag{5.117}$$

and the partition function (5.85) becomes

$$Z_{\text{v}}^{\text{hy}} = \int_{-\infty}^{\infty} \mathcal{D}\mathbf{a} \, \Phi_T[\mathbf{a}] \sum_{\{S\}} \Phi[\boldsymbol{\theta}^{\text{v}}] e^{-\beta H_{\text{avc}}}. \tag{5.118}$$

In terms of the Hamiltonian (5.116), the partition function becomes a sum over vortex lines L:

$$Z_{\text{v}}^{\text{hy}} = \int_{-\infty}^{\infty} \mathcal{D}\mathbf{a} \, \Phi_T[\mathbf{a}] \sum_{\{L\}} \Phi_T[\mathbf{j}^{\text{v}}] e^{-\beta H'_{\text{avc}}}, \tag{5.119}$$

where

$$\Phi_T[\mathbf{j}^{\text{v}}] = \delta[\boldsymbol{\nabla} \cdot \mathbf{j}^{\text{v}}] \tag{5.120}$$

ensures the closure of the vortex lines.

Note that if the energies $H_{\text{vc}}^{\text{hy}}$ or H_{v}^{hy} in (5.32) and (5.67) contain an explicit θ-dependent term, such as the mass term in the Hamiltonian (5.75), there exists no reformulation of the θ-fluctuations in terms of a gauge field $\mathbf{a}$. For a mass term, the formula (5.94) turns into

$$\int_{-\infty}^{\infty} \mathcal{D}\theta \, e^{-\int d^3x \left[\beta m^2 \theta^2(\mathbf{x})/2 - if(\mathbf{x})\theta(\mathbf{x}) \right]} = \delta_m[f(\mathbf{x})], \tag{5.121}$$

where $\delta_m[f(\mathbf{x})]$ denotes the softened δ-functional

$$\delta_m[f(\mathbf{x})] \propto e^{-\int d^3x \, f^2(\mathbf{x})/2\beta m^2}. \tag{5.122}$$

For $f(\mathbf{x}) = \boldsymbol{\nabla} \cdot \mathbf{b}(\mathbf{x})$ this implies that $\mathbf{b}(\mathbf{x})$ is no longer purely transverse, as in (5.95). Hence it no longer possesses a curl representation (5.113).

5.1.10 Disorder Field Theory

In order to understand the thermal behavior of the partition function (5.119) it is useful to study separately the sum over all vortex line configurations at a fixed vector potential **a**. Thus we consider the **a**-dependent vortex partition function

$$Z^v[\mathbf{a}] = \sum_{\{L\}} \delta[\nabla \cdot \mathbf{j}^v] \exp\left[-\int d^3x \left(\frac{\beta\epsilon_c}{2}\mathbf{j}^{v2} - i\,\mathbf{a}\cdot\mathbf{j}^v\right)\right]. \qquad (5.123)$$

It is possible to re-express this with the help of an auxiliary fluctuating vortex gauge field $\tilde{\boldsymbol{\theta}}^v(\mathbf{x})$, which is singular on surfaces $\tilde{S}$, as a sum over auxiliary surface configurations $\tilde{S}$ as follows

$$Z^v[\mathbf{a}] = \sum_{\{\tilde{S}\}} \int \mathcal{D}\mathbf{j}^v \delta[\nabla \cdot \mathbf{j}^v] \exp\left\{-\int d^3x \left[\frac{\beta\epsilon_c}{2}\mathbf{j}^{v2} - i\,\mathbf{j}^v\cdot\left(\tilde{\boldsymbol{\theta}}^v + \mathbf{a}\right)\right]\right\}. \qquad (5.124)$$

In this expression, $\mathbf{j}^v$ is an ordinary field. The sum over all $\tilde{S}$-configuration ensures via formula (5.98) that the functional integral over $\mathbf{j}^v$ really represents a sum over δ-functions on lines $\tilde{L}$, so that (5.124) is the same as (5.123), up to an irrelevant overall factor.

Next we introduce an auxiliary field $\tilde{\theta}$, and rewrite the δ functional of the divergence of $\mathbf{j}^v$ as a functional Fourier integral, so that we obtain the identity

$$Z^v[\mathbf{a}] = \sum_{\{\tilde{S}\}} \int \mathcal{D}\mathbf{j}^v \int \mathcal{D}\tilde{\theta} \exp\left\{-\int d^3x \left[\frac{\beta\epsilon_c}{2}\mathbf{j}^{v2} + i\,\mathbf{j}^v\cdot\left(\nabla\tilde{\theta} - \tilde{\boldsymbol{\theta}}^v - \mathbf{a}\right)\right]\right\}. \qquad (5.125)$$

Now $\mathbf{j}^v$ is a completely unrestricted ordinary field. It can therefore be integrated to yield

$$Z^v[\mathbf{a}] = \sum_{\{\tilde{L}\}} \int \mathcal{D}\tilde{\theta} \exp\left[-\frac{1}{2\beta\epsilon_c}\int d^3x (\nabla\tilde{\theta} - \tilde{\boldsymbol{\theta}}^v - \mathbf{a})^2\right]. \qquad (5.126)$$

Remembering the derivation of the Hamiltonian (5.26) from the hydrodynamic limit of the Ginzburg-Landau $|\phi|^4$ theory (5.6), we may interprete (5.126) as the partition function of the hydrodynamic limit of another U(1)-invariant Ginzburg-Landau theory whose partition function is given by the functional integral

$$\tilde{Z}^v[\mathbf{a}] = \int \mathcal{D}\psi\mathcal{D}\psi^* \exp\left\{-\frac{1}{2\beta}\int d^3x \left[|(\nabla - i\mathbf{a})\,\psi|^2 + m^2\,|\psi|^2 + \frac{g}{2}\,|\psi|^4\right]\right\}, \qquad (5.127)$$

where $\psi(\mathbf{x})$ is another complex field ψ with a $|\psi|^4$ interaction. Inserting this into (5.119) we obtain the combined partition function

$$Z_v^{\text{hy}} = \int_{-\infty}^{\infty} \mathcal{D}\mathbf{a}\, \Phi_T[\mathbf{a}]\tilde{Z}^v[\mathbf{a}] \exp\left\{-\int d^3x \left[\frac{1}{2\beta}(\nabla\times\mathbf{a})^2\right]\right\}, \qquad (5.128)$$

which defines the desired *disorder field theory*.

The representation of ensembles of lines in terms of a single disorder field is the Euclidean version of what is known as *second quantization* in the quantum mechanics of many-particle systems.

At high temperature, the mass term m^2 of the ψ-field is negative and the disorder field acquires a nonzero expectation value $\psi_0 = \sqrt{-m^2/g}$. Setting, as in (5.10),

$$\psi(\mathbf{x}) = \tilde{\rho}(\mathbf{x})e^{i\tilde{\theta}(\mathbf{x})} \tag{5.129}$$

and freezing out the fluctuations of ρ leads directly to the partition function (5.126).

The disorder field theory possesses similar vortex lines as the original Ginzburg-Landau theory with Hamiltonian (5.6), or its hydrodynamic limit (5.32). But in contrast to it, the fluctuations of the disorder field are "frozen out" at high temperature, as we can see from the prefactor $1/\beta$ in the exponents of (5.126) and (5.127), and the partition function (5.127) reduces to (5.126) in the hydrodynamic limit. As before in (5.66) we may perform the functional integral over $\tilde{\theta}$. This removes the longitudinal part of $\tilde{\boldsymbol{\theta}}^{\mathrm{v}} - \mathbf{a}$, and (5.126) becomes

$$Z^{\mathrm{v}}[\mathbf{a}] = \exp\left[-\frac{m_a^2}{2\beta}\int d^3x \left(\tilde{\boldsymbol{\theta}}^{\mathrm{v}} - \mathbf{a}\right)_T^2\right], \tag{5.130}$$

where

$$m_a^2 = \frac{1}{\epsilon_c}, \tag{5.131}$$

and $\mathbf{v}_T$ denotes the transverse part of the vector field $\mathbf{v}$. This and the longitudinal part $\mathbf{v}_L$ are defined by

$$v_{Ti} \equiv \left(\delta_{ij} - \frac{\nabla_i \nabla_j}{\nabla^2}\right)v_j, \qquad v_{Li} \equiv \frac{\nabla_i \nabla_j}{\nabla^2}v_j. \tag{5.132}$$

At high temperatures, where the disorder field ψ has no vortex lines $\tilde{L}$ (while the order field ϕ has many vortex lines L), the partition function (5.130) becomes

$$Z^{\mathrm{v}}[\mathbf{a}] \approx \exp\left(-\frac{m_a^2}{2\beta}\int d^3x\, \mathbf{a}_T^2\right), \tag{5.133}$$

and the exponent gives a mass to the transverse part $\mathbf{a}_T$ of the gauge field of superflow. Recalling the gradient term $(1/\beta)(\boldsymbol{\nabla} \times \mathbf{a})^2$ of the $\mathbf{a}$-field in (5.128) we see that the mass has the value m_a.

Having obtained this result we go once more back to the expression (5.123) and realize that the same mass can also be obtained from $Z^{\mathrm{v}}[\mathbf{a}]$ by simply ignoring the δ-function nature of $\mathbf{j}^{\mathrm{v}}(\mathbf{x}) = 2\pi\boldsymbol{\delta}(\mathbf{x}, L)$ and integrating $\mathbf{j}^{\mathrm{v}}(\mathbf{x})$ out using the Gaussian formula (5.72). With such an approximate treatment, the partition function (5.123) yields, for the vortex density, the simple correlation function

$$\langle j_i^{\mathrm{v}}(\mathbf{x})j_j^{\mathrm{v}}(\mathbf{x}')\rangle = \frac{1}{\epsilon_c}\left(\delta_{ij} - \frac{\nabla_i \nabla_j}{\nabla^2}\right)\delta^{(3)}(\mathbf{x} - \mathbf{x}'). \tag{5.134}$$

The reason why this simplification is applicable in the high-temperature phase is easy to understand. On a lattice, the sums over lines L in (5.123) correspond to Gaussian sums of the form $\sum_{n_i=-\infty}^{\infty} e^{-\beta\epsilon_c 4\pi n_i^2/2}$ at each $\mathbf{x}, i$. At high temperatures where β is small, the sum over n_i can obviously be replaced by $1/\sqrt{\beta}$ times an integral over the quasi-continuous variable $\nu_i \equiv \sqrt{\beta} n_i$. In general, if lines or surfaces of volumes are prolific, the statistical mechanics of fields which are proportional to the corresponding δ-functions $\boldsymbol{\delta}(\mathbf{x}; L)$, $\boldsymbol{\delta}(\mathbf{x}; S)$, $\delta(\mathbf{x}; V)$ can be treated as if they were ordinary fields. The sums of the geometric configurations turn into functional integrals.

The same mass generation can, of course, be observed in the complex disorder field theory (5.127). At high temperature, the mass term m^2 of the ψ-field is negative and the disorder field acquires a nonzero expectation value $\psi_0 = \sqrt{-m^2/g}$. This produces again the mass term (5.130) with $m_a^2 = \psi_0^2$.

Let us now look at the low-temperature phase. There the δ-function nature of the density $\mathbf{j}^{\mathrm{v}}(\mathbf{x}) = 2\pi\boldsymbol{\delta}(\mathbf{x}; L)$ cannot be ignored in the partition function (5.123). At low temperatures, vortex lines appear only as small loops. An infinitesimal loop gives a simple curl contribution [22]

$$Z^{\mathrm{v}}[\mathbf{a}] \sim \exp\left[-\frac{1}{2\beta} \int d^3x \left(\boldsymbol{\nabla} \times \mathbf{a}\right)^2\right], \qquad (5.135)$$

whereas larger loops contribute

$$Z^{\mathrm{v}}[\mathbf{a}] \sim \exp\left[-\frac{1}{2} \int d^3x \left(\boldsymbol{\nabla} \times \mathbf{a}\right) f(-i\boldsymbol{\nabla})(\boldsymbol{\nabla} \times \mathbf{a})\right], \qquad (5.136)$$

with $f(\mathbf{k})$ being some smooth function of $\mathbf{k}$ starting out with a constant, the so-called *stiffness* of the $\mathbf{a}$-field. Hence the contributions of small vortex loops change only the *dispersion* of the gauge fields of superflow. Infinitely long vortex lines in $\boldsymbol{\theta}^{\mathrm{v}}$ are necessary to produce a mass term. These appear when the temperature is raised above the critical point, in particular at high temperatures, where the correlation function of the vortex densities is approximately given by (5.134), and (5.123) leads directly to the mass term in (5.133).

With the help of the disorder partition function $Z^{\mathrm{v}}[\mathbf{a}]$, the partition function (5.33) can be replaced by the completely equivalent dual partition function

$$\tilde{Z}_{\mathrm{v}}^{\mathrm{hy}} = \int_{-\infty}^{\infty} \mathcal{D}\mathbf{a}\Phi_T[\mathbf{a}] \sum_{\{\tilde{S}\}} \Phi[\tilde{\boldsymbol{\theta}}^{\mathrm{v}}] \int_{-\infty}^{\infty} \mathcal{D}\tilde{\theta}\, e^{-\beta\tilde{H}_{\mathrm{v}}^{\mathrm{hy}}} \qquad (5.137)$$

with the exponent

$$\beta\tilde{H}_{\mathrm{v}}^{\mathrm{hy}} = \frac{1}{2\beta} \int d^3x \left[(\boldsymbol{\nabla} \times \mathbf{a})^2 + m_a^2 \left(\boldsymbol{\nabla}\tilde{\theta} - \tilde{\boldsymbol{\theta}}^{\mathrm{v}} - \mathbf{a}\right)^2\right]. \qquad (5.138)$$

This energy is invariant under the following two gauge transformations. First, there is invariance under the gauge transformations of superflow (5.115), if it is accompanied by a compensating transformation of the angular field $\tilde{\theta}$:

$$\mathbf{a}(\mathbf{x}) \to \mathbf{a}(\mathbf{x}) + \boldsymbol{\nabla}\Lambda(\mathbf{x}), \qquad \tilde{\theta}(\mathbf{x}) \to \tilde{\theta}(\mathbf{x}) + 2\pi\Lambda(\mathbf{x}). \qquad (5.139)$$

Second, there is gauge invariance under the vortex gauge transformations of the form (5.28), here applied to the disorder field:

$$\tilde{\theta}^{\text{v}}(\mathbf{x}) \to \tilde{\theta}^{\text{v}}(\mathbf{x}) + \boldsymbol{\nabla}\tilde{\Lambda}_{\delta}^{\text{v}}(\mathbf{x}), \qquad \tilde{\theta}(\mathbf{x}) \to \tilde{\theta}(\mathbf{x}) + \tilde{\Lambda}_{\delta}^{\text{v}}(\mathbf{x}), \qquad (5.140)$$

with gauge functions

$$\tilde{\Lambda}^{\text{v}}(\mathbf{x}) = 2\pi\delta(\mathbf{x}; \tilde{V}). \qquad (5.141)$$

At high temperatures, the vortex lines in $\tilde{\theta}^{\text{v}}$ are frozen out and the energy (5.138) shows again the mass term (5.130).

The mass term implies that at high temperatures, the gauge field of super-flow possesses a finite range. At some critical temperature superfluidity has been destroyed. This is the disorder analog of the famous Meissner effect in superconduc-tors [23], to be discussed in Section 5.2.1. Without the gauge field of superflow $\mathbf{a}$, the field $\tilde{\theta}$ would be of long range, i.e., massless. The gauge field of superflow absorbs this massless mode and the system has only short-range excitations. More pre-cisely, it can be shown that all correlation functions involving local gauge-invariant observable quantities must be of short range in the high-temperature phase.

Take, for instance, the local gauge-invariant current operator of the disorder field

$$\mathbf{j}_s \equiv \boldsymbol{\nabla}\tilde{\theta} - \mathbf{a}. \qquad (5.142)$$

Choosing $\tilde{\theta}$ to absorb the longitudinal part of $\mathbf{a}$, only the transverse part of $\mathbf{a}$ re-mains in (5.142), which becomes $\mathbf{j}^s = -\mathbf{a}_T$. [24]. From the Hamiltonian (5.138) we immediately find the free correlation function of superflow:

$$\langle j_i{}^s(\mathbf{x}_1)j_j{}^s(\mathbf{x}_2)\rangle \propto \int \frac{d^3k}{(2\pi)^3} \frac{\delta_{ij} - k_ik_j/m_a^2}{\mathbf{k}^2 + m_a^2} e^{i\mathbf{k}(\mathbf{x}_1 - \mathbf{x}_2)}, \qquad (5.143)$$

which has *no* zero-mass pole.

5.2   Phase Transition in Superconductor

The specific heat of a superconductor is shown in Fig. 5.8. It looks quite different from that of helium on p. 131. It starts out with a behavior typical for an activation process, which is governed by a Boltzmann factor $c_s \propto e^{-\Delta(0)/k_BT}$, where k_B is the Boltzmann constant. The activation energy $\Delta(0)$ shows the *energy gap* in the electron spectrum at $T = 0$. It is equal to the binding energy of the *Cooper pairs* formed from electrons of opposite momentum near the Fermi sphere. At the critical temperature T_c, the specific heat drops down to the specific heat of a free electron gas

$$c_n = \frac{2}{3}\pi^2\mathcal{N}(0)T, \qquad (5.144)$$

where [25]

$$\mathcal{N}(0) = \frac{3n_e}{4\epsilon_F} = \frac{3mn_e}{2p_F^2} = \frac{3n_e}{2mv_F^2} \qquad (5.145)$$

is the density of electrons of mass m at the surface of the Fermi sphere of energy ϵ_F and momentum p_F, velocity $v_F = p_F/m$, and n_e is the density of electrons of both spin directions. The Fermi velocity v_F is typically of the order 10^8 (cm/sec)($\sim$ c/300).

According to the theory of Bardeen, Cooper, and Schrieffer (BCS) [26], the jump is given by the universal law [to be derived in Eq. (7A.24)]

$$\frac{c_s - c_n}{c_n} \equiv \frac{\Delta c}{c_n} = \frac{3}{2}\frac{8}{7\zeta(3)} \equiv 1.4261\ldots, \tag{5.146}$$

where $\zeta(3)$ is Riemann's zeta function $\zeta(z) \equiv \sum_{n=1}^{\infty} n^{-z}$, with $\zeta(3) = 1.202057\ldots$. This jump agrees perfectly with the experiment.

In the BCS-theory there exists a universal ratio between the gap $\Delta(0)$ and T_c:

$$\frac{\Delta(0)}{T_c} = \pi e^{-\gamma} \approx 1.76, \tag{5.147}$$

where $\gamma \approx 0.577\ldots$ is the *Euler-Mascheroni constant*. This ratio is also observed in Fig. 5.8.

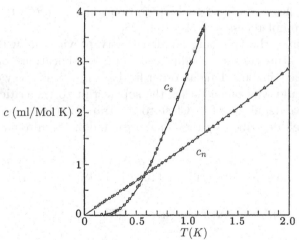

FIGURE 5.8 Specific heat of superconducting aluminum [N.E. Phillips, Phys. Rev. **114**, 676 (1959)]. For very small T, it shows the typical power behavior $e^{-\Delta(0)/k_B T}$ instead of the power behavior in superfluid helium. From the curve we extract $\Delta(0) \equiv 2.04\,\mathrm{K}$. At the critical temperature $T_c \approx 1.2\,\mathrm{K}$, there is a jump down to the linear behavior characteristic for a free electron gas. The ratios $\Delta c/c_n = 1.43$ and $T_c \approx \Delta(0)/1.76$, agree well with the BCS results (5.146) and (5.147) [26]. A normal metal shows only the linear behavior labeled by c_s.

5.2.1   Ginzburg-Landau Theory

The BCS theory can be used to derive the *Ginzburg-Landau Hamiltonian* for the superconducting phase transition [5, 27]

$$H_{\text{HL}}[\psi, \psi^*, \mathbf{A}] = \frac{1}{2} \int d^3x \left\{ |(\boldsymbol{\nabla} - iq\mathbf{A})\psi|^2 + \tau|\psi|^2 + \frac{g}{2}|\psi|^4 + (\boldsymbol{\nabla} \times \mathbf{A})^2 \right\} \quad (5.148)$$

governing the neighborhood of the critical point. The parameter q is the charge of the ψ-field, and τ may be identified with $\tilde{T}/\tilde{T}_c^{\text{MF}} - 1$, the relative temperature distance from the critical point. It is positive in the normal state and negative in the superconducting state. The field $\psi(\mathbf{x})$ is a so-called *collective field* describing the Cooper pairs of electrons of opposite momenta slightly above and below the Fermi sphere [28]. The Cooper pairs carry a charge twice the electron charge, $q = 2e$. In (5.148), they are coupled minimally to the vector potential $\mathbf{A}(\mathbf{x})$. For simplicity, we have set the light velocity c equal to unity. The size of ψ is equal to the energy gap in the electron spectrum, and as such to the binding energy of the electrons to Cooper pairs.

Ginzburg and Landau [29] found the Hamiltonian (5.148) by a formal expansion of the energy in powers of the energy gap which they considered as an order parameter. They convinced themselves that for small τ only the terms up to ψ^4 would be important. To this truncated expansion they added a gradient term to allow for spatial variations of the order parameter, making it an *order field* denoted by $\psi(\mathbf{x})$. There exists an elegant derivation of the Ginzburg-Landau Hamiltonian (5.148) from the BCS theory via functional integration which is briefly recapitulated in Appendix 7A, for completeness (see also Ref. [28]).

In the critical regime, the Ginzburg-Landau theory provides us with a simple explanation of many features of superconductors. In most applications, one may neglect fluctuations of the Ginzburg-Landau order field $\phi(\mathbf{x})$, which is why one speaks of mean-field results, and why one attaches the superscript to the critical temperature $\tilde{T}_c^{\text{MF}}$ in the above definition of τ. Close to the transition, the properties of a superconductor are well described by the Ginzburg-Landau Hamiltonian [compare (5.127)].

The Ginzburg-Landau Hamiltonian (5.148) possesses a conserved supercurrent which is found by applying Noether's rule (3.116) to (5.148). The current density is

$$\mathbf{j}(\mathbf{x}, t) \equiv \frac{1}{2i} \left(\psi^\dagger(\mathbf{x}, t)[\boldsymbol{\nabla} - iq\mathbf{A}(\mathbf{x}, t)]\psi(\mathbf{x}, t) - \{[\boldsymbol{\nabla} - iq\mathbf{A}(\mathbf{x}, t)]\,\psi(\mathbf{x}, t)\}^\dagger \psi \right)$$

$$= \frac{i}{2}\psi^*(\mathbf{x}, t) \overleftrightarrow{\boldsymbol{\nabla}} \psi(\mathbf{x}, t) - q\mathbf{A}\,\psi^*(\mathbf{x}, t)\psi(\mathbf{x}, t). \quad (5.149)$$

This differs from the Schrödinger current density (3.118) by the use of natural units $m = 1$, $c = 1$, and by the fact that the charge q is equal to $2e$ for the Cooper pairs.

Let us now proceed as in (5.10) and (5.129) and decompose the field ψ as

$$\psi(\mathbf{x}) = \tilde{\rho}(\mathbf{x})\,e^{i\tilde{\theta}(\mathbf{x})}. \quad (5.150)$$

Inserting this into (5.148) and remembering (5.22), we find

$$H_{\mathrm{GL}}[\tilde{\rho},\tilde{\theta},\tilde{\boldsymbol{\theta}}^{\mathrm{v}},\mathbf{A}] = \int d^3x \left[\frac{\tilde{\rho}^2}{2}(\boldsymbol{\nabla}\tilde{\theta} - \tilde{\boldsymbol{\theta}}^{\mathrm{v}} - q\mathbf{A})^2 + \frac{1}{2}(\boldsymbol{\nabla}\tilde{\rho})^2 + V(\tilde{\rho}) + \frac{1}{2}(\boldsymbol{\nabla}\times\mathbf{A})^2 \right], \quad (5.151)$$

where $V(\tilde{\rho})$ is the potential of the $\tilde{\rho}$-field:

$$V(\tilde{\rho}) = \frac{\tau}{2}\tilde{\rho}^2 + \frac{g}{4}\tilde{\rho}^4. \qquad (5.152)$$

In the low-temperature phase we go to the hydrodynamic limit by setting $\tilde{\rho}(\mathbf{x})$ equal to its value $\tilde{\rho}_0 = \sqrt{-\tau/g}$ at the minimum of the energy (5.148). The resulting hydrodynamic or London energy of the superconductor is

$$H_{\mathrm{SC}}^{\mathrm{hy}}[\tilde{\theta},\tilde{\boldsymbol{\theta}}^{\mathrm{v}},\mathbf{A}] = \int d^3x \left[\frac{m_A^2}{2q^2}(\boldsymbol{\nabla}\tilde{\theta} - \tilde{\boldsymbol{\theta}}^{\mathrm{v}} - q\mathbf{A})^2 + \frac{1}{2}(\boldsymbol{\nabla}\times\mathbf{A})^2 \right]. \qquad (5.153)$$

where we have introduced a mass parameter

$$m_A^2 = n_0 q^2 \qquad (5.154)$$

proportional to the density of superfluid particles

$$n_0 = \tilde{\rho}_0^2. \qquad (5.155)$$

At very low temperatures where vortices are absent, the first term in (5.153) makes the transverse part of the vector field massive. This causes a finite *penetration depth* $\lambda = 1/m_A$ of the magnetic field in a superconductor, thus explaining the famous Meissner effect of superconductivity [30].

This mechanism is imitated in the standard model of electromagnetic and weak interactions to give the vector mesons $W^{+,0,-}$ and Z a finite mass, thereby explaining the strong suppression of weak with respect to electromagnetic interactions. There the Meissner effect is called *Higgs effect*.

In the same limit, the current density of superfluid particles becomes

$$\mathbf{j}_s = n_0(\boldsymbol{\nabla}\tilde{\theta} - \tilde{\boldsymbol{\theta}}^{\mathrm{v}} - q\mathbf{A}). \qquad (5.156)$$

The partition function reads

$$Z_{\mathrm{SC}}^{\mathrm{hy}} = \int \mathcal{D}\mathbf{A}\,\Phi_T[\mathbf{A}] \sum_{\{\tilde{S}\}} \Phi[\tilde{\boldsymbol{\theta}}^{\mathrm{v}}] \int_{-\infty}^{\infty} \mathcal{D}\theta\, e^{-\tilde{\beta}H_{\mathrm{SC}}^{\mathrm{hy}}[\tilde{\theta},\tilde{\boldsymbol{\theta}}^{\mathrm{v}},\mathbf{A}]}. \qquad (5.157)$$

To distinguish this discussion from the previous one of superfluid helium we call the temperature of the superconductor $\tilde{T}$ and its inverse $\tilde{\beta}$.

The energy (5.153) has the same form as the energy in the disorder representation (5.138) of superfluid ^{4}He. The role of the gauge field of superflow is now played by the vector potential $\mathbf{A}$ of magnetism. The energy has the following two types of gauge symmetries: the magnetic invariance

$$\mathbf{A}(\mathbf{x}) \to \mathbf{A}(\mathbf{x}) + q^{-1}\boldsymbol{\nabla}\Lambda(\mathbf{x}), \qquad \tilde{\theta}(\mathbf{x}) \to \tilde{\theta}(\mathbf{x}) + \Lambda(\mathbf{x}), \qquad (5.158)$$

and the vortex gauge invariance

$$\tilde{\boldsymbol{\theta}}^{\mathrm{v}}(\mathbf{x}) \to \tilde{\boldsymbol{\theta}}^{\mathrm{v}}(\mathbf{x}) + \partial_i\tilde{\Lambda}_\delta(\mathbf{x}), \qquad \tilde{\theta}(\mathbf{x}) \to \tilde{\theta}(\mathbf{x}) + \tilde{\Lambda}_\delta(\mathbf{x}), \qquad (5.159)$$

with gauge functions

$$\tilde{\Lambda}_\delta(\mathbf{x}) = 2\pi\delta(\mathbf{x}; \tilde{V}). \qquad (5.160)$$

As in the description of superfluid ^{4}He with the partition function (5.33), the partition function (5.157) gives us the statistical behavior of the superconductor not only at zero temperature, where the energy (5.153) was constructed, but at all not too large temperatures. The fluctuating vortex gauge field $\tilde{\boldsymbol{\theta}}^{\mathrm{v}}$ ensures the validity through the phase transition.

5.2.2   Disorder Theory of Superconductor

We shall now derive the disorder representation of this partition function in which the vortex lines of the superconductor play a central role in describing the phase transition [23].

At low temperatures, the vortices are frozen, and the $\tilde{\theta}$-fluctuations in the partition function (5.157) can be integrated out. This reduces the energy (5.153) to

$$H_{\mathrm{SC}}^{\mathrm{hy}}[\mathbf{A}] \approx \int d^3x \left[\frac{m_A^2}{2}\mathbf{A}_T^2 + \frac{1}{2}(\boldsymbol{\nabla} \times \mathbf{A})^2 \right]. \qquad (5.161)$$

i.e., to a free vector potential $\mathbf{A}$ with a transverse mass term. This is the famous Meissner effect in a superconductor, which limits the range of a magnetic field to a finite penetration depth $\lambda = 1/m_A$. The effect is completely analogous to the one observed previously in the disorder description of the superfluid where the superfluid acquired a finite range in the normal phase.

To derive the disorder theory of the partition function (5.157), we supplement the energy (5.153) by a core energy of the vortex lines

$$H_{\mathrm{c}} = \frac{\tilde{\epsilon}_c}{2} \int d^3x \, (\boldsymbol{\nabla} \times \tilde{\boldsymbol{\theta}}^{\mathrm{v}})^2. \qquad (5.162)$$

As in the partition function (5.85), an auxiliary $\tilde{b}_i$ field can be introduced to bring the exponent in (5.153) to the canonical form

$$\tilde{\beta}H_{\mathrm{SC}}^{\mathrm{hy}} = \int d^3x \left[\frac{1}{2\tilde{\beta}m_A^2}\tilde{\mathbf{b}}^2 + i\tilde{\mathbf{b}}\left(\boldsymbol{\nabla}\tilde{\theta} - \tilde{\boldsymbol{\theta}}^{\mathrm{v}} - q\mathbf{A}\right) + \frac{\tilde{\beta}}{2}\left(\boldsymbol{\nabla} \times \mathbf{A}\right)^2 + \frac{\tilde{\beta}\tilde{\epsilon}_c}{2}(\boldsymbol{\nabla} \times \tilde{\boldsymbol{\theta}}^{\mathrm{v}})^2 \right]. (5.163)$$

By integrating out the $\tilde{\theta}$-fields in the associated partition function, we obtain the conservation law

$$\boldsymbol{\nabla} \cdot \tilde{\mathbf{b}} = 0, \qquad (5.164)$$

which is fulfilled by expressing $\tilde{\mathbf{b}}$ as a curl of the gauge field $\tilde{\mathbf{a}}$ of superflow in the superconductor

$$\tilde{\mathbf{b}} = \boldsymbol{\nabla} \times \tilde{\mathbf{a}}. \tag{5.165}$$

This brings the energy to the form

$$\tilde{\beta} H_{\mathrm{SC}}^{\mathrm{hy}} = \int d^3x \left[\frac{1}{2\tilde{\beta}m_A^2} (\boldsymbol{\nabla} \times \tilde{\mathbf{a}})^2 - iq\tilde{\mathbf{a}} \cdot (\boldsymbol{\nabla} \times \mathbf{A}) + \frac{\tilde{\beta}}{2} (\boldsymbol{\nabla} \times \mathbf{A})^2 - i\tilde{\mathbf{a}} \cdot \tilde{\mathbf{j}}^{\mathrm{v}} + \frac{\tilde{\beta}\tilde{\epsilon}_c}{2} \tilde{\mathbf{j}}^{\mathrm{v}2} \right], \tag{5.166}$$

where

$$\tilde{\mathbf{j}}^{\mathrm{v}} = \boldsymbol{\nabla} \times \tilde{\boldsymbol{\theta}}^{\mathrm{v}} \tag{5.167}$$

is the vortex density in the superconductor. At low temperatures where $\tilde{\beta}$ is large and the vortex lines are frozen out, the last two terms in the Hamiltonian can be neglected. Integrating out the $\tilde{\mathbf{a}}$-field we re-obtain the transverse mass term (5.161) of the Meissner effect. At high temperatures, on the other hand, the vortex lines are prolific and the vortex density $\tilde{\mathbf{j}}^{\mathrm{v}}$ can be integrated out in the associated partition function like an ordinary field using the analog of the correlation function (5.134). This produces the transverse mass term

$$\frac{1}{2\tilde{\beta}m_A^2} \int d^3x\, m_{\tilde{a}}^2\, \tilde{\mathbf{a}}_T^2 \tag{5.168}$$

where the mass $m_{\tilde{a}}$ of the $\tilde{\mathbf{a}}_T$-field is given by

$$m_{\tilde{a}}^2 = q^2 m_A^2 / \tilde{\epsilon}_c. \tag{5.169}$$

Such a mass term can immediately be seen to destroy the Meissner effect in the superconductor at high temperature. Indeed, inserting the curl (5.165) into the energy (5.163), and using the result (5.168), we obtain at high $\tilde{T}$:

$$\tilde{\beta}\tilde{H}_{\mathrm{SC}}^{\mathrm{hy}} = \int d^3x \left[\frac{1}{2\tilde{\beta}m_A^2} \left[(\boldsymbol{\nabla} \times \tilde{\mathbf{a}})^2 + m_{\tilde{a}}^2\, \tilde{\mathbf{a}}_T^2 \right] - i\tilde{\mathbf{a}} \cdot (\boldsymbol{\nabla} \times \mathbf{A}) + \frac{\tilde{\beta}}{2} (\boldsymbol{\nabla} \times \mathbf{A})^2 \right]. \tag{5.170}$$

If we integrate out the massive $\tilde{\mathbf{a}}$-field in the partition function, the Hamiltonian of the vector potential becomes

$$H_{\mathbf{A}} = \frac{1}{2} \int d^3x\, \boldsymbol{\nabla} \times \mathbf{A} \left(1 + \frac{m_A^2}{-\boldsymbol{\nabla}^2 + m_{\tilde{a}}^2} \right) \boldsymbol{\nabla} \times \mathbf{A}. \tag{5.171}$$

Expanding the denominator in powers of $-\boldsymbol{\nabla}^2$ we see that only gradient energies appear, but no mass term. Thus, the vector potential $\mathbf{A}$ maintains its long range and yields Coulomb-like forces at large distances. Only its dispersion is modified to a more complicated $\mathbf{k}$-dependence of the energy.

In the low-temperature phase, on the other hand, the mass $m_{\tilde{a}}$ is zero, and the m_A^2-term in (5.171) produces again the transverse mass Hamiltonian (5.161) which is responsible for the Meissner effect.

We can represent the fluctuating vortices in the superconductor by a disorder field theory in the same way as we did for the vortices in the superfluid, by repeating the transformations in Eqs. (5.123)–(5.126). The angular field variable of disorder will now be denoted by $\theta(\mathbf{x})$, the vortex lines in the disorder theory by $\boldsymbol{\theta}^{\mathrm{v}}(\mathbf{x})$. The disorder action reads

$$\tilde{\beta}\tilde{H}_{\mathrm{SC}}^{\mathrm{hy}} = \int d^3x \left[\frac{1}{2\tilde{\beta}m_A^2}(\boldsymbol{\nabla} \times \tilde{\mathbf{a}})^2 - iq\tilde{\mathbf{a}} \cdot (\boldsymbol{\nabla} \times \mathbf{A}) + \frac{\tilde{\beta}}{2}(\boldsymbol{\nabla} \times \mathbf{A})^2 \right.$$
$$\left. + \frac{m_{\tilde{a}}^2}{2\tilde{\beta}m_A^2}(\boldsymbol{\nabla}\theta - \boldsymbol{\theta}^{\mathrm{v}} - \tilde{\mathbf{a}})^2 \right]. \quad (5.172)$$

Near the phase transition, this is equivalent to a disorder field energy

$$\tilde{\beta}\tilde{H}_{\mathrm{SC}}^{\mathrm{hy}} \sim \int d^2x \left[\frac{1}{2\tilde{\beta}m_A^2}(\boldsymbol{\nabla} \times \tilde{\mathbf{a}})^2 - iq\tilde{\mathbf{a}} \cdot (\boldsymbol{\nabla} \times \mathbf{A}) + \frac{\tilde{\beta}}{2}(\boldsymbol{\nabla} \times \mathbf{A})^2 \right.$$
$$\left. + \frac{1}{2}[(\boldsymbol{\nabla} - i\tilde{\mathbf{a}})\,\phi]^2 + \frac{\tau}{2}|\phi|^2 + \frac{g}{4}|\phi|^4 \right]. \quad (5.173)$$

The complex disorder field $\phi(\mathbf{x})$ has a phase $\theta(\mathbf{x})$, and its size $|\phi(\mathbf{x})|$ is fixed by the parameters $\tau < 0$ and g to have $|\phi(\mathbf{x})|^2 = m_{\tilde{a}}^2/\tilde{\beta}m_A^2$. The vector potential $\mathbf{A}$ fluctuates harmonically in such a way that the associated magnetic field is on the average equal to $q\tilde{\mathbf{a}}/\tilde{\beta}$. Integrating out $\mathbf{A}$, we obtain from (5.173)

$$\tilde{\beta}\tilde{H}_{\mathrm{SC}}^{\mathrm{hy}} \sim \int d^2x \left[\frac{1}{2\tilde{\beta}m_A^2}[(\boldsymbol{\nabla} \times \tilde{\mathbf{a}})^2 + m_{\tilde{a}}^2\tilde{\mathbf{a}}_T^2] + \frac{1}{2}[(\boldsymbol{\nabla} - i\tilde{\mathbf{a}})\,\phi]^2 + \frac{\tau}{2}|\phi|^2 + \frac{g}{4}|\phi|^4 \right].$$
$$(5.174)$$

This Hamiltonian is invariant under the gauge transformations

$$\phi(\mathbf{x}) \to e^{i\tilde{\Lambda}(\mathbf{x})}\phi(\mathbf{x}), \quad \tilde{\mathbf{a}}(\mathbf{x}) \to \tilde{\mathbf{a}}(\mathbf{x}) + \boldsymbol{\nabla}\tilde{\Lambda}(\mathbf{x}). \quad (5.175)$$

The partition function is

$$Z_{\mathrm{SC}}^{\mathrm{dual}} = \int \mathcal{D}\phi \int \mathcal{D}\phi^* \, \mathcal{D}\tilde{\mathbf{a}} \, \Phi[\tilde{\mathbf{a}}] \, e^{-\tilde{\beta}\tilde{H}_{\mathrm{SC}}^{\mathrm{hy}}}, \quad (5.176)$$

where $\Phi[\tilde{\mathbf{a}}]$ is some gauge-fixing functional.

This partition function can be evaluated perturbatively as a power series in g. The terms of order g^n consist of *Feynman integrals* which can be pictured by *Feynman diagrams* with $n + 1$ loops [31]. These loops are pictures of the topology of vortex loops in the superconductor.

The disorder field theory for the superconductor was for a long time the only formulation which has led to a determination of the critical and tricritical properties of the superconducting phase transition [23, 32]. Within the Ginzburg-Landau theory, an explanation was found only recently [33].

In the hydrodynamic Hamiltonian (5.172), the elimination of $\mathbf{A}(\mathbf{x})$ leads to the Hamiltonian

$$\tilde{\beta}\tilde{H}_{\mathrm{SC}}^{\mathrm{hy}} = \int d^3x \left[\frac{1}{2\tilde{\beta}m_A^2}[(\boldsymbol{\nabla} \times \tilde{\mathbf{a}})^2 + m_a^2\tilde{\mathbf{a}}_T^2] + \frac{m_a^2}{2\tilde{\beta}m_A^2}(\boldsymbol{\nabla}\theta - \boldsymbol{\theta}^{\mathrm{v}} - \tilde{\mathbf{a}})^2 \right], \quad (5.177)$$

which is gauge-invariant under

$$\theta(\mathbf{x}) \to \theta(\mathbf{x}) + \tilde{\Lambda}(\mathbf{x}), \quad \tilde{\mathbf{a}}(\mathbf{x}) \to \tilde{\mathbf{a}}(\mathbf{x}) + \boldsymbol{\nabla}\tilde{\Lambda}(\mathbf{x}). \tag{5.178}$$

5.3 Order versus Disorder Parameter

Since Landau's 1947 work [2], phase transitions are characterized by an order parameter which is nonzero in the low-temperature ordered phase, and zero in the high-temperature disordered phase. In the 1980s, this characterization has been enriched by the dual disorder field theory of various phase transitions [7]. The expectation value of the disorder field provides us with the disorder parameter which has the opposite temperature behavior, being nonzero in the high-temperature and zero in the low-temperature phase. Let us identify the order and disorder fields in superfluids and superconductors, and study their expectation values in the hydrodynamic theories of the two systems.

5.3.1 Superfluid ^{4}He

In Landau's original description of the superfluid phase transition with the Hamiltonian (5.6), the role of the order parameter $\mathcal{O}$ is played by the expectation value of the complex order field $\mathcal{O}(\mathbf{x}) = \phi(\mathbf{x})$:

$$\mathcal{O} \equiv \langle \mathcal{O}(\mathbf{x}) \rangle = \langle \phi(\mathbf{x}) \rangle. \tag{5.179}$$

Its behavior can be extracted from the large-distance limit of the correlation function of two order fields $\mathcal{O}(\mathbf{x})$:

$$G_{\mathcal{O}}(\mathbf{x}_2, \mathbf{x}_1) \equiv \langle \mathcal{O}(\mathbf{x}_2)\mathcal{O}^*(\mathbf{x}_1) \rangle = \langle \phi(\mathbf{x}_2)\phi^*(\mathbf{x}_1) \rangle. \tag{5.180}$$

This is done by taking advantage of the cluster property of the correlation functions of arbitrary local operators

$$\langle O_1(\mathbf{x}_2)O_2(\mathbf{x}_1) \rangle \xrightarrow[|\mathbf{x}_2 - \mathbf{x}_1| \to \infty]{} \langle O_1(\mathbf{x}_2) \rangle \langle O_2(\mathbf{x}_1) \rangle. \tag{5.181}$$

Hence we obtain the large-distance limit of the correlation function (5.180)

$$G_{\mathcal{O}}(\mathbf{x}_2, \mathbf{x}_1) \xrightarrow[|\mathbf{x}_2 - \mathbf{x}_1| \to \infty]{} |\mathcal{O}|^2. \tag{5.182}$$

If we go to the hydrodynamic limit of the theory where the size of $\phi(\mathbf{x})$ is frozen and the order field reduces to $O(\mathbf{x}) = e^{i\theta(\mathbf{x})}$, the order parameter becomes

$$\mathcal{O} \equiv \langle O(\mathbf{x}) \rangle = \langle e^{i\theta(\mathbf{x})} \rangle. \tag{5.183}$$

This is extracted from the large-distance limit of the correlation function

$$G_{\mathcal{O}}(\mathbf{x}_2, \mathbf{x}_1) - \langle e^{i\theta(\mathbf{x}_2)} e^{-i\theta(\mathbf{x}_1)} \rangle. \tag{5.184}$$

If we want to use (5.183) as an order parameter to replace (5.179), it is important that the correlation function (5.184) is vortex-gauge-invariant under the transformations (5.28). This is not immediately obvious. A quantity where the invariance is obvious is the expectation value

$$G_{\mathcal{O}}(\mathbf{x}_2, \mathbf{x}_1) = \left\langle e^{i \int_{\mathbf{x}_1}^{\mathbf{x}_2} d\mathbf{x} [\boldsymbol{\nabla}\theta(\mathbf{x}) - \boldsymbol{\theta}^{\mathrm{v}}(\mathbf{x})]} \right\rangle. \tag{5.185}$$

The transformations (5.28) do not change the exponent. We have, however, achieved vortex gauge invariance at the price of an apparent dependence of (5.185) on the shape of the path from $\mathbf{x}_1$ to $\mathbf{x}_2$. Fortunately it is possible to show that this shape dependence is not really there, so that the vortex gauge-invariant correlation function is uniquely defined, and that it is in fact the same as (5.184), thus ensuring the vortex gauge invariance of (5.184).

In order to prove this, let us rewrite (5.185) in the form

$$G_{\mathcal{O}}(\mathbf{x}_2, \mathbf{x}_1) = \left\langle e^{i \int d^3x\, \mathbf{b}^{\mathrm{m}}(\mathbf{x}) [\boldsymbol{\nabla}\theta(\mathbf{x}) - \boldsymbol{\theta}^{\mathrm{v}}(\mathbf{x})]} \right\rangle, \tag{5.186}$$

where the field

$$\mathbf{b}^{\mathrm{m}}(\mathbf{x}) = \boldsymbol{\delta}(\mathbf{x}; \tilde{L}_{\mathbf{x}_1}^{\mathbf{x}_2}) \tag{5.187}$$

is a δ-function on an arbitrary line $\tilde{L}_{\mathbf{x}_1}^{\mathbf{x}_2}$ running from $\mathbf{x}_1$ to $\mathbf{x}_2$. This field satisfies [recall (4.10) and (4.11)]

$$\boldsymbol{\nabla} \cdot \mathbf{b}^{\mathrm{m}}(\mathbf{x}) = q(\mathbf{x}), \tag{5.188}$$

where

$$q(\mathbf{x}) = \delta^{(3)}(\mathbf{x} - \mathbf{x}_1) - \delta^{(3)}(\mathbf{x} - \mathbf{x}_2). \tag{5.189}$$

It is now easy to see that the expression (5.186) is invariant under deformations of $\tilde{L}_{\mathbf{x}_1}^{\mathbf{x}_2}$. Indeed, let $\tilde{L}'^{\mathbf{x}_2}_{\mathbf{x}_1}$ be a different path running from $\mathbf{x}_1$ to $\mathbf{x}_2$. Then the difference between the two is a closed path $\tilde{L}$, and the exponents in (5.186) differ by an integral

$$i \int d^3x\, \boldsymbol{\delta}(\mathbf{x}; \tilde{L}) \left[\boldsymbol{\nabla}\theta(\mathbf{x}) - \boldsymbol{\theta}^{\mathrm{v}}(\mathbf{x}) \right]. \tag{5.190}$$

The first term vanishes after a partial integration due to Eq. (4.9). The second term becomes, after inserting (5.24),

$$- 2\pi i \int d^3x\, \boldsymbol{\delta}(\mathbf{x}; \tilde{L})\, \boldsymbol{\delta}(\mathbf{x}; S) = -2\pi i k, \qquad k = \text{integer}. \tag{5.191}$$

The integer k counts how many times the line $\tilde{L}$ pierces the surface S. Since $-2\pi i k$ appears in the exponential, it does not contribute to the correlation function (5.186). Thus we have proved that the expectation value (5.185) is independent of the path along which the integral runs from $\mathbf{x}_1$ to $\mathbf{x}_2$.

We recognize the analogy to the discussion of magnetic monopoles in Section 4.4. For this reason we shall speak of $q(\mathbf{x})$ as a *charge density of a monopole-antimonopole pair* located at $\mathbf{x}_2$ and $\mathbf{x}_1$, respectively. In the description of monopoles in Section 4.4, a monopole at $\mathbf{x}_2$ is attached to a Dirac string $L^{\mathbf{x}_2}$ along which the flux is imported from infinity, whereas an antimonopole at $\mathbf{x}_1$ carries a Dirac string $L_{\mathbf{x}_1}$ along which the flux is exported to infinity. Since the shape of the two strings is irrelevant, we may distort them to a single line connecting $\mathbf{x}_1$ with $\mathbf{x}_2$ along an arbitrary path. This is the line $\tilde{L}^{\mathbf{x}_2}_{\mathbf{x}_1}$ in (5.187).

The field $\mathbf{b}^m(\mathbf{x})$ is a gauge field with the same properties as the *monopole gauge field* in Section 4.4 (see also Ref. [34]). A change of the shape of the line $\tilde{L}^{\mathbf{x}_2}_{\mathbf{x}_1}$ is achieved by a *monopole gauge transformation* [recall (4.64)]

$$\mathbf{b}^m(\mathbf{x}) \to \mathbf{b}^m(\mathbf{x}) + \boldsymbol{\nabla} \times \boldsymbol{\delta}(\mathbf{x}; \tilde{S}). \tag{5.192}$$

Note that the invariant field strength of this gauge field is the divergence (5.188) rather than a curl [recall Eq. (5.30) for a vortex gauge field].

In this way, the independence of the manifestly vortex gauge-invariant correlation function (5.186) on the shape of the line connecting $\mathbf{x}_1$ with $\mathbf{x}_2$ is expressed as an additional invariance under monopole gauge transformations. The correlation function is thus a double-gauge-invariant object.

After this discussion we are able to define a manifestly vortex gauge-invariant formulation of the order parameter (5.183). It is given by the expectation value

$$\mathcal{O} = \langle \mathcal{O}(\mathbf{x}) \rangle = \left\langle e^{i \int^{\mathbf{x}} d\mathbf{x}' [\boldsymbol{\nabla}\theta(\mathbf{x}') - \boldsymbol{\theta}^v(\mathbf{x}')]} \right\rangle = \left\langle e^{i \int d^3x' \, \boldsymbol{\delta}(\mathbf{x}'; L^{\mathbf{x}})[\boldsymbol{\nabla}\theta(\mathbf{x}') - \boldsymbol{\theta}^v(\mathbf{x}')]} \right\rangle, \tag{5.193}$$

where $\boldsymbol{\delta}(\mathbf{x}; L^{\mathbf{x}})$ is the δ-function on an arbitrary line as defined in Eq. (4.58). It comes from infinity along an arbitrary path ending at $\mathbf{x}$.

Let us now study the large-distance behavior (5.181) of the correlation function (5.185) at low and high temperatures. At low temperature where vortices are rare, the $\theta(\mathbf{x})$-field fluctuates almost harmonically. By *Wick's theorem*, according to which harmonically fluctuating variables θ_1, θ_2 satisfy the equation [35]

$$\langle e^{i\theta_1} e^{i\theta_2} \rangle = e^{-\frac{1}{2}\langle \theta_1 \theta_2 \rangle}, \tag{5.194}$$

we can approximate

$$G_{\mathcal{O}}(\mathbf{x}_2, \mathbf{x}_1) \underset{T \approx 0}{\approx} e^{-\frac{1}{2}\langle [\theta(\mathbf{x}_2) - \theta(\mathbf{x}_1)]^2 \rangle} = e^{\langle [\theta(\mathbf{x}_2)\theta(\mathbf{x}_1) - \frac{1}{2}\theta^2(\mathbf{x}_1) - \frac{1}{2}\theta^2(\mathbf{x}_2)] \rangle}. \tag{5.195}$$

The correlation function of two $\theta(\mathbf{x})$-fields is

$$\langle \theta(\mathbf{x}_2)\theta(\mathbf{x}_1) \rangle \approx T v_0(|\mathbf{x}_2 - \mathbf{x}_1|), \tag{5.196}$$

where $v_0(r)$ is the Coulomb potential (5.48) which goes to zero for $r \to \infty$. The correlation function (5.195) is then equal to

$$G_{\mathcal{O}}(\mathbf{x}_2, \mathbf{x}_1) \approx e^{-T v_0(\mathbf{0})} e^{T v_0(|\mathbf{x}_2 - \mathbf{x}_1|)}. \tag{5.197}$$

This is finite only after remembering that we are studying the superfluid in the hydrodynamic limit, which is correct only for length scales larger than the coherence length ξ. In He this is of the order of a few Å. Thus we should perform all wave vector integrals only for $|\mathbf{k}| \leq \Lambda \equiv 1/\xi$, which makes $v_0(\mathbf{0})$ a finite quantity

$$v_0(\mathbf{0}) = 1/2\xi\pi^2. \tag{5.198}$$

As a result, the correlation function (5.195) has a nonzero large-distance limit

$$G_{\mathcal{O}}'(\mathbf{x}_2, \mathbf{x}_1) \xrightarrow[|\mathbf{x}_2 - \mathbf{x}_1| \to \infty]{} \text{const}, \tag{5.199}$$

implying via Eq. (5.181) that the order parameter $\mathcal{O} = \langle e^{i\theta(\mathbf{x})} \rangle$ is nonzero.

Let us now calculate the large-distance behavior in the high-temperature phase. To find the correlation function $G_{\mathcal{O}}(\mathbf{x}_2, \mathbf{x}_1)$, we insert the extra source term

$$e^{i\theta(\mathbf{x}_2)} e^{-i\theta(\mathbf{x}_1)} = e^{-i \int d^3x \, q(\mathbf{x})\theta(\mathbf{x})} \tag{5.200}$$

into the partition function (5.33). This term enters the canonical representation (5.86) of the energy as follows:

$$\beta H = \int d^3x \left[\frac{1}{2\beta} \mathbf{b}^2 - i\mathbf{b} \left(\boldsymbol{\nabla}\theta - \boldsymbol{\theta}^{\mathrm{v}} \right) + \frac{\beta \epsilon_c}{2} (\boldsymbol{\nabla} \times \boldsymbol{\theta}^{\mathrm{v}})^2 + iq(\mathbf{x})\theta(\mathbf{x}) \right], \tag{5.201}$$

where we have allowed for an extra core energy, for the sake of generality. Integrating out the θ-field in the partition function gives the constraint

$$\boldsymbol{\nabla} \cdot \mathbf{b}(\mathbf{x}) = -q(\mathbf{x}). \tag{5.202}$$

The constraint is solved by the negative of the monopole gauge field (5.187), and the general solution is

$$\mathbf{b}(\mathbf{x}) = \boldsymbol{\nabla} \times \mathbf{a}(\mathbf{x}) - \mathbf{b}^{\mathrm{m}}(\mathbf{x}), \tag{5.203}$$

so that the energy (5.201) can be replaced by [using once more (5.191)]

$$\beta H = \int d^3x \left[\frac{1}{2\beta} \left(\boldsymbol{\nabla} \times \mathbf{a} - \mathbf{b}^{\mathrm{m}} \right)^2 - i\mathbf{a} \cdot \mathbf{j}^{\mathrm{v}} + \frac{\beta \epsilon_c}{2} \mathbf{j}_c^{\mathrm{v2}} \right]. \tag{5.204}$$

Under a monopole gauge transformation (5.192), this remains invariant if the gauge field of superflow is simultaneously transformed as

$$\mathbf{a}(\mathbf{x}) \to \mathbf{a}(\mathbf{x}) + \boldsymbol{\delta}(\mathbf{x}; \tilde{S}). \tag{5.205}$$

The correlation function (5.186) is now calculated from the functional integral over the Boltzmann factor with the Hamiltonian (5.204). The presence of the source term (5.200) is accounted for in the functional integral over $e^{-\beta H}$ by the $\mathbf{b}^{\mathrm{m}}$-dependent integrand

$$e^{-i \int d^3x q(\mathbf{x})\theta(\mathbf{x})} \; \hat{=} \; e^{-\frac{1}{\beta} \int d^3x \left\{ \frac{1}{2} \mathbf{b}^{\mathrm{m}}(\mathbf{x})^2 - \mathbf{b}^{\mathrm{m}}(\mathbf{x})[\boldsymbol{\nabla} \times \mathbf{a}(\mathbf{x})] \right\}}. \tag{5.206}$$

It is instructive to calculate the large-distance behavior (5.199) of the correlation function in the low-temperature phase once more in this canonical formulation. At low temperatures, the vortex lines are frozen out and we can omit the last two terms in (5.204). We integrate out the gauge field $\mathbf{a}$ of superflow in the associated partition function and find that the partition function contains $\mathbf{b}^m$ in the form of a factor

$$e^{-\frac{1}{2\beta}\int d^3x\left\{\mathbf{b}^m(\mathbf{x})^2-[\nabla\times\mathbf{b}^m(\mathbf{x})]\frac{1}{-\nabla^2}[\nabla\times\mathbf{b}^m(\mathbf{x})]\right\}} = e^{-\frac{1}{2\beta}\int d^3x\,\nabla\cdot\mathbf{b}^m(\mathbf{x})\frac{1}{-\nabla^2}\nabla\cdot\mathbf{b}^m(\mathbf{x})}. \qquad (5.207)$$

From this we obtain the correlation function

$$G_O(\mathbf{x}_1,\mathbf{x}_2) = e^{-\frac{1}{2\beta}\int d^3x\,q(\mathbf{x})\frac{1}{-\nabla^2}q(\mathbf{x})} = e^{-\frac{1}{2\beta}\int d^3x d^3x'\,q(\mathbf{x})v_0(\mathbf{x}-\mathbf{x}')q(\mathbf{x}')}. \qquad (5.208)$$

Inserting (5.189), this becomes

$$G_O(\mathbf{x}_1,\mathbf{x}_2) = e^{-v_0(\mathbf{0})/\beta}e^{v_0(\mathbf{x}_1-\mathbf{x}_2)/\beta}, \qquad (5.209)$$

in agreement with the previous result (5.197).

The canonical formulation (5.204) of the energy enables us to calculate the large-distance behavior of the correlation function in the high-temperature phase. The prolific vortex fluctuations produce a transverse mass term $m_a^2\mathbf{a}^2$ which changes (5.207) to (see also Ref. [36])

$$e^{-\frac{1}{2\beta}\int d^3x\left\{\mathbf{b}^m(\mathbf{x})^2-[\nabla\times\mathbf{b}^m(\mathbf{x})]\frac{1}{-\nabla^2+m_a^2}[\nabla\times\mathbf{b}^m(\mathbf{x})]\right\}}$$
$$= e^{-\frac{1}{2\beta}\int d^3x\left[\nabla\cdot\mathbf{b}^m(\mathbf{x})\frac{1}{-\nabla^2+m_a^2}\nabla\cdot\mathbf{b}^m+\mathbf{b}^m\frac{m_a^2}{-\nabla^2+m_a^2}\mathbf{b}^m(\mathbf{x})\right]}. \qquad (5.210)$$

Using (5.188), we factorize this as

$$e^{-\frac{1}{2\beta}\int d^3x\,q(\mathbf{x})\frac{1}{-\nabla^2+m_a^2}q(\mathbf{x})} \times e^{-\frac{1}{2\beta}\int d^3x\,\mathbf{b}^m(\mathbf{x})\frac{m_a^2}{-\nabla^2+m_a^2}\mathbf{b}^m(\mathbf{x})}. \qquad (5.211)$$

The first exponent contains the Yukawa potential

$$v_{m_a}(r) \equiv \int \frac{d^3k}{(2\pi)^3}e^{i\mathbf{k}\mathbf{x}}\frac{1}{\mathbf{k}^2+m_a^2} = \frac{e^{-m_ar}}{4\pi r} \qquad (5.212)$$

between the monopole-antimonopole pair at $\mathbf{x}_2$ and $\mathbf{x}_1$, respectively, in the same form as in (5.209), $e^{-v_{m_a}(0)/\beta}e^{v_{m_a}(|\mathbf{x}_1-\mathbf{x}_2|)/\beta}$. The potential $v_{m_a}(|\mathbf{x}_1-\mathbf{x}_2|)$ goes to zero for large distances, so that the exponential tends towards a constant. The second factor in (5.211) has the form [recall (5.187)]

$$e^{-\frac{1}{2\beta}\int d^3x d^3x'\,\delta(\mathbf{x};\tilde{L}_{\mathbf{x}_1}^{\mathbf{x}_2})v_{m_a}(|\mathbf{x}-\mathbf{x}'|)\delta(\mathbf{x}';\tilde{L}_{\mathbf{x}_1}^{\mathbf{x}_2})}. \qquad (5.213)$$

This is the Yukawa self-energy of the line $\tilde{L}_{\mathbf{x}_1}^{\mathbf{x}_2}$ connecting $\mathbf{x}_1$ and $\mathbf{x}_2$. For $|\mathbf{x}_1-\mathbf{x}_2|$ much larger than the range of the Yukawa potential $1/m_a$, this is proportional to $|\mathbf{x}_1-\mathbf{x}_2|$. Hence the second exponential in (5.211) vanishes in this limit, and so does the correlation function:

$$G_O(\mathbf{x}_1,\mathbf{x}_2) \sim e^{-\text{const}\cdot|\mathbf{x}_1-\mathbf{x}_2|} \xrightarrow[|\mathbf{x}_1-\mathbf{x}_2|\to\infty]{} 0. \qquad (5.214)$$

Due to the cluster property (5.181) of correlation functions, this shows that at high temperatures, the expectation value $\mathcal{O} = \langle O(\mathbf{x}) \rangle = \langle e^{i\theta(\mathbf{x})} \rangle$ vanishes, so that O is indeed a good order parameter.

The mechanism which gives an energy to the initially irrelevant line $\tilde{L}^{\mathbf{x}_2}_{\mathbf{x}_1}$ connecting monopole and antimonopole is completely analogous to the generation of surface energy in the previous Eq. (5.82). There the energy arose from a mass of the θ-fluctuations, here from a mass of the $\mathbf{a}$-field fluctuations which was caused by the proliferation of infinitely long vortex lines in the high-temperature phase.

Note that an exponential falloff is also found within Landau's complex order field theory where

$$\langle \psi(\mathbf{x}_1)\psi(\mathbf{x}_2) \rangle \propto \int \frac{d^3k}{(2\pi)^3} e^{ik(\mathbf{x}_1 - \mathbf{x}_2)} \frac{1}{\mathbf{k}^2 + m^2} = \frac{1}{4\pi} \frac{e^{-m|\mathbf{x}_1 - \mathbf{x}_2|}}{|\mathbf{x}_1 - \mathbf{x}_2|}. \tag{5.215}$$

However, here the finite range arises in a different way. In calculating (5.215), the size fluctuations of the order field play an essential role. In the partition function (5.33), their role is taken over by the fluctuations of the vortex gauge field $\boldsymbol{\theta}^{\mathrm{v}}(\mathbf{x})$, as pointed out at the end of Section 5.1.4. The proliferation of the vortex lines produces the finite range $1/m_a$ of the Yukawa potential and the exponential falloff (5.214).

5.3.2   Superconductor

In contrast to the expectation value (5.183) for superfluid helium, the expectation value of the order field $\psi(\mathbf{x})$ of the Ginzburg-Landau Hamiltonian (5.148) cannot be used as an order parameter since it is not invariant under ordinary magnetic gauge transformations (5.158). The expectation of all non-gauge-invariant quantities vanishes for all temperatures. This intuitively obvious fact is known as *Elitzur's theorem* [37]. The theorem applies also to the hydrodynamic limit of $\psi(\mathbf{x})$, so that the expectation value of the exponential $e^{i\tilde{\theta}(\mathbf{x})}$ cannot serve as an order parameter. Let us search for other possible candidates to be extracted from the large-distance limit of various gauge-invariant correlation functions.

a) Schwinger Candidate for Order Parameter

As a first possible candidate, consider the following gauge-invariant version of the expectation value of $\langle e^{i\tilde{\theta}(\mathbf{x}_2)} e^{-i\tilde{\theta}(\mathbf{x}_1)} \rangle$:

$$G_{\tilde{O}}(\mathbf{x}_2, \mathbf{x}_1) = \langle e^{i\tilde{\theta}(\mathbf{x}_2)} e^{-i \int_{\mathbf{x}_1}^{\mathbf{x}_2} d\mathbf{x} \, \mathbf{A}(\mathbf{x})} e^{-i\tilde{\theta}(\mathbf{x}_1)} \rangle, \tag{5.216}$$

which can also be written as

$$G_{\tilde{O}}(\mathbf{x}_2, \mathbf{x}_1) = \langle e^{i\tilde{\theta}(\mathbf{x}_2)} e^{-i \int d^3x \, \mathbf{b}^{\mathrm{m}}(\mathbf{x}) \mathbf{A}(\mathbf{x})} e^{-i\tilde{\theta}(\mathbf{x}_1)} \rangle, \tag{5.217}$$

where $\mathbf{b}^{\mathrm{m}}(\mathbf{x})$ is the δ-function (5.187) along the line $L^{\mathbf{x}_2}_{\mathbf{x}_1}$ connecting $\mathbf{x}_1$ with $\mathbf{x}_2$. This expression is obviously invariant under magnetic gauge transformations (5.158), due to Eqs. (5.188) and (5.189).

As before, we must make the correlation function (5.217) manifestly invariant under vortex gauge transformations (5.159). This can be done by adding, as in (5.200), a vortex gauge field:

$$G_{\tilde{O}}(\mathbf{x}_2, \mathbf{x}_1) = \langle e^{i \int d^3 x \, \mathbf{b}^m(\mathbf{x}) [\nabla \tilde{\theta}(\mathbf{x}) - \mathbf{A}(\mathbf{x}) - \tilde{\theta}^v(\mathbf{x})]} \rangle. \tag{5.218}$$

The associated order parameter would be [compare (5.193)]

$$\tilde{\mathcal{O}} \equiv \langle \tilde{O}(\mathbf{x}) \rangle = \langle e^{i \int d^3 x' \, \delta(\mathbf{x}; L^{\mathbf{x}}) [\nabla \tilde{\theta}(\mathbf{x}') - \mathbf{A}(\mathbf{x}') - \tilde{\theta}^v(\mathbf{x}')]} \rangle. \tag{5.219}$$

We now observe that in contrast to the correlation function in the superfluid (5.184), this is not invariant under deformations of the shape of the line $\tilde{L}_{\mathbf{x}_1}^{\mathbf{x}_2}$ connecting the points $\mathbf{x}_1$ and $\mathbf{x}_2$. Indeed, if we apply the associated monopole gauge transformation (5.192) to (5.217), we see that

$$e^{-i \int d^3 x \, \mathbf{b}^m(\mathbf{x}) \mathbf{A}(\mathbf{x})} \rightarrow e^{-i \int d^3 x \{\mathbf{b}^m(\mathbf{x}) \mathbf{A}(\mathbf{x}) + [\nabla \times \delta(\mathbf{x}; \tilde{S})] \mathbf{A}(\mathbf{x})\}} = e^{-i \int d^3 x \{\mathbf{b}^m(\mathbf{x}) \mathbf{A}(\mathbf{x}) + \mathbf{B}(\mathbf{x}) \delta(\mathbf{x}; \tilde{S})\}}, \tag{5.220}$$

where $\tilde{S}$ is the surface over which $\tilde{L}_{\mathbf{x}_1}^{\mathbf{x}_2}$ has swept. Thus, the correlation function (5.217) changes under monopole gauge transformations by a phase

$$G_{\tilde{O}}(\mathbf{x}_2, \mathbf{x}_1) \rightarrow e^{-i \int d^3 x \, \mathbf{B}(\mathbf{x}) \delta(\mathbf{x}; \tilde{S})} G_{\tilde{O}}(\mathbf{x}_2, \mathbf{x}_1), \tag{5.221}$$

which depends on the fluctuating magnetic flux through the surface $\tilde{S}$. For this reason, we must first remove the freedom of choosing the shape of $\tilde{L}_{\mathbf{x}_1}^{\mathbf{x}_2}$ which connects $\mathbf{x}_1$ with $\mathbf{x}_2$. The simplest choice made by Schwinger [38] is the straight path from $\mathbf{x}_1$ to $\mathbf{x}_2$.

Still, the correlation function (5.218) does not supply us with an order parameter when taking the limit of large $|\mathbf{x}_2 - \mathbf{x}_1|$. In order to verify this, we go to the partition function with the canonical representation (5.163) of the Hamiltonian, and insert the expression (5.218). Then we change field variables from $\tilde{\mathbf{b}}$ to $\tilde{\mathbf{b}} - \mathbf{b}^m$, and use (5.165) to obtain (5.166), with $(\nabla \times \tilde{\mathbf{a}})^2$ replaced by $(\nabla \times \tilde{\mathbf{a}} - \mathbf{b}^m)^2$:

$$\tilde{\beta} H_{\mathrm{SC}}^{\mathrm{hy}} = \int d^3 x \left[\frac{1}{2\tilde{\beta} m_A^2} (\nabla \times \tilde{\mathbf{a}} - \mathbf{b}^m)^2 - i\tilde{\mathbf{a}} \cdot (\nabla \times \mathbf{A}) + \frac{\tilde{\beta}}{2} (\nabla \times \mathbf{A})^2 - i\tilde{\mathbf{a}} \cdot \tilde{\mathbf{j}}^v + \frac{\tilde{\beta} \tilde{\epsilon}_c}{2} \tilde{\mathbf{j}}_c^{v2} \right]. \tag{5.222}$$

This is quadratic in the magnetic vector potential $\mathbf{A}$ which can be integrated out in the associated partition function, leading to the Hamiltonian

$$\tilde{\beta} H_{\mathrm{SC}}^{\mathrm{hy}} = \int d^3 x \, \frac{1}{2\tilde{\beta} m_A^2} \left[(\nabla \times \tilde{\mathbf{a}} - \mathbf{b}^m)^2 + m_A^2 \tilde{\mathbf{a}}^2 - i\tilde{\mathbf{a}} \cdot \tilde{\mathbf{j}}^v + \frac{\tilde{\beta} \tilde{\epsilon}_c}{2} \tilde{\mathbf{j}}_c^{v2} \right]. \tag{5.223}$$

With this Hamiltonian, the correlation function (5.217) can be calculated from the expectation value [compare (5.206)]:

$$G_{\tilde{O}}(\mathbf{x}_2, \mathbf{x}_1) = \left\langle e^{-\frac{1}{\tilde{\beta} m_A^2} \int d^3 x \{\frac{1}{2} \mathbf{b}^m(\mathbf{x})^2 - \mathbf{b}^m(\mathbf{x}) [\nabla \times \tilde{\mathbf{a}}(\mathbf{x})]\}} \right\rangle. \tag{5.224}$$

Consider first the low-temperature phase where the vortices in the superconductor are frozen out, and we may omit the last two terms in (5.222). Then the massive field $\tilde{\mathbf{a}}$ can be integrated out trivially in the partition function, leading to

$$G_{\tilde{O}}(\mathbf{x}_1, \mathbf{x}_2) \sim e^{-\frac{\tilde{\beta}m_A^2}{2}\int d^3x \left[\mathbf{b}^{m\,2} - (\nabla\times\mathbf{b}^m)\frac{1}{-\nabla^2+m_A^2}(\nabla\times\mathbf{b}^m)\right]}. \tag{5.225}$$

This is the same expression as in the high-temperature phase of the superfluid in Eq. (5.210), except that the relevant mass is now the Meissner mass m_A of the superconductor rather than m_a. The mass m_A makes the line $\tilde{L}_{\mathbf{x}_1}^{\mathbf{x}_2}$ between $\mathbf{x}_1$ and $\mathbf{x}_2$ in $\mathbf{b}^m(\mathbf{x})$ energetic, and leads to the same type of exponential long-distance falloff of the correlation function as in Eq. (5.214):

$$G_{\tilde{O}}(\mathbf{x}_1, \mathbf{x}_2) \sim e^{-\text{const}\cdot|\mathbf{x}_1-\mathbf{x}_2|} \xrightarrow[|\mathbf{x}_1-\mathbf{x}_2|\mapsto\infty]{} 0. \tag{5.226}$$

This implies a vanishing of the candidate (5.219) for the order parameter:

$$\tilde{O} = \langle \tilde{O}(\mathbf{x}) \rangle = 0. \tag{5.227}$$

Thus $\tilde{O}$ fails to indicate the order of the low-temperature phase.

Could $\tilde{O}$ be a disorder parameter? To see this we go to the high-temperature phase where the vortex lines are prolific. In the Hamiltonian (5.222), this corresponds to integrating out $\tilde{\mathbf{j}}_c^v$ like an ordinary Gaussian variable, producing a Hamiltonian

$$\tilde{\beta}H_{SC}^{hy} = \int d^3x \left[\frac{1}{2\tilde{\beta}m_A^2}\left[(\nabla\times\tilde{\mathbf{a}} - \mathbf{b}^m)^2 + m_{\tilde{a}}^2\,\tilde{\mathbf{a}}^2\right] - i\tilde{\mathbf{a}}\cdot(\nabla\times\mathbf{A}) + \frac{\tilde{\beta}}{2}(\nabla\times\mathbf{A})^2\right]. \tag{5.228}$$

If we now integrate out the magnetic vector potential $\mathbf{A}$, the mass term changes from $m_{\tilde{a}}^2$ to $m_{\tilde{a}}^2 + m_A^2$, causing the correlation function to fall off even faster than in (5.226). Hence $\tilde{O}$ is again zero and does not distinguish the different phases.

b) Dirac Candidate for Order Parameter

As an alternative to Schwinger's choice of a straight line connection from $\mathbf{x}_1$ to $\mathbf{x}_2$ in Eq. (5.216) we may choose a different monopole gauge field in Eq. (5.218) which possesses the same divergence $\nabla\cdot\mathbf{b}^m(\mathbf{x}) = q(\mathbf{x})$ as $\mathbf{b}^m(\mathbf{x})$ in Eq. (5.218), but has a *longitudinal* gauge [39, 40, 41]:

$$\nabla\times\mathbf{b}^m(\mathbf{x}) = 0. \tag{5.229}$$

Such a choice exists. We simply take

$$\mathbf{b}^m(\mathbf{x}) = \nabla\frac{1}{\nabla^2}q(\mathbf{x}) = -\frac{1}{4\pi}\nabla\left[\frac{1}{|\mathbf{x}-\mathbf{x}_1|} - \frac{1}{|\mathbf{x}-\mathbf{x}_2|}\right]. \tag{5.230}$$

The monopole gauge field (5.230) is the associated Coulomb field which is longitudinal. Now the exponent in (5.225) simplifies and we obtain the limit

$$G_{\tilde{O}}(\mathbf{x}_1, \mathbf{x}_2) \sim e^{-\frac{\tilde{\beta} m_A^2}{2} \int d^3x \, \mathbf{b}^{m2}} \sim e^{-\frac{\tilde{\beta} m_A^2}{2} \int d^3x \, q \frac{1}{-\nabla^2} q} = e^{-\tilde{\beta} m_A^2 / 8\pi |\mathbf{x}_1 - \mathbf{x}_2|} \xrightarrow[|\mathbf{x}_1 - \mathbf{x}_2| \mapsto \infty]{} 1. \quad (5.231)$$

Actually, this result could have been deduced directly from the energy (5.223). In the longitudinal gauge, $\mathbf{b}^m$ is orthogonal to the purely transversal field $\nabla \times \tilde{\mathbf{a}}$, so that it decouples:

$$(\nabla \times \tilde{\mathbf{a}} - \mathbf{b}^m)^2 = (\nabla \times \tilde{\mathbf{a}})^2 + \mathbf{b}^{m\,2}, \quad (5.232)$$

thus leading directly to (5.231).

The nonzero long-distance limit (5.231) is what we expect in the ordered phase. Thus we may have the hope that (5.219), with $\mathbf{b}^m(\mathbf{x})$ replaced by the field (5.230) restricted to a single monopole at $\mathbf{x}$, i.e.,

$$\mathbf{b}^m_{\mathbf{x}}(\mathbf{x}') \equiv -\nabla' \frac{1}{\nabla'^2} \delta^{(3)}(\mathbf{x}' - \mathbf{x}) = \frac{1}{4\pi} \nabla' \frac{1}{|\mathbf{x}' - \mathbf{x}|}, \quad (5.233)$$

can supply us with an order parameter:

$$\tilde{O} = \langle \tilde{O}(\mathbf{x}) \rangle = \left\langle \exp \left\{ i\tilde{\theta}(\mathbf{x}) - \int d^3x' \, \mathbf{b}^m_{\mathbf{x}}(\mathbf{x}') \cdot \left[\mathbf{A}(\mathbf{x}') - \tilde{\boldsymbol{\theta}}^{\mathrm{v}}(\mathbf{x}') \right] \right\} \right\rangle. \quad (5.234)$$

The important question is whether this is zero in the high-temperature disordered phase of the superconductor [40, 41]. Unfortunately, the answer is negative. We have observed before that the vortex lines merely change the mass square in (5.225) from m_A^2 to $m_A^2 + m_{\tilde{a}}^2$. This does not modify the expression (5.231). Hence the correlation function has the same type of large-distance limit as before in (5.231), implying that (5.234) is again nonzero and thus capable of distinguishing the disordered from the ordered phase.

The reason why (5.231) is the same in both phases is very simple: It lies in the decoupling of the transverse $\nabla \times \tilde{\mathbf{a}}$ from the longitudinal field $\mathbf{b}^m$ in Eq. (5.231), so that the asymptotic behavior is unaffected by a change in the mass of $\tilde{\mathbf{a}}$.

c) Disorder Parameter

The only way to judge the order of the superconductor is to use the disorder field theory and define a disorder parameter whose expectation value is zero for the low-temperature ordered phase and nonzero for the high-temperature disordered phase. For a superconductor, the disorder Hamiltonian was written down in (5.173). Recalling (5.180) we might at first consider extracting the disorder parameter from a large-distance limit of the correlation function

$$G_{\tilde{D}}(\mathbf{x}_2, \mathbf{x}_1) = \langle \phi(\mathbf{x}_2) \phi^*(\mathbf{x}_1) \rangle. \quad (5.235)$$

This, however, would not possess the gauge invariance (5.175) of the disorder Hamiltonian (5.173). An invariant expression is obtained by inserting a factor of the type used in (5.217)

$$G_{\tilde{D}}(\mathbf{x}_2, \mathbf{x}_1) = \langle \phi(\mathbf{x}_2) e^{-i \int d^3x \, \mathbf{b}^m(\mathbf{x}) \tilde{\mathbf{a}}(\mathbf{x})} \phi^*(\mathbf{x}_1) \rangle, \quad (5.236)$$

where $\mathbf{b}^m(\mathbf{x})$ is again the δ-function (5.187) along the line $L_{\mathbf{x}_1}^{\mathbf{x}_2}$ connecting $\mathbf{x}_1$ with $\mathbf{x}_2$. The phase factor ensures the gauge invariance under (5.175). In the hydrodynamic limit, (5.236) becomes

$$G_{\tilde{\mathcal{D}}}(\mathbf{x}_2, \mathbf{x}_1) = \langle e^{i\theta(\mathbf{x}_2)} e^{-i\int d^3x\, \mathbf{b}^m(\mathbf{x})\tilde{\mathbf{a}}(\mathbf{x})} e^{-i\theta(\mathbf{x}_1)} \rangle, \tag{5.237}$$

where $\mathbf{b}^m(\mathbf{x})$ is again the δ-function (5.187) along the line $L_{\mathbf{x}_1}^{\mathbf{x}_2}$ connecting $\mathbf{x}_1$ with $\mathbf{x}_2$. This can be rewritten like (5.218) as

$$G_{\tilde{\mathcal{D}}}(\mathbf{x}_2, \mathbf{x}_1) = \langle e^{i\int d^3x\, \mathbf{b}^m(\mathbf{x})[\nabla\theta(\mathbf{x}) - \tilde{\mathbf{a}}(\mathbf{x}) - \boldsymbol{\theta}^v(\mathbf{x})]} \rangle, \tag{5.238}$$

which now defines a disorder parameter of the superconductor [compare (5.193)]

$$\tilde{\mathcal{D}} \equiv \langle \tilde{D}(\mathbf{x}) \rangle = \langle e^{i\int d^3x'\, \delta(\mathbf{x}';L^{\mathbf{x}})[\nabla\theta(\mathbf{x}') - \tilde{\mathbf{a}}(\mathbf{x}') - \boldsymbol{\theta}^v(\mathbf{x}')]} \rangle, \tag{5.239}$$

where the line L imports the flux from infinity to $\mathbf{x}$.

Thus we must study the energy

$$\begin{aligned}
\beta \tilde{H}_{\mathrm{SC}}^{\mathrm{hy},\mathcal{D}} = \int d^3x \Bigg\{ &\frac{1}{2\tilde{\beta}m_A^2}(\nabla \times \tilde{\mathbf{a}})^2 - i\tilde{\mathbf{a}}\cdot(\nabla \times \mathbf{A}) + \frac{\tilde{\beta}}{2}(\nabla \times \mathbf{A})^2 \\
&+ \frac{m_a^2}{2\tilde{\beta}m_A^2}(\nabla\theta - \boldsymbol{\theta}^v - \tilde{\mathbf{a}})^2 + \mathbf{b}^m \cdot (\nabla\theta - \tilde{\mathbf{a}} - \boldsymbol{\theta}^v) \Bigg\}.
\end{aligned} \tag{5.240}$$

Integrating out the $\mathbf{A}$-field in (5.240) makes the $\tilde{\mathbf{a}}$-field massive and the Hamiltonian becomes

$$\begin{aligned}
\tilde{\beta} \tilde{H}_{\mathrm{SC}}^{\mathrm{hy},\mathcal{D}} = \int d^3x \Bigg\{ &\frac{1}{2\tilde{\beta}m_A^2}\left[(\nabla \times \tilde{\mathbf{a}})^2 + m_A^2\tilde{\mathbf{a}}^2\right] \\
&+ \frac{m_{\tilde{a}}^2}{2\tilde{\beta}m_A^2}(\nabla\theta - \boldsymbol{\theta}^v - \tilde{\mathbf{a}})^2 + \mathbf{b}^m \cdot (\nabla\theta - \tilde{\mathbf{a}} - \boldsymbol{\theta}^v) \Bigg\},
\end{aligned} \tag{5.241}$$

where $m_{\tilde{a}}$ is the mass parameter in Eq. (5.169), although it does not coincide with the mass of the $\tilde{\mathbf{a}}$-field as it did there.

As done before, we introduce an auxiliary field $\mathbf{b}$ to rewrite the last two terms of (5.241) in the form

$$\int d^3x \left[\frac{\tilde{\beta}m_A^2}{2m_{\tilde{a}}^2}(\mathbf{b} - \mathbf{b}^m)^2 + i\,\mathbf{b}\cdot(\nabla\theta - \tilde{\mathbf{a}} - \boldsymbol{\theta}^v) \right], \tag{5.242}$$

and further as

$$\int d^3x \left[\frac{\tilde{\beta}m_A^2}{2m_{\tilde{a}}^2}(\nabla \times \mathbf{a} - \mathbf{b}^m)^2 + i\,\mathbf{a}\cdot(\nabla \times \tilde{\mathbf{a}} + \mathbf{j}^v) + \frac{\tilde{\beta}\epsilon_c}{2}\mathbf{j}^{v2} \right]. \tag{5.243}$$

We have added a core energy to simplify the following discussion.

In the low-temperature phase, there are no vortices in the superconductor but prolific vortices in the dual formulation whose vortex density is $\mathbf{j}^v$, so that we can integrate out $\mathbf{j}^v$ in (5.243) as if it were an ordinary Gaussian field. This gives rise to a mass term for the $\mathbf{a}$-field, so that the Hamiltonian (5.241) becomes

$$
\tilde{\beta}\tilde{H}_{SC}^{hy,\mathcal{D}'} = \int d^3x \left\{ \frac{1}{2\tilde{\beta}m_A^2} \left[(\boldsymbol{\nabla} \times \tilde{\mathbf{a}})^2 + m_A^2 \tilde{\mathbf{a}}^2 \right] \right.
$$
$$
\left. + \frac{\tilde{\beta}m_A^2}{2m_{\tilde{a}}^2} \left[(\boldsymbol{\nabla} \times \mathbf{a} - \mathbf{b}^m)^2 + m_a^2 \mathbf{a}^2 \right] + i\,\tilde{\mathbf{a}} \cdot \boldsymbol{\nabla} \times \mathbf{a} \right\}. \qquad (5.244)
$$

Upon integrating out the $\tilde{\mathbf{a}}$-field, we obtain

$$
\tilde{\beta}\tilde{H}_{SC}^{hy,\mathcal{D}'} = \int d^3x \left\{ \frac{1}{2\tilde{\beta}m_A^2} \left[(\boldsymbol{\nabla} \times \mathbf{a} - \mathbf{b}^m)^2 + m_a^2 \mathbf{a}^2 \right] \right\} + \Delta H, \qquad (5.245)
$$

where

$$
\Delta H = \frac{\tilde{\beta}m_A^2}{2} \int d^3x \, \boldsymbol{\nabla} \times \mathbf{a} \frac{1}{-\boldsymbol{\nabla}^2 + m_A^2} \boldsymbol{\nabla} \times \mathbf{a}. \qquad (5.246)
$$

If we forget this term for a moment we derive from the Hamiltonian (5.245) the correlation function

$$
G_{\tilde{D}}(\mathbf{x}_1, \mathbf{x}_2) \sim e^{-\frac{\tilde{\beta}m_a^2}{2} \int d^3x \left[\mathbf{b}^{m2} - (\boldsymbol{\nabla} \times \mathbf{b}^m) \frac{1}{-\boldsymbol{\nabla}^2 + m_a^2} (\boldsymbol{\nabla} \times \mathbf{b}^m) \right]}. \qquad (5.247)
$$

As in Eq. (5.210), the mass of $\mathbf{a}$ provides the line $L_{\mathbf{x}_1}^{\mathbf{x}_2}$ in $\mathbf{b}^m(\mathbf{x})$ with an energy proportional to its length, so that the disorder correlation function (5.238) goes to zero at large distances.

This result is unchanged by the omitted term (5.246). By expanding it in powers of $\boldsymbol{\nabla}^2$, it consists of a sum of gradient terms

$$
\Delta H = \frac{\tilde{\beta}}{2} \int d^3x \, \boldsymbol{\nabla} \times \mathbf{a} \left[1 + \sum_{1=0}^{\infty} \left(\frac{\boldsymbol{\nabla}^2}{m_A^2} \right)^n \right] \boldsymbol{\nabla} \times \mathbf{a}, \qquad (5.248)
$$

which change only the dispersion of the $\mathbf{a}$-field, but not its mass.

In the high-temperature phase, there are no dual vortices so that the $\mathbf{a}$-field remains massless, and the correlation function is given by an expression like (5.207):

$$
G_{\tilde{D}}(\mathbf{x}_2, \mathbf{x}_1) \approx e^{-\frac{\tilde{\beta}m_a^2}{2} \int d^3x \left[\mathbf{b}^{m2} - (\boldsymbol{\nabla} \times \mathbf{b}^m) \frac{1}{-\boldsymbol{\nabla}^2} (\boldsymbol{\nabla} \times \mathbf{b}^m) \right]} = e^{-\frac{\tilde{\beta}m_a^2}{2} \int d^3x \, \boldsymbol{\nabla} \cdot \mathbf{b}^m \frac{1}{-\boldsymbol{\nabla}^2} \boldsymbol{\nabla} \cdot \mathbf{b}^m}. \qquad (5.249)
$$

This has the same constant large-distance behavior as (5.208) which is independent of the shape of $L_{\mathbf{x}_1}^{\mathbf{x}_2}$, implying a nonzero disorder parameter (5.239). The monopole gauge invariance is unbroken in this phase.

Thus (5.239) is a good disorder parameter for the superconducting phase transition.

c) Alternative Disorder Parameter

At this point we realize that the large-distance behaviors of (5.247) and (5.249) are also found in the correlation function

$$G_{\tilde{D}}(\mathbf{x}_2, \mathbf{x}_1) = \left\langle e^{-\frac{1}{\tilde{\beta}m_A^2} \int d^3x \left\{ \frac{1}{2}\mathbf{b}^m(\mathbf{x})^2 - \mathbf{b}^m(\mathbf{x})[\nabla \times \mathbf{A}(\mathbf{x})] \right\}} \right\rangle, \tag{5.250}$$

where the singular line $L_{\mathbf{x}_1}^{\mathbf{x}_2}$ in $\mathbf{b}^m(\mathbf{x})$ [recall (5.187)] is taken to be the straight line connecting $\mathbf{x}_1$ with $\mathbf{x}_2$ [as in (5.224)]. In the low-temperature ordered phase the vector potential $\mathbf{A}(\mathbf{x})$ has a nonzero Meissner mass m_A and thus the same long-distance behavior as (5.247). In the high-temperature disordered phase, $\mathbf{A}(\mathbf{x})$ is massless and (5.250) behaves like (5.249).

The Hamiltonian leading to the correlation function (5.250) looks like (5.222), but with the magnetic gauge field $\mathbf{b}^m(\mathbf{x})$ inserted into the magnetic gradient term rather than the gradient term of the field $\tilde{\mathbf{a}}(\mathbf{x})$:

$$\tilde{\beta}H_{SC}^{hy} = \int d^3x \left[\frac{1}{2\tilde{\beta}m_A^2}(\nabla \times \tilde{\mathbf{a}})^2 - i\tilde{\mathbf{a}} \cdot (\nabla \times \mathbf{A}) + \frac{\tilde{\beta}}{2}(\nabla \times \mathbf{A} - \mathbf{b}^m)^2 - i\tilde{\mathbf{a}} \cdot \tilde{\mathbf{j}}^v + \frac{\tilde{\beta}\tilde{\epsilon}_c}{2}\tilde{\mathbf{j}}_c^{v2} \right]. \tag{5.251}$$

The disorder parameter associated with the correlation function (5.250) is the expectation value

$$\mathcal{D} = \langle \mathcal{D}(\mathbf{x}) \rangle = \left\langle e^{-\frac{1}{\tilde{\beta}m_A^2} \int d^3x \left\{ \frac{1}{2}\mathbf{b}_{\mathbf{x}}^m(\mathbf{x})^2 - \mathbf{b}_{\mathbf{x}}^m(\mathbf{x})[\nabla \times \mathbf{A}(\mathbf{x})] \right\}} \right\rangle, \tag{5.252}$$

where

$$\mathbf{b}_{\mathbf{x}}^m(\mathbf{x}) = \boldsymbol{\delta}(\mathbf{x}; \tilde{L}_{\mathbf{x}}) \tag{5.253}$$

is singular on any straight line $\tilde{L}_{\mathbf{x}}$ from $\mathbf{x}$ to infinity.

5.4   Order of Superconducting Phase Transition and Tricritical Point

Since the discovery superconductivity by Kamerlingh Onnes in 1908, most experimental data of the phase transition are fitted very well by the BCS theory (recall Fig. 5.8). In the neighborhood of the critical point, the BCS theory is approximated by the Ginzburg-Landau Hamiltonian (5.151) [28]. Fluctuations of the Ginzburg-Landau order field $\phi(\mathbf{x})$ are usually so small that they can be neglected, i.e., mean-field results provide us with good approximations to the data. In the immediate vicinity of the transition, however, they become important.

5.4.1   Fluctuation Regime

The reason why mean-field results are so accurate was first explained by Ginzburg [42] who found a criterion estimating the temperature interval ΔT_G around T_c for which fluctuations can be important. Strictly speaking, his criterion is inapplicable to superconductors, as has often been done, but only to systems with a single real order parameter. Since superconductors have a complex order parameter, a different criterion is relevant which has only recently been found [43]. If the order parameter is not a single real quantity but has a symmetry $O(N)$, the true fluctuation interval ΔT_{GK} has turned out to be by a factor N^2 larger than Ginzburg's estimate $\dot{T}_{\mathrm{G}}$.

The fluctuations in the corrected Ginzburg interval cause a divergence in the specific heat at T_c very similar to the divergence observed in the λ-transition of superfluid helium (recall Fig. 5.1). This interval is in all traditional superconductors too small to be resolved [42, 43], so that it was not astonishing that no fluctuations were observed [44, 45]. The situation has changed only recently with the advent of high-temperature superconductivity where ΔT_{G} is large enough to be experimentally resolved [46, 47].

5.4.2 First- or Second-Order Transition?

In 1972, the order of the superconducting phase transition became a matter of controversy after a theoretical paper by Halperin, Lubensky, and Ma [48] predicted that the transition should really be of first order. The argument was based on an application of renormalization group methods [49] to the partition function

$$Z_{\mathrm{GL}} = \int \mathcal{D}\psi \mathcal{D}\psi^* \int \mathcal{D}\mathbf{A}\, \Phi_T[\mathbf{A}]\, e^{-\tilde{\beta} H_{\mathrm{GL}}[\psi,\psi^*,\mathbf{A}]} \tag{5.254}$$

associated with the Ginzburg-Landau Hamiltonian (5.148) in $4 - \epsilon$ dimensions. The technical signal for the first-order transition was the nonexistence of an infrared-stable fixed point in the renormalization group flow [50] of the coupling constants e and g as a function of the renormalization scale. The fact that all experimental observations indicated a second-order transition was explained by the fact that the fluctuation interval ΔT_{GK} was too small to be detected. Since then there has been much work [51] trying to find an infrared-stable fixed point by going to higher loop orders or by different resummations of the divergent perturbation expansions, but with little success. The controversy was resolved only 10 years later by the author [52] who demonstrated that superconductors can have first- *and* second-order transitions, separated by a *tricritical point*, a result confirmed recently by Monte Carlo simulations [53].

With the advent of modern high-T_c superconductors, the experimental situation has been improved. The temperature interval of large fluctuations is now broad enough to observe critical properties beyond the mean-field approximation. Several experiments have found a critical point of the XY universality class [46]. In addition, there seems to be recent evidence for an additional critical behavior associated with the so-called charged fixed point [47]. In future experiments it will be important to understand the precise nature of the critical fluctuations.

Starting point of the theoretical discussion is the Ginzburg-Landau Hamiltonian (5.148). It contains the field $\psi(x)$ describing the Cooper pairs, and the vector potential $\mathbf{A}(x)$. Near the critical temperature, but outside the narrow interval ΔT_{GK} of large fluctuations, the energy (5.148) describes well the second-order phase transition of the superconductor. It takes place when τ drops below zero where the pair field $\psi(\mathbf{x}) = \rho(\mathbf{x})e^{i\theta(\mathbf{x})}$ acquires the nonzero expectation value $\rho_0 = \sqrt{-\tau/g}$. The properties of the superconducting phase are approximated well by the energy (5.153). The Meissner-Higgs mass term in (5.153) gives rise to a finite penetration depth of the magnetic field $\lambda = 1/m_A = 1/\rho_0 q$.

By expanding the Hamiltonian (5.151) around the hydrodynamic limit (5.153) in powers of the fluctuations $\delta\rho \equiv \tilde{\rho} - \rho_0$, we find that the ρ-fluctuations have a quadratic energy

$$H_{\delta\rho} = \frac{1}{2}\int d^3x \left[(\boldsymbol{\nabla}\delta\rho)^2 - 2\tau(\delta\rho)^2 \right], \qquad (5.255)$$

implying that these have a finite *coherence length* $\xi = 1/\sqrt{-2\tau}$.

The ratio of the two length scales λ and ξ:

$$\kappa \equiv \lambda/\sqrt{2}\xi, \qquad (5.256)$$

which for historic reasons [54] carries a factor $\sqrt{2}$, is the so-called *Ginzburg parameter* whose mean field value is $\kappa_{\mathrm{MF}} \equiv \sqrt{g/q^2}$. Type-I superconductors have small values of κ, type-II superconductors have large values. At the mean-field level, the dividing line lies at $\kappa = 1/\sqrt{2}$.

5.4.3   Partition Function of Superconductor with Vortex Lines

The higher operating temperatures in the new high-T_c superconductors make field fluctuations important. These can be taken into account either by calculating the partition function and field correlation functions from the functional integral [compare (5.157)] or, after the field decomposition (5.150), from

$$Z_{\mathrm{GL}} = \int \mathcal{D}\tilde{\rho}\,\tilde{\rho} \int \mathcal{D}\mathbf{A}\,\Phi_T[\mathbf{A}] \sum_{\{S\}} \Phi[\tilde{\boldsymbol{\theta}}^{\mathrm{v}}] \int \mathcal{D}\tilde{\theta}\, e^{-\tilde{\beta}H_{\mathrm{GL}}[\tilde{\rho},\tilde{\theta},\mathbf{A}]}. \qquad (5.257)$$

This can be approximated by the hydrodynamic formulation (5.157). From now on we use natural temperature units where $k_B T = 1$ and omit all tildes on top of $\tilde{\rho}$, $\tilde{\theta}$, T, etc., for brevity, so that we shall rewrite (5.157):

$$Z_{\mathrm{SC}}^{\mathrm{hy}} = \int \mathcal{D}\mathbf{A}\,\Phi_T[\mathbf{A}] \sum_{\{S\}} \Phi[\boldsymbol{\theta}^{\mathrm{v}}] \int_{-\infty}^{\infty} \mathcal{D}\theta\, e^{-\tilde{\beta}H_{\mathrm{SC}}^{\mathrm{hy}}}. \qquad (5.258)$$

It is instructive recapitulate the basic difficulties in explaining the order of the superconducting phase transition. The simplest argument suggesting a first-order transition is based on performing a mean-field approximation in the pair field ρ and ignore the effect of vortex fluctuations. Thus one sets $\boldsymbol{\theta}^{\mathrm{v}} \equiv 0$ in the Hamiltonian (5.151), and considers

$$H_{\mathrm{GL}}^{\mathrm{app}} \approx \int d^3x \left[\frac{\rho^2}{2}(\boldsymbol{\nabla}\theta - q\mathbf{A}_L)^2 + \frac{1}{2}(\boldsymbol{\nabla}\rho)^2 + V(\rho) + \frac{1}{2}(\boldsymbol{\nabla}\times\mathbf{A})^2 + \frac{\rho^2 q^2}{2}\mathbf{A}_T^2 \right], \qquad (5.259)$$

where the wiggles on top of ρ, θ, and $\boldsymbol{\theta}^{\mathrm{v}}$ have been omitted to simplify the notation. The right-hand side is only an approximation for the following reason. We have separated $\mathbf{A}$ into longitudinal and transverse parts $\mathbf{A}_L$ and $\mathbf{A}_T$ as defined in Eq. (5.132). If ρ were a constant and not a field this separation would lead exactly to (5.259). Due to the $\mathbf{x}$-dependence, however, there will be corrections proportional to the gradient of $\rho(\mathbf{x})$ which we shall ignore, assuming the field $\rho(\mathbf{x})$ to be sufficiently smooth [55].

After these approximations we can integrate out the Gaussian phase fluctuations $\theta(\mathbf{x})$ in the partition function (5.254) and obtain

$$Z_{\text{GL}}^{\text{app}'} = \text{Det}^{-1/2}[-\boldsymbol{\nabla}^2] \int \mathcal{D}\rho\,\mathcal{D}\mathbf{A}\,\Phi_T[\mathbf{A}]\,e^{-\tilde{\beta}H_{\text{GL}}^{\text{app}'}}, \qquad (5.260)$$

with the Hamiltonian

$$H_{\text{GL}}^{\text{app}'} = \int d^3x \left[\frac{1}{2}(\boldsymbol{\nabla}\rho)^2 + V(\rho) + \frac{1}{2}(\boldsymbol{\nabla}\times\mathbf{A})^2 + \frac{\rho^2 q^2}{2}\mathbf{A}_T^2\right]. \qquad (5.261)$$

The fluctuations of the vector potential are also Gaussian, and can be integrated out in (5.260) to yield

$$\bar{Z}_{\text{GL}}^{\text{app}'} = \text{Det}^{-1/2}[-\boldsymbol{\nabla}^2]\,\text{Det}^{-1}[-\boldsymbol{\nabla}^2 + \rho^2 q^2] \int \mathcal{D}\rho\,e^{-\tilde{\beta}\bar{H}_{\text{GL}}^{\text{app}'}}, \qquad (5.262)$$

where

$$\bar{H}_{\text{GL}}^{\text{app}'} = \int d^3x \left[\frac{1}{2}(\boldsymbol{\nabla}\rho)^2 + V(\rho)\right]. \qquad (5.263)$$

5.4.4 First-Order Regime

Assuming again that ρ is smooth, the functional determinant $\text{Det}^{-1}[-\boldsymbol{\nabla}^2 + \rho^2 q^2]$ may be done in the Thomas-Fermi approximation [56] where it yields

$$\text{Det}^{-1}[-\boldsymbol{\nabla}^2 + \rho^2 q^2] = e^{-\text{Tr}\log[-\boldsymbol{\nabla}^2 + \rho^2 q^2]} \approx e^{-V\int[d^3k/(2\pi)^3](k^2 + \rho^2 q^2)} = e^{\rho^3 q^3/6\pi}. \quad (5.264)$$

From now on we shall use natural units for the energy and measure energies in units of $k_B T$. Then we can set $\tilde{\beta} = 1$ in the Boltzmann factors (5.261) and (5.263), and the result (5.264) implies that $\mathbf{A}$-fluctuations contribute a cubic term to the potential $V(\rho)$ in Eq. (5.152), changing it to

$$\bar{V}(\rho) = \frac{\tau}{2}\rho^2 + \frac{g}{4}\rho^4 - \frac{c}{3}\rho^3, \qquad c \equiv \frac{q^3}{2\pi}. \qquad (5.265)$$

The cubic term generates, for $\tau < c^2/4g$, a second minimum in the potential $\tilde{V}(\rho)$ at

$$\tilde{\rho}_0 = \frac{c}{2g}\left(1 + \sqrt{1 - \frac{4\tau g}{c^2}}\right), \qquad (5.266)$$

as illustrated in Fig. 5.9.

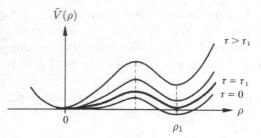

FIGURE 5.9 Potential for the order parameter ρ with cubic term. A new minimum develops around ρ_1 causing a first-order transition for $\tau = \tau_1$.

If τ decreases below

$$\tau_1 = 2c^2/9g, \qquad (5.267)$$

the new minimum drops *below* the minimum at the origin, so that the order parameter jumps from $\rho = 0$ to

$$\rho_1 = 2c/3g \qquad (5.268)$$

in a phase transition. At this point, the coherence length of the ρ-fluctuations $\xi = 1/\sqrt{\tau + 3g\rho^2 - 2c\rho}$ has the finite value

$$\xi_1 = \frac{3}{c}\sqrt{\frac{g}{2}}, \qquad (5.269)$$

which happens to be the same as for the fluctuations around $\rho = 0$. The fact that the transition occurs at a finite $\xi = \xi_1 \neq 0$ indicates that the phase transition is of first order. In a second-order transition, ξ would go to infinity as T approaches T_c.

This conclusion is reliable only if the jump of ρ_0 is sufficiently large. For small jumps, the mean-field discussion of the energy density (5.265) cannot be trusted. At a certain small ρ_0, the transition becomes second-order. The change of the order is caused by the neglected vortex fluctuations in (5.261). We must calculate the partition function (5.260) including the sum over vortex gauge fields $\boldsymbol{\theta}^v(\mathbf{x})$, with a Hamiltonian equal to (5.151) but with omitted wiggles:

$$H_{\mathrm{GL}} = \int d^3x \left[\frac{\rho^2}{2} (\boldsymbol{\nabla}\theta - \boldsymbol{\theta}^v - q\mathbf{A})^2 + \frac{1}{2}(\boldsymbol{\nabla}\rho)^2 + V(\rho) + \frac{1}{2}(\boldsymbol{\nabla}\times\mathbf{A})^2 \right]. \qquad (5.270)$$

If we now integrate out the θ-fluctuations, and assume smooth ρ-fields, we obtain the partition function (5.260) extended by the sum over vortex gauge fields $\boldsymbol{\theta}^v(\mathbf{x})$, and with the Hamiltonian

$$H_{\mathrm{GL}}^{\mathrm{app}'} = \int d^3x \left[\frac{1}{2}(\boldsymbol{\nabla}\rho)^2 + V(\rho) + \frac{1}{2}(\boldsymbol{\nabla}\times\mathbf{A})^2 + \frac{\rho^2}{2}(q\mathbf{A} + \boldsymbol{\theta}^v)_T^2 \right]. \qquad (5.271)$$

We may now study the vortex fluctuations separately by defining a partition function of vortex lines in the presence of a fluctuating $\mathbf{A}$-field for smooth $\rho(\mathbf{x})$:

$$Z_{\boldsymbol{\theta}^v,\mathbf{A}}[\rho] \equiv \int \mathcal{D}\boldsymbol{\theta}_T^v \mathcal{D}\mathbf{A}_T \exp\left\{ -\frac{1}{2}\int d^3x \left[(\boldsymbol{\nabla}\times\mathbf{A})^2 + \frac{\rho^2}{2}(q\mathbf{A} + \boldsymbol{\theta}^v)_T^2 \right] \right\}. \qquad (5.272)$$

The transverse part of the vortex gauge field $\boldsymbol{\theta}^{\mathrm{v}}$ is defined as in (5.132). We have abbreviated the sum over the jumping surfaces S with vortex gauge fixing, $\sum_{\{S\}} \Phi[\boldsymbol{\theta}^{\mathrm{v}}]$, defined in (5.35) by $\int \mathcal{D}\boldsymbol{\theta}_T^{\mathrm{v}}$. In addition, we have fixed of the vector potential to be transverse and indicated this by the functional integration symbol $\mathcal{D}\mathbf{A}_T$.

5.4.5 Vortex Line Origin of Second-Order Transition

The important observation is now that for smooth ρ-fields, this partial partition function possesses a second-order transition of the XY-model type if the average value of ρ drops below a critical value ρ_c. To see this we integrate out the **A**-field and obtain

$$Z_{\boldsymbol{\theta}^{\mathrm{v}},\mathbf{A}}[\rho] \approx \exp\left[\int d^3x \frac{q^3\rho^3}{6\pi}\right] \int \mathcal{D}\boldsymbol{\theta}_T^{\mathrm{v}} \exp\left[\frac{\rho^2}{2}\int d^3x \left(\frac{1}{2}\boldsymbol{\theta}_T^{\mathrm{v}\,2} - \boldsymbol{\theta}_T^{\mathrm{v}} \frac{\rho^2 q^2}{-\boldsymbol{\nabla}^2 + \rho^2 q^2}\boldsymbol{\theta}_T^{\mathrm{v}}\right)\right]. \tag{5.273}$$

The first factor yields, once more, the cubic term of the potential (5.265). The second factor accounts for the vortex loops. The integral in the exponent can be rewritten as

$$\frac{\rho^2}{2}\int d^3x \left(\boldsymbol{\theta}_T^{\mathrm{v}} \frac{-\boldsymbol{\nabla}^2}{-\boldsymbol{\nabla}^2 + \rho^2 q^2}\boldsymbol{\theta}_T^{\mathrm{v}}\right). \tag{5.274}$$

Integrating this by parts, and using the identity

$$\int d^3x \, \nabla_i\mathbf{A}\,\nabla_i\mathbf{B} = \int d^3x\,[(\boldsymbol{\nabla}\times\mathbf{A})(\boldsymbol{\nabla}\times\mathbf{B}) + (\boldsymbol{\nabla}\cdot\mathbf{A})(\boldsymbol{\nabla}\cdot\mathbf{B})], \tag{5.275}$$

together with the transversality property $\boldsymbol{\nabla}\cdot\boldsymbol{\theta}_T^{\mathrm{v}} = 0$ and the curl relation $\boldsymbol{\nabla}\times\boldsymbol{\theta}_T^{\mathrm{v}} = \mathbf{j}^{\mathrm{v}}$ of Eq. (5.30), the partition function (5.273) without the prefactor takes the form

$$\bar{Z}_{\boldsymbol{\theta}^{\mathrm{v}},\mathbf{A}}[\rho] \approx \int \mathcal{D}\boldsymbol{\theta}_T^{\mathrm{v}} \exp\left[-\frac{\rho^2}{2}\int d^3x \left(\mathbf{j}^{\mathrm{v}}\frac{1}{-\boldsymbol{\nabla}^2 + \rho^2 q^2}\mathbf{j}^{\mathrm{v}}\right)\right]. \tag{5.276}$$

This is the partition function of a grand-canonical ensemble of closed fluctuating vortex lines. The interaction between them is of the Yukawa type, with a finite range equal to the penetration depth $\lambda = 1/\rho q$.

It is well-known how to compute pair and magnetic fields of the Ginzburg-Landau theory for a single straight vortex line from the extrema of the energy density [44]. In an external magnetic field, there exist triangular and various other regular arrays of vortex lines, such as vortex lattices. In the presence of impurities, the vortex lattices may turn into vortex glasses. The study of such phases and the transitions between them is an active field of research [57].

In the core of each vortex line, the pair field ρ goes to zero over a distance ξ. If we want to sum over a grand-canonical ensemble of fluctuating vortex lines of any shape in the partition function (5.276), the space dependence of ρ causes complications. These can be avoided by an approximation, in which the system is placed on a simple-cubic lattice of spacing $a = \alpha\,\xi$, with α of the order of unity, and a *fixed* value $\rho = \tilde{\rho}_0$ given by Eq. (5.266). Thus we replace the partial partition function (5.276) approximately by

$$\bar{Z}_{\boldsymbol{\theta}^{\mathrm{v}},\mathbf{A}}[\tilde{\rho}_0] \approx \sum_{\{\mathbf{l};\nabla\cdot\mathbf{l}=0\}} \exp\left[-\frac{4\pi^2\tilde{\rho}_0^2 a}{2}\sum_{\mathbf{x}}\mathbf{l}(\mathbf{x})v_{\tilde{\rho}_0 q}(\mathbf{x}-\mathbf{x}')\mathbf{l}(\mathbf{x}')\right]. \qquad (5.277)$$

The sum runs over the discrete versions of the vortex density $\mathbf{j}^{\mathrm{v}}$ in (5.276). Recalling (5.34) and (5.46), these are 2π times the integer-valued vectors $\mathbf{l}(\mathbf{x}) = (l_1(\mathbf{x}), l_2(\mathbf{x}), l_3(\mathbf{x})) = \nabla\times\mathbf{n}(x)$, where ∇ denotes the lattice derivative (5.38). Being lattice curls of the integer vector field $\mathbf{n}(\mathbf{x}) = (n_1(\mathbf{x}), n_2(\mathbf{x}), n_3(\mathbf{x}))$, they satisfy $\nabla\cdot\mathbf{l}(\mathbf{x}) = 0$. This condition restricts the sum over $\mathbf{l}(\mathbf{x})$-configurations in (5.277) to all non-selfbacktracking integer-valued closed loops. The partition function (5.272) has precisely the form discussed before in Eq. (5.52) with $\rho_0 q$ playing the role of the Yukawa mass m in (5.52). The lattice partition function (5.277) has therefore a second-order phase transition The transition temperature was plotted in Fig. 5.7.

5.4.6   Tricritical Point

We are now prepared to locate the position of the tricritical point where the order of the superconducting phase transition changes from second to first. Consider the superconductor in the low-temperature phase where the size of the order parameter lies in the right-hand minimum of the potential $\tilde{V}(\rho)$ in Fig. 5.9. Upon heating, the minimum moves closer to the origin. The decrease of ρ^2 in the Boltzmann factor of (5.276) increases the number and the length of the vortex lines. If the critical value of vortex condensation is reached before the right-hand minimum arrives at the same height as the left-hand minimum at the origin in Fig. 5.9. Then the superconductor undergoes a second-order phase transition of the XY-model type. If equal height is reached before this, the order parameter ρ jumps to zero in a first-order transition. The first-order transition takes certainly place for large q, i.e., for small mean-field Ginzburg parameter $\kappa_{\mathrm{MF}} = \sqrt{g/q^2}$ [recall (5.256)], where the jump is large. In the opposite limit, the vortex loops will condense before the order parameter jumps to zero. The above considerations permit us to determine at which value of κ_{MF} the order changes.

Comparing (5.277) with the partition function (5.106) of the Yukawa loop gas, we conclude that the vortex condensation takes place when [compare (5.108)]

$$4\pi^2 a\tilde{\rho}_0^2 v_{\tilde{\rho}_0 q}(\mathbf{0}) \approx T_c \approx 3. \qquad (5.278)$$

Using the analytic approximation (5.111), we may write this as

$$4\pi^2 a v_0(0)\frac{\tilde{\rho}_0^2}{\sigma a^2\tilde{\rho}_0^2 q^2/6 + 1} \approx T_c \approx 3, \qquad (5.279)$$

or

$$\frac{\tilde{\rho}_0^2 a}{\sigma a^2\tilde{\rho}_0^2 q^2/6 + 1} \approx \frac{r}{3}, \qquad (5.280)$$

where $r = 9/4\pi^2 v_0(0) \equiv 0.90$. The solution is

$$\tilde{\rho}_0 \approx \frac{1}{\sqrt{3a}} \sqrt{\frac{r}{1 - \sigma r a q^2/18}}. \tag{5.281}$$

Replacing here a by $\alpha\xi_1 = \alpha(3/c)\sqrt{g/2}$ of Eq. (5.269), and $\tilde{\rho}_0$ by $\rho_1 = 2c/3g$ of Eq. (5.268), and inserting further $c = q^3/2\pi$ of Eq. (5.265), we find the equation for the mean-field Ginzburg parameter $\kappa_{\text{MF}} = \sqrt{g/q^2}$ [recall (5.256)]:

$$\kappa_{\text{MF}}^3 + \alpha^2 \sigma \frac{\kappa_{\text{MF}}}{3} - \frac{\sqrt{2}\alpha}{\pi r} = 0. \tag{5.282}$$

For the best value $\sigma \approx 1.6$ in the approximation (5.111), the parameter $r \approx 0.9$, and the rough estimate $\alpha \approx 1$, the solution of this equation yields the tricritical value

$$\kappa_{\text{MF}}^{\text{tric}} \approx 0.82/\sqrt{2}. \tag{5.283}$$

In spite of the roughness of the approximations, this result is very close to the value

$$\kappa_{\text{MF}}^{\text{tric}} = \frac{3\sqrt{3}}{2\pi} \sqrt{1 - \frac{4}{9}\left(\frac{\pi}{3}\right)} \approx \frac{0.80}{\sqrt{2}} \tag{5.284}$$

derived from the dual theory in [23]. The approximation (5.283) has three uncertainties. First, the identification of the effective lattice spacing $a = \alpha\xi$ with $\alpha \approx 1$; second the associated neglect of the $\mathbf{x}$-dependence of ρ and its fluctuations, and third the localization of the critical point of the XY-model type transition in Eqs. (5.112) and the ensuing (5.280).

5.4.7 Disorder Theory

In the disorder theory (5.174) it is much easier to prove that superconductors can have a first- and a second-order phase transition, depending on the size of the Ginzburg parameter κ defined in Eq. (5.256). Before we start let us rewrite the disorder theory in a more convenient way. As before, we decompose the complex disorder field ϕ as $\phi = \rho e^{i\theta}$. In the partition function (5.176), this changes the measure of functional integration from $\int \mathcal{D}\phi \int \mathcal{D}\phi^*$ to $\int \mathcal{D}\rho\rho \int \mathcal{D}\theta$. Now we fix the gauge by absorbing the phase θ of the field into $\tilde{\mathbf{a}}$ by a gauge transformation (5.175). This brings the Hamiltonian (5.174) to the form

$$\tilde{\beta}\tilde{H}_{\text{SC}}^{\text{hy}} \sim \int d^3x \left[\frac{1}{2\tilde{\beta}m_A^2}[(\nabla \times \tilde{\mathbf{a}})^2 + m_A^2\tilde{\mathbf{a}}_T^2] + \frac{\rho^2}{2}\left(\tilde{\mathbf{a}}_T^2 + \tilde{\mathbf{a}}_L^2\right) + \frac{1}{2}(\nabla\rho)^2 + \frac{\tau}{2}\rho^2 + \frac{g}{4}\rho^4 \right], \tag{5.285}$$

where we have again assumed a smooth ρ-field to separate $\rho^2\tilde{\mathbf{a}}^2$ into $\rho^2\tilde{\mathbf{a}}_T^2 + \rho^2\tilde{\mathbf{a}}_L^2$. The partition function (5.176) reads now

$$Z_{\text{SC}}^{\text{dual}} = \int \mathcal{D}\rho\rho \int \mathcal{D}\tilde{\mathbf{a}}\, e^{-\tilde{\beta}\tilde{H}_{\text{SC}}^{\text{hy}}}. \tag{5.286}$$

We may integrate out $\tilde{\mathbf{a}}_L$ to obtain a factor $\mathrm{Det}[\rho^2]^{-1/2}$ which removes the factor ρ in the measure of path integration $\mathcal{D}\rho\,\rho$. This leads to the partition function

$$Z_{\mathrm{SC}}^{\mathrm{dual}} = \int \mathcal{D}\rho\, \mathrm{Det}[-\boldsymbol{\nabla}^2 + m_A^2(1 + \tilde{\beta}\rho^2)]e^{-\tilde{\beta}\tilde{H}_{\mathrm{SC}}^{\mathrm{hy}}}, \qquad (5.287)$$

with

$$\tilde{\beta}\tilde{H}_{\mathrm{SC}}^{\mathrm{hy}} = \int d^3x \left[\frac{1}{2}\,(\boldsymbol{\nabla}\rho)^2 + \frac{\tau}{2}\rho^2 + \frac{g}{4}\rho^4\right] + \mathrm{Tr}\log[-\boldsymbol{\nabla}^2 + m_A^2(1 + \tilde{\beta}\rho^2)]. \quad (5.288)$$

In the superconducting phase, there are only a few vortex lines and the disorder field ρ of vortex lines fluctuates around zero. In this phase we may expand the Tracelog into powers of ρ^2. The first expansion term is proportional to ρ^2 and renormalizes τ in the Hamiltonian (5.288), corresponding to a shift in the critical temperature.

The second expansion term is approximately given by

$$-\tilde{\beta}^2 m_A^4 \int d^3x \int \frac{d^3k}{(2\pi)^3}\frac{1}{(k^2 + m_A^2)^2}\rho^4 \propto -m_A^3 \int d^3x\, \rho^4. \qquad (5.289)$$

This term *lowers* the interaction term $(g/4)\rho^4$ in the Hamiltonian (5.288). An increase in m_A corresponds to a decrease of the penetration depth in the superconductor, i.e., to materials moving towards the type-I regime. At some larger value of m_A, the ρ^4-term vanishes and the disorder field theory requires a ρ^6-term to stabilize the fluctuations of the vortex lines. In such materials, the superconducting phase transition turns from second to first order.

A more quantitative version of this argument was used in Ref. [52] to show the existence of the tricritical point and its location at the Ginzburg parameter $\kappa \equiv g/q^2$ in Eq. (5.284). This agrees well with a recent Monte Carlo value $(0.76 \pm 0.04)/\sqrt{2}$ of Ref. [53].

5.5 Vortex Lattices

The model action (5.26) represents the gradient energy in superfluid ^{4}He correctly only in the long-wavelength limit. The neutron scattering data yield the energy spectrum $\omega = \epsilon(\mathbf{k})$ shown in Fig. 5.2.

To account for this, the energy should be taken as follows:

$$H_{\mathrm{NG}} = \frac{1}{2}\int d^3x(\boldsymbol{\nabla}\theta - \boldsymbol{\theta}^{\mathrm{v}})\frac{\epsilon^2(-i\boldsymbol{\nabla})}{-\boldsymbol{\nabla}^2}(\boldsymbol{\nabla}\theta - \boldsymbol{\theta}^{\mathrm{v}}). \qquad (5.290)$$

The *roton peak* near 2Å^{-1} gives rise to a repulsion between opposite vortex line elements at the corresponding distance. If a layer of superfluid ^{4}He is diluted with ^{3}He, the core energy of the vortices decreases, the fugacity y and the average vortex number increases. For a sufficiently high average spacing, a vortex lattice forms. In this regime, the superfluid has three transitions when passing from zero temperature

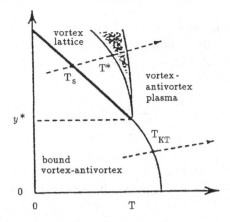

FIGURE 5.10 Phase diagram of a two-dimensional layer of superfluid ^{4}He. At a higher fugacity $y > y^*$, an increase in temperature causes the vortices to first condense to a lattice. This has a melting transition to a phase which possesses a Kosterlitz-Thouless vortex unbinding transition (after Ref. [58]).

to the normal phase. There is first a condensation process to a vortex lattice, then a melting transition of this lattice into a fluid of bound vortex-antivortex pairs, and finally a pair-unbinding transition of the Kosterlitz-Thouless type. The latter two transitions have apparently been seen experimentally (see Fig. 5.10).

Appendix 5A Single Vortex Line in Superfluid

Here we derive some properties of an individual vortex line obtained by extremizing the Ginzburg-Landau Hamiltonian (5.6). For simplicity, we shall focus attention only upon a straight line. Such a line can be obtained as a cylindrical solution to the field equation

$$-\boldsymbol{\nabla}^2\phi + \tau\phi + \lambda|\phi|^2\phi = 0, \qquad (5A.1)$$

which minimizes the energy (5.6).

Decomposing ϕ into its polar components as in (5.10), $\phi(\mathbf{x}) = \rho e^{i\theta(\mathbf{x})}$, and ignoring for a moment the vortex gauge field, the real and imaginary parts of this equation read

$$\left[-\boldsymbol{\nabla}^2 + (\boldsymbol{\nabla}\theta)^2 + \tau + \lambda\rho^2\right]\rho = 0 \qquad (5A.2)$$

and

$$\boldsymbol{\nabla}\mathbf{j}_s(\mathbf{x}) = 0, \qquad \mathbf{j}_s(\mathbf{x}) \equiv \rho^2\boldsymbol{\nabla}\theta(\mathbf{x}) = 0. \qquad (5A.3)$$

The latter equation is the statement of current conservation for the current density of superflow $\mathbf{j}_s(\mathbf{x})$.

Current conservation can be ensured by a purely circular flow in which ρ depends on the distance r from the cylindrical axis and the phase θ is an integer multiple of the azimuthal angle in space, $\theta = n \arctan(x_2/x_1)$. Then (5A.2) reduces to the radial differential equation

$$-\left(\partial_r^2 + \frac{1}{r}\partial_r - \frac{n^2}{2}\right)\rho + \lambda(\rho^2 - \rho_0^2)\rho = 0, \tag{5A.4}$$

where $\rho_0 = \sqrt{-\tau/\lambda} = \sqrt{(1 - T/T_c)/\lambda}$ [compare (5.7), (5.8)].

In order to solve this equation it is convenient to go to reduced quantities $\bar{r}, \bar{\rho}$, which measure the distance r in units of the coherence length [recall (5.13)]

$$\xi = \xi_0 \frac{1}{\sqrt{2(1 - T/T_c)}} \tag{5A.5}$$

and the size of the order parameter ρ in units of ρ_0, i.e., we introduce

$$\bar{x} = x/\xi, \quad \bar{r} = r/\xi \tag{5A.6}$$

$$\bar{\rho} = \rho/\rho_0. \tag{5A.7}$$

Then (5A.4) takes the form

$$\left[-\left(\partial_{\bar{r}}^2 + \frac{1}{\bar{r}}\partial_{\bar{r}} - \frac{n^2}{\bar{r}^2}\right) + (\bar{\rho}^2 - 1)\right]\bar{\rho}(\bar{r}) = 0. \tag{5A.8}$$

Multiplying this with the phase factor

$$e^{in\theta} = e^{in\tan^{-1}(x_2/x_1)}, \tag{5A.9}$$

we see that the complex field $\phi(\mathbf{x})$ has the following small $|\mathbf{x}|$ behavior

$$\phi(\mathbf{x}) \propto \bar{r}^n e^{in\tan^{-1}(x_2/x_1)} = (x_1 + ix_2)^n, \tag{5A.10}$$

which corresponds to a zero of n-th order in $\phi(\mathbf{x})$.

For large $\bar{r} \gg 1$, $\bar{\rho}(\bar{r})$ approaches the asymptotic value $\bar{\rho} = 1$. In fact, Eq. (5A.8) is solved by a large-$\bar{r}$ expansion:

$$\bar{\rho}_n(\bar{r}) = 1 - \frac{n^2}{2\bar{r}^2} - \left(n^2 + \frac{1}{8}n^4\right)\frac{1}{\bar{r}^4} - \left(8 + 2n^2 + \frac{1}{16}n^4\right)\frac{n}{\bar{r}^6} - \mathcal{O}\left(\frac{1}{\bar{r}^8}\right). \tag{5A.11}$$

Integrating the differential equation numerically inward, we find the solution displayed in Fig. 5.11.

Let us now study the energy of these vortex lines. The calculation can be simplified by a scaling argument: if $\phi(\mathbf{x})$ is the solution of the differential equation (5A.1), the rescaled solution

$$\phi_\delta(\mathbf{x}) \equiv e^\delta \cdot \phi(\mathbf{x}) \tag{5A.12}$$

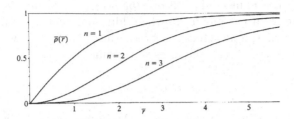

FIGURE 5.11 Order parameter $\bar{\rho} = |\phi|/|\phi_0|$ around a vortex line of strength $n = 1, 2, 3, \ldots$ as a function of the reduced distance $\bar{r} = r/\xi$, where r is the distance from the axis and ξ the healing length.

must extremize the energy for $\delta = 0$. Inserting (5A.12) into (5.6), we calculate

$$E = \frac{1}{2} \int d^3x \left[e^{2\delta} |\nabla \phi|^2 + e^{2\delta} \tau |\phi|^2 + \frac{\lambda}{2} e^{4\delta} |\phi|^4 \right]. \tag{5A.13}$$

Setting the derivative with respect to δ equal to zero gives at $\delta = 0$

$$\frac{1}{2} \int d^3x \left[|\nabla \phi|^2 + \tau |\phi|^2 + \lambda |\phi|^4 \right] = 0. \tag{5A.14}$$

Subtracting this from (5A.13) for $\delta = 0$ we see that the energy of a solution of the field equation is simply given by

$$E = -\frac{\lambda}{4} \int d^3x |\phi|^4. \tag{5A.15}$$

Most of this energy is due to the asymptotic regime where $\phi \to \phi_0 = \rho_0 = \sqrt{-\tau/\lambda}$. Inserting this into (5A.16) yields the condensation energy (5.15)

$$E_c = -V \frac{\lambda}{4} \rho_0^4 = -V f_c, \qquad f_c = \frac{\tau^2}{4\lambda}. \tag{5A.16}$$

Subtracting this from (5A.16) we find the energy due to the presence of the vortex line

$$E_v = -\frac{\lambda}{4} \int d^3x \left(|\phi|^4 - |\phi_0|^4 \right). \tag{5A.17}$$

In terms of natural units introduced in (5A.6) and (5A.7), this is simply

$$E_v = -f_c \xi^3 \int d^3\bar{x} \left[1 - \bar{\rho}^4(\mathbf{x}) \right]. \tag{5A.18}$$

Going over to cylindrical coordinates $\bar{r}, \theta, z$, the integral becomes, for a line of reduced length $\bar{L}$ along the z-direction:

$$2\pi \bar{L} \int_0^\infty d\bar{r} \, \bar{r} \left[1 - \bar{\rho}^4(\bar{r}) \right]. \tag{5A.19}$$

Before inserting the numerical solutions for $\bar{\rho}(\bar{r})$ plotted in Fig. 5.11 we note that due to the factor $\bar{r}$ in the integrand, the additional energy comes mainly from the large-$\bar{r}$ regime, i.e., the far zone around the line. In fact, if we insert the leading asymptotic behavior (5A.11), the integral becomes

$$2\pi n^2 \bar{L} \int^{\infty} d\bar{r}/\bar{r}, \tag{5A.20}$$

which diverges logarithmically for large $\bar{r}$. An immediate conclusion is that a single vortex line can have a finite energy only in a finite container. If this container is cylindrical of radius R, the integral is finite and becomes $4\pi n^2 \bar{L} \log(R/\xi)$. In an infinite container, straight vortex lines can only exist in pairs of opposite circulation.

Consider now the small-$\bar{r}$ behavior. From (5A.8) we see that close to the origin, $\bar{\rho}(\bar{r})$ behaves like $\bar{r}^n$. Hence $1 - \bar{\rho}^4$, and the energy of a thin cylindrical section of radius $\bar{r}$ grows like $\bar{r}^2$. For increasing $\bar{r}$, the rate of growth rapidly slows down and settles at the asymptotic rate $4\pi n^2 \bar{L} \times \log(r/\xi)$, where ξ is the coherence length. The proper inclusion of the nonasymptotic behavior gives a finite correction to this asymptotic energy. In a container of radius R, the integral (5A.19) becomes $4\pi n^2 \bar{L}[\log(R/\xi) + c]$. The precise numerical evaluation of the differential equation (5A.8) and the integral (5A.19) shows that for the lowest vortex line, c has the value

$$c = 0.385. \tag{5A.21}$$

Hence the energy of the vortex line per unit length becomes

$$\frac{E_v}{L} \approx f_c \xi^2 4\pi n^2 [\log(R/\xi) + 0.385]. \tag{5A.22}$$

The same result would have been obtained by replacing the integrand $\bar{r}[1 - \bar{\rho}^4(\bar{r})]$ by its asymptotic form $2n^2/\bar{r}$ and integrating from a radius

$$r_c = \xi e^{-c} \tag{5A.23}$$

to R. The quantity r_c is called the *core radius* of the vortex line.

The logarithmic divergence of the energy has a simple physical meaning. In order to see this let us calculate this energy once more using the original expression (5.6), i.e., without invoking the property (5A.14). It reads, for a cylindrical solution:

$$\frac{E_v}{L} = f_c \xi^2 4\pi \int d\bar{r}\, \bar{r} \left\{ (\partial_{\bar{r}} \bar{\rho})^2 + \frac{1}{2}(1 - \bar{\rho}^2)^2 + \frac{n^2}{\bar{r}^2} \bar{\rho}^2 \right\}. \tag{5A.24}$$

The first two terms are rapidly converging. Thus the energy of the far-zone resides completely in the last term

$$\frac{n^2}{\bar{r}^2} \bar{\rho}^2 \approx \frac{n^2}{\bar{r}^2}. \tag{5A.25}$$

This energy is a consequence of the angular behavior of the condensate phase $\theta = n \tan^{-1}(x_2/x_1)$ around a vortex line. In fact, the term (5A.26) is entirely due to the *azimuthal part* of the gradient energy $(1/2)\rho^2(\nabla\theta)^2$, i.e., the term which describes the *Nambu-Goldstone modes*.

The dominance of the energy carried by the phase gradient can also be described in a different and more physical way. In Eq. (5A.3) we have seen that $\mathbf{j}_s(\mathbf{x}) = \rho^2\nabla\theta(\mathbf{x})$ is the current density of superflow, and we may identify the gradient $\nabla\theta(\mathbf{x})$ as the superflow velocity $\mathbf{v}_s(\mathbf{x})$. In physical units, it is given by $\hbar\nabla\theta(\mathbf{x})$. Far away from the line it reads explicitly

$$\mathbf{v}_s = \frac{\hbar}{M}n\nabla\arctan\left(\frac{x_2}{x_1}\right) = \frac{\hbar}{M}\frac{n}{r^2}(-x_2,x_1,0) = \frac{\hbar}{M}n\frac{1}{r}\mathbf{e}_\phi, \tag{5A.26}$$

where $\mathbf{e}_\phi$ is the unit vector in azimuthal direction. Thus, around every vortex line, there is a circular flow of the superfluid whose velocity decreases like the inverse distance from the line. With the notation (5.17) for the superfluid density, the hydrodynamic energy density of the velocity field (5A.26) is

$$\mathcal{E}(\mathbf{x}) = \frac{\rho_s}{2}\mathbf{v}_s^2(\mathbf{x}) = \frac{\rho_s}{2}\frac{\hbar^2}{M^2}\frac{n^2}{r^2}. \tag{5A.27}$$

This is precisely the dominant third Nambu-Goldstone term in the energy integral (5A.24). Thus the energy of the vortex line is indeed mainly due to the hydrodynamic energy of the superflow around the line.

For the major portion of the fluid, the limiting hydrodynamic expressions (5A.26), (5A.27) give an excellent approximation to these quantities. Only in the neighborhood of the line, i.e., for small radii $r \leq \xi$, the energy density differs from (5A.27) due to gradients of the size of the field $|\phi|$. It is therefore suggestive to *idealize* the superfluid and *assume* the validity of the pure gradient energy density

$$\mathcal{E}(\mathbf{x}) = \frac{\rho_s}{2}\mathbf{v}_s^2(\mathbf{x}) = \frac{\hbar^2\rho_s}{2M^2}[\nabla\theta(\mathbf{x})]^2 \tag{5A.28}$$

everywhere in space.

The deviations from this law, which become significant only very close to a vortex line, i.e., at distances of the order of the coherence length ξ, are treated *approximately* by simply cutting off the energy integration at the core radius r_c (5A.23) around the vortex line. In other words, we pretend as though there is no superflow at all within the thin tubes of radius ξ, and assume a sudden onset of idealized flow outside r_c, moving with the limiting velocity (5A.26).

Although the internal part of the thin tube carries no superflow, it nevertheless carries rotational energy. Within the present approximation, this energy is associated with the number $c = 0.385$ in (5A.22). This piece will be called the *core energy*. The core energy has a physical interpretation. At distances smaller than the core radius, the different parts of the liquid can no longer slip past each other freely. Hence the core of a vortex line is expected to rotate roughly like a solid rod, rather than with the diverging velocity $v_s \sim 1/r$. Indeed, if we use the approximation

$$v_s \approx n\begin{cases} r/\xi, & r \leq \xi, \\ 1/r, & r > \xi \end{cases} \tag{5A.29}$$

for a line of vortex strength n, the energy density has, for small r, precisely the behavior proportional to r^2 observed before in the exact expressions (5A.18). Moreover, the energy integration gives

$$n^2 \left[\int_1^{R/\xi} d\bar{r}\, \bar{r}\, \frac{1}{\bar{r}^2} + \int_0^1 d\bar{r}\, \bar{r}\, \bar{r}^2 \right] = n^2 [\log(R/\xi) + 0.25], \qquad (5A.30)$$

and we see that the number c for the core energy emerges with this approximation as 0.25, which is of the correct order of magnitude.

To complete our discussion of the hydrodynamic picture, let us calculate the circulation of the superfluid velocity field around the vortex lines:

$$\kappa \equiv \oint_B dx_i v_{s_i} = \frac{\hbar}{M} n \oint_B dx_i \partial_i \theta = \frac{\hbar}{M} 2\pi n = n \frac{h}{M} = n\kappa_1. \qquad (5A.31)$$

This integral is the same for any size and shape of the circuit B around the vortex line. Thus the circulation is quantized and always appears in multiplets of $\kappa_1 = h/M \approx 10^{-3}$ cm^2/sec. The number n is called *vortex strength*.

The integral (5A.31) can be transformed into a surface integral via Stokes' theorem (4.21):

$$\int_{S^B} d^2x\, \boldsymbol{\nabla} \times \mathbf{v}_s = \frac{\hbar}{M} 2\pi n, \qquad (5A.32)$$

where S^B is some surface spanned by the circuit B in (5A.31). This integral is the same for any size and shape of S^B. From this result we conclude that the third component of the curl of $\mathbf{v}_s$ must vanish everywhere, except at the origin. There it must have a singularity of such a strength that the two-dimensional integral gives the correct vortex strength. Hence

$$\boldsymbol{\nabla} \times \mathbf{v}_s = \frac{\hbar}{M} 2\pi n\, \delta^{(2)}(\mathbf{x}_\perp)\hat{\mathbf{z}}, \qquad (5A.33)$$

are the coordinates orthogonal to the vortex line. This is the two-dimensional version of (5.30).

If the nonlinearities of the field are taken into account, the $\delta^{(2)}$-function is really smeared out over a circle whose radius is of order ξ. Typically, the term $1/r^2$ in (5A.27) will be softened to $1/(r^2 + \varepsilon^2)$, in which case the curl of the superfluid velocity becomes

$$\boldsymbol{\nabla} \times \mathbf{v}_s = \frac{\hbar}{M} n \frac{2\varepsilon^2}{(r^2 + \varepsilon^2)^2}\hat{\mathbf{z}}. \qquad (5A.34)$$

The right-hand side is nonzero only within a small radius $r \geq \varepsilon$, where it diverges with the total strength

$$\int d^2x\, \frac{2\varepsilon^2}{(r^2 + \varepsilon^2)^2} = 2\pi\varepsilon^2 \int dr\, \frac{2r}{(r^2 + \varepsilon^2)^2} = 2\pi. \qquad (5A.35)$$

This shows that (5A.34) is, indeed, a smeared-out version of the singular relation (5A.33).

Due to their rotational properties, vortex lines can be generated experimentally by rotating a vessel with an angular velocity Ω. Initially, the lack of friction will cause the superfluid part of the liquid to remain at rest. This situation cannot, however, persist forever since it is not in a state of thermal equilibrium. After some time, vortex lines form on the walls which migrate into the liquid and distribute evenly. This goes on until their total number is such that the rotational Helmholtz free energy

$$E_\Omega = H - \mathbf{\Omega} \cdot \mathbf{L} \approx \int d^3x \left(\frac{\rho_s}{2} \mathbf{v}_s^2 - \mathbf{\Omega} \cdot \mathbf{x} \times \rho_s \mathbf{v}_s \right) \qquad (5A.36)$$

is minimal. This equilibration process has been observed in the laboratory and has even been photographed. This was done using the property that vortex lines trap ions which can be accelerated against a photographic plate.

If we evaluate the energy (5A.36) with the circular velocity field $\mathbf{v}_s(\mathbf{x})$ of a single vortex of Eq. (5A.26), we find that in a cylindrical vessel of radius R, the first vortex line $n = 1$ appears at a critical angular velocity

$$\Omega_c = \frac{\kappa_1}{\pi R^2} \log \frac{R}{\xi} \qquad (5A.37)$$

and settles on the axis of rotation. It is useful to observe that the vortex lines of higher n are all unstable. Since the energy increases quadratically with n, it is favorable for a single line with $n > 1$ to decay into n lines with $n = 1$. When generating vortex lines by stirring a vessel, one may nevertheless be able to create, for a short time, such an unstable line, and to observe its decay.

After the seminal work by Onsager [5] and Feynman [6] the properties of vortex lines in rotating superfluid helium were calculated by Hess in [59] in 1967. His results were improved in many later papers. In 1969, vortex lines were produced experimentally by Packard's group in Berkeley [60]. Since the discovery of Bose-Einstein condensation by Eric Cornell and Carl Wieman at Cornell university and by Wolfgang Ketterle at MIT in 1995, experimentalist are able to produce them in well-controlled experiments at ultralow temperatures. This has led to a flurry of activity in this field. A small selection of the publications is cited in Ref. [61]. The reader can trace the development from the references therein.

Notes and References

[1] P. Debye, *Zur Theorie der spezifischen Wärmen*, Annalen der Physik 39(4), 789 (1912).

[2] L.D. Landau, J. Phys. U.S.S.R. **11**, 91 (1947) [see also Phys. Rev. **75**, 884 (1949)].

[3] C.A. Jones and P.H. Roberts, J. Phys. A: Math. Gen. **15**, 2599 (1982).

[4] R.A. Cowley and A.D. Woods, Can. J. Phys **49**, 177 (1971).

[5] L. Onsager, Nuovo Cimento Suppl. **6**, 249 (1949).

[6] R.P. Feynman, in *Progress in Low Temperature Physics*, ed. by C. J. Gorter, North-Holland, Amsterdam, 1955.

[7] H. Kleinert, *Gauge Fields in Condensed Matter*, Vol. I: *Superflow and Vortex Lines, Disorder Fields, Phase Transitions*, World Scientific, Singapore, 1989 (`kl/b1`), where `kl` is short for the www address `http://www.physik.fu-berlin.de/~kleinert`.

[8] W. Shockley, in *L'Etat Solide*, Proc. Neuvi'eme Conseil de Physique, Brussels, ed. R. Stoops (Inst. de Physique, Solvay, Brussels, 1952).

[9] H. Kleinert, *Gauge Fields in Condensed Matter*, Vol. II: *Stresses and Defects, Differential Geometry, Crystal Defects*, World Scientific, Singapore, 1989 (`kl/b2`).

[10] Note that this configurational entropy cannot be properly accounted for by a model restricted only to circular vortex lines proposed by G. Williams, Phys. Rev. Lett. **59**, 1926 (1987).

[11] This relation was first stated on p. 517 of the textbook [7] and confirmed experimentally by Y.J. Uemura, Physica B **374**, 1 (2006) (cond-mat/0512075).

[12] See pp. 529-530 in textbook [7] (`kl/b1/gifs/v1-530s.html`).

[13] L.P. Pitaevskii, Zh. Eksp. Teor. Fiz. **40**, 646 (1961) [Sov. Phys.-JETP **13**, 451 (1961)].

[14] H. Kleinert, *Theory of Fluctuating Nonholonomic Fields and Applications: Statistical Mechanics of Vortices and Defects and New Physical Laws in Spaces with Curvature and Torsion*, in: Proceedings of NATO Advanced Study Institute on Formation and Interaction of Topological Defects, Plenum Press, New York, 1995, pp. 201–232 (`kl/227`).

[15] See pp. 136–145 in textbook [7] (`kl/b1/gifs/v1-136s.html`).

[16] See p. 468 in textbook [7] (`kl/b1/gifs/v1-468s.html`).

[17] See p. 241 in textbook [7] (`kl/b1/gifs/v1-241s.html`).

[18] J. Villain, J. Phys. (Paris) **36**, 581 (1977). See also the textbook discussion at `kl/b1/gifs/v1-489s.html`.

[19] See p. 503 in textbook [7] (`kl/b1/gifs/v1-503s.html`). The high-temperature expansions of the partition function (5.36) and the associated free energy are given in Eqs. (7.42a) and (7.42b), the low-temperature expansion of the free energy in Eq. (7.43).

[20] H. Kleinert, *Path Integrals in Quantum Mechanics, Statistics, Polymer Physics, and Financial Markets,* 4th ed., World Scientific, Singapore 2006 (`kl/b5`).

[21] Similar canonical representations for the defect ensembles in a variety of physical systems are given in
H. Kleinert, J. Phys. **44**, 353 (1983) (Paris) (`kl/102`).

[22] See textbook [7], Part 2, Sections 9.7 and 11.9, in particular Eq. (11.133);
H. Kleinert and W. Miller, Phys. Rev. Lett. **56**, 11 (1986) (`kl/130`);
Phys. Rev. D **38**, 1239(1988) (`kl/156`).

[23] H. Kleinert, Lett. Nuovo Cimento **35**, 405 (1982) (`kl/97`). The tricritical value $\kappa \approx 0.80/\sqrt{2}$ derived in this paper was confirmed only recently by Monte Carlo simulations [53].

[24] The equation $\mathbf{j}^s = -\mathbf{a}_T$ is the disorder version of the famous *first London equation* for the superconductor $\mathbf{j}^s = -(q^2 n_0/Mc)\mathbf{A}_T$ to be discussed further in Section 7.2.

[25] Recall that the density of states per spin direction in the energy interval $E, E + dE$ is given, in proper physical units, by $\mathcal{N} = \int [d^3p/(2\pi\hbar)^3]\delta(p^2/2m - E) = (2mE/\hbar^2)^{1/2}m/2\pi^2\hbar^2$. On the surface of the Fermi sphere it becomes $mp_F/2\pi^2\hbar^2$. The total density of electrons n_e is $2\int_0^{E_F} dE\, \mathcal{N} = p_F^3/3\pi^2$, so that we obtain indeed (5.145).

[26] J. Bardeen, L.N. Cooper, J.R. Schrieffer, Phys. Rev. **108**, 1175 (1957);
M. Tinkham, *Introduction to Superconductivity,* McGraw-Hill, New York, 1975.

[27] Gorkov's derivation was valid only at the mean field level. A modern derivation based on functional integrals [28] permits the inclusion of fluctuations to all orders.

[28] H. Kleinert, *Collective Quantum Fields,* Lectures presented at the First Erice Summer School on Low-Temperature Physics, 1977, Fortschr. Physik **26**, 565-671 (1978) (`kl/55`). See Eq. (4.118).

[29] L.D. Landau, Zh. Eksp. Teor. Fiz. **7**, 627 (1937);
V.L. Ginzburg and L.D. Landau, ibid. **20**, 1064 (1950).

[30] The effect was discovered in collaboration with R. Ochsenfeld in
     W. Meissner and R. Ochsenfeld, *Ein neuer Effekt bei Eintritt der
     Supraleitfähigkeit*, Naturwissenschaften **21**, 787, (1933).
     Due to the lengthy name of the second author, it has become a (certainly bad)
     habit to call it Meissner effect, for brevity.

[31] This property of the disorder theory was demonstrated in detail in Ref. [7].

[32] M. Kiometzis, H. Kleinert, and A.M.J. Schakel, Phys. Rev. Lett. **73**, 1975
     (1994) (cond-mat/9503019).

[33] H. Kleinert, Europhys. Letters **74**, 889 (2006) (cond-mat/0509430).

[34] See Ref. [9], Vol. I, Part 2, Chapter 10.

[35] For a derivation see Section 3.10 of textbook [20].

[36] This is an analog of the Meissner effect in the dual description of superfluid
     helium.

[37] S. Elitzur, Phys. Rev. D **12**, 3978 (1975).

[38] J. Schwinger, Phys. Rev. **115**, 721 (1959); **127**, 324 (1962).

[39] P.A.M. Dirac, *Principles of Quantum Mechanics*, 4th ed., Clarendon, Cam-
     bridge 1981, Section 80; *Gauge-Invariant Formulation of Quantum Electrody-
     namics*, Can. J. of Physics **33**, 650 (1955). See, in particular, his Eqs. (16)
     and (19).

[40] T. Kennedy and C. King, Phys. Rev. Lett. **55**, 776 (1985); Comm. Math.
     Phys. **104**, 327 (1986).

[41] M. Kiometzis and A.M.J. Schakel, Int. J. Mod. Phys. B **7**, 4271 (1993).

[42] V.L. Ginzburg, Fiz. Tverd. Tela **2**, 2031 (1960) [Sov. Phys. Solid State **2**, 1824
     (1961)].
     See also the detailed discussion in Chapter 13 of textbook
     L.D. Landau and E.M. Lifshitz, *Statistical Physics*, 3rd edition, Pergamon
     Press, London, 1968.

[43] H. Kleinert, *Criterion for Dominance of Directional over Size Fluctuations in
     Destroying Order*, Phys. Rev. Lett. **84**, 286 (2000) (cond-mat/9908239).
     The Ginzburg criterion estimates the energy necessary for hopping over an
     energy barrier for a real order parameter. For a superconductor, however,
     the size of directional fluctuations is relevant which gives rise to vortex loop
     proliferation. In general, fluctuations with symmetry $O(N)$ are important in
     an N^2-times larger temperature interval than Ginzburg's. See also pp. 18–23
     in textbook [50].

[44] D. Saint-James, G. Sarma, and E.J. Thomas, *Type II Superconductivity*, Pergamon, Oxford, 1969;

[45] M. Tinkham, *Introduction to Superconductivity*, 2nd ed., Dover, New York, 1996.

[46] T. Schneider, J. M. Singer, *A Phase Transition Approach to High Temperature Superconductivity: Universal Properties of Cuprate Superconductors*, World Scientific, Singapore, 2000; *Evidence for 3D-XY Critical Properties in Underdoped YBa2Cu3O7+x* (cond-mat/0610289).

[47] T. Schneider, R. Khasanov, and H. Keller, Phys. Rev. Lett. **94**, 77002 (2005); T. Schneider, R. Khasanov, K. Conder, E. Pomjakushina, R. Bruetsch, and H. Keller, J. Phys. Condens. Matter **16**, L1 (2004) (cond-mat/0406691).

[48] B.I. Halperin, T.C. Lubensky, and S. Ma, Phys. Rev. Lett. **32**, 292 (1972).

[49] L.P. Kadanoff, Physics **2**, 263 (1966); K.G. Wilson, Phys. Rev. B **4**, 3174, 3184 (1971); K.G. Wilson and M.E. Fisher, Phys. Rev. Lett. **28**, 240 (1972); and references therein.

[50] For a general treatment of the renormalization group see the textbooks
J. Zinn-Justin, *Quantum Field Theory and Critical Phenomena*, 4th ed., Oxford Science Publications, Oxford 2002;
H. Kleinert and V. Schulte-Frohlinde, *Critical Phenomena in ϕ^4-Theory*, World Scientific, Singapore, 2001 (`kl/b8`).

[51] A small selection of papers on this subject is
J. Tessmann, *Two Loop Renormalization of Scalar Electrodynamics*, MS thesis 1984 (the pdf file is available on the internet at (`kl/MS-Tessmann.pdf`), where kl is short for `www.physik.fu-berlin.de/~kleinert`;
S. Kolnberger and R. Folk, *Critical Fluctuations in Superconductors*, Phys. Rev. B **41**, 4083 (1990);
R. Folk and Y. Holovatch, *On the Critical Fluctuations in Superconductors*, J. Phys. A **29**, 3409 (1996);
I.F. Herbut and Z. Tešanović, *Critical Fluctuations in Superconductors and the Magnetic Field Penetration Depth*, Phys. Rev. Lett. **76**, 4588 (1996);
H. Kleinert and F.S. Nogueira, *Charged Fixed Point in the Ginzburg-Landau Superconductor and the Role of the Ginzburg Parameter κ*, Nucl. Phys. B **651**, 361 (2003) (cond-mat/0104573).

[52] H. Kleinert, *Disorder Version of the Abelian Higgs Model and the Superconductive Phase Transition*, Lett. Nuovo Cimento **35**, 405 (1982). See also the more detailed discussion in Ref. [7], Part 2, Chapter 13 (`kl/b1/gifs/v1-716s.html`), where the final disorder theory was derived [see, in particular, Eq. (13.30)].

[53] J. Hove, S. Mo, and A. Sudbø, *Vortex Interactions and Thermally Induced Crossover from Type-I to Type-II Superconductivity*, Phys. Rev. B **66**, 64524 (2002) (cond-mat/0202215);
S. Mo, J. Hove, and A. Sudbø, *Order of the Metal to Superconductor Transition*, Phys. Rev. B **65**, 104501 (2002) (cond-mat/0109260).
The Monte Carlo simulations of these authors yield the tricritical value $(0.76 \pm 0.04)/\sqrt{2}$ for the Ginzburg parameter $\kappa = \sqrt{g/q^2}$.

[54] There is also a good physical reason for the factor $\sqrt{2}$: In the high-temperature disordered phase the fluctuations of real and imaginary parts of the order field $\psi(\mathbf{x})$ have equal range for $\kappa = 1/\sqrt{2}$.

[55] L. Marotta, M. Camarda, G.G.N. Angilella, F. Siringo, *A General Interpolation Scheme for Thermal Fluctuations in Superconductors*, Phys. Rev. B **73**, 104517 (2006).

[56] See Section 4.10 of textbook [35].

[57] For a theoretical discussion and a detailed list of references see
J. Dietel and H. Kleinert, *Defect-Induced Melting of Vortices in High-T_c Superconductors: A Model Based on Continuum Elasticity Theory*, Phys. Rev. B **74**, 024515 (2006) (cond-mat/0511710); *Phase Diagram of Vortices in High-Tc Superconductors from Lattice Defect Model with Pinning* (cond-mat/0612042).

[58] M. Gabay and A. Kapitulnik, Phys. Rev. Lett. **71**, 2138 (1993).
See also
S.-C. Zhang, Phys. Rev. Lett. **71**, 2142 (1993).
M.T. Chen, J.M. Roesler, and J.M. Mochel, J. Low Temp. Phys. **89**, 125 (1992).

[59] G.B. Hess, *Angular Momentum of a Superfluid Helium in a Rotating Cylinder*, Phys. Rev. **161**, 189 (1967);
D. Stauffer and A.L. Fetter, *Distribution of Vortices in Rotating Helium II*, Phys. Rev. **168**, 156 (1968).
See also the textbook
R.J. Donnelly, *Quantized Vortices in Helium II*, Cambridge University Press, 1991;
and the review article
A.L. Fetter, in *The Physics of Solid and Liquid Helium*, ed. by K.H. Bennemann and J.B. Ketterson, Wiley, New York, Springer, Berlin, 1976.

[60] R.E. Packard and T.M. Sanders, Jr., *Detection of Single Quantized Vortex Lines in Rotating He II*, Phys. Rev. Lett. **22**, 823 (1969).
See also
K. DeConde and R.E. Packard, *Measurement of Equilibrium Critical Velocities for Vortex Formation in Superfluid Helium*, Phys. Rev. Lett. **35**, 732 (1975).

[61] A.L. Fetter and A.A. Svidzinsky, *Vortices in a Trapped Dilute Bose-Einstein Condensate*, J. Phys.: Condensed Matter **13**, R135 (2001)

A.L. Fetter, *Rotating Vortex Lattice in a Condensate Trapped in Combined Quadratic and Quartic Radial Potentials*, Phys. Rev. A **64**, 3608 (2001);

A.L. Fetter, B. Jackson, and S. Stringari, *Rapid Rotation of a Bose-Einstein Condensate in a Harmonic Plus Quartic Trap*, Phys. Rev. A **71**, 013605 (2005)

A.A. Svidzinsky and A.L. Fetter, *Normal Modes of a Vortex in a Trapped Bose-Einstein Condensate*, Phys. Rev. A **58**, 3168 (1998);

J. Tempere, M. Wouters, and J.T. Devreese, *The Vortex State in the BEC to BCS Crossover: A Path-Integral Description*, Phys. Rev. A **71**, 033631 (2005) (cond-mat/0410252);

V.N. Gladilin, J. Tempere, I.F. Silvera, J.T. Devreese, V.V. Moshchalkov, *Vortices on a Superconducting Nanoshell: Phase Diagram and Dynamics*, (arxiv:0709.0463).

6

Dynamics of Superfluids

It has been argued by Feynman [1] that at zero temperature, the time dependence of the ϕ-field in the Hamiltonian (5.6) is governed by the action

$$A = \int dt \int d^3x \, \mathcal{L} = \int dt \left\{ \int d^3x \, i\hbar \phi^* \partial_t \phi - H[\phi] \right\}, \tag{6.1}$$

so that the Lagrangian density is

$$\mathcal{L} = i\hbar \phi^* \partial_t \phi - \frac{1}{2} \left\{ |\nabla \phi|^2 + \tau |\phi|^2 + \frac{\lambda}{2} |\phi|^4 \right\}. \tag{6.2}$$

In the superfluid phase where $\tau < 0$ and $\phi = \phi_0 = \sqrt{-\tau/\lambda} e^{i\alpha}$, this can be written more explicitly, using proper physical units rather than natural units, i.e., including the Planck constant $\hbar$ and the mass M of the superfluid particles, as

$$\mathcal{L} = i\hbar \phi^* \partial_t \phi - \frac{\hbar^2}{2M} |\nabla \phi|^2 - \frac{c_0^2 M}{2n_0} (\phi^* \phi - n_0)^2 + \frac{c_0^2 M n_0}{2}, \tag{6.3}$$

where $n_0 = |\phi_0|^2 = -\tau/\lambda$ is the density $\phi^* \phi$ of the superfluid particles in the ground state, i.e., the superfluid density (5.17) which we name $n(x)$ to avoid confusion with the field size $\rho(x) = |\phi(x)|$ in Eq. (5.10)–(5.25). The last term is the negative condensation energy density $-c_0^2 M n_0 / 2$ in the superfluid phase. The interaction strength λ in (6.2) has been reparametrized as $2c_0^2 M / n_0$ and τ as $-2c_0^2 M$, for reasons to be understood below.

The equation of motion of the time-dependent field $\phi(t, \mathbf{x}) \equiv \phi(x)$ is

$$i\hbar \partial_t \phi(x) = \left[-\frac{\hbar}{2M} \nabla^2 - c_0^2 M + \frac{c_0^2 M}{n_0} \phi^*(x) \phi(x) \right] \phi(x). \tag{6.4}$$

6.1   Hydrodynamic Description of Superfluid

After substituting $\phi(x)$ by $\rho(x) e^{i\theta(x)}$ as in Eq. (5.10), and further $\rho(x)$ by $\sqrt{n(x)}$, the Lagrangian density in (6.3) becomes

$$\mathcal{L} = n(x) \left\{ -\hbar [\partial_t \theta(x) + \theta_t^v(x)] - \frac{\hbar^2}{2M} [\nabla \theta(x) - \theta^v(x)]^2 - e_{\nabla n}(x) - e_n(x) \right\} \tag{6.5}$$

196

where

$$e_n(x) \equiv \frac{c_0^2 M}{2n_0 n(x)} \left\{ [n(x) - n_0]^2 - n_0^2 \right\}$$ (6.6)

is the internal energy per particle in the fluctuating condensate, and

$$e_{\nabla n}(x) \equiv \frac{\hbar^2}{8M} \frac{[\boldsymbol{\nabla} n(x)]^2}{n^2(x)}$$ (6.7)

the gradient energy of the condensate. This energy may also be written with

$$e_{\nabla n}(x) = \frac{\mathbf{p}^{\mathrm{os}2}(x)}{2M}$$ (6.8)

where

$$\mathbf{p}^{\mathrm{os}}(x) \equiv M\mathbf{v}^{\mathrm{os}}(x) \equiv \frac{\hbar}{2} \frac{\boldsymbol{\nabla} n(x)}{n(x)}$$ (6.9)

is i times the quantum-mechanical momentum associated with the expansion of the condensate, the so-called *osmotic momentum*. The vector $\mathbf{v}^{\mathrm{os}}(x)$ is the associated osmotic velocity.

If the particles move in an external trap potential $V(x)$, this is simply added to $e(x)$. Examples are the Bosons in a Bose-Einstein condensate [9] or in an optical lattice [10]. Then the two last terms in (6.5) are replaced by

$$e_{\mathrm{tot}}(x) = e_{\nabla n}(x) + e_n(x) + V(x).$$ (6.10)

We may conveniently chose the axial gauge of the vortex gauge field where the time component $\theta_t^{\mathrm{v}}(x)$ vanishes and only the spatial part $\boldsymbol{\theta}^{\mathrm{v}}(x)$ is nonzero. After this, the field $\theta(x)$ runs from $-\infty$ to ∞ rather than $-\pi$ to π [recall the steps leading from the partition function (5.36) to (5.44)].

We now introduce the velocity field with vortices

$$\mathbf{v}(x) \equiv \hbar[\boldsymbol{\nabla}\theta(x) - \boldsymbol{\theta}^{\mathrm{v}}(x)]/M,$$ (6.11)

and the local deviation of the particle density from the ground-state value $\delta n(x) \equiv n(x) - n_0$, so that (6.5) can be written as

$$\mathcal{L} = -n(x)\left[\hbar\partial_t\theta(x) + \frac{M}{2}\mathbf{v}^2(x) + e_{\mathrm{tot}}(x)\right].$$ (6.12)

The Lagrangian density (6.12) is invariant under changes of $\theta(x)$ by an additive constant Λ. According to Noether's theorem, this implies the existence of a conserved current density. We can calculate the charge and particle current densities from the rule (3.102) as

$$n(x) = -\frac{1}{\hbar}\frac{\partial\mathcal{L}}{\partial\partial_t\theta(x)}, \quad \mathbf{j}(x) = -\frac{1}{\hbar}\frac{\partial\mathcal{L}}{\partial\boldsymbol{\nabla}\theta(x)} = n(x)\mathbf{v}(x).$$ (6.13)

To find the second expression we must remember (6.11). The prefactor $1/\hbar$ is chosen to have the correct physical dimensions. The associated conservation law reads

$$\partial_t n(x) = -\boldsymbol{\nabla} \cdot [n(x)\mathbf{v}(x)], \tag{6.14}$$

which is the *continuity equation* of hydrodynamics. This equation is found from the Lagrangian density (6.12) by extremizing the associated action with respect to $\theta(x)$.

Functional extremization with respect to $\delta n(x)$ yields

$$\hbar\partial_t\theta(x) + \frac{M}{2}\mathbf{v}^2(x) + V(x) + h_{\nabla n}(x) + h_n(x) = 0, \tag{6.15}$$

where we have included a possible external potential $V(x)$ as in Eq. (6.10). The last term is the *enthalpy* per particle associated with the energy per particle $e_n(x)$. It is defined by

$$h_n(x) \equiv \frac{\partial[n(x)e_n(x)]}{\partial n(x)} = e_n(x) + n(x)\frac{\partial e_n(x)}{\partial n(x)} = e_n(x) + \frac{p_n(x)}{n(x)}, \tag{6.16}$$

where $p_n(x)$ is the pressure due to the energy $e_n(x)$:

$$p_n(x) \equiv n^2(x)\frac{\partial}{\partial n}e_n(x) = \left(n\frac{\partial}{\partial n} - 1\right)[n(x)e_n(x)]. \tag{6.17}$$

For $e_n(x)$ from Eq. (6.6), and allowing for an external potential $V(x)$ as in (6.10), we find

$$h_n(x) = \frac{c_0^2 M}{n_0}\delta n(x), \quad p_n(x) = \frac{c_0^2 M}{2n_0}n^2(x). \tag{6.18}$$

The term $h_{\nabla n}(x)$ is the so-called *quantum enthalpy*. It is obtained from the energy density $e_{\nabla n}(x)$ as a contribution from the Euler-Lagrange equation:

$$h_{\nabla n}(x) \equiv \frac{\partial[n(x)e_{\nabla n}(x)]}{\partial n(x)} - \boldsymbol{\nabla}\frac{\partial[n(x)e_{\nabla n}(x)]}{\partial\boldsymbol{\nabla}n(x)}. \tag{6.19}$$

This can be written as

$$h_{\nabla n}(x) = e_{\nabla n}(x) + \frac{p_{\nabla n}(x)}{n(x)}, \tag{6.20}$$

where

$$p_{\nabla n}(x) = n^2(x)\left[\frac{\partial}{\partial n} - \boldsymbol{\nabla}\frac{\partial}{\partial\boldsymbol{\nabla}n}\right]e_{\nabla n}(x) = \left\{n(x)\left[\frac{\partial}{\partial n} - \boldsymbol{\nabla}\frac{\partial}{\partial\boldsymbol{\nabla}n}\right] - 1\right\}[n(x)e_{\nabla n}(x)] \tag{6.21}$$

is the so-called *quantum pressure*.

Inserting (6.7) yields

$$h_{\nabla n}(x) = \frac{\hbar^2}{8M}\left\{\frac{[\boldsymbol{\nabla}n(x)]^2}{n(x)} - 2\boldsymbol{\nabla}^2 n(x)\right\}, \quad p_{\nabla n}(x) = -\frac{\hbar^2}{4M}\boldsymbol{\nabla}^2 n(x). \tag{6.22}$$

The two equations (6.14) and (6.15) were found by Madelung in 1926 [2].

The gradient of (6.15) yields the equation of motion

$$M\partial_t \mathbf{v}(x) + \hbar\partial_t\boldsymbol{\theta}^{\mathrm{v}} + \frac{M}{2}\boldsymbol{\nabla}\,\mathbf{v}^2(x) = -\boldsymbol{\nabla}V_{\mathrm{tot}}(x) - \boldsymbol{\nabla}h_{\nabla n}(x) - \boldsymbol{\nabla}h_n(x), \qquad (6.23)$$

where

$$V_{\mathrm{tot}}(x) \equiv V(x) \equiv h_{\nabla n}(x) + \boldsymbol{\nabla}h_n(x). \qquad (6.24)$$

We now use the vector identity

$$\frac{1}{2}\boldsymbol{\nabla}\,\mathbf{v}^2(x) = \mathbf{v}(x) \times [\boldsymbol{\nabla}\times\mathbf{v}(x)] + [\mathbf{v}(x)\cdot\boldsymbol{\nabla}]\mathbf{v}(x), \qquad (6.25)$$

and rewrite Eq. (6.23) as

$$M\partial_t\mathbf{v}(x) + M[\mathbf{v}(x)\cdot\boldsymbol{\nabla}]\mathbf{v}(x) = -\boldsymbol{\nabla}V_{\mathrm{tot}}(x) + \mathbf{f}^{\mathrm{v}}(x) \qquad (6.26)$$

where

$$\mathbf{f}^{\mathrm{v}}(x) \equiv -\hbar\partial_t\boldsymbol{\theta}^{\mathrm{v}}(x) - M\mathbf{v}(x)\times[\boldsymbol{\nabla}\times\mathbf{v}(x)] = -\hbar\partial_t\boldsymbol{\theta}^{\mathrm{v}}(x) + \hbar\mathbf{v}(x)\times[\boldsymbol{\nabla}\times\boldsymbol{\theta}^{\mathrm{v}}(x)] \qquad (6.27)$$

is a force due to the vortices. The classical contribution to the second term is the important Magnus force [3] acting upon a rotating fluid:

$$\mathbf{f}^{\mathrm{v}}_{\mathrm{Magnus}}(x) \equiv -M\mathbf{v}(x) \times [\boldsymbol{\nabla}\times\mathbf{v}(x)]. \qquad (6.28)$$

The important observation is now that the force (6.27) is in fact zero in the superfluid,

$$\mathbf{f}^{\mathrm{v}}(x) = 0, \qquad (6.29)$$

implying that the time dependence of the vortex gauge field is driven by the Magnus force (6.28).

Let us prove this. Consider first the two-dimensional situation with a point-like vortex which lies at the origin at a given time t. This can be described by a vortex gauge field

$$\theta_1^{\mathrm{v}}(\mathbf{x}) = 0, \qquad \theta_2^{\mathrm{v}}(\mathbf{x}) = 2\pi\Theta(x_1)\delta(x_2), \qquad (6.30)$$

where $\Theta(x_1)$ is the Heaviside step function which is zero for negative and unity for positive x_1. The curl of (6.30) is the vortex density, which is proportional to a δ-function at the origin:

$$\boldsymbol{\nabla}\times\boldsymbol{\theta}^{\mathrm{v}}(\mathbf{x}) = \nabla_1\theta_2^{\mathrm{v}}(\mathbf{x}) - \nabla_2\theta_1^{\mathrm{v}}(\mathbf{x}) = 2\pi\delta^{(2)}(\mathbf{x}), \qquad (6.31)$$

in agreement with the general relation (5.30). Suppose that the vortex moves, after a short time Δt, to the point $\mathbf{x} + \Delta\mathbf{x} = (\Delta x_1, 0)$, where

$$\theta_1^{\mathrm{v}}(\mathbf{x}) = 0, \qquad \theta_2^{v}(\mathbf{x}) = \Theta(x_1 + \Delta x_1)\delta(x_2), \qquad \boldsymbol{\nabla}\times\boldsymbol{\theta}^{\mathrm{v}}(\mathbf{x}) = 2\pi\delta^{(2)}(\mathbf{x} + \Delta\mathbf{x}). \quad (6.32)$$

Since $\Theta(x_1 + \Delta x_1) = \Theta(x_1) + \Delta x_1\delta(x_1)$, we see that $\Delta\boldsymbol{\theta}^{\mathrm{v}}(\mathbf{x}) = \Delta\mathbf{x}\times[\boldsymbol{\nabla}\times\boldsymbol{\theta}^{\mathrm{v}}(\mathbf{x})]$ which becomes

$$\partial_t\boldsymbol{\theta}^{\mathrm{v}}(x) = \mathbf{v}(x)\times[\boldsymbol{\nabla}\times\boldsymbol{\theta}^{\mathrm{v}}(x)] \qquad (6.33)$$

after dividing it by Δt and taking the limit $\Delta t \to 0$, thus proving the vanishing of $\mathbf{f}^{\mathrm{v}}(x)$.

The result can easily be generalized to a line with wiggles by approximating it as a sequence of points in closely stacked planes orthogonal to the line elements. As long as the line is smooth, the change in the direction is of higher order in $\Delta \mathbf{x}$ and does not influence the result in the limit $\Delta t \to 0$. Thus we can omit the last term in (6.26).

Equation (6.33) is the equation of motion for the vortex gauge field. The time dependence of this field is governed by quantum analog of the Magnus force (6.27).

Note that for a vanishing force $\mathbf{f}^{\mathrm{v}}(x)$ and quantum pressure $p_{\nabla n}(x)$, Eq. (6.26) coincides with the classical *Euler equation of motion* for an ideal fluid

$$M\frac{d}{dt}\mathbf{v}(x) = M\partial_t \mathbf{v}(x) + M[\mathbf{v}(x) \cdot \boldsymbol{\nabla}]\mathbf{v}(x) = -\boldsymbol{\nabla}V(x) - \frac{\boldsymbol{\nabla}p_n(x)}{n(x)}. \qquad (6.34)$$

The last term is initially equal to $-\boldsymbol{\nabla}h_n(x)$. However, since $e_n(x)$ depends only on $n(x)$, in which case the system is referred to as *barytropic*, we see that (6.16) implies

$$\boldsymbol{\nabla}h_n(x) = \boldsymbol{\nabla}e_n(x) - \frac{p_n(x)}{n^2(x)}\boldsymbol{\nabla}n(x) + \frac{\boldsymbol{\nabla}p_n(x)}{n(x)} = \left[\frac{\partial e_n(x)}{\partial n(x)} - \frac{p_n(x)}{n^2(x)}\right]\boldsymbol{\nabla}n(x) + \frac{\boldsymbol{\nabla}p_n(x)}{n(x)}$$

$$= \frac{\boldsymbol{\nabla}p_n(x)}{n(x)}. \qquad (6.35)$$

There are only two differences between (6.26) with $\mathbf{f}^{\mathrm{v}}(x) = 0$ and the classical equation (6.34). One is the extra quantum part $-\boldsymbol{\nabla}h_{\nabla n}(x)$ in (6.26). The other lies in the nature of the vortex structure. In a classical fluid, the *vorticity*[1]

$$\mathbf{w}(x) \equiv \boldsymbol{\nabla} \times \mathbf{v}(x) \qquad (6.36)$$

can be an arbitrary function of x. For instance, a velocity field $\mathbf{v}(x) = (0, x_1, 0)$ has the constant vorticity $\boldsymbol{\nabla} \times \mathbf{v}(x) = 1$. In a superfluid, such vorticities do not exist. If one performs the integral over any closed contour $M\oint d\mathbf{x} \cdot \mathbf{v}(x)$, one must always find an integer multiple of $\hbar$ to ensure the uniqueness of the wave function around the vortex line. This corresponds to the Sommerfeld quantization condition $\oint d\mathbf{x} \cdot \mathbf{p}(x) = \hbar n$. In a superfluid, there exists no continuous regions of nonzero vorticity, only infinitesimally thin lines. This leaves only vorticities which are superpositions of δ-functions $2\pi\hbar\delta(\mathbf{x}; L)$, which is guaranteed here by the expression (6.11) for the velocity.

By taking the curl of the right-hand side of the vanishing force $\mathbf{f}^{\mathrm{v}}(x)$, we obtain an equation for the time derivative of the vortex density

$$\partial_t[\boldsymbol{\nabla} \times \boldsymbol{\theta}^{\mathrm{v}}(x)] = \boldsymbol{\nabla} \times \left[\mathbf{v}(x) \times [\boldsymbol{\nabla} \times \boldsymbol{\theta}^{\mathrm{v}}(x)]\right]. \qquad (6.37)$$

---

[1]The letter $\mathbf{w}$ stems from the German word for vorticity="Wirbelstärke".

Such an equation was first found in 1942 by Ertel [4] for the vorticity $\mathbf{w}(x)$ of a classical fluid, rather than $\nabla \times \boldsymbol{\theta}^{\mathrm{v}}(x)$. Using the vector identity

$$\nabla \times \mathbf{v}(x) \times \mathbf{w}(x) = -\mathbf{w}(x)[\nabla \cdot \mathbf{v}(x)] - [\mathbf{v}(x) \cdot \nabla]\mathbf{w}(x)$$
$$+\mathbf{v}(x)[\nabla \cdot \mathbf{w}(x)] + [\mathbf{w}(x) \cdot \nabla]\mathbf{v}(x), \qquad (6.38)$$

and the identity $\nabla \cdot \mathbf{w}(x) = \nabla \cdot [\nabla \times \mathbf{v}(x)] \equiv 0$, we can rewrite the classical version of (6.37) in the form

$$\frac{d}{dt}\mathbf{w}(x) = \partial_t \mathbf{w}(x) + [\mathbf{v}(x) \cdot \nabla]\mathbf{w}(x) = -\mathbf{w}(x)[\nabla \cdot \mathbf{v}(x)] + [\mathbf{w}(x) \cdot \nabla]\mathbf{v}(x), \quad (6.39)$$

which is the form stated by Ertel. This and Eq. (6.34) are the basis for deriving the famous Helmholtz-Thomson theorem of an ideal perfect classical fluid which states that the vorticity is constant along a vortex line if the forces possess a potential.

Equation (6.37) is the quantum version of Ertel's equation where the vorticity occurs only in infinitesimally thin lines satisfying the quantization condition $\oint dx \cdot \mathbf{p}(x) = \hbar n$.

Inserting the vortex density (5.30) into Eq. (6.37), we obtain for a line $L(t)$ moving in a fluid with a velocity field $\mathbf{v}(x)$ the equation

$$\partial_t \delta(\mathbf{x}; L(t)) = \nabla \times [\mathbf{v}(x) \times \delta(\mathbf{x}; L(t))] . \qquad (6.40)$$

It has been argued by L. Morati [5, 6] on the basis of a stochastic approach to quantum theory by E. Nelson [7] that the force $\mathbf{f}^{\mathrm{v}}(x)$ is not zero but equal to quantum force

$$\mathbf{f}^{\mathrm{qu}}(x) \equiv -\frac{\hbar}{2}\left[\frac{\nabla n(x)}{n(x)} + \nabla\right] \times [\nabla \times \mathbf{v}(x)] . \qquad (6.41)$$

Our direct derivation from the superfluid Lagrangian density (6.2) does not produce such a term.

6.2   Velocity of Second Sound

Consider the Lagrangian density (6.12) and omit the trivial constant condensation energy density $-c_0^2 M n_0/2$ as well as external potential $V(x)$. The result is

$$\mathcal{L} = -[n_0 + \delta n(x)]\left[\hbar \partial_t \theta(x) + \frac{M}{2}\mathbf{v}^2(x)\right] - \frac{\hbar^2}{8M}\frac{[\nabla \delta n(x)]^2}{n(x)} - \frac{c_0^2 M}{2n_0}[\delta n(x)]^2 . \quad (6.42)$$

For small $\delta n(x) \ll n_0$, this is extremal at

$$\delta n(x) = \frac{n_0}{c_0^2 M}\frac{1}{1 - \xi^2 \nabla^2}\left[\hbar \partial_t \theta(x) + \frac{M}{2}\mathbf{v}^2(x)\right], \qquad (6.43)$$

where

$$\xi \equiv \frac{1}{2}\frac{\hbar}{c_0 M} = \frac{1}{2}\frac{c}{c_0}\lambda_M \qquad (6.44)$$

is the range of the $\delta n(x)$-fluctuations, i.e., the coherence length of the superfluid, and $\lambda_M = \hbar/Mc$ the Compton wavelength of the particles of mass M.

Reinserting (6.43) into (6.42) leads to the alternative Lagrangian density

$$
\begin{aligned}
\mathcal{L} = & - n_0 \left[\hbar \partial_t \theta(x) + \frac{M}{2} \mathbf{v}^2(x) \right] \\
& + \frac{n_0}{2c_0^2 M} \left[\hbar \partial_t \theta(x) + \frac{M}{2} \mathbf{v}^2(x) \right] \frac{1}{1 - \xi^2 \boldsymbol{\nabla}^2} \left[\hbar \partial_t \theta(x) + \frac{M}{2} \mathbf{v}^2(x) \right].
\end{aligned} \tag{6.45}
$$

The first term is an irrelevant surface term and can be omitted. The quadratic fluctuations of $\theta(x)$ are governed by the Lagrangian density

$$
\mathcal{L}_0 = \frac{n_0 \hbar^2}{2M} \left\{ \frac{1}{c_0^2} [\partial_t \theta(x) - \theta_t^{\mathrm{v}}(x)] \frac{1}{1 - \xi \boldsymbol{\nabla}^2} [\partial_t \theta(x) - \theta_t^{\mathrm{v}}(x)] - [\boldsymbol{\nabla}\theta(x) - \boldsymbol{\theta}^{\mathrm{v}}(x)]^2 \right\}. \tag{6.46}
$$

For the sake of manifest vortex gauge invariance we have reinserted the time-component $\theta_t^{\mathrm{v}}(x)$ of the vortex gauge field which was omitted in (6.12) where we used the axial gauge.

In the absence of vortices, and in the long-wavelength limit, the Lagrangian density (6.46) leads to the equation of motion

$$
(-\partial_t^2 + c_0^2 \boldsymbol{\nabla}^2)\theta(x) = 0. \tag{6.47}
$$

This is a Klein-Gordon equation for $\theta(x)$ which shows that the parameter c_0 is the propagating velocity of phase fluctuations, which form the second sound in the superfluid.

Note the remarkable fact that although the initial equation of motion (6.4) is nonrelativistic, the sound waves follow a Lorentz-invariant equation in which the sound velocity c_0 is playing the role of the light velocity. If there is a potential, the velocity of second sound will no longer be a constant but depend on the position.

6.3   Vortex-Electromagnetic Fields

It is useful to carry the analogy between the gauge fields of electromagnetism further and define $\theta_0^{\mathrm{v}}(x) \equiv \theta_t^{\mathrm{v}}(x)/c$, and vortex-electric and vortex-magnetic fields as

$$
\mathbf{E}^{\mathrm{v}}(x) \equiv - \left[\boldsymbol{\nabla}\theta_0^{\mathrm{v}}(x) + \frac{1}{c}\partial_t \boldsymbol{\theta}^{\mathrm{v}}(x) \right], \quad \mathbf{B}^{\mathrm{v}}(x) \equiv \boldsymbol{\nabla} \times \boldsymbol{\theta}^{\mathrm{v}}(x), \tag{6.48}
$$

These are analogs of the electromagnetic fields (2.74) and (2.75). The zeroth component $\theta_0^{\mathrm{v}}(x)$ is defined from $\theta_t(x)$ in the same way as $A_0(x)$ is from $A_t(x) = cA_0(x)$, to make the two alternative expression for $dx^\mu A_\mu$ equal:

$$
dx^\mu A_\mu \equiv dx^0 A_0 - d\mathbf{x} \cdot \mathbf{A} = dt\, A_t - d\mathbf{x} \cdot \mathbf{A}. \tag{6.49}
$$

The dimension of both $\boldsymbol{\theta}^{\mathrm{v}}(x)$ and $\theta_0^{\mathrm{v}}(x)$ is 1/length, that of the vortex-electromagnetic fields $\mathbf{E}^{\mathrm{v}}$ and $\mathbf{B}^{\mathrm{v}}$ have the dimensions 1/length². The fields satisfy the same type of Bianchi identities as the electromagnetic fields in (1.189) and (1.190):

$$\boldsymbol{\nabla} \cdot \mathbf{B}^{\mathrm{v}}(x) \;=\; 0, \tag{6.50}$$

$$\boldsymbol{\nabla} \times \mathbf{E}^{\mathrm{v}}(x) + \frac{1}{c}\partial_t \mathbf{B}^{\mathrm{v}}(x) \;=\; 0. \tag{6.51}$$

With these fields, the vortex force (6.27) becomes

$$\mathbf{f}^{\mathrm{v}}(x) = \hbar c \left[\mathbf{E}^{\mathrm{v}}(x) + \frac{\mathbf{v}(x)}{c} \times \mathbf{B}^{\mathrm{v}}(x) \right], \tag{6.52}$$

which has the same form as the electromagnetic force upon a moving particle of unit charge in Eq. (1.186). According to Eq. (6.29), the corresponding vortex force vanishes, so that

$$\mathbf{E}^{\mathrm{v}}(x) = -\frac{\mathbf{v}(x)}{c} \times \mathbf{B}^{\mathrm{v}}(x). \tag{6.53}$$

By substituting Eq. (6.48) for $\mathbf{B}^{\mathrm{v}}(x)$ into (6.37), we find the following equation of motion for the vortex magnetic field:

$$\partial_t \mathbf{B}^{\mathrm{v}}(x) = \boldsymbol{\nabla} \times [\mathbf{v}(x) \times \mathbf{B}^{\mathrm{v}}(x)]. \tag{6.54}$$

Note that due to Eq. (6.11),

$$\boldsymbol{\nabla} \times \mathbf{v}(x) = -\frac{\hbar}{M}\boldsymbol{\nabla} \times \boldsymbol{\theta}^{\mathrm{v}}(x) = -\frac{\hbar}{M}\mathbf{B}^{\mathrm{v}}(x). \tag{6.55}$$

Using Eq. (6.53), and Eq. (6.55), we can rewrite the divergence of $\mathbf{E}^{\mathrm{v}}(x)$ as

$$\boldsymbol{\nabla} \cdot \mathbf{E}^{\mathrm{v}}(x) = -\boldsymbol{\nabla} \cdot \left[\frac{\mathbf{v}(x)}{c} \times \mathbf{B}^{\mathrm{v}}(x) \right] = \frac{1}{c}\left\{ -[\boldsymbol{\nabla} \times \mathbf{v}(x)]\mathbf{B}^{\mathrm{v}}(x) + \mathbf{v}(x) \cdot [\boldsymbol{\nabla} \times \mathbf{B}^{\mathrm{v}}(x)] \right\}$$

$$= \frac{\hbar}{Mc}[\mathbf{B}^{\mathrm{v}}(x)]^2 + \frac{\mathbf{v}(x)}{c} \cdot [\boldsymbol{\nabla} \times \mathbf{B}^{\mathrm{v}}(x)]. \tag{6.56}$$

6.4 Simple Example

As a simple example illustrating the above extension of Madelung's theory consider a harmonic oscillator in two dimensions with the Schrödinger equation in cylindrical coordinates (r, φ) with $r \in (0, \infty)$ and $\varphi \in (0, 2\pi)$:

$$\left(-\frac{1}{2}\boldsymbol{\nabla}^2 + \frac{1}{2}r^2 \right) \psi_{nm}(r, \varphi) = E_{nm}\psi_{nm}(r, \varphi), \tag{6.57}$$

where n, m are the principal quantum number the azimuthal quantum numbers, respectively. For simplicity, we have set $M = 1$ and $\hbar = 1$. In particular, we shall focus on the state

$$\psi_{11}(r, \theta) - \pi^{-1/2}\, r\, e^{-r^2/2}e^{i\varphi}. \tag{6.58}$$

The Hamiltonian of the two-dimensional oscillator corresponding to the field formulation (5.25) reads

$$H[\phi] \;=\; \frac{1}{2}\int d^2x\,\phi^*(-\boldsymbol{\nabla}^2+\mathbf{x}^2)\phi, \tag{6.59}$$

where we have done a nabla integration to replace $|\boldsymbol{\nabla}\phi|^2$ by $-\phi^*\boldsymbol{\nabla}^2\phi$. The wave function (6.58) corresponds to the specific field configuration

$$\rho(r)=\pi^{-1/2}\,r, \qquad \theta=\arctan(x_2/x_1). \tag{6.60}$$

Thus we calculate the energy (6.62) in cylindrical coordinates, where $\phi^*(-\boldsymbol{\nabla}^2)\phi = -\phi^*(r^{-1}\partial_r r\partial_r - r^{-2}\partial_\varphi^2)\phi$ becomes for $\phi(x)=\psi_{11}(r,\varphi)$:

$$-\phi^*(-\boldsymbol{\nabla}^2)\phi=\frac{1}{\pi}\left(4-r^2\right)r^2e^{-r^2}, \tag{6.61}$$

so that we find

$$E_{11} \;=\; \pi\int_0^\infty dr\,r\left[\frac{1}{\pi}(4-r^2)e^{-r^2}+\frac{1}{\pi}r^4e^{-r^2}\right]=1+1=2. \tag{6.62}$$

Let us now calculate the same energy from the hydrodynamic expression for the energy which we read directly off the Lagrangian density (6.5) as

$$H=\int d^2x\,\mathcal{H}=\int d^2x\,n(x)\left\{\frac{1}{2}[\boldsymbol{\nabla}\theta(x)-\boldsymbol{\theta}^{\mathrm{v}}(x)]^2+\frac{\mathbf{p}^{\mathrm{os}\,2}(x)}{2}+\frac{\mathbf{x}^2}{2}\right\}. \tag{6.63}$$

The gradient of $\theta(x)=\arctan(x_2/x_1)$ has the jump at the cut of $\arctan(x_2/x_1)$, which runs here from zero to infinity in the x_1,x_2-plane:

$$\boldsymbol{\nabla}_1\arctan(x_2/x_1)=-x_2/r^2, \qquad \boldsymbol{\nabla}_2\arctan(x_2/x_1)=x_1/r^2+2\pi\Theta(x_1)\delta(x_2). \tag{6.64}$$

The vortex gauge field is the same as in the example (6.30).

When forming the superflow velocity (6.11), the second term in $\boldsymbol{\nabla}_2\arctan(x_2/x_1)$ is removed by the vortex gauge field (6.30), and we obtain simply

$$v_1(x)=-x_2/r^2, \qquad v_2(x)=x_1/r^2. \tag{6.65}$$

Since the wave function has $n(x)=r^2e^{-r^2}/\pi$, the osmotic momentum (6.9) is

$$\mathbf{p}^{\mathrm{os}}=\frac{1}{2}\frac{\boldsymbol{\nabla}n(x)}{n(x)}=\frac{1}{2}\frac{\boldsymbol{\nabla}(r^2e^{-r^2})}{r^2e^{-r^2}}\frac{\mathbf{x}}{r}=\left(\frac{1}{r}-r\right)\frac{\mathbf{x}}{r}. \tag{6.66}$$

Inserting this into (6.63) yields the energy

$$H=\pi\int_0^\infty dr\,r\,\frac{r^2}{\pi}e^{-r^2}\left[\frac{1}{r^2}+\left(\frac{1}{r}-r\right)^2+r^2\right], \tag{6.67}$$

which gives the same value 2 as in the calculation (6.62).

Let us check the validity of the equation of motion (6.54) in this example. The vortex magnetic field $\mathbf{B}^v(x)$ is according to Eq. (6.48) in the present natural units

$$\mathbf{B}^v(x) = 2\pi\delta^{(2)}(x). \tag{6.68}$$

This is time-independent, so that the right-hand side of Eq. (6.54) must vanish. Indeed, from (6.65) we see that

$$\mathbf{v}(x) \times \mathbf{B}^v(x) = 2\pi \left(\frac{x_1}{r}, \frac{x_2}{r}\right) \delta^{(2)}(x), \tag{6.69}$$

so that its curl gives

$$2\pi\boldsymbol{\nabla} \times \left(\frac{x_1}{r}, \frac{x_2}{r}\right) \delta^{(2)}(x) = 2\pi \left(\nabla_1 \frac{x_2}{r} - \nabla_2 \frac{x_1}{r}\right) \delta^{(2)}(x). \tag{6.70}$$

This vanishes identically due to the rotational symmetry of the δ-function in two dimensions

$$\delta^{(2)}(x) = \frac{1}{2\pi r}\delta(r). \tag{6.71}$$

After applying the chain rule of differentiation to (6.70) one obtains zero.

It is interesting to note that the extra quantum force (6.41) happens to vanish as well in this atomic state. In terms of the vortex magnetic field it reads

$$\mathbf{f}^{\text{qu}}(x) \equiv \frac{\hbar^2}{2M} \left[\frac{\boldsymbol{\nabla} n(x)}{n(x)} + \boldsymbol{\nabla}\right] \times \mathbf{B}^v(x). \tag{6.72}$$

The $\mathbf{B}^v(x)$-field (6.68) has a curl

$$\boldsymbol{\nabla} \times \mathbf{B}^v = 2\pi \left(\nabla_2 \delta^{(2)}(x), -\nabla_1 \delta^{(2)}(x)\right). \tag{6.73}$$

With the help of the rotationally symmetric expression (6.71) for $\delta^{(2)}(x)$, this is rewritten as

$$\boldsymbol{\nabla} \times \mathbf{B}^v = \left(\frac{x_2}{r}, -\frac{x_1}{r}\right) \left[\frac{1}{r}\delta(r)\right]' = -\left(\frac{x_2}{r}, -\frac{x_1}{r}\right) \left[\frac{1}{r^2}\delta(r) - \frac{1}{r}\delta'(r)\right]. \tag{6.74}$$

The osmotic term adds to this:

$$\frac{\boldsymbol{\nabla} n(x)}{n(x)} \times \mathbf{B}^v(x) = 2 \left(\frac{1}{r} - r\right) \frac{\mathbf{x}}{r} \times \mathbf{B}^v(x) = 2 \left(\frac{1}{r} - r\right) 2\pi \left(\frac{x_2}{r}, -\frac{x_1}{r}\right) \frac{1}{r}\delta(r). \tag{6.75}$$

The two distributions (6.74) and (6.75) are easily shown to cancel each other. Since they both point in the same direction, we remove the unit vectors $(x_2, -x_1)/r$ and compare the two contributions in the force (6.72) which are proportional to

$$-\left[\frac{1}{r^2}\delta(r) - \frac{1}{r}\delta'(r)\right], \quad 2\left(\frac{1}{r^2} - 1\right)\delta(r). \tag{6.76}$$

Multiplying both expressions by an arbitrary smooth rotation-symmetric test function $f(r)$ and integrating we obtain

$$2\pi \int_0^\infty dr\, r\, f(r) \left\{ -\left[\frac{1}{r^2}\delta(r) - \frac{1}{r}\delta'(r) \right], \quad 2\pi \int_0^\infty dr\, r\, f(r)\, 2\left(\frac{1}{r^2} - 1 \right)\delta(r) \right\}. \quad (6.77)$$

These integrals are finite only if $f(0) = 0$, $f'(0) = 0$, so that $f(r)$ must have the small-r behavior $f''(0)r^2/2! + f^{(3)}(0)r^3/3! + \ldots$. Inserting this into the two integrals and using the formula $\int_0^\infty dr\, r^n\, \delta'(r) = -\delta_{n,1}$ for $n \geq 1$, we obtain the values $-2\pi f''(0)/2$ and $2\pi f''(0)/2$, respectively, so that the force (6.72) is indeed equal to zero.

6.5 Eckart Theory of Ideal Quantum Fluids

It is instructive to compare the above equations with those for an ideal isentropic quantum fluid without vortices which is described by a Lagrangian density due to Eckart [8]:

$$\mathcal{L} = n(x)\frac{M}{2}\mathbf{v}^2(x) + \lambda(x)M\{\partial_t n(x) + \boldsymbol{\nabla} \cdot [n(x)\mathbf{v}(x)]\} - n(x)e_{\text{tot}}(x), \quad (6.78)$$

where $e_{\text{tot}}(x)$ is the internal energy (6.10) per particle, and $\lambda(x)$ a Lagrange multiplyer $\lambda(x)$. If we extremize the action (6.78) with respect to $\lambda(x)$, we obtain once more the continuity equation (6.14). Extremizing (6.78) with respect to $\mathbf{v}(x)$, we see that the velocity field is given by the gradient of the Lagrange multiplyer:

$$\mathbf{v}(x) = \boldsymbol{\nabla}\lambda(x). \quad (6.79)$$

Reinserting this into (6.78), the Lagrangian density of the fluid becomes

$$\mathcal{L} = n(x)\frac{M}{2}[\boldsymbol{\nabla}\lambda(x)]^2 - n(x)e(x) + \lambda(x)\{\partial_t M n(x) + M\boldsymbol{\nabla}\cdot[n(x)\boldsymbol{\nabla}\lambda(x)]\} - n(x)e(x),$$
$$(6.80)$$

or, after a partial integration in the associated action,

$$\mathcal{L} = -n(x)\left\{ M\partial_t\lambda(x) + \frac{M}{2}[\boldsymbol{\nabla}\lambda(x)]^2 + e_{\text{tot}}(x) \right\}. \quad (6.81)$$

Since the gradient of a scalar field has no curl, this implies that these actions describe only a vortexless flow.

Comparing (6.79) with (6.11) in the absence of vortices, we identify the velocity potential as

$$\lambda(x) \equiv \hbar\theta(x)/M. \quad (6.82)$$

6.6   Rotating Superfluid

If we want to study a superfluid in a vessel which rotates with a constant angular velocity $\boldsymbol{\Omega}$, we must add to the Lagrangian density (6.3) a source term

$$\mathcal{L}_{\Omega} = -\mathbf{l}(\mathbf{x}) \cdot \boldsymbol{\Omega} = [\mathbf{x} \times \mathbf{j}(x)] \cdot \boldsymbol{\Omega} = \frac{\hbar}{2Mi} \hbar \, \phi^*(x)[\mathbf{x} \times \overset{\leftrightarrow}{\boldsymbol{\nabla}}]\phi(x) \cdot \boldsymbol{\Omega}, \qquad (6.83)$$

where $\mathbf{j}(x)$ is the current density [compare (2.64)], and $\mathbf{l}(\mathbf{x}) \equiv \mathbf{x} \times \mathbf{j}(x)$ the density of angular momentum of the fluid. After substituting $\phi(x)$ by $\sqrt{n(x)}e^{i\theta(x)}$, this becomes

$$\mathcal{L}_{\Omega} = -i\hbar \nabla_{\varphi} n(x) - Mn(x)\mathbf{v}(x) \cdot \mathbf{v}_{\Omega}, \qquad (6.84)$$

where $\nabla_{\varphi} n(x)$ denotes the azimuthal derivative of the density around the direction of the rotation axis $\boldsymbol{\Omega}$, and

$$\mathbf{v}_{\Omega}(x) \equiv \boldsymbol{\Omega} \times \mathbf{x} \qquad (6.85)$$

is the velocity which the particles at $\mathbf{x}$ would have if the fluid would rotate as a whole like a solid. The action associated with the first term vanishes by partial integration since $n(x)$ is periodic around the axis $\boldsymbol{\Omega}$. Adding this to the hydrodynamic Lagrangian density (6.5) and performing a quadratic completion in $\mathbf{v}_{\Omega}(x)$, we obtain

$$\mathcal{L} = n(x)\left\{-\hbar[\partial_t \theta(x) + \theta_t^{\text{v}}(x)] - \frac{\hbar^2}{2M}\left[\boldsymbol{\nabla}\theta(x) - \boldsymbol{\theta}^{\text{v}}(x) - \frac{M}{\hbar}\mathbf{v}_{\Omega}(x)\right]^2 - e_{\text{tot}}(x) - n(x)V_{\Omega}(x)\right\}, \qquad (6.86)$$

where $V_{\Omega}(x)$ is the harmonic potential

$$V_{\Omega}(x) \equiv -\frac{M}{2}\mathbf{v}_{\Omega}^2(x) = -\frac{M}{2}\Omega^2 r_{\perp}^2, \qquad (6.87)$$

depending quadratically on the distance $r_{\perp}$ from the rotation axis.

The velocity $\mathbf{v}_{\Omega}(x)$ has a constant curl

$$\boldsymbol{\nabla} \times \mathbf{v}_{\Omega}(x) = 2\boldsymbol{\Omega}. \qquad (6.88)$$

It can therefore not be absorbed into the wave function by a phase transformation $\phi(x) \to e^{i\alpha(x)}\phi(x)$, since this would make the wave function multivalued. The energy of the rotating superfluid can be minimized only by a triangular lattice of vortex lines. Their total number N is such that the total circulation equals that of a solid body rotation with $\boldsymbol{\Omega}$. Thus, if we integrate along a circle C of radius R around the rotation axis, the number of vortices enclosed is given by

$$M \oint_C d\mathbf{x} \cdot \mathbf{v} = 2\pi\hbar N. \qquad (6.89)$$

In this way the average of the vortex gauge field $\boldsymbol{\theta}^{\text{v}}(x)$ cancels the rotation field $\mathbf{v}_{\Omega}(x)$ of constant vorticity.

Triangular vortex lattices have been observed in rotating superfluid ^{4}He [11], and recently in Bose-Einstein condensates [12]. The theory of these lattices was developed in the 1960's by Tkachenko and others for superfluid ^{4}He [13], and recently by various authors for Bose-Einstein condensates [14].

Notes and References

[1] R.P. Feynman, *Statistical Mechanics*, Addison Wesley, New York, 1972, Sec. 10.12.

[2] E. Madelung, Z. Phys. **40**,322 (1926).
See also
T.C. Wallstrom, Phys. Rev. A **49**, 1613 (1994).

[3] The Magnus force is named after the German physicist Heinrich Magnus who described it in 1853. According to
J. Gleick, *Isaac Newton*, Harper Fourth Estate, London (2004),
Newton observed the effect 180 years earlier when watching tennis players in his Cambridge college. The Magnus force makes airplanes fly due to a circulation of air around the wings. The circulation forms at takeoff, leaving behind an equal opposite circulation at the airport. The latter has caused crashes of small planes starting too close to a jumbo jet. The effect was used by the German engineer Anton Flettner in the 1920's to drive ships by a rotor rather than a sail. His ship *Baden-Baden* crossed the Atlantic in 1926. Presently, only the French research ship *Alcyone* built in 1985 uses such a drive with two rotors shaped like an airplane wing.

[4] H. Ertel, *Ein neuer hydrodynamischer Wirbelsatz*, Meteorol. Z. **59**, 277 (1942); Naturwissenschaften **30**, 543 (1942); *Über hydrodynamische Wirbelsätze*, Physik. Z. **43**, 526 (1942); *Über das Verhältnis des neuen hydrodynamischen Wirbelsatzes zum Zirkulationssatz von V. Bjerknes*, Meteorol. Z. **59**, 385 (1942).

[5] M. Caliari, G. Inverso, and L.M. Morato, *Dissipation caused by a vorticity field and generation of singularities in Madelung fluid*, New J. Phys. **6**, 69 (2004).

[6] M.I. Ioffredo and L.M. Morato, *Lagrangian Variational Principle in Stochastic Mechanics: Gauge Structure and Stability*, J. Math. Phys. **30**, 354 (1988).

[7] E. Nelson, *Quantum Fluctuations*, Princeton University Press, Princeton, NJ, 1985.

[8] C. Eckart, Phys. Rev. **54**, 920 (1938); W. Yourgrau and S. Mandelstam, *Variational Principles in Dynamics and Quantum Theory*, Pitman, London, 1968; A.M.J. Schakel, *Boulevard of Broken Symmetries*, (cond-mat/9805152).

[9] For detailed reviews and references see
C.J. Pethick and H. Smith, *Bose-Einstein Condensation in Dilute Gases*, Cambridge University Press, Cambridge, 2001;
L.P. Pitaevskii and S. Stringari, *Bose-Einstein Condensation*, Oxford University Press (Intern. Ser. Monog. on Physics), Oxford, 2003.

[10] For a detailed review and references see
I. Bloch, *Ultracold Quantum Gases in Optical Lattices*, Nat. Phys. **1**, 23 (2005).

[11] E.J. Yarmchuk, M.J.V. Gordon, and R.E. Packard, Phys. Rev. Lett. **43**, 214 (1979);
E.J. Yarmchuk and R.E. Packard, J. Low Temp. Phys. **46**, 479 (1982).

[12] M.R. Matthews, B.P. Anderson, P.C. Haljan, D.S. Hall, C.E. Wieman, and E.A. Cornell, *Vortices in a Bose-Einstein Condensate*, Phys. Rev. Lett. **83**, 2498 (1999);
J.R. Abo-Shaeer, C. Raman, J.M. Vogels, and W. Ketterle, Science **292**, 476 (2001);
V. Bretin, S. Stock, Y. Seurin, and J. Dalibard, *Fast Rotation of a Bose-Einstein Condensate*, Phys. Rev. Lett. **92**, 050403 (2004);
S. Stock, V. Bretin, F. Chevy, and J. Dalibard, *Shape Oscillation of a Rotating Bose-Einstein Condensate*, Europhys. Lett. **65**, 594 (2004).

[13] V.K. Tkachenko, Zh. Eksp. Teor. Fiz. **49**, 1875 (1965) [Sov. Phys.–JETP **22**, 1282 (1966)]; Zh. Eksp. Teor. Fiz. **50**, 1573 (1966) [Sov. Phys.–JETP **23**, 1049 (1966)]; Zh. Eksp. Teor. Fiz. **56**, 1763 (1969) [Sov. Phys.–JETP **29**, 945 (1969)].
D. Stauffer and A.L. Fetter, *Distribution of Vortices in Rotating Helium II*, Phys. Rev. **168**, 156 (1968);
G. Baym, *Stability of the Vortex Lattice in a Rotating Superfluid*, Phys. Rev. B **51**, 11697 (1995).

[14] A.L. Fetter, *Rotating Vortex Lattice in a Condensate Trapped in Combined Quadratic and Quartic Radial Potentials*, Phys. Rev. A **64**, 3608 (2001);
A.L. Fetter and A.A. Svidzinsky, *Vortices in a Trapped Dilute Bose-Einstein Condensate*, J. Phys.: Condensed Matter **13**, R135 (2001);
A.L. Fetter, B. Jackson, and S. Stringari, *Rapid Rotation of a Bose-Einstein Condensate in a Harmonic Plus Quartic Trap*, Phys. Rev. A **71**, 013605 (2005);
A.A. Svidzinsky and A.L. Fetter, *Normal Modes of a Vortex in a Trapped Bose-Einstein Condensate*, Phys. Rev. A **58**, 3168 (1998);
K. Kasamatsu, M. Tsubota, and M. Ueda, *Vortex lattice formation in a rotating Bose-Einstein condensate* Phys. Rev. A **65**, 023603 (2002); *Nonlinear dynamics of vortex lattice formation in a rotating Bose-Einstein condensate*, Phys. Rev. A **67**, 033610 (2003).

7

Dynamics of Charged Superfluid and Superconductor

In the presence of electromagnetism, we extend the derivatives in the Lagrangian density (6.3) by covariant derivative containing the minimally coupled vector potential $A^\mu(x) = (A^0(x), \mathbf{A}(x)) = (A_t(x)/c, \mathbf{A}(x))$:

$$\partial_t \phi(x) \quad \rightarrow \quad D_t \phi(x) \equiv [\partial_t + i\frac{q}{\hbar c}A_t(x)]\phi(x), \tag{7.1}$$

$$\boldsymbol{\nabla} \phi(x) \quad \rightarrow \quad \mathbf{D}\phi(x) \equiv [\boldsymbol{\nabla} - i\frac{q}{\hbar}\mathbf{A}(x)]\phi(x), \tag{7.2}$$

where q is the charge of the particles in the superfluid. In the Lagrangian density (6.5), the vector potential appears in the following form:

$$\mathcal{L} = n(x)\left\{-\hbar\left[\partial_t\theta(x) + \theta_t^{\mathrm{v}}(x) + \frac{q}{\hbar c}A_t(x)\right] - \frac{\hbar^2}{2M}\left[\boldsymbol{\nabla}\theta(x) - \boldsymbol{\theta}^{\mathrm{v}}(x) - \frac{q}{\hbar c}\mathbf{A}(x)\right]^2 - e_{\mathrm{tot}}(x)\right\}. \tag{7.3}$$

Thus the vector potential is simply added to the vortex-gauge field:

$$\theta_t^{\mathrm{v}}(x) \rightarrow \theta_t^{\mathrm{v}}(x) + \frac{q}{\hbar c}A_t(x), \quad \boldsymbol{\theta}^{\mathrm{v}}(x) \rightarrow \boldsymbol{\theta}^{\mathrm{v}}(x) + \frac{q}{\hbar c}\mathbf{A}(x), \tag{7.4}$$

This Lagrangian density has to be supplemented by Maxwell's electromagnetic Lagrangian density (2.85).

Conversely, we may take the electromagnetically coupled Lagrangian density (6.5) with the covariant derivatives (7.1) and (7.2), and replace the vector potential by

$$A_t(x) \rightarrow \tilde{A}_t(x) \equiv A_t(x) + q_m\theta_t^{\mathrm{v}}(x), \quad \mathbf{A}(x) \rightarrow \tilde{\mathbf{A}}(x) \equiv \mathbf{A}(x) + q_m\boldsymbol{\theta}^{\mathrm{v}}(x), \tag{7.5}$$

where we have introduced a magnetic charge associated with the electric charge q:

$$q_m = \frac{\hbar c}{q}. \tag{7.6}$$

Then we can rewrite (7.3) in the short form

$$\mathcal{L} = n(x)\left\{-\left[\hbar\partial_t\theta(x) + \frac{q}{c}\tilde{A}_t(x)\right] - \frac{1}{2M}\left[\hbar\boldsymbol{\nabla}\theta(x) - \frac{q}{c}\tilde{\mathbf{A}}(x)\right]^2 - e_{\text{tot}}(x)\right\}. \quad (7.7)$$

The equation of motion of the time-dependent field $\phi(t, \mathbf{x}) \equiv \phi(x)$ is

$$i\left[\hbar\partial_t + \frac{q}{c}\tilde{A}_t(x)\right]\phi(x) = \left\{-\frac{1}{2M}\left[\hbar\boldsymbol{\nabla}\theta(x) - \frac{q}{c}\tilde{\mathbf{A}}\right]^2 - c_0^2 M + \frac{c_0^2 M}{n_0}\phi^*(x)\phi(x)\right\}\phi(x). \quad (7.8)$$

7.1 Hydrodynamic Description of Charged Superfluid

For a charged superfluid, the velocity field is given by

$$\mathbf{v}(x) \equiv \frac{1}{M}\left[\hbar\boldsymbol{\nabla}\theta(x) - \frac{q}{c}\tilde{\mathbf{A}}(x)\right] = \frac{\hbar}{M}\left[\boldsymbol{\nabla}\theta(x) - \boldsymbol{\theta}^{\text{v}}(x) - \frac{q}{\hbar c}\mathbf{A}(x)\right]. \quad (7.9)$$

It is invariant under both magnetic and vortex gauge transformations. In terms of the local deviation of the particle density from the ground-state value $\delta n(x) \equiv n(x) - n_0$, the hydrodynamic Lagrangian density (7.7) can be written as

$$\mathcal{L} = -n(x)\left[\hbar\partial_t\theta(x) + \hbar\theta_t^{\text{v}}(x) + \frac{q}{c}A_t(x) + \frac{M}{2}\mathbf{v}^2(x) + e_{\text{tot}}(x)\right]. \quad (7.10)$$

The electric charge- and current densities are simply q times the particle- and current densities (6.13). They can now be derived alternatively from the Noether rule (3.117):

$$\rho(x) = -\frac{1}{c}\frac{\partial\mathcal{L}}{\partial A_t(x)} = qn(x), \quad \mathbf{J}(x) = \frac{1}{c}\frac{\partial\mathcal{L}}{\partial\mathbf{A}(x)} = qn(x)\mathbf{v}(x). \quad (7.11)$$

These satisfy the *continuity equation*

$$q\partial_t n(x) = -\boldsymbol{\nabla}\cdot\mathbf{J}(x) = 0, \quad (7.12)$$

which can again be found by extremizing the associated action with respect to $\theta(x)$.

Functional extremization of the action with respect to $n(x)$ yields

$$\hbar\partial_t\theta(x) + \hbar\theta_t^{\text{v}}(x) + \frac{q}{c}A_t(x) + \frac{M}{2}\mathbf{v}^2(x) + h_{\text{tot}}(x) = 0, \quad (7.13)$$

where $p^{\text{qu}}(x)$ is the quantum pressure defined in Eq. (6.21). These are the extensions of the Madelung equations (6.14) and (6.15) by vortices and electromagnetism. The last term may, incidentally, be replaced by the enthalpy per particle $h(x)$ of Eq. (6.16).

The gradient of (7.13) yields the equation of motion

$$M\partial_t\mathbf{v}(x) + q\left[\frac{1}{c}\partial_t\mathbf{A}(x) + \boldsymbol{\nabla}A_0(x)\right] + \hbar\left[\partial_t\boldsymbol{\theta}^{\text{v}}(x) + \boldsymbol{\nabla}\theta_t^{\text{v}}(x)\right] + \frac{M}{2}\boldsymbol{\nabla}\,\mathbf{v}^2(x) = -\boldsymbol{\nabla}h_{\text{tot}}(x). \quad (7.14)$$

Inserting here the identity (6.25) we obtain

$$M\partial_t \mathbf{v}(x) + M[\mathbf{v}(x) \cdot \boldsymbol{\nabla}]\mathbf{v}(x) = -\boldsymbol{\nabla} h_{\text{tot}}(x) + \mathbf{f}^{\text{tot}}(x), \tag{7.15}$$

where

$$\mathbf{f}^{\text{tot}}(x) = -q\left[\frac{1}{c}\partial_t \mathbf{A}(x) + \boldsymbol{\nabla} A_0(x)\right] - \hbar\left[\partial_t \boldsymbol{\theta}^{\text{v}}(x) + \boldsymbol{\nabla}\theta_t^{\text{v}}(x)\right] - M\mathbf{v}(x) \times \left[\boldsymbol{\nabla} \times \mathbf{v}(x)\right].\,(7.16)$$

On the right-hand side we may insert the velocity (7.9), further the defining equation (2.74) and (2.80) for the electromagnetic fields, and finally Eqs. (6.48) for the vortex electromagnetic fields. Then we see that $\mathbf{f}^{\text{tot}}(x) = \mathbf{f}^{\text{em}}(x) + \mathbf{f}^{\text{v}}(x)$ is the sum of the vortex force $\mathbf{f}^{\text{v}}(x)$ of Eq. (6.52), and the electromagnetic Lorentz force (1.186) upon the charged moving particles. Recall that according to Eq. (6.33) the vortex version $\mathbf{f}^{\text{v}}(x)$ of the Lorentz force happened to be zero, so that it can be omitted from $\mathbf{f}^{\text{tot}}(x)$.

The additional Maxwell action adds the equations for the electromagnetic field

$$\boldsymbol{\nabla} \cdot \mathbf{E} = \rho \quad \text{(Coulomb's law)}, \tag{7.17}$$

$$\boldsymbol{\nabla} \times \mathbf{B} - \frac{1}{c}\frac{\partial \mathbf{E}}{\partial t} = \frac{1}{c}\mathbf{J} \quad \text{(Ampère's law)}, \tag{7.18}$$

$$\boldsymbol{\nabla} \cdot \mathbf{B} = 0 \quad \text{(absence of magnetic monopoles)}, \tag{7.19}$$

$$\boldsymbol{\nabla} \times \mathbf{E} + \frac{1}{c}\frac{\partial \mathbf{B}}{\partial t} = 0 \quad \text{(Faraday's law)}. \tag{7.20}$$

The classical limit of Eq. (7.15) together with (7.17)–(7.20) are the well-known equations of motion of magnetohydrodynamics [1].

7.2   London Theory of Charged Superfluid

If we ignore the vortex gauge field in Eq. (7.9), the current density (7.11) is

$$\mathbf{J}(x) \equiv qn(x)\mathbf{v}(x) = \frac{qn(x)}{M}\left[\hbar\boldsymbol{\nabla}\theta(x) - \frac{q}{c}\mathbf{A}(x)\right\}. \tag{7.21}$$

The charge q is equal to $-2e$ since the charge carriers in the superconductor are Cooper pairs of electrons.

The brothers Heinz and Fritz London [2] were studying superconductors with constant density $n(x) \equiv n_0$, which is the reason for calling the hydrodynamic limit in Eq. (5.14) also the London limit. These authors absorbed the phase variable $\theta(x)$ into the vector potential $\mathbf{A}(x)$ by a gauge transformation

$$A_\mu(x) \to A_\mu'(x) = A_\mu(x) - \frac{c}{q}\hbar\partial_\mu\theta, \tag{7.22}$$

so that the supercurrent became directly proportional to the vector potential:

$$\mathbf{J}(x) \equiv qn_0\mathbf{v}(x) = -\frac{q^2 n_0}{cM}\mathbf{A}'(x). \tag{7.23}$$

This vector potential satisfies the transversal gauge condition

$$\nabla \cdot \mathbf{A}'(x) = 0, \tag{7.24}$$

which makes (7.23) compatible with the current conservation law (7.12).

Taking the time derivative of this and using the defining equation (2.74) for the electric field in terms of the vector potential, the London brothers obtained the equation of motion for the current

$$\partial_t \mathbf{J}(x) = \frac{q^2 n_0}{M}[\mathbf{E}(x) + \nabla A_0'(x)]. \tag{7.25}$$

At this place they postulated that the electric potential $A_0'(x)$ vanishes in a superconductor, which led them to their famous *first London equation*:

$$\partial_t \mathbf{J}(x) = \frac{q^2 n_0}{M}\mathbf{E}(x). \tag{7.26}$$

In a second step they formed, at a constant $n(x) = n_0$, the curl of the current (7.21), and obtained the *second London equation*:

$$\nabla \times \mathbf{J}(x) + \frac{q^2 n_0}{Mc}\mathbf{B}(x) = 0. \tag{7.27}$$

To check the compatibility of the two London equations one may take the curl of (7.25) and use Faraday's law of induction (7.20) to find

$$\partial_t \left[\nabla \times \mathbf{J}(x) + \frac{q^2 n_0}{Mc}\mathbf{B}(x)\right] = 0, \tag{7.28}$$

in agreement with (7.27).

From the second London equation (7.27) one derives immediately the Meissner effect. First one recalls how the electromagnetic waves are derived from the combination of Ampère's and Faraday's laws (1.188) and (1.190), in combination with the magnetic source condition (1.189):

$$\nabla \times \nabla \times \mathbf{B}(x) + \frac{1}{c^2}\partial_t^2\mathbf{B}(x) = -\nabla^2\mathbf{B}(x) + \frac{1}{c^2}\partial_t^2\mathbf{B}(x) = \frac{1}{c}\nabla \times \mathbf{J}(x). \tag{7.29}$$

In the absence of currents, this equation describes electromagnetic waves propagating with light velocity c. In a superconductor, the right-hand side is replaced by the second London equation (7.27), leading to the wave equation

$$\left[\frac{1}{c^2}\partial_t^2\mathbf{B}(x) - \nabla^2 + \lambda_{\mathrm{L}}^{-2}\right]\mathbf{B}(x) = 0, \tag{7.30}$$

with

$$\lambda_{\mathrm{L}} = \sqrt{\frac{Mc^2}{n_0 q^2}} = \frac{1}{2}\sqrt{\frac{Mc^2}{n_0 e^2}} = \frac{1}{2\sqrt{n_0 \lambda_M 4\pi\alpha}}, \tag{7.31}$$

where $\alpha \approx 1/137.0359\ldots$ is the fine-structure constant (1.145), and $\lambda_M = \hbar/Mc$ the Compton wavelength of the particles of mass M.

Equation (7.30) shows that inside a superconductor, the magnetic field has a finite *London penetration depth* λ_{L}.

7.3   Including Vortices in London Equations

The development in the last section allows us to correct the London equations. First we add the vortex gauge field, so that (7.21) becomes

$$\mathbf{J}(x) \equiv qn(x)\mathbf{v}(x) = \frac{\hbar qn(x)}{M}[\boldsymbol{\nabla}\theta(x) - \boldsymbol{\theta}^{\mathrm{v}}(x)] - \frac{q^2 n(x)}{cM}\mathbf{A}(x). \tag{7.32}$$

In the London limit where $n(x) \approx n_0$, we take again the time derivative of (7.32) and, recalling Eq. (6.48), we obtain

$$\partial_t \mathbf{J}(x) = \frac{\hbar qn_0}{M}[\boldsymbol{\nabla}\partial_t\theta(x) + \mathbf{E}^{\mathrm{v}}(x) + \boldsymbol{\nabla}\theta_t^{\mathrm{v}}(x)] + \frac{q^2 n_0}{M}[\mathbf{E}(x) + \boldsymbol{\nabla}A_0(x)]. \tag{7.33}$$

As before, we fix the vortex gauge to have $\theta_t^{\mathrm{v}}(x) = 0$, and absorb the phase variable $\theta(x)$ in the vector potential $\mathbf{A}$ by a gauge transformation (7.22). Thus we remain with the vortex-corrected first London equation

$$\partial_t \mathbf{J}(x) = \frac{q^2 n_0}{M}[\mathbf{E}(x) + q_m\mathbf{E}^{\mathrm{v}}(x) + \boldsymbol{\nabla}A_0(x)], \tag{7.34}$$

where q_m is the magnetic charge (7.6) associated with the electric charge q.

Taking the curl of Eq. (7.32) in the London limit with the same fixing of vortex and electromagnetic gauge, we obtain the vortex-corrected second London equation

$$\boldsymbol{\nabla} \times \mathbf{J}(x) + \frac{q^2 n_0}{Mc}[\mathbf{B}(x) + q_m\mathbf{B}^{\mathrm{v}}(x)] = 0. \tag{7.35}$$

The compatibility with (7.34) is checked by forming the curl of (7.34) and inserting Faraday's law of induction (7.20) and its vortex analog (6.51). The result is the statement that the time derivative of (7.35) vanishes, which is certainly true.

Inserting (7.35) into the combined Maxwell equation (7.29) yields the vortex-corrected Eq. (7.30):

$$\left[\frac{1}{c^2}\partial_t^2 - \boldsymbol{\nabla}^2 + \lambda_{\mathrm{L}}^{-2}\right]\mathbf{B}(x) = -\lambda_{\mathrm{L}}^{-2}q_m\,\mathbf{B}^{\mathrm{v}}(x). \tag{7.36}$$

From this we can directly deduce the interaction between vortex lines

$$\mathcal{A}_{\mathrm{int}} = -\frac{q_m^2}{2}\int d^4x\,d^4x'\,\mathbf{B}^{\mathrm{v}}(x)G_{\lambda_{\mathrm{L}}}^R(x - x')\mathbf{B}^{\mathrm{v}}(x'), \tag{7.37}$$

where $G_{\lambda_{\mathrm{L}}}^R(x - x')$ is the retarded Yukawa Green function

$$G_{\lambda_{\mathrm{L}}}^R(x - x') = \frac{1}{-c^{-2}\partial_t^2 + \boldsymbol{\nabla}^2 - \lambda_{\mathrm{L}}^{-2}}(x, x') = -\Theta(t - t')\frac{e^{-R/\lambda_{\mathrm{L}}}}{4\pi c^2 R}\delta(t - t' - R/c), \tag{7.38}$$

in which R denotes the spatial distance $R \equiv |\mathbf{x} - \mathbf{x}'|$.

In the limit $\lambda_{\mathrm{L}} \to \infty$ this goes over to the Coulomb version which is the origin of the well-known Liènard-Wiechert potential of electrodynamics.

For slowly moving vortices, the retardation can be neglected and, after inserting q_m from (7.6) and $\mathbf{B}^v(x)$ from (6.48), and performing the time derivatives in (7.37), we find

$$\mathcal{A}_{\text{int}} = -\frac{\hbar^2 c^2}{2q^2} \int dt \int d^3x \, \mathbf{j}^v(\mathbf{x}, t) \frac{1}{-\boldsymbol{\nabla}^2 + \lambda_{\mathrm{L}}^{-2}} \mathbf{j}^v(\mathbf{x}, t). \tag{7.39}$$

This agrees with the previous static interaction energy in the partition function (5.276), if we go to natural units $\hbar = c = M = 1$.

7.4 Hydrodynamic Description of Superconductor

For a superconductor, the above theory of a charged superfluid is not applicable since the initial Ginzburg-Landau Lagrangian density can be derived [3] only near the phase transition where it has, moreover, a purely damped temporal behavior. Hence there is no time derivative term as in Eq. (6.2). At zero temperature, however, the superflow can be described by simple hydrodynamic equations. From the BCS theory, one can derive a Lagrangian of the type (6.46) in the harmonic approximation [4]

$$\mathcal{L}_0 = -n_0 \hbar \partial_t \theta(x) + \frac{n_0 \hbar^2}{2M} \left\{ \frac{1}{c_0^2} [\partial_t \theta(x) + \theta_t^v(x)]^2 - [\boldsymbol{\nabla}\theta(x) - \boldsymbol{\theta}^v(x)]^2 \right\} + \dots, \tag{7.40}$$

with the second sound velocity [3]

$$c_0 = \frac{v_F}{\sqrt{3}}, \tag{7.41}$$

where $v_F = p_F/M = \sqrt{2ME_F}$ is the velocity of electrons on the surface of the Fermi sphere, which is calculated from the density of electrons (which is twice as big as the density of Cooper pairs n_0):

$$n_{\text{int}} = 2 \int \frac{d^3p}{(2\pi\hbar)^3} = \frac{p_F^3}{3\hbar^3\pi^2} = \frac{v_F^3}{3\hbar^3 M^3\pi^2}. \tag{7.42}$$

The dots in (7.40) indicate terms which can be ignored in the long-wavelength limit. These are different from those of the Bose Lagrangian density in (6.46). There we see that the energy spectrum of second sound excitations has a first correction term of the form

$$\epsilon(\mathbf{k}) = c_0 |\mathbf{k}|(1 - \gamma \mathbf{k}^2 + \dots), \tag{7.43}$$

with a *negative* $\gamma = -\xi < 0$. In a superconductor at $T = 0$, on the other hand, the BCS theory yields a *positive* γ:

$$\gamma = \frac{\hbar^2 v_F^2}{45\Delta^2} = \frac{1}{45} l^2, \quad l \equiv \hbar\sqrt{1/M\Delta}, \tag{7.44}$$

where Δ is the energy gap of the quasiparticle excitations of the superconductor, which is of the order of the transition temperature (times k_B) (see Appendix 7A). The length scale l is of the order of the zero-temperature coherence length.

The positivity of γ ensures the stability of the long-wavelength excitations against decay since it makes $|\mathbf{k}_1 + \mathbf{k}_2|[1 - \gamma(\mathbf{k}_1 + \mathbf{k}_2)^2] < |\mathbf{k}_1|(1 - \gamma\mathbf{k}_2^1) + |\mathbf{k}_2|(1 - \gamma\mathbf{k}_2^2)$.

We now add the electromagnetic fields by minimal coupling [compare (7.7)], and find

$$\mathcal{L}_0 = -n_0\hbar\left[\partial_t\theta(x) + \frac{q}{\hbar c}\tilde{A}_t(x)\right] + \frac{n_0\hbar^2}{2M}\left\{\frac{1}{c_0^2}\left[\partial_t\theta(x) + \frac{q}{\hbar c}\tilde{A}_t(x)\right]^2 - [\mathbf{\nabla}\theta(x) - \frac{q}{\hbar c}\tilde{\mathbf{A}}(x)]^2\right\},$$
(7.45)

to be supplemented by the Maxwell Lagrangian density (2.85).

The derivative of $\mathcal{L}_0$ with respect to $-\mathbf{A}(x)/c$ yields the current density [recall (3.117)]:

$$\mathbf{J}(x) = qn_0\mathbf{v}(x) = \frac{qn_0\hbar}{M}[\mathbf{\nabla}\theta(x) - \mathbf{\theta}^{\mathrm{v}}(x)] - \frac{q^2n_0}{cM}\mathbf{A}(x).$$
(7.46)

From the derivative of $\mathcal{L}_0$ with respect to $-A_t(x)/c$ we obtain the charge density:

$$q[n(x) - n_0] = \frac{qn_0\hbar}{M}\frac{1}{c_0^2}[\partial_t\theta(x) + \theta_t^{\mathrm{v}}(x)] - \frac{q^2n_0}{c_0^2cM}A_t(x).$$
(7.47)

If we absorb the field $\theta(x)$ in the vector potential, we find the same supercurrent as in (7.23):

$$\mathbf{J}(x) = -\frac{q^2n_0}{cM}\tilde{\mathbf{A}}(x),$$
(7.48)

whereas the charge density becomes

$$qn(x) = -\frac{q^2n_0}{c_0^2cM}\tilde{A}_t(x).$$
(7.49)

The current conservation law implies that

$$\mathbf{\nabla}\cdot\mathbf{A}(x) + \frac{c^2}{c_0^2}\frac{1}{c^2}\partial_t A_t = 0.$$
(7.50)

Note the difference by the large factor c^2/c_0^2 of the time derivative term with respect to the Lorentz gauge (2.106):

$$\partial_a A^a(x) = \mathbf{\nabla}\cdot\mathbf{A}(x) + \partial_0 A^0(x) = \mathbf{\nabla}\cdot\mathbf{A}(x) + \frac{1}{c^2}\partial_t A^t(x) = 0.$$
(7.51)

Since the velocity $c_0 = v_F/\sqrt{3}$ is much smaller than the light velocity c, typically by a factor $1/100$, the ratio c^2/c_0^2 is of the order of 10^4.

At this point we recall that according to definition (5.256), the ratio of the penetration depth λ_{L} of Eq. (7.31) and the coherence length ξ of Eq. (6.44) define the Ginzburg parameter $\kappa \equiv \lambda_{\mathrm{L}}/\sqrt{2}\xi$. This allows us to express the ratio c_0/c in terms of κ as follows:

$$\frac{c_0}{c} = \frac{\kappa}{\sqrt{2}}\sqrt{n_0 \lambda_M^3 q^2}. \tag{7.52}$$

If the current density (7.23) is inserted into the combined Maxwell equation (7.29), we obtain once more the field equation (7.30) for the screened magnetic field $\mathbf{B}$ and its vortex-corrected version (7.36).

The field equation for A_0, however, has quite different wave propagation properties, due to the factor c^2/c_0^2. It is obtained by varying the action $\mathcal{A} = \int dt d^3x \, [\mathcal{L}^{\mathrm{em}}] + \mathcal{L}_0$, with respect to $-A_0(x) = -A_t(x)/c$, which yields

$$\boldsymbol{\nabla} \cdot \mathbf{E}(x) = qn(x). \tag{7.53}$$

Inserting $\mathbf{E}(x)$ from (2.74), and $qn(x)$ from Eq. (7.49) in the axial vortex gauge, we find

$$-\boldsymbol{\nabla}^2 A^0(x) - \frac{1}{c}\partial_t \boldsymbol{\nabla} \cdot \mathbf{A}(x) = -\frac{q^2 n_0}{c_0^2 M} A^0(x). \tag{7.54}$$

Eliminating $\boldsymbol{\nabla} \cdot \mathbf{A}(x)$ with the help of Eq (7.50), we obtain

$$\left(-\boldsymbol{\nabla}^2 + \lambda_{\mathrm{L}0}^{-2}\right) A^0(x) - \frac{c^2}{c_0^2}\frac{1}{c^2}\partial_t^2 A^0 = 0. \tag{7.55}$$

This equation shows that the field $A^0(x)$ penetrates a superconductor over the distance

$$\lambda_{\mathrm{L}0} = \frac{c_0}{c}\lambda_{\mathrm{L}} = \frac{c_0}{c}\frac{1}{\sqrt{n_0 \lambda_M q^2}}, \tag{7.56}$$

which is typically two orders of magnitude smaller than the penetration depth λ_{L} of the magnetic field. Moreover, the propagation velocity of $A^0(x)$ is not the light velocity c but the much smaller velocity $c_0 = v_F/\sqrt{3}$.

Note that Eq. (7.55) for $A^0(x)$ has no gauge freedom left, the gauge being fixed by Eq. (7.50). This is best seen by expressing $A^0(x)$ in terms of the charge density $qn(x)$ via Eq. (7.49) which yields

$$\left[-\frac{1}{c^2}\partial_t^2 + \frac{c_0^2}{c^2}\left(-\boldsymbol{\nabla}^2 + \lambda_{\mathrm{L}0}^{-2}\right)\right] n(x) = 0. \tag{7.57}$$

Using the vanishing of $\mathbf{f}^{\mathrm{v}}(x)$ of Eq. (6.52), and Eq. (6.55), we can rewrite the divergence of $\mathbf{E}^{\mathrm{v}}(x)$ as

$$\boldsymbol{\nabla} \cdot \mathbf{E}^{\mathrm{v}}(x) = -\boldsymbol{\nabla} \cdot [\mathbf{v}(x) \times \mathbf{B}^{\mathrm{v}}(x)] = -[\boldsymbol{\nabla} \times \mathbf{v}(x)]\mathbf{B}^{\mathrm{v}}(x) + \mathbf{v}(x) \cdot [\boldsymbol{\nabla} \times \mathbf{B}^{\mathrm{v}}(x)]$$
$$= \frac{\hbar}{M}[\mathbf{B}^{\mathrm{v}}(x)]^2 + \mathbf{v}(x) \cdot [\boldsymbol{\nabla} \times \mathbf{B}^{\mathrm{v}}(x)]. \tag{7.58}$$

The Lagrangian density of the vector field $A^\mu(x)$ can also be written as

$$\mathcal{L} = \frac{1}{2}A^a(x)\left(\partial^2 g_{ab} - \partial_a \partial_b\right)A^b(x) + \frac{m_A^2}{2}\{a[A^0(x)]^2 - \mathbf{A}^2(x)\}, \tag{7.59}$$

where $m_A^2 = \lambda_L^{-2}$ and $u^2 = c^2/c_0^2$. The field equation in the presence of an external source $j^a(x)$ is

$$\left[(\partial^2 g_{ab} - \partial_a \partial_b) + m_A^2 g_{ab} + m_A^2(u^2 - 1)h_{ab}\right] A^b(x) = j_a(x), \tag{7.60}$$

where h_{ab} has only one nonzero matrix element $h_{00} = 1$. By contracting (7.60) with ∂_a, and using current conservation $\partial_a j^a(x) = 0$, we obtain

$$\partial_a A^a(x) + (u^2 - 1)\partial_0 A^0(x) = 0, \tag{7.61}$$

which is the divergence equation (7.50). Reinserting this into (7.60) yields

$$\left(u^2 \partial_0^2 - \nabla^2 + u^2 m_A^2\right) A^0(x) = j^0(x), \tag{7.62}$$

$$(\partial^2 + m_A^2)\mathbf{A}(x) + \frac{u^2 - 1}{u^2}\nabla[\nabla \cdot \mathbf{A}(x)] = \mathbf{j}(x). \tag{7.63}$$

To check the consistency, we take the divergence of the second equation and use the current conservation law to write

$$-(\partial^2 + m_A^2)[\nabla \cdot \mathbf{A}(x)] + \frac{u^2 - 1}{u^2}\nabla^2[\nabla \cdot \mathbf{A}(x)] = \partial_0 j^0(x). \tag{7.64}$$

Then we replace $\nabla \cdot \mathbf{A}(x)$ by $-u^2 \partial_0 A^0(x)$ and obtain

$$u^2 \partial_0(\partial_0^2 - \nabla^2 + m_A^2)A^0(x) + (u^2 - 1)\partial_0 \nabla^2 A^0(x) = \partial_0 j^0(x), \tag{7.65}$$

which is the time derivative of Eq. (7.62).

The inverse of Eq. (7.62) is

$$A^0(x) = \frac{1}{u^2}\left(\partial_0^2 - u^{-2}\nabla^2 + m_A^2\right)^{-1} j^0(x). \tag{7.66}$$

To invert Eq. (7.63) we rewrite it in terms of spatial transverse and longitudinal projection matrices

$$P_{ij}^t = \delta_{ij} - \frac{\nabla_i \nabla_j}{\nabla^2}, \qquad P_{ij}^l = \frac{\nabla_i \nabla_j}{\nabla^2}, \tag{7.67}$$

as

$$\left[(\partial^2 + m_A^2)P^t + (\partial_0^2 - u^{-2}\nabla^2 + m_A^2)P^l\right] \mathbf{A}(x) = \mathbf{j}(x). \tag{7.68}$$

This is immediately inverted to

$$\mathbf{A}(x) = \left[(\partial^2 + m_A^2)^{-1}P^t + (\partial_0^2 - u^{-2}\nabla^2 + m_A^2)^{-1}P^l\right] \mathbf{j}(x). \tag{7.69}$$

From these equations we derive the interaction between external currents

$$\mathcal{A}^{\text{int}} = \int d^4x \left[\frac{1}{u^2}j^0(x)\frac{1}{\partial_0^2 - u^{-2}\nabla^2 + m_A^2}j^0(x) - \mathbf{j}^l(x)\frac{1}{\partial_0^2 - u^{-2}\nabla^2 + m_A^2}\mathbf{j}^l(x) \right.$$
$$\left. - \mathbf{j}^t(x)\frac{1}{\partial_0^2 - \nabla^2 + m_A^2}\mathbf{j}^t(x) \right]. \tag{7.70}$$

Only the transverse currents interact with the relativistic retarded interaction. Due to current conservation the first two terms can be combined and we obtain

$$\mathcal{A}^{\text{int}} = \int d^4x \left[j^0(x) \frac{u^{-2} - \boldsymbol{\nabla}^{-2}\partial_0^2}{\partial_0^2 - u^{-2}\boldsymbol{\nabla}^2 + m_A^2} j^0(x) - \mathbf{j}^t(x) \frac{1}{\partial_0^2 - \boldsymbol{\nabla}^2 + m_A^2} \mathbf{j}^t(x) \right]. \quad (7.71)$$

For $u^2 = c^2/c_0^2 = 1$, this reduces to the usual relativistic interaction

$$\mathcal{A}^{\text{int}} = \int d^4x\, j_\mu(x) \frac{1}{\partial^2 + m_A^2} j^\mu(x). \quad (7.72)$$

Appendix 7A Excitation Spectrum of Superconductor

For understanding the time dependence of the hydrodynamic equations of a super-conductor it is important to know the spectrum of the low-energy excitations. This is derived here from the theory of Bardeen, Cooper, and Schrieffer (BCS) [4]. The electrons are bound to Cooper pairs, and the quasiparticle energies have the form

$$E(\mathbf{p}) = \sqrt{\xi^2(\mathbf{p}) + \Delta^2}, \quad (7A.1)$$

where

$$\xi(\mathbf{p}) \equiv \frac{\mathbf{p}^2}{2M} - \mu \quad (7A.2)$$

are the free-electron energies measured from the chemical potential μ. At zero temperature, this is equal to the Fermi energy $\epsilon_F = M v_F^2/2$.

7A.1 Gap Equation

The quasiparticle energies have a gap Δ which is determined by the *gap equation*

$$\frac{1}{g} = \frac{T}{V} \sum_{\omega_m, \mathbf{p}} \frac{1}{\omega_m^2 + E^2(\mathbf{p})} = \frac{1}{V} \sum_{\mathbf{p}} \frac{1}{2E(\mathbf{p})} \tanh \frac{E(\mathbf{p})}{2T}, \quad (7A.3)$$

where g is the attractive short-range interaction between electrons near the surface of the Fermi sea caused by the electron-phonon interaction. The sum over ω_m runs over the *Matsubara frequencies* $\omega_m = 2\pi k_B T m$, for $m = 0, \pm 1, \pm 2, \ldots$. The equality of the second and third expression in (7A.3) follows from the summation formula [3]

$$T \sum_{\omega_m} \frac{e^{i\omega_m \eta}}{i\omega_m - E} = n(E), \quad (7A.4)$$

where $0 < \eta \ll 1$ is an infinitesimal parameter to make the sum convergent, and $n(E)$ is the Fermi distribution function

$$n(E) \equiv \frac{1}{e^{E/T} + 1} = \frac{1}{2}\left(1 - \tanh \frac{E}{2T}\right). \quad (7A.5)$$

By combining the sums (7A.4) with E and $-E$ we obtain the important formula

$$T \sum_{\omega_m} \frac{1}{\omega_m^2 + E^2} = \frac{1}{2E} T \sum_{\omega_m} \left(\frac{e^{i\omega_m \eta}}{i\omega_m + E} - \frac{e^{i\omega_m \eta}}{i\omega_m - E} \right) = \frac{1}{2E}[n(-E) - n(E)]$$

$$= \frac{1}{2E} \tanh \frac{E}{2T} = \frac{1}{2E}[1 - 2n(E)]. \tag{7A.6}$$

In terms of $n(E)$ of (7A.5), the gap equation (7A.3) reads

$$\frac{1}{g} = \frac{1}{V} \sum_{\mathbf{p}} \frac{1}{2E(\mathbf{p})}[1 - 2n(E)]. \tag{7A.7}$$

The momentum sums in (7A.3) are conveniently performed in an approximation which is excellent for small electron-phonon interaction, where only the neighborhood of the Fermi surface contributes significantly, as follows:

$$\frac{1}{V} \sum_{\mathbf{p}} \rightarrow \int \frac{d^3 p}{(2\pi)^3} = \int d\hat{\mathbf{p}} \int \frac{dp^2}{2} p \approx \frac{M^2 v_F}{2\pi^2} \int \frac{d\hat{\mathbf{p}}}{4\pi} \int d\xi. \tag{7A.8}$$

The integral $\int d\hat{\mathbf{p}}$ runs over all momentum directions $\hat{\mathbf{p}} = \mathbf{p}/|\mathbf{p}|$. The prefactor

$$\mathcal{N}(0) \equiv \frac{m^2 v_F}{2\pi^2} \tag{7A.9}$$

is the density of states of one spin orientation on the surface of the Fermi sea. This brings the gap equation (7A.3) to the form

$$\frac{1}{g} = \mathcal{N}(0) \int_0^{\omega_D} \frac{d\xi}{\sqrt{\xi^2 + \Delta^2}} \tanh \frac{\sqrt{\xi^2 + \Delta^2}}{2T}. \tag{7A.10}$$

The integral is logarithmically divergent. Since the attraction is due to the phonons in a crystal whose spectrum is limited by the Debye frequency ω_D (which in conventional superconductors is much smaller than the Fermi energy ϵ_F), we have cut off the integral at ω_D.

The critical temperature T_c is the place where the gap Δ vanishes, and (7A.19) reduces to

$$\frac{1}{g} = \mathcal{N}(0) \int_0^{\omega_D} \frac{d\xi}{\xi} \tanh \frac{\xi}{2T_c}. \tag{7A.11}$$

The integral is done as follows. It is performed first by parts to yield

$$\int_0^{\omega_D} \frac{d\xi}{\xi} \tanh \frac{\xi}{2T_c} = \log \frac{\xi}{T_c} \tanh \frac{\xi}{2T_c} \Big|_0^{\omega_D} - \frac{1}{2} \int_0^{\infty} d\frac{\xi}{T_c} \log \frac{\xi}{T_c} \cosh^{-2} \frac{\xi}{2T_c}. \tag{7A.12}$$

Since $\omega_D/\pi T_c \gg 1$, the first term is equal to $\log(\omega_D/2T_c)$, with exponentially small corrections from the hyperbolic tangens, which can be ignored. In the second integral, we have taken the upper limit of integration to infinity since it converges. We may use the integral formula[1]

$$\int_0^{\infty} dx \frac{x^{\mu-1}}{\cosh^2(ax)} = \frac{4}{(2a)^\mu} \left(1 - 2^{2-\mu} \right) \Gamma(\mu) \zeta(\mu - 1), \tag{7A.13}$$

[1]See, for instance, I.S. Gradshteyn and I.M. Ryzhik, *Table of Integrals, Series, and Products*, Academic Press, New York, 1980, Formula 3.527.3.

set $\mu = 1 + \delta$, expand the formula to order δ, and insert the special values

$$\Gamma'(1) = -\gamma, \quad \zeta'(0) = -\frac{1}{2}\log(2\pi)\log(4e^{\gamma}/\pi), \tag{7A.14}$$

where γ is Euler's constant

$$\gamma = -\Gamma'(1)/\Gamma(1) \approx 0.577, \tag{7A.15}$$

so that $e^{\gamma}/\pi \approx 1.13$. Thus we find from the linear terms in δ:

$$\int_0^{\infty} dx \frac{\log x}{\cosh^2(x/2)} = -2\log(2e^{\gamma}/\pi), \tag{7A.16}$$

and Eq. (7A.11) becomes

$$\frac{1}{g} = \mathcal{N}(0)\log\left(\frac{\omega_D}{T_c}\frac{2e^{\gamma}}{\pi}\right), \tag{7A.17}$$

which determines T_c in terms of the coupling strength g as

$$T_c = \omega_D \frac{2e^{\gamma}}{\pi}e^{-1/g\mathcal{N}(0)}. \tag{7A.18}$$

In order to find the T-dependence of the gap, we may expand the hyperbolic tangens in Eq. (7A.19) in powers of $e^{-E(\mathbf{p})/T}$ and obtain

$$\frac{1}{g} = \mathcal{N}(0)\int_0^{\omega_D}\frac{d\xi}{\sqrt{\xi^2 + \Delta^2}}\left[1 + 2\sum_{n=1}^{\infty}(-1)^n\exp(-n\sqrt{\xi^2 + \Delta^2}/T)-\right], \tag{7A.19}$$

where $K_0(z)$ are the modified Bessel functions of the second kind. The cutoff is needed only in the first integral, in the others it can be moved to infinity, and we obtain

$$\frac{1}{g} = \mathcal{N}(0)\left[\log\frac{2\omega_D}{\Delta} + 2\sum_{n=1}^{\infty}(-1)^n K_0(n\Delta/T)\right]. \tag{7A.20}$$

Replacing $1/g$ by (7A.18), we find

$$\log\left(\frac{\Delta}{T_c}\frac{e^{\gamma}}{\pi}\right) = 2\sum_{n=1}^{\infty}(-1)^n K_0(n\Delta/T). \tag{7A.21}$$

For small T, K_0 vanishes exponentially fast:

$$2K_0\left(\frac{\Delta}{T}\right) \to \frac{1}{\Delta}\sqrt{2\pi T\Delta}e^{-\Delta/T}. \tag{7A.22}$$

Hence we find the $T = 0$ -gap

$$\Delta(0) = 2\omega_D e^{-1/g\mathcal{N}(0)}. \tag{7A.23}$$

Combining this with (7A.18) we obtain the important universal relation between critical temperature T_c and energy gap at zero-temperature $\Delta(0)$:

$$\Delta(0)/T_c = \pi e^{-\gamma} \approx 1.76. \tag{7A.24}$$

This value is approached exponentially as $T \to 0$, since from (7A.19)

$$\log \frac{\Delta(T)}{\Delta(0)} \approx \frac{\Delta(T)}{\Delta(0)} - 1 \approx -\frac{1}{\Delta(0)} \sqrt{2\pi T \Delta(0)} e^{-\Delta(0)/T}. \tag{7A.25}$$

With $\Delta(0)$ determined by (7A.24), we may replace the left-hand side of (7A.21) by $\log[\Delta/\Delta(0)]$.

For $T \approx T_c$, the gap is calculated most efficiently by combining the gap equation (7A.19) with its $T = T_c$-version to obtain

$$\int_0^{\omega_D} \frac{d\xi}{\xi} \left(\tanh \frac{\xi}{2T_c} - \tanh \frac{\xi}{2T} \right) = \int_0^{\omega_D} d\xi \left(\frac{1}{\sqrt{\xi^2 + \Delta^2}} \tanh \frac{\sqrt{\xi^2 + \Delta^2}}{2T} - \frac{1}{\xi} \tanh \frac{\xi}{2T} \right). \tag{7A.26}$$

The integrals on both sides are now convergent so that the cutoff frequency ω_D can be removed. If the integrals on the left-hand side are performed individually as in Eqs. (7A.11)–(7A.17), they yield

$$\log \left(\frac{\omega_D}{T_c} \frac{2e^{\gamma}}{\pi} \right) - \log \left(\frac{\omega_D}{T} \frac{2e^{\gamma}}{\pi} \right) = \log \frac{T}{T_c}. \tag{7A.27}$$

On the right-hand side we replace each hyperbolic tangens by a Matsubara sum according to Eq. (7A.6), and arrive at

$$\log \frac{T}{T_c} = T \sum_{\omega_m} \int_0^{\infty} d\xi \left(\frac{1}{\omega_m^2 + \xi^2 + \Delta^2} - \frac{1}{\omega_m^2 + \xi^2} \right). \tag{7A.28}$$

This can be integrated over ξ to yield the gap equation

$$\log \frac{T}{T_c} = 2\pi T \sum_{\omega_m > 0} \left(\frac{1}{\sqrt{\omega_m^2 + \Delta^2}} - \frac{1}{\omega_m} \right). \tag{7A.29}$$

It is convenient to introduce the reduced gap

$$\delta \equiv \frac{\Delta}{T} \tag{7A.30}$$

and a reduced version of the Matsubara frequencies:

$$x_n \equiv (2n + 1)\pi/\delta. \tag{7A.31}$$

Then the gap equation (7A.29) takes the form

$$\log \frac{T}{T_c} = \frac{2\pi}{\delta} \sum_{n=0}^{\infty} \left(\frac{1}{\sqrt{x_n^2 + 1}} - \frac{1}{x_n} \right). \tag{7A.32}$$

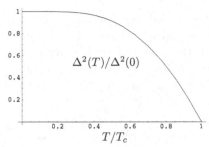

FIGURE 7.1 Energy gap (squared) of superconductor as a function of temperature.

It can be used to calculate T/T_c as a function of δ, from which we obtain $\Delta(T)/\Delta(0) = (e^\gamma/\pi)\Delta(T)/T_c = (e^\gamma\delta/\pi)T/T_c$ as a function of T/T_c, as shown in Fig. 7.1.

The behavior in the vicinity of the critical temperature T_c can be extracted from Eq. (7A.32) by expanding the sum under the assumption of small δ and large x_n. The leading term gives

$$\log\frac{T}{T_c} \approx \frac{2\pi}{\delta}\sum_{n=0}^{\infty}\frac{1}{2x_n^3} = -\frac{\delta^2}{\pi^2}\sum_{n=0}^{\infty}\frac{1}{(2n+1)^2} = -\frac{\delta^2}{\pi^2}\frac{7}{8}\zeta(3) \qquad (7A.33)$$

so that

$$\delta^2 \approx \frac{8\pi^2}{7\zeta(3)}\left(1 - \frac{T}{T_c}\right) \qquad (7A.34)$$

and

$$\frac{\Delta}{T_c} = \delta_c = \pi\sqrt{\frac{8}{7\zeta(3)}}\left(1 - \frac{T}{T_c}\right)^{1/2} \approx 3.063 \times \left(1 - \frac{T}{T_c}\right)^{1/2}. \qquad (7A.35)$$

7A.2 Action of Quadratic Fluctuations

The small fluctuations $\delta\Delta(x)$ of the complex field of Cooper pairs are governed by the quadratic action

$$\mathcal{A}_2[\delta\Delta^*, \delta\Delta] = -\frac{i}{2}\mathrm{Tr}\left[\mathbf{G}_\Delta\begin{pmatrix}0 & \delta\Delta \\ \delta\Delta^* & 0\end{pmatrix}\mathbf{G}_\Delta\begin{pmatrix}0 & \delta\Delta \\ \delta\Delta^* & 0\end{pmatrix}\right] - \frac{1}{g}\int dx|\delta\Delta(x)|^2, \qquad (7A.36)$$

where $\delta\Delta(x)$ is a small fluctuation of the complex gap field around the real background value $\Delta(T)$ given by the gap equation (7A.3). The gap equation is determined from the extremum of the action which ensures that the fluctuation expansion $\delta\Delta(x)$ has no linear term and is dominated by (7A.36). The matrix $\mathbf{G}_\Delta(x, x')$ denotes the free correlation functions of the electrons in a constant background pair field $\Delta(x) = \Delta$:

$$\mathbf{G}_\Delta(x, x') = i\begin{pmatrix}[i\partial_t - \xi(-i\boldsymbol{\nabla})]\delta & -\Delta \\ -\Delta & \mp i[\partial_t - \xi(i\boldsymbol{\nabla})]\delta\end{pmatrix}^{-1}(x, x'). \qquad (7A.37)$$

At finite temperature we go to Fourier space with Matsubara frequencies $\omega_m = 2\pi(m + \frac{1}{2})T$ for the electrons and $\nu_n = 2\pi nT$ for the pair field which guarantee the antiperiodicity of the Fermi and the periodicity of the Cooper pair field in the imaginary time interval $\tau \in (0, 1/T)$. If we use a four-vector notation for the Euclidean electron momenta $p \equiv (\omega_m, \mathbf{p})$ and pair momenta $k = (\nu_n, \mathbf{k})$, the action (7A.36) reads

$$
A_2[\delta\Delta^*, \delta\Delta] = \frac{1}{2}\frac{T}{V}\sum_{p,k}\left\{\left[\left(\omega_m + \frac{\nu_n}{2}\right)^2 + E^2\left(\mathbf{p} + \frac{\mathbf{k}}{2}\right)\right]\left[\left(\omega_m - \frac{\nu_n}{2}\right)^2 + E^2\left(\mathbf{p} - \frac{\mathbf{k}}{2}\right)\right]\right\}^{-1}
$$

$$
\times \left\{\left[\omega_m^2 - \frac{\nu_n^2}{4} + \xi\left(\mathbf{p} + \frac{\mathbf{k}}{2}\right)\xi\left(\mathbf{p} - \frac{\mathbf{k}}{2}\right)\right][\Delta'^*(k)\delta\Delta(k) + \delta\Delta(-k)\delta\Delta^*(-k)]\right.
$$

$$
\left. - |\Delta_0|^2\left[\delta\Delta^*(k)\delta\Delta^*(-k) + \delta\Delta(k)\delta\Delta(-k)\right]\right\} - \frac{1}{g}\sum_k \delta\Delta^*(k)\delta\Delta(k). \quad (7A.38)
$$

This has the generic quadratic form

$$
A_2[\delta\Delta^*, \delta\Delta] = \frac{1}{2}\frac{T}{V}\sum_k [\delta\Delta^*(k)L_{11}(k)\delta\Delta(k) + \delta\Delta(-k)L_{22}(k)\delta\Delta(-k)
$$

$$
+ \delta\Delta^*(k)L_{12}(k)\delta\Delta^*(-k) + \delta\Delta(-k)L_{21}(k)\delta\Delta(k)], \quad (7A.39)
$$

with coefficients

$$
L_{11}(k) = L_{22}(k) = \int\frac{d^3p}{(2\pi)^3}T\sum_{\omega_m}\frac{\omega_m^2 - \nu_n^2/4 + \xi_+\xi_-}{\left[\left(\omega_m + \frac{\nu_n}{2}\right)^2 + E_+^2\right]\left[\left(\omega_m - \frac{\nu_n}{2}\right)^2 + E_-^2\right]} - \frac{1}{g}, \quad (7A.40)
$$

$$
L_{12}(k) = [L_{21}(k)]^* = -\Delta^2\int\frac{d^3p}{(2\pi)^3}T\sum_{\omega_m}\frac{1}{\left[\left(\omega_m + \frac{\nu_n}{2}\right)^2 + E_+^2\right]\left[\left(\omega_m - \frac{\nu_n}{2}\right)^2 + E_-^2\right]}.
$$

$$
\quad (7A.41)
$$

The quantities $\xi_\pm$ and $E_\pm$ are defined in terms of $\xi(\mathbf{p})$ and $E(\mathbf{p})$ of Eqs. (7A.1) and (7A.2) as follows:

$$
\xi_\pm = \xi(\mathbf{p} \pm \mathbf{k}/2) = \frac{\mathbf{p}^2}{2M} \pm \frac{1}{2}\mathbf{v}\mathbf{k} + \frac{\mathbf{k}^2}{8M} - \mu, \qquad E_\pm = E(\mathbf{p} \pm \mathbf{k}/2). \quad (7A.42)
$$

The combinations in Eqs. (7A.40) and (7A.41) are

$$
\xi_+\xi_- = \xi^2 - \frac{1}{4}(\mathbf{v}\mathbf{k})^2 + \xi\frac{\mathbf{k}^2}{2m} + \frac{\mathbf{k}^4}{64m^2}, \qquad \mathbf{v} \equiv \frac{\mathbf{p}}{m}, \quad (7A.43)
$$

$$
\left\{\begin{matrix} E_+^2 \\ E_-^2 \end{matrix}\right\} = E^2 \pm \xi\mathbf{v}\mathbf{k} + \frac{1}{4}(\mathbf{v}\mathbf{k})^2 + \xi\frac{\mathbf{k}^2}{4m} \pm \mathbf{v}\mathbf{k}\frac{\mathbf{k}^2}{8m} + \frac{\mathbf{k}^4}{64m^2}. \quad (7A.44)
$$

After some straightforward algebra and using the sum formula (7A.4), we replace the Euclidean pair energy ν_n by $-i$ times a real continuous energy ϵ, and obtain

$$
L_{11}(\epsilon, k) = \int\frac{d^3p}{(2\pi)^3}\left\{\frac{E_+E_- + \xi_+\xi_-}{2E_+E_-}\frac{E_+ + E_-}{(E_+ + E_-)^2 - \epsilon^2}[1 - n(E_+) - n(E_-)]\right. \quad (7A.45)
$$

$$
\left. - \frac{E_+E_- - \xi_+\xi_-}{2E_+E_-}\frac{E_+ - E_-}{(E_+ - E_-)^2 - \epsilon^2}[n(E_+) - n(E_-)]\right\} - \int\frac{d^3p}{(2\pi)^3}\frac{1}{2E}[1 - 2n(E)],
$$

and

$$L_{12}(\epsilon, \mathbf{k}) = -\Delta^2 \int \frac{d^3p}{(2\pi)^3} \frac{1}{2E_-E_+} \tag{7A.46}$$

$$\times \left\{ \frac{E_+ + E_-}{(E_+ + E_-)^2 - \epsilon^2} [1 - n(E_+) - n(E_-)] + \frac{E_+ - E_-}{(E_+ - E_-)^2 - \epsilon^2} [n(E_+) - n(E_-)] \right\}.$$

In the last term of (7A.45) we have used the gap equation in the form (7A.7) to eliminate $1/g$.

The excitation spectrum is determined by the vanishing of the fluctuation determinant in the quadratic form (7A.39), i.e., from

$$L_{11}(k)L_{22}(k) - L_{12}(k)L_{12}^*(k) = 0. \tag{7A.47}$$

This equation has two solutions:

$$L_{11}(k) = \pm L_{12}(k), \tag{7A.48}$$

the first giving the low-, the second the high-energy excitations.

7A.3 Long-Wavelength Excitations at Zero Temperature.

At zero temperature, $n(E)$ vanishes and the functions (7A.45), (7A.46) reduce to

$$L_{11}(\epsilon, \mathbf{k}) = \frac{1}{V} \sum_{\mathbf{p}} \left\{ \frac{E_+E_- + \xi_+\xi_-}{2E_+E_-} \frac{E_+ + E_-}{(E_+ + E_-)^2 - \epsilon^2} - \frac{1}{2E} \right\}, \tag{7A.49}$$

$$L_{12}(\epsilon, \mathbf{k}) = -\frac{1}{V} \sum_{\mathbf{p}} \frac{\Delta^2}{2E_+E_-} \frac{E_+ + E_-}{(E_+ + E_-)^2 - \epsilon^2}, \tag{7A.50}$$

where the momentum sums are evaluated as integrals in the large-volume limit (7A.8). Inserting here (7A.43) and (7A.44), and expanding the integrands in powers of $\mathbf{k}$, we can perform the integrals over all momentum directions $\hat{\mathbf{p}} = \mathbf{p}/|\mathbf{p}|$ using the formula

$$\int \frac{d\hat{\mathbf{p}}}{4\pi} (\mathbf{v}\mathbf{k})^n = \mathbf{v}^{2n/2} \int \frac{d\cos\theta}{2} \cos^n \theta = \mathbf{v}^{2n/2} \begin{cases} 1, & n = \text{even}, \\ \frac{1}{n+1} & 0, & n = \text{odd}, \end{cases} \tag{7A.51}$$

and obtain

$$L_{11}(\epsilon, \mathbf{k}) = -\frac{\mathcal{N}(0)}{2} \left(1 - \frac{\epsilon^2}{3\Delta^2} + \frac{v_F^2 k^2}{9\Delta^2} + \frac{v_F^2 \epsilon^2 k^2}{30\Delta^4} - \frac{\epsilon^4}{20\Delta^4} - \frac{v_F^4 k^4}{100\Delta^4} + \ldots \right), \tag{7A.52}$$

$$L_{12}(\epsilon, \mathbf{k}) = -\frac{\mathcal{N}(0)}{2} \left(1 + \frac{\epsilon^2}{6\Delta^2} - \frac{v_F^2 k^2}{18\Delta^2} - \frac{v_F^2 \epsilon^2 k^2}{45\Delta^4} + \frac{\epsilon^4}{30\Delta^4} + \frac{v_F^4 k^4}{150\Delta^4} + \ldots \right). \tag{7A.53}$$

We have ignored terms such as $\mathbf{k}^4/m^2\Delta^2$ compared to $v_F^2 k^4/\Delta^4$ since the Fermi energy is much larger than the gap in a superconductor, i.e., $mv_F^2/2 \gg \Delta$. This limit is most easily accommodated by replacing $\mathbf{k}$ by $\eta\mathbf{k}$, and v_F by v_F/η, and taking the limit $\eta \to 0$. In the expressions (7A.43) and (7A.44) this procedure eliminates the last term in comparison to the term before it.

The spectrum of the long-wavelength excitation is found from the equation $L_{11}(\epsilon, \mathbf{k}) = L_{12}(\epsilon, \mathbf{k})$ which leads to the small-$\mathbf{k}$ expansion (7.43) of the energy. For higher k we must solve the equation $L_{11}(\epsilon, \mathbf{k}) = L_{12}(\epsilon, \mathbf{k})$ numerically. The result is shown in Fig. 7.2.

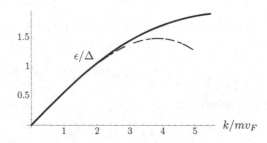

FIGURE 7.2 Energies of the low-energy excitations in superconductor. The curve approaches the energy $\epsilon = 2\Delta$ for large k. The dashed curve shows the analytic small-k expansion (7.43).

The fluctuations of the size of the order parameter are found by solving the equation $L_{11}(\epsilon, \mathbf{k}) = -L_{12}(\epsilon, \mathbf{k})$. Since ϵ remains large we may perform only a small-$\mathbf{k}$ expansion which leads to the energies [3, 6]

$$\epsilon^{(n)}(\mathbf{k}) = 2\Delta + \Delta \left(\frac{v_F\mathbf{k}}{2\Delta}\right)^2 z_n, \tag{7A.54}$$

where z_n are the solutions of the integral equation

$$\int_{-1}^{1} dx \int_{-\infty}^{\infty} dy \, \frac{x^2 - z}{x^2 + y^2 - z} = 0. \tag{7A.55}$$

Setting $e^t = \left(\sqrt{1-z}+1\right) / \left(\sqrt{1-z}-1\right)$ this turns into the algebraic equation

$$\frac{\pi}{2\sinh^2(t/2)} (t + \sinh t) = 0, \tag{7A.56}$$

which has infinitely many solutions t_n. The lowest is

$$t_1 = 2.25073 + 4.21239\, i, \tag{7A.57}$$

the higher ones tend asymptotically to

$$t_n \approx \log[\pi(4n - 1)] + i \left(2\pi n - \frac{\pi}{2}\right). \tag{7A.58}$$

See the contour plot in Fig. 7.3.

The excitation energies are

$$\epsilon^{(n)}(\mathbf{k}) = 2\Delta - \frac{v_F^2}{4\Delta}\mathbf{k}^2\frac{1}{\sinh^2 t_n/2}. \tag{7A.59}$$

Of these, only the first at $\epsilon^{(1)}(\mathbf{k}) \approx 2\Delta + (0.2369 - 0.2956\,i)v_F^2/4\Delta^2\mathbf{k}^2$ lies in the second sheet and may have observable consequences. The others are hiding under lower and lower Riemann sheets below the two-particle branch cut from 2Δ to ∞. This cut is logarithmic due to the dimensionality $D = 2$ of the surface of the Fermi sea at $T = 0$.

7A.4   Long-Wavelength Excitations at Nonzero Temperature

Consider now the case of nonzero temperature where the energy gap is calculated from Eq. (7A.3).

Let us first study the static case and consider only the long-wavelength limit of small $\mathbf{k}$. Hence, we set $\epsilon = 0$ and keep only the lowest orders in $\mathbf{k}$. At $\mathbf{k} = 0$ we find from (7A.45) and (7A.8):

$$L_{11}(0,0) = \mathcal{N}(0)\int_{-\infty}^{\infty} d\xi \left\{\frac{E^2 + \xi^2}{4E^3}[1 - 2n(E)] - \frac{E^2 - \xi^2}{2E^2}n'(E) - \frac{1}{2E}[1 - 2n(E)]\right\}. \tag{7A.60}$$

Inserting $E = \sqrt{\Delta^2 + \xi^2}$, and using the reduced variable $\delta \equiv \Delta/T$ of Eq. (7A.30), this becomes

$$L_{11}(0,0) = -\frac{1}{2}\mathcal{N}(0)\phi(\delta), \tag{7A.61}$$

where we have introduced the so-called *Yoshida function*

$$\phi(\delta) \equiv \Delta^2\left\{\int_0^{\infty} d\xi\left[\frac{1}{E^3}[1 - 2n(E)] + 2\frac{1}{E^2}n'(E)\right]\right\}. \tag{7A.62}$$

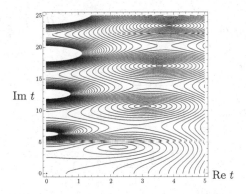

FIGURE 7.3 Contour plot of zeros for energy eigenvalues following from Eq. (7A.56) which are approximately given by Eq. (7A.58). The contour lines indicate fixed values of $\mathrm{Abs}[(t + \sinh t)/\sinh(t/2)]$.

Here we observe that

$$\partial_\xi \left[\frac{\xi}{\Delta^2 E} n(E) \right] = \frac{1}{E^3} n(E) + \left(\frac{1}{\Delta^2} - \frac{1}{E^2} \right) n'(E), \qquad (7A.63)$$

to bring (7A.62) to the form

$$\phi(\delta) \equiv \Delta^2 \int_0^\infty d\xi \left\{ \frac{1}{E^3} + \frac{2}{\Delta^2} n'(E) - 2\partial_\xi \left[\frac{\xi}{|\Delta|^2 E} n(E) \right] \right\}. \qquad (7A.64)$$

The surface term vanishes, and the first integral in Eq. (7A.64) can be done, so that we arrive at the more convenient form

$$\phi(\delta) = 1 + 2 \int_0^\infty d\xi \, n'(E) = 1 - \frac{1}{2T} \int_0^\infty d\xi \frac{1}{\cosh^2(E/2T)}. \qquad (7A.65)$$

For $T \approx 0$, this function approaches zero exponentially.

For the function $L_{12}(\epsilon, \mathbf{k})$ in Eq. (7A.46) we find at $\epsilon = 0$, $\mathbf{k} = 0$:

$$L_{12}(0, 0) = -\mathcal{N}(0)\Delta^2 \int_{-\infty}^\infty d\xi \left\{ \frac{1}{4E^3} [1 - 2n(E)] + \frac{1}{2E^2} n'(E) \right\}, \qquad (7A.66)$$

which can again be expressed in terms of the Yoshida function (7A.65) as

$$L_{12}(0, 0) = -\frac{1}{2}\mathcal{N}(0)\phi(\delta). \qquad (7A.67)$$

This implies that the modes following from the equation $L_{11}(\epsilon, \mathbf{k}) = L_{12}(\epsilon, \mathbf{k})$, which at $T = 0$ have zero energy for $\mathbf{k} = 0$, maintain this property also for nonzero T. This is a consequence of the Nambu-Goldstone theorem.

The full temperature behavior is best calculated by using the Matsubara sum (7A.6). Taking the derivative of this with respect to the energy we see that (7A.65) can be rewritten as

$$
\begin{aligned}
\phi(\delta) &= 2T \sum_{\omega_m} \int d\xi \frac{\Delta^2}{(\omega_m^2 + E^2)^2} = -2\Delta^2 T \sum_{\omega_m} \frac{\partial}{\partial \Delta^2} \int d\xi \frac{1}{\omega_m^2 + \xi^2 + \Delta^2} \\
&= -2\Delta^2 T \sum_{\omega_m} \frac{\partial}{\partial \Delta^2} \frac{\pi}{\sqrt{\omega_m^2 + \Delta^2}} = 2T\pi \sum_{\omega_m > 0} \frac{\Delta^2}{\sqrt{\omega_m^2 + \Delta^2}^3}.
\end{aligned} \qquad (7A.68)
$$

Using the reduced variable (7A.31), this becomes

$$\phi(\delta) = \frac{2\pi}{\delta} \sum_{n=0}^\infty \frac{1}{\sqrt{x_n^2 + 1}^3}. \qquad (7A.69)$$

For T near T_c where δ is small [see Eq. (7A.34)], we can approximate

$$\phi(\Delta) \approx 2\frac{\delta^2}{\pi^2} \sum_{n=0}^\infty \frac{1}{(2n+1)^3} = 2\frac{\delta^2}{\pi^2} \frac{7\zeta(3)}{8} \approx 2\left(1 - \frac{T}{T_c} \right). \qquad (7A.70)$$

In the limit $T \to 0$, the sum turns into an integral. Using the formula

$$\int_0^\infty dx \frac{x^{\mu-1}}{(x^2+1)^\nu} = \frac{1}{2} B(\mu/2, \nu_n - \mu/2) \tag{7A.71}$$

with $B(x,y) = \Gamma(x)\Gamma(y)/\Gamma(x+y)$ we verify that

$$\phi(\Delta)|_{T=0} = 1. \tag{7A.72}$$

7A.5 Bending Energies of Order Field

Let us now calculate the bending energies of the collective field $\Delta(x)$. For this, we expand $L_{11}(0,\mathbf{k})$ and $L_{12}(0,\mathbf{k})$ in powers of the momentum $\mathbf{k}$ up to $\mathbf{k}^2$. We start from the Matsubara sums (7A.40) and (7A.41):

$$L_{11}(0,\mathbf{k}) = \frac{T}{V} \sum_{\omega_m, \mathbf{p}} \frac{\omega_m^2 + \xi_+ \xi_-}{(\omega_m^2 + E_+^2)(\omega_m^2 + E_-^2)} - \frac{1}{g}, \tag{7A.73}$$

$$L_{12}(0,\mathbf{k}) = -\frac{T}{V} \sum_{\omega_m, \mathbf{p}} \frac{\Delta^2}{(\omega_m^2 + E_+^2)(\omega_m^2 + E_-^2)}. \tag{7A.74}$$

Inserting the expansions (7A.43) and (7A.44), these become

$$L_{11}(0,\mathbf{k}) - L_{12}(0,\mathbf{k}) \approx \int \frac{d^3 p}{(2\pi)^3} T \sum_{\omega_m} \frac{\omega_m^2 + \Delta^2 + \xi^2 - \frac{1}{4}(\mathbf{vk})^2}{(\omega_m^2 + E^2)^2 \left[1 + \frac{1}{2}(\mathbf{vk})^2 \frac{\omega_m^2 - \xi^2 + \Delta^2}{(\omega_m^2 + E^2)^2}\right]} - \frac{1}{g} + \cdots$$

$$= \int \frac{d^3 p}{(2\pi)^3} \left\{ \left(T \sum_{\omega_m} \frac{1}{\omega_m^2 + E^2} - \frac{1}{g} \right) \right.$$

$$\left. + T \sum_{\omega_m} \left[\frac{1}{4} \frac{1}{(\omega_m^2 + E^2)^2} - \frac{\omega_m^2 + \Delta^2}{(\omega_m^2 + E^2)^3} \right] (\mathbf{vk})^2 \right\} + \cdots \tag{7A.75}$$

Due to the gap equation (7A.3), the first term in the curly brackets vanishes, and using the directional integral (7A.51), we find

$$L_{11}(0,\mathbf{k}) - L_{12}(0,\mathbf{k}) \approx \mathcal{N}(0) \frac{v_F^2 \mathbf{k}^2}{3}$$

$$\times \sum_{\omega_m} \int_{-\infty}^\infty d\xi \left[\frac{1}{4} \frac{1}{(\omega_m^2 + \xi^2 + \Delta^2)^2} - \frac{\omega_m^2 + \Delta^2}{(\omega_m^2 + \xi^2 + \Delta^2)^3} \right]. \tag{7A.76}$$

Similarly we obtain

$$L_{12}(0,\mathbf{k}) \approx -\mathcal{N}(0) \sum_{\omega_m} \int_{-\infty}^\infty d\xi \left\{ \frac{\Delta^2}{(\omega_m^2 + \xi^2 + \Delta^2)^2} \right.$$

$$\left. + \frac{v_F^2 \mathbf{k}^2 \Delta^2}{3} \left[\frac{1}{2} \frac{1}{(\omega_m^2 + \xi^2 + \Delta^2)^3} - \frac{\omega_m^2 + \Delta^2}{(\omega_m^2 + \xi^2 + \Delta^2)^4} \right] \right\}. \tag{7A.77}$$

Using the integrals

$$\int_{-\infty}^{\infty} d\xi \frac{1}{(\omega_m^2 + \xi^2 + \Delta^2)^{2,3,4}} = \left(\frac{1}{2}, \frac{3}{8}, \frac{5}{16}\right) \frac{\pi}{\sqrt{\omega_m^2 + \Delta^2}^{3,5,7}}, \qquad (7A.78)$$

we find with the help of (7A.68):

$$L_{11}(0, \mathbf{k}) - L_{12}(0, \mathbf{k}) \approx -\frac{\mathcal{N}(0)}{4\Delta^2} \frac{v_F^2 \mathbf{k}^2}{3} \phi(\delta), \qquad (7A.79)$$

$$L_{12}(0, \mathbf{k}) \approx -\frac{\mathcal{N}(0)}{2} \phi(\delta) + \frac{\mathcal{N}(0)}{8\Delta^2} \frac{v_F^2 \mathbf{k}^2}{3} \frac{2}{3} \bar{\phi}(\delta), \qquad (7A.80)$$

where $\phi(\delta)$ is the Yoshida function (7A.69), while $\bar{\phi}(\delta)$ is a further gap function:

$$\bar{\phi}(\delta) \equiv 3\Delta^4 \pi T \sum_{\omega_m > 0} \frac{1}{\sqrt{\omega_m^2 + \Delta^2}^5} = \frac{3\pi}{\delta} \sum_{n=0}^{\infty} \frac{1}{\sqrt{x_n^2 + 1}^5}. \qquad (7A.81)$$

In the limit $T \to 0$, the sum turns into an integral whose value is, by formula (7A.71),

$$\bar{\phi}(\delta)\big|_{T=0} = 1. \qquad (7A.82)$$

Together with (7A.72) we see that (7A.79) and (7A.80) reproduce correctly the $\mathbf{k}^2$-terms of Eqs. (7A.52) and (7A.53).

For $T \approx T_c$, where $\delta \to 0$, we find

$$\bar{\phi}(\delta) \approx 3\frac{\delta^4}{\pi^4} \sum_{n=0}^{\infty} \frac{1}{(2n+1)^5} = 3\frac{\delta^4}{\pi^4} \frac{31\zeta(5)}{32}, \qquad (7A.83)$$

and thus, by (7A.34),

$$\bar{\phi}(\delta) \approx \frac{3}{\pi^4} \frac{31\zeta(5)}{32} \left(\frac{8\pi^2}{7\zeta(3)}\right)^2 \left(1 - \frac{T}{T_c}\right)^2 \approx 2.7241 \times \left(1 - \frac{T}{T_c}\right)^2. \qquad (7A.84)$$

Inserting this together with (7A.70) into (7A.79) and (7A.80), and considering only long-wavelength excitations with $v_F^2 \mathbf{k}^2 \leq \Delta^2$, so that $(v_F^2 \mathbf{k}^2/\Delta^2)\tilde{\phi}(\delta)$ is of the order of $(1 - T/T_c)^2$, we obtain to lowest order in $1 - T/T_c$:

$$L_{11}(0, \mathbf{k}) - L_{12}(\epsilon, \mathbf{k}) \approx -\mathcal{N}(0)\frac{v_F^2}{\pi^2 T_c^2} \frac{7\zeta(3)}{48} \mathbf{k}^2, \qquad (7A.85)$$

$$L_{12}(0, \mathbf{k}) \approx -\mathcal{N}(0)\left(1 - \frac{T}{T_c}\right). \qquad (7A.86)$$

Using (5.145), we see that

$$\mathcal{N}(0)v_F^2 = \frac{3}{2}\frac{n_e}{m} = \frac{3}{2}\frac{n_\Delta}{M} = \frac{3}{2}\frac{\rho}{m^2} = 6\frac{\rho}{M^2}, \qquad (7A.87)$$

where n_Δ is the density of Cooper pairs of mass $M = 2m$, and $\rho \equiv Mn_\Delta = 2mn_\Delta = mn_e$ is their mass density. Their mass density is, of course, equal to the electron mass density: $\rho \equiv 2Mn_e = Mn_\Delta$. Thus we may eliminate $\mathcal{N}(0)$ in favor of the pair density and obtain

$$L_{11}(0, \mathbf{k}) - L_{12}(0, \mathbf{k}) \approx -\frac{\rho}{2M^2\Delta^2}\mathbf{k}^2\phi(\delta), \tag{7A.88}$$

$$L_{12}(0, \mathbf{k}) \approx -\frac{6\rho}{2M^2v_F^2}\phi(\delta) + \frac{\rho}{4M^2\Delta^2}\mathbf{k}^2\frac{2}{3}\bar{\phi}(\delta). \tag{7A.89}$$

Returning to x-space we decompose the collective field $\Delta(x)$ into a real size field $|\Delta(x)|$ fluctuating around Δ and a phase field $\theta(x)$ fluctuating around zero as

$$\Delta(x) = |\Delta(x)|e^{i\theta(x)}, \tag{7A.90}$$

and extract from the action (7A.39), the energy for static small fluctuations as

$$\mathcal{E}(x) = \frac{1}{2M^2\Delta^2}\left\{\rho^{11}\boldsymbol{\nabla}\Delta^*(x)\boldsymbol{\nabla}\Delta(x) + \mathrm{Re}\left[\rho^{12}\boldsymbol{\nabla}\Delta^*(x)\boldsymbol{\nabla}\Delta^*(x) + a^{12}\delta\Delta^*(x)\delta\Delta^*(x)\right]\right\}. \tag{7A.91}$$

From (7A.88) and (7A.89) we identify the coefficients as:

$$\rho^{11} - \rho^{12} = \rho\phi(\delta), \quad \rho^{12} = -\frac{\rho}{2}\frac{2}{3}\bar{\phi}(\delta), \quad a^{12} = 6\rho\frac{\Delta^2}{v_F^2}\phi(\delta). \tag{7A.92}$$

Decomposing the collective field $\Delta(x)$ into a real size field $\Delta(x)$ and a phase field $\theta(x)$ as

$$\Delta(x) = \Delta(x)e^{i\theta(x)}, \tag{7A.93}$$

the energy density reads

$$\mathcal{E}(x) = \frac{1}{2M^2}\left\{\rho_s(\boldsymbol{\nabla}\theta - \boldsymbol{\theta}^v)^2 + \rho_\Delta(\boldsymbol{\nabla}|\Delta(x)|)^2/\Delta^2 + 2a^{12}[\delta|\Delta(x)|]^2\right\}. \tag{7A.94}$$

The factor before the first gradient term is the *superfluid density*:

$$\rho_s \equiv \rho^{11} - \rho^{12} = \rho\phi(\delta). \tag{7A.95}$$

By analogy we have introduced the quantities

$$\rho_\Delta \equiv \rho^{11} + \rho^{12} = \rho_s + 2\rho^{12} = \rho_s - \frac{2}{3}\bar{\rho}_s, \quad \bar{\rho}_s \equiv \rho\bar{\phi}(\delta). \tag{7A.96}$$

The behavior of ρ_s and $\bar{\rho}_s$ for all $T \leq T_c$ is shown in Fig. 7.4.

The phase fluctuations are of infinite range, the size fluctuations have a finite range characterized by the temperature-dependent *coherence length* (with reinserted $\hbar$ to have proper physical units)

$$\xi(T) = \frac{\hbar v_F}{\Delta}\sqrt{\frac{1}{12\Delta^2}\frac{\rho_s - 2\bar{\rho}_s/3}{\rho_s}}. \tag{7A.97}$$

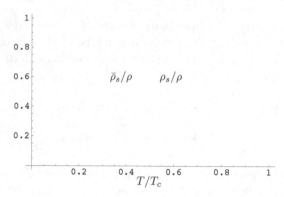

FIGURE 7.4 Temperature behavior of superfluid density ρ_s/ρ (Yoshida function) and the gap function $\bar{\rho}_s/\rho$ of Eqs. (7A.95) and (7A.96).

Near T_c, we insert $\rho_s \approx 2(1 - T/T_c)$, $\bar{\rho}_s \approx (1 - T/T_c)^2$, and Δ from (7A.35) to find the limit (in physical units)

$$\xi(T) = \frac{\xi_0}{\sqrt{2}} \left(1 - \frac{T}{T_c}\right)^{-1/2}, \quad \xi_0 \equiv \sqrt{\frac{7\zeta(3)}{48}} \frac{\hbar v_F}{\pi k_B T_c} \approx 0.42 \frac{\hbar v_F}{\pi k_B T_c} \sim 0.27 \frac{T_F}{T_c} l_F. \quad (7A.98)$$

In the last expression, we have introduced the *Fermi length* $l_F = \hbar/p_F$, and the *Fermi temperature* $T_F \equiv p_F^2/2Mk_B$. In old-fashioned superconductors, l_F is of the order of the lattice spacing, while T_F is usually larger than T_c by factors 10^3–10^4, so that the ratio of the coherence length with respect to the Fermi length is quite large. In high-temperature superconductors, however, ξ_0 can shrink to only a few times l_F, which greatly increases the effect of fluctuations.

At zero temperature we obtain, with the help of (7A.24):

$$\xi(0) = \frac{1}{6} \frac{v_F}{\Delta(0)} = \frac{1}{6e^\gamma} \frac{v_F}{\pi T_c} \approx 0.0935 \frac{v_F}{\pi T_c}, \quad (7A.99)$$

which is about a sixth of the length parameter ξ_0 in the $T \approx T_c$ -equation (7A.98).

7A.6   Kinetic Terms of Pair Field at Nonzero Temperature

At nonzero temperature, we shall extract the dynamics of the kinetic term of a slowly varying pair field by calculating the excitation energies from Eqs. (7A.45) and (7A.46) at small $\mathbf{k}$. We begin with $\mathbf{k} = 0$ where we obtain, instead of (7A.60) and (7A.66):

$$L_{11}(\epsilon, \mathbf{0}) - L_{11}(\epsilon, \mathbf{0}) = \mathcal{N}(0) \frac{1}{2} \int_{-\infty}^{\infty} d\xi \left[\frac{E}{E^2 - \epsilon^2/4} - \frac{1}{E}\right] [1 - 2n(E)], \quad (7A.100)$$

$$L_{12}(\epsilon, \mathbf{0}) - L_{12}(0, \mathbf{0}) = -\mathcal{N}(0) \frac{\epsilon^2}{4\Delta^2} \frac{1}{4} \int_{-\infty}^{\infty} d\xi \frac{\Delta^4}{E^3(E^2 - \epsilon^2/4)} [1 - 2n(E)]. \quad (7A.101)$$

They can be rewritten as

$$L_{11}(\epsilon, \mathbf{0}) - L_{12}(\epsilon, \mathbf{0}) = \mathcal{N}(0) \frac{\epsilon^2}{4\Delta^2} \gamma(\delta, \epsilon), \tag{7A.102}$$

$$L_{12}(\epsilon, \mathbf{0}) - L_{12}(0, \mathbf{0}) = -\mathcal{N}(0) \frac{\epsilon^2}{4\Delta^2} \frac{1}{3} \bar{\gamma}(\delta, \epsilon), \tag{7A.103}$$

where

$$\gamma(\delta, \epsilon) \equiv \frac{1}{2} \int_{-\infty}^{\infty} d\xi \, \frac{\Delta^2}{E(E^2 - \epsilon^2/4)} [1 - 2n(E)], \tag{7A.104}$$

$$\bar{\gamma}(\delta, \epsilon) \equiv \frac{3}{4} \int_{-\infty}^{\infty} d\xi \, \frac{\Delta^4}{E^3(E^2 - \epsilon^2/4\Delta^2)} [1 - 2n(E)]. \tag{7A.105}$$

At zero temperature where $n(E) = 0$, both $\gamma(\delta, 0)$ and $\bar{\gamma}(\delta, 0)$ start out with the value 1, so that the results (7A.102) and (7A.103) reproduce the ϵ^2-terms of (7A.52) and (7A.53). The full temperature behavior of $\gamma(\delta, 0)$ and $\bar{\gamma}(\delta, 0)$ is plotted in Fig. 7.5.

For arbitrary temperature we calculate (7A.104) and (7A.105) most conveniently by expanding $n(E)$ into a Matsubara sum via Eq. (7A.6), so that it takes the form

$$\gamma(\delta, \epsilon) = 2T \sum_{\omega_m > 0} \int_{-\infty}^{\infty} d\xi \, \frac{\Delta^2}{E^2 - \epsilon^2/4} \frac{1}{\omega_m^2 + E^2}, \tag{7A.106}$$

$$\bar{\gamma}(\delta, \epsilon) = 3T \sum_{\omega_m > 0} \int_{-\infty}^{\infty} d\xi \, \frac{\Delta^4}{E^2(E^2 - \epsilon^2/4)} \frac{1}{\omega_m^2 + E^2}. \tag{7A.107}$$

Performing the integrals over ξ yields

$$\gamma(\delta, \epsilon) = 2\pi \frac{T}{\Delta} \sum_{\omega_m > 0} \left(\frac{1}{\sqrt{1 - \epsilon^2/4\Delta^2}} \frac{\Delta^2}{\omega_m^2 + \epsilon^2/4} - \frac{\Delta^2}{\omega_m^2 + \epsilon^2/4} \frac{\Delta}{\sqrt{\omega_m^2 + \Delta^2}} \right), \tag{7A.108}$$

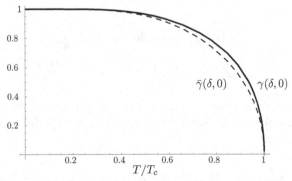

FIGURE 7.5 Temperature behavior of the functions $\gamma(\delta, 0)$ and $\bar{\gamma}(\delta, 0)$ of Eqs. (7A.104) and (7A.105).

$$\bar{\gamma}(\delta, \epsilon) = 3\pi \frac{T}{\Delta} \sum_{\omega_m > 0} \left[\frac{4\Delta^2}{\epsilon^2} \frac{\Delta^2}{\omega_m^2 + \epsilon^2/4} \left(\frac{1}{\sqrt{1 - \epsilon^2/4\Delta^2}} - 1 \right) \right.$$
$$\left. + \left(\frac{\Delta}{\sqrt{\omega_m^2 + \Delta^2}} - 1 \right) \frac{\Delta^2}{\omega_m^2} \frac{\Delta^2}{\omega_m^2 + \epsilon^2/4} \right]. \tag{7A.109}$$

For $\epsilon = 0$, these become

$$\gamma(\delta, 0) = 2\pi \frac{T}{\Delta} \sum_{\omega_m > 0} \left(\frac{\Delta^2}{\omega_m^2} - \frac{\Delta^3}{\omega_m^2 \sqrt{\omega_m^2 + \Delta^2}} \right), \tag{7A.110}$$

and

$$\bar{\gamma}(\delta, 0) = 3\pi \frac{T}{\Delta} \sum_{\omega_m > 0} \left[\frac{1}{2} \frac{\Delta^2}{\omega_m^2} + \frac{\Delta^4}{\omega_m^4} \left(\frac{\Delta}{\sqrt{\omega_m^2 + \Delta^2}} - 1 \right) \right]. \tag{7A.111}$$

Here we can replace

$$\sum_{\omega_m > 0} \omega_m^{-k} = (1 - 2^{-k}) \zeta(k) (\pi T)^{-k}, \tag{7A.112}$$

which is equal to $1/8T^2$ for $k = 2$, and $1/96T^4$ for $k = 4$. In the limit $T \to 0$, the Matsubara sums $T \sum_{\omega_m > 0}$ become integrals $\int_0^\infty d\omega_m/2\pi$ and we recover the limits $\gamma(\delta) \to 1$, $\bar{\gamma}(\delta) \to 1$ obtained before from (7A.104) and (7A.105).

In the limit $T \to T_c$ where $\Delta \to 0$, the functions (7A.108) and (7A.109) have the limit

$$\gamma(\delta, \epsilon) \to 2\pi \frac{T\Delta}{\sqrt{1 - \epsilon^2/4\Delta^2}} \sum_{\omega_m > 0} \frac{1}{\omega_m^2} = \frac{\pi\Delta}{4T\sqrt{1 - \epsilon^2/4\Delta^2}} \to \frac{\pi\Delta^2}{2T\sqrt{-\epsilon^2}}, \tag{7A.113}$$

$$\frac{\epsilon^2}{4\Delta^2} \bar{\gamma}(\delta, \epsilon) \to 3\pi \frac{T\Delta}{\sqrt{1 - \epsilon^2/4\Delta^2}} \sum_{\omega_m > 0} \frac{1}{\omega_m^2} = \frac{3\pi\Delta}{8T\sqrt{1 - \epsilon^2/4\Delta^2}} \to \frac{3\pi\Delta^2}{4T\sqrt{-\epsilon^2}}, \tag{7A.114}$$

so that

$$L_{11}(\epsilon, \mathbf{k}) - L_{12}(\epsilon, \mathbf{k}) \approx \mathcal{N}(0) \frac{i\pi\epsilon}{8T}, \tag{7A.115}$$

$$L_{12}(\epsilon, 0) - L_{12}(0, 0) \approx -\mathcal{N}(0) \frac{i\pi\Delta^2}{4T\epsilon}. \tag{7A.116}$$

For $\epsilon \gg \Delta^2$, the second function can be ignored in comparison with the first.

The same results could have been derived directly from Eqs. (7A.102) and (7A.101) for $\Delta = 0$:

$$L_{11}(\epsilon, 0) - L_{12}(\epsilon, 0) \approx \mathcal{N}(0) \int_{-\infty}^\infty d\xi \left[\frac{\xi}{2(\xi^2 - \epsilon^2/4)} - \frac{1}{2\xi} \right] \tanh \frac{\xi}{2T}, \tag{7A.117}$$

$$L_{12}(\epsilon, 0) \approx 0. \tag{7A.118}$$

Together with (7A.85), (7A.86), and (7A.99), this yields

$$L_{11}(\epsilon, \mathbf{k}) - L_{12}(\epsilon, \mathbf{k}) \equiv -\mathcal{N}(0)\left(-i\frac{\pi\epsilon}{8T} + \xi_0^2\mathbf{k}^2 + \ldots\right), \qquad (7A.119)$$

$$L_{12}(\epsilon, \mathbf{k}) = -\mathcal{N}(0)\left(1 - \frac{T}{T_c}\right) + \ldots . \qquad (7A.120)$$

This shows that for $T \approx T_c$, the excitations are purely damped with a decay rate

$$\Gamma = 2\frac{8T}{\pi}\xi_0^2\mathbf{k}^2. \qquad (7A.121)$$

The above results provide us with all information to set up a Ginzburg-Landau action for describing a superconductor in the neighborhood of the critical temperature. This action is a low-order expansion in powers of the Cooper pair field and its gradients

$$\mathcal{A}_2[\Delta, \Delta^*] \approx \mathcal{N}(0)\int dt \int d^3x \left\{\Delta^*(x)\left[-\frac{\pi}{8T}\partial_t + \xi_0^2\boldsymbol{\nabla}^2 - a_2\right]\Delta(x) - \frac{a_4}{2}|\Delta(x)|^4 + \ldots\right\}, \qquad (7A.122)$$

where the gradient terms follow directly from (7A.119). The dots indicate higher expansion terms which contain more powers of the field such as $|\Delta(x)|^6$, $|\Delta(x)|^8$, $\ldots$, or higher derivative terms such as $|\boldsymbol{\nabla}^2\Delta(x)|^2$, $|\partial_t\Delta(x)|^2$, $\ldots$. For the study of the phase transition these are all irrelevant.

To determine a_2 and a_4 we insert the decomposition $\Delta(x) = \Delta + \delta\Delta(x)$, into (7A.122) and find that for $a_2 < 0$, the action has an extremum at $\Delta = \sqrt{-a_2/a_4}$. The quadratic fluctuations $\delta\Delta(x)$ possess the same gradient terms as in (7A.122), while the potential terms $-a_2|\Delta(x)|^2 - a_4|\Delta(x)|^4/2$ contribute

$$\Delta\mathcal{A}_2 \approx -\int dt \int d^3x \left[\left(a_2 + 2a_4\Delta^2\right)\delta\Delta^*(x)\delta\Delta(x) + \frac{a_4}{2}\left(\Delta^2\left\{[\delta\Delta(x)]^2 + [\delta\Delta^*(x)]^2\right\}\right)\right]. \qquad (7A.123)$$

At the extremum, this becomes

$$\Delta\mathcal{A}_2 \approx a_2\int dt \int d^3x \left(\delta\Delta^*(x)\delta\Delta(x) + \frac{1}{2}\left\{[\delta\Delta(x)]^2 + [\delta\Delta^*(x)]^2\right\}\right). \qquad (7A.124)$$

The imaginary part of $\delta\Delta(x)$ drops out ensuring that its static infinite-wavelength fluctuations have an infinite range, in accordance with the Nambu-Goldstone theorem.

Comparing (7A.124) with (7A.39) we identify $a_4\Delta^2$ with $L_{12}(0,\mathbf{0})$ in Eq. (7A.80) for small Δ, i.e.,

$$a_4 \approx \frac{\mathcal{N}(0)}{2\Delta^2}\phi(\delta) \approx \mathcal{N}(0)\frac{1}{\pi^2 T^2}\frac{7\zeta(3)}{8} = \mathcal{N}(0)\frac{6\xi_0^2}{\hbar^2 v_F^2}. \qquad (7A.125)$$

The constant a_2 is then [recalling (7A.35) and (7A.98)]

$$a_2 = -\Delta^2 a_4 \approx \mathcal{N}(0) \left(\frac{T}{T_c} - 1 \right). \tag{7A.126}$$

Inserting this into (7A.122) we see that the fluctuations of $\Delta(x)$ around Δ have the coherence lengths

$$\xi(T) = \xi_0 \left(\frac{T}{T_c} - 1 \right)^{-1/2}, \quad T > T_c, \tag{7A.127}$$

$$\xi_{\text{size}}(T) = \frac{\xi_0}{\sqrt{2}} \left(\frac{T}{T_c} - 1 \right)^{-1/2}, \quad T < T_c. \tag{7A.128}$$

For critical temperatures of the order of 1 to $10\,\mathrm{K}$, Fermi temperatures of the order of 10^4 to $10^5\,\mathrm{K}$, and Fermi momenta of the order $\hbar/\mathring{\mathrm{A}}$, we obtain quite a large coherence length of the order of 10^3–$10^4\,\mathring{\mathrm{A}}$.

The energy in the action (7A.122) coincides with the Ginzburg-Landau energy (5.148) if we identify:[2]

$$\psi(x) = \sqrt{2\mathcal{N}(0)}\xi_0 \,\Delta(x), \quad \tau = \frac{1}{\xi_0^2} \left(\frac{T}{T_c} - 1 \right). \tag{7A.129}$$

Then a_4 of Eq. (7A.124) implies that the coupling constant g in (5.148) has the BCS-value

$$g = \frac{3}{\mathcal{N}(0)\hbar^2 v_F^2 \xi_0^2}. \tag{7A.130}$$

The condensation energy density of the superconductor is given by

$$\mathcal{E}_c = \frac{\tau^4}{4g} = \frac{1}{4\xi_0^4} \frac{1}{3} \frac{(k_B T_c)^2 \xi_0^4}{7\xi(3)/48\pi^2} \mathcal{N}(0) \left(1 - \frac{T}{T_c} \right)^2$$

$$= \frac{1}{7\xi(3)} \left(\frac{p_F}{\hbar} \right)^3 \frac{T_c}{T_F} k_B T_c \left(1 - \frac{T}{T_c} \right)^2, \tag{7A.131}$$

which is of the order of

$$\mathcal{E}_c \approx 10^{-4} k_B \left(1 - \frac{T}{T_c} \right)^2 k_B \mathrm{K}/\mathring{\mathrm{A}}^3 \approx 10^4 \left(1 - \frac{T}{T_c} \right)^2 \mathrm{erg/cm}^3. \tag{7A.132}$$

In order to obtain a better idea of the size of interaction strength, it is useful to go to natural units used in the fluctuation discussion after Eq. (5.262), and measure energies in units of $k_B T_c$. In addition, we measure distances in units of ξ_0. Then, taking a factor $\sqrt{k_B T_c/\xi_0}$ out of ψ and A, and ξ_0 out of x (i.e.,

---

[2]Note that the dimensions of ψ, $\mathcal{N}(0)$, $\boldsymbol{\Delta}$, g are (energy/length) $^{1/2}$, energy$^{-1}\cdot$ length^{-3}, energy, and (energy $\cdot$ length)$^{-1}$, respectively.

$\sqrt{\xi_0/k_BT_c}\psi$, $A_{\text{new}} = \sqrt{\xi_0/k_BT_c}A$, $x_{\text{new}} = x/\xi_0$), we arrive at the dimensionless Ginzburg-Landau Hamiltonian

$$\mathcal{H}_{\text{GL}} = \int d^3x \left\{ \frac{1}{2}|(\boldsymbol{\nabla} - iq\mathbf{A})\psi|^2 + \frac{1}{2}\left(\frac{T}{T_c}-1\right)|\psi|^2 + \frac{g}{4}|\psi|^4 + \frac{1}{2}(\boldsymbol{\nabla}\times\mathbf{A})^2 \right\}. \quad (7A.133)$$

Here the coupling g and q are dimensionless and their magnitudes are

$$g = \frac{3\xi_0 k_B T_c}{N(0)\hbar^2 v_F^2 \xi_0^2} = \frac{3}{2}\pi^2 \sqrt{\frac{7\xi(3)}{48\pi^2}}^{-1}\left(\frac{T_c}{T_F}\right)^2 \sim 111.08\left(\frac{T_c}{T_F}\right)^2, \quad (7A.134)$$

$$q = \frac{2e}{\hbar c}\sqrt{k_B T_c \xi_0} = 2\sqrt{4\pi\alpha\frac{v_F}{c}\sqrt{\frac{7\xi(3)}{48\pi^2}}} \sim 2.59\sqrt{\alpha\frac{v_F}{c}}, \quad (7A.135)$$

where $\alpha = (e^2/4\pi)/\hbar c = 1/137$ is the fine structure constant. Since $T_c/T_F \sim 10^{-4}$ and $\alpha(v_F/c) \sim 10^{-4}$, both couplings are extremely small, i.e., $g \sim 10^{-6}$, $q \sim 10^{-2}$.

Gorkov's original derivation [5] was valid only for perfect crystals. In dirty materials, the mean free path of the electron has finite value, say ℓ. In that case, the length scale ξ_0^2 in front of the gradient term of (7A.122) receives a correction factor

$$r = \sum_{n=1}^{\infty} \frac{1}{(2n+1)^2(2n+1+\xi_0/2\pi\cdot 0.18\ell)} \Big/ \sum_{n=1}^{\infty} \frac{1}{(2n+1)^3}, \quad (7A.136)$$

with the other terms remaining the same. In the clean limit $\ell = \infty$, r is equal to 1. In very dirty materials, however, $\ell \ll \xi_0$ and r becomes $\sim 0.18(\ell\xi_0)$ which can be quite small. If $\xi_0' = r^{1/2}\xi_0$ is used as a new length scale and $\sqrt{r^{1/2}\xi_0/k_BT_c}$ is taken out of the fields, instead of ξ_0/k_BT, the correction factor r changes the constants in the reduced energy as follows:

$$g \to gr^{-3/2}, \qquad q \to qr^{1/4}. \quad (7A.137)$$

Note that the reduced condensation energy density

$$\beta_c \mathcal{E}_c = \frac{1}{4g\xi_0^3}\left(1 - \frac{T}{T_c}\right)^2 \quad (7A.138)$$

remains unchanged since ξ_0 and g are modified by r oppositely. This is the content of a theorem by Abrikosov which states that dirt does not change the global thermodynamics of the superconductor.

Appendix 7B Properties of Ginzburg-Landau Theory of Superconductivity

Let us discuss some properties of the Ginzburg-Landau field theory with Hamiltonian (5.148). The field equations are

$$\left[-(\boldsymbol{\nabla} - iq\mathbf{A})^2 + \tau + g|\psi|^2\right]\psi = 0, \quad (7B.1)$$

and

$$\boldsymbol{\nabla} \times \boldsymbol{\nabla} \times \mathbf{A} = q\,\mathbf{j}_s, \qquad (7\text{B}.2)$$

with the supercurrent (5.149), which is conserved as a consequence of Eq. (7B.1):

$$\boldsymbol{\nabla} \cdot \mathbf{j}_s = 0. \qquad (7\text{B}.3)$$

In order to show this explicitly, observe that the product rule of differentiation holds also for the covariant derivative of a product of complex fields:

$$
\begin{aligned}
\boldsymbol{\nabla}(a^\dagger b) &= (\boldsymbol{\nabla}a^\dagger)b + a^\dagger(\boldsymbol{\nabla}b) = (\boldsymbol{\nabla} + iq\mathbf{A})a^\dagger b + a^\dagger(\boldsymbol{\nabla} - iq\mathbf{A})b \\
&= (\mathbf{D}a)^\dagger b + a^\dagger \mathbf{D}b.
\end{aligned}
\qquad (7\text{B}.4)
$$

Thus, in each term of the product rule, we may directly replace the ordinary derivative $\boldsymbol{\nabla}$ by the covariant one $\mathbf{D} = \boldsymbol{\nabla} - iq\mathbf{A}$. Applying this rule to (7B.3), we see that

$$
\begin{aligned}
\boldsymbol{\nabla} \cdot \mathbf{j}_s &= \frac{1}{2i}\left\{(\mathbf{D}\psi)^\dagger(\mathbf{D}\psi) + \psi^\dagger \mathbf{D}^2\psi - (\mathbf{D}\psi)^\dagger(\mathbf{D}\psi) - \left(\mathbf{D}^2\psi\right)^\dagger \psi\right\} \\
&= \frac{1}{2i}\left\{\psi^\dagger \mathbf{D}^2\psi - \left(\mathbf{D}^2\psi\right)^\dagger \psi\right\},
\end{aligned}
\qquad (7\text{B}.5)
$$

which vanishes indeed due to the field equation (7B.1).

The invariance of the Ginzburg-Landau equations (7B.1) and (7B.1) under the gauge transformations

$$\mathbf{A}(\mathbf{x}) \to \mathbf{A}(\mathbf{x}) + \boldsymbol{\nabla}\Lambda(\mathbf{x}), \qquad \psi(\mathbf{x}) \to e^{iq\Lambda(\mathbf{x})}\psi(\mathbf{x}) \qquad (7\text{B}.6)$$

can be used to transform away the phase of the ψ-field. As in (5.150), we shall parametrize it in terms of size and phase angle as $\psi(\mathbf{x}) = \rho(\mathbf{x})e^{i\theta(\mathbf{x})}$, but omit the wiggles for brevity. We may choose $q\Lambda(\mathbf{x}) = -\theta(\mathbf{x})$, and the field equations become

$$\left[-\left(\boldsymbol{\nabla} - iq\mathbf{A}\right)^2 + \tau + g\rho^2\right]\rho = 0, \qquad (7\text{B}.7)$$

$$\boldsymbol{\nabla} \times \boldsymbol{\nabla} \times \mathbf{A}(\mathbf{x}) = q\,\mathbf{j}_s = -q^2\rho^2\mathbf{A}. \qquad (7\text{B}.8)$$

Separating real and imaginary parts, the first equation decomposes into an equation for $\rho(\mathbf{x})$:

$$(-\boldsymbol{\nabla}^2 + q^2 A^2 + \tau + g\rho^2)\rho = 0, \qquad (7\text{B}.9)$$

and another one for $\mathbf{A}(\mathbf{x})$;

$$\rho\boldsymbol{\nabla} \cdot \mathbf{A} + 2\mathbf{A} \cdot \boldsymbol{\nabla}\rho = 0. \qquad (7\text{B}.10)$$

The latter expresses the current conservation law (7B.3) in terms of size and phase fields where (5.149) reads

$$\mathbf{j}_s(\mathbf{x}) = \rho^2\left[\boldsymbol{\nabla}\theta(\mathbf{x}) - q\mathbf{A}(\mathbf{x})\right]. \qquad (7\text{B}.11)$$

The Hamiltonian in these field variables was written down in Eq. (5.151). In the present notation without wiggles, the Hamiltonian density without vortices reads:

$$\mathcal{H}_{\rm GL} = \int d^3x \int d^3x \left\{ \frac{1}{2} (\nabla \rho)^2 + \frac{\tau}{2} \rho^2 + \frac{g}{4} \rho^4 + \frac{1}{2} q^2 \rho^2 \mathbf{A}^2 + \frac{1}{2} (\nabla \times A)^2 \right\}. \quad (7B.12)$$

Let us extract from this a few experimental properties. The derivation of the finite penetration depth λ in the superconducting phase was described in Section 5.2.1. The result was, in the present natural units,

$$\lambda = 1/q\rho. \quad (7B.13)$$

Here we exhibit a few more important properties.

7B.1   Critical Magnetic Field

The Ginzburg-Landau equations explain the existence of a critical external magnetic field $\mathbf{H}_c$ at which the Meissner effect breaks down and the field invades into the superconductor, thereby destroying all supercurrents. This is most easily derived by studying the *magnetic enthalpy*, whose density is

$$\mathcal{E}_H = \mathcal{H}_{\rm GL} - \mathbf{H} \cdot \mathbf{H}^{\rm ext}. \quad (7B.14)$$

We can then see that for $H^{\rm ext} < H_c = (1/\sqrt{2})|\tau|/\sqrt{g}$ the enthalpy $\mathcal{E}_H$ is minimized by $\rho_0 = \sqrt{-\tau/g}$, $\mathbf{A} = 0$ with a minimal density

$$\mathcal{E}_H = -\frac{\tau^2}{4g} = \mathcal{E}_c. \quad (7B.15)$$

For $H^{\rm ext} > H_c$, however, the minimum of (7B.14) lies at $\rho = 0$, $H = H^{\rm ext}$, where it has the value

$$\mathcal{E}_H = -\frac{(\mathbf{H}^{\rm ext})^2}{2}. \quad (7B.16)$$

Since the order parameter vanishes, this state is no longer superconducting.

For $H^{\rm ext} = H_c$ the system can be in either state. In cgs units, the critical field is given by $(H_c^{\rm ext})^2/2 = \mathcal{E}_c$ so that its order of magnitude lies, according to (7A.131), (7A.132), in the range of a few gauss.

The interesting consequence of the Ginzburg-Landau equations is that it allows for both the superconducting and the normal phase in one and the same sample separated by domain walls. This *mixed state*, also called the *Shubnikov phase*, is experimentally of particular importance and deserves some discussion.

7B.2   Two Length Scales and Type I or II Superconductivity

In the superconducting phase with field expectation value $\rho = \rho_0 = \sqrt{-\tau/g}$, the ρ fluctuations $\delta\rho = \rho - \rho_0$ have a coherence length given by Eq. (7A.128) in natural units:

$$\xi_{\text{size}}(T) = \frac{1}{\sqrt{-2\tau}}. \tag{7B.17}$$

By comparison with Eq. (7B.13) we find the ratio of penetration depth and coherence length, the Ginzburg parameter κ [recall (5.256)]

$$\kappa \equiv \frac{\lambda}{\sqrt{2}\xi_{\text{size}}} = \sqrt{\frac{g}{q^2}}. \tag{7B.18}$$

For $\kappa > 1/\sqrt{2}$ or $< 1/\sqrt{2}$, the magnetic penetration depth is larger or smaller than the coherence length of the order parameter. These two cases are called type-II and type-I superconductivity, respectively.

Inserting Eq. (7A.135) we estimate

$$\kappa \equiv \frac{\lambda}{\sqrt{2}\xi_{\text{size}}} \approx 4.06 \frac{1}{\sqrt{\alpha v_F/c}} \frac{T_c}{T_F}, \tag{7B.19}$$

which is of the order of $1/10$. Thus, a clean superconductor is usually of type-I.

In a dirty superconductor the result is modified by a factor r^{-1} as a consequence of Eq. (7A.137).

Hence impurities can bring κ into the type-II zone. In aluminum, for instance, 0.1% of impurities are sufficient to achieve this.

Let us now study types of domain walls between normal and superconducting materials; they differ significantly for the two types of superconductors. It will be convenient to go to a further reduced field variable $\hat{\rho} = \rho/(-\tau/g)^{1/2}$ which, in the superconductive state, fluctuates around unity instead of $\rho_0 = (-\tau/g)^{1/2}$. Similarly we shall define a reduced vector potential $\hat{A} = A/\kappa(-\tau/g)^{1/2}$ and measure lengths in units of the temperature dependent coherence length $r^{1/2}\xi_0/(-\tau)^{1/2}$, rather than $r^{1/2}\xi_0$. Then the Hamiltonians (7A.133) and (7B.12) become, for $\tau < 0$,

$$\hat{H}_{\text{GL}} \equiv \frac{g}{\tau^2} H_{\text{red}} = \frac{1}{2} \int d^3x \left\{ |(\boldsymbol{\nabla} - i\mathbf{A})\psi|^2 - |\psi|^2 + \frac{1}{2}|\psi|^4 + \frac{\kappa^2}{2}(\boldsymbol{\nabla} \times \mathbf{A})^2 \right\}, \tag{7B.20}$$

and in size and phase angle fields

$$\hat{H}_{\text{GL}} = \frac{1}{2} \int d^3x \left\{ (\boldsymbol{\nabla}\rho)^2 - \rho^2 + \frac{1}{2}\rho^4 + \left[\rho^2\mathbf{A}^2 + \kappa^2(\boldsymbol{\nabla} \times \mathbf{A})^2 \right] \right\}, \tag{7B.21}$$

where we have dropped the hats on top of the fields, for brevity. The associated supercurrent density is

$$\mathbf{j}_s = \frac{1}{2i}\psi^\dagger\boldsymbol{\nabla}\psi - \mathbf{A}|\psi|^2 = -\rho^2\mathbf{A}. \tag{7B.22}$$

We also define a reduced magnetic field

$$\mathbf{H} \equiv \kappa \, \mathbf{\nabla} \times \mathbf{A}, \tag{7B.23}$$

such that the magnetic field energy takes the usual form $\mathbf{H}^2/2$, and the critical magnetic field H_c is equal to $1/\sqrt{2}$. In these units, the field equations (7B.7) and (7B.8) become simply

$$\left(-\mathbf{\nabla}^2 + \mathbf{A}^2 - 1 + \rho^2\right)\rho = 0, \tag{7B.24}$$

$$\kappa^2 \, \mathbf{\nabla} \times (\mathbf{\nabla} \times \mathbf{A}) = \kappa \mathbf{\nabla} \times \mathbf{H} = -\rho^2 \mathbf{A}. \tag{7B.25}$$

They can be solved for an H and a ρ field varying, say, along the x-direction with $\mathbf{H}$ pointing in the y-direction. Accordingly, we choose a potential along the z-direction

$$\mathbf{A}(\mathbf{x}) = (0, 0 - A(x)), \tag{7B.26}$$

so that [with $' \equiv \partial_x$]

$$H(x) = \kappa A'(x). \tag{7B.27}$$

The field equations are

$$-\rho''(x) + A^2\rho(x) = \rho(x) - \rho^3(x), \tag{7B.28}$$

$$\kappa^2 A''(x) = \kappa H'(x) = \rho^2 A(x). \tag{7B.29}$$

Differentiating the second equation, it reduces to an equation for the magnetic field

$$\rho^2 H = \kappa^2(H'' - 2H'\rho'/\rho) = \kappa^2\rho^2 \left(\frac{1}{\rho^2}H'\right)'. \tag{7B.30}$$

In the first equation, we can eliminate A in favor of the magnetic field by writing the second equation as

$$A = \kappa^2 A''/\rho^2 = \kappa H'/\rho^2 \tag{7B.31}$$

so that

$$-\rho'' + \kappa^2 H'^2/\rho^3 = \rho - \rho^3. \tag{7D.32}$$

Now we observe that for the value $\kappa = 1/\sqrt{2}$, where magnetic and size fluctuations have equal length scales, these equations become particularly simple. For, if we make a trial ansatz

$$H = \frac{1}{\sqrt{2}}(1 - \rho^2) \tag{7B.33}$$

and insert it into Eq. (7B.30) to obtain

$$\frac{1}{\sqrt{2}} \left(1 - \rho^2\right) \rho^2 = -\frac{1}{\sqrt{2}} \left(\rho\rho'' - \rho'^2\right). \tag{7B.34}$$

But this happens to coincide with the second field equation (7B.32). Moreover, introducing $\sigma = 2 \log \rho$ we see that (7B.34) reduces to a differential equation of the Liouville type

$$\frac{\sigma''}{2} = e^\sigma - 1. \tag{7B.35}$$

This can be integrated to yield

$$\frac{\sigma'^2}{4} = e^\sigma - 1 - \sigma, \tag{7B.36}$$

or

$$x = \frac{1}{2} \int_{-1}^{\sigma} \frac{d\zeta'}{\sqrt{e^\zeta - 1 - \zeta}}. \tag{7B.37}$$

For $x \to -\infty$, σ goes to zero, like $e^{x/\sqrt{2}}$, so that $\rho \sim \exp\left(e^{x/\sqrt{2}}/2\right) \to 1$. For $x \to -\infty$, there is superconducting order and no magnet field; for $x \to \infty$ there is no order, $\rho = 0$, and the critical magnetic field $H = H_c = 1/\sqrt{2}$. The important point about a domain wall for $\kappa = 1/\sqrt{2}$ is that it can be formed in an external magnetic field $H^{\text{ext}} = 1/\sqrt{2}$, without any cost in energy. In order to see this we calculate, in reduced units, the magnetic enthalpy (for any κ)

$$\begin{aligned}
\hat{E}_H &= \int d^3x \, \hat{\mathcal{H}}_{\text{GL}} - \int d^3x \, \mathbf{H} \cdot \mathbf{H}^{\text{ext}} \\
&= \frac{1}{2} \int d^3x \left[(\nabla\rho)^2 - \rho^2 + \frac{1}{2}\rho^4 + \left(\rho^2 \mathbf{A}^2 + \mathbf{H}^2\right) \right] - \int d^3x \, \mathbf{H} \cdot \mathbf{H}^{\text{ext}}. \tag{7B.38}
\end{aligned}$$

Subtracting from this the condensation energy $\hat{E}_c = -(1/4) \int d^3x$, and inserting the field equations (7B.24), we find

$$\hat{E}_H - \hat{E}_c = \int d^3x \left[\left(1 - \rho^4\right) + \frac{1}{2}\mathbf{H}^2 - \mathbf{H} \cdot \mathbf{H}^{\text{ext}} \right]. \tag{7B.39}$$

This is the energy of a domain wall. At the critical field strength $H^{\text{ext}} = H_c = 1/\sqrt{2}$ pointing in the y-direction it becomes

$$\hat{E}_H - \hat{F}_c = \frac{1}{2} \int d^3x \left[-\frac{\rho^4}{2} + \left(H - \frac{1}{\sqrt{2}}\right)^2 \right]. \tag{7B.40}$$

Inserting $\kappa = 1/\sqrt{2}$ into Eq. (7B.33) we indeed obtain zero. For $\kappa = 1/\sqrt{2}$, a domain wall costs no energy.

Assuming that the wall energy is a monotonic function of κ, we expect the regions $\kappa > 1/\sqrt{2}$ and $\kappa < 1/\sqrt{2}$ to have wall energies of the opposite sign. Indeed, a numerical discussion of the different equations confirms this expectation. The solutions to the field equations are shown in Fig. 7.6. Inserting them into (7B.40) shows that the energy $\hat{F}_H - \hat{F}_c$ is positive for $\kappa < 1/\sqrt{2}$ and negative for $\lambda < 1/\sqrt{2}$. Hence we can conclude that type-I superconductors prefer a uniform state, type-II superconductors a mixed state.

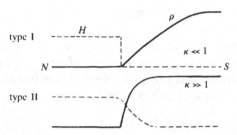

FIGURE 7.6 Spatial variation of order parameter ρ and magnetic field H in the neighborhood of a planar domain wall between normal and superconducting phases N and S, respectively. The magnetic field points parallel to the wall.

Actually, the planar domain walls calculated above are not the most energetically favorable way of forming a mixed state. A much better configuration is given by a bundle of magnetic vortex lines. In order to see this, let us study the properties of a solution corresponding to a single vortex line.

7B.3   Single Vortex Line and Critical Field H_{c_1}

In a type-II superconductor, the mixed state begins to form for much lower fields than the critical magnetic field $H_c = 1/\sqrt{2}$. The reason lies in the fact that there exists a solution in which only a very small magnetic flux invades into the superconductor, namely, the flux

$$\Phi_0 = \frac{ch}{q} = \pi\frac{c\hbar}{e} \approx 2 \times 10^{-7} \text{ gauss} \cdot \text{cm}^2. \tag{7B.41}$$

This solution has the form of a vortex line. Such a vortex line may be considered as a line-like defect in a uniform superconducting state. In this respect, it is a relative of a vortex line in superfluid ^{4}He. The two are, however, objects with quite different physical properties, as we shall now see.

Suppose the system is in the superconducting state without an external voltage so that there is no current j. Let us introduce a vortex line along the z-axis. Then we can use the current formula (7B.22) to find the vector potential,

$$\mathbf{A} = -\frac{\mathbf{j}_s}{|\psi|^2} + \frac{1}{2i}\frac{1}{|\psi|^2}\psi^\dagger \overset{\leftrightarrow}{\nabla} \psi. \tag{7B.42}$$

Far away from the vortex line, the state is undisturbed, i.e., $\mathbf{j}_s$ vanishes, and we have the relation

$$\mathbf{A} = \frac{1}{2i} \frac{1}{|\psi|^2} \psi^\dagger \overset{\leftrightarrow}{\boldsymbol{\nabla}} \psi. \tag{7B.43}$$

In a polar decomposition, $\psi(\mathbf{x}) = \rho(\mathbf{x}) e^{i\theta(\mathbf{x})}$, the derivative of $\rho(\mathbf{x})$ cancels and $A_i(\mathbf{x})$ depends only on the phase of the order parameter,

$$\mathbf{A}(\mathbf{x}) = \boldsymbol{\nabla}\theta(\mathbf{x}). \tag{7B.44}$$

Here we can establish contact with the discussion in superfluid ^{4}He. There the superflow velocity was proportional to the gradient of a phase angle variable θ. The multivaluedness of θ led to the quantization rule that any integral over $d\theta(\mathbf{x})$ along a closed circuit around the vortex line has to be an integer multiple of 2π. The same rule now applies here:

$$\oint_B d\theta(\mathbf{x}) = \oint_B d\mathbf{x} \cdot \boldsymbol{\nabla}\theta(\mathbf{x}) = 2\pi n. \tag{7B.45}$$

Expressing $\boldsymbol{\nabla}\theta(\mathbf{x})$ in terms of $\mathbf{A}(\mathbf{x})$ via 7B.44) and applying Stokes' theorem (4.21), this is equal to the magnetic flux through the area of the circuit [recall Eq. (7B.23)]

$$\Phi = \int_{S_B} d\mathbf{S} \cdot \mathbf{H} = \kappa \int_{S_B} d\mathbf{S} \cdot [\boldsymbol{\nabla} \times \mathbf{A}] = \kappa \oint_B d\mathbf{x} \cdot \mathbf{A} = 2\pi n \kappa. \tag{7B.46}$$

This holds in natural units. The quantization condition in physical units follows by applying the same argument to the original current (5.149) associated with the energy (5.148), which leads to

$$\Phi = n\Phi_0, \tag{7B.47}$$

with Φ_0 of Eq. (7B.41) [7].

Observe that when performing the integral along a circle closer to the vortex axis, we cannot ignore $\mathbf{j}_s(\mathbf{x})$, and Eq. (7B.44) becomes

$$\mathbf{A}(\mathbf{x}) + \frac{\mathbf{j}_s}{|\psi|^2} = \boldsymbol{\nabla}\theta(\mathbf{x}). \tag{7B.48}$$

The angular integral $\oint d\mathbf{x} \cdot \boldsymbol{\nabla}\theta$ still remains quantized and equal to $2\pi n$. Thus we find the quantization rule

$$\oint_B d\mathbf{x} \cdot \left(\mathbf{A} + \frac{\mathbf{j}_s}{|\psi|^2} \right) = 2\pi n, \tag{7B.49}$$

or

$$\Phi = -\frac{1}{|\psi|^2} \oint_B d\mathbf{x} \cdot \mathbf{j}_s + 2\pi n \kappa. \tag{7B.50}$$

This equation shows that a smaller circuit contains less magnetic flux. Part of the quantized flux $2\pi n\kappa$ is destroyed by the magnetic field of the supercurrent flowing around the vortex line.

Quantitatively, we can deduce the properties of a vortex line by solving the field equations (7B.24), (7B.25) in cylindrical coordinates. Inserting the second into the first equation, we find

$$-\frac{1}{r}\frac{d}{dr}r\frac{d\rho}{dr} + \frac{\kappa^2}{\rho^3}\left(\frac{d}{dr}H\right)^2 - (1-\rho^2)\rho = 0. \tag{7B.51}$$

Forming the curl of the second gives the cylindrical analog of (7B.30), i.e.,

$$H = \kappa^2 \frac{1}{r}\frac{d}{dr}\frac{f}{\rho^2}\frac{d}{dr}H. \tag{7B.52}$$

For $r \to \infty$ we have the boundary condition $\rho = 1$, $H = 0$ (superconducting state with Meissner effect) and $\mathbf{j}_s = 0$ (no current). Since for stationary supercurrents, Ampère's law (1.188) tells us that $\mathbf{j}_s \propto \nabla \times \mathbf{H}$, the last condition amounts to

$$H'(r) = 0, \quad r \to \infty. \tag{7B.53}$$

In cylindrical coordinates, flux quantization can be written in the form

$$\Phi = 2\pi \int_0^\infty dr r H = 2\pi n\kappa. \tag{7B.54}$$

Inserting Eq. (7B.52) into this result gives

$$\Phi = 2\pi\kappa^2 \left[\frac{r}{\rho^2}H'\right]_0^\infty = -2\pi\kappa^2 \left[\frac{r}{\rho^2}H'\right]_{r=0}, \tag{7B.55}$$

so that the quantization condition turns into a boundary condition at the origin:

$$H' \to -\rho^2 \frac{n}{\kappa}\frac{1}{r}, \quad r \to 0. \tag{7B.56}$$

Inserting this condition into (7B.51) we see that, close to the origin, $\rho(r)$ satisfies the equation

$$-\frac{1}{r}\frac{d}{dr}r\frac{d}{dr}\rho(r) + \frac{n^2}{r^2}\rho - (1-\rho^2)\rho \sim 0, \tag{7B.57}$$

which amounts to a behavior

$$\rho(r) = c_n \left(\frac{r}{\kappa}\right)^n + O(r^{n+1}). \tag{7B.58}$$

Putting this back into (7B.56) we have

$$H(r) = H(0) - \frac{c_n^2}{2\kappa}\left(\frac{r}{\kappa}\right)^{2n} + O(r^{2n+1}). \tag{7B.59}$$

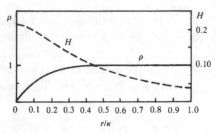

FIGURE 7.7 Order parameter ρ and magnetic field H for a vortex line with $n = 1$ fundamental flux units in a deep type-II superconductor ($\kappa = 10$). The radius where ρ approaches unity is the core radius r_c (here $r_c \approx 0.5\kappa$).

For large r, where $\rho \to 1$, Eq. (7B.52) is solved by the modified Bessel function K_0, with some factor α, namely[3]

$$H(r) \to \alpha K_0 \left(\frac{r}{\kappa}\right), \quad r \to \infty. \tag{7B.60}$$

For large $\kappa \gg 1/\sqrt{2}$ (i.e. deep type-II regime) ρ goes rapidly to 1 as compared to the length scale over which H changes (which is κ). Therefore, the behavior (7B.60) holds very close to the origin. We can determine α by matching (7B.60) to (7B.56) which fixes

$$\alpha \approx \frac{n}{\kappa}, \tag{7B.61}$$

where we have used the small-r behavior $K_0' = K_1 \sim -1/r$. In general, $H(r)$ and $\rho(r)$ have to be found numerically. A typical solution for $n = 1$ is shown in Fig. 7.7 for $\kappa = 10$. The energy of a vortex line can be calculated by using (7B.21). Inserting the equations of motion, and subtracting the condensation energy $\hat{E}_c = -(1/4) \int d^3x$ as in (7B.39), we find

$$\hat{E}_v = \hat{E}_H - \hat{E}_c = \frac{1}{2} \int d^3x \left[\frac{1}{2}\left(1 - \rho^4\right) + \mathbf{H}^2\right]. \tag{7B.62}$$

For $\kappa \gg 1/\sqrt{2}$, we may neglect the small radius $r \leq 1$, over which ρ increases quickly from zero to its asymptotic value $\rho = 1$. Beyond $r \geq 1$, but for $r \leq \kappa$, H is given by (7B.60). Inserting this into (7B.51) with (7B.58), we find

$$\rho(r) \sim 1 - \frac{n^2}{2r^2}. \tag{7B.63}$$

For the region $1 \leq r \leq \kappa$ this gives for the energy per length of a vortex line

$$\frac{1}{L}\hat{E}_v = \frac{1}{2}2\pi \int_1^\kappa dr\, r \left[\frac{1}{2}\left(1 - \rho^4\right) + H^2\right] = \pi n^2 \int_1^\kappa dr\, r \left[\frac{1}{r^2} + \frac{1}{\kappa^2}K_0^2\left(\frac{r}{\kappa}\right)\right]. \tag{7B.64}$$

---

[3] For very large r, this has the limit $\sqrt{\pi\kappa/2r}\, e^{-r/k}$.

For $\kappa \to \infty$, the second integral goes toward a constant [since $\int_0^\infty dx\, x K_0^2(x) = \frac{1}{2}$]. The first integral, however, has a logarithmic divergence so that we find the energy of a vortex line

$$\frac{1}{L}\hat{E}_v \approx \pi n^2 \left[\log \kappa + \text{const.}\right]. \tag{7B.65}$$

A more careful estimate gives $\pi n^2 (\log \kappa + 0.08)$.

Let us now see at which external magnetic field such a vortex line can form. For this we consider again the magnetic enthalpy (7B.39) and subtract from $(1/L)\hat{E}_v$ the magnetic $\hat{E}_v/L$ coupling $H H^{\text{ext}}$ so that, per length unit,

$$\frac{1}{L}\hat{E}_H = \pi n^2 (\log \kappa + 0.08) - 2\pi \int_0^\infty dr\, r H H^{\text{ext}}. \tag{7B.66}$$

The integral over H is simply the flux quantum (7B.46) associated with the vortex line, i.e.,

$$\frac{1}{L}\hat{E} = \pi n^2 (\log \kappa + 0.08) - 2\pi n\kappa H^{\text{ext}}. \tag{7B.67}$$

When this drops below zero, a vortex line invades the superconductor along the z-axis. The associated critical magnetic field is

$$H_{c_1} = \frac{n}{2\kappa} (\log \kappa + 0.08). \tag{7B.68}$$

For large κ this field can be quite small. The asymptotic result is compared with a numerical solution of the differential equation for $n = 1, 2, 3, \ldots$ in Fig. 7.8. For a

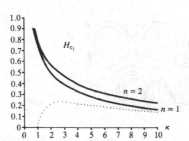

FIGURE 7.8 Critical field H_{c_1}/n at which a vortex line of strength n forms when it first invades a type-II superconductor as a function of the parameter κ. The dotted line indicates the asymptotic result $(1/2\kappa) \log \kappa$ of Eq. (7B.68). The magnetic field H_{c_1} is measured in units of $\sqrt{2}H_c$ where H_c is the magnetic field at which the magnetic energy equals the condensation energy.

comparison with experimental data one expresses this field in terms of the critical magnetic field $H_c = 1/\sqrt{2}$ and measures the ratio

$$\frac{H_{c_1}}{H_c} = \frac{n}{\sqrt{2}\kappa} (\log \kappa + 0.08). \tag{7B.69}$$

As an example, pure lead is a type-I superconductor with $H_{c_1} = H_c \simeq 550$ gauss. An admixture of 15% Iridium or 30% Thallium brings H_c up to 650 or 430, and H_{c_1} down to 250 or 145, respectively (see Table (7.1)).

TABLE 7.1 Critical magnetic fields in gauss for Pb and Nb with various impurities.

material	H_c	H_{c_1}	H_{c_2}	T_c/K
Pb	550	550	550	4.2
0.850 Pb, 0.150 Ir	650	250	3040	4.2
0.750 Pb, 0.250 In	570	200	3500	4.2
0.700 Pb, 0.300 Tl	430	145	2920	4.2
0.976 Pb, 0.042 Hg	580	340	1460	4.2
0.916 Pb, 0.088 Bi	675	245	3250	4.2
Nb	1608	1300	2680	4.2
0.500 Nb, 0.500 Ta	252	–	1470	5.6

7B.4   Critical Field H_{c_2} where Superconductivity is Destroyed

As the field increases above H_{c_1}, more and more vortex lines invade the superconductor. For $H \sim H_c$, they form a hexagonal array as shown in (7.9). If the field increases even more, the superconducting regions separating the vortex lines become thinner and thinner until, finally, the whole material is filled with the magnetic field, and superconductivity is destroyed. The field where this happens is denoted by H_{c_2}. Its value can be estimated quite simply following Abrikosov. He noticed that close to H_{c_2}, the order parameter is so small that the nonlinear terms can be forgotten, and the Ginzburg-Landau equation reads

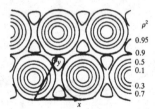

FIGURE 7.9 Lines of equal size of order parameter $\rho(\mathbf{x})$ in a typical mixed state in which the vortex lines form a hexagonal lattice [8].

$$\left[\left(\frac{1}{i}\nabla - \mathbf{A}\right)^2 - 1\right]\psi(\mathbf{x}) = 0. \tag{7B.70}$$

For H along the z direction one may choose

$$\mathbf{A}(\mathbf{x}) = \left(0, \frac{1}{\kappa}Hx, 0\right) \tag{7B.71}$$

and the following equation emerges:

$$\left[-\partial_x^2 - \left(-i\partial_y - \frac{1}{\kappa}Hx\right)^2 - \partial_z^2 - 1\right]psi(\mathbf{x}) = 0. \tag{7B.72}$$

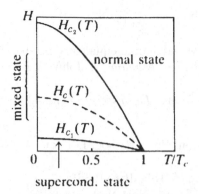

supercond. state

FIGURE 7.10 Temperature behavior of the critical magnetic fields of a type-II supercon-ductor: H_{c_1} (when the first vortex line invades), H_{c_2} (when superconductivity is destroyed) and H_c (when the magnetic field energy is equal to the condensation energy).

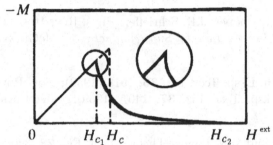

FIGURE 7.11 Magnetization curve as a function of the external magnetic field H^{ext}. The dashed curve shows how a type-I superconductor would behave.

The lowest nontrivial eigenstate is

$$\psi(\mathbf{x}) = \text{const.} \; e^{-(1/\kappa)H(x - p_y\kappa/H)^2/2} e^{ip_y y}. \tag{7B.73}$$

For this solution to occur, the energy eigenvalue $H/\kappa - 1$ must be negative. This happens for $H < H_{c_2} = \kappa$. The field H_{c_2} is larger than the critical field H_c by a factor $\sqrt{2}\kappa$, which can be a large factor in the deep type-II regime. As an example, pure lead has $H_{c_2} = H_c = 550$ gauss. An admixture of 15% Indium or 30% Thallium which changes H_c to 650 or 430, increases H_{c_2} to 3040 or 2920 gauss, respectively. The typical behavior of the critical fields H_c, H_{c_1}, H_{c_2} as a function of T is shown in Fig. 7.10.

The invasion of vortex lines becomes apparent from the curve depicted in Fig. 7.11 which shows the behavior of the magnetization curve as a function of H^{ext}

$$-M = H^{\text{ext}} - H \tag{7B.74}$$

in a type-II superconductor as compared with a type-I superconductor.

Notes and References

[1] The classical magnetohydrodynamic equations are discussed in
    J.D. Jackson, *Classical Electrodynamics*, John Wiley and Sons, New York,
    1975, Sects. 6.12-6.13;
    W.F. Hughes and F.J. Young, *The Electromagnetodynamics of Fluids*, Wiley,
    New York, 1966.

[2] F. London and H. London, Proc. R. Soc. London, A **149**, 71 (1935); Physica
    A **2**, 341 (1935);
    H. London, Proc. R. Soc. A **155**, 102 (1936);
    F. London, *Superfluids*, Dover, New York, 1961.

[3] H. Kleinert, *Collective Quantum Fields*, Lectures presented at the First Erice
    Summer School on Low-Temperature Physics, 1977, in Fortschr. Physik **26**,
    565-671 (1978) (**kl/55**). See Eq. (4.118).

[4] J. Bardeen, L.N. Cooper, J.R. Schrieffer, Phys. Rev. **108**, 1175 (1957);
    M. Tinkham, *Introduction to Superconductivity*, McGraw-Hill, New York,
    1975.

[5] L.P. Gorkov, Zh. Eksp. Teor. Fiz. **36**, 1918 (1959); [Sov. Phys.-JETP **9**, 1364
    (1959)]; Zh. Eksp. Teor. Fiz. **37**, 1407 (1959); [Sov. Phys.-JETP **10**, 998
    (1960)].

[6] V.A. Adrianov and V.N. Popov, Theor. Math. Fiz. **28**, 340 (1976).

[7] As a cross check, we calculate once more the value of the flux by transforming
    the twice reduced field variables to physical ones. Thus we insert into the physi-
    cal flux $\Phi^{\text{phys}} = \oint d\mathbf{x} \cdot \mathbf{A}^{\text{phys}}$ the physical field $\mathbf{A}^{\text{phys}} = \sqrt{k_B T_c/\xi_0}\sqrt{-\tau/g}\,\kappa\mathbf{A}$ and
    $x^{\text{phys}}/x = \xi_0/\sqrt{-\tau}$ to find $\Phi^{\text{phys}} = \sqrt{\xi_0 k_B T_c}(1/q)\int d\mathbf{x}\cdot\mathbf{A} = \sqrt{\xi_0 k_B T_c}\,2\pi n/q =$
    $n(ch/2e)$, where we have used Eq. (7A.135), according to which $q = (2e/\pi e)\sqrt{k_B T_c \xi_0}$.

[8] A.A. Abrikosov, Zh. Eksp. Teor. Fiz. **32**, 1442 (1957) [Sov. Phys.-JETP **5**,
    1174 (1957)];
    W.M. Kleiner, L.M. Roth, and S.H. Autler, Phys. Rev. A **133**, 1225 (1964);
    E.H. Brandt, Phys. Status Solidi B **51**, 345 (1972);
    J.L. Harden and V. Arp, Cryogenics **3**, 105 (1963).

Do not confine your children to your own learning,
for they were born in another time.
HEBREW PROVERB

8

Relativistic Magnetic Monopoles and Electric Charge Confinement

The theory of multivalued fields in magnetism in Chapter 4 can easily be extended to a full relativistic theory of charges and monopoles [1, 2]. For this we go over to four spacetime dimensions, which are assumed to be Euclidean with a fourth spatial component $dx^4 = icdt$, to avoid factors of i.

8.1 Monopole Gauge Invariance

The covariant extension of the Maxwell equation (4.54) is a natural modification of Eq. (1.200) and reads, by complete analogy with the Maxwell equation (1.196):

$$\partial_b \tilde{F}^{ab} = -\frac{1}{c}\tilde{j}_b, \tag{8.1}$$

where $\tilde{j}_a$ is the magnetic current density

$$\tilde{j}_a = (c\rho_{\mathrm{m}}, \mathbf{j}_{\mathrm{m}}). \tag{8.2}$$

Eq. (8.1) Equation (8.1) implies that the magnetic current density is conserved:

$$\partial_a \tilde{j}_a = 0. \tag{8.3}$$

The zeroth component of (8.1) reproduces the divergence equation (4.54) for the magnetic field [recall the identification of the field components (1.172)]. The spatial components yield the extension of the Maxwell equation (1.18):

$$\nabla \times \mathbf{E} + \frac{1}{c}\frac{\partial \mathbf{B}}{\partial t} = -\frac{1}{c}\mathbf{j}_{\mathrm{m}} \qquad \text{(extended Faraday law)}. \tag{8.4}$$

For a single monopole of strength g moving along a world line L parametrized by $q_a(\sigma)$, the magnetic current density $\tilde{j}_a$ can be expressed in terms of a δ-function on the world line,

$$\delta_a(x; L) = \int d\sigma \frac{d\bar{x}_a(\sigma)}{d\sigma}\delta^{(4)}(x - \bar{x}(\sigma)), \tag{8.5}$$

as follows

$$\tilde{j}_a = g\,c\,\delta_a(x;L). \tag{8.6}$$

This satisfies the magnetic current conservation law (10.16) as a consequence of the divergence property of the δ-function on a closed world line:

$$\partial_a \delta_a(x;L) = 0, \tag{8.7}$$

which is the four-dimensional version of (4.10).

The spacetime components of the magnetic current density are [compare (1.211) and (1.212)]

$$c\rho_m(\mathbf{x},t) = g\,c \int_{-\infty}^{\infty} d\tau\,\gamma c\,\delta^{(4)}\left(x - \bar{x}(\tau)\right), \tag{8.8}$$

$$\mathbf{j}_m(\mathbf{x},t) = g\,c \int_{-\infty}^{\infty} d\tau\,\gamma\mathbf{v}\,\delta^{(4)}\left(x - \bar{x}(\tau)\right). \tag{8.9}$$

With this notation, the electric current density (1.213) of a particle on the worldline L reads

$$j_a = e\,c\,\delta_a(x;L), \tag{8.10}$$

and satisfies the electric current conservation law $\partial_a j^a = 0$, again due to the identity (8.7).

Equation (8.1) shows that F_{ab} cannot be represented as a curl of a single-valued vector potential A_a, since left-hand side is equal to $\frac{1}{2}\epsilon_{abcd}(\partial_a\partial_c - \partial_c\partial_a)A_d$. The non-vanishing of the commutator of derivatives implies a violation of Schwarz's integrability condition and the multivaluedness of A_d. As in the magnetostatic discussion in Section 4.6, the simplest way to incorporate the monopole worldline into the electromagnetic field theory is by the introduction of an extra monopole gauge field. In four spacetime dimensions, this has the form

$$F_{ab}^M \equiv g\,\tilde{\delta}_{ab}(x;S), \tag{8.11}$$

where $\tilde{\delta}_{ab}(x;S)$ is the *dual tensor*

$$\tilde{\delta}_{ab}(x;S) \equiv \frac{1}{2}\epsilon_{abcd}\delta_{cd}(x;S), \tag{8.12}$$

of the δ-function $\delta_{cd}(x;S)$ that is singular on the world surface S:

$$\delta_{ab}(x;S) \equiv \int d\sigma d\tau \left[\frac{\partial \bar{x}_a(\sigma,\tau)}{\partial \sigma}\frac{d\bar{x}_b(\sigma,\tau)}{\partial \tau} - (a \leftrightarrow b)\right]\delta^{(4)}(x - \bar{x}(\sigma,\tau)). \tag{8.13}$$

This δ-function has the divergence

$$\partial_a \delta_{ab}(x;S) = \delta_b(x;L), \tag{8.14}$$

where L is the boundary line of the surface. This follows directly from the simple calculation:

$$\partial_a \delta_{ab}(x;S) = \int d\tau \left[\frac{d\bar{x}_b(\sigma_b, \tau)}{\partial \tau} \delta^{(4)}(x - \bar{x}(\sigma_b, \tau)) - \frac{d\bar{x}_b(\sigma_a, \tau)}{\partial \tau} \delta^{(4)}(x - \bar{x}(\sigma_a, \tau)) \right]$$
$$- \int d\sigma \left[\frac{d\bar{x}_b(\sigma, \tau_b)}{\partial \tau} \delta^{(4)}(x - \bar{x}(\sigma, \tau_b)) + \frac{d\bar{x}_b(\sigma, \tau_a)}{\partial \tau} \delta^{(4)}(x - \bar{x}(\sigma, \tau_a)) \right],$$

where $\sigma_{a,b}$ and $\tau_{a,b}$ are the lower and upper values of the surface parameters, respectively, so that $\bar{x}(\sigma_a, \tau)$, $\bar{x}(\sigma, \tau_a)$, $\bar{x}(\sigma_b, \tau)$, and $\bar{x}(\sigma, \tau_b)$ run around the boundary line of the surface. For the dual δ-function (8.12) the relation (8.14) reads

$$\frac{1}{2} \epsilon_{abcd} \partial_b \tilde{\delta}_{cd}(x;S) = \delta_a(x;L), \tag{8.15}$$

due to the identity (1A.24). We recognize the four-dimensional version of the local formulation (4.24) of Stokes' theorem.

Expressing the δ-functions in terms of the monopole gauge field (8.11) and the magnetic current density (8.6), the relation (8.15) implies

$$\frac{1}{2} \epsilon_{abcd} \partial_a F_{cd}^M = \frac{1}{c} \tilde{j}_b. \tag{8.16}$$

The surface S is the worldsheet of the Dirac string. For any given line L, there are many possible surfaces S. We can go over from one S to another, say S', with a fixed boundary L, as follows

$$\tilde{\delta}_{cd}(x;S) \rightarrow \tilde{\delta}_{cd}(x;S') = \tilde{\delta}_{cd}(x;S) + \partial_a \tilde{\delta}_b(x;V) - \partial_b \tilde{\delta}_a(x;V), \tag{8.17}$$

where $\tilde{\delta}_a(x;V)$ is the δ-function

$$\tilde{\delta}_a(x;V) \equiv \epsilon_{abcd} \delta_{bcd}(x;V), \tag{8.18}$$

with

$$\delta_{abc}(x;V) \equiv \int d\sigma d\tau d\lambda \left[\sum_{P(abc)} \epsilon_P \frac{\partial \bar{x}_a}{\partial \sigma} \frac{\partial \bar{x}_b}{\partial \tau} \frac{\partial \bar{x}_c}{\partial \lambda} \right] \delta^{(4)}(x - \bar{x}(\sigma, \tau, \lambda)). \tag{8.19}$$

The sum runs over all 6 permutations P of the indices and ϵ_P denotes their parity ($\epsilon_P = +1$ for even and -1 for odd permutation P). These δ-functions are singular on the three-dimensional volume V in four-space swept out when the surface S moves through four-space. The transformation law (8.17) is the obvious generalization of (4.29).

An ensemble of many noninteracting monopoles is represented by a gauge field (8.11) with a superposition of many different surfaces S.

We are now ready to set up the electromagnetic action in the presence of an arbitrary number of monopoles. By analogy with Eq. (4.86) and (5.26) it depends only on the difference between the total field strength $F_{ab} = \partial_a A_b - \partial_b A_a$ of the integrable vector potential A_a and the monopole gauge field F_{ab}^M of (8.11), i.e., it is given by [3, 4, 5, 6]

$$\mathcal{A}_0 + \mathcal{A}_{\text{mg}} \equiv \mathcal{A}_{0,\text{mg}} = \int d^4x \, \frac{1}{4c} \left(F_{ab} - F_{ab}^M \right)^2. \tag{8.20}$$

The subtraction of F_{ab}^M is essential in avoiding an infinite energy density in the Maxwell action

$$\mathcal{A}_0 \equiv \int d^4x \, \frac{1}{4c} F_{ab}^2, \tag{8.21}$$

that would arise from the flux tube in F_{ab} inside the Dirac string. The difference

$$F_{ab}^{\text{obs}} \equiv F_{ab} - F_{ab}^M \tag{8.22}$$

is the nonsingular observable field strength. Since only fields with finite action are physical, the action contains no contributions from squares of δ-functions, as it might initially appear.

The action (8.20) exhibits two types of gauge invariances. First, the original electromagnetic one under [compare (2.104)]

$$A_a(x) \longrightarrow A_a'(x) = A_a(x) + \partial_a \Lambda(x), \tag{8.23}$$

where $\Lambda(x)$ is any smooth field which satisfies the integrability condition

$$(\partial_a \partial_b - \partial_b \partial_a)\Lambda(x) = 0, \tag{8.24}$$

under which F_{ab}^M is trivially invariant. Second, there is gauge invariance under *monopole gauge transformations*

$$F_{ab}^M \rightarrow F_{ab}^M + \partial_a \Lambda_b^M - \partial_b \Lambda_a^M, \tag{8.25}$$

with integrable vector functions $\Lambda_a^M(x)$. By Eq. (8.17) they have the general form

$$\Lambda_a^M(x) = g \, \delta_a(x; V), \tag{8.26}$$

with arbitrary choices of three-volumes V. If the monopole gauge field (8.11) contains many jumping surfaces S, the function $\Lambda_a^M(x)$ will contain a superposition of many volumes V.

To have invariance of the action (8.20), the transformation (8.25) must be accompanied by a shift in the electromagnetic gauge field [3, 4, 5, 6]

$$A_a \rightarrow A_a + \Lambda_a^M. \tag{8.27}$$

From Eqs. (8.11), (8.15), and (8.17) we see that the physical significance of the part (8.25) of the monopole gauge transformation is to change the Dirac world surface without changing its boundary, the monopole world line. An exception are vortex gauge transformations (8.27) of the gradient type, in which Λ_a^M is g times the gradient $\partial_a \Lambda^M$ of the δ-function on the four-volume V_4:

$$\delta(x; V_4) \equiv \epsilon_{abcd} \int d\sigma d\tau d\lambda d\kappa \, \frac{\partial \bar{x}_a}{\partial \sigma} \frac{\partial \bar{x}_b}{\partial \tau} \frac{\partial \bar{x}_c}{\partial \lambda} \frac{\partial \bar{x}_d}{\partial \kappa} \delta^{(4)} \left(x - \bar{x}(\sigma, \tau, \lambda, \kappa) \right), \qquad (8.28)$$

i.e.,

$$A_a \to A_a + g \partial_a \delta(x; V_4). \qquad (8.29)$$

These do not give any change in F_{ab}^M since they are particular forms of the original electromagnetic gauge transformations (8.23).

The field strength F_{ab} is, of course, changed when distorting the Dirac string in spacetime, only the observable field strength $F_{ab}^{\text{obs}} = F_{ab} - F_{ab}^M$ remains invariant.

The part (8.27) of the monopole gauge transformations expresses the fact that, in the presence of monopoles, the gauge field A_a is necessarily a *cyclic variable* for which $A_a(x)$ and $A_a(x) + gn$ are identical at each point x for any integer n.

At this place we should emphasize once more (what, is of course, clear to the reader from the previous discussions) that the vortex gauge transformations bear no relation to the original gauge transformation (8.23). One sometimes finds in the literature the misconception [7, 8] that, since the movement of the Dirac string can be achieved by a transformation,

$$A_a \to A_a + g \partial_a \Omega, \qquad (8.30)$$

where Ω is the spherical angle over which the string has swept, the invisibility of the string may be related to the electromagnetic gauge invariance (together with the single-valuedness of wave functions). After all, (8.30) looks precisely like (8.23), with $\Lambda = g\Omega$. This argument, however, is invalid since the spherical angle is a multivalued function which fails to satisfy the integrability condition (8.24). This is why (8.30) is *not* a gauge transformation in spite of its suggestive appearance. It cannot possibly be since it changes the magnetic field along the Dirac string. Sometimes, (8.30) is referred to as a "singular gauge transformation" or "general gauge transformation". This terminology is misleading and must be rejected. After all, if we were to allow for such "singular" (i.e. nonintegrable) transformations Λ in (8.23) we could reach an arbitrary field F_{ab} starting from $F_{ab} \equiv 0$, and the physics would certainly not be invariant under this [8].

The complete conceptual independence of the two gauge invariances is most easily seen in the model discussion in Subsection 4.6, where the ordinary gauge invariance is absent while the invariance under deformations of the surface S is essential for obtaining the correct magnetic fields around current loops.

The partition function of magnetic monopoles with their electromagnetic interactions is given by the functional integral

$$Z = \int \mathcal{D} A_a^T \int \mathcal{D} F_{ab}^M \, e^{-\mathcal{A}_{0,\text{mg}}^E}. \qquad (8.31)$$

where $\mathcal{A}_{0,\text{mg}}^E$ is the Euclidean version of the action (8.20). We have used the same short notation for the measure as in (5.272), indicating by the symbol $\int \mathcal{D} F_{ab}^{MT}$ the gauge-fixed sum $\sum_{\{S\}} \Phi[F_{ab}^M]$ over fluctuating jumping surfaces S, the world sheets of the Dirac strings. The symbol $\int \mathcal{D}A_a^T$ denotes the gauge-fixed functional integral over A_a in the Lorentz gauge $\partial_a A_a = 0$.

8.2   Charge Quantization

Let us now introduce electrically charged particles into the action (8.20). This is done via the Euclidean version of the current interaction in (2.83):

$$\mathcal{A}_{\text{int}}^E = \frac{i}{c^2} \int d^4 x_E \, j_a(x) A_a(x), \tag{8.32}$$

where $j_a(x)$ is the electric current of the world line L of a charged particle

$$j_a = e \, \delta_a(x; L). \tag{8.33}$$

For brevity, we shall omit in this chapter all superscripts E.

Due to electric current conservation

$$\partial_a j_a = 0, \tag{8.34}$$

the action (8.32) is trivially invariant under electromagnetic gauge transformations (8.23). In contrast, it can remain invariant under monopole gauge transformations (8.25) and (8.27) only if the monopole charge satisfies the famous *Dirac quantization condition* derived before in Eq. (4.109). Let us see how this comes about in the present four-dimensional theory.

Under the monopole gauge transformation (8.27), only the part $\mathcal{A}_0 + \mathcal{A}_{\text{mg}}$ of the total action

$$\mathcal{A}_{\text{tot}} \equiv \mathcal{A}_0 + \mathcal{A}_{\text{mg}} + \mathcal{A}_{\text{int}} = \int d^4 x \, \frac{1}{4c} \left(F_{ab} - F_{ab}^M \right)^2 + \frac{i}{c^2} \int d^4 x \, j_a(x) A_a(x) \tag{8.35}$$

is manifestly invariant. The electric part $\mathcal{A}_{\text{int}}$, and thus the total action, changes by

$$\Delta \mathcal{A}_{\text{tot}} = \Delta \mathcal{A}_{\text{int}} = i \frac{eg}{c} I, \tag{8.36}$$

where I denotes the integral

$$I \equiv \int d^4 x \, \delta_a(L) \delta_a(V). \tag{8.37}$$

This is an integer number if L passes through V, and zero if it misses V. In the former case, the string in the operation (8.17) sweeps across L, in the other case it does not. To prove this we let L run along the first axis and let V be the entire volume in 234-subspace. Then $\delta_a(x; L)$ and $\delta_a(x; V)$ have nonzero components only in the 1-direction; $\delta_1(x; L) = \delta(x_2)\delta(x_3)\delta(x_4)$, and $\delta_1(x; V) = \delta(x_1)$. Inserting these into the integral (8.37) yields $I = 1$.

Dirac's charge quantization rule is a consequence of quantum theory. As explained in Section 2.8, probability amplitudes are determined from a sum over all possible field configurations

$$\text{Amplitude} = \sum_{\text{field configurations}} e^{i\mathcal{A}/\hbar}, \tag{8.38}$$

where $\mathcal{A}$ is the classical action of the system. This amplitude is invariant under jumps of the classical action by $2\pi\hbar \times$ integer since these do not contribute to the phase factors $e^{i\mathcal{A}/\hbar}$.

In the Euclidean formulation, a corresponding sum yields the partition function of the system in Eq. (2.162). This function is invariant under jumps of the Euclidean action by $2\pi i\hbar \times$ integer since these do not contribute to the Boltzmann factors $e^{-\mathcal{A}^E/\hbar}$ in the sum (2.162). The property (8.36) of the Euclidean action implies therefore that if all electric charges satisfy the relation

$$\frac{eg}{\hbar c} = 2\pi \times \text{integer}, \tag{8.39}$$

physics is invariant under arbitrary shape deformations of the Dirac string. This invariance makes it invisible to all electric particles.

The result may be stated in a dimensionless way by expressing e in terms of the fine-structure constant $\alpha \approx 1/137.0359\ldots$ of Eq. (1.145) as $e^2 = 4\pi hbarc\,\alpha$, so that the charge quantization condition becomes[1]

$$g/e = \text{integer}/2\alpha. \tag{8.40}$$

It must be emphasized that the above derivation of (8.40) requires much less quantum mechanical input than most derivations in the literature, which involve the wave functions for the charged particle in a monopole field [1, 2, 7]. In the above derivation, however, the particle orbits remain fixed, and only the worldsheets of the Dirac strings are moved around by monopole gauge transformation. The quantization follows only from the requirement of invariance under these transformations.

Observe that after the quantization of the charge, the total action (8.35) is double-gauge invariant — it is invariant under the ordinary electromagnetic gauge transformations (8.23) *and* the monopole gauge transformations (8.25).

It should be mentioned that the Dirac quantization condition (8.39) guarantees the invisibility of the Dirac string only for electric charges of integer spin. For electrons, and all particles of half-integer spin, the wave function is double-valued since it returns to its original value only after rotating it by 4π. For these particles the electric charge must be twice as big as for integer spins, and satisfy the *Schwinger quantization condition* [9]

$$\frac{eg}{\hbar c} = 2 \times 2\pi \times \text{integer}. \tag{8.41}$$

---

[1]In many textbooks, the action (8.20) has a prefactor $1/4\pi$, leading to Dirac's charge quantization condition in the form $2eg/\hbar c =$ integer. In these conventions, $e^2 = \hbar c\,\alpha$, so that the condition (8.40) is the same.

8.3   Electric and Magnetic Current-Current Interactions

If we integrate out the A_a-field in the partition function associated with the action (8.35) we obtain the interaction

$$
\mathcal{A}_{\mathrm{int}} = \int d^4x \left\{ \frac{1}{4c} \left[(F_{ab}^M)^2 + 2\partial_a F_{ab}^M (-\partial^2)^{-1} \partial_c F_{cb}^M \right] \right.
$$

$$
\left. + \frac{1}{2c^3} j_a (-\partial^2)^{-1} j_a + \frac{i}{2c^2} \partial_a F_{ab}^M (-\partial^2)^{-1} j_b \right\}. \tag{8.42}
$$

The second term

$$
\mathcal{A}_{jj} = \frac{1}{2c^3} \int d^4x j_a (-\partial^2)^{-1} j_a \tag{8.43}
$$

is the usual electric current-current interaction, where $(-\partial^2)^{-1}$ denotes the Euclidean version of *retarded Green function* of the vector potential $A^a(x)$

$$
(-\partial^2)^{-1}(x, x') = G(x - x') \equiv \int \frac{d^4k}{(2\pi)^4} \frac{e^{ik(x-x')}}{k^2}. \tag{8.44}
$$

Indeed, inserting the components of the four-component current density $j^a = (c\rho, \mathbf{j})$ [recall (1.197)], the interaction (8.43) reads

$$
\mathcal{A}_{jj} = \frac{1}{2c} \int d^4x d^4x' \, \rho(x) G(x, x') \rho(x') + \frac{1}{2c^3} \int d^4x d^4x' \, \mathbf{j}(x) G(x, x') \mathbf{j}(x'). \tag{8.45}
$$

For the static charges and currents in Minkowski spacetime, this becomes [compare (4.98)]

$$
\mathcal{A}_{jj} = \frac{1}{2} \int dt d^3x d^3x' \, \rho(t, \mathbf{x}) \frac{1}{|\mathbf{x}-\mathbf{x}'|} \rho(t, \mathbf{x}') - \frac{1}{2c^2} \int dt d^3x d^3x' \, \mathbf{j}(t, \mathbf{x}) \frac{1}{|\mathbf{x}-\mathbf{x}'|} \mathbf{j}(t, \mathbf{x}'). \tag{8.46}
$$

The first term is the *Coulomb interaction*, the second the *Biot-Savart interaction* of an arbitrary current distribution.

The first two terms in the interaction (8.42) reduce to the magnetic current-current interaction [compare (4.100)]

$$
\mathcal{A}_{\tilde{j}\tilde{j}} = \frac{1}{2c^3} \int d^4x \, \tilde{j}_a (-\partial^2)^{-1} \tilde{j}_a. \tag{8.47}
$$

This follows from (8.16) and the simple calculation with the help of the tensor identity (1A.23):

$$
\tilde{j}^2 = \left(\frac{c}{2} \epsilon_{abcd} \partial_b F_{cd}^M \right)^2 = c^2 \left[\partial^2 (F_{cd}^M)^2 - 2(\partial_a F_{ab}^M)^2 \right]. \tag{8.48}
$$

The magnetic interaction (8.47) can be decomposed into time- and space-like components in the same way as in Eq. (8.45), but with magnetic and electric charge and current densities.

The last term in (8.42)

$$A_{\bar{j}j} = \int d^4x \, \frac{i}{2c^2} \partial_a F_{ab}^M (-\partial^2)^{-1} j_b \qquad (8.49)$$

specifies the interaction between electric and magnetic currents. It is the relativistic version of the interaction (4.102).

All three interactions are invariant under monopole gauge transformations (8.25). For the electric and magnetic current-current interactions (8.43) and (8.47) this is immediately obvious since they depend only on the world lines of electric and magnetic charges. Only for the mixed interaction (8.49), the invariance is not obvious. In fact, performing a monopole gauge transformation (8.25), and using the world line representation (8.10) of the electric four-vector current, this interaction is changed by

$$\Delta A_{\bar{j}j} = i \int d^4x \, \frac{g}{c^2} \partial^2 \delta_b(x; V)(-\partial^2)^{-1} j_b = i\frac{ge}{c} \int d^4x \delta_b(x; V)\delta(x; L) = i\frac{ge}{c} I, \quad (8.50)$$

with I of (8.37). This is nonzero, but the theory is still invariant since $(ge/c)I$ is equal to $2\pi\hbar$ times a number which is integer due to Dirac's charge quantization condition (8.39). Thus $e^{-\Delta A_{\bar{j}j}/\hbar}$ is equal to one and the theory invariant. The reader will recognize the analogy with the three-dimensional situation in the mixed interaction (4.102).

8.4   Dual Gauge Field Representation

It is instructive to subject the total action (8.35) to a similar duality transformation as the Hamiltonian (4.86) by which we derived the dual gauge formulation (4.91). Thus we introduce an independent fluctuating field f_{ab} and replace the action (8.20) by the equivalent expression [the four-dimensional analog of (4.87)]

$$\tilde{A}_{0,\text{mg}} = \int d^4x \left[\frac{1}{4c} f_{ab}^2 + \frac{i}{2c} f_{ab} \left(F_{ab} - F_{ab}^M \right) \right], \qquad (8.51)$$

with two independent fields A_a and f_{ab}. Inserting $F_{ab} \equiv \partial_a A_b - \partial_a A_a$, the partition function (8.31) becomes

$$Z = \int \mathcal{D}A_T^a \int \mathcal{D}f_{ab} \int \mathcal{D}F_{ab}^M \, e^{-\tilde{A}_{0,\text{mg}}}. \qquad (8.52)$$

Here we may integrate out the vector potential A_u to obtain the constraint

$$\partial_b f_{ab} = 0. \qquad (8.53)$$

This can be satisfied identically (as a Bianchi identity) by introducing a dual magnetoelectric vector potential $\tilde{A}_a$ and writing [compare (4.89)]

$$f_{ab} \equiv \epsilon_{abcd} \partial_c \tilde{A}_d. \qquad (8.54)$$

If we also introduce a dual field tensor

$$\tilde{F}_{ab} \equiv \partial_a \tilde{A}_b - \partial_b \tilde{A}_a, \tag{8.55}$$

the action (8.51) takes the dual form

$$\tilde{\mathcal{A}}_{0,\mathrm{mg}} \equiv \tilde{\mathcal{A}}_0 + \tilde{\mathcal{A}}_{\mathrm{mg}} = \int d^4 x \left(\frac{1}{4c} \tilde{F}_{ab}^2 + \frac{i}{c^2} \tilde{A}_a \tilde{\jmath}_a \right), \tag{8.56}$$

with the magnetoelectric source

$$\tilde{\jmath}_a \equiv \frac{c}{2} \epsilon_{abcd} \partial_b F_{cd}^M. \tag{8.57}$$

By inserting (8.11) and (8.15), we see that $\tilde{\jmath}_a$ is the magnetic current density (8.6). Since (8.57) satisfied trivially the current conservation law $\partial_a \tilde{\jmath}_a = 0$ [recall (10.16)], the action (8.56) allows for an additional set of gauge transformations which are the magnetoelectric ones

$$\tilde{A}_a \to \tilde{A}_a + \partial_a \tilde{\Lambda} \tag{8.58}$$

with arbitrary integrable functions $\tilde{\Lambda}$,

$$(\partial_a \partial_b - \partial_b \partial_a) \tilde{\Lambda} = 0. \tag{8.59}$$

If we include the electric current (8.33) into the dual form of the action (8.56) it becomes

$$\tilde{\mathcal{A}}_{\mathrm{tot}} = \int d^4 x \left[\frac{1}{4c} f_{ab}^2 + \frac{i}{2c} f_{ab} \left(F_{ab} - F_{ab}^M \right) + \frac{i}{c^2} \jmath_a A_a \right]. \tag{8.60}$$

Extremizing this with respect to the field A_a gives now

$$\partial_a f_{ab} = -\frac{1}{c} \jmath_b, \tag{8.61}$$

rather than (8.53). The solution of this requires the introduction of a gauge field analog to (8.11), the *charge* gauge field

$$\tilde{F}_{ab}^E = e\, \tilde{\delta}_{ab}(x; S'). \tag{8.62}$$

Then (8.61) is solved by

$$f_{ab} \equiv \frac{1}{2} \epsilon_{abcd} (\tilde{F}_{ab} - \tilde{F}_{ab}^E). \tag{8.63}$$

The identity (8.15) ensures (8.61).

When inserting (8.63) into (8.60), we find

$$\tilde{\mathcal{A}}_{\mathrm{tot}} = \int d^4 x \left[\frac{1}{4c} (\tilde{F}_{ab} - \tilde{F}_{ab}^E)^2 - \frac{i}{4c} \tilde{F}_{ab} \, \epsilon_{abcd} \tilde{F}_{cd}^M + \frac{i}{4c} \tilde{F}_{ab}^E \, \epsilon_{abcd} \tilde{F}_{cd}^M \right]. \tag{8.64}$$

Integrating the second term by parts and using Eq. (8.16) we obtain

$$\tilde{\mathcal{A}}_{\mathrm{tot}} = \int d^4 x \left[\frac{1}{4c} (\tilde{F}_{ab} - \tilde{F}_{ab}^E)^2 + \frac{i}{c^2} \tilde{A}_a \tilde{\jmath}_a \right] + \Delta \mathcal{A}, \tag{8.65}$$

where

$$\Delta\mathcal{A} = \frac{i}{4c} \int d^4x \; \tilde{F}^E_{ab} \, \epsilon_{abcd} F^M_{cd}. \tag{8.66}$$

Remembering Eqs. (8.11) and (8.62), this can be shown to be an integer multiple of eg/c:

$$\Delta\mathcal{A} = eg \frac{i}{4c} \int d^4x \; \tilde{\delta}_{ab}(x; S) \, \epsilon_{abcd} \tilde{\delta}_{cd}(x; S') = i \frac{eg}{c} \, n, \quad n = \text{integer}. \tag{8.67}$$

To see this we simply choose the surface S to be the 12-plane and S' to be the 34-plane. Then $\tilde{\delta}_{12}(x; S) = -\tilde{\delta}_{21}(x; S) = \delta(x_1)\delta(x_2)$ and $\tilde{\delta}_{34}(x; S') = -\tilde{\delta}_{43}(x; S') = \delta(x_3)\delta(x_4)$, and all other components vanish, so that $\int d^4x \, \epsilon_{abcd}\tilde{\delta}_{ab}(x; S)\tilde{\delta}_{ab}(x; S') = 4 \int d^4x \, \delta(x_1)\delta(x_2)\delta(x_3)\delta(x_4) = 4$. This proof can easily be generalized to arbitrary S, S' configurations.

We now impose Dirac's quantization condition (8.39) to guarantee the invariance under monopole gauge transformations (8.25), thereby ensuring the invariance under string deformations (8.17). This makes the phase factor $e^{-\Delta\mathcal{A}/\hbar}$ equal to unity, so that it has no influence upon any quantum process.

The dual version of the total action (8.35) of monopoles and charges is therefore

$$\tilde{\mathcal{A}}_{\text{tot}} \equiv \tilde{\mathcal{A}}_0 + \tilde{\mathcal{A}}_{\text{int}} + \tilde{\mathcal{A}}_{\text{mg}} = \int d^4x \left[\frac{1}{4c} (\tilde{F}_{ab} - \tilde{F}^E_{ab})^2 + \frac{i}{c^2} \tilde{A}_a \tilde{j}_a \right]. \tag{8.68}$$

It describes the same physics as the action (8.35). Here the magnetic monopole is coupled locally whereas the world line of the charged particle is represented by the charge gauge field (8.62). With the predominance of electric charges in nature, however, this dual action is only of academic interest.

Note that the dual magnetoelectric action (8.68) has the same double-gauge invariance as the electromagnetic action with monopoles in Eq. (8.35). With the Dirac quantization of the charge, it is invariant under the magnetoelectric gauge transformations (8.58) and the deformations of the surface S monopole gauge transformations (8.25).

8.5 Monopole Gauge Fixing

First we should eliminate the superfluous monopole gauge transformation (8.29) with the special gauge functions $\Lambda^M_a = g\partial_a \sum_{V_4} \delta(x; V_4)$ which do not give any change in F^M_{ab}. They may be removed from Λ^M_a by a gauge-fixing condition such as

$$n_a \Lambda^M_a \equiv 0, \tag{8.69}$$

where n_a is an arbitrary fixed unit vector.

The remaining monopole gauge freedom can be used to bring all Dirac strings to a standard shape so that $F^M_{ab}(x)$ becomes only a function of the boundary lines L. In fact, for any choice of the above unit vector n_a, we may always go to the *axial monopole gauge* defined by

$$n_a F^M_{ab} = 0. \tag{8.70}$$

To see this we take n_a along the 4-axis and consider the gauge fixing equations

$$F_{4i} + \partial_4 \Lambda_i^M - \partial_i \Lambda_4^M = 0, \quad i = 1, 2, 3. \tag{8.71}$$

Equation (8.69) requires that $\Lambda_4^M \equiv 0$. The spatial components Λ_i^M can be determined from (8.71). If they were ordinary real functions, this would be trivial. The fact that they are superpositions of δ-functions makes the proof more subtle, but still possible, so that the axial gauge (8.70) can indeed be reached.

This is most easily shown by proceeding as in Subsection 5.1.4. We approximate four-space by a fine-grained hypercubic lattice of spacing a, and imagining the $d(d-1)/2$ components of F_{ab}^M to be functions defined on the plaquettes. Then the above δ-functions (8.28), (8.26), (8.5), and (8.13) correspond to integer-valued functions on sites $\delta(x; V_4) \hat{=} N(x)$, on links $\delta_a(x; V) \hat{=} N_a/a$, on plaquettes $\delta_{ab}(x; S) \hat{=} N_{ab}/a^2$, and on links $\delta_a(x; L) \hat{=} N_a/a^3$, respectively. The derivatives ∂_a correspond to lattice derivatives ∇_a across links as defined in Eq. (5.38). Thus F_{ab}^M can be written as $g N_{ab}(x)/a^2$ with integer $N_{ab}(x)$. The gauge fixing in (8.71) with the restricted gauge functions amounts then to solving a set of integer-valued equations of the type

$$N_{4i} + \nabla_4 N_i^M - \nabla_i N_4 = 0, \quad i = 1, 2, 3, \tag{8.72}$$

with $N_4 \equiv 0$. This is always possible as has been shown with similar equations in Ref. [10].

Having fixed the gauge in this way we can solve Eq. (8.1) uniquely by the monopole gauge field

$$F_{ab}^M = -2\epsilon_{abcd} n_c (n\partial)^{-1} \tilde{j}_d. \tag{8.73}$$

With this, the interaction between electric and magnetic currents in the last term of (8.42) becomes

$$\mathcal{A}_{j\tilde{j}} = \epsilon_{abcd} \int d^4x \, j_a (n\partial \, \partial^2)^{-1} n_b \partial_c \tilde{j}_d. \tag{8.74}$$

This interaction can be found in textbooks [11].

8.6   Quantum Field Theory of Spinless Electric Charges

The full Euclidean quantum field theory of electrically and magnetically charged particles is obtained from the functional integral over the Boltzmann factors $e^{-\mathcal{A}_{\text{tot}}/\hbar}$ with the action (8.35). The functional integral has to be performed over all electromagnetic fields A_a and over all electric and magnetic world line configurations L and L'. These, in turn, can be replaced by fluctuating disorder fields which account for grand-canonical ensembles of world lines [12]. This replacement, the Euclidean analog of *second quantization*, in many-body quantum mechanics, was explained in the last chapter.

Let us assume that only a few fixed worldlines L of monopoles are present. The electric world lines, on the other hand will be assumed to consist of a few fixed world lines L' plus a fluctuating grand-canonical ensemble of closed world lines L''. The latter are converted into a single complex disorder field ψ_e whose Feynman diagrams are pictures of the lines L'' [13]. The technique was explained in Subsection 5.1.10 and corresponds to the second quantization of many-body quantum mechanics. In other words, we shall study the partition function of the *disorder field theory*

$$Z = \int \mathcal{D}A_a^T e^{-A_{tot}} \int \mathcal{D}\psi_e \int \mathcal{D}\psi_e^* \, e^{-A_{\psi_e}}, \tag{8.75}$$

with the field action of the fluctuating electric orbits

$$A_{\psi_e} = \int d^4x \, \frac{1}{2}\left[|D\psi_e|^2 + m^2|\psi_e|^2 + \lambda|\psi_e|^4\right], \tag{8.76}$$

where D_a denotes the covariant derivative involving the gauge field A_a:

$$D_a \equiv \partial_a - i\frac{e}{c}A_a. \tag{8.77}$$

When performing a perturbation expansion of this functional integral in powers of the coupling constant e, the Feynman loop diagrams of the ψ_e field provide direct pictures of the different ways in which the fluctuating closed charged worldlines interact in the ensemble.

8.7 Theory of Magnetic Charge Confinement

The field action of fluctuating electric charges is the four-dimensional version of the Ginzburg-Landau Hamiltonian (5.148). We have learned in the previous chapter that this Hamiltonian allows for a phase transition as a function of the mass parameter m^2 in (8.76). There exists a critical value of m^2 where the system changes from an ordered to a disordered phase. At the mean-field level, the critical value is zero. For $m^2 > 0$, only a few vortex loops are excited. In this phase, the field has a vanishing expectation value $\langle\psi_e\rangle$. For $e < e_c$, on the other hand, the configurational entropy wins over the Boltzmann suppression and infinitely long vortex are thermally created. The mass parameter m^2 becomes negative and the disorder field ψ_e develops a nonzero expectation value $\langle\psi_e\rangle$ whose absolute value is equal to $\sqrt{-m^2/2\lambda}$. This is a condensed phase where the charge worldlines are infinitely long and prolific. The passage of e through e_c is a *phase transition*. From the derivative term $|D\psi_e|^2$, the vector field A_a receives a mass term $(m_A^2/2c)A_a^2$ with m_A^2 equal to $q^2c|m^2|/\lambda$. For very small $e \ll e_c$, the penetration depth $1/m_A$ of the vector potential is much larger than the coherence length $1/m$ of the disorder field, and the system behaves like a superconductor of type II.

Between magnetic monopoles of opposite sign, the magnetic field lines are squeezed into the four-dimensional analogs of the Abrikosov flux tubes. Within the present functional integral, the initially irrelevant surfaces S enclosed by the charge worldlines L acquire, via the phase transition, an energy proportional to their area which removes the charge gauge invariance of the action. They become physical fluctuating objects and generate the linearly rising static potential between the charges, thus causing magnetic charge confinement.

The confinement mechanism is most simply described in the hydrodynamic or London limit. In this limit, the magnitude $|\psi_e|$ of the field is frozen so it can be replaced by a constant $|\psi_e|$ multiplied by a spacetime-dependent phase factor $e^{i\theta(x)}$. The functional integral over ψ_e and ψ_e^* in (8.75) reduces therefore to

$$\sum_{\{V\}} \int_{-\infty}^{\infty} \mathcal{D}\theta \exp\left\{-\frac{m_A^2 c}{2q^2} \int d^4x \left[\partial_a\theta - \theta_a^v(x) - \frac{e}{c}A_a\right]^2\right\}, \tag{8.78}$$

where $\theta_a^v(x)$ is the four-dimensional vortex gauge field

$$\theta_a^v(x) \equiv 2\pi\delta_a(x; V). \tag{8.79}$$

This may be chosen in a specific gauge, for instance in the axial gauge with $\delta_4(x; V) = 0$, so that V is uniquely fixed by its surface S, the worldsheet of a vortex line in the ϕ_e-field. Thus the action in (8.75) reads, in the hydrodynamic limit,

$$\mathcal{A}^{\text{hy}} = \int d^4x \left\{\frac{1}{4c}(F_{ab} - F_{ab}^M)^2 + \frac{i}{c^2}\int d^4x\, j_a(x)A_a(x) + \frac{m_A^2 c}{2q^2}\left[\partial_a\theta - \theta_a^v(x) - \frac{e}{c}A_a\right]^2\right\}. \tag{8.80}$$

If we ignore the vortices and eliminate the θ-fluctuations from the functional integral, we generate a transverse mass term

$$\frac{m_A^2}{2c}A_a^{T2} \tag{8.81}$$

where $A_a^T \equiv (g_{ab} - \partial_a\partial_b/\partial^2)A_b$. This causes the celebrated Meissner effect in the superconductor. The action becomes therefore

$$A^{\text{hy}} = \int d^4x \left[\frac{1}{4c}(F_{ab} - F_{ab}^M)^2 + \frac{m_A^2}{2c}A_a^{T2}\right]. \tag{8.82}$$

If we now integrate out the A_a-fields in the partition function (8.75), we obtain the interaction between the worldlines of electric charges L and the surfaces S whose boundaries are the monopole worldlines:

$$A_{\text{int}}^{\text{hy}} = \int d^4x \left\{\frac{1}{4c}\left[(F_{ab}^M)^2 - 2\partial_a F_{ab}^M(-\partial^2 + m_A^2)^{-1}\partial_c F_{cb}^M\right]\right.$$
$$\left. + \frac{1}{2c^3}\partial_a F_{ab}^M(-\partial^2 + m_A^2)^{-1}j_b + \frac{1}{2c^3}j_a(-\partial^2 + m_A^2)^{-1}j_a\right\}. \tag{8.83}$$

This is a generalization of the previous current-current interaction (8.42), to which it reduces for $m_A = 0$. Applying the right-hand part of relation (8.48) to the first two terms of (8.83), and introducing the massive correlation function

$$(-\partial^2 + m_A^2)^{-1}(x, x') = G_{m_A}(x - x') \equiv \int \frac{d^4k}{(2\pi)^4}e^{-ik(x-x')}\frac{1}{k^2 + m_A^2}, \tag{8.84}$$

we obtain

$$\mathcal{A}_{\text{int}}^{\text{hy}} = \int d^4x \int d^4x' \left[\frac{m_A^2}{16\pi} F_{ab}^M(x) G_{m_A}(x-x') F_{ab}^M(x') \right.$$
$$\left. + \frac{1}{2} \partial_a F_{ab}^M G_{m_A}(x-x') j_a(x-x') + \frac{4\pi}{2} j_a(x) G_{m_A}(x-x') j_a(x') \right]. \quad (8.85)$$

The presence of the mass m_A gives the interaction between the electric charges in the last term of (8.85) a short range of the Yukawa type. The second term is a short-range interaction between the surfaces and the boundary lines.

The first term is most interesting. It gives an energy to the previously irrelevant surfaces S enclosed by the magnetic worldlines L. The energy covers S and a neighborhood of it up to a distance $1/m_A$. It converts S into a worldsheet of thickness $1/m_A$. This is the world surface of a flux tube connecting the magnetic charges. To leading order in the thickness, this causes a surface tension, giving rise to a linearly rising potential between magnetic charges, and thus to confinement. To next order, it causes a curvature stiffness [14].

The fact that the energy of the surface S enclosed by a monopole world line causes confinement can be phrased as a criterion for confinement due to Wilson. In the duality transformation of the monopole part of the action (8.35) to (8.56) we have observed that a surface S in the monopole gauge field F_{ab}^M corresponds to a local coupling $(i/c) \int d^4x \, \tilde{A}_a \tilde{j}_a$ in the dual action. This implies that the expectation value of the exponential,

$$\left\langle \exp\left(\frac{i}{c} \oint_L d^4x \, \tilde{A}_a \tilde{j}_a \right) \right\rangle, \quad (8.86)$$

falls off like $\exp\left(-\text{area enclosed by } L\right)$ in the confined phase where the interaction is given by (8.85), but only like $\exp\left(-\text{length of } L\right)$ in the unconfined phase where the interaction is given by (8.42).

If the charged particles are electrons, the field $\psi_e(x)$ must consist of four anti-commuting Grassmann components and the action must be of the Dirac type which has the form in Minkowski spacetime:

$$\mathcal{A}_e^{\text{Dirac}} = \int d^4x \left\{ \bar{\psi}_e(x) \left[\gamma^a \left(i\hbar \partial_a - \frac{e}{c} A_a \right) \psi_e(x) - m_e c^2 \bar{\psi}_e(x) \psi_e(x) \right] \right\}, \quad (8.87)$$

where m_e is the mass of the electron and $\psi_e(x)$ the standard Dirac field. Fermi fields cannot form a condensate, so that there is no confinement of monopoles. The second quantization of this theory leads to the standard quantum field theory of electromagnetism (QED) with the minimal electromagnetic interaction:

$$\mathcal{A}_{\text{int}} = \frac{i}{c^2} \int d^4x \, A_a j_a, \qquad j_a = e c \bar{\psi} \gamma_a \psi. \quad (8.88)$$

8.8   Second Quantization of the Monopole Field

For monopoles described by the action (8.76), second quantization seems at first impossible since the partition function contains sum over a grand-canonical ensemble of surfaces S rather than lines. Up to now, there exists no satisfactory second-quantized field theory which could replace such a sum. According to present belief, the vacuum fluctuations of some nonabelian gauge theory is able to do this, but a convincing theoretical formulation is still missing.

Fortunately, the monopole gauge invariance of the action (8.35) under (8.25) makes most configurations of the surfaces S' physically irrelevant and allows us to return to a worldline description of the monopoles after all. We simply fix the gauge as described above, which makes the monopole gauge field unique. It is given by Eq. (8.73) and thus completely specified by the worldlines L' of the monopoles. With this we can rewrite the action as [15]

$$
\begin{aligned}
\mathcal{A} &= \mathcal{A}'_1 + \mathcal{A}_{\text{int}} + \mathcal{A}_{\lambda 1} + \mathcal{A}_{\lambda 2} \\
&= \int d^4x \left\{ \left[\frac{1}{4c}(F_{ab} - f^M_{ab})^2 + \frac{i}{c^2} A_a j_a \right] + \frac{i}{c^2}\lambda_{ab}\left(n_\sigma \partial_\sigma f^M_{ab} + 2\epsilon_{abcd} n_c \tilde{j}_d \right) \right\},
\end{aligned}
\tag{8.89}
$$

where f^M_{ab} and λ_{ab} are now two *arbitrary* fluctuating fields, i.e., f^M_{ab} is no longer of the restricted δ-function type implied by (8.11). This form restriction is enforced by the fluctuating λ_{ab}-field. The two terms in the action containing λ_{ab} have been denoted by $\mathcal{A}$ and $\mathcal{A}_{\lambda 2}$. The monopole worldline appears only in the magnetic current coupling

$$
\mathcal{A}_{\text{mg}} \equiv \mathcal{A}_{\lambda 2} = \frac{i}{c^2}\int d^4x\, \tilde{A}^n_d\, \tilde{j}_d,
\tag{8.90}
$$

where $\tilde{A}^n_a$ is short for the vector field

$$
\tilde{A}^n_d \equiv 2\lambda_{ab}\epsilon_{abcd} n_c.
\tag{8.91}
$$

In the partition function associated with this action we may now sum over a grand-canonical ensemble of monopole worldlines L by converting it into a functional integral over a single fluctuating monopole field ϕ_g as in the derivation of (8.75). If monopoles carry no spin, this obviously replaces the sum over all fluctuating monopole world lines with the magnetic interaction (8.90) by a functional integral

$$
\int \mathcal{D}\phi_g \mathcal{D}\phi_g^*\, e^{-\mathcal{A}^n_g},
\tag{8.92}
$$

where $\mathcal{A}_g$ is the action of the complex monopole field

$$
\mathcal{A}^n_g = \int d^4x\, \frac{1}{2}\left[|D^n_a \phi_g|^2 + m^2_g|\phi_g|^2 + \lambda|\phi_g|^4 \right],
\tag{8.93}
$$

and D^n_a the covariant derivative

$$
D^n_a \equiv \partial_a - \frac{g}{c}\tilde{A}^n_a.
\tag{8.94}
$$

By allowing all fields ψ_e, ϕ_g, A_a, f^M_{ab}, and λ_{ab}, to fluctuate with a Euclidean amplitude $e^{-\mathcal{A}/\hbar}$ we obtain the desired quantum field theory of electric charges and Dirac monopoles [16]. The total field action of charged spin-1/2 particles and spin-zero monopoles is therefore

$$\mathcal{A} = \int d^4x \left[\frac{1}{4c}(F_{ab} - f_{ab}^M)^2 + \frac{i}{c^2}\lambda_{ab}n_\sigma\partial_\sigma f_{ab}^M \right] + \mathcal{A}_e^{\text{Dirac}} + \mathcal{A}_g^n. \tag{8.95}$$

Note that the effect of monopole gauge invariance is much more dramatic than that of the ordinary gauge invariance in pure QED. The electromagnetic gauge transformation $A_a \rightarrow A_a + \partial_a\Lambda$ eliminated only the longitudinal polarization of the photons. The monopole gauge transformations, in contrast, (8.25) reduce the *dimensionality* of the fluctuations from surfaces S to lines L, which is crucial for setting up the disorder field theory (8.92).

It is obvious that there exists a dual formulation of this theory with the action

$$\mathcal{A} = \int d^4x \left[\frac{1}{4c}(\tilde{F}_{ab} - \tilde{f}_{ab}^E)^2 + \frac{i}{c^2}\lambda_{ab}n_\sigma\partial_\sigma \tilde{f}_{ab}^E \right] + \mathcal{A}_g + \mathcal{A}_e^{\text{Diracn}}, \tag{8.96}$$

where $\mathcal{A}_g$ is the action (8.92) with the covariant derivative

$$D_a \equiv \partial_a - \frac{g}{c}\tilde{A}_a, \tag{8.97}$$

and $\mathcal{A}_e^{\text{Diracn}}$ is the Dirac action coupled minimally to the vector potential (8.91):

$$\mathcal{A}_e^{\text{Diracn}} = \int d^4x \left\{ \bar{\psi}_e(x) \left[\gamma^a \left(i\hbar\partial_a - \frac{e}{c}A_a^n \right) \psi_e(x) - m_e c^2 \bar{\psi}_e(x)\psi_e(x) \right] \right\}. \tag{8.98}$$

8.9 Quantum Field Theory of Electric Charge Confinement

It has long been known that quantum electrodynamics on a lattice with a cyclic vector potential (called *compact* QED) shows quark confinement for a sufficiently strong electric charge e. The system contains a grand-canonical ensemble of magnetic monopoles which condense at some critical value e_c. The condensate squeezes the electric field lines emerging from any charge into a thin tube giving rise to a confining potential [17, 18, 19]. In three spacetime dimensions, confinement is permanent. This is due to the magnetic version of the Debye screening which always generates a mass term in the dual vector potential $\tilde{A}_a$, thus causing a physical flux tube between electric charges.

It is possible to transform the partition function to the dual version of a standard Higgs model coupled minimally to the dual vector potential $\tilde{A}_a$ [20]. The Higgs field is the *disorder field* [10] of the magnetic monopoles, i.e., its Feynman graphs are the direct pictures of the monopole worldlines in the ensemble. Two electric charges in this model are connected by Abrikosov vortices producing the linearly rising potential between the charges, thus leading to confinement. The system is a perfect dielectric. While there is no problem in taking the dual Higgs field description of quark confinement to the continuum limit [20], the same thing has apparently never been done in the original formulation in terms of the gauge field A_a. The reason was a lack of an adequate continuum description of the integer-valued jumps in the electromagnetic gauge field A_a across the worldsheets spanned by the worldlines of the magnetic monopoles. After the development of the previous sections we can easily construct a simple quantum field theory which exhibits electric charge confinement. It is based on a slight modification of the dual magnetoelectric action (8.68). The modification gives rise to the formation of thin electric flux tubes between opposite electric charges.

For a fixed set of electric and magnetic charges, the Euclidean action reads [recall (8.35)]

$$\mathcal{A}_{\text{tot}} = \frac{1}{4c} \int d^4x \left[F_{ab}(x) - F^M_{ab}(x) \right]^2 + \frac{i}{c^2} \int d^4x\, j_a(x) A_a(x), \qquad (8.99)$$

where $F_{ab} = \partial_a A_b - \partial_b A_a$ is the electromagnetic field tensor,

$$j_a(x) \equiv e\, \delta_a(x; L) \qquad (8.100)$$

the electric current density (8.10), and

$$F^M_{ab}(x) = \tilde{\delta}_{ab}(x; S) \qquad (8.101)$$

the *gauge field of monopoles* (8.11).

In Section 8.4 we have derived the completely equivalent dual action [see Eq. (8.68)]

$$\tilde{\mathcal{A}}_{\text{tot}} = \frac{1}{4c} \int d^4x \left[\tilde{F}_{ab}(x) - \tilde{F}^E_{ab}(x) \right]^2 + \frac{i}{c^2} \int d^4x\, \tilde{j}_a(x) \tilde{A}_a(x) \qquad (8.102)$$

where $\tilde{F}_{ab} = \partial_a \tilde{A}_b - \partial_b \tilde{A}_a$ is the dual field tensor $\tilde{F}_{ab} \equiv (1/2)\epsilon_{abcd} F_{ab}$ and

$$\tilde{j}_a(x) = g\, \delta_a(x; \tilde{L}) \qquad (8.103)$$

the dual current density (8.6) associated with the magnetic monopole worldlines $\tilde{L}$. Now the electric charges are described by a *charge gauge field* $\tilde{F}^E_{ab}$ of Eq. (8.62), that is singular on some worldsheets S enclosed by the electric worldlines L:

$$\tilde{F}^E_{ab}(x) = e\, \tilde{\delta}_{ab}(x; S). \qquad (8.104)$$

The Euclidean quantum partition function of the system is found by summing, in a functional integral, the Boltzmann factor $e^{-\tilde{\mathcal{A}}_{\text{tot}}}$ over all field configurations $\tilde{A}_a$, all line configurations $\tilde{L}$ in $\tilde{j}^a$, and all surface configurations S in $\tilde{F}^E_{ab}$. This action (8.102) is invariant under the *magnetoelectric gauge transformations*

$$\tilde{A}_a \to \tilde{A}_a + \partial_a \tilde{\Lambda}, \qquad (8.105)$$

and under the discrete-valued *charge gauge transformations*

$$\begin{aligned} \tilde{A}_a &\to \tilde{A}_a + \tilde{\Lambda}^E_a, \\ \tilde{F}^E_{ab} &\to \tilde{F}^E_{ab} + \partial_a \tilde{\Lambda}^E_b - \partial_b \tilde{\Lambda}^E_a. \end{aligned} \qquad (8.106)$$

We proceed as in the derivation of the disorder theory (8.75), but now in the dual form. The resulting second-quantized action is

$$\tilde{\mathcal{A}}_{\mathrm{tot}} = \int d^4x \frac{1}{4c}(\tilde{F}_{ab} - \tilde{F}^E_{ab})^2 + \int d^4x \left[|\tilde{D}\phi_g|^2 + m^2|\phi_g|^2 + \lambda|\phi_g|^4 \right], \qquad (8.107)$$

where

$$\tilde{D}_a \equiv \partial_a - \frac{g}{c}\tilde{A}_a. \qquad (8.108)$$

If we choose the mass parameter m^2 of the monopole field ϕ_g to be negative, then ϕ_g acquires a nonzero expectation value $(-m^2/2\lambda)^{1/2}$, which generates a Meissner mass $m^2_{\tilde{A}}$ for the dual vector potential $\tilde{A}_a$. The energy has again the form (8.85), but with electric and magnetic sources exchanged:

$$\mathcal{A}^{\mathrm{hy}}_{\mathrm{int}} = \int d^4x \int d^4x' \left[\frac{m^2_{\tilde{A}}}{16\pi} F^E_{ab}(x) G_{m_{\tilde{A}}}(x - x') F^E_{ab}(x') \right.$$
$$\left. + \frac{1}{2}\partial_a F^E_{ab} G_{m_{\tilde{A}}}(x - x')\tilde{j}_a(x - x') + \frac{4\pi}{2}\tilde{j}_a(x)G_{m_{\tilde{A}}}(x - x')\tilde{j}_a(x') \right]. (8.109)$$

This gives the surfaces S' enclosed by the electric world lines an energy with the properties discussed above, causing now the confinement of electric charges. The surface has tension and, due to its finite thickness $1/m_{\tilde{A}}$, a nonzero curvature stiffness. The consequences of this for world sheets of hadronic strings have been calculated independently by Polyakov [22] and the author [21].

In the case of electric charge confinement, the expectation value of the dual of the Wilson loop (8.86)

$$\left\langle \exp\left(\frac{i}{c}\oint_L d^4x\, A_a \tilde{j}_a \right) \right\rangle \qquad (8.110)$$

behaves like $\exp(-\text{area enclosed by } L)$.

It goes without saying that in order to apply the model to quarks, the action (8.87) has to be replaced by a Dirac action with three colors and six flavors in a gauge-invariant coupling

$$\mathcal{A}_D = \int d^4x\, \bar{\psi}(\slashed{D} - \mathcal{M})\psi, \qquad (8.111)$$

where $D_\mu + iG_\mu$ is a covariant derivative in color space, and G_μ a traceless 3×3-matrix color-electric gauge field with the field action

$$\mathcal{A}_{G_\mu} = -\frac{1}{4}\int d^4x\, \mathrm{tr}\left(\partial_\mu G_\nu - \partial_\nu G_\mu - [G_\mu, G_\nu] \right)^2. \qquad (8.112)$$

The symbol $\mathcal{M}$ denotes a mass matrix in the six-dimensional flavor space of u, d, c, s, t, b.

If one applies the above model interaction (8.109) to quarks, one may study low-energy phenomena by approximating it roughly by a four-Fermi interaction. This can be converted into a chirally invariant effective action for pseudoscalar, scalar, vector, and axial-vector mesons by functional integral technique (*hadronization*) [23]. The effective action reproduces qualitatively many of the low-energy properties of these particles, in particular their chiral symmetry, its spontaneous breakdown, and the difference between the observed quark masses and the masses in the action (8.111) (*current quark masses*). It also explains why the quarks u, d in a nucleon have approximately a third of a nucleon mass while their masses $\mathcal{M}_u$, $\mathcal{M}_d$ in the action (8.111) are very small.[2]

The technique of hadronization developed in [23] has been generalized in various ways, in particular by including the color degree of freedom [24]. It has also been used to describe the low-lying baryons and the restoration of chiral symmetry by thermal effects [25].

An interesting aspect of (8.85) is that the local part of the four-Fermi interaction, which is proportional to $1/\tilde{m}_A^2$, arises by the same mechanism as the confining potential, whose tension is proportional to $\tilde{m}_A^2 \log(\Lambda^2/\tilde{m}_A^2)$, with Λ being some ultraviolet cutoff parameter. One would therefore predict that at an increased temperature of the order of $\tilde{m}_A$ the spontaneous symmetry breakdown, which is caused by the four-Fermi interaction, takes place at the same temperature at which the potential looses its deconfinement properties. This initially surprising coincidence has long been observed in Monte Carlo simulations of lattice gauge theories.

It is an important open problem to generalize the above hydrodynamic discussion to the case of colored gluons. In particular, the existence of three- and four-string vertices must be accounted for in a simple way. A promising intermediate solution was suggested by 't Hooft's [26] hypothesis of dominance of abelian monopoles [27].

Notes and References

[1] Magnetic monopoles were first introduced by
P.A.M. Dirac, Proc. Roy. Soc. A **133**, 60 (1931); Phys. Rev. **74**, 817 (1948).

[2] M.N. Saha, Ind. J. Phys. **10**, 145 (1936);
J. Schwinger, *Particles, Sources, and Fields*, Vols. 1 and 2, Addison Wesley, Reading, Mass., 1970 and 1973;
G. Wentzel, Progr. Theor. Phys. Suppl. **37**, 163 (1966);
E. Amaldi, in *Old and New Problems in Elementary Particles*, ed. by G. Puppi, Academic Press, New York (1968);
D. Villaroel, Phys. Rev. D **14**, 3350 (1972);
Y.D. Usachev, Sov. J. Particles Nuclei **4**, 92 (1973);
A.O. Barut, J. Phys. A **11**, 2037 (1978);
J.D. Jackson, *Classical Electrodynamics*, John Wiley and Sons, New York, 1975, Sects. 6.12-6.13.

---

[2]The quark masses in the action (8.111) are $\mathcal{M}_u \approx 4\,\text{MeV}$, $\mathcal{M}_d \approx 8\,\text{MeV}$, $\mathcal{M}_c \approx 1.5\,\text{GeV}$, $\mathcal{M}_s \approx 0.150\,\text{GeV}$, $\mathcal{M}_t \approx 176\,\text{GeV}$, $\mathcal{M}_b \approx 4.7\,\text{GeV}$).

[3] H. Kleinert, *Defect Mediated Phase Transitions in Superfluids, Solids, and Relation to Lattice Gauge Theories*, Lecture presented at the 1983 Cargèse Summer School on Progress in Gauge Theories, publ. in *Gauge Theories*, ed. by G. 't Hooft et al., Plenum Press, New York, 1984, pp 373-401 (`kl/118`), where `kl` is short for the www address `http://www.physik.fu-berlin.de/~kleinert.`;

[4] H. Kleinert, *The Extra Gauge Symmetry of String Deformations in Electromagnetism with Charges and Dirac Monopoles*, Int. J. Mod. Phys. A **7**, 4693 (1992) (`kl/203`).

[5] H. Kleinert, *Double-Gauge Invariance and Local Quantum Field Theory of Charges and Dirac Magnetic Monopoles*, Phys. Lett. B **246**, 127 (1990) (`kl/205`).

[6] H. Kleinert, *Abelian Double-Gauge Invariant Continuous Quantum Field Theory of Electric Charge Confinement*, Phys. Lett. B **293**, 168 (1992) (`kl/211`).

[7] See, most notably, the textbook by Jackson in Ref. [1], p. 258, where it is stated that "a choice of different string positions is equivalent to different choices of (electromagnetic) gauge". See also his Eq. (6.162) and the lines below it. In the unnumbered equation on p. 258 Jackson observes that the physical monopole field is $F_{ab}^{\mathrm{monop}} = F_{ab} - F_{ab}^{M}$. However, he does not notice the independent gauge properties of F_{ab}^{M} and the need to use the action (8.20) rather than (8.21).

[8] Compare the review article by
P. Goddard and D. Olive, Progress in Physics **41**, 1357 (1978),
who use such "general gauge transformations" [their Eq. (2.46)]. They observe that the field tensor $F_{ab}^{\mathrm{obs}} = F_{ab} + (1/a^3 a)\mathbf{\Phi} \cdot (\partial_a \mathbf{\Phi} \times \partial_b \mathbf{\Phi})$ introduced by 't Hooft into his $SU(2)$ gauge theory with Higgs fields $\mathbf{\Phi}$ to describe magnetic monopoles can be brought to the form $F_{ab}^{\mathrm{obs}} \equiv F_{ab} - F_{ab}^{M}$ by a gauge transformation within the SU(2) gauge group. This moves magnetic fields from F_{ab} to F_{ab}^{M} without changing F_{ab}^{obs} [see their Eq. (4.30) and the last two equations in their Section 4.5]. These are *not* permissible gauge transformation of the electromagnetic type, but have a similar effect as our *monopole gauge transformations* of the form (8.27), although with more general transformation functions than (8.26) due to the more general SU(2) symmetry.

[9] J. Schwinger, Phys. Rev. **144**, 1087 (1966).

[10] H. Kleinert, *Gauge Fields in Condensed Matter*, Vol. I, *Superflow and Vortex Lines*, World Scientific, Singapore, 1989 (`kl/b1`).

[11] J. Schwinger, *Particles, Sources and Fields*, Vol. 1, Addison Wesley, Reading, Mass., 1970, p. 235.

[12] H. Kleinert, *Path Integrals in Quantum Mechanics, Statistics, Polymer Physics, and Financial Markets*, 4th ed., World Scientific, Singapore, 2006 (kl/b5).

[13] K. Symanzik, Varenna Lectures 1986, in *Euclidean Quantum Field Theory*, ed. R. Jost, Academic Press, New York, 1969.

[14] Surface tension and stiffness coming from interactions like the first term in (8.85) have been calculated for biomembranes in
H. Kleinert, *Dynamical Generation of String Tension and Stiffness in Strings and Membranes*, Phys. Lett. B **211**, 151 (1988) (kl/177).

[15] See pp. 570–578 in textbook [10] (kl/b1/gifs/v1-570s.html).

[16] The construction of a quantum field theory of monopoles is impossible in the theory of "exorcized" Dirac strings proposed by
T.T. Wu and C.N. Yang, Phys. Rev. D **14**, 437 (1976); Phys. Rev. D **12**, 3845, (1075); D **16**, 1018 (1977); Nuclear Physics B **107**, 365 (1976);
C.N. Yang, Lectures presented at the 1976 Erice summer school, in *Gauge Interactions*, Plenum Press, New York 1978, ed. by A. Zichichi and at the 1982 Erice summer school, in *Gauge Interactions*, Plenum Press, New York 1984, ed. by A. Zichichi.

[17] Y. Nambu, Phys. Rev. D **10**, 4262 (1974).

[18] S. Mandelstam, Phys. Rep. C **23**, 245 (1976); Phys. Rev. D **19**, 2391 (1979).

[19] G. 't Hooft, Nucl. Phys. B **79**, 276 (1974); and in *High Energy Physics*, ed. by A. Zichichi, Editrice Compositori, Bologna, 1976.

[20] H. Kleinert, *Higgs Particles from Pure Gauge Fields*, Lecture presented at the Erice Summer School, 1982, in *Gauge Interactions*, ed. by A. Zichichi, Plenum Press, New York, 1984 (kl/117).

[21] H. Kleinert, Phys. Lett. B **174**, 335 (1986) (kl/149).
This work was inspired by the author's previous results on biomembranes.

[22] A.M. Polyakov, Nucl. Phys. B **268**, 406 (1986); see also his monograph *Gauge Fields and Strings*, Harwood Academic, Chur, 1987.

[23] H. Kleinert, *On the Hadronization of Quark Theories*, Lectures presented at the Erice Summer Institute 1976, *Understanding the Fundamental Constituents of Matter*, Plenum Press, New York, 1978, A. Zichichi (ed.), pp. 289-390 (kl/53). See also *Hadronization of Quark Theories and a Bilocal form of QED*, Phys. Lett. B **62**, 429 (1976).

[24] See for example
K. Rajagopal and F. Wilczek, *The Condensed Matter Physics of QCD*, (hep-ph/0011333);
S.B. Ruster, V. Werth, M. Buballa, I.A. Shovkovy, D.H. Rischke, *Phase Diagram of Neutral Quark Matter at Moderate Densities*, Phys. Rev. D **73**, 034025 (2006) (nucl-th/0602018),
and references therein.

[25] R. Cahill, C.D. Roberts, and J. Praschifka, Aust. J. Phys. **42**, 129 (1989);
R. Cahill, J. Praschifka, and C.J. Burden, Aust. J. Phys. **42**, 161 (1989);
R. Cahill, Aust. J. Phys. **42**, 171 (1989);
H. Reinhardt, Phys. Lett. B **244**, 316 (1990);
V. Christov, E. Ruiz-Arriola, K. Goeke, Phys. Lett. B **243**, 191 (1990);
R. Cahill, Nucl. Phys. A **543**, 63c (1992).

[26] G. 't Hooft, Nucl. Phys. B **190**, 455 (1981).

[27] A.S. Kronfeld, G. Schierholz, and U.-J. Wiese, Nucl. Phys. B **293**, 461 (1987);
A.S. Kronfeld, M.L. Laursen, G. Schierholz, and U.-J. Wiese, Phys. Lett. B **198**, 516 (1987);
F. Brandstaeter, G. Schierholz, and U.-J. Wiese, DESY preprint 91-040 (1991);
T. Suzuki and I. Yotsuyanagi, Phys. Rev. D **42**, 4257 (1991);
S. Hioki, S. Kitahara, S. Kiura, Y. Matsubara, O. Miyamura, S. Ohuo, and T. Suzuki, Phys. Lett. **B** 272, 326 (1991);
J. Smit and A.J. Van der Sijs, Nucl. Phys. B **355**, 603 (1991);
V.G. Bornyakov, E.M. Ilgenfritz, M.L. Laursen, V.K. Mitrijushkin, M. Müller-Preussker, A.J. Van der Sijs, and A.M. Zadorozhyn, Phys. Lett. B **261**, 116 (1991);
J. Greensite and J. Winchester, Phys. Rev. D **40**, 4167 (1989);
J. Greensite and J. Iwasaki, Phys. Lett. B **255**, 415 (1991);
A. Di Giacomo, M. Maggiore, and Š. Olejník, Nucl. Phys. B **347**, 441 (1990);
M. Campostrini, A. Di Giacomo, H. Panagopoulos, and E. Vicari, Nucl. Phys. B **329**, 683 (1990);
M. Campostrini, A. Di Giacomo, M. Maggiore, Š. Olejník, H. Panagopoulos, and E. Vicari, Nucl. Phys. B (Proc. Suppl.) **17**, 563 (1990);
L. Del Debbio, A. Di Giacomo, M. Maggiore, and S. Olejník, Phys. Lett. B **267**, 254 (1991);
L. Polley and U.-J. Wiese, Nucl. Phys. B **356**, 629 (1991);
H.G. Evertz, K. Jansen, J. Jersak, C.B. Lang, and T. Neuhaus, Nucl. Phys. B **285**, 590 (1987);
T.L. Ivanenko, A.V. Pochinsky, and M.I. Polikarpov, Phys. Lett. **B** 302, 458 (1993);
P. Cea and L. Cosmai, Nuovo Cim. **A** 106, 1361 (1993).

Better die than live mechanically a life
that is a repetition of repetitions.
D. H. LAWRENCE (1885–1930)

9

Multivalued Mapping from Ideal Crystals to Crystals with Defects

In the last chapter we have learned how multivalued gauge transformations allow us to transform theories in field-free space into theories coupled to electromagnetism. By analogy, we expect that multivalued coordinate transformations can be used to transform theories in flat space into theories in spaces with curvature and torsion. This is indeed possible. The mathematical methods have been developed in the theory of line-like defects in crystals [1, 2, 3]. Let us briefly review those parts of the theory which will be needed for our purposes.

9.1 Defects

No crystal produced in the laboratory is perfect. It always contains a great number of defects. These may be chemical, electrical, or structural in character, involving foreign atoms. They may be classified according to their space dimensionality. The simplest type of defect is the point defect. It is characterized by the fact that, within a certain finite neighborhood only one cell shows a drastic deviation from the perfect crystal symmetry. The most frequent origin of such point defects is irradiation or an isotropic mechanical deformation under strong shear stresses. There are two types of intrinsic point defects. Either an atom may be missing from its regular lattice site (vacancy) or there may be an excess atom (interstitial) (see Fig. 9.1). Vacancies and interstitials are mobile defects. A vacancy can move if a neighboring atom moves into its place, leaving a vacancy at its own former position. An interstitial atom can move in two ways. It may hop directly from one interstitial site to another. This happens in strongly anisotropic materials such as graphite but also in some cubic materials like Si or Ge. Or it may move in a way more similar to the vacancies by replacing atoms, i.e., by pushing a regular atom out of its place into an interstitial position which, in turn affects the same change on its neighbor, etc.

Intrinsic point defects have the property that the total energy of a bunch of them is smaller than the sum of the individual energies. The reason for this is easily seen. If two vacancies in a simple cubic lattice come to lie side by side, there are only 10 broken valencies compared to 12 when they are separated. If a larger set

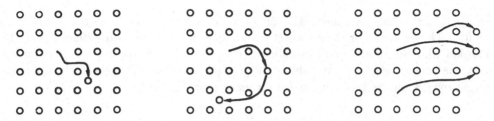

FIGURE 9.1 Intrinsic point defects in a crystal. An atom may become interstitial, leaving behind a vacancy. It may perform random motion via interstitial places until it reaches another vacancy where it recombines. The exterior of the crystal may be seen as a reservoir of vacancies.

of vacancies comes to lie side by side forming an entire disc of missing atoms, the crystal planes can move together and make the disc disappear (see Fig. 9.2). In this way, the crystal structure is repaired. Close to the boundary line, however, such a repair is impossible. The boundary line forms a line-like defect.

Certainly, line-like defects can arise also in an opposite process of clustering of interstitial atoms. If they accumulate side by side forming an interstitial disc, the crystal planes move apart and accommodate the additional atoms in a regular atomic array, again with the exception of the boundary line. Line-like defects of this type are called *dislocation lines*.

It is obvious that a dislocation line may also be the result of several discs of missing or excessive atoms stacked on top of each other. Their boundary forms a dislocation line of higher strength. The energy of such a higher dislocation line increases roughly with the square of the strength. Dislocations are created and set into motion if stresses exceed certain critical values. This is why they were first

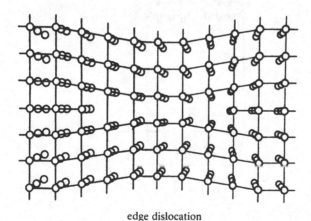

edge dislocation

FIGURE 9.2 Formation of a dislocation line (of the edge type) from a disc of missing atoms. The atoms above and below the missing ones have moved together and repaired the defect, except at the boundary.

seen in plastic deformation experiments of the nineteenth century in the form of slip bands. The grounds for their theoretical understanding were laid much later by Y. Frenkel who postulated the existence of crystalline defects in order to understand why materials yield to plastic shear about a thousand times more easily than one might expect on the basis of a naive estimate (see Fig. 9.3).

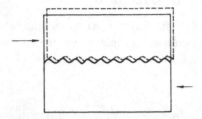

FIGURE 9.3 Naive estimate of maximal stress supported by a crystal under shear stress as indicated by the arrows. The two halves tend to slip against each other.

Assuming a periodic behavior $\sigma = \sigma_{max} \sin(2\pi x/a)$, this reduces to $\sigma \sim \sigma_{max} 2\pi(x/a) \sim \mu(x/a)$. Hence $\sigma_{max} = \mu/2\pi$. Experimentally, however, $\sigma_{max} \sim 10^{-3}\mu$ to $10^{-4}\mu$.

The large discrepancy was explained by Frenkel by noting that the plastic slip would not proceed by the two halves moving against each other as a whole but stepwise, by means of defects. In 1934, Orowan, Polany, and Taylor identified these defects as dislocation lines. The presence of a single moving edge dislocation allows for a plastic shear movement of the one crystal half against the other. The movement proceeds in the same way as that of a caterpillar. This is pictured in Fig. 9.4. One leg is always in the air breaking translational invariance. This is step by step exchanged

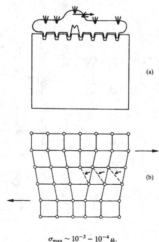

$\sigma_{max} \sim 10^{-3} - 10^{-4}\mu.$

FIGURE 9.4 Dislocation line permitting two crystal pieces to move across each other in the same way as a caterpillar moves across the ground. The bonds can flip direction successively, a rather easy process.

against the one in front of it, etc. In the crystal shown in the lower part of Fig. 9.4, the single leg corresponds to the lattice plane of excess atoms. Under stress along the arrows, this leg moves to the right. After a complete sweep across the crystal, the upper half is shifted against the lower by precisely one lattice spacing.

If many discs of missing or excess atoms come to lie close together, a further cooperative phenomenon can be observed. This is illustrated in Fig. 9.5. On the

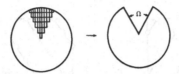

FIGURE 9.5 Formation of a disclination from a stack of layers of missing atoms (cf. Fig. 9.2). Equivalently, one may cut out an entire section of the crystal. In a real crystal, the section has to conform with the symmetry angles. In the continuum approximation, the angel Ω is meant to be very small.

left-hand side, an infinite number of atomic half planes (discs of semi-infinite size) has been removed from an ideal crystal. If the half planes themselves form a regular crystalline array, they can fit smoothly into the original crystal. Only the origin shows a breakdown of crystal symmetry. Everywhere else, the crystal is only slightly distorted. What has been formed is again a line-like defect called a *disclination*. Dislocations and disclinations will play a central role in our further discussion.

Before entering into the detailed discussion, let us complete the dimensional classification of two-dimensional defects. They are of three types.

First, there are *grain boundaries* where two regular lattice parts meet, with the lattice orientations being different on both sides of the interface (see Fig. 9.6). They

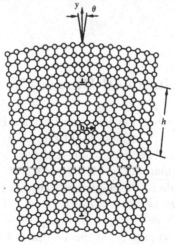

FIGURE 9.6 Grain boundary where two crystal pieces meet with different orientations in such a way that not every atomic layer matches (here only every other one does).

may be considered as arrays of dislocation lines in which half planes of point defects are stacked on top of each other with some spacing, having completely regular lattice planes between them.

The second type of planar defects are *stacking faults*. They contain again completely regular crystal pieces on both sides of the plane, but instead of being oriented differently they are shifted against each other (see Fig. 9.7).

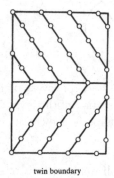
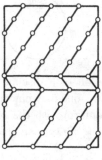

twin boundary   stacking fault

FIGURE 9.7 Two typical stacking faults. The first is called growth-stacking fault or twin boundary, the second deformation-stacking fault.

The third, always unavoidable, type of defects consists of the surfaces of the crystal.

The defects which will play an important role in this textbook are the line-like defects. Their presence will be seen to equip space with a discrete version of a Riemann-Cartan geometry. Its continuous limit can be used as a basis for a theory of gravitation.

9.2   Dislocation Lines and Burgers Vector

Let us first see how a dislocation line can be characterized mathematically. For this we look at Fig. 9.8 in which a closed circuit in the ideal crystal is mapped into the disturbed crystal. The orientation is chosen arbitrarily to be anticlockwise. The prescription for the mapping is that for each step along a lattice direction, a corresponding step is made in the disturbed crystal. If the original lattice sites are denoted by $\mathbf{x}_n$, the image points are given by $\mathbf{x}'_n = \mathbf{x}_n + \mathbf{u}(\mathbf{x}_n)$, where $\mathbf{u}(\mathbf{x}_n)$ is the displacement field: At each step, the image point moves in a slightly different original point. After the original point has completed a closed circuit, call it B, the image point will not have arrived at the point of departure. The image of the closed contour B is no longer closed. This closure failure is given precisely by a lattice

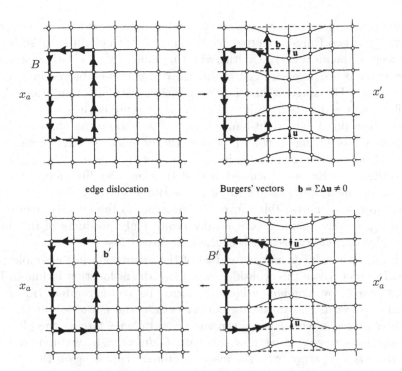

edge dislocation   Burgers' vectors $\mathbf{b} = \Sigma\Delta\mathbf{u} \neq 0$

FIGURE 9.8 Definition of Burgers vector. In the presence of a dislocation line the image of a Burgers circuit B which is closed in the ideal crystal fails to close in the defected crystal. The opposite is also true. The closure failure is measured by a lattice vector, called *Burgers vector* $\mathbf{b}$. The dislocation line in the figure is of the edge type, and the Burgers vector points orthogonally to the line.

vector $\mathbf{b}(\mathbf{x})$ called a *local Burgers vector*, which points from the beginning to the end of the circuit.[1] Thus, the dislocation line is characterized by the equation

$$\sum_B \Delta\mathbf{u}(\mathbf{x}) = \mathbf{b}, \tag{9.1}$$

where $\Delta\mathbf{u}(\mathbf{x}_n)$ are the increments of the displacement vector from step to step. If we consider the same process in the continuum limit, we can write

$$\int_B d\mathbf{u}(\mathbf{x}) = \mathbf{b}. \tag{9.2}$$

The closed circuit B is called *Burgers circuit*.

Equivalently, we can consider a closed circuit in the disturbed crystal, call it B', and find that its counter image in the ideal crystal does not close by a vector $\mathbf{b}'$

---

[1]Our sign convention is the opposite of Bilby *et al.* and agrees with Read's (see Notes and References). Note that in contrast to the local Burgers vector, the true Burgers vector is defined on a perfect lattice.

called the *true Burgers vector* which now points from the end to the beginning of the circuit. The two Burgers vectors are the same if both circuits are so large that they lie deep in the ideal crystal. Otherwise they differ by an elastic distortion.

A few remarks are necessary concerning the convention employed in defining the Burgers vector. The singular line L is in principle without orientation. We may arbitrarily assign a direction to it. The Burgers circuit is then taken to encircle this chosen direction in the right-handed way. If we choose the opposite direction, the Burgers vector changes sign. However, the product $\mathbf{b} \cdot d\mathbf{x}$, where $d\mathbf{x}$ is the infinitesimal tangent vector to L, is invariant under this change. Note that this is similar to the magnetic case discussed in Part II, where the direction of the current was defined by the flow of *positive charge*. The Burgers circuit gives $\oint du = I$. One could, however, also reverse this convention referring to the negative charge. Then $\oint du$ would give $-I$. Again, $I\,d\mathbf{x}$ is an invariant. Only products of this invariant appear in physical observables such as the Biot-Savart law.

The invariance of $\mathbf{b} \cdot d\mathbf{x}$ under reversal of the orientation has a simple physical meaning. In order to see this, consider once more the dislocation in line in Fig. 9.8 which was created by *removing* a layer of atoms. Its vector product $\mathbf{b} \times d\mathbf{x}$ points *inwards*, i.e., towards the vacancies. Consider now the opposite case in which a layer of new atoms is inserted between the crystal planes forcing the planes apart to relax the local stress. If we now calculate $\oint_B du(\mathbf{x}) = \mathbf{b}$, we find that $\mathbf{x} \times d\mathbf{x}$ points *outwards*, i.e., away from the inserted atoms. This is again the direction in which there are fewer atoms. Both statements are independent of the choice of the orientation of the Burgers circuit. Since the second case has extra atoms inside the circle, where the previous one had vacancies, the two can be considered as antidefects of one another. If the boundary lines happen to fall on top of each other, they can annihilate each other and yield back a perfect crystal. This annihilation can happen piece-wise in which case a large dislocation decomposes into several disjoint sections.

In the above examples, the Burgers vectors are everywhere orthogonal to the dislocation line, and one speaks of a pure *edge dislocation* (see Fig. 9.2).

There is no difficulty in constructing another type of dislocation: one cuts a crystal along a lattice half-plane up to some straight line L, and translates one of the lips against the other along the direction of L. In this way one arrives at the so-called *screw dislocation* shown in Fig. 9.9 in which the Burgers vector points parallel to the line L.

When drawing crystals out of a melt, it always contains a certain fraction of dislocations. Even in clean samples, at least one in 10^6 atoms is dislocated. Their boundaries run in all directions through space. We shall see very soon that their Burgers vector is a topological invariant for any closed dislocation loop. Therefore, the character "edge" versus "screw" of a dislocation line is not an invariant. It changes according to the direction of the line with respect to the invariant Burgers vector $\mathbf{b}_i$. It is obvious from the Figs. 9.2 to 9.9 that a dislocation line destroys the translational invariance of the crystal by multiples of the lattice vectors. If there are only a few lines this destruction is not very drastic. Locally, i.e., in any small subspecimen which does not lie too close to the dislocation line, the crystal can still

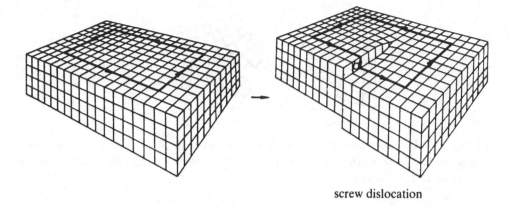

screw dislocation

FIGURE 9.9 Screw disclination which arises when tearing a crystal. The Burgers vector is parallel to the vertical line.

be described by a periodic array of atoms whose order is disturbed only slightly by a smooth displacement field $u_i(\mathbf{x})$.

9.3 Disclination Lines and Frank Vector

Since the crystal is not only invariant under discrete translations but also under certain discrete rotations we expect the existence of another type of defects which is capable of destroying the global rotational order, while maintaining it locally. These are the disclination lines of which an example was given in Fig. 9.5. It arose as a superposition of stacks of layers of missing atoms. In the present context, it is useful to construct it by means of the following Gedanken experiment. Take a regular crystal in the form of a round cheese and remove a section subtending an angle Ω (see Fig. 9.10). The free surfaces can be forced together. For large Ω this requires considerable energy. Still, if the atomic layers on the free surfaces match perfectly, and the crystal re-establishes locally its periodic structure. This happens for all symmetries of the crystal. In a simple cubic crystal, Ω can be 90^0, 180^0, 270^0.

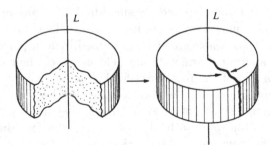

FIGURE 9.10 Volterra cutting and welding process leading to a wedge disclination.

The 90^0 case is displayed in Fig. 9.11. In that figure we have also gone through the

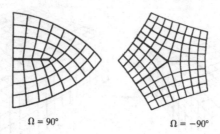

$$\Omega = 90° \qquad\qquad \Omega = -90°$$

FIGURE 9.11 Lattice structure at a wedge disclination in a simple cubic lattice. The Frank angle Ω is equal to the symmetry angles 90^0 or -90^0. The crystal is locally perfect, except close to the disclination line.

opposite procedure of going from the right in Fig. 9.10 to the left. The crystal is cut and the lips are opened by Ω to insert new undistorted crystalline matter matching the atoms in the free surfaces. These are the disclinations of negative angles Ω, here -90^0.

The local crystal structure is destroyed only along the singular line along the axis of the cheese. The rotation which has to be imposed upon the free surfaces in order to force them together may be represented by a rotation vector Ω which, in the present example, points parallel to L and to the cut. This is called a *wedge disclination*. It is not difficult to construct other rotational defects. The three possibilities are shown in Fig. 9.12. Each case is characterized by a vector $\boldsymbol{\Omega}$. In the first case, $\boldsymbol{\Omega}$ points parallel to the line L and the cut. Now, in the second case, $\boldsymbol{\Omega}$ points orthogonal to the line L, and parallel to the cut. This is a *splay disclination*. In the third case, $\boldsymbol{\Omega}$ points orthogonal to the line and cut. This is a *twist disclination*.

The vector $\boldsymbol{\Omega}$ is referred to as the *Frank vector* of the disclination. As in the construction of dislocations, the interface at which the material is joined together does not have any physical reality. For example, in Fig. 9.12a we could have cut out the piece along any other direction which is merely rotated with respect to the first around L by a discrete symmetry irregular piece as long as the faces fit together smoothly (recall Fig. 9.10). Only the singular line is a physical object.

Disclinations were first observed and classified by F.C. Frank in 1958 in the context of liquid crystals. Liquid crystals are mesophases. They are liquids consisting of rod-like molecules. Thus, they cannot be described by a displacement field $u_i(\mathbf{x})$ alone but require an additional orientational field $n_i(\mathbf{x})$ for their description. This orientation is *independent* of the rotational field $\omega_i(\mathbf{x}) = \frac{1}{2}\epsilon_{ijk}\partial_j u_k(\mathbf{x})$. The disclination lines defined by Frank are the rotational defect lines with respect to this *independent* orientational degree of freedom. Thus, they are *a priori* unrelated to the disclination lines in the rotation field $\omega_i(\mathbf{x}) = \frac{1}{2}\epsilon_{ijk}\partial_j u_k(\mathbf{x})$. In fact, a liquid is always filled with dislocations and ω-disclinations, even if the orientation field $n_j(\mathbf{x})$ is completely ordered.

In his book on dislocations, J.P. Friedel called these n_j-disclinations *rotation dislocations* (see the Notes and References at the end of chapter). But later the name disclinations became customary (see Kléman's article cited in Notes and References). In general, there is little danger of confusion, if one knows which system and phase one is talking about.

The Gedanken experiments of cutting a crystal, removing or inserting slices or sections, and joining the free faces smoothly together were first performed by Volterra in 1907. For this reason one speaks of the creation of a defect line as a *Volterra process* and calls the cutting surfaces, where the free faces are joined together, *Volterra surfaces*.

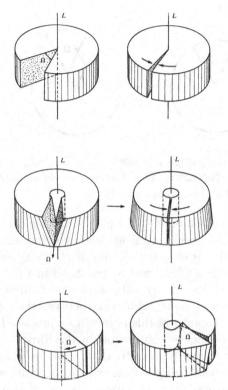

FIGURE 9.12 Three different possibilities of constructing disclinations: (a) wedge, (b) splay, and (c) twist disclinations.

9.4 Interdependence of Dislocation and Disclinations

It must be pointed out that dislocation and disclination lines are not completely independent. We have seen before in Fig. 9.5 that a disclination line was created by removing stacks of atomic layers from a crystal. But each layer can be considered as a dislocation line running along the boundary. Thus a disclination line is apparently indistinguishable from a stack of dislocation lines, placed with equal spacing on top of each other. Conversely, a dislocation line is very similar to a pair of disclination lines running in opposite directions close to each other. This is illustrated in Fig. 9.13. What we have here is a pair of opposite Volterra processes of disclination lines. We have cut out a section of angle Ω, but instead of removing it completely we have displaced it merely by one lattice spacing a. This is equivalent to generating a disclination of the Frank vector Ω and another one with the opposite Frank vector $-\Omega$ whose rotation axis is displaced by a. It is obvious from the figure that the result is a dislocation line with Burgers vector **b**. Due to this interdependence of dislocations and disclinations, the defect lines occurring in a real crystal will, in general, be of a mixed nature.

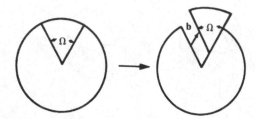

FIGURE 9.13 Generation of dislocation line from a pair of disclination lines running in opposite directions at a fixed distance b. The Volterra process amounts to cutting out a section and reinserting it, but shifted by the amount **b**.

The interdependence discussed here is of a purely topological character. It does not imply that the elastic energy of equivalent defect configurations, such as those in Fig. 9.13, is the same. This is only true for linear elasticity with first gradients of the displacement field. Such a model will be presented in Eq. (10.9).

In a realistic crystal, the elastic energy will always contain higher-gradients of the displacement fields, and these will make a difference between the topologically equivalent defect configurations. A model of this type will be proposed in Eq. (10.29).

Remarkably, the interdependence plays a role in the Einstein-Cartan theory of gravitation. As will be shown in Chapter 21, there exists a reformulation of Einstein's theory of relativity in which all gravitational effects come from torsion rather than curvature. This is the so-called teleparallel theory of gravity. In the next chapter we shall see that curvature and torsion may be viewed as being due to disclinations and dislocations in spacetime, the teleparallel theory is based precisely on the above interdependence, the possibility of obtaining curvature from a combination of torsion fields.

9.5 Defect Lines with Infinitesimal Discontinuities in Continuous Media

The question arises as to how one can properly describe the wide variety of line-like defects which can exist in a crystal. In general, this is a rather difficult task due to the many possible different crystal symmetries. For the sake of gathering some insight it is useful to restrict oneself to continuous isotropic media. Then defects may be created with arbitrarily small Burgers and Frank vectors. Such infinitesimal defects have the great advantage of being accessible to differential analysis. This is essential for a simple treatment of rotational defects. It permits a characterization of disclinations in a way very similar to that of dislocations via a Burgers circuit integral. Consider, for example, the wedge disclination along the line L (shown in Figs. 9.5, 9.10, 9.11 or 9.12a), and form an integral over a closed circuit B enclosing L.

Just as in the case of dislocations this measures the thickness of the material section removed in the Volterra process. Unlike the situation for dislocations, this thickness increases with distance from the line. If Ω is very small, the displacement field across the cut has a discontinuity which can be calculated from an *infinitesimal* rotation

$$\Delta u_i = (\Omega \times \mathbf{x})_i \,, \tag{9.3}$$

where $\mathbf{x}$ is the vector pointing to the place where the integral starts and ends. In order to turn this statement into a circuit integral it is useful to remove the explicit dependence on $\mathbf{x}$ and consider, instead of the displacement field $u_i(\mathbf{x})$, the *local rotation field* accompanying the displacement instead. This is given by the antisymmetric tensor field

$$\omega_{ij}(\mathbf{x}) \equiv \frac{1}{2}\left[\partial_i u_j(\mathbf{x}) - \partial_j u_i(\mathbf{x})\right]. \tag{9.4}$$

The rotational character of this tensor field is obvious when looking at the change of an infinitesimal distance vector under a distortion

$$dx_i' \quad dx_i \;=\; (\partial_j u_i)\, dx_j = u_{ij}\, dx_j - \omega_{ij}\, dx_j. \tag{9.5}$$

The tensor field ω_{ij} is associated with a vector field ω_i as follows:

$$\omega_{ij}(\mathbf{x}) = \epsilon_{ijk}\omega_k(\mathbf{x}), \tag{9.6}$$

i.e.,

$$\omega_{ij}(\mathbf{x}) \equiv \frac{1}{2}\epsilon_{ijk}\omega_{jk}(\mathbf{x}) = \frac{1}{2}\left(\nabla \times \mathbf{u}\right)_i. \tag{9.7}$$

The right-hand side of (9.6) separates the local distortion into a sum of a local change of shape and a local rotation. Now, when looking at the wedge disclination in Fig. 9.12a, we see that due to (9.3), the field $\omega_i(\mathbf{x})$ has a constant discontinuity Ω across the cut. This can be formulated as a circuit integral

$$\Delta \omega_i = \oint_B d\omega_i = \Omega_i. \tag{9.8}$$

The value of this integral is the same for any choice of the circuit B as long as it encloses the disclination line L.

This simple characterization depends essentially on the infinitesimal size of the defect. If Ω were finite, the differential expression (9.3) would not be a rotation and the discontinuity across the cut could not be given in the form (9.8) without specifying the circuit B. The difficulties for finite angles are a consequence of the nonabelian nature of the rotation group.

Only infinitesimal local rotations have additive rotation angles, since the quadratic and higher-order corrections can be neglected.

9.6 Multivaluedness of Displacement Field

The displacement field is a prime example of a *multivalued* field. In a perfect crystal, in which the atoms deviate little from their equilibrium positions $\mathbf{x}$, it is natural to draw the displacement vector from the lattice places $\mathbf{x}$ to the *nearest* atom. In principle, however, the identity of the atoms makes such a specific assignment impossible. Due to thermal fluctuations, the atoms exchange positions from time to time by a process called *self-diffusion*. After a very long time, the displacement vector, even in a regular crystal, will run through the entire lattice. Thus, if we describe a regular crystal initially by very small displacement vectors $u_i(\mathbf{x})$, then, after a very long time, these will have changed to a permutation of lattice vectors, each of them occurring precisely once, plus some small fluctuations around them. Hence the displacement vectors are intrinsically multivalued, with $u_i(\mathbf{x})$ being indistinguishable from $u_i(\mathbf{x}) + aN_i(\mathbf{x})$, where $N_i(\mathbf{x})$ are integer numbers and a is the lattice spacing.

It is interesting to realize that this property puts the displacement fields on the same footing with the phase variable $\gamma(\mathbf{x})$ of superfluid ^{4}He. There the indistinguishability of $\gamma(\mathbf{x})$ and $\gamma(\mathbf{x}) + 2\pi N(\mathbf{x})$ has an entirely different reason: it follows directly from the fact that the physical field is the complex field $\psi(\mathbf{x}) = |\psi(\mathbf{x})|e^{i\gamma(\mathbf{x})}$, which is invariant under the exchange $\gamma(\mathbf{x}) \to \gamma(\mathbf{x}) + 2\pi N(\mathbf{x})$.

Thus, in spite of the different physics described by the variables $\gamma(\mathbf{x}), u_i(\mathbf{x})$, they both share the characteristic multivaluedness. It is just as if the rescaled $u_i(\mathbf{x})$ variables $\gamma_i(\mathbf{x}) = (2\pi/a)u_i(\mathbf{x})$ were phases of three complex fields

$$\psi_i(\mathbf{x}) = |\psi_i(\mathbf{x})|e^{i\gamma_i(\mathbf{x})},$$

which serve to describe the positions of the atoms in a crystal.

In a regular crystal, the multivaluedness of $u_i(\mathbf{x})$ has no important physical consequence. The atoms are strongly localized and the exchange of positions occurs very rarely. The exchange is made irrelevant by the identity of the atoms and symmetry of the many-body wave function. This is why the natural assignment of $u_i(\mathbf{x})$ to the nearest equilibrium position $\mathbf{x}$ presents no problems. As soon as defects are present, however, the full ambiguity of the assignment comes up: When removing a layer of atoms, the result is a dislocation line along the boundary of the layer. Across the layer, the positions $u_i(\mathbf{x})$ jump by a lattice spacing. This means that the atoms on both sides are interpreted as having moved towards each other. Figure 9.14 shows that the same dislocation line could have been constructed by removing a completely different layer of atoms, say S', just as long as it has the same boundary line. Physically, there is no difference. There is only a difference in the *descriptions* which amounts to a difference in the assignment of the equilibrium positions from where to count the displacement field $u_i(\mathbf{x})$. In contrast to regular crystals there now exists no choice of the nearest equilibrium point. It is this multivaluedness which will form the basis for the geometric description of line-like defects in solids.

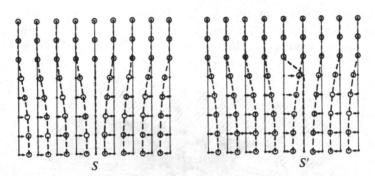

FIGURE 9.14 In the presence of a dislocation line, the displacement field is defined only modulo lattice vectors. The ambiguity is due to the fact that the surface S, from which the atoms have been removed in the Volterra process, is arbitrary as long as the boundary line stays fixed. Shifting S implies shifting the reference positions, from which to count the displacements $u_i(\mathbf{x})$.

9.7 Smoothness Properties of Displacement Field and Weingarten's Theorem

In order to be able to classify a general defect line we must first give a characterization of the smoothness properties of the displacement field away from the singularity. In physical terms, we have to make sure that the crystal matches properly together when cutting and rejoining the free faces.

In the gradient representation of magnetism in Subsection 4.2, the presence of a magnetic field was signalized by a violation of the integrability condition [recall (4.37)]

$$\left(\partial_i\partial_j - \partial_j\partial_i\right)\Omega(\mathbf{x}) = 0. \tag{9.9}$$

In a crystal, this property will hold away from the cutting surface S, where $u_i(\mathbf{x})$ is perfectly smooth and satisfies the corresponding integrability condition

$$\left(\partial_i\partial_j - \partial_j\partial_i\right)u_k(\mathbf{x}) = 0. \tag{9.10}$$

Across the surface, $u_i(\mathbf{x})$ is discontinuous. However, the open faces of the crystalline material must fit properly to each other. This implies that the strain as well as its first derivatives should have the same values on both sides of the cutting surface S:

$$\Delta u_{ij} = 0, \tag{9.11}$$
$$\Delta \partial_k u_{ij} = 0. \tag{9.12}$$

These conditions restrict severely the discontinuities of $u_i(\mathbf{x})$ across S. In order to see this let $\mathbf{x}(1), \mathbf{x}(2)$ be two different crystal points slightly above and below S, and C^+, C^- two curves connecting the two points (see Fig. 9.15). We can then calculate

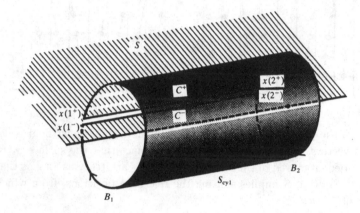

FIGURE 9.15 Geometry used in the derivation of Weingarten's theorem. See Eqs. (9.13)–(9.22).

the difference of the discontinuities as follows:

$$\Delta u_i(1) - \Delta u_i(2) = \left[u_i(1^-) u_i(1^+) \right] - \left[u_i(2^-) - u_i(2^+) \right]$$
$$= \int_{\substack{1^+ \\ C^+}}^{2^+} dx_j \partial_j u_i - \int_{\substack{1^- \\ C^-}}^{2^-} dx_j \partial_j u_i. \tag{9.13}$$

Using the local rotation field $\omega_{ij}(\mathbf{x})$ we can rewrite this as

$$\Delta u_i(1) - \Delta u_i(2) = \int_{\substack{1^+ \\ C^+}}^{2^+} dx_j dx_j (u_{ij} - \omega_{ij} - \int_{\substack{1^- \\ C^-}}^{2^-} dx_j (u_{ij} - \omega_{ij}). \tag{9.14}$$

The ω_{ij} pieces may be integrated by parts:

$$-\left(x_j - x_j(1^+) \right) \omega_{ij} \Big|_{1^+}^{2^+} + \int_{1^+}^{2^+} dx_k \left(x_j - x_j(1^+) \right) \partial_k \omega_{ij}$$

$$+ \left(x_j - x_j(1^-) \right) \omega_{ij} \Big|_{1^-}^{2^-} dx_k \left(x_j - x_j(1^-) \right) \partial_k \omega_{ij} \tag{9.15}$$

$$= \left[-\left(x_j(2^+) - x_j(1^+) \right) \omega_{ij}(2^+) + \int_{1^+}^{2^+} dx_k \left(x_j - x_j(1^+) \right) \partial_k \omega_{ij} \right] - [+ \rightarrow -].$$

Since
$$x_j(1^+) = x_j(1^-), \quad x_j(2^+) = x_j(2^-),$$
we arrive at the relation
$$\Delta u_i(1) - \Delta u_i(2) = -\left(x_j(1) - x_j(2)\right)\left(\omega_{ij}(2^-) - \omega_{ij}(2^+)\right)$$
$$+ \oint_{C^{+-}} dx_k \left\{u_{ik} + (x_j - x_j(1))\partial_k\omega_{ij}\right\}, \tag{9.16}$$
where C^{+-} is the closed contour consisting of C^+ followed by $-C^-$. Since C^+ and $-C^-$ are running back and forth on top of each other, the closed contour integral can be rewritten as a single integral along $-C^-$ with u_{ik} and $\partial_k\omega_{ij}$ replaced by their discontinuities across the sheet S. Moreover, the discontinuity of $\partial_k\omega_{ij}$ can be decomposed in the following manner:

$$\Delta\left(\partial_k\omega_{ij}\right) = \frac{1}{2}\partial_k\left(\partial_i u_j - \partial_j u_i\right)(\mathbf{x}^-) - \left(\mathbf{x}^- \to \mathbf{x}^+\right)$$
$$= \partial_i u_{kj}(\mathbf{x}^-) - \partial_j u_{ki}(\mathbf{x}^-) + \frac{1}{2}\left(\partial_k\partial_i - \partial_i\partial_k\right)u_j(\mathbf{x}^-) \tag{9.17}$$
$$- \frac{1}{2}\left(\partial_k\partial_j - \partial_j\partial_k\right)u_i(\mathbf{x}^-) + \frac{1}{2}\left(\partial_j\partial_i - \partial_i\partial_j\right)u_k(\mathbf{x}^-) - \left(\mathbf{x}^- \to \mathbf{x}^-\right).$$

Since the displacement field is smooth above and below the sheet, the two derivatives in front of $\mathbf{u}(x^\pm)$ commute. Hence the integral in (9.16) becomes
$$-\int_{C^-} dx_k \left\{\Delta u_{ij} + (x_j - x_j)(1)\Delta(\partial_i u_{kj} - \partial_j u_{ki})\right\}. \tag{9.18}$$

This expression vanishes due to the physical requirements (9.11) and (9.12). As a result we find that the discontinuities between two arbitrary points 1 and 2 on the sheet have the simple relation
$$\Delta u_i(2) = \Delta u_i(1) - \Omega_{ij}\left(x_j(2) - x_j(1)\right), \tag{9.19}$$
where Ω_{ij} is a fixed infinitesimal rotation matrix given by
$$\Omega_{ij} = \Delta\omega_{ij}(2) = \omega_{ij}(2^-) - \omega_{ij}(2^+). \tag{9.20}$$
We now define the rotation vector $\mathbf{\Omega}$ with components
$$\Omega_k = \frac{1}{2}\epsilon_{ijk}\Omega_{ij} \tag{9.21}$$
in terms of which (9.19) reads
$$\Delta\mathbf{u}(2) = \Delta\mathbf{u}(1) + \mathbf{\Omega} \times (\mathbf{x}(2) - \mathbf{x}(1)). \tag{9.22}$$

This is the content of *Weingarten's theorem* which states that the discontinuity of the displacement field across the cutting surface can only be a constant vector plus a fixed rotation.

Note that these are precisely the symmetry elements of a solid continuum. When looking back at the particular dislocation and disclination lines in Figs. 9.2–9.12 we see that all discontinuities obey this theorem, as they should. The vector $\mathbf{\Omega}$ is the Frank vector of the disclination lines. For a pure disclination line, $\mathbf{\Omega} = \mathbf{0}$ and $\Delta\mathbf{u}(1) = \Delta\mathbf{u}(2) = \mathbf{b}$ is the Burgers vector.

9.8   Integrability Properties of Displacement Field

The rotation field $\omega_{ij}(\mathbf{x})$ has also nontrivial integrability properties. Taking Weingarten's theorem (9.19) and forming derivatives, we see that the jump of the $\omega_{ij}(\mathbf{x})$ field is necessarily a constant, namely Ω_{ij}. Hence ω_{kl} also satisfies everywhere the integrability condition

$$\left(\partial_i \partial_j - \partial_j \partial_i\right) \omega_{kl} = 0, \tag{9.23}$$

except on the defect line. The argument is the same as that for the vortex lines. We simply observe that the contour integral over a Burgers circuit

$$\Delta \omega_{ij} = \oint_B d\omega_{ij} = \oint_B dx_k \partial_k \omega_{ij} \tag{9.24}$$

can be cast, by Stokes' theorem (4.21), in the form

$$\Delta \omega_{ij} = \int_{S^B} dS_m \epsilon_{mkl} \partial_k \partial_l \omega_{ij}, \tag{9.25}$$

where S^B is some surface enclosed by the Burgers circuit. Since the result is independent of the size, shape, and position of the Burgers circuit as long as it encloses the defect line L, this implies that everywhere away from L

$$\epsilon_{mkl} \partial_k \partial_l \omega_{ij}(\mathbf{x}) = 0, \tag{9.26}$$

which is what we wanted to show.

In fact, the constancy of the jump in ω_{ij} could have been derived somewhat more directly, without going through (9.23)–(9.26), by taking again the curves C^+, C^- on Fig. 9.15 and calculating

$$\Delta \omega_{ij}(1) - \Delta \omega_{ij}(2) = \underset{C^+}{\int_{1^+}^{2^+}} dx_k \partial_k \omega_{ij} - \underset{C^-}{\int_{1^-}^{2^-}} dx_k \partial_k \omega_{ij} = - \underset{C^-}{\int_{1^-}^{2^-}} dx_k \Delta \left(\partial_k \omega_{ij}\right). \tag{9.27}$$

From the assumptions (9.11) and (9.12), together with (9.18), we see that $\omega_{ij}(\mathbf{x})$ does not jump across the Volterra surface S. But then (9.27) shows us that $\Delta \omega_{ij}$ is a constant.

Let us now consider the displacement field itself. As a result of Weingarten's theorem, the integral over the Burgers circuit B_2 in Fig. 9.15 gives

$$\Delta u_i(2) = \oint_{B_2} du_i = \Delta u(1) - \Omega_{ij} \left[x_j(2) - x_j(1)\right], \tag{9.28}$$

$$\Delta u_i(1) - \Omega_{ij} \left[x_j(2) - x_j(1)\right] = \oint_{B_2} dx_k \left\{u_{ik} + \left[x_j - x_j(2)\right] \partial_k \omega_{ij}\right\}. \tag{9.29}$$

Here we observe that the factors of $x_i(2)$ can be dropped on both sides by (9.24) and $\Delta \omega_{ij} = \Omega_{ij}$. By Stokes' theorem (4.21), the remaining equation then becomes an equation for the surface integral over S^{B_2},

$$\begin{aligned} \Delta u_i(1) + \Omega_{ij} x_j(1) &= \int_{S^{B_2}} dS_l \, \epsilon_{lmk} \partial_m \left(u_{ik} + x_j \partial_k \omega_{ij}\right) \\ &= \int_{S^{B_2}} dS_l \, \epsilon_{lmk} \left[(\partial_m u_{ik} + \partial_k \omega_m) + x_j \partial_m \partial_k \omega_{ij}\right]. \end{aligned} \tag{9.30}$$

This must hold for any size, shape, and position of the circuit B_2 as long as it encircles the defect line L. For all these different configurations, the left-hand side of (9.30) is a constant. We can therefore conclude that

$$\int_S dS_l \left[\epsilon_{lmk} \left(\partial_m u_{ik} + \partial_k \omega_{im} \right) + x_j \epsilon_{lmk} \partial_k \omega_{ij} \right] = 0 \qquad (9.31)$$

for any surface S which does *not* enclose L. Moreover, from (9.23) we see that the last term cannot contribute. The first two terms, on the other hand, can be rewritten, using the same decomposition of $\partial_k \omega_{im}$ as in (9.18), in the form

$$- \int_S dS_l \, \epsilon_{lmk} \left(S_{kmi} - S_{mik} + S_{ikm} \right) = \int_S dS_l \, \epsilon_{lmk} S_{mki}, \qquad (9.32)$$

where we have abbreviated

$$S_{kmi}(\mathbf{x}) \equiv \frac{1}{2} \left(\partial_k \partial_m - \partial_m \partial_k \right) u_i(\mathbf{x}). \qquad (9.33)$$

Since this has to vanish for any S we conclude that, at some distance from the defect line, the displacement field $u_i(\mathbf{x})$ also satisfies the integrability condition

$$\left(\partial_k \partial_m - \partial_m \partial_k \right) u_i(\mathbf{x}) = 0. \qquad (9.34)$$

On the line L, the integrability conditions for u_i and ω_{ij} are, in general, both violated. Let us first consider ω_{ij}. In order to give the constant result $\Delta_{ij}(\mathbf{x}) \equiv \Omega_{ij}$ in (9.25) the integrability condition must be violated by a singularity in the form of a δ-function along the line L (4.10), namely:

$$\epsilon_{lmk} \partial_k \omega_{ij} = \delta_l(\mathbf{x}; L). \qquad (9.35)$$

Then (9.25) gives $\Delta \omega_{ij} = \Omega_{ij}$ via the formula

$$\int_{S^B} dS_l \, \delta_l(\mathbf{x}; L) = 1. \qquad (9.36)$$

In order to see how the integrability condition is violated for $u_i(\mathbf{x})$, consider now the integral (9.30) and insert the result (9.32). This gives

$$\Delta u_i(1) + \Omega_{ij} x_j(1) = \int_{S^{B_2}} dS_l \, \epsilon_{lmk} \left(S_{mki} + x_j \partial_m \partial_k \omega_{ij} \right). \qquad (9.37)$$

The right hand side is a constant independent of the position of the surface S^{B_2}. This implies that the singularity along L is of the form

$$\epsilon_{lmk} \left(\partial_m \partial_k u_i + x_j \partial_m \partial_k \omega_{ij} \right) = b_i \delta_l(\mathbf{x}; L), \qquad (9.38)$$

where we have introduced the quantity

$$b_i \equiv \Delta u_i(1) + \Omega_{ij} x_j(1). \qquad (9.39)$$

Inserting (9.35) into (9.38) leads to the following violation of the integrability condition for $u_i(\mathbf{x})$ along L:

$$\epsilon_{lmk} \partial_m \partial_k u_i = \left(b_i - \Omega_{ij} x_j \right) \delta_l(\mathbf{x}; L). \qquad (9.40)$$

In terms of the tensor (9.33), this reads

$$\epsilon_{lmk} S_{mki} = \left(b_i - \Omega_{ij} x_j \right) \delta_l(\mathbf{x}; L). \qquad (9.41)$$

9.9   Dislocation and Disclination Densities

The violation of the integrability condition for displacement and rotation fields proportional to δ-functions along lines L is analogous to the situation in the multivalued description of the magnetic field in Chapter 4. The analogy can be carried further. Consider, for example, the current density of magnetism in Eq. (4.1), which by Eqs. (4.36), (4.23), and (4.37), can be rewritten in the multivalued description as

$$j_i(\mathbf{x}) = \epsilon_{ijk}\partial_j B_k(\mathbf{x}) = \frac{I}{4\pi}\epsilon_{ijk}\partial_j\partial_k\Omega(\mathbf{x}) = I\delta_i(\mathbf{x}; L). \qquad (9.42)$$

Here Ω is the solid angle (4.25) under which the loop L is seen from the point $\mathbf{x}$. By analogy, we introduce *densities* for dislocations and disclinations, respectively, as follows:

$$\alpha_{ij}(\mathbf{x}) \equiv \epsilon_{ikl}\partial_k\partial_l u_j(\mathbf{x}), \qquad (9.43)$$
$$\theta_{ij}(\mathbf{x}) \equiv \epsilon_{ikl}\partial_k\partial_l \omega_j(\mathbf{x}), \qquad (9.44)$$

where we have used the vector form of the rotation field $\omega_i = (1/2)\epsilon_{ijk}\omega_{jk}$, in order to save one index. For the general defect line along L, these densities have the form

$$\alpha_{ij}(\mathbf{x}) = \delta_i(\mathbf{x}; L)\left(b_i - \Omega_{jk}x_k\right), \qquad (9.45)$$
$$\theta_{ij}(\mathbf{x}) = \delta_i(\mathbf{x}; L)\Omega_j, \qquad (9.46)$$

where $\Omega_i = (1/2)\epsilon_{ijk}\Omega_{jk}$ are the components of the Frank vector.

Note that in terms of the tensor field $S_{ijk}(\mathbf{x})$ of Eq. (9.33), the dislocation density (9.43) reads

$$\alpha_{ij}(\mathbf{x}) \equiv \epsilon_{ikl}S_{lkj}(\mathbf{x}). \qquad (9.47)$$

In (9.45) and (9.46) the rotation by Ω is performed around the origin. Obviously, the position of the rotation axis can be changed to any other point $\mathbf{x}_0$ by a simple shift in the constant $b_j \rightarrow b'_j + (\Omega \times \mathbf{x}_0)_j$. Then $\alpha_{ij}(\mathbf{x}) = \delta_i(\mathbf{x}; L)\left\{b'_j + (\Omega \times (\mathbf{x} - \mathbf{x}_0))_j\right\}$. Note that due to the identity $\partial_i\delta_i(\mathbf{x}; L) = 0$ for closed lines L [recall (4.12)], the disclination density satisfies the conservation law

$$\partial_i\theta_{ij}(\mathbf{x}) = 0, \qquad (9.48)$$

which implies that disclination lines are always closed. This is not true for media with a directional field, e.g., for nematic liquid crystals. Such media are not considered here since they cannot be described by a displacement field alone. Differentiating (9.45) we find the conservation law for disclination lines $\partial_i\alpha_{ij}(\mathbf{x}) = -\Omega_{ij}\delta_i(\mathbf{x}; L)$ which, in turn, can be expressed in the form

$$\partial_i\alpha_{ij}(\mathbf{x}) = -\epsilon_{jkl}\theta_{kl}(\mathbf{x}). \qquad (9.49)$$

Expressed terms of the tensor S_{ijk} via (9.47), it reads

$$\epsilon_{jkl}\left(\partial_i S_{kli} + \partial_k S_{lnm} - \partial_l S_{knn}\right) = -\epsilon_{jkl}\theta_{kl}. \qquad (9.50)$$

Indeed, inserting $S_{klj} = (1/2)\epsilon_{kli}\alpha_{ij}$ from (9.47), this reduces to the conservation law (9.49) for the dislocation density.

From the linearity of the relations (9.43) and (9.44) in u_j and ω_j, respectively, it is obvious that these conservation laws remain true for any ensemble of infinitesimal defect lines. The conservation law (9.49) may, in fact, be derived by purely differential techniques from the first smoothness assumption (9.11). Using Stokes' theorem (4.21), Δu_{ij} can be expressed in the same way as $\Delta\omega_{ij}$ in (9.25). By the same argument as the one used for ω_{ij} we conclude that the strain is an integrable function *in all space* and satisfies

$$\left(\partial_i\partial_k - \partial_k\partial_i\right) u_{lj}(\mathbf{x}) = 0. \qquad (9.51)$$

Now we take α_{ij} in the general definition (9.43) and rewrite it as

$$\begin{aligned}
\alpha_{ij} &= \epsilon_{ikl}\partial_k\partial_l u_j = \epsilon_{ikl}\partial_k\left(u_{lj} + \omega_{lj}\right) \\
&= \epsilon_{ikl}\partial_k u_{lj} + \delta_{ij}\partial_k\omega_k - \partial_j\omega_i.
\end{aligned} \qquad (9.52)$$

Applying to this the derivative ∂_i, and using (9.51) we find directly (9.49).

In a similar way, the conservation law (9.48) can be derived by combining the two smoothness assumptions (9.11) and (9.12). The first can be restated, via Stokes' theorem (4.21), as an integrability condition for the derivative of strain, i.e.,

$$\left(\partial_l\partial_n - \partial_n\partial_l\right) \partial_k u_{ij}(\mathbf{x}) = 0. \qquad (9.53)$$

Let us recall that from the assumption (9.11) has led in (9.18) to the conclusion that $\partial_k\omega_{ij}(\mathbf{x})$ is also a completely smooth function across the surface S. Hence, $\partial_k\omega_{ij}$ must also satisfy the integrability condition

$$\left(\partial_l\partial_n - \partial_n\partial_l\right) \partial_k\omega_{ij}(\mathbf{x}) = 0.$$

Together with (9.53) this implies that $\partial_k\partial_i u_j(\mathbf{x})$ is integrable:

$$\left(\partial_l\partial_n - \partial_n\partial_l\right) \partial_k\partial_i u_j(\mathbf{x}) = 0. \qquad (9.54)$$

If we write down this relation three times, each time with l, n, k exchanged cyclically, we find

$$\partial_l R_{nkij} + \partial_n R_{klij} + \partial_k R_{lnij} = 0, \qquad (9.55)$$

where R_{nkij} is an abbreviation for the expression,

$$R_{nkij} = \left(\partial_n\partial_k - \partial_k\partial_n\right) \partial_i u_j(\mathbf{x}). \qquad (9.56)$$

Contracting k with i and l with j gives

$$\partial_j R_{niij} + \partial_n R_{ijij} + \partial_i R_{jnij} = 0. \tag{9.57}$$

Now we observe that, because of (9.51), R_{nkij} is anti-symmetric not only in n and k but also in i and j so that

$$2\partial_j R_{inji} - \partial_n R_{ijji} = 0. \tag{9.58}$$

With the help of the identity (1A.16) for ϵ-tensors, this may be rewritten as [see Eq. (1A.16)]

$$2\partial_j \left(\frac{1}{4}\epsilon_{jpq}\epsilon_{nkl} R_{pqkl} \right) = 0, \tag{9.59}$$

Recalling now the definition (9.56) of the curvature tensor, and using $\omega_n = (1/2)\epsilon_{nkl}\partial_k u_l$, Eq. (9.59) becomes

$$2\epsilon_{ipq}\partial_i\partial_p\partial_q\omega_m = 0. \tag{9.60}$$

This is precisely the conservation law $\partial_i\theta_{ik} = 0$ for disclinations in Eq. (9.48) which we wanted to prove.

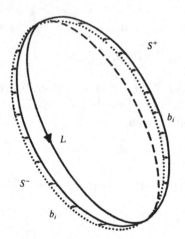

FIGURE 9.16 Illustration of Volterra process in which an entire volume piece is moved with the vector b_i.

Note the appearance of torsion and curvature in Eqs. (9.33) and (9.56). In fact, Eqs. (9.50) and (9.55) will turn out to linearized versions of the famous fundamental identity and Bianchi identity, to be discussed in detail in Sections 12.1 and 12.5 [see Eqs. (12.103) and (12.115), respectively].

9.10 Mnemonic Procedure for Constructing Defect Densities

There exists a simple mnemonic procedure for constructing the defect densities and their conservation laws.

Suppose we perform the Volterra cutting procedure on a *closed surface* S, dividing it mentally into two parts, joined along some line L (see Fig. 9.16). On one part of S, say S^+, we remove material of thickness b_i and on the other we add the same material. This corresponds to a simple translational movement of crystalline material by b_i, i.e., to a displacement field

$$u_l(\mathbf{x}) = -\delta(\mathbf{x}; V)b_l, \tag{9.61}$$

with the δ-function on a volume V defined in Eq. (4.30). By this transformation the elastic properties of the material are unchanged.

Consider now the distortion field $\partial_k u_l(\mathbf{x})$. Under (9.61), it changes by

$$\partial_k u_l(\mathbf{x}) \to \partial_k u_l(\mathbf{x}) - \partial_k \delta(\mathbf{x}; V)b_l. \tag{9.62}$$

The derivative of the δ-function is singular only on the surface of the volume V. In fact, in Eq. (4.35) we derived the formula

$$\partial_k \delta(\mathbf{x}; V) = -\delta_k(\mathbf{x}; S), \tag{9.63}$$

so that (9.62) reads

$$\partial_k u_l(\mathbf{x}) \to \partial_k u_l(\mathbf{x}) + \delta_k(\mathbf{x}; S)b_l. \tag{9.64}$$

From this trivial transformation we can now construct a proper dislocation line by assuming S to be no longer a closed surface but an open one, i.e. we may restrict S to the shell S^+ with a boundary L. Then we can form the dislocation density

$$\alpha_{il}(\mathbf{x}) = \epsilon_{ijk}\partial_j \partial_k u_l(\mathbf{x}) = \epsilon_{ijk}\partial_j \delta_k(\mathbf{x}; S)b_l. \tag{9.65}$$

The superscript $+$ was dropped. Using Stokes' theorem for the function $\delta_k(\mathbf{x}; S)$ in the form (4.24), this becomes simply

$$\alpha_{il}(\mathbf{x}) = \delta_i(\mathbf{x}; L)b_l. \tag{9.66}$$

For a closed surface, (9.65) vanishes.

For a general defect line, the starting point is the trivial Volterra operation of *translating and rotating* a piece of crystalline volume. This corresponds to a displacement field

$$u_l(\mathbf{x}) = -\delta(\mathbf{x}; V)\left(b_l + \epsilon_{lqr}\Omega_q x_r\right). \tag{9.67}$$

If we now form the distortion, we find using (9.63):

$$\partial_k u_l(\mathbf{x}) = \delta_k(\mathbf{x}; S)\left(b_l + \epsilon_{lqr}\Omega_q x_r\right) - \delta(\mathbf{x}; V)\epsilon_{lqk}\Omega_q. \tag{9.68}$$

In this expression it is still impossible to assume S to be an open surface, since this would not have a well-defined enclosed volume V. However, if we form the symmetric combination, the volume term cancels, and the *strain tensor* becomes

$$u_{kl} = \frac{1}{2}\left(\partial_k u_l + \partial_l u_k\right) = \frac{1}{2}\left[\delta_k(\mathbf{x}; S)\left(b_l + \epsilon_{lqr}\Omega_q x_r\right) + (k \leftrightarrow l)\right], \qquad (9.69)$$

It contains only the surface S which can be assumed to be open. In this case we refer to u_{kl} as the *plastic strain* and denote it by u_{kl}^p. The field

$$\beta_{kl}^p \equiv \delta_k(\mathbf{x}; S)\left(b_l + \epsilon_{lqr}\Omega_q x_r\right) \qquad (9.70)$$

plays the role of a dipole density of the defect line across the surface S. It is called the *plastic distortion*. It is a single valued field, i.e., derivatives in front of it commute. In terms of β_{kl}^p, the plastic strain is simply

$$u_{kl}^p = \frac{1}{2}\left(\beta_{kl}^p + \beta_{kl}^p\right). \qquad (9.71)$$

The full displacement field (9.67) is not defined for an open surface since it contains the volume V. The dislocation density, however, is single valued. We can easily calculate with the help of (9.63) and Stokes theorem (4.24):

$$\begin{aligned} \alpha_{il} &= \epsilon_{ijk}\partial_j\partial_k u_l(\mathbf{x}) = \epsilon_{ijk}\partial_j\left[\delta_k(\mathbf{x}; S)\left(b_l + \epsilon_{lqr}\Omega_q x_r\right) - \delta(\mathbf{x}; V)\epsilon_{lqk}\Omega_q\right] \\ &= \delta_i(\mathbf{x}; L)\left(b_l + \epsilon_{lqr}\Omega_q x_r\right), \end{aligned} \qquad (9.72)$$

and see that this coincides with (9.45).

Let us now turn to the disclination density $\theta_{pj} = \epsilon_{pmn}\partial_m\partial_n\omega_j$. From (9.67) we find the gradient of the rotation field

$$\begin{aligned} \partial_n\omega_j &= \frac{1}{2}\epsilon_{jkl}\partial_n\partial_k u_l \\ &= \frac{1}{2}\epsilon_{jkl}\partial_n\left[\delta_k(\mathbf{x}; S)\left(b_l + \epsilon_{lqr}\Omega_q x_r\right) - \delta(\mathbf{x}; V)\epsilon_{lqk}\Omega_q\right] \\ &= \frac{1}{2}\epsilon_{jkl}\partial_n\beta_{kl}^p + \delta_n(\mathbf{x}; S)\Omega_j. \end{aligned} \qquad (9.73)$$

This gradient is defined for an open surface S, and is called the field of *plastic bend-twist*, denoted by $\kappa_{nj}^p \equiv \partial_n\omega_j^p$.

It is useful to define the *plastic rotation*

$$\phi_{nj}^p \equiv \delta_n(\mathbf{x}; S)\Omega_j, \qquad (9.74)$$

which plays the role of a dipole density for disclinations. With this, the plastic gradient of ω_j is given by

$$\kappa_{nj}^p = \partial_n\omega_j^p = \frac{1}{2}\epsilon_{jkl}\partial_n\beta_{kl}^p + \phi_{nj}^p. \qquad (9.75)$$

We can now easily calculate the disclination density:

$$\theta_{pj} = \epsilon_{pmn}\partial_m\partial_n\omega_j = \epsilon_{pmn}\partial_m\kappa^p_{nj} = \frac{1}{2}\epsilon_{jkl}\epsilon_{pmn}\partial_m\partial_n\beta^p_{kl} + \epsilon_{pmn}\partial_m\phi^p_{nj}.$$

The derivatives in front of β^p_{kl} commute [see (9.70)], so that the first term vanishes. Applying Stokes' theorem (4.24) to the second term gives

$$\theta_{pj} = \epsilon_{pmn}\partial_m\phi^p_{nj} = \delta_p(\mathbf{x}; L)\Omega_j, \tag{9.76}$$

in agreement with (4.12).

Note that according to the second line of (9.72), the dislocation density can also be expressed in terms of β^p_{kl} and ϕ^p_{li} as

$$\alpha_{il} = \epsilon_{ijk}\partial_j\beta^p_{kl} + \delta_{il}\phi^p_{pp} - \phi^p_{li}. \tag{9.77}$$

In fact, this is a direct consequence of the decomposition (9.52), which can be written in terms of plastic strain and bend-twist fields as

$$\alpha_{ij} = \epsilon_{ikl}\partial_k u^p_{lj} + \delta_{ij}\kappa^p_{qq} - \kappa^p_{ji}. \tag{9.78}$$

Expressing u^p_{li} in terms of β^p_{li}, and κ^p_{ij} in terms of ϕ^p_{ij} [see (9.71) and (9.75)], we find

$$\alpha_{ij} = \frac{1}{2}\epsilon_{ikl}\partial_k\beta^p_{lj} + \delta_{ij}\phi^p_{qq} - \phi^p_{qq} - \phi^p_{ji} + \frac{1}{2}\left(\epsilon_{ijk}\partial_k\beta^p_{jl} + \delta_{ij}\epsilon_{qkl}\partial_q\beta^p_{kl} - \epsilon_{ikl}\partial_i\beta^p_{kl}\right).$$

The quantity inside the parentheses is equal to $\frac{1}{2}\epsilon_{ikl}\partial_k\beta^p_{lj}$, as can be seen by applying to $\partial_q\beta_{kl}$ the identity (1A.20). Thus α_{ij} takes again the form (9.77).

9.11   Defect Gauge Invariance

A given defect distribution can be derived from many different plastic strains and rotations. This is an obvious consequence of the freedom in choosing Volterra surfaces S for the construction of defect lines L. These lines run along the boundary lines of surfaces S, whose shape is irrelevant. From the discussion of the gradient representation of magnetic fields of current loops in Subsection 4.6, and from the theory of vortex lines in superfluids in Chapter 5 we know that this freedom can be formulated mathematically as a gauge symmetry. Recall that a magnetic field caused by a line-like current density was represented as a curl of a δ-function on a surface [see Eq. (4.92)]:

$$\mathbf{j}(\mathbf{x}) = I\,\mathbf{\nabla} \times \boldsymbol{\delta}(\mathbf{x}; S). \tag{9.79}$$

This representation was invariant under a gauge transformation

$$\boldsymbol{\delta}(\mathbf{x}; S) \rightarrow \boldsymbol{\delta}(\mathbf{x}; S') = \boldsymbol{\delta}(\mathbf{x}; S) - \mathbf{\nabla}\delta(\mathbf{x}; V), \tag{9.80}$$

[see Eq. (4.29)], which shifts the surface S to a new position S' with the same boundary line. The same gauge invariance was found for vortex lines [recall (5.28)].

The gauge field (9.80) enabled us to construct a gauge-invariant gradient of a multivalued field. In the simplest case of a superfluid, this was the cyclic phase angle $\theta(\mathbf{x})$ with period 2π, and the gauge-invariant gradient was the superfluid velocity (5.27):

$$\mathbf{v}(\mathbf{x}) \equiv \boldsymbol{\nabla}\theta(\mathbf{x}) - \boldsymbol{\theta}^{\mathrm{v}}(\mathbf{x}), \tag{9.81}$$

In the light of the discussion in the previous section we may set up a simple construction rule for this invariant. We perform a trivial "Volterra" operation on the cyclic field $\theta(\mathbf{x})$, shifting it -2π on an arbitrary volume V:

$$\theta(\mathbf{x}) \rightarrow \theta(\mathbf{x}) - 2\pi\delta(\mathbf{x}, V), \tag{9.82}$$

Since $\theta(\mathbf{x})$ and $\theta(\mathbf{x}){-}2\pi$ are physically indistinguishable, this is certainly a symmetry operation. Under this transformation, the gradient changes by

$$\boldsymbol{\nabla}\theta(\mathbf{x}) \rightarrow \boldsymbol{\nabla}\theta(\mathbf{x}) - 2\pi\boldsymbol{\nabla}\delta(\mathbf{x}, V) = \boldsymbol{\nabla}\theta(\mathbf{x}) + 2\pi\boldsymbol{\delta}(\mathbf{x}, S), \tag{9.83}$$

where S is the surface of V, and we have used the identity (9.63). The additional term $2\pi\boldsymbol{\delta}(\mathbf{x}, S)$ is not only defined for surfaces S which are boundaries of a volume V which are closed, but also for open surfaces. In this case it is a plastic distortion of the gradient

$$[\boldsymbol{\nabla}\theta(\mathbf{x})]^p \equiv 2\pi\boldsymbol{\delta}(\mathbf{x}, S). \tag{9.84}$$

The vortex gauge-invariant (9.81) is obtained from the difference

$$\mathbf{v}(\mathbf{x}) \equiv \boldsymbol{\nabla}\theta(\mathbf{x}) - [\boldsymbol{\nabla}\theta(\mathbf{x})]^p. \tag{9.85}$$

If we now perform a shift in the surface S to S', the plastic distortion (9.84) changes, due to the gauge property (4.29), in precisely the same way as under a trivial "Volterra" operation (9.82), except for the sign:

$$[\boldsymbol{\nabla}\theta(\mathbf{x})]^p \rightarrow [\boldsymbol{\nabla}\theta(\mathbf{x})]^p + 2\pi\boldsymbol{\nabla}\delta(\mathbf{x}, V). \tag{9.86}$$

The only difference is that V is now the volume over which th open surface S has swept when being shifted to S'. This change can therefore be compensated by the "Volterra" operation

$$\theta(\mathbf{x}) \rightarrow \theta(\mathbf{x}) + 2\pi\delta(\mathbf{x}, V). \tag{9.87}$$

This ensures vortex gauge-invariance under the simultaneous transformations (9.86) and (9.87).

If we want to set up a theory of elasticity for crystals with defects, as will be done in Chapter 10, we must find analogs of the invariant gradient (9.85). By analogy with (9.85), these are simply the differences between elastic and plastic distortions obtained from the mnemonic procedure in the previous section. For the strain tensor u_{kl}, we subtract from u_{kl} the plastic strain tensor (9.71), and form the defect gauge-invariant strain tensor

$$u_{ij}^{\mathrm{inv}}(\mathbf{x}) \equiv u_{ij}(\mathbf{x}) - u_{ij}^p(\mathbf{x}). \tag{9.88}$$

Indeed, under a shift of the Volterra surface from S to S', plastic strain tensor (9.72) changes due to the transformation property (9.80) of $\delta(\mathbf{x}; S)$ like

$$\beta_{kl}^p \to \beta_{kl}^p - \partial_k \delta(\mathbf{x}; V) \left(b_l + \epsilon_{lqr} \Omega_q x_r \right). \qquad (9.89)$$

This change is precisely compensated by the combined translation and rotation (9.67), which produced the tensor (9.69) via the identity (9.61).

Similarly we can form the invariant combination

$$\kappa_{nj}^{\text{inv}} \equiv \partial_n \omega_j - \kappa_{nj}^p = \partial_n \omega_j - \frac{1}{2}\epsilon_{jkl}\partial_n \beta_{kl}^p - \phi_{nj}^p, \qquad (9.90)$$

which is defect gauge-invariant since the transformation (9.89) is compensated by the trivial Volterra operation (9.67) which produced the plastic bend-twist (9.75) via (9.61).

Given the general form of plastic distortions in Eq. (9.70) we are able to characterize the interdependence of dislocations and disclinations observed in Section 9.4 by an extra type of gauge invariance. Obviously, β_{kl}^p remains invariant under the transformations

$$b_l \to b_l + \epsilon_{lqr}\Delta\Omega_q x_r, \qquad \Omega_q \to \Omega_q - \Delta\Omega_q. \qquad (9.91)$$

Since x_r are all integer numbers on a simple cubic lattice, there exists a choice of $\Delta\Omega_q$ which makes either b_l of Ω_q equal to zero.

9.12 Branching Defect Lines

Recall that, from the geometric point of view, the defect conservation laws imply that disclination lines never end and dislocations end at most at a disclination line. Consider, for example, a branching configuration where a line L splits into two lines L and L'. Assign an orientation to each line and suppose that their disclination density is

$$\theta_{ij}(\mathbf{x}) = \Omega_i \delta_j(\mathbf{x}; L) + \Omega_i' \delta_j(\mathbf{x}; L') + \Omega_i'' \delta_j(\mathbf{x}; L''), \qquad (9.92)$$

with their dislocation density being

$$\begin{aligned}
\alpha_{ij}(\mathbf{x}) &= \delta_i(\mathbf{x}; L)\left\{ b_j + [\mathbf{\Omega} \times (\mathbf{x} - \mathbf{x}_0)]_j \right\} + \delta_i(\mathbf{x}; L')\left\{ b_j' + [\Omega' \times (\mathbf{x} - \mathbf{x}_0')]_j \right\} \\
&\quad \delta_i(\mathbf{x}; L')\left\{ b_j'' + [\Omega'' \times (\mathbf{x} - \mathbf{x}_0'')]_j \right\}.
\end{aligned} \qquad (9.93)$$

The conservation law $\partial_i \theta_{ij} = 0$ then implies that the Frank vectors satisfy the equivalent of Kirchhoff's law for currents

$$\Omega_j + \Omega_i'' = \Omega_i'. \qquad (9.94)$$

This follows directly from the identity for lines

$$\partial_i \delta_i(\mathbf{x}; L) = \int ds \frac{d\bar{x}_i}{ds}\delta^{(3)}\left(\mathbf{x} - \bar{\mathbf{x}}(3) \right) = \delta^{(3)}(\mathbf{x} - \mathbf{x}_i) - \delta^{(3)}(\mathbf{x} - \mathbf{x}_f),$$

where $\mathbf{x}_i$ and $\mathbf{x}_f$ are the initial and final points of the curve L. The conservation law $\partial_i \alpha_{ij} = \epsilon_{ikl}\theta_{kl}$, on the other hand, gives

$$b_i - [\boldsymbol{\Omega} \times (\mathbf{x} - \mathbf{x}_0)]_i + b_i'' - \left[\boldsymbol{\Omega}'' \times \left(\mathbf{x} - \mathbf{x}_0''\right)\right]_i = b_i' - \left[\boldsymbol{\Omega} \times \left(\mathbf{x}' - \mathbf{x}_0'\right)\right]_i. \quad (9.95)$$

If the same position is chosen for all rotation axes, the Burgers vectors b_i satisfy again a Kirchhoff-like law:

$$b_i + b_i'' = b_i'. \quad (9.96)$$

But Burgers vectors can be modified by different rotation axes, for example, L' and L'' could be pure disclination lines with different axes through $\mathbf{x}_0', \mathbf{x}_0''$, while L' a pure dislocation line through $\mathbf{x}_0' = -\Omega' \times (\mathbf{x}_0' - \mathbf{x}_0'')$, and end on L', L''. Equation (9.94) renders different choices equivalent.

9.13   Defect Density and Incompatibility

As far as classical linear elasticity is concerned, the information contained in α_{ij} and θ_{ij} can be combined efficiently in a single symmetric tensor, called the *defect density* $\eta_{ij}(\mathbf{x})$. It is defined as the double curl of the strain tensor resulting from the trivial Volterra operation (9.61):

$$\eta_{ij}(\mathbf{x}) \equiv \epsilon_{ikl}\epsilon_{jmn}\partial_k\partial_m u_{ln}(\mathbf{x}). \quad (9.97)$$

If the surface S around V is opened, the incompatibility is nonzero and strain tensor $u_{ln}(\mathbf{x})$ on the right-hand side can be replaced by the plastic strain tensor $u_{ln}^p(\mathbf{x})$.

In order to see its relation with α_{ij} and θ_{ij}, we take (9.43) and contract the indices i and j, obtaining

$$\alpha_{ii} = 2\partial_i\omega_i. \quad (9.98)$$

Using this, (9.52) can be written in the form

$$\epsilon_{ikl}\partial_k u_{ln} = \partial_n\omega_i - \left(-\alpha_{in} + \frac{1}{2}\delta_{in}\alpha_{kk}\right). \quad (9.99)$$

The expression in parentheses was first introduced by Nye and called *contortion*[2]

$$K_{ni} \equiv -\alpha_{in} + \frac{1}{2}\delta_{in}\alpha_{kk}. \quad (9.100)$$

The inverse relation is

$$\alpha_{ij} = -K_{ji} + \delta_{ij}K_{kk}. \quad (9.101)$$

---

[2]In terms of the plastic quantities introduced in the last section the plastic part of K_{ij} reads

$$K_{ij}^p = -\epsilon_{ikl}\partial_k\beta_{li}^p + \frac{1}{2}\delta_{ij}\epsilon_{nkl}\partial_k\beta_{ln}^p + \phi_{ij}^p.$$

Multiplying (9.97) by $\epsilon_{jmn}\partial_m$, we find with (9.44)

$$
\begin{aligned}
\eta_{ij} &= \epsilon_{jmn}\epsilon_{ikl}\partial_m\partial_k u_{ln} = \epsilon_{jmn}\partial_m\partial_n\omega_i - \epsilon_{jmn}\partial_m K_{ni} \\
&= \theta_{ij} - \epsilon_{jmn}\partial_m K_{ni}.
\end{aligned}
\tag{9.102}
$$

Although it is not at all obvious, the final expression is symmetric in ij. Indeed, if it is contracted with the antisymmetric tensor ϵ_{lij}, it yields $\epsilon_{lij}\theta_{ij}\partial_l K_{ii} - \partial_i K_{li} = \epsilon_{lij}\theta_{ij}\theta_{ij} + \partial_i\alpha_{il}$, which vanishes due to the conservation law (9.49) for the dislocation density.

There is yet another version of the decomposition (9.102) which is obtained after applying the identity (1A.17) to $\partial_m\alpha_{qn}$ giving

$$
\epsilon_{njm}\partial_m\left(\alpha_{in} - \frac{1}{2}\delta_{in}\alpha_{kk}\right) = -\frac{1}{2}\partial_m\left[\epsilon_{mjn}\alpha_{in} + (i \leftrightarrow j) + \epsilon_{ijn}\alpha_{mn}\right].
\tag{9.103}
$$

Hence

$$
\eta_{ij} = \theta_{ij} - \frac{1}{2}\partial_m\left[\epsilon_{min}\alpha_{jn} + (i \leftrightarrow j) - \epsilon_{ijn}\alpha_{mn}\right].
\tag{9.104}
$$

This type of decomposition will be encountered in the context of general relativity later in Eqs. (17.156) and (18.53) in the Belinfante construction of symmetric energy-momentum tensors.

The double curl operation is a useful generalization of the curl operation on vector fields to symmetric tensor fields. Recall that the vanishing of a curl of a vector field $\mathbf{E}$ implies that $\mathbf{E}$ can be written as the gradient of a scalar potential $\phi(\mathbf{x})$ which satisfies the integrability condition $(\partial_i\partial_j - \partial_j\partial_i)\phi(\mathbf{x}) = 0$:

$$
\nabla \times \mathbf{E} = 0 \quad \Rightarrow \quad E_i = \partial_i\phi(\mathbf{x}).
\tag{9.105}
$$

The double curl operation implies a similar property for the symmetric tensor, as was shown a century ago by Riemann and by Christoffel. If the double curl of a symmetric tensor field vanishes everywhere in space, his field can be written as the strain of some displacement field $u_i(\mathbf{x})$ which is integrable in all space [i.e., it satisfies (9.34)]. We may state this conclusion briefly as follows:

$$
\epsilon_{ikl}\epsilon_{jmn}\partial_k\partial_m u_{ln}(\mathbf{x}) = 0 \quad \Rightarrow \quad u_{ij} = \frac{1}{2}\left(\partial_i u_j + \partial_j u_i\right).
\tag{9.106}
$$

If the double curl of $u_{ln}(\mathbf{x})$ is zero one says that $u_{ln}(\mathbf{x})$ is *compatible* with a displacement field. A nonzero double curl

$$
(\text{inc } u)_{ij} \equiv \epsilon_{ikl}\epsilon_{jmn}\partial_k\partial_m u_{ln}
\tag{9.107}
$$

measures the *incompatibility* of the displacement field. The proof of statement (9.106) follows from (9.105) for a vector field: we simply observe that every vector field $V_k(\mathbf{x})$ vanishing at infinity and satisfying the integrability condition

$\left(\partial_i\partial_j - \partial_j\partial_i\right) V_k(\mathbf{x}) = 0$ can be decomposed into transverse and longitudinal pieces, namely, a gradient whose curl vanishes and a curl whose gradient vanishes,

$$V_i = \partial_i\varphi + \epsilon_{ijk}\partial_j A_k, \tag{9.108}$$

both fields φ and A_k being integrable. Explicitly these are given by

$$\varphi = \frac{1}{\partial^2}\partial_i V_i, \tag{9.109}$$

$$A_k = -\frac{1}{\partial^2}\epsilon_{klm}\partial_l V_m + \partial_k C, \tag{9.110}$$

where $1/\partial^2$ is a short notation for the Coulomb Green function $(1/\partial^2)(\mathbf{x}, \mathbf{x}')$ which acts on an arbitrary function in the usual way:

$$-\frac{1}{\partial^2}f(\mathbf{x}) \equiv \int d^3x \frac{1}{4\pi|\mathbf{x} - \mathbf{x}'|}f(\mathbf{x}'). \tag{9.111}$$

Note that the field A_k is determined by (9.110) only up to an arbitrary pure gradient $\partial_k C$.

By repeated application of this formula, we find the decompositions of an arbitrary, not necessarily symmetric, tensor u_{il}:

$$u_{il} = \partial_i\varphi'_l + \epsilon_{ijk}\partial_j A'_{kl} = \partial_i\varphi'_l + \epsilon_{ijk}\partial_j\left(\partial_l\varphi_k + \epsilon_{lmn}\partial_m A_{kn}\right). \tag{9.112}$$

Setting

$$\varphi''_i \equiv \epsilon_{ijk}\partial_j\varphi_k, \tag{9.113}$$

this may be cast as

$$u_{il} = \partial_i\varphi'_l + \partial_l\varphi''_i + \epsilon_{ijk}\epsilon_{lmn}\partial_j\partial_m A_{kn}. \tag{9.114}$$

For the special case of a symmetric tensor u_{il} we can symmetrize this result and decompose it as

$$u_{il} = \partial_i u_j + \partial_j u_i + \epsilon_{ijk}\epsilon_{lmn}\partial_j\partial_m A^S_{kn}, \tag{9.115}$$

where

$$u_i = \frac{1}{2}\left(\varphi'_i + \varphi''_i\right), \tag{9.116}$$

and A^S_{kn} is the symmetric part of A_{kn}, both being integrable fields. The first term in (9.115) has zero incompatibility, the second has zero divergence when applied to either index.

In the general case, i.e., when there is no symmetry, we can use the formulas (9.109), (9.110) twice and determine the fields $\varphi'_l, \varphi''_i, A_{kn}$ as follows:

$$\varphi'_l = \frac{1}{\partial^2}\partial_k u_{kl}, \tag{9.117}$$

$$A'_{kl} = -\frac{1}{\partial^2}\epsilon_{kpq}\partial_p u_{ql} + \partial_k C_l, \tag{9.118}$$

$$\varphi_k = -\frac{1}{\partial^4}\epsilon_{kpq}\partial_p\partial_l u_{ql} + \frac{1}{\partial^2}\partial_k\partial_l C_l, \tag{9.119}$$

$$A_{kn} = -\frac{1}{\partial^4}\epsilon_{klm}\epsilon_{npq}u_{mq} + \partial_k\left(-\frac{1}{\partial^2}\epsilon_{njl}\partial_j C_l\right) + \partial_n D_k, \tag{9.120}$$

so that from (9.112):

$$\varphi''_i = -\frac{1}{\partial^4}\partial_i\partial_p\partial_q u_{pq} + \frac{1}{\partial^2}\partial_l u_{il}. \tag{9.121}$$

Reinserting this into decomposition (9.114) we find the identity

$$
\begin{aligned}
u_{il} &= \frac{1}{\partial^2}\left(\partial_i\partial_k u_{kl} + \partial_l\partial_k u_{ik}\right) - \frac{1}{\partial^4}\partial_i\partial_l\left(\partial_p\partial_q u_{pq}\right) \\
&+ \frac{1}{\partial^4}\epsilon_{ijk}\epsilon_{lmn}\partial_j\partial_m\left(\epsilon_{kpr}\epsilon_{nqs}\partial_p\partial_q u_{rs}\right),
\end{aligned}
\tag{9.122}
$$

which is valid for any tensor of rank two. This may be verified by working out the contractions of the ϵ tensors.

While the statements (9.105) and (9.106) for vector and tensor fields are completely analogous to each other, it is important to realize that there exists an important difference between the two. For a vector field with no curl, the potential can be calculated uniquely (up to boundary conditions) from

$$\varphi = \frac{1}{\partial^2}\partial_i E_i. \tag{9.123}$$

This is no longer true, however, for the compatible tensor field u_{il}. The point of departure lies in the nonuniqueness of functions φ'_l and φ''_i in the decomposition (9.114). They are determined only modulo a common arbitrary local rotation field $\omega_i(\mathbf{x})$. In order to see this we perform the replacements

$$\partial_i\varphi'_l(\mathbf{x}) \rightarrow \partial_i\varphi'_l(\mathbf{x}) + \epsilon_{ilq}\omega_q(\mathbf{x}), \tag{9.124}$$
$$\partial_l\varphi''_i(\mathbf{x}) \rightarrow \partial_l\varphi''_i(\mathbf{x}) + \epsilon_{liq}\omega_q(\mathbf{x}), \tag{9.125}$$

and see that (9.114) is still true. The field (9.116) is only *a particular example* of a displacement field which has the strain tensor equal to the given u_{kl}:

$$u^0_{kl} = \frac{1}{2}\left(\partial_k u^0_l + \partial_l u^0_k\right) = u_{kl}. \tag{9.126}$$

This displacement field may not, however, be the true displacement field $u_l(\mathbf{x})$ in the crystal, which also satisfies

$$\frac{1}{2}\left(\partial_k u_l + \partial_l u_k\right) = u_{kl}. \tag{9.127}$$

In order to find the latter, we need additional information on the rotation field

$$\omega_{kl} = \frac{1}{2}\left(\partial_k u_l - \partial_l u_k\right). \tag{9.128}$$

We must know both $u_{kl}(\mathbf{x})$ and $\omega_{kl}(\mathbf{x})$ to calculate

$$\partial_k u_l(\mathbf{x}) = u_{kl}(\mathbf{x}) + \omega_{kl}(\mathbf{x}) \tag{9.129}$$

and solve this equation for $u_l(\mathbf{x})$.

In order to make use of this observation we have to be sure that $\omega_i = \frac{1}{2}\epsilon_{ijk}\omega_{jk}$ can be written as the curl of a displacement field $u_i(\mathbf{x})$. This is possible if

$$\partial_i \omega_i = \epsilon_{ijk}\partial_i\partial_j u_k = 0, \tag{9.130}$$

which implies that [see (9.98)]

$$\alpha_{ii}(\mathbf{x}) = 0. \tag{9.131}$$

In later discussions we shall be confronted with the situation in which u_{kl} and $\partial_i\omega_j$ are both given. In order to obtain ω_i from the latter we have to make sure that ω_i is an integrable field, which is assured by the constraint

$$\theta_{ij} = \epsilon_{ikl}\partial_k\partial_l\omega_j = 0. \tag{9.132}$$

Thus we can state the following important result: Suppose a crystal is subject to a strain $u_{kl}(\mathbf{x})$ and a rotational distortion $\omega_{kl}(\mathbf{x})$. There exists an associated single-valued displacement field $u_l(\mathbf{x})$, if and only if the crystal possesses a vanishing defect density $\eta_{ij}(\mathbf{x})$, a vanishing disclination density $\theta_{ij}(\mathbf{x})$, and a vanishing $\alpha_{ii} = 0$, i.e.,

$$\eta_{ij}(\mathbf{x}) = 0, \quad \theta_{ij}(\mathbf{x}) = 0, \quad \alpha_{ii}(\mathbf{x}) = 0. \tag{9.133}$$

Relation (9.104) implies that this is only true if two of these densities vanish, for example

$$\eta_{ij}(\mathbf{x}) = 0, \quad \alpha_{ij}(\mathbf{x}) = 0, \tag{9.134}$$

or

$$\theta_{ij}(\mathbf{x}) = 0, \quad \alpha_{ij}(\mathbf{x}) = 0. \tag{9.135}$$

Note that it is possible to introduce nonzero rotational and translational defects into a given elastically distorted crystal in such a way that θ_{ij} and α_{ij} in (9.104) cancel each other. Then the elastic distortions remain unchanged. The local rotation field, however, can be changed. In particular, it may no longer be integrable.

Notes and References

[1] The first observations and theoretical studies of dislocations are found in
O. Mügge, Neues Jahrb. Min. **13** (1883);
A. Ewing and W. Rosenhain, Phil. Trans. Roy. Soc. A **193**, 353 (1899);
J. Frenkel, Z. Phys. **37**, 572 (1926);
E. Orowan, Z. Phys. **89**, 605, 634 (1934);
M. Polany, Z. Phys. **89**, 660 (1934);
G.I. Taylor, Proc. Roy. Soc. A **145**, 362 (1934);
J.M. Burgers, Proc. Roy. Soc. A **145**, 362 (1934);
F. Kroupa and P.B. Price, Phil. Mag. **6**, 234 (1961);
W.T. Read, *Disclinations in Crystals*, McGraw-Hill, New York, 1953;
F.C. Frank and W.T. Read, Phys. Rev. **79**, 722 (1950);
B.A. Bilby, R. Bullough, and E. Smith, Proc. Roy. Soc. A **231**, 263 (1955).
Disclination lines were described by
F.C. Frank, Disc. Farad. Soc. **25**, 1 (1958).
More details can be found in the books quoted below.
Weingarten's theorem is derived in
F.R.N. Nabarro, *Theory of Dislocations*, Clarendon Press, Oxford, 1967;
C. Truesdell and R. Toupin, in *Handbook of Physics*, Vol III(1), ed. S. Flügge,
Springer, Berlin 1960.
The plastic strain described in Section 2.9 was used efficiently by
J.D. Eshelby, Brit. J. Appl. Phys. **17**, 1131 (1966)
to calculate elastic field configurations. The plastic distortions and rotations
were introduced by
T. Mura, Phil. Mag. **8**, 843 (1963), Arch. Mech. **24**, 449 (1972),
Phys. Stat. Sol. **10**, 447 (1965), **11**, 683 (1965).
See also his book *Micromechanics of Defects in Solids*, Noordhoof, Amsterdam, 1987.
For an excellent review see
E. Kröner, Les Houches lectures, publ. in *The Physics of Defects*,
eds. R. Balian et al., North-Holland, Amsterdam, 1981, p. 264.
Further useful literature can be found in
J.P. Friedel, *Disclinations*, Pergamon Press, Oxford, 1964.
See also his 1980 Les Houches lectures, op. cit.;
M. Kléman, *The General Theory of Disclinations*, in *Dislocations in Solids*,
ed. F.R.N. Nabarro, North-Holland, Amsterdam, 1980;
W. Bollmann, *Crystal Defects and Crystalline Interfaces*, Springer, Berlin, 1970.
B. Henderson, *Defects in Crystalline Solids*, Edward Arnold, London, 1972.

[2] We follow closely the theory presented in
H. Kleinert, *Gauge Fields in Condensed Matter*, World Scientific, Singapore,
1989, Vol. II, Part IV, *Differential Geometry of Defects and Gravity with Torsion*, p. 1432 (kl/b2).

[3] For details on vortex lines see
    H. Kleinert, *Gauge Fields in Condensed Matter*, World Scientific, Singapore,
    1989, Vol. I, *Superflow and Vortex Lines*, pp. 1–742 (kl/b2).

10

Defect Melting

In Chapter 5 we have seen that the phase transitions in superfluid helium and in superconductors can be understood as a consequence of the proliferation of vortex lines at the critical temperature. A similar proliferation mechanism of dislocation and disclination lines will now be shown to lead to the melting of crystals.

10.1 Specific Heat

The specific heat of solids has several parallels with the specific heat of the λ-transition. For low temperature it starts out like T^3 [see Fig. 10.1], the typical signal for the existence of massless excitations in a system. In solids, these are the longitudinal and transverse phonons. They are the Goldstone modes caused by the spontaneous breakdown of translational symmetry in the crystalline ground state. For higher temperatures the specific heat saturates at a value $6 \times k_B N/2$, in accordance with the *Dulong-Petit rule* of classical statistics. Recall that this rule assigns a specific heat $k_B/2$ to each harmonic degree of freedom. In the solid, these are three potential and three kinetic degrees of freedom per particle.

The transition between the two regimes lies at the *Debye temperature* Θ_D which is determined by the longitudinal and transversal sound velocities c_s^L, c_s^T and the particle density $n \equiv N/V$. For one atom per lattice cell, and three equal sound velocities, it is given by

$$\Theta_D = 2\pi \frac{\hbar c_s}{k_B} \left(\frac{2n}{4\pi} \right)^{1/3}. \tag{10.1}$$

The internal energy is given by the universal Debye function

$$D(z) \equiv \frac{3}{z^3} \int_0^z \frac{x^3}{e^x - 1} \tag{10.2}$$

as

$$U = 3Nk_B T D(\Theta_D/T). \tag{10.3}$$

The specific heat follows from this:

$$C = \frac{\partial U}{\partial T} = 3Nk_B T \left[D(\Theta_D/T) - (\Theta_D/T) D'(\Theta_D/T) \right]. \tag{10.4}$$

Using the limiting behavior

$$D(z) = \begin{cases} \dfrac{\pi^2}{5z^3} - 3e^{-z} + \dots & \text{for } z \gg 1, \\ 1 - \dfrac{3}{8}z^2 + \dots & \text{for } z \ll 1, \end{cases} \tag{10.5}$$

we find

$$C = 3Nk_B \begin{cases} \dfrac{4\pi^4}{5}\left(\dfrac{T}{\Theta_D}\right)^3, & \text{for } T \ll \Theta_D, \\ 1 & \text{for } T \gg \Theta_D. \end{cases} \tag{10.6}$$

The result agrees well with experiments as shown in Fig. 10.1.

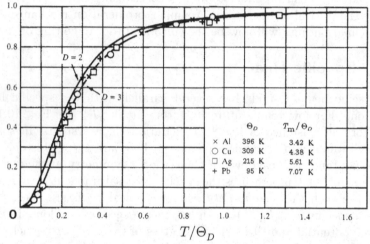

FIGURE 10.1 Specific heat of various solids. By plotting the data against the ratio T/Θ_D, where Θ_D is the Debye temperature (10.1), the data fall on a universal curve. The insert lists Θ_D-values and melting temperature T_{m}.

10.2   Elastic Energy of Solid with Defects

In solids, the elastic energy of long-wavelength distortions is usually expressed in terms of a material displacement field $u_i(\mathbf{x})$ as

$$E = \int d^3x \left[\mu u_{ij}^2(\mathbf{x}) + \frac{\lambda}{2}u_{ii}^2(\mathbf{x}) \right], \tag{10.7}$$

where μ is the shear module, λ the Lamé constant, and

$$u_{ij}(\mathbf{x}) = \frac{1}{2}[\partial_i u_j(\mathbf{x}) + \partial_j u_i(\mathbf{x})] \tag{10.8}$$

the strain tensor (9.69). The elastic energy goes to zero for infinite wave length since in this limit $u_i(\mathbf{x})$ reduces to a pure translation and the energy of the system is translationally invariant. The crystallization process causes a spontaneous breakdown of the translational symmetry of the system. The elastic distortions describe the Nambu-Goldstone-modes resulting from this symmetry breakdown.

As discussed in the previous chapter, a crystalline material always contains defects which make the displacement fields $u_i(\mathbf{x})$ multivalued. The elastic energy (10.8) is therefore incorrect. It must necessarily depend on the defect gauge-invariant strains u_{ij}^{inv} of Eq. (9.88). The correct version of (10.8) is therefore

$$\mathcal{E} = \int d^3x \left[\mu(u_{ij} - u_{ij}^{\text{p}})^2 + \frac{\lambda}{2}(u_{ii} - u_{ii}^{\text{p}})^2 \right],$$ (10.9)

where u_{ij}^{p} is the *plastic strain tensor* (9.71), with the latter describing the defects.

The above energy is the continuum limit of the energy of a crystal lattice. If we want to study the statistical mechanics of elastic and defect fluctuations, we may discretize the energy (10.9) for mathematical simplicity, on a simple cubic lattice of spacing 2π. Then the energy of $u_i(\mathbf{x})$ and $u_i(\mathbf{x}) + 2\pi N_i(\mathbf{x})$ are indistinguishable for any integer-valued field $N_i(\mathbf{x})$, which correspond to permutations of the lattice sites.

If we include only dislocation lines, the plastic strain tensor u_{ij}^{p} contains three types of surfaces where the displacement field $u_i(\mathbf{x})$ jumps by 2π, one for each lattice direction. They are characterized by the three Burgers vectors $\mathbf{b}^{(1),(2),(3)}$, and the plastic distortion (9.70) has the simpler form

$$\beta_{il}^{\text{p}}(\mathbf{x}) = \delta_i(\mathbf{x}; S)b_l,$$ (10.10)

where b_l are the components of any of the three Burgers vectors $\mathbf{b}^{(i)}$. The irrelevant surfaces S are the Volterra surfaces of the dislocation lines. The lattice version of (10.10) is

$$\beta_{ij}^p = 2\pi n_i(\mathbf{x})b_j,$$ (10.11)

where $n_i(\mathbf{x})$ are integer numbers. The lattice version of $u_{ij}^p(\mathbf{x})$ is, of course [recall (9.71)]

$$u_{ij}^p = 2\pi[n_i(\mathbf{x})b_j + n_j(\mathbf{x})b_j].$$ (10.12)

By analogy with the superfluid in Eq. (5.183), we may define the expectation value

$$\mathcal{O}_i \equiv \langle O_i(\mathbf{x}) \rangle = \langle e^{u_i(\mathbf{x})} \rangle$$ (10.13)

as an order parameter of the system. It will be nonzero in the crystalline phase since $\mathbf{u}_i(\mathbf{x})$ fluctuates around zero, and will vanish in the molten state in which $\mathbf{u}_i(\mathbf{x})$ fluctuate through the entire crystal.

We can now calculate the partition function of lattice fluctuations governed by the energy (10.9) from the functional integral and the sum over all Volterra surfaces

$$Z \equiv e^{-\beta F} = \int \mathcal{D}\mathbf{u} \sum_{S} e^{-\overline{cal}E}, \tag{10.14}$$

where $\beta \equiv 1/k_B T$. This can be done approximately with the help of low- and high-temperature approximations [1], and more precisely by Monte-Carlo simulations. The resulting specific heat near the melting transition is shown in Fig. 10.2.

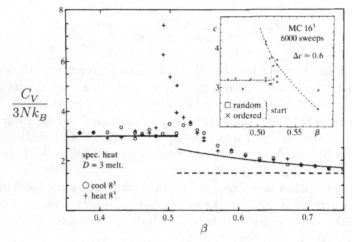

FIGURE 10.2 Specific heat of melting model (10.14). The solid lines are obtained from lowest-order high- and low-temperature expansions. The symbols + and ○ show results of Monte Carlo simulations. The insert resolves the jump of the specific heat at the transition temperature. For more details see Chapter 12 in the textbook [1].

The effect of defect lines upon the statistical mechanics of the crystal can be exhibited by proceeding along the lines of Section 5.1.7 and transforming the energy to the conjugate form corresponding to Eq. (5.86). Here the conjugate variable is the strain tensor $\sigma_{ij}(\mathbf{x})$ playing the role of the supercurrent $\mathbf{b}(\mathbf{x})$ in (5.86).

For simplicity, let us ignore the term proportional to the Lamè constant λ. Then we can easily see that the energy $\beta\mathcal{E}$ of Eq. (10.9) can be replaced by

$$\beta\mathcal{E} = \int d^3x \left[\frac{1}{4\mu}\sigma_{ij}^2 + i\sigma_{ij}(u_{ij} - u_{ij}^P) \right], \tag{10.15}$$

with an additional functional integral over the stress tensor σ_{ij}. Indeed, by a quadratic completion of this integral we recover (10.9).

If we now perform the integral over the displacement field we obtain the analog of the conservation law (5.95):

$$\partial_i \sigma_{ij} = 0. \tag{10.16}$$

This allows us to represent σ_{ij} with the help of a stress gauge field χ_{ij} as a double-curl:

$$\sigma_{ij} = \epsilon_{ikl}\epsilon_{jmn}\partial_k\partial_m\chi_{ln}, \tag{10.17}$$

in terms of which the energy (10.15) takes the form

$$\beta\mathcal{E} = \int d^3x \left[\frac{1}{4\mu}(\epsilon_{ikl}\epsilon_{jmn}\partial_k\partial_m\chi_{ln})^2 - i\epsilon_{ikl}\epsilon_{jmn}\partial_k\partial_m\chi_{ln}u_{ij}^p \right]. \tag{10.18}$$

This action is double-gauge invariant in a similar way as the superfluid energy (5.114). It is invariant under defect gauge transformations associated with the shift of S to S':

$$u_{ij}^p \rightarrow u_{ij}^p + \partial_i\delta(\mathbf{x};V)b_j + \partial_j\delta(\mathbf{x};V)b_i \tag{10.19}$$

where V is the volume over which S has swept on the way to S', if u_i is simultaneously transformed as

$$u_i \rightarrow u_i + 2\pi\delta(\mathbf{x},V). \tag{10.20}$$

And it is manifestly invariant under stress gauge transformations

$$\chi_{ij} \rightarrow \chi_{ij} + \partial_i\Lambda_j + \partial_j\Lambda_i. \tag{10.21}$$

A partial integration of the energy (10.18) leads to the form [compare (5.114)]

$$\beta\mathcal{E} = \int d^3x \left[\frac{1}{4\mu}(\epsilon_{ikl}\epsilon_{jmn}\partial_k\partial_m\chi_{ln})^2 - i\chi_{ij}\eta_{ij} \right], \tag{10.22}$$

where η_{ij} is the defect tensor defined in Eq. (9.97) [the analog of the magnetic current (4.92), or of the vortex density (5.30)]. Note that the strain tensor u_{ij} of Eq. (10.8) has no incompatibility so that only u_{ij}^p contributes to the defect tensor η_{ij}.

Let us simplify the expression (10.17) for the stress field as follows. First we use the identity (1A.16) to obtain

$$\sigma_{ij} = -(\partial^2\chi_{ij} + \partial_i\partial_j\chi_k{}^k - \partial_i\partial_k\chi_i{}^k - \partial_j\partial_k\chi_i{}^k) + \eta_{ij}(\partial^2\chi_k{}^k - \partial_k\partial_l\chi^{kl}). \tag{10.23}$$

Then we introduce the modified gauge field

$$\phi_{ij} \equiv \chi_{ij} - \tfrac{1}{2}\delta_{ij}\chi_{kk}, \tag{10.24}$$

and finally we go to the so-called *Hilbert gauge* in which

$$\partial_i\phi_{ij} = 0. \tag{10.25}$$

As a result, the stress field becomes simply

$$\sigma_{ij} = -\partial^2\phi_{ij}, \tag{10.26}$$

and the energy (10.22) takes the form

$$\beta\mathcal{E} = \int d^3x \left[\frac{1}{4\mu}(\partial^2\phi_{ij})^2 - i\phi_{ij}(\eta_{ij} - \frac{1}{2}\delta_{ij}\eta_{kk}) \right]. \tag{10.27}$$

Extremization with respect to the field ϕ_{ij} yields the interaction of an arbitrary distribution of defects [the analog of (4.93) and (5.82)]:

$$\beta\mathcal{E} = \mu \int d^3x \left(\eta_{ij} - \tfrac{1}{2}\delta_{ij}\eta_{kk} \right) \frac{1}{(\partial^2)^2} \left(\eta_{ij} - \tfrac{1}{2}\delta_{ij}\eta_{kk} \right). \tag{10.28}$$

Inserting for η_{ij} the decomposition (9.104) into dislocation and disclination densities, we find that dislocation lines interact with Biot-Savart type forces with parallel line elements having a repulsive $1/r$-interaction. Disclination lines, on the other hand repel each other with a linearly rising potential. For more details see Ref. [1].

In the interaction (10.28), the interplay between dislocations and disclinations discussed in Section 9.4 is perfect. Dislocations can be replaced freely by adjacent pairs of disclinations, and disclinations by a string of dislocations. The partition function containing the energy (10.28) can therefore be done only over dislocations. Otherwise it diverges due to overcounting. The interchangeability between the defects is a kind of gauge freedom which must be fixed in the partition function.

In a real crystal, however, the two kinds of defects can be distinguished, and for this reason we must extend the elastic energy (10.9) by higher gradient terms in the displacement field. An extended energy contains the invariant bend-twist (9.90) and reads, in the simplest possible version [1, 2]

$$\beta\mathcal{E} = \mu \int d^3x \left[\left(u_{ij} - u_{ij}^{\mathrm{p}} \right)^2 + \ell^2 \left(\partial_i\omega_j - \kappa_{ij}^{\mathrm{p}} \right)^2 \right]. \tag{10.29}$$

The parameter ℓ is the length scale over which the crystal is rotationally stiff.

By analogy with the treatment of the above energy (10.9) we reformulate the energy (10.29) in a canonical representation of the type (10.15). In addition to the stress field σ_{ij} there is now a rotational stress field τ_{ij}, and we may rewrite the elastic action of defect lines as [1, 2]

$$\beta\mathcal{E} = \int d^3x \left[\frac{1}{4\mu} \left(\sigma_{ij} + \sigma_{ji} \right)^2 + \frac{1}{8\mu\ell^2} \tau_{ij}^2 \right. $$
$$\left. + i\sigma_{ij} \left(\partial_i u_j - \epsilon_{ijk}\omega_k - \beta_{ij}^{\mathrm{p}} \right) + i\tau_{ij} \left(\partial_i\omega_j - \phi_{ij}^{\mathrm{p}} \right) \right]. \tag{10.30}$$

We have found it convenient to introduce an independent variable ω_i. The partition function is defined by integrating the Boltzmann factor $e^{-\beta\mathcal{E}}$ functionally over $\sigma_{ij}, \tau_{ij}, u_i, \omega_j$ and summing over all jumping surfaces S in the plastic fields. The functional integral over the antisymmetric part of σ_{ij} fixes the independent variable ω_i to satisfy $\omega_i = \tfrac{1}{2}\epsilon_{ijk}(\partial_j u_k + \beta_{ij}^{\mathrm{p}})$. Reinserting this into (10.30) and integrating out the remaining σ_{ij} and τ_{ij} we recover the original expression (10.29).

Alternatively, we may integrate out ω_j and u_i in the partition function. This leads to the conservation laws, generalizing (10.16),

$$\partial_i\sigma_{ij} = 0, \quad \partial_i\tau_{ij} = -\epsilon_{jkl}\sigma_{kl}. \tag{10.31}$$

These are dual to the conservation laws for disclination and dislocation densities (9.48) and (9.49), respectively.

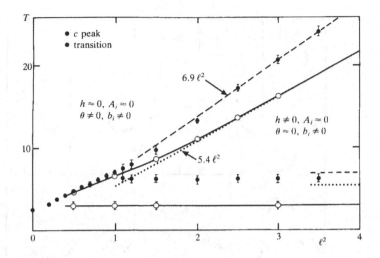

FIGURE 10.3 Phase diagram in the T-ℓ-plane in two-dimensional melting. Theoretical curves are from Ref. [2], Monte Carlo data from [3]. Detailed discussion is in textbook [1].

The conservation laws are guaranteed as Bianchi identities by introducing the stress gauge fields A_{ij} and h_{ij} and writing

$$
\begin{aligned}
\sigma_{ij} &= \epsilon_{ikl}\partial_k A_{lj} \\
\tau_{ij} &= \epsilon_{ikl}\partial_k h_{lj} + \delta_{ij}A_{ll} - A_{ji}.
\end{aligned}
\tag{10.32}
$$

This allows us to re-express the energy (10.30) as

$$
\beta\mathcal{E} = \int d^3x \left[\frac{1}{4\mu}\left(\sigma_{ij} + \sigma_{ji}\right)^2 + \frac{1}{8\mu l^2}\tau_{ij}^{\;2} + A_{ij}\alpha_{ij} + h_{ij}\theta_{ij} \right].
\tag{10.33}
$$

The stress gauge fields couple locally to the defect densities, and these are singular on the boundary lines of the Volterra surfaces. In the limit of a vanishing length scale ℓ, τ_{ij} is forced to be identically zero and (10.32) allows us to express A_{ij} in terms of h_{ij} which may be identified with the stress gauge field χ_{ij} in Eq. (10.17). Then the energy (10.30) reduces to (10.22).

Depending on the length parameter ℓ of rotational stiffness, the defect system was predicted in Ref. [2] to have either a single first-order transition (for small ℓ), or two successive continuous melting transitions. In the first transitions, dislocation lines proliferate and destroy the translational order, in the second transition, disclination lines proliferate and destroy the rotational order (see Figs. 10.4 and 10.3).

The existence of two successive continuous transitions was conjectured a long time ago [4, 5, 6, 7] for two-dimensional melting, where these transitions should be of the Kosterlitz-Thouless type. However, the simplest lattice defect models constructed to illustrate this behavior displayed only a single first-order transition [8].

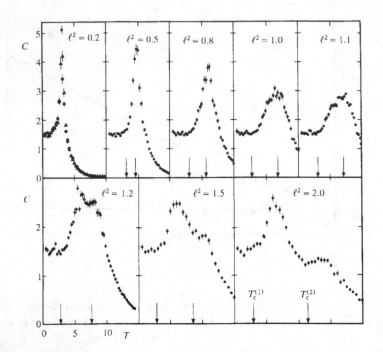

FIGURE 10.4 Separation of first-order melting transition into two successive Kosterlitz-Thouless transitions in two dimensions when increasing the length scale ℓ of rotational stiffness of the defect model. Monte Carlo data are from Ref. [3]. See also textbook [1].

Only after introducing the angular stiffness ℓ in Ref. [2] was it possible to separate the first-order melting transition into two successive Kosterlitz-Thouless transitions. The dependence on ℓ is shown in Figs. 10.4 and 10.3.

Notes and References

[1] H. Kleinert, *Gauge Fields in Condensed Matter*, Vol. II: *Stresses and Defects, Differential Geometry, Crystal Defects*, World Scientific, Singapore, 1989 (kl/b2).

[2] H. Kleinert, Phys. Lett. A **130**, 443 (1988) (kl/174).

[3] W. Janke and H. Kleinert, Phys. Rev. Lett. **61**, 2344 (1988) (kl/179).

[4] D. Nelson, Phys. Rev. B **18**, 2318 (1978).

[5] D. Nelson and B.I. Halperin, Phys. Rev. B **19**, 2457 (1979).

[6] A.P. Young, ibid., 1855 (1979).

[7] D. Nelson, Phys. Rev. B **26**, 269 (1982).

[8] W. Janke and H. Kleinert, Phys. Lett. A **105**, 134 (1984); (kl/120); Phys. Lett. A **114**, 255 (1986) (kl/135).

11

Relativistic Mechanics in Curvilinear Coordinates

The basic idea which led Einstein to his formulation of the theory of gravitation in terms of curved spacetime was the observation by Galileo Galilei (1564-1642) that, in the absence of air friction, all bodies would fall with equal velocity. This observation was confirmed with higher accuracy by C. Huygens (1629-1695). In 1889, R. Eötvös found a simple trick to remove the air friction completely [1]. This enabled him to limit the relative difference between the falling speeds of wood and platinum to one part in 10^9. This implies that the *inertial mass* m which governs the acceleration of a body if subjected to a force $\mathbf{f}(t)$ in Newton's equation of motion

$$m\,\ddot{\mathbf{x}}(t) = \mathbf{f}(t), \tag{11.1}$$

which appears on the left-hand side of the equations of motion (1.2), and the *gravitational mass* on the right-hand side of Eq. (1.2) cannot differ by more than this extremely small amount.

11.1 Equivalence Principle

Einstein considered the result of the Eötvös experiment as evidence that inertial and gravitational masses are *exactly* equal. From this he concluded that the motion of all point particles under the influence of a gravitational field can be described completely in geometric terms. The basic thought experiment which led him to this conclusion consisted in imagining an elevator in a large sky scraper to fall freely. Since all bodies in it would fall with the same speed, they would appear weightless. Thus, for an observer inside the cabin, the gravitational attraction to the earth would have disappeared. Einstein concluded that gravitational forces can be removed by acceleration. This is the content of the *equivalence principle*.

Mathematically, the cabin is just an accelerating coordinate frame. If the original spacetime coordinates with gravity are denoted by x^μ, the coordinates of the small cabin are given by a function $x^\mu(x^a)$. Hence the equivalence principle states that the behavior of particles under the influence of gravitational forces can be found by going to a new coordinate frame $x^a(x^\mu)$ in which the motion within the cabin proceeds without gravity.

316

There is a converse way of stating this principle. Given an inertial frame x^a, we can simulate a gravitational field at a point by going to a small cabin with x^μ, which is accelerated with respect to the inertial frame x^a. In the coordinates x^μ, the motion of the particle looks the same as if a gravitational field were present.

This suggests a simple way of finding the equations of motion of a point particle in a gravitational field: one must simply transform the known equations of motion in an inertial frame to arbitrary curvilinear coordinates x^μ. When written in general coordinates x^μ, the flat-spacetime equations must be valid also in the presence of gravitational fields.

In formulating the equivalence principle it must be realized that by a coordinate transformation the gravitational field can only be removed at a single point. In a falling cabin, a point particle will remain at the same place only if it resides at the center of mass of the cabin. Particles in the neighborhood of this point will move slowly away from this point. The force causing this are called tidal forces. They are the same forces which give rise to the tidal waves of the oceans. Earth and moon circle around each other and their center of mass circles around the sun. The center of mass is in "free fall", the gravitational attraction proportional to the gravitational mass being canceled by the centrifugal force proportional to the inertial mass. Any point on the earth which lies farther from the sun than the center of mass is pulled outwards by the centripetal force, those which lie closer are pulled inwards by the gravitational force.

It is important to realize that the existence of tidal forces makes it impossible to simulate gravitational forces by coordinate transformation in quantum mechanics. Due to Heisenberg's uncertainty relation, quantum particles can never be localized to a point but always occur in the form of wave packets. These flow apart and are therefore increasingly sensitive to the tidal forces. If one wants to remove the gravitational forces for a wave packet, *multivalued* coordinate transformations will be necessary of the type used in the last chapter to create defects. These will supply us with a *quantum equivalence principle* to be derived in Chapter 12.

11.2 Free Particle in General Coordinates Frame

As a first application of Einstein's equivalence principle, consider the action (2.19) of a free massive point particle in Minkowski spacetime:

$$\overset{m}{\mathcal{A}} = -mc \int_{s_a}^{s_b} ds = \int_{\sigma_a}^{\sigma_b} d\sigma \ \overset{m}{\mathcal{L}} \tag{11.2}$$

with the Lagrangian density

$$\overset{m}{\mathcal{L}} = -mc \, ds/d\sigma = -mc \sqrt{g_{ab} \, \dot{x}^a(\sigma) \dot{x}^b(\sigma)}. \tag{11.3}$$

The parameter s denotes the invariant length along the path $x(\sigma)$ defined in Eq. (1.139), which is proportional to the proper time $s = c\tau$ [recall (1.141)].

A free particle moves along a straight line

$$\ddot{x}^a(\tau) = 0, \tag{11.4}$$

which is the shortest spacetime path between initial and final points. This path extremizes the action:

$$\delta \overset{m}{\mathcal{A}} = 0. \tag{11.5}$$

When going to an arbitrary curvilinear description of the same Minkowski spacetime in terms of coordinates x^μ carrying latin indices

$$x^\mu = x^\mu(x^a), \tag{11.6}$$

the invariant length ds is given by

$$ds = d\sigma \sqrt{g_{\mu\nu}(x(\sigma))\dot{x}^\mu(\sigma)\dot{x}^\nu(\sigma)}. \tag{11.7}$$

The 4×4 spacetime-dependent matrix

$$g_{\mu\nu}(q) = g_{ab}\frac{\partial x^a}{\partial x^\mu}\frac{\partial x^b}{\partial x^\nu} \tag{11.8}$$

plays the role of a *metric* in the spacetime. Note that the inverse metric is given by

$$g^{\mu\nu}(x) = g^{ab}\frac{\partial x^\mu}{\partial x^a}\frac{\partial x^\nu}{\partial x^b}. \tag{11.9}$$

Since spacetime has not really changed, only its parametrization, the path is still straight. The equation of motion in the new curvilinear coordinates x^μ can be found in two ways. One is to simply transform the free equation of motion in Minkowski spacetime (11.4) to curvilinear coordinates. This will be done at the end of this section.

To begin we derive the equation of motion by extremizing the action written in general coordinates:

$$\overset{m}{\mathcal{A}} = \int_{\sigma_a}^{\sigma_b} d\sigma \overset{m}{L}(\dot{x}^\mu(\sigma)), \tag{11.10}$$

with the transformed Lagrangian [compare (2.19)]

$$\overset{m}{L}(\dot{x}^\mu(\sigma)) = -mc\frac{ds}{d\sigma} = -mc\left[g_{\mu\nu}(x(\sigma))\dot{x}^\mu(\sigma)\dot{x}^\nu(\sigma)\right]^{\frac{1}{2}}. \tag{11.11}$$

As observed in Subsection 2.2, the action (11.10) is invariant under arbitrary reparametrizations

$$\sigma \to \sigma' = f(\sigma). \tag{11.12}$$

Variation of the action yields

$$\begin{aligned}
\delta \overset{m}{\mathcal{A}} &= \int_{\sigma_a}^{\sigma_b} d\sigma\, \delta L(\dot{x}^\mu(\sigma)) \\
&= -m^2 c^2 \frac{1}{2}\int_{\sigma_a}^{\sigma_b}\frac{d\sigma}{L(\dot{x}^\mu(\sigma))}\left[(\partial_\lambda g_{\mu\nu})\,\delta x^\lambda\,\dot{x}^\mu(\sigma)\dot{x}^\nu(\sigma) + 2g_{\lambda\nu}\frac{d\delta x^\lambda}{d\sigma}\dot{x}^\nu(\sigma)\right].
\end{aligned} \tag{11.13}$$

On the right-hand side we have used the property (2.7) that the variation of the derivative is equal to the derivative of the variation.

The factor before the bracket is equal to $d\sigma/(-mc\,ds/d\sigma) = -(d\sigma/ds)^2 ds/mc$. Thus, if we choose σ to be the proper time τ for which $d\sigma/ds = d\tau/ds = 1/c$, we may rewrite the variations as

$$\delta \overset{m}{\mathcal{A}} = -m\frac{1}{2}\int_{\tau_a}^{\tau_b} d\tau \left[(\partial_\lambda g_{\mu\nu})\,\delta x^\lambda\,\dot{x}^\mu(\tau)\dot{x}^\nu(\tau) + 2g_{\lambda\nu}\frac{d\delta x^\lambda}{d\tau}\dot{x}^\nu(\tau) \right]. \tag{11.14}$$

This shows that if we use the proper time τ to parameterize the paths, the equations of motion can alternatively be derived from the simpler action

$$\overset{m}{\mathcal{A}} = \int_{\tau_a}^{\tau_b} d\tau\, \overset{m}{L}\left(\dot{x}^\mu(\tau) \right), \tag{11.15}$$

where

$$\overset{m}{L}\left(\dot{x}^\mu(\tau) \right) \equiv -\frac{m}{2}\, g_{\mu\nu}(x(\tau))\dot{x}^\mu(\tau)\dot{x}^\nu(\tau). \tag{11.16}$$

This has the same form as the action of a nonrelativistic point particle in four-dimensional spacetime parameterized by a pseudotime τ. Note that although the action (11.15) has the same extrema as (11.10), it has only half the size.

The second integral in (11.14) can be performed by parts to yield

$$2g_{\lambda\nu}(x(\tau))\delta x^\lambda(\tau)\dot{x}^\nu(\tau)\Big|_{\tau_a}^{\tau_b} - 2\int_{\tau_a}^{\tau_b} d\tau\, \delta x^\lambda(\tau)\frac{d}{d\tau}\left[g_{\lambda\nu}(x(\tau))\dot{x}^\nu(\tau) \right]. \tag{11.17}$$

According to the extremal principle of classical mechanics, we derive the equations of motion by varying the action with vanishing variations of the paths δq^μ at the endpoints [recall (2.3), which leads to the equation

$$\frac{1}{2}\int_{\tau_a}^{\tau_b} d\tau \left[\left(\partial_\lambda g_{\mu\nu} - 2\partial_\mu g_{\lambda\nu} \right)\dot{x}^\mu(\tau)\dot{x}^\nu(\tau) - 2g_{\lambda\nu}\ddot{x}^\nu(\tau) \right]\delta x^\lambda(\tau) = 0. \tag{11.18}$$

This is valid for all $\delta x^\mu(\tau)$ vanishing at the endpoints, in particular for the infinitesimal local spikes:

$$\delta x^\mu(\tau) = \epsilon\delta(\tau - \tau_0). \tag{11.19}$$

Inserting these into (11.18) we obtain the equations of motion

$$g_{\lambda\nu}\ddot{x}^\nu(\tau) + \left(\partial_\mu g_{\lambda\nu} - \frac{1}{2}\partial_\lambda g_{\mu\nu} \right)\dot{x}^\mu(\tau)\dot{x}^\nu(\tau) = 0. \tag{11.20}$$

It is convenient to introduce a quantity called the *Riemann connection*, or *Christoffel symbol*:

$$\bar{\Gamma}_{\mu\nu\lambda} \equiv \{\mu\nu, \lambda\} = \frac{1}{2}\left(\partial_\mu g_{\nu\lambda} + \partial_\nu g_{\mu\lambda} - \partial_\lambda g_{\mu\nu} \right). \tag{11.21}$$

Then the equation of motion can be written as

$$g_{\lambda\nu}\ddot{x}^\nu(\tau) + \bar{\Gamma}_{\mu\nu\lambda}\,\dot{x}^\mu(\tau)\dot{x}^\nu(\tau) = 0. \tag{11.22}$$

By further introducing the *modified Christoffel symbol*

$$\bar{\Gamma}_{\mu\nu}{}^{\kappa} \equiv \left\{ \begin{matrix} \kappa \\ \mu\nu \end{matrix} \right\} = g^{\kappa\lambda}\bar{\Gamma}_{\mu\nu\lambda} = g^{\kappa\lambda}\{\mu\nu\lambda\}\,, \tag{11.23}$$

we can bring Eq. (11.20) to the form

$$\ddot{x}^{\lambda}(\tau) + \bar{\Gamma}_{\mu\nu}{}^{\lambda}\dot{x}^{\mu}(\tau)\dot{x}^{\nu}(\tau) = 0. \tag{11.24}$$

A path $x^{\lambda}(\tau)$ satisfying this differential equation of shortest length is called a *geodesic trajectory*. It is Einstein's postulate, that this equation describes correctly the motion of a point particle in the presence of a gravitational field.

Now we turn to the simpler direct derivation of the equation of motion applying the coordinate transformation $x^{a}(x^{\mu})$ to the straight-line equation of motion (11.4) in Minkowski spacetime:

$$\ddot{x}^{a}(\tau) = \frac{d}{dt}\left[\frac{\partial x^{a}}{\partial x^{\mu}}\dot{x}^{\mu}(\tau)\right] = \frac{\partial x^{a}}{\partial x^{\mu}}\ddot{x}^{\mu}(\tau) + \left(\frac{d}{dt}\frac{\partial x^{a}}{\partial x^{\mu}}\right)\dot{x}^{\mu}(\tau) = 0, \tag{11.25}$$

where we have written $\partial x^{a}/\partial x^{\mu}$ for the coordinate transformation matrix evaluated on the trajectory $x(\tau)$. Multiplying this by $\partial x^{\lambda}/\partial x^{a}$ and summing over repeated indices a yields

$$\ddot{x}^{\lambda}(\tau) + \frac{\partial x^{\lambda}}{\partial x^{a}}\left(\frac{d}{dt}\frac{\partial x^{a}}{\partial x^{\mu}}\right)\dot{x}^{\mu}(\tau) = \ddot{x}^{\lambda}(\tau) + \frac{\partial x^{\lambda}}{\partial x^{a}}(\partial_{\mu}\partial_{\nu}x^{a})\,\dot{x}^{\mu}(\tau)\dot{x}^{\nu}(\tau) = 0. \tag{11.26}$$

The second term can be processed using (11.9) as follows:

$$\frac{\partial x^{\lambda}}{\partial x^{a}}(\partial_{\mu}\partial_{\nu}x^{a}) = g^{\lambda\sigma}(\partial_{\sigma}x_{a})(\partial_{\mu}\partial_{\nu}x^{a}). \tag{11.27}$$

It takes a little algebra to verify that this is equal to $\bar{\Gamma}_{\lambda\kappa}{}^{\mu}$, so that the transformed equation of motion (11.26) coincides, indeed, with the geodesic equation (11.24).

11.3 Minkowski Geometry formulated in General Coordinates

In Einstein's theory, all gravitational effects can be completely described by a non-trivial geometry of spacetime. As a first step towards developing this theory it is important to learn to distinguish between inessential properties of the geometry which are merely due to the formulation in terms of general coordinates, as in the last section, and those which are caused by the presence of gravitational forces. For this purpose we study in more detail the mathematics of coordinate transformation in Minkowski spacetime.

11.3.1   Local Basis tetrads

As in Eq. (1.25) we use coordinates x^a ($a = 0, 1, 2, 3$) to specify the points in Minkowski spacetime. From now on it will be convenient to use fat latin letters to denote four-vectors in spacetime. Thus we shall denote the four-dimensional basis vectors by $\mathbf{e}_a$, and an arbitrary four-dimensional vector with coordinates x^a by $\mathbf{x} = \mathbf{e}_a x^a$. The basis vectors are orthonormal with respect to the Minkowski metric g_{ab} of Eq. (1.29):

$$\mathbf{e}_a \mathbf{e}_b = g_{ab}. \qquad (11.28)$$

The basis vectors $\mathbf{e}_a$ define an inertial frame of reference.

Let us now reparametrize this Minkowski spacetime by a new set of coordinates x^μ whose values are given by a mapping

$$x^a \to x^\mu = x^\mu(x^a). \qquad (11.29)$$

Since x^μ still labels the same spacetime we shall assume the function $x^\mu(x^a)$ to possess an inverse $x^a = x^a(x^\mu)$ and to be sufficiently smooth so that $x^\mu(x^a)$ and $x^a(x^\mu)$ have at least two smooth derivatives. These will always commute with each other. In other words, the general coordinate transformation (11.29) and their inverse $x^a(x^\mu)$ will satisfy the integrability conditions of Schwartz:

$$\left(\partial_\mu \partial_\nu - \partial_\nu \partial_\mu \right) x^a(x^\kappa) = 0, \qquad (11.30)$$

$$\left(\partial_\mu \partial_\nu - \partial_\nu \partial_\mu \right) \partial_\lambda x^a(x^\kappa) = 0. \qquad (11.31)$$

The conditions $x^\mu(x^a) = $ const. define a network of new coordinate hypersurfaces whose normal vectors are given by (see Fig. 11.1)

$$\mathbf{e}_\mu(x) \equiv \mathbf{e}_a e^a{}_\mu(x) = \mathbf{e}_a \frac{\partial x^a}{\partial x^\mu}. \qquad (11.32)$$

These are called *local basis vectors*. Their components $e^a{}_\mu(x)$ are called local *basis tetrads*. The difference vector between two points $\mathbf{x}'$ and $\mathbf{x}$ has, in the inertial frame

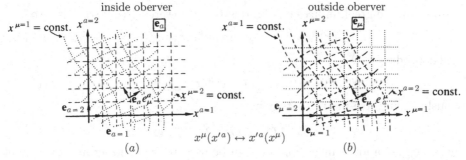

FIGURE 11.1 Illustration of crystal planes ($x^\mu = $ const.) before and after elastic distortion, once seen from within the crystal (a) and once from outside (b).

of reference, the description $\Delta \mathbf{x} = \mathbf{e}_a \left(x'^a - x^a \right)$. When going to coordinates x'^μ, x^μ, this becomes

$$\Delta \mathbf{x} = \mathbf{e}_a \int_x^{x'} e^a{}_\mu(x) dx^\mu. \tag{11.33}$$

The length of an infinitesimal vector $d\mathbf{x}$ is given by

$$ds = \sqrt{d\mathbf{x}^2} = \sqrt{\left(\mathbf{e}_\mu dx^\mu \right)^2} = \sqrt{\mathbf{e}_\mu \mathbf{e}_\nu dx^\mu dx^\nu}. \tag{11.34}$$

The right-hand side shows that the metric in the curvilinear coordinates can be expressed as a scalar product of the local basis vectors:

$$g_{\mu\nu}(x) = \mathbf{e}_\mu(x) \mathbf{e}_\nu(x). \tag{11.35}$$

Indeed, inserting here (11.32) and using (11.28) leads back to Eq. (11.8):

$$g_{\mu\nu}(x) = g_{ab} e^a{}_\mu(x) e^b{}_\nu(x) = \mathbf{e}_a \mathbf{e}_b \frac{\partial x^a}{\partial x^\mu} \frac{\partial x^b}{\partial x^\nu} = g_{ab} \frac{\partial x^a}{\partial x^\mu} \frac{\partial x^b}{\partial x^\nu}. \tag{11.36}$$

In the sequel, we shall freely raise and lower the latin index using the Minkowski metric $g_{ab} = g^{ab}$, and define

$$e^{a\mu} \equiv g^{ab} e_b{}^\mu, \qquad e_{a\mu} \equiv g_{ab} e^b{}_\mu. \tag{11.37}$$

Then we can rewrite (11.35) as

$$g_{\mu\nu}(x) = \mathbf{e}_\mu(x) \mathbf{e}_\nu(x) = e^a{}_\mu(x) e_{a\nu}(x). \tag{11.38}$$

Since the general coordinate formulation (11.28) was assumed to have an inverse, we can also calculate the derivatives $\partial x^\mu / \partial x^a$. These are orthonormal to the derivatives $\partial x^a / \partial x^\mu$ in two ways:

$$\frac{\partial x^a}{\partial x^\mu} \frac{\partial x^\mu}{\partial x^b} = \delta^a{}_b, \qquad \frac{\partial x^\mu}{\partial x^a} \frac{\partial x^a}{\partial x^\nu} = \delta^\mu{}_\nu. \tag{11.39}$$

It is useful to denote the inverse derivatives $\partial x^\mu / \partial x^a$ by $e_a{}^\mu$ and introduce the vectors $\mathbf{e}^\mu = \mathbf{e}_a g^{ab} e_b{}^\mu$, called the *reciprocal multivalued basis tetrads* (see also Fig. 11.2). With this notation, the equations in (11.39) become orthonormality and completeness relations of the tetrads:

$$e^a{}_\mu(x) e_b{}^\mu(x) = \delta^a{}_b, \qquad e_a{}^\mu(x) e^a{}_\nu(x) = \delta^\mu{}_\nu. \tag{11.40}$$

The position of the greek indices can be raised and lowered freely so that also

$$e^{a\mu}(x) e_{b\mu}(x) = \delta^a{}_b, \qquad e_{a\mu}(x) e^{a\nu}(x) = \delta_\mu{}^\nu. \tag{11.41}$$

The scalar product

$$g^{\mu\nu}(x) = \mathbf{e}^\mu(x) \mathbf{e}^\nu(x) = e^{a\mu}(x) e_a{}^\nu(x) \tag{11.42}$$

is obviously the inverse metric, satisfying

$$g^{\mu\nu}(x) g_{\nu\lambda}(x) = \delta^\mu{}_\lambda. \tag{11.43}$$

The metric $g_{\mu\nu}(x)$ and its inverse $g^{\mu\nu}(x)$ can be used to freely lower and raise greek indices on any tensor, and to form invariants under coordinate transformations by contraction of all indices.

11.3.2 Vector- and Tensor Fields in Minkowski Coordinates

In Section 1.4 we have analyzed physical quantities according to their transformation properties under Lorentz transformations. These transformations change the coordinates of an inertial frame by multiplication with a matrix $\Lambda^a{}_b$,

$$x^a \to x'^a \equiv (\Lambda x)^a = \Lambda^a{}_b x^b. \tag{11.44}$$

This is done in such a way that the scalar products (1.78) and length elements (1.142) are the same in both sets of coordinates x^a and x'^a. The transformation matrices $\Lambda^a{}_b$ satisfy the pseudo-orthogonality relation (1.28), and have the infinitesimal representation (1.103):

$$\Lambda^a{}_b = \delta^a{}_b + \omega^a{}_b, \qquad \left(\Lambda^{-1}\right)^a{}_b = \delta^a{}_b - \omega^a{}_b, \tag{11.45}$$

where $\omega_{ab} = -\omega_{ba}$ has the six independent matrix elements rotation angle φ^k and rapidity ζ^i of Eqs. (1.55) and (1.56).

Since the physical points are the same before and after a Lorentz transformation, the basis vectors $\mathbf{e}_a$ change according to the law

$$\mathbf{e}_a \to \mathbf{e}'_a \equiv \mathbf{e}_b \left(\Lambda^{-1}\right)^b{}_a. \tag{11.46}$$

This gives

$$\mathbf{x} \equiv \mathbf{e}_a x^a \longrightarrow \mathbf{e}'_a x'^a = \mathbf{e}_b \left(\Lambda^{-1}\right)^b{}_a \Lambda^a{}_c x^c = \mathbf{e}_a x^a = \mathbf{x}, \tag{11.47}$$

showing that the vectors in the inertial frame are the same before and after the transformation. So far the discussion is only a reminder of Section 1.4.

Consider now a *vector field* $v^a(x)$. It assigns to every point P a vector

$$\mathbf{v}(P) = \mathbf{e}_a v^a(x). \tag{11.48}$$

After a Lorentz transformation of the coordinates x^a and the basis vectors $\mathbf{e}_a$, the observable vector $\mathbf{v}(P)$ specifies the same point, i.e.,

$$\mathbf{v}'(P) = \mathbf{v}(P). \tag{11.49}$$

Writing this as

$$\mathbf{v}'(P) = \mathbf{e}'_a v'^a(x') = \mathbf{v}(P) = \mathbf{e}_a v^a(x) \tag{11.50}$$

we see that the components of the vector in the two sets of coordinates have to be related in the same way as the coordinates x'^a and x^a, i.e.,

$$v'^a(x') = \Lambda^a{}_b v'^b(x), \tag{11.51}$$

or, after replacing $x' \to x$ and $x \to x^{-1}x$,

$$v'^a(x) = \Lambda^a{}_b v^b \left(\Lambda^{-1}x\right). \tag{11.52}$$

For infinitesimal transformations (1.74 of the coordinates

$$\Lambda^a{}_b x^b = \left(\delta^a{}_b + \omega^a{}_b\right) x^b, \qquad \left(\Lambda^{-1} x\right)^a = x^a - \omega^a{}_b x^b, \tag{11.53}$$

the contravariant vector field $v^a(x)$ goes over into

$$v'^a(x) = v^a(x) + \omega^a{}_b v^b(x) - \omega^{b'}{}_b x^b \partial_{b'} v^a(x). \tag{11.54}$$

The infinitesimal version of the transformation law (11.54) is recognized to be a substantial variation δ_s as defined in (3.6), which we shall write for Lorentz transformations as

$$\begin{aligned}
\delta_\Lambda v^a(x) &= v'^a(x) - v^a(x) \\
&= \omega^a{}_b v^b(x) - \omega^b{}_{b'} x^{b'} \partial_b v^a(x). \tag{11.55}
\end{aligned}$$

Let us find the infinitesimal transformation laws also for the covariant vector field $v_a(x) = g_{ab}(x) v^b(x)$. The substantial variation of $v_a(x)$ analogous to (11.55) is

$$\begin{aligned}
\delta_\Lambda v_a(x) &= \omega_a{}^b v_\Lambda(x) - \omega^{b'}{}_b x^b \partial_{b'} v_a(x) \\
&= \omega_a{}^b v_b(x) + \omega_b{}^{b'} x^b \partial_{b'} v_a(x) \tag{11.56}
\end{aligned}$$

where we have introduced the matrix elements of ω_{ab}:

$$\omega_b{}^{b'} = g_{ab} g^{a'b'} \omega^a{}_{b'} = g^{a'b'} \omega_{ba}. \tag{11.57}$$

The derivatives of a contravariant vector field v_a with respect to changes of x^a are higher tensor fields. Infinitesimally, derivatives transform via the *sum* of operations (11.56), one applied to each index. This follows directly from (11.56) and the commutation rule $[\partial_a, x_b] = g_{ab}$:

$$\begin{aligned}
\delta_\Lambda \partial_b v_a &= \partial_b \delta_\Lambda v_a \\
&= \partial_b \left(\omega_a{}^{a'} v_{a'} + \omega_a{}^{c'} x^c \partial_{c'} v_a \right) \tag{11.58} \\
&= \omega_a{}^{a'} \partial_b v_{a'} + \omega_b{}^{b'} \partial_{b'} v_a + \omega_c{}^{c'} x^c \partial_{c'} \partial_b v_a.
\end{aligned}$$

This simple rule can easily be extended to arbitrary higher derivatives thereby forming higher tensor fields. Note that since the arguments in f and f' in (3.6) are the same, the operation "substantial variation" commutes with the derivative.

11.3.3   Vector- and Tensor Fields in General Coordinates

Consider now the same physical objects but described in terms of curvilinear coordinates $x^\mu(x^a)$. Then the components of $\mathbf{v}(P)$ are not measured with respect to the basis $\mathbf{e}_a$ but with respect to the *local* basis $\mathbf{e}_\mu(x) = \mathbf{e}_a e^a{}_\mu(x)$ of Eq. (11.32). It is then natural to specify $\mathbf{v}(P)$ in terms of its local components $v^\mu(x) = v^a(x) e_a{}^\mu(x)$.

On the fields $v^\mu(x)$ one cannot only perform Lorentz transformations but any general coordinate transformation $x^\mu \to x'^\mu(x^\mu)$ which will be referred to as *Einstein transformations*.

Under Einstein transformations the vectors $e_a{}^\mu(x)$, being derivatives of the coordinate transformation functions $x^\mu(x^a)$, undergo the following changes

$$
\begin{aligned}
e_a{}^\mu(x) \to e'_a{}^\mu(x') &\equiv \frac{\partial x'^\mu(x)}{\partial x^a} = \frac{\partial x'^\mu(x)}{\partial x^\nu}\frac{\partial x^\nu}{\partial x^a} \\
&= \alpha^\mu{}_\nu(x)e_a{}^\nu(x) \\
e^a{}_\mu(x) \to e'^a{}_\mu(x') &\equiv \frac{\partial x^a(x)}{\partial x'^\mu} = \frac{\partial x^\nu(x')}{\partial x'^\mu}\frac{\partial x^a}{\partial x^\nu} \\
&= \alpha_\mu{}^\nu(x)e^a{}_\nu(x).
\end{aligned}
\tag{11.59}
$$

The matrices

$$
\alpha^\mu{}_\nu(x) \equiv \frac{\partial x'^\mu}{\partial x^\nu}, \qquad \alpha_\mu{}^\nu(x) \equiv \frac{\partial x^\nu}{\partial x'^\mu}
\tag{11.60}
$$

are orthogonal to each other

$$
\alpha^\nu{}_\lambda \alpha_\nu{}^\mu = \delta_\lambda{}^\mu, \qquad \alpha_\nu{}^\mu \alpha^\lambda{}_\mu = \delta_\nu{}^\lambda,
\tag{11.61}
$$

i.e.,

$$
\left(\alpha^{-1}\right)^\nu{}_\lambda = \alpha_\lambda{}^\nu
\tag{11.62}
$$

is a right- as well as a left-inverse of the matrix $\alpha_\nu{}^\mu$.

A tensor under Einstein transformations has, by analogy with Eq. (1.77), the transformation property

$$
t'^{\mu'\nu'}(x') = \alpha^{\mu'}{}_\mu(x)\alpha^{\nu'}{}_\nu(x)t^{\mu\nu}, \quad t'^{\mu'\nu'\lambda'}(x') = \alpha^{\mu'}{}_\mu(x)\alpha^{\nu'}{}_\nu(x)\alpha^{\lambda'}{}_\lambda(x)t^{\mu\nu\lambda}(x).
\tag{11.63}
$$

Similarly we have, by analogy with Eq. (1.93),

$$
t'_{\mu'\nu'}(x') = \alpha_{\mu'}{}^\mu(x)\alpha_{\nu'}{}^\nu(x)t_{\mu\nu}, \quad t'_{\mu'\nu'\lambda'}(x') = \alpha_{\mu'}{}^\mu(x)\alpha_{\nu'}{}^\nu(x)\alpha_{\lambda'}{}^\lambda(x)t_{\mu\nu\lambda}(x).
\tag{11.64}
$$

It will be convenient to write the transformation $x^\mu \to x'^\mu(x^\mu)$ also as

$$
x^\mu \to x'^\mu \equiv x^\mu - \xi^\mu(x),
\tag{11.65}
$$

which shows that Einstein transformations can be interpreted as *local* translations. The transformation matrices are

$$
\alpha^\lambda{}_\nu(x) = \delta_\nu{}^\lambda - \partial_\nu \xi^\lambda(x), \qquad \alpha_\mu{}^\nu(x) = \delta_\mu{}^\nu + \partial_\mu \xi^\nu(x).
\tag{11.66}
$$

Let us now study the substantial variations δ_s under Einstein transformations, to be denoted by δ_E [recall again the definition in Eq. (3.6)]. Thus we consider infinitesimal translations $\xi^\lambda(x)$ and find for the basis tetrads $e_a{}^\mu(x)$ and $e^a{}_\mu(x)$:

$$
\begin{aligned}
\delta_E e_a{}^\mu(x) &\equiv e'_a{}^\mu(x) - e_a{}^\mu(x) = e'_a{}^\mu(x') - e_a{}^\mu(x') \\
&= e_a{}^\mu(x) - e_a{}^\mu(x') + e'_a{}^\mu(x') - e_a{}^\mu(x) \\
&= \xi^\lambda \partial_\lambda e_a{}^\mu(x) - \partial_\lambda \xi^\mu e_a{}^\lambda(x),
\end{aligned}
\tag{11.67}
$$

$$
\delta_E e^a{}_\mu(x) = \xi^\lambda \partial_\lambda e^a{}_\mu(x) + \partial_\mu \xi^\lambda e^a{}_\lambda(x).
\tag{11.68}
$$

Analogous substantial variations can be derived for the components of the vector fields $v^\mu(x)$ and $v_\mu(x)$. These follow from the fact that the components $v^a(x^b), v_a(x^b)$ remain the same under changes of the general coordinates from x^μ to $x^{\mu'}$. Thus we have the obvious relation

$$v'^a(x^b) = v^a(x^b). \tag{11.69}$$

When reparametrizing the point x^b in two different coordinates x'^μ and x^μ, this relation takes the form

$$v'^a(x') = v^a(x), \tag{11.70}$$

where we have omitted the greek superscripts of x' and x. With these arguments, the substantial variations, i.e., the changes at the same values of the general coordinates x^μ, are

$$\delta_E v^a(x) = v'^a(x) - v^a(x) = \xi^\lambda \partial_\lambda v^a(x). \tag{11.71}$$

Using this and (11.68), we derive from (11.70)

$$v'^\mu(x') = \alpha^\mu{}_\nu v^\nu(x), \qquad v'_\mu(x') = \alpha_\mu{}^\nu v_\nu(x), \tag{11.72}$$

with the substantial variations

$$
\begin{aligned}
\delta_E v^\mu(x) &= v'^\mu(x) - v^\mu(x) = \xi^\lambda \partial_\lambda v^\mu - \partial_\lambda \xi^\mu v^\lambda & (11.73)\\
\delta_E v_\mu(x) &= v'_\mu(x) - v_\mu(x) = \xi^\lambda \partial_\lambda v_\mu \partial_\mu \xi^\lambda v_\lambda. & (11.74)
\end{aligned}
$$

Any four-component field with these transformation properties is called *Einstein vector* or *world vector*.

This definition can trivially be extended to higher *Einstein-* or *world tensors*. We merely apply separately the transformation matrices (11.66) to each index. In particular, the metric $g_{\mu\nu}(x)$ transforms as

$$g'_{\lambda\kappa}(x') = \alpha_\lambda{}^\mu \alpha_\kappa{}^\nu g_{\mu\nu}(x), \qquad g'^{\lambda\kappa}(x') = \alpha^\lambda{}_\mu \alpha^\kappa{}_\nu g^{\mu\nu}(x), \tag{11.75}$$

or, infinitesimally, as

$$
\begin{aligned}
\delta_E g_{\mu\nu} &= \xi^\lambda \partial_\lambda g_{\mu\nu} + \partial_\mu \xi^\lambda g_{\lambda\nu} + \partial_\nu \xi^\lambda g_{\mu\lambda}, & (11.76)\\
\delta_E g^{\mu\nu} &= \xi^\lambda \partial_\lambda g^{\mu\nu} - \partial_\lambda \xi^\mu g^{\lambda\nu} - \partial_\lambda \xi^\nu g^{\mu\lambda}. & (11.77)
\end{aligned}
$$

This can be rewritten in a manifestly covariant form as follows:

$$
\begin{aligned}
\delta_E g_{\mu\nu} &= \bar{D}_\mu \xi_\nu + \bar{D}_\nu \xi_\mu, & (11.78)\\
\delta_E g^{\mu\nu} &= \bar{D}^\mu \xi^\nu + \bar{D}^\nu \xi^\mu. & (11.79)
\end{aligned}
$$

It is now obvious from (11.61) that one can multiply any set of world tensors with each other by a simple contraction of upper and lower indices. The contracted object transforms again like a world tensors. In particular, one obtains an *Einstein-* or *world invariants* if the contraction is complete, i.e., if no uncontracted index is left.

11.3.4   Affine Connections and Covariant Derivatives

The multiplication rules for world tensors are completely analogous to those for Lorentz tensors. There is, however, one important difference. Contrary to the Lorentz case, derivatives of world tensors are no longer tensors. In curvilinear coordinates, certain modifications of the derivatives are required in order to make them proper tensors. It is quite easy to find these modifications and construct objects analogous to the derivative tensors in the Minkowski coordinates. For this we rewrite the derivative tensors in terms of the general curvilinear components. Take, for example, the tensor $\partial_b v_a(x)$. Going over to curvilinear components x^μ we can write this as

$$\partial_b v_a = \partial_b \left(e_a{}^\mu v_\mu \right). \tag{11.80}$$

If we take the derivative ∂_b past the basis tetrad $e_a{}^\mu$, we find

$$\partial_b v_a = e_a{}^\mu \partial_b v_\mu + \partial_b e_a{}^\mu v_\mu. \tag{11.81}$$

Using the relation

$$\partial_b = e_b{}^\lambda \partial_\lambda \tag{11.82}$$

we see that

$$\partial_b v_a = e_a{}^\mu e_b{}^\nu \partial_\nu v_\mu + \left(e_b{}^\nu \partial_\nu e_a{}^\lambda \right) v_\lambda. \tag{11.83}$$

The right-hand side can be rewritten as

$$\partial_b v_a \equiv e_a{}^\mu e_b{}^\nu D_\nu v_\mu, \tag{11.84}$$

where the symbol D_ν stands for the modified derivative

$$D_\nu v_\mu = \partial_\nu v_\mu - e_c{}^\lambda \partial_\nu e_\mu{}^c v_\lambda \equiv \partial_\nu v_\mu - \Gamma_{\nu\mu}{}^\lambda v_\lambda. \tag{11.85}$$

The explicit form on the right-hand side follows from the simple relation

$$\partial_\nu e_a{}^\lambda = -e_a{}^\mu \left(e_c{}^\lambda \partial_\nu e^c{}_\mu \right) \tag{11.86}$$

$$\partial_\nu e^a{}_\lambda = -e^a{}_\mu \left(e^c{}_\lambda \partial_\nu e_c{}^\mu \right) \tag{11.87}$$

which, in turn, is a consequence of differentiating the orthogonality relation $e_a{}^\lambda e^b_\lambda = \delta_a{}^b$. Similarly, we can find the Einstein version of the derivative of a contravariant vector field $\partial_b v^a(x)$, which can be rewritten as

$$\partial_b v^a = \partial_b \left(e^a{}_\mu v^\mu \right) = e^a{}_\mu e_b{}^\nu \partial_\nu v^\mu + (e_b{}^\nu \partial_\nu e^a{}_\lambda) v^\lambda \tag{11.88}$$

and brought to the form

$$e^a{}_\mu e_b{}^\nu D_\nu v^\mu, \tag{11.89}$$

with a covariant derivative

$$D_\nu v^\mu \;=\; \partial_\nu v^\mu - e^c{}_\lambda \partial_\nu e_c{}^\mu v^\lambda = \partial_\nu v^\mu + e_c{}^\mu \partial_\nu e_{c\lambda} v^\lambda \equiv \partial_\nu v^\mu + \Gamma_{\nu\lambda}{}^\mu v^\lambda. \quad (11.90)$$

The extra term appearing in (11.85) and (11.90):

$$\Gamma_{\mu\nu}{}^\lambda \equiv e_a{}^\lambda \partial_\mu e^a{}_\nu \equiv -e^a{}_\nu \partial_\mu e_a{}^\lambda \quad (11.91)$$

is called the *affine connection* of the spacetime under consideration. In general, a spacetime with a metric $g_{\mu\nu}$ and an affine connection $\Gamma_{\mu\nu}{}^\lambda$ defining covariant derivatives is called an *affine spacetime*, and the associated geometry is referred to as a *metric-affine geometry*. Note that by definition, the covariant derivatives of $e^a{}_\nu$ and $e_a{}^\nu$ vanish:

$$D_\mu e^a{}_\nu \;=\; \partial_\mu e^a{}_\nu - \Gamma_{\mu\nu}{}^\lambda e_\lambda{}^a = 0, \quad (11.92)$$

$$D_\mu e_a{}^\nu \;=\; \partial_\mu e_a{}^\nu + \Gamma_{\mu\lambda}{}^\nu e_a{}^\lambda = 0. \quad (11.93)$$

Since $g_{\mu\nu} = e^a{}_\mu e_{a\nu}$, the same property holds for the metric tensor[1].

$$D_\lambda g_{\mu\nu} \;=\; \partial_\lambda g_{\mu\nu} - \Gamma_{\lambda\mu}{}^\sigma g_{\sigma\nu} - \Gamma_{\lambda\nu}{}^\sigma g_{\mu\sigma} = 0, \quad (11.94)$$

$$D_\lambda g^{\mu\nu} \;=\; \partial_\lambda g^{\mu\nu} + \Gamma_{\lambda\sigma}{}^\mu g^{\sigma\nu} + \Gamma_{\lambda\sigma}{}^\nu g^{\mu\sigma} = 0. \quad (11.95)$$

It is worth noting that the metric satisfies once more relations like (11.95), in which the connections are replaced by Christoffel symbols. In fact, from the definition (11.21) we can verify directly that

$$\bar{D}_\lambda g_{\mu\nu} \;=\; \partial_\lambda g_{\mu\nu} - \bar{\Gamma}_{\lambda\mu}{}^\sigma g_{\sigma\nu} - \bar{\Gamma}_{\lambda\nu}{}^\sigma g_{\mu\sigma} = 0, \quad (11.96)$$

$$\bar{D}_\lambda g^{\mu\nu} \;=\; \partial_\lambda g^{\mu\nu} + \bar{\Gamma}_{\lambda\sigma}{}^\mu g^{\sigma\nu} + \bar{\Gamma}_{\lambda\sigma}{}^\nu g^{\mu\sigma} = 0. \quad (11.97)$$

The left-hand sides of (11.84) and (11.88) are tensors with respect to Lorentz transformations. Hence the covariant derivatives $D_\nu v_\mu$ and $D_\nu v^\mu$ in Eqs. (11.85) and (11.90) must be tensors with respect to general coordinate transformation, i.e., world tensors. In fact, one can easily verify that they transform covariantly:

$$\partial'_{\mu'} v_{\nu'}(x') = \alpha_{\mu'}{}^\mu \alpha_{\nu'}{}^\nu \partial_\mu v_\nu(x). \quad (11.98)$$

Working out the derivative on the left-hand side we obtain

$$\partial'_{\mu'} v_{\nu'}(x') \;=\; \alpha_{\mu'}{}^\mu \partial_\mu \left[\alpha_{\nu'}{}^\nu v_\nu(x) \right]$$

$$=\; \alpha_{\mu'}{}^\mu \alpha_{\nu'}{}^\nu \partial_\mu v_\nu(x) + \alpha_{\mu'}{}^\mu \partial_\mu \alpha_{\nu'}{}^\nu. \quad (11.99)$$

The last term is an obstacle to covariance. It is compensated by a similar term in nontensorial behavior of $\Gamma_{\mu\nu}{}^\lambda$:

$$\Gamma'_{\mu'\nu'}{}^{\lambda'}(x') \;=\; e_a{}^{\lambda'} \partial'_{\mu'} e'^a{}_{\nu'} = \alpha^{\lambda'}{}_\lambda \alpha_{\mu'}{}^\mu e_a{}^\lambda \partial_\mu \left(x_{\nu'}{}^\nu e^a{}_\nu \right)$$

$$=\; \alpha_{\mu'}{}^\mu \left[\alpha_{\nu'}{}^\nu \alpha^{\lambda'}{}_\lambda \Gamma_{\mu\nu}{}^\lambda(x) + \alpha^{\lambda'}{}_\nu \partial_\mu \alpha_{\nu'}{}^\nu \right], \quad (11.100)$$

---

[1] There exist more general geometries where this is no longer true [2]

$$\Gamma'_{\mu'\nu'}{}^{\lambda'}(x') = -e'^a_{\nu'}\partial_{\mu'}e_a{}^{\lambda'} = -\alpha_{\nu'}{}^{\nu}\alpha_{\mu'}{}^{\mu}e^a{}_{\nu}\partial_{\mu}\left(\alpha^{\lambda'}{}_{\lambda}e_a{}^{\lambda}\right)$$

$$= \alpha_{\mu'}{}^{\mu}\left[\alpha_{\nu'}{}^{\nu}\alpha^{\lambda'}{}_{\lambda}\Gamma^{\lambda}_{\mu\nu}(x) - \alpha_{\nu'}{}^{\nu}\partial_{\mu}\alpha^{\lambda'}{}_{\nu}\right]. \tag{11.101}$$

Infinitesimally, the transformation matrices are $\alpha_{\mu}{}^{\nu} = \delta_{\mu}{}^{\nu} + \partial_{\mu}\xi^{\nu}$ and $\alpha^{\mu}{}_{\nu} = \delta^{\mu}{}_{\nu} - \partial_{\nu}\xi^{\mu}$, and we easily verify that the covariant derivatives $D_{\mu}v_{\nu}, D_{\mu}v^{\nu}$ have the correct substantial transformation properties of world tensors:

$$\delta_E D_{\mu}v_{\nu} = \xi^{\lambda}\partial_{\lambda}D_{\mu}v_p + \partial_{\mu}\xi^{\lambda}D_{\lambda}v_{\nu} + \partial v_{\nu}\xi^{\lambda}D_{\mu}v_{\lambda},$$

$$\delta_E D_{\mu}v^{\nu} = \xi^{\lambda}\partial_{\lambda}D_{\mu}v^{\nu} + \partial_{\mu}\xi^{\lambda}D_{\lambda}v_{\nu} - \partial_{\nu}\xi^{\nu}D_{\mu}v^{\lambda}. \tag{11.102}$$

The last noncovariant piece in

$$\delta_E\partial_{\mu}v_{\nu} = \partial_{\mu}\delta_E v_{\mu} = \partial_{\mu}\left(\xi^{\lambda}\partial_{\lambda}v_{\nu} + \partial_{\nu}\xi^{\lambda}v_{\lambda}\right)$$

$$= \xi^{\lambda}\partial_{\lambda}\partial_{\mu}v_{\nu} + \partial_{\mu}\xi^{\lambda}\partial_{\lambda}v_{\nu} + \partial_{\nu}\xi^{\lambda}\partial_{\mu}v_{\lambda} + \partial_{\mu}\partial_{\nu}\xi^{\lambda}v_{\lambda} \tag{11.103}$$

is canceled by the last nontensorial piece in $\delta_E\Gamma^{\kappa}_{\mu\nu}$:

$$\delta_E\Gamma_{\mu\nu}{}^{\kappa} = \xi^{\lambda}\partial_{\lambda}\Gamma_{\mu\nu}{}^{\kappa} + \partial_{\mu}\xi^{\lambda}\Gamma_{\mu\nu}{}^{\kappa} + \partial_{\nu}\xi^{\lambda}\Gamma_{\mu\nu}{}^{\kappa} + \partial_{\mu}\partial_{\nu}\xi^{\kappa}. \tag{11.104}$$

One can easily check that the same cancellation occurs in the covariant derivative of an arbitrary tensor field, defined as

$$D_{\mu}t_{\nu_1\ldots\nu_n}{}^{\nu'_1\ldots\nu'_{n'}} \equiv \partial_{\mu}t_{\nu_1\ldots\nu_n}{}^{\nu'_1\ldots\nu'_{n'}} - \sum_i \Gamma_{\mu\nu_i}{}^{\lambda_i}\, t_{\nu_1\ldots\lambda_i\ldots\nu_n}{}^{\nu'_1\ldots\nu'_{n'}}$$

$$+ \sum_{i'}\Gamma_{\mu\lambda'_{i'}}{}^{\nu'_{i'}}\, t_{\nu_1\ldots\nu_n}{}^{\nu'_1\ldots\lambda'_{i'}\ldots\nu'_{n'}}. \tag{11.105}$$

11.4 Torsion tensor

As long as the coordinate transformations $x^{\mu}(x^a)$ and $x^a(x^{\mu})$ are integrable, the derivatives of the infinitesimal local translation field $\xi^{\mu}(x)$ commute with each other:

$$\left(\partial_{\mu}\partial_{\nu} - \partial_{\nu}\partial_{\mu}\right)\xi^{\lambda}(x) = 0. \tag{11.106}$$

For multivalued coordinate transformations, this is no longer true. Then there exists a nonzero antisymmetric part of the connection

$$S_{\mu\nu}{}^{\lambda} \equiv \frac{1}{2}\left(\Gamma_{\mu\nu}{}^{\lambda} - \Gamma_{\nu\mu}{}^{\lambda}\right) = e_a{}^{\lambda}\partial_{\mu}e^a{}_{\nu} - e_a{}^{\lambda}\partial_{\nu}e^a{}_{\mu}, \tag{11.107}$$

whose linear approximation is

$$S_{\mu\nu}{}^{\lambda} = \left(\partial_{\mu}\partial_{\nu} - \partial_{\nu}\partial_{\mu}\right)\xi^{\lambda}(x). \tag{11.108}$$

Remarkably, this transforms like a proper tensor,

$$\delta_E S_{\mu\nu}{}^{\kappa} = \xi^{\lambda}\partial_{\lambda}S_{\mu\nu}{}^{\kappa} + \partial_{\mu}\xi^{\lambda}S_{\mu\nu}{}^{\kappa} + \partial_{\nu}\xi^{\lambda}S_{\mu\nu}{}^{\kappa}, \tag{11.109}$$

as follows directly from the transformation law (11.104). The additional derivative term $\partial_\mu\partial_\nu\xi^\kappa$ arising in the transformation of $\Gamma_{\mu\nu}{}^\kappa$ is symmetric in $\mu\nu$, and thus disappears after antisymmetrization. For this reason, $S_{\mu\nu}$ the *torsion tensor*.

It is useful to realize that with the help of the torsion tensor, the connection can be decomposed into tow parts: a Christoffel part (11.23), which depends only on the metric $g_{\mu\nu}(x)$, and a second part called the *contortion tensor*, which is a combination of torsion tensors.

To derive this decomposition, which is valid in spacetimes with torsion, let us define the modified connection

$$\Gamma_{\mu\nu\lambda} \equiv \Gamma_{\mu\nu}{}^\kappa g_{\kappa\lambda} = e_{a\lambda}\partial_\mu e^a{}_\nu,$$

and decompose this trivially as

$$\Gamma_{\mu\nu\lambda} = \overset{e}{\Gamma}_{\mu\nu\lambda} + \overset{e}{K}_{\mu\nu\lambda}, \tag{11.110}$$

where

$$\overset{e}{\Gamma}_{\mu\nu\lambda} \equiv \frac{1}{2}\left\{e_{a\lambda}\partial_\mu e^a{}_\nu + \partial_\mu e_{a\lambda}e^a{}_\nu + e_{a\mu}\partial_\nu e^a{}_\lambda + e_{a\lambda}\partial_\nu e^a{}_\mu - e_{a\mu}\partial_\lambda e^a{}_\nu - \partial_\lambda e_{a\mu}e^a{}_\nu\right\}, \tag{11.111}$$

$$\overset{e}{K}_{\mu\nu\lambda} \equiv \frac{1}{2}\left\{e_{a\lambda}\partial_\mu e^a{}_\nu - e_{a\lambda}\partial_\nu e^a{}_\mu - e_{a\mu}\partial_\nu e^a{}_\lambda + e_{a\mu}\partial_\lambda e^a{}_\nu + e_{a\nu}\partial_\lambda e^a{}_\mu - e_{a\nu}\partial_\mu e^a{}_\lambda\right\}. \tag{11.112}$$

The terms in the first expression can be combined to

$$\overset{e}{\Gamma}_{\mu\nu\lambda} = \frac{1}{2}\left\{\partial_\mu\left(e_{a\lambda}e^a{}_\nu\right) + \partial_\nu\left(e_{a\mu}e^a{}_\lambda\right) - \partial_\lambda(e_{a\mu}e^a{}_\nu)\right\} = \frac{1}{2}\left(\partial_\mu g_{\lambda\nu} + \partial_\nu g_{\mu\lambda} - \partial_\lambda g_{\mu\nu}\right), \tag{11.113}$$

which shows that $\overset{e}{\Gamma}_{\mu\nu\lambda}$ is equal to the Riemann connection $\bar{\Gamma}_{\mu\nu\lambda}$ in Eq. (11.21), i.e., to the Christoffel symbol. The second expression, the contortion tensor *contortion tensor* $K_{\mu\nu\lambda}$, is a combination of three torsion tensors (11.107). Defining an associated torsion tensor $S_{\mu\nu\lambda} \equiv S_{\mu\nu}{}^\kappa g_{\kappa\lambda}$, we see that

$$\overset{e}{K}_{\mu\nu\lambda} \equiv K_{\mu\nu\lambda} \equiv S_{\mu\nu\lambda} - S_{\nu\lambda\mu} + S_{\lambda\mu\nu}. \tag{11.114}$$

The order of the indices of the three torsion terms are easy to remember: The first starts out with the same indices as $K_{\mu\nu\lambda}$. The second and third terms are shifted cyclically to the left with alternating signs. Note that the antisymmetry of $S_{\mu\nu\lambda}$ in the first two indices makes the contortion tensor $K_{\mu\nu\lambda}$ antisymmetric in the last two indices.

Summarizing, we have found that the full affine connection $\Gamma_{\mu\nu\lambda}$ can be decomposed into a sum of a Riemann connection and a contortion tensor:

$$\Gamma_{\mu\nu\lambda} = \bar{\Gamma}_{\mu\nu\lambda} + K_{\mu\nu\lambda}. \tag{11.115}$$

Since torsion transforms like a tensor, also the contortion is a tensor. As a consequence we may omit the contortion part in the covariant derivatives (11.85) and (11.90), and define the *Riemann-covariant derivatives*

$$\bar{D}_\nu v_\mu \equiv \partial_\nu v_\mu - \bar{\Gamma}_{\nu\mu}{}^\lambda v_\lambda, \quad \bar{D}_\nu v^\mu \equiv \partial_\nu v^\mu + \Gamma_{\nu\lambda}{}^\mu v^\lambda, \tag{11.116}$$

which contain only the Christoffel part of the affine connection.

11.5 Covariant Time Derivative and Acceleration

It is useful to introduce the concept of a *covariant time derivative* of an arbitrary vector field $v^\mu(x)$ in spacetime along a trajectory $x^\mu(\tau)$, where it has the time dependence $v^\mu(\tau) \equiv v^\mu(q(\tau))$. The four-velocity $u^\mu(\tau) = \dot{x}^\mu(\tau)$ transforms like a four-vector. By analogy with the covariant derivative of a vector field $v^\mu(x)$ in Eqs. (11.85) and (11.90), we define the covariant time derivatives of $v^\mu(\tau)$ and $v_\mu(\tau) = g_{\mu\nu}(x(\tau))v^\mu(\tau)$ as

$$\frac{D}{d\tau}v^\mu(\tau) \equiv u^\kappa D_\kappa v^\mu(\tau), \quad \frac{D}{d\tau}v_\mu(\tau) \equiv u^\kappa D_\kappa v_\mu(\tau), \tag{11.117}$$

which become, by Eqs. (11.85) and (11.90):

$$\frac{D}{d\tau}v^\mu(\tau) \equiv \frac{d}{d\tau}v^\mu(\tau) + \Gamma_{\lambda\kappa}{}^\mu v^\lambda(\tau)u^\kappa(\tau), \quad \frac{D}{d\tau}v_\mu(\tau) \equiv \frac{d}{d\tau}v_\mu(\tau) - \Gamma_{\lambda\mu}{}^\kappa u^\lambda(\tau)v^\kappa(\tau). \tag{11.118}$$

If the vector trajectory is the velocity trajectory of a point particle, we replace $v^\mu(\tau)$ and $v_\mu(\tau)$ by $u^\mu(\tau)$ and $u_\mu(\tau)$, and obtain the *covariant accelerations*.

We may also define the Riemann-covariant time derivatives

$$\frac{\bar{D}}{d\tau}v^\mu(\tau) \equiv \frac{d}{d\tau}v^\mu(\tau) + \bar{\Gamma}_{\lambda\kappa}{}^\mu v^\lambda(\tau)u^\kappa(\tau), \quad \frac{\bar{D}}{d\tau}v_\mu(\tau) \equiv \frac{d}{d\tau}v_\mu(\tau) - \bar{\Gamma}_{\lambda\mu}{}^\kappa u^\lambda(\tau)v^\kappa(\tau), \tag{11.119}$$

and the corresponding accelerations.

The above equations serve to define covariant derivatives and accelerations along any curve in a metric-affine spacetime. From the variation of the action (11.2) of a point particle we have learned in Section 11.2 that the particle trajectories in a curved space are geodesics with the equation of motion (11.24). Thus we may also say that the particle orbit has a zero Riemann-covariant acceleration (11.119).

If a particle trajectory has a vanishing acceleration (11.118) involving the total metric-affine connection $\Gamma_{\mu\nu}{}^\lambda$, it is called an *autoparallel trajectory*. This will play an important role in Chapter 14.

The differences between the two covariant time derivatives (11.118) and (11.119) are found with the help of the decomposition (11.115) as:

$$\frac{D}{d\tau}v^\mu(\tau) = \frac{\bar{D}}{d\tau}v^\mu(\tau) + K_{\lambda\kappa}{}^\mu v^\lambda(\tau)u^\kappa(\tau), \quad \frac{D}{d\tau}v_\mu(\tau) = \frac{\bar{D}}{d\tau}v_\mu(\tau) - K_{\lambda\mu}{}^\kappa u^\lambda(\tau)v^\kappa(\tau). \tag{11.120}$$

11.6 Curvature Tensor as Covariant Curl of Affine Connection

In the last section we have seen that even though the connection $\Gamma_{\mu\nu}{}^\lambda$ is not a tensor, its antisymmetric part, the torsion $S^\lambda_{\mu\nu}$, is a tensor. The question arises whether it is possible to form a covariant object which contains information on the content of gravitational forces in the symmetric Christoffel part of the connection. Such a tensor does indeed exist.

When looking back at the transformation properties (11.104) of the connection we see that the tensor character is destroyed by the last term which is additive in the derivative of an arbitrary function $\partial_\mu \partial_\nu \xi^\kappa(x)$. Such additive derivative terms were encountered before in Subsection 2.4.4 in gauge transformations of electromagnetism. Recall that the gauge field of magnetism transform with such an additive derivative term [recall (2.104)]

$$\delta A_a(x) = \partial_a \Lambda(x), \tag{11.121}$$

where $\Lambda(x)$ are arbitrary gauge functions with commuting derivatives [recall (2.105)]. The experimentally measurable physical fields are given by the gauge invariant antisymmetric combination of derivatives (2.81):

$$F_{ab} = \partial_a A_b - \partial_b A_a. \tag{11.122}$$

The additional derivative terms (11.121) disappear in the antisymmetric combination (11.122). This suggests that a similar antisymmetric construction exists also for the connection. The construction is slightly more complicated since the transformation law (11.104) contains also contributions which are linear in the connection.

In a nonabelian gauge theory associated with an internal symmetry which is independent of the spacetime coordinate x, the covariant field strength F_{ab} is a matrix. If g are the elements of the gauge group and $D(g)$ a representation of g in this matrix space, the field strength transforms like a tensor

$$F_{ab} \to F'_{ab} = D(g) F_{ab} D^{-1}(g). \tag{11.123}$$

The gauge field A_a behaves under such transformations as

$$A_a(x) \to A'_a(x) = D(g) A_a(x) D^{-1}(g) + [\partial_a D(g)] D^{-1}(g), \tag{11.124}$$

which is the generalization of the gauge transformations (2.104). The covariant field strength with the transformation property (11.123) is obtained from this by forming the nonabelian curl

$$F_{ab} = \partial_a A_b - \partial_b A_a - [A_a, A_b]. \tag{11.125}$$

This kind of gauge transformations and covariant field strengths appear in nonabelian gauge theories used to describe the vector bosons $W^{0,\pm}$ and Z^0 of weak interactions, where the gauge group is SU(2). They are also needed to describe the octet of gluons $G^{1,\dots,8}$ in the theory of strong interactions, where the gauge group is SU(3). In either case, the representation matrices $D(g)$ belong to the adjoint representation of the gauge group.

Now we observe that the transformation law (11.100) of the affine connection can be written in a way completely analogous to the transformation law (11.124) of a nonabelian gauge field. For this we consider $\Gamma^\lambda_{\mu\nu}$ as the matrix elements of four 4×4 matrix Γ_μ matrices:

$$\Gamma_{\mu\nu}{}^\lambda = \left(\Gamma_\mu\right)_\nu{}^\lambda. \tag{11.126}$$

Then (11.101) may be viewed as a matrix equation

$$\mathbf{\Gamma}'_{\mu'}(x') = \alpha_{\mu'}{}^\mu \left[\alpha\mathbf{\Gamma}_\mu(x)\alpha^{-1} + (\partial_\mu\alpha)\alpha^{-1}\right]. \tag{11.127}$$

This equation is a direct generalization of Eq. (11.124) to the case that the symmetry group acts also on the spacetime coordinates. To achieve covariance, the vector index μ of the gauge field must be transformed accordingly.

Actually, this observation comes as no surprise if we remember the original purpose of introducing the connection $\Gamma_{\mu\nu}{}^\lambda$. It served to form the covariant derivatives (11.85) and (11.90). Equation (11.127) shows that the connection may be viewed as a nonabelian gauge field of the group of local general coordinate transformations $\alpha^\mu{}_\nu(x)$. Einstein vectors and tensors in curvilinear coordinates are the associated gauge covariant quantities.

By analogy with the field strength (16.15), we can immediately write down a covariant curl of the matrix field $\mathbf{\Gamma}_\mu$:

$$\mathbf{R}_{\mu\nu} \equiv \partial_\mu\mathbf{\Gamma}_\nu - \partial_\nu\mathbf{\Gamma}_\mu - \left[\mathbf{\Gamma}_\mu, \mathbf{\Gamma}_\nu\right], \tag{11.128}$$

which should transform like a tensor under general coordinate transformations. In component form, this tensor reads

$$R_{\mu\nu\lambda}{}^\sigma = \partial_\mu\Gamma_{\nu\lambda}{}^\sigma - \partial_\nu\Gamma_{\mu\lambda}{}^\sigma - \Gamma_{\mu\lambda}{}^\delta\Gamma_{\nu\delta}{}^\sigma + \Gamma_{\nu\lambda}{}^\delta\Gamma_{\mu\delta}{}^\sigma. \tag{11.129}$$

The covariance properties of $R_{\mu\nu\lambda}{}^\kappa$ follow most easily by realizing that, in terms of the basic tetrads $e_a{}^\mu$, the covariant curl has the simple representation

$$R_{\mu\nu\lambda}{}^\sigma = e_a{}^\sigma\left(\partial_\mu\partial_\nu - \partial_\nu\partial_\mu\right)e^a{}_\lambda = -e^a{}_\lambda\left(\partial_\mu\partial_\nu - \partial_\nu\partial_\mu\right)e_a{}^\sigma. \tag{11.130}$$

The first line is obtained directly by inserting $\Gamma_{\mu\nu}{}^\lambda = e_a{}^\lambda\partial_\mu e^a{}_\nu$ into (11.128), and executing the derivatives

$$\left[\partial_\mu\Gamma_{\nu\lambda}{}^\kappa - \left(\Gamma_\mu\Gamma_\nu\right)_\lambda{}^\kappa\right] - [\mu \leftrightarrow \nu]$$
$$= \left(\partial_\mu e_a{}^\kappa\partial_\nu e^a{}_\lambda + e_a{}^\kappa\partial_\mu\partial_\nu e^a_\lambda + e_b{}^\rho\partial_\mu e^b{}_\lambda e^a{}_\rho\partial_\nu e_a{}^\kappa\right) - (\mu \leftrightarrow \nu)$$
$$= e_a{}^\kappa\left(\partial_\mu\partial_\nu - \partial_\nu\partial_\mu\right)e^a{}_\lambda. \tag{11.131}$$

The second line in (11.131) is obtained from the first by inserting $\Gamma_{\mu\nu}{}^\lambda = -e^a{}_\nu\partial_\mu e_a{}^\lambda$ or $\Gamma_{\nu\rho}{}^\kappa = e_a{}^\kappa\partial_\nu e^a_\rho$.

We are now ready to realize another property of Minkowski spacetime. Just as this spacetime had a vanishing torsion tensor for any curvilinear parametrization, it also has a vanishing curvature tensor. The representation (11.130) shows that a spacetime x^μ can have curvature only if the derivatives of the mapping functions $x^a \to x^\mu$ are not integrable in the Schwarz sense. Expressed differently, the vanishing of $R_{\mu\nu\lambda}{}^\kappa$ follows from the obvious fact that

$$R_{\mu\nu\lambda}{}^\kappa = e_a{}^\kappa \left(\partial_\mu \partial_\nu - \partial_\nu \partial_\mu \right) e^a{}_\lambda \equiv 0 \tag{11.132}$$

for the trivial choice of the basis tetrad $e_a{}^\kappa = \delta_a{}^\kappa$. Together with the tensor transformation law (11.64) we find that $R_{\mu\nu\lambda}{}^\kappa$ remains identically zero in any curvilinear parametrization of Minkowski spacetime.

From the tetrad expression for $R_{\mu\nu\lambda}{}^\kappa$ the tensor transformation law is easily found [using (11.60)]

$$\begin{aligned}
R_{\mu\nu\lambda}{}^\kappa(x) &\to R'_{\mu'\nu'\lambda'}{}^{\kappa'}(x') \\
&= e_a{}^{x'}(x') \left(\partial'_{\mu'} \partial'_{\nu'} - \partial'_{\nu'} \partial'_{\mu'} \right) e'^a{}_{\lambda'}(x) \\
&= \alpha^{\kappa'}{}_\kappa \alpha_{\mu'}{}^\mu e_a{}^\kappa(x) \left(\partial_\mu \partial_\nu - \partial_\nu \partial_\mu \right) \left(\alpha_{\lambda'}{}^\lambda e^a{}_\lambda \right) \\
&= \alpha_{\mu'}{}^\mu \alpha_{\nu'}{}^\nu \alpha_{\lambda'}{}^\lambda \alpha^{\kappa'}{}_\kappa R_{\mu\nu\lambda}{}^\kappa(x) \\
&\quad + \alpha_{\mu'}{}^\mu \alpha_{\nu'}{}^\nu \alpha^{\kappa'}{}_\lambda \left[\left(\partial_\mu \partial_\nu - \partial_\nu \partial_\mu \right) \alpha_{\lambda'}{}^\lambda \right]. \tag{11.133}
\end{aligned}$$

Since general coordinate transformations are assumed to be smooth, the derivatives in front of $\alpha_{\lambda'}{}^\lambda$ commute and $R_{\mu\nu\lambda}{}^\kappa$ is a proper tensor. It is called the *curvature tensor*.

By construction, this curvature tensor is antisymmetric in the first index pair. A property that is not so easy to see is the antisymmetry with respect to the second index pair, if the last index is lowered to $R_{\mu\nu\lambda\kappa} \equiv R_{\mu\nu\lambda}{}^\sigma g_{\kappa\sigma}$:

$$R_{\mu\nu\lambda\kappa} = -R_{\mu\nu\kappa\lambda} \tag{11.134}$$

Indeed, if we calculate the difference between the two sides using the definition (11.130) we find

$$\begin{aligned}
R_{\mu\nu\lambda\kappa} + R_{\mu\nu\kappa\lambda} &= e_{a\kappa} \left(\partial_\mu \partial_\nu - \partial_\nu \partial_\mu \right) e^a{}_\lambda + e_{a\lambda} \left(\partial_\mu \partial_\nu - \partial_\nu \partial_\mu \right) e^a{}_\kappa \\
&= \partial_\mu \partial_\nu \left(e_{a\kappa} e^a{}_\lambda \right) - \partial_\nu \partial_\mu \left(e_{a\kappa} e^a{}_\lambda \right) \\
&= \left(\partial_\mu \partial_\nu - \partial_\nu \partial_\mu \right) g_{\lambda\kappa}. \tag{11.135}
\end{aligned}$$

The physical observability requires the metric

$$g_{\lambda\kappa}(x) = \frac{\partial x^a}{\partial x^\lambda} \frac{\partial x_a}{\partial x^\kappa} \tag{11.136}$$

to be a smooth single-valued function, so that it satisfies the integrability condition

$$\left(\partial_\mu \partial_\nu - \partial_\nu \partial_\mu \right) g_{\lambda\kappa} = 0. \tag{11.137}$$

Inserting this into Eq. (11.135) proves that the Riemannian-Cartan curvature tensor is indeed antisymmetric in the last two indices [3]. According to the definition given after Eq. (2.89), this antisymmetry is therefore a Bianchi identity.

An integrability assumption of the type (11.137) must also be imposed upon the affine connection to make it a physically observable field:

$$\left(\partial_\mu \partial_\nu - \partial_\nu \partial_\mu \right) \Gamma_{\lambda\kappa}{}^\delta = 0. \tag{11.138}$$

This integrability condition gives rise to the famous Bianchi identity of Riemann-Cartan spacetimes to be derived in Section 12.5.

It should be pointed out that a nonvanishing curvature tensor has the consequence that covariant derivatives no longer commute. If we form

$$D_\mu D_\mu v_\lambda - D_\nu D_\mu v_\lambda \tag{11.139}$$

we find using (11.85), (11.90), and (11.129):

$$
\begin{aligned}
D_\nu D_\mu v_\lambda - D_\mu D_\nu v_\lambda &= -R_{\nu\mu\lambda}{}^\kappa v_\kappa - 2S_{\nu\mu}{}^\rho R_\rho v_\lambda, \\
D_\nu D_\mu v^\kappa - D_\mu D_\nu v^\kappa &= R_{\nu\mu\lambda}{}^\kappa v^\lambda - 2S_{\nu\mu}{}^\rho D_\rho v^\kappa.
\end{aligned}
\tag{11.140}
$$

Since $R_{\mu\nu\lambda}{}^\kappa$ is a tensor, it can be contracted with the metric tensor to form covariant quantities of lower rank. There are two possibilities

$$R_{\mu\nu} \equiv R_{\kappa\mu\nu}{}^\kappa \tag{11.141}$$

called the *Ricci tensor* and

$$R = R_{\mu\nu} g^{\mu\nu} \tag{11.142}$$

called the *scalar curvature*. A combination of both

$$G_{\mu\nu} \equiv R_{\mu\nu} - \frac{1}{2} g_{\mu\nu} R \tag{11.143}$$

was introduced by Einstein and is therefore called the *Einstein curvature tensor*. It can also be written as

$$G^{\nu\mu} = \frac{1}{4} e^{\mu\alpha\beta\gamma} e^\nu{}_\alpha{}^{\delta\tau} R_{\beta\gamma\delta\tau}, \tag{11.144}$$

where $e^{\mu\nu\lambda\kappa}$ is the contravariant version of the Levi-Civita tensor defined in Appendix 11A. The equality between (11.144) and (11.143) follows directly from the curved-spacetime version of the identity (1A.24).

11.7 Riemann Curvature Tensor

Actually, Einstein worked with a related tensor which deals exclusively with the Riemann part of the connection and the curvature tensor. Since the contortion $K_{\mu\nu}{}^\lambda$ is a tensor, the Riemann part $\bar{\Gamma}_{\mu\nu}{}^\lambda$ of $\Gamma_{\mu\nu}{}^\lambda$ has the same transformation properties (11.104) as $\Gamma^\lambda_{\mu\nu}$, and we can form the *Riemann curvature tensor*

$$\bar{R}_{\mu\nu\lambda}{}^\sigma = \partial_\mu \bar{\Gamma}_{\nu\lambda}{}^\sigma - \partial_\nu \bar{\Gamma}_{\mu\lambda}{}^\sigma - \bar{\Gamma}_{\mu\lambda}{}^\rho \bar{\Gamma}_{\nu\rho}{}^\sigma + \bar{\Gamma}_{\nu\lambda}{}^\rho \bar{\Gamma}_{\mu\rho}{}^\sigma. \tag{11.145}$$

Contrary to $R_{\mu\nu\lambda}{}^\kappa$ in Eq. (11.129), this curvature tensor can be expressed completely in terms of derivatives of the metric [recall (11.21), (11.23)]. The difference between the two tensors is the following function of the contortion tensor

$$R_{\mu\nu\lambda}{}^\kappa - \bar{R}_{\mu\nu\lambda}{}^\kappa = \bar{D}_\mu K_{\nu\lambda}{}^\kappa - \bar{D}_\nu K_{\mu\lambda}{}^\kappa - \left(K_{\mu\lambda}{}^\rho K_{\nu\rho}{}^\kappa - K_{\nu\lambda}{}^\rho K_{\mu\rho}{}^\kappa \right), \quad (11.146)$$

where $\bar{D}_\mu$ denotes the Riemann-covariant derivative (11.116) formed with the Christoffel part of the connection.

Note that Eq. (11.146) is compatible with the antisymmetry of $R_{\mu\nu\lambda\kappa}$ and $\bar{R}_{\mu\nu\lambda\kappa}$ in the first and second index pairs. For the first index pair, where the antisymmetry is implied by the definitions (11.129) and (11.145), the difference in (11.146) has obviously the same antisymmetry. The antisymmetry of the difference in the second index pair $\lambda\kappa$ is a consequence of the antisymmetry of the contortion tensor $K_{\nu\lambda\kappa}$ in the last two indices.

In addition the curvature tensor $\bar{R}_{\mu\nu\lambda\kappa}$ is symmetric under the exchange of the first and the second index pair

$$\bar{R}_{\mu\nu\lambda\kappa} = \bar{R}_{\lambda\kappa\mu\nu}. \quad (11.147)$$

This can be shown by expressing the first two terms in (11.145) as derivatives of the metric tensor

$$\bar{R}_{\mu\nu\lambda\kappa} = \left[g_{\kappa\delta} \partial_\mu \frac{g^{\delta\sigma}}{2} \left(\partial_\nu g_{\lambda\sigma} + \partial_\lambda g_{\nu\sigma} - \partial_\sigma g_{\nu\lambda} \right) \right] - [\mu \leftrightarrow \nu] - g_{\kappa\delta} \left(\bar{\Gamma}_{\mu\lambda}{}^\rho \bar{\Gamma}_{\nu\rho}{}^\delta - \bar{\Gamma}_{\nu\lambda}{}^\rho \bar{\Gamma}_{\mu\rho}{}^\delta \right),$$

$$(11.148)$$

and using (11.97) to express $\partial_\mu g^{\delta\sigma}$ in terms of Christoffel symbols,

$$g_{\kappa\delta} \partial_\mu g^{\delta\sigma} = -\left(\partial_\mu g_{\kappa\delta} \right) g^{\delta\sigma}$$

$$= -\left(\bar{\Gamma}_{\mu\kappa}{}^\tau g_{\tau\delta} + \bar{\Gamma}_{\mu\delta}{}^\tau g_{\kappa\tau} \right) g^{\delta\sigma} = -\bar{\Gamma}_{\mu\kappa}{}^\sigma - \bar{\Gamma}_{\mu\delta\kappa} g^{\delta\sigma}. \quad (11.149)$$

In this way we obtain

$$\bar{R}_{\mu\nu\lambda\kappa} = \frac{1}{2} \left[\left(\partial_\mu \partial_\lambda g_{\nu\kappa} - \partial_\mu \partial_\kappa g_{\nu\lambda} \right) - (\mu \leftrightarrow \nu) \right]$$

$$- \left[\left(\bar{\Gamma}_{\mu\kappa}{}^\sigma + \bar{\Gamma}_{\mu\kappa'\kappa} g^{\lambda\sigma} \right) \bar{\Gamma}_{\nu\lambda\sigma} - (\mu \leftrightarrow \nu) \right] - \left(\bar{\Gamma}_{\mu\lambda}{}^\rho \bar{\Gamma}_{\nu\rho\kappa} - \bar{\Gamma}_{\nu\lambda}{}^\rho \bar{\Gamma}_{\mu\rho\kappa} \right). (11.150)$$

A further use of relation (11.97) brings the second line to

$$-\frac{1}{2} \left\{ \left(\bar{\Gamma}_{\mu\kappa}{}^\sigma + g^{\delta\sigma} \bar{\Gamma}_{\mu\delta\kappa} \right) \left[\left(\bar{\Gamma}_{\nu\lambda\sigma} + \bar{\Gamma}_{\nu\sigma\lambda} \right) + (\lambda \leftrightarrow \nu) - \bar{\Gamma}_{\sigma\nu\lambda} - \bar{\Gamma}_{\sigma\lambda\nu} \right] \right\} - \{\mu \leftrightarrow \nu\},$$

and we find that almost all terms cancel, due to the symmetry of $\bar{\Gamma}_{\mu\nu\lambda}$ in $\mu\nu$. Only

$$- \left(\bar{\Gamma}_{\mu\kappa}{}^\sigma \bar{\Gamma}_{\nu\lambda\sigma} + \bar{\Gamma}_{\mu\delta\kappa} \bar{\Gamma}_{\nu\lambda}{}^\delta \right) + (\mu \leftrightarrow \nu)$$

survives, whose second term cancels the third line in (11.150), bringing $\bar{R}_{\mu\nu\lambda\kappa}$ to the form

$$\bar{R}_{\mu\nu\lambda\kappa} = \frac{1}{2}\left[\left(\partial_\mu\partial_\lambda g_{\nu\kappa} - \partial_\mu\partial_\kappa g_{\nu\lambda}\right) - (\mu \leftrightarrow \nu)\right] - \left(\bar{\Gamma}_{\mu\kappa}{}^\sigma \bar{\Gamma}_{\nu\lambda\sigma} - \bar{\Gamma}_{\nu\kappa}{}^\sigma \bar{\Gamma}_{\mu\lambda\sigma}\right). \quad (11.151)$$

This expression shows manifestly the symmetry $\mu\nu \leftrightarrow \lambda\kappa$ as a consequence of the integrability property $\left(\partial_\mu\partial_\nu - \partial_\nu\partial_\mu\right) g_{\lambda\kappa} = 0$. The same property makes $\bar{R}_{\mu\nu\lambda\kappa}$ antisymmetric under the exchange $\mu \leftrightarrow \nu$, as follows from Eqs. (11.135) and (11.146).

By contracting (11.151) with $g^{\nu\lambda} g^{\mu\kappa}$, we can derive the following compact expression for the curvature scalar

$$\sqrt{-g}\bar{R} = \partial_\lambda\left[\left(g^{\mu\nu}\sqrt{-g}\right)\left(\bar{\Gamma}_{\mu\nu}{}^\lambda - \delta_\mu{}^\lambda \bar{\Gamma}_{\nu\kappa}{}^\kappa\right)\right] + \sqrt{-g}g^{\mu\nu}\left(\bar{\Gamma}_{\mu\lambda}{}^\kappa\bar{\Gamma}_{\nu\kappa}{}^\lambda - \bar{\Gamma}_{\mu\nu}{}^\lambda\bar{\Gamma}_{\lambda\kappa}{}^\kappa\right). \quad (11.152)$$

It is instructive to check this equation. For this we use the identity (11.149) in the form

$$\partial_\kappa g_{\mu\nu} = -g_{\mu\sigma}g_{\nu\tau}\partial_\kappa g^{\sigma\tau} = \bar{\Gamma}_{\kappa\mu\nu} + \bar{\Gamma}_{\kappa\nu\mu}. \quad (11.153)$$

In addition, we employ the identity

$$\partial_\lambda\sqrt{-g} = \frac{1}{2}\sqrt{-g}g^{\sigma\tau}\partial_\lambda g_{\sigma\tau} = \bar{\Gamma}_{\lambda\mu}{}^\mu, \quad (11.154)$$

which follows directly from Eq. (11A.23). Combining these we derive

$$\partial_\lambda(g^{\mu\nu}\sqrt{-g}) = \sqrt{-g}\left[-g^{\mu\sigma}\bar{\Gamma}_{\lambda\sigma}{}^\nu - g^{\nu\sigma}\bar{\Gamma}_{\lambda\sigma}{}^\mu + g^{\mu\nu}\bar{\Gamma}_{\lambda\sigma}{}^\sigma\right]. \quad (11.155)$$

This allows us to rewrite the first term in (11.152) as

$$\partial_\lambda(g^{\mu\nu}\sqrt{-g})\left(\bar{\Gamma}_{\mu\nu}{}^\lambda - \delta_\mu{}^\lambda \bar{\Gamma}_{\nu\kappa}{}^\kappa\right) \quad (11.156)$$
$$= \sqrt{-g}\left[-g^{\mu\sigma}\bar{\Gamma}_{\sigma\lambda}{}^\nu - g^{\nu\sigma}\bar{\Gamma}_{\sigma\lambda}{}^\mu + g^{\mu\nu}\bar{\Gamma}_{\sigma\lambda}{}^\sigma\right]\left(\bar{\Gamma}_{\lambda\mu\nu} - \delta_\mu{}^\lambda\bar{\Gamma}_{\kappa\nu}{}^\kappa\right)$$
$$= -2\sqrt{-g}\left[\bar{\Gamma}_{\mu\nu\sigma}g^{\mu\lambda}g^{\nu\kappa}\bar{\Gamma}_{\lambda\kappa}{}^\sigma - \bar{\Gamma}_{\lambda\sigma}{}^\sigma g^{\lambda\kappa}\bar{\Gamma}_{\kappa\mu}{}^\mu\right].$$

As a consequence, Eq. (11.152) becomes

$$\sqrt{-g}R = \sqrt{-g}\left[g^{\mu\nu}g^{\lambda\kappa}\left(\partial_\lambda\Gamma_{\kappa\mu\nu} - \partial_\mu\Gamma_{\nu\lambda\kappa} - g^{\sigma\tau}\bar{\Gamma}_{\lambda\sigma\mu}\bar{\Gamma}_{\kappa\tau\nu} + \bar{\Gamma}_{\sigma\lambda}{}^\sigma\bar{\Gamma}_{\kappa\mu\nu}\right)\right], \quad (11.157)$$

which is the contraction of the defining equation (11.145) for the Riemann curvature tensor with $\delta^\mu{}_\sigma g^{\nu\lambda}$.

Appendix 11A   Curvilinear Versions of Levi-Civita Tensor

In Appendix 1A we have listed the properties of the Levi-Civita tensor $\epsilon^{a_1\cdots a_D}$ in Euclidean space as well as Minkowski spacetime. These properties acquire little change if the spacetimes are reparametrized with curvilinear coordinates. To be specific, we consider only a four-dimensional Minkowski spacetime whose metric arises from a coordinate transformation of (1A.21). The same formulas hold also if the spacetime is curved. The curvilinear Levi-Civita tensor is

$$e^{\mu_1 \cdots \mu_D} = \frac{1}{\sqrt{-g}} \epsilon^{\mu_1 \cdots \mu_D}, \tag{11A.1}$$

where

$$-g \equiv \det\left(-g_{\mu\nu}\right) \tag{11A.2}$$

is the positive determinant of $-g_{\mu\nu}$, and $\sqrt{-g}$ is the positive square root of it. Just as $\epsilon^{a_1 \cdots a_D}$ was a pseudotensor under Lorentz transformations [recall (1A.11)], $e^{\mu_1 \cdots \mu_D}$ is a pseudotensor under general coordinate transformations, which transform

$$\epsilon^{\mu_1 \cdots \mu_D} \quad \rightarrow \quad \alpha^{\mu_1}{}_{\nu_1} \cdots \alpha^{\mu_D}{}_{\nu_D} \epsilon^{\nu_1 \cdots \nu_D} = \det\left(\alpha\right) \epsilon^{\mu_1 \cdots \mu_D}. \tag{11A.3}$$

Since $g_{\mu\nu}$ is transformed as

$$g_{\mu\nu} \rightarrow \alpha_\mu{}^\lambda \alpha_\nu{}^\kappa g_{\lambda\kappa}, \tag{11A.4}$$

its determinant behaves like

$$g \rightarrow \det\left(\alpha_\mu{}^\lambda\right)^2 g = \det\left(\alpha^\mu{}_\nu\right)^{-2} g. \tag{11A.5}$$

Hence

$$e^{\mu_1 \cdots \mu_D} \quad \rightarrow \quad \frac{\det\left(\alpha^\mu{}_\nu\right)}{|\det\left(\alpha^\mu{}_\nu\right)|} e^{\mu_1 \cdots \mu_D}, \tag{11A.6}$$

showing the pseudotensor property.

The same thing holds for the tensor

$$e_{\mu_1 \cdots \mu_D} = \sqrt{-g}\, \epsilon_{\mu_1 \cdots \mu_D}. \tag{11A.7}$$

It arises from $e^{\nu_1 \cdots \mu_D}$ by multiplication with $g_{\mu_1\nu_1} \cdots g_{\mu_D\nu_D}$, as it should.

The co- and contravariant antisymmetric tensors $e_{\mu_1 \cdots \mu_D}, e^{\mu_1 \cdots \mu_D}$ share an important property with the symmetric tensors $g_{\mu_1\mu_2}, g^{\mu_1\mu_2}$. All of them are invariant under covariant differentiation:

$$D_\lambda e_{\mu_1 \cdots \mu_D} = 0, \qquad D_\lambda e^{\mu_1 \cdots \mu_D} = 0. \tag{11A.8}$$

Indeed, since $e_{\mu_1 \cdots \mu_D}$ is a tensor, we can write this equation explicitly as

$$\partial_\lambda e_{\mu_1 \cdots \mu_D} = \Gamma_{\lambda\mu_1}{}^{\nu_1} e_{\nu_1\mu_2 \cdots \mu_D} + \Gamma_{\lambda\mu_2}{}^{\nu_2} e_{\mu_1\nu_2 \cdots \mu_D} + \cdots + \Gamma_{\lambda\mu_D}{}^{\nu_D} e_{\mu_1\mu_2 \cdots \nu_D}. \tag{11A.9}$$

Using $e_{\mu_1 \cdots \mu_D} = \sqrt{-g}\,\varepsilon_{\mu_1 \cdots \mu_D}$, the left-hand side is equal to

$$\frac{1}{\sqrt{-g}} \left(\partial_\lambda \sqrt{-g}\right) e_{\mu_1 \cdots \mu_D}, \tag{11A.10}$$

from which the equality follows from the covariant version of the identity (1A.27):

$$e_{\mu_1\ldots\mu_D}g_{\sigma\tau} = e_{\tau\mu_2\ldots\mu_D}g_{\sigma\mu_1} + e_{\mu_1\tau\ldots\mu_D}g_{\sigma\mu_2} + \ldots + e_{\mu_1\mu_2\ldots\tau}g_{\sigma\mu_D}, \quad (11A.11)$$

after multiplying it by $g^{\sigma\delta}\Gamma_{\lambda\delta}{}^{\tau}$.

An important consequence of the vanishing covariant derivative of the antisymmetric tensors in Eq. (11A.8) is that antisymmetric products satisfy the covariant version of the chain rule of differentiation without an extra term. For instance, the vector product in three curved dimensions

$$(\mathbf{x} \times \mathbf{w})_{\mu} = e_{\mu\lambda\kappa}x^{\lambda}w^{\kappa} \quad (11A.12)$$

has the covariant derivative

$$D_{\sigma}(\mathbf{x} \times \mathbf{w}) = D_{\sigma}\mathbf{x} \times \mathbf{w} + \mathbf{x} \times D_{\sigma}\mathbf{w}, \quad (11A.13)$$

just as in flat spacetime. The same rule applies, of course, to the scalar product

$$\mathbf{x} \cdot \mathbf{w} = g_{\mu\nu}v^{\mu}\omega^{\lambda}, \quad (11A.14)$$

as a consequence of the vanishing covariant derivative of the metric in Eq. (11.94):

$$D_{\sigma}(\mathbf{x} \cdot \mathbf{w}) = D_{\sigma}\mathbf{x} \cdot \mathbf{w} + \mathbf{x} \cdot D_{\sigma}\mathbf{w}. \quad (11A.15)$$

The determinant of an arbitrary tensor $t_{\mu\nu}$ is given by a formula similar to (1A.9)

$$\det\left(t_{\mu\nu}\right) = \frac{1}{D!}\epsilon^{\mu_1\ldots\mu_D}\epsilon^{\nu_1\ldots\nu_D}t_{\mu_1\nu_1}\ldots t_{\mu_D\nu_D} = -\frac{g}{D!}e^{\mu_1\ldots\mu_D}e^{\nu_1\ldots\nu_D}t_{\mu_1\nu_1}\ldots t_{\mu_D\nu_D}. \quad (11A.16)$$

The determinant of $t_{\mu}{}^{\nu}$, on the other hand, is equal to

$$\det\left(t_{\mu}{}^{\nu}\right) = -\frac{1}{D!}\epsilon^{\mu_1\ldots\mu_D}\epsilon_{\nu_1\ldots\nu_D}t_{\mu_1}{}^{\nu_1} = -\frac{1}{D!}e^{\mu_1\ldots\mu_D}e_{\nu_1\ldots\nu_D}t_{\mu_1}{}^{\nu_1}\ldots t_{\mu_D}{}^{\nu_D}, \quad (11A.17)$$

in agreement with the relation $\det\left(t_{\mu}{}^{\nu}\right) = \det\left(t_{\mu\lambda}g^{\lambda\nu}\right) = \det\left(t_{\mu\nu}\right)g^{-1}$.

The covariant tensors $e^{\nu_1\ldots\nu_D}$ are useful for writing down explicitly the cofactors $M_{\nu}{}^{\mu}$ in the expansion of a determinant.

$$\det\left(t_{\mu}{}^{\nu}\right) = \frac{1}{D}t_{\mu}{}^{\nu}M_{\nu}{}^{\mu}. \quad (11A.18)$$

By comparison with Eq. (11A.17) we identify:

$$M_{\nu_1}{}^{\mu_1} = -\frac{1}{(D-1)!}\epsilon^{\mu_1\ldots\mu_D}\epsilon_{\nu_1\ldots\nu_D}t_{\mu_2}{}^{\nu_2}\ldots t_{\mu_D}{}^{\nu_D}. \quad (11A.19)$$

The inverse of the matrix $t_{\mu}{}^{\nu}$ has then the explicit form

$$\left(t^{-1}\right)_{\nu}{}^{\mu} = \frac{1}{\det\left(t_{\mu}{}^{\nu}\right)}M_{\nu}{}^{\mu}. \quad (11A.20)$$

For a determinant $\det\left(t_{\mu\nu}\right)$ we find, similarly,

$$\det\left(t_{\mu\nu}\right) = \frac{1}{D}t_{\mu\nu}M^{\mu\nu}, \tag{11A.21}$$

with

$$
\begin{aligned}
M^{\mu_1\nu_1} &= \frac{1}{(D-1)!}\epsilon^{\mu_1\cdots\mu_D}\epsilon^{\nu_1\cdots\nu_D}t_{\mu_2\nu_2}\cdots t_{\mu_D\nu_D}\\
&= \det\left(t_{\mu\nu}\right)\left(t^{-1}\right)^{\mu_1\nu_1}.
\end{aligned} \tag{11A.22}
$$

This equation is useful for calculating variations of the determinant g upon variations of the metric $g_{\mu\nu}$, which will be needed later in Eq. (15.24). Inserting $g_{\mu\nu}$ into (11A.16) and using the first line of (11A.22), we find immediately

$$
\begin{aligned}
\delta g &= \frac{1}{D!}\epsilon^{\mu_1\cdots\mu_D}\epsilon^{\nu_1\cdots\nu_D}\delta\left(g_{\mu_1\nu_1}g_{\mu_2\nu_2}\cdots g_{\mu_D\nu_D}\right)\\
&= \frac{1}{(D-1)!}\epsilon^{\mu_1\cdots\mu_D}\epsilon^{\nu_1\cdots\nu_D}\delta g_{\mu_1\nu_1}g_{\mu_2\nu_2}\cdots g_{\mu_D\nu_D} = \delta g_{\mu\nu}M^{\mu\nu} = \det\left(g_{\mu\nu}\right)g^{\mu\nu}\delta g_{\mu\nu}\\
&= gg^{\mu\nu}\delta g_{\mu\nu}.
\end{aligned} \tag{11A.23}
$$

The identity $g^{\mu\nu}g_{\nu\lambda} = \delta^\mu{}_\lambda$ implies opposite signs of co- and contravariant variations:

$$g^{\lambda\mu}\delta g_{\mu\nu} = -g_{\nu\kappa}\delta g^{\lambda\kappa}, \tag{11A.24}$$

so that $\delta g^{\lambda\kappa} = -g^{\lambda\mu}g^{\kappa\nu}\delta g_{\mu\nu}$ and

$$\delta g = gg^{\mu\nu}\delta g_{\mu\nu} = -gg_{\mu\nu}\delta g^{\mu\nu}. \tag{11A.25}$$

Another way of deriving this result employs the identity valid for any nonsingular matrix A:

$$\det A = e^{\operatorname{tr}\log A}, \tag{11A.26}$$

from which we find

$$\delta\det A = \det A\,\delta(\operatorname{tr}\log A) = \det A\,\operatorname{tr}(A^{-1}\delta A). \tag{11A.27}$$

Replacing A by the metric gives directly (11A.23).

Notes and References

For more details see Chapters 10 and 11 of
H. Kleinert, *Path Integrals in Quantum Mechanics, Statistics, Polymer Physics, and Financial Markets*, World Scientific, Singapore, 4th edition, 2006 (kl/b5), where kl is short for the www address http://www.physik.fu-berlin.de/~kleinert, and Part IV of
H. Kleinert, *Gauge Fields in Condensed Matter*, Vol. II, *Stresses and Defects*, World

Scientific, Singapore, 1989 (`kl/b2`).

The geometric aspects of general relativity is discussed in great detail in
L.D. Landau and E.M. Lifshitz, *Classical Field Theory* Addison-Wesley, Reading, Mass., 1958);
C.W. Misner, K.S. Thorne, and J.A. Wheeler, *Gravitation* , Freeman and Co., New York, 1973;
E. Schmutzer, *Relativistische Physik*, Akad. Verlagsg. Geest und Portig, Leipzig, 1968;
S. Weinberg, *Gravitation and Cosmology*, J. Wiley and Sons, New York, 1972.

For the mathematics of metric-affine spaces see
J.A. Schouten, *Ricci Calculus*, Springer, Berlin, 1954.
We use the same notation.

The particular citations in this chapter refer to the publications

[1] R. Eötvös, Math. Nat. Ber. Ungarn **8**, 65 (1890).
See also
J. Renner, Hung. Acad. Sci. *53*, Part II (1935).

[2] In general relativity there have been theories based on spaces in which this is not satisfied. Then the covariant derivative of the metric tensor $D_\lambda g_{\mu\nu} = -Q_{\lambda\mu\nu}$ becomes a dynamical field to be determined from field equations. See T. De Donder, , *La gravitation de Weyl-Eddington-Einstein*, Gauthier-Villars, Paris, 1924;
H. Weyl, Phys. Z. **22**, 473 (1921); Ann. Phys. 59, 101 (1919); 65, 541 (1921);
A.S. Eddington, Proc. Roy. Soc. **99**, 104 (1921) and *The Mathematical Theory of Relativity*, Springer, Berlin 1925.
In such spaces, the correction is defined as

$$\Gamma_{\mu\nu}{}^{\lambda} \equiv e_a{}^{\lambda}\left(\partial_\mu - D_\mu\right)e^a{}_\nu$$

and can be decomposed as

$$\Gamma_{\mu\nu}{}^{\lambda} \equiv e_a{}^{\lambda}\left(\partial_\mu - D_\lambda\right)e^a_\nu = \bar{\Gamma}_{\mu\nu}{}^{\lambda} - \left(S_{\mu\nu}^{\lambda} - S_{\nu\mu}^{\lambda} + S_{\nu\mu}^{\lambda}\right) + \frac{1}{2}(Q_\mu{}^\lambda{}_\nu - Q^\lambda{}_{\nu\mu} + Q_{\nu\mu}^\lambda),$$

with $S_{\mu\nu}{}^{\lambda}$ from Eq. (11.107).

[3] In the more general geometries of the previous remark, there can also exist a nonzero symmetric part

$$R_{\mu\nu\lambda\kappa} + R_{\mu\nu\kappa\lambda} = \left[D_\mu Q_{\nu\lambda\kappa} - (\nu \leftrightarrow \mu)\right] + 2S_{\mu\nu}{}^{\rho}Q_{\rho\lambda\kappa}.$$

*Get your facts first,
and then you can distort them as much as you please.*
MARK TWAIN (1835–1910)

12

Torsion and Curvature from Defects

In the last chapter we have seen that Minkowski spacetime has neither torsion nor curvature. The absence of torsion follows from its tensor property, which was a consequence of the commutativity of derivatives in front of the infinitesimal translation field

$$\left(\partial_\mu\partial_\nu - \partial_\nu\partial_\mu\right)\xi^\kappa(x) = 0. \tag{12.1}$$

The absence of curvature, on the other hand, was a consequence of the integrability condition (11.31) of the transformation matrices

$$\left(\partial_\mu\partial_\nu - \partial_\nu\partial_\mu\right)\alpha^\kappa{}_\lambda(x) = 0. \tag{12.2}$$

Infinitesimally, this implies that

$$\left(\partial_\mu\partial_\nu - \partial_\nu\partial_\mu\right)\partial_\lambda\xi^\kappa(x) = 0, \tag{12.3}$$

i.e., that derivatives commute in front of *derivatives*of the infinitesimal translation field.

The situation is similar to those in electromagnetism in Chapter 4. Arbitrary gauge transformations (2.104) whose gauge functions $\Lambda(x)$ have commuting derivatives [see (2.105)] do not change the electromagnetic fields in spacetime. In particular, a field-free spacetime remains field-free. In Subsection 4.3 we have seen however, that it is possible to generate thin nonzero magnetic field tubes in a field-free spacetime by performing multivalued gauge transformations which violate Schwarz' integrability conditions. It is useful to imagine these coordinate transformations as being plastic distortions of a *world crystal*. Ordinary single-valued coordinate transformations correspond to elastic distortions of the world crystal which do not change the geometry represented by defects.

In Chapter 9 we have shown that the theoretical description of crystals with defects is very similar to that of electromagnetism in terms of a multivalued scalar field. This suggests a simple way of constructing general affine spacetimes with torsion or curvature or both from a Minkowski spacetime by performing *multivalued* coordinate transformations which do not satisfy (12.1) and (12.3).

12.1   Multivalued Infinitesimal Coordinate Transformations

Let us study the properties of spacetimes which can be reached from basis tetrads $e_a{}^\mu = \delta_a{}^\mu, e^a{}_\mu = \delta^a{}_\mu$ by applying *infinitesimal multivalued* coordinate transformations $\xi^\kappa(x)$. According to (11.68), the new basis tetrads are

$$e_a{}^\mu = \delta_a{}^\mu - \partial_a \xi^\mu, \quad e^a{}_\mu = \delta^a{}_\mu + \partial_\mu \xi^a, \tag{12.4}$$

and the metric is

$$g_{\mu\nu} = e^a{}_\mu e_{a\nu} = \eta_{\mu\nu} + \left(\partial_\mu \xi_\nu + \partial_\nu \xi_\mu \right), \tag{12.5}$$

where $\eta_{\mu\nu}$ denotes the Minkowski metric (1.29). The different notation with respect to (1.29) is necessary in order to adhere to our convention that Greek subscripts refer to curvilinear coordinates (otherwise we would have had to write somewhat clumsily $g_{ab}|_{a=\mu, b=\nu}$).

Inserting the basis tetrads (12.4) into Eq. (11.91) we find the affine connection

$$\Gamma_{\mu\nu}{}^\lambda = \partial_\mu \partial_\nu \xi^\lambda, \tag{12.6}$$

and from this the torsion and curvature tensors

$$S_{\mu\nu}{}^\lambda = \frac{1}{2} \left(\partial_\mu \partial_\nu - \partial_\nu \partial_\mu \right) \xi^\lambda, \quad R_{\mu\nu\lambda}{}^\kappa = \left(\partial_\mu \partial_\nu - \partial_\nu \partial_\mu \right) \partial_\lambda \xi^\kappa. \tag{12.7}$$

Since ξ^λ are infinitesimal displacements, we can lower the last index in both equations with an error quadratic in ξ^κ, and thus negligible for small ξ^κ, so that

$$\Gamma_{\mu\nu\lambda} = \partial_\mu \partial_\nu \xi_\lambda, \quad S_{\mu\nu\lambda} = \frac{1}{2} \left(\partial_\mu \partial_\nu - \partial_\nu \partial_\mu \right) \xi_\lambda, \quad R_{\mu\nu\lambda\kappa} = \left(\partial_\mu \partial_\nu - \partial_\nu \partial_\mu \right) \partial_\lambda \xi_\kappa. \tag{12.8}$$

The curvature tensor is trivially antisymmetric in the first two indices [as in (11.134)].

For singular $\xi(x)$, the metric and the connection are, in general, also singular. This would cause difficulties in performing consistent length measurements and parallel displacements. To avoid such difficulties, Einstein postulated that the metric $g_{\mu\nu}$ and the connection $\Gamma_{\mu\nu}{}^\lambda$ should be smooth enough to permit two differentiations which commute which each other, as stated earlier in (11.137) and (11.138). For the infinitesimal expressions (12.4) and (12.5), these properties imply that we must consider only such singular coordinate transformations which satisfy the condition

$$\left(\partial_\mu \partial_\nu - \partial_\nu \partial_\mu \right) \left(\partial_\lambda \xi_\kappa + \partial_\kappa \xi_\lambda \right) = 0, \tag{12.9}$$

$$\left(\partial_\mu \partial_\nu - \partial_\nu \partial_\mu \right) \partial_\sigma \partial_\lambda \xi_\kappa = 0. \tag{12.10}$$

The integrability conditions (12.9) show again, now for the linearized metric, that the curvature tensor (12.12) is antisymmetric in the last two indices [recall (11.135)].

For completeness, let us also write down the pure Christoffel part of the connection obtained by inserting (12.5) into (11.23):

$$\bar{\Gamma}_{\mu\nu\kappa} = \frac{1}{2}\{\mu\nu, \kappa\} = \frac{1}{2}\left[\partial_\mu\left(\partial_\nu\xi_\kappa + \partial_\kappa\xi_\nu\right) + \partial_\nu\left(\partial_\mu\xi_\kappa + \partial_\kappa\xi_\mu\right) - \partial_\kappa\left(\partial_\mu\xi_\nu + \partial_\nu\xi_\mu\right)\right]$$

(12.11)

For completeness, let us also write down the decomposition (11.115) of the connection into the Christoffel part and the contortion tensor obtained by inserting (12.5) into (11.23):

$$\Gamma_{\mu\nu\kappa} = \{\mu\nu, \kappa\} + K_{\mu\nu\kappa}$$

(12.12)

with

$$
\begin{aligned}
\{\mu\nu, \kappa\} &= \frac{1}{2}\partial_\mu\left(\partial_\nu\xi_\kappa\xi_\nu\right) + \frac{1}{2}\partial_\nu\left(\partial_\mu\xi_\kappa + \partial_\kappa\xi_\mu\right) - \frac{1}{2}\partial_\kappa\left(\partial_\mu\xi_\nu + \partial_\nu\xi_\mu\right) \\
K_{\mu\nu\lambda} &= \frac{1}{2}\left(\partial_\mu\partial_\nu - \partial_\nu\partial_\mu\right)\xi_\lambda - \frac{1}{2}\left(\partial_\nu\partial_\lambda - \partial_\lambda\partial_\nu\right)\xi_\mu + \frac{1}{2}\left(\partial_\lambda\partial_\mu - \partial_\mu\partial_\lambda\right)\xi_\nu \\
&= \frac{1}{2}\partial_\mu\left(\partial_\nu\xi_\lambda - \partial_\lambda\xi_\nu\right) + \frac{1}{2}\partial_\lambda\left(\partial_\nu + \partial_\mu\xi_\nu\right) - \frac{1}{2}\partial_\nu\left(\partial_\lambda\xi_\mu + \partial_\mu\xi_\lambda\right).
\end{aligned}
$$

(12.13)

From the Christoffel symbol we find the Riemann curvature tensor

$$
\begin{aligned}
\bar{R}_{\mu\nu\lambda\kappa} &= \frac{1}{2}\partial_\mu[\partial_\nu\left(\partial_\lambda\xi_\kappa + \partial_\kappa\xi_\lambda\right) + \partial_\lambda\left(\partial_\nu\xi_\kappa + \partial_\kappa\xi_\nu\right) - \partial_\kappa\left(\partial_\nu\xi_\lambda + \partial_\lambda\xi_\nu\right)] \\
&= -\frac{1}{2}\partial_\nu\left[\partial_\mu\left(\partial_\lambda\xi_\kappa + \partial_\kappa\xi_\lambda\right) + \partial_\lambda\left(\partial_\nu\xi_\kappa + \partial_\kappa\xi_\nu\right) - \partial_\kappa\left(\partial_\mu\xi_\lambda + \partial_\lambda\xi_\lambda\right)\right].
\end{aligned}
$$

(12.14)

Due to the integrability condition (12.10) the first terms in each line cancel and this becomes

$$\bar{R}_{\mu\nu\lambda\kappa} = \frac{1}{2}\left\{\left[\partial_\mu\partial_\lambda\left(\partial_\nu\xi_\kappa + \partial_\kappa\xi_\nu\right) - (\mu \leftrightarrow \nu)\right] - (\lambda\kappa)\right\}.$$

(12.15)

The geometry generated in this way coincides with the geometry generated in crystals by infinitesimal multivalued displacements of the atoms. The infinitesimal singular transformations of spacetime

$$x^a \to x^\mu = \left[x^a - \xi^a\left(x^b\right)\right]\delta_a{}^\mu$$

(12.16)

correspond roughly to the infinitesimal displacements of atoms of Section 9.2:

$$\mathbf{x}_n \to \mathbf{x}'_n = \mathbf{x}_n + \mathbf{u}(\mathbf{x}_n),$$

(12.17)

where $\mathbf{x}'_n$ are the shifted positions, as seen from an ideal reference crystal. If we change the point of view to an intrinsic description, i.e., if we measure coordinates by counting the number of atomic steps *within* the distorted crystal, then the atoms of the ideal reference crystal are displaced by

$$\mathbf{x}_n \to \mathbf{x}'_n = \mathbf{x}_n - \mathbf{u}(\mathbf{x}_n),$$

(12.18)

This corresponds now precisely to the transformations (12.16). Hence the noncommutativity of derivatives in front of singular coordinate changes $\xi^a\left(x^\lambda\right)$ is completely analogous to that in front of crystal displacements $u_i(\mathbf{x})$. In crystals this was a signal for the presence of defects. For the purpose of a better visualization, let us restrict our consideration to the three-dimensional flat sub-spacetime of the Minkowski spacetime. Then we have to identify the physical coordinates of material points x^a for $a = 1, 2, 3$ with the previous spatial coordinates[1] x_i for $i = 1, 2, 3$, and $\partial_a = \partial/\partial x^a (a = i)$ with the previous derivatives ∂_i. The infinitesimal translations $\xi^{a=i}(\mathbf{x})$ in (11.143) are equal to the displacements $u_i(\mathbf{x})$. The associated basis tetrads are, as in (12.4),

$$e'_a = \delta_a^i - \partial_a u_i, \quad e^a{}_i = \delta^a{}_i + \partial_i u_a, \tag{12.19}$$

and the metric becomes, as in (12.5),

$$g_{ij} = e_{ai}e^a{}_j = \delta_{ij} + \partial_i u_j + \partial_j u_i = \delta_{ij} + 2u_{ij}. \tag{12.20}$$

The connection is simply

$$\Gamma_{ijk} = \partial_i\partial_j u_k \tag{12.21}$$

with torsion and curvature tensors

$$S_{ijk} = \frac{1}{2}\left(\partial_i\partial_j - \partial_j\partial_i\right)u_k, \quad R_{ijkl} = \left(\partial_i\partial_j - \partial_j\partial_i\right)\partial_k u_l. \tag{12.22}$$

The integrability conditions (12.9) and (12.10) can be combined to the three relations

$$\left(\partial_i\partial_j - \partial_j\partial_i\right)\left(\partial_k u_l + \partial_l u_k\right) = 0, \tag{12.23}$$

$$\left(\partial_i\partial_j - \partial_j\partial_i\right)\partial_n\left(\partial_k u_i + \partial_1 u_k\right) = 0, \tag{12.24}$$

$$\left(\partial_i\partial_j - \partial_j\partial_i\right)\partial_k\left(\partial_k u_i - \partial_l u_k\right) = 0. \tag{12.25}$$

They state that the strain tensor, its derivative, and the derivative of the local rotation field are all twice-differentiable single-valued functions everywhere. It was argued that this should be true in a crystal. We can take advantage of the first condition and write the curvature tensor alternatively as

$$R_{ijkl} = \left(\partial_i\partial_j - \partial_j\partial_i\right)\frac{1}{2}\left(\partial_k u_l - \partial_l u_k\right). \tag{12.26}$$

The antisymmetry in ij and kl suggests, in three dimensions, the introduction of a tensor of second rank analogous to (11.144)

$$G_{ji} \equiv \frac{1}{4}e_{ikl}e_{jmn}R^{klmn}, \tag{12.27}$$

---

[1]When working with four-vectors it is conventional to consider the upper indices as physical components. In purely three-dimensional calculations we employ the metric $g_{ab} = \delta_{ab}$ such that $x^{a=i}$ and x_i are the same.

where e_{ijk} is the covariant Levi-Civita tensor defined in Eq. (11A.7). The tensor G_{ji} coincides with the Einstein tensor (11.143) due to the identity (1A.17). Inserting here R_{ijkl} of Eq. (12.22), we find in linear approximation

$$G_{ij} = \epsilon_{ikl}\partial_k\partial_l \left(\frac{1}{2}\epsilon_{jmn}\partial_m u_n\right).$$
(12.28)

The expression in parentheses is the local rotation $\omega_j = \frac{1}{2}\epsilon_{jmn}\partial_m u_n$, implying that the Einstein curvature tensor can be written as

$$G_{ji} = \epsilon_{ikl}\partial_k\partial_l\omega_j.$$
(12.29)

Let us also form the Einstein tensor $\bar{G}_{ij}$ associated with the Riemann curvature tensor $\bar{R}_{ijkl}$. Using (12.15) we find

$$\bar{G}_{ji} = \epsilon_{ikl}\epsilon_{jmn}\partial_k\partial_m \frac{1}{2}\left(\partial_l u_n + \partial_n u_l\right).$$
(12.30)

In the discussion of crystal defects we have introduced the following measures for the noncommutativity of derivatives. The dislocation density

$$\alpha_{ij} = \epsilon_{ikl}\partial_k\partial_l u_j,$$
(12.31)

the disclination density

$$\theta_{ij} = \epsilon_{ikl}\partial_k\partial_l,\omega_j$$
(12.32)

and the defect density

$$g_{ij} = \epsilon_{ikl}\epsilon_{jmn}\partial_k\partial_m u_{lm}.$$
(12.33)

Comparison with (12.15) shows that α_{ij} is directly related to the torsion tensor $S_{kl}{}^i = \frac{1}{2}\left(\Gamma_{kl}{}^i - \Gamma_{lk}{}^i\right)$:

$$\alpha_{ij} \equiv \epsilon_{ikl}\Gamma_{klj} \equiv \epsilon_{ikl}S_{klj}.$$
(12.34)

Hence torsion is a measure of the translational defects contained in the multivalued coordinate transformations, which may be pictured as combinations of elastic plus plastic distortions of our world crystal.

We can also use the decomposition (11.115) and write, due to the symmetry of the Christoffel symbol $\{kl, j\}$ in kl:

$$\alpha_{ij} = \epsilon_{ikl}K_{klj},$$
(12.35)

where K_{klj} is the contortion tensor. In terms of the displacement field $u(\mathbf{x})$,

$$
\begin{aligned}
K_{ijk} &= \frac{1}{2}\partial_j\left(\partial_j u_k - \partial_k u_j\right) - \frac{1}{2}\left[\partial_j\left(\partial_k u_j + \partial_i u_k\right) - (j \leftrightarrow k)\right] \\
&= \partial_i\omega_{jk} - \left(\partial_j u_{ki} - \partial_k u_{ji}\right).
\end{aligned}
$$
(12.36)

Since K_{ijk} is antisymmetric in jk, it is useful to introduce the tensor of second rank called *Nye contortion tensor*:

$$K_{ln} = \frac{1}{2}K_{klj}\epsilon_{ljn}.$$ (12.37)

Inserting this into (12.35) we see that

$$\alpha_{ij} = -K_{ji} + \delta_{ij}K_{ll}.$$ (12.38)

In terms of the displacement and rotation fields, one has

$$K_{il} = \partial_i\omega_l - \epsilon_{lkj}\partial_j u_{kj}.$$ (12.39)

Consider now the disclination density θ_{ij}. Comparing (12.33) with (12.29) we see that it coincides exactly with the Einstein tensor G_{jl} formed from the full curvature tensor

$$\theta_{ij} \equiv G_{ji}.$$ (12.40)

The defect density (12.33), finally, coincides with the Einstein tensor formed from the Riemann curvature tensor:

$$g_{ij} = \bar{G}_{ij}.$$ (12.41)

Hence we can conclude: Spacetime with small torsion and curvature can be generated from a Minkowski spacetime via singular coordinate transformations. It is completely equivalent to a crystal filled with dislocations and disclinations after a plastic deformation.

In Minkowski spacetime, the trajectories of free particles are straight lines. In spacetime with defects, this is no longer possible and particles run along the straightest possible path. In Einstein's theory, the motion of mass points in a gravitational field is governed by the principle of shortest path in the geometry defined by metric $g_{\mu\nu}$. There the metric contains all gravitational effects in the world crystal. The motion of particles in the world crystal will be discussed later in Chapter 14.

The natural length scale of gravitation is the *Planck length* which is the following combination of Newton's gravitational constant $G_N \approx 6.673 \times 10^{-8}$ cm^3/g s^2 [recall (1.3)] with the light velocity c ($\approx 3 \times 10^{10}$ cm/s) and Planck's constant $\hbar$ ($\approx 1.05459 \times 10^{-27}$ erg/s):

$$l_{\rm P} = \left(\frac{c^3}{G_N\hbar}\right)^{-1/2} \approx 1.616 \times 10^{-33}{\rm cm}.$$ (12.42)

The Planck length is an extremely small quantity. It is by a factor 10^{-25} smaller than an atom, which is roughly the ratio between the radius of an atom ($\approx 10^{-8}$ cm) and the radius of the solar system ($\approx 10^{10}$ km). Such small distances are at present beyond any experimental resolution, and will probably be so in the distant future.

Particle accelerators are presently able to probe distances which are still 10 orders of magnitude larger than l_P. Considering the fast growing costs of accelerators with higher energy, it is unimaginable, that they will be able to probe distances near the Planck length for many generations to come. This length may therefore be considered as the shortest length accessible to experimental physics. Thus the Planck length l_P may easily be imagined as the lattice constant of our world crystal with defects, without running into experimental contradictions.

The mass whose Compton wavelength is l_P,

$$
\begin{aligned}
m_P &= \frac{\hbar}{c l_P} = \sqrt{\frac{\hbar c}{G_N}} = 1.221\,047(79) \times 10^{19}\,\text{GeV} \\
&= 0.021\,7671(14)\,\text{mg} = 1.30138(6) \times 10^{19} m_{\text{proton}}, \tag{12.43}
\end{aligned}
$$

is called the *Planck mass*. Being 19 orders of magnitude larger than the proton mass, it is much larger than any elementary particle mass.

12.2 Examples for Nonholonomic Coordinate Transformations

It may be useful to give a few explicit examples of multivalued mappings $x^\mu(x^a)$ leading from a flat spacetime to a spacetime with curvature and torsion. We shall do so by appealing to actual physical situations. For simplicity, we consider two dimensions. Imagine an ideal crystal with atoms placed at $x^a = (n^1, n^2, n^3) \cdot b$ with infinitesimal lattice constant b.

12.2.1 Dislocation

The simplest example for a crystalline defect is the *edge dislocation* and the *edge disclination* shown in Fig. 12.1. The mapping transforms the lattice points to new

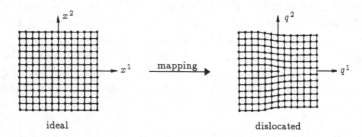

FIGURE 12.1 Edge dislocation in a crystal associated with a missing semi-infinite plane of atoms. The multivalued mapping from the ideal crystal to the crystal with the dislocation introduces a δ-function type torsion in the image space.

distorted positions of which $x^\mu(x^a)$ are the Cartesian coordinates. There exists no one-to-one mapping between the two figures since the excessive atoms in the middle horizontal layer $x^a < 0, x^2 = 0$ have no correspondence in x^a spacetime. In the continuum limit of an infinitesimally small Burgers vector, the mapping can be described by the *multivalued* function

$$\bar{x}^1 = x^1, \qquad \bar{x}^2 = x^2 - \frac{b}{2\pi}\phi, \qquad (12.44)$$

where

$$\phi(x) = \arctan \frac{x^2}{x^1} \qquad (12.45)$$

with the multivalued definition of arctg. On the physical Riemann sheet it is equal to $\pm\pi$ for $x^1 = 0, x^2 = \pm\epsilon$. Its differential version is

$$d\bar{x}^1 = dx^1 \qquad (12.46)$$

$$d\bar{x}^2 = dx^2 + \frac{b}{2\pi}\frac{1}{(x^1)^2 + (x^2)^2}\left(x^2 dx^1 - x^1 dx^2\right) \qquad (12.47)$$

with the basis diads $e^a{}_\mu = \partial\bar{x}^a/\partial x^\mu$

$$e^a{}_\mu = \left(\begin{array}{cc} 1 & 0 \\ \dfrac{b}{2\pi}\dfrac{x^2}{(x^1)^2 + \left(x^2\right)^2} & -\dfrac{b}{2\pi}\dfrac{x^1}{(x^1)^2 + \left(x^2\right)^2} \end{array}\right). \qquad (12.48)$$

We have used the notation $\bar{x}^a \equiv x^a$ in order to distinguish $x^{a=1,2}$ from $x^{\mu=1,2}$.

Let us now integrate dx^μ over a Burgers circuit which consists of a closed circuit $C\left(x^\mu\right)$ in x^μ-space around the origin,

$$b^a = \int_{C(x^\mu)} d\bar{x}^a = \int_{C(x^\mu)} dx^\mu \frac{\partial\bar{x}^a}{\partial x^\mu} = \int_{C(x^\mu)} dx^\mu e^a{}_\mu. \qquad (12.49)$$

Inserting (12.46) and (12.47) we see that

$$b^1 = \oint_{C(x^\mu)} dx^{\bar{1}} = \int_{C(x^\mu)} dx^\mu \frac{\partial x^{\bar{1}}}{\partial x^\mu} = \int_{C(x^\mu)} dx^\mu e^1{}_\mu = 0, \qquad (12.50)$$

$$b^2 = \oint_{C(x^\mu)} dx^{\bar{2}} = \int_{C(x^\mu)} dx^\mu \frac{\partial x^{\bar{2}}}{\partial x^\mu} = \int_{C(x^\mu)} dx^\mu e^2{}_\mu = -b. \qquad (12.51)$$

It is easy to calculate the torsion tensor $S^a{}_{\mu\nu}$ associated with the multivalued mapping (12.46) and (12.47). Because of its antisymmetry, only $S_{12}{}^1$ and $S_{12}{}^2$ are independent. These become

$$S_{12}{}^2 = \partial_1 e^2_2 - \partial_2 e^2_1 = \partial_1 \frac{\partial\bar{x}^?}{\partial x^2} - \partial_2 \frac{\partial\bar{x}^?}{\partial x^1} = -b\delta^{(2)}(\mathbf{x}),$$

$$S_{12}^1 = \partial_1 e^1_2 - \partial_2 e^1_1 = \partial_1 \frac{\partial\bar{x}^1}{\partial x^2} - \partial_2 \frac{\partial\bar{x}^2}{\partial x^1} = 0. \qquad (12.52)$$

We may write this result with the Burgers vector $b^a = (0, b)$ in the form

$$S^a{}_{\mu\nu} = b^a\delta^{(2)}(\mathbf{x}). \qquad (12.53)$$

Let us now calculate the curvature tensor for this defect which is

$$R_{\mu\nu\lambda\kappa} = e_{a\kappa}\left(\partial_\mu\partial_\nu - \partial_\nu\partial_\mu\right)e^a_\lambda.$$ (12.54)

Since $e^a_{\ \mu}$ in (12.48) is single-valued, derivatives in front of it commute. Hence $R_{\mu\nu\lambda\kappa}$ vanishes identically,

$$R_{\mu\nu\lambda\kappa} \equiv 0.$$ (12.55)

A pure dislocation gives rise to torsion but not to curvature.

12.2.2   Disclination

As a second example for a multivalued mapping, we generate curvature by the transformation

$$x^{\bar{i}} = \delta^i_{\ \mu}[x^\mu + \Omega\epsilon^\mu_{\ \nu}x^\nu\phi(x)],$$ (12.56)

with the multi-valued function (12.45). The symbol $\epsilon_{\mu\nu}$ denotes the antisymmetric Levi-Civita tensor. The transformed metric

$$g_{\mu\nu} = \delta_{\mu\nu} - \frac{2\Omega}{x^\sigma x_\sigma}\epsilon_{\mu\nu}\epsilon^\mu_{\ \lambda}\epsilon^\nu_{\ \kappa}x^\lambda x^\kappa.$$ (12.57)

is single-valued and has commuting derivatives. The torsion tensor vanishes since $(\partial_1\partial_2 - \partial_2\partial_1)x^{1,2}$ is proportional to $x^{2,1}\delta^{(2)}(x) = 0$. The local rotation field $\omega(x) \equiv \frac{1}{2}(\partial_1 x^2 - \partial_2 x^1)$, on the other hand, is equal to the multi-valued function $-\Omega\phi(x)$, thus having the noncommuting derivatives:

$$(\partial_1\partial_2 - \partial_2\partial_1)\omega(x) = -2\pi\Omega\delta^{(2)}(x).$$ (12.58)

To lowest order in Ω, this determines the curvature tensor, which in two dimensions possesses only one independent component, for instance R_{1212}. Using the fact that $g_{\mu\nu}$ has commuting derivatives, R_{1212} can be written as

$$R_{1212} = (\partial_1\partial_2 - \partial_2\partial_1)\omega(x).$$ (12.59)

In defect physics, the mapping (12.56) is associated with a disclination which corresponds to an entire section of angle α missing in an ideal atomic array (see Fig. 10.2).

12.3   Differential Geometric Properties of Affine Spaces

Up to now we have studied only affine spacetimes obtained from a Minkowski spacetime by infinitesimal defects. In reality, defects can pile up, and the full affine spacetime requires a nonlinear formulation, which will now be developed.

12.3.1   Integrability of Metric and Affine Connection

The general, affine spacetime will be characterized by the same type of integrability conditions as the spacetime with infinitesimal defects stated in Eqs. (12.9) and (12.10). In the nonlinear formulation, these conditions are imposed upon metric and affine connection:

$$\left(\partial_\mu \partial_\nu - \partial_\nu \partial_\mu\right) g_{\lambda\kappa} = 0, \tag{12.60}$$

$$\left(\partial_\mu \partial_\nu - \partial_\nu \partial_\mu\right) \Gamma^\kappa_{\sigma\lambda} = 0. \tag{12.61}$$

Remember that the first condition ensures the antisymmetry of the curvature tensor in the last two indices [see (11.135)]. By antisymmetrizing the second condition in $\sigma\lambda$ it can also be replaced by an integrability condition for the torsion

$$\left(\partial_\mu \partial_\nu - \partial_\nu \partial_\mu\right) S_{\sigma\lambda}{}^\kappa = 0. \tag{12.62}$$

Moreover, using the decomposition (11.115), the Christoffel symbol is seen to be integrable as well:

$$\left(\partial_\mu \partial_\nu - \partial_\nu \partial_\mu\right) \left\{ {\kappa \atop \sigma\lambda} \right\} = 0. \tag{12.63}$$

There exists also the nonlinear version of Eq. (12.24):

$$\left(\partial_\mu \partial_\nu - \partial_\nu \partial_\mu\right) \partial_\sigma g_{\lambda\kappa} = 0. \tag{12.64}$$

To prove this we note that products of integrable functions f and g are themselves integrable since

$$\left(\partial_\mu \partial_\nu - \partial_\nu \partial_\nu\right)(fg) = \left[\left(\partial_\mu \partial_\nu - \partial_\nu \partial_\mu\right) f\right] g + f \left(\partial_\mu \partial_\nu - \partial_\nu \partial_\mu\right) g = 0. \tag{12.65}$$

Since the derivatives of $g_{\lambda\kappa}$ can be expressed as sums of products of Christoffel symbols and metric tensors, which are integrable, Eq. (12.64) is indeed true. Thus, given the integrability property (12.60) of the metric tensor, the two equations (12.63) and equivalent (12.64) are equivalent.

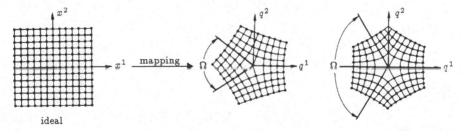

FIGURE 12.2 Edge disclination in a crystal associated with a missing semi-infinite section of atoms of angle Ω. The multivalued mapping from the ideal crystal to the crystal with the disclination introduces a δ-function type curvature in the image spacetime.

12.3.2   Local Parallelism

In order to understand the geometric properties of such a general affine spacetime let us first introduce the concept of *local parallelism*.

Consider a vector field $\mathbf{v}(x) = \mathbf{e}_a v^a(x)$ which is parallel in the inertial frame in the naive sense that all vectors point in the same direction. This simply means $\partial_b \mathbf{v}(x) = \mathbf{e}_a \partial_b v^a = 0$. But when changing to the coordinates x^μ we find

$$\partial_b v^a = \partial_b e^a_\mu v^\mu = e_b^{\ \nu} \partial_\nu \left(e^a_\mu v^\mu \right) = e_b^\nu e^a_\mu D_\nu v^\mu = 0. \tag{12.66}$$

Thus parallel vector fields have their local components v^μ change in such a way that their covariant derivatives vanish:

$$D_\nu v^\mu = \partial_\nu v^\mu + \Gamma_{\nu\lambda}{}^\mu v^\lambda = 0. \tag{12.67}$$

Similarly we find:

$$D_\nu v_\mu = \partial_\nu v_\mu - \Gamma_{\nu\mu}{}^\lambda v_\lambda = 0. \tag{12.68}$$

Note that the basis tetrads e^ν_a, e^a_ν are parallel vector fields, by construction [see (11.93)].

Let us study this type of situation in general: Given an arbitrary connection $\Gamma_{\mu\nu}{}^\lambda$ we first ask the question under what condition it is possible to find a parallel vector field in the whole spacetime. For this we consider the vector field $v^\mu(x)$ at a point x_0 where it has the value $v^\mu(x_0)$. Let us now move to the neighboring position $x_0 + dx$. There the field has the components

$$v^\mu (x_0 + dx) = v^\mu(x_0) + \partial_\nu v^\mu(x_0) d^\nu_x. \tag{12.69}$$

If $v^\mu(x)$ is a parallel vector field with $D_\nu v^\mu = 0$ the derivative satisfies

$$\partial_\nu v^\mu = -\Gamma_{\nu\kappa}{}^\mu v^\kappa. \tag{12.70}$$

This differential equation is integrable over a finite region of spacetime if and only if Schwarz's criterion is fulfilled:

$$(\partial_\lambda \partial_\nu - \partial_\nu \partial_\lambda) v^\mu = 0. \tag{12.71}$$

If we calculate

$$(\partial_\lambda \partial_\nu - \partial_\nu \partial_\lambda) v^\mu = -\partial_\lambda (\Gamma^\mu_{\nu\kappa} v^\kappa) + \partial_\nu (\Gamma_{\lambda\kappa}{}^\mu v^\kappa), \tag{12.72}$$

we find

$$- (\partial_\lambda \Gamma_{\nu\kappa}{}^\mu - \partial_\nu \Gamma_{\lambda\kappa}{}^\mu) v^\kappa - \Gamma^\mu_{\nu\kappa} \partial_\lambda v^\kappa + \Gamma^\mu_{\lambda\kappa} \partial_\nu v^\kappa \tag{12.73}$$

which becomes after using once more (12.70):

$$(\partial_\lambda \partial_\nu - \partial_\nu \partial_\lambda) v^\mu = -R_{\lambda\nu\kappa}{}^\mu v^\kappa. \tag{12.74}$$

Thus the parallel field $v^\mu(x)$ exists in the whole spacetime if and only if the curvature tensor vanishes everywhere.

If $R_{\lambda\nu\kappa}$ is nonzero, the concept of parallel vectors cannot be carried over from Minkowski spacetime to the general affine spacetime over any finite distance. Such spacetimes are called *curved*. One says that in curved spacetimes there exists no *teleparallelism*.

We have illustrated before, that this is the case in the presence of disclinations. Disclinations generate curvature, i.e., a crystal containing disclinations is curved in the differential-geometric sense.

This is in accordance with the previous observation that the disclination density θ_{ij} coincides with the Einstein curvature tensor G_{ij}.

In Fig. 12.3 we also see that even in the presence of a disclination it still is meaningful to define a vector field as *locally parallel*. The condition for this is that the covariant derivatives vanish at the point x_0 : $D_\nu v^\mu(x_a) = 0$. If this condition is satisfied, neighboring vectors $v^\mu(x)$ differ from $v^\mu(x_0)$ at most by terms of the order $(x - x_0)^2$, rather than $(x - x_0)$ for nonparallel vectors. In order to see this in more detail let us draw an infinitesimal quadrangle $ABCD$ in the coordinates x^μ spanned by $AB = dx^\mu = DC$ and $BC = dx_2^\mu = AD$ (see Fig. 12.3). Now we compare the directions of $v^\mu(x)$ before and after going around the circumference. When passing from A at x^μ to B at $x^\mu + dx_1^y$ the vector components change from $v_1^\mu = v^\mu(x)$ to

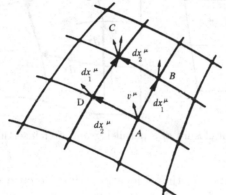

FIGURE 12.3 Illustration of parallel transport of a vector around a closed circuit ABCD.

$$v_B^\mu = v^\mu\left(x^\mu + x_1^\nu\right) = v_1^\mu + \partial_\nu v^\mu d x_1^\nu v_A^\mu - \overset{A}{\Gamma}_{\nu\lambda}{}^\mu v^\lambda d x_1^\nu. \qquad (12.75)$$

When continuing to C at $x^\mu \mid d\overset{\mu}{x} \mid d\overset{\mu}{x}$ we have

$$\begin{aligned}
v_C{}^\mu &= v_B{}^\mu - \overset{B}{\Gamma}_{\tau\kappa}{}^\mu v_B{}^\kappa d x_2^\tau \\
&= v_A{}^\mu - \overset{A}{\Gamma}_{\nu\lambda}{}^\mu v^\lambda d x_1^\nu - \overset{B}{\Gamma}_{\tau\kappa}{}^\mu v_A{}^\kappa d x_2^\tau + \overset{B}{\Gamma}_{\tau\kappa}{}^\mu \overset{A}{\Gamma}_{\nu\lambda}{}^\kappa v_A d\overset{\nu}{x}_1 d x_2^\tau \\
&= v_A^\mu - \overset{A}{\Gamma}_{\nu\lambda}{}^\mu v^\lambda \left(d x_1^\nu + d x_2^\nu\right) - \partial_\nu \overset{A}{\Gamma}_{\tau\kappa}{}^\mu v_A^\kappa d x_1^\nu d x_2^\tau +
\end{aligned}$$

$$+ \overset{A}{\Gamma}_{\tau\kappa}{}^{\mu} \overset{A}{\Gamma}_{\nu\lambda}{}^{\kappa} v_A^{\lambda} d\underset{1}{x}{}^{\nu} d\underset{2}{x}{}^{\tau} + \mathcal{O}\left(dx^3\right). \tag{12.76}$$

We can now repeat the same procedure along the line ADC, and we find the same result with interchanged $d\underset{1}{x}$ and $d\underset{2}{x}$. The difference between the two results is

$$v^{\mu}_{ABC} - v^{\mu}_{ADC} = -\frac{1}{2} R_{\nu\tau\kappa}{}^{\mu} v_a^{\kappa} ds^{\nu\tau} + \mathcal{O}\left(dx^3\right) \tag{12.77}$$

where $ds^{\nu\tau} = \left(d\underset{1}{x}{}^{\nu} d\underset{2}{x}{}^{\tau} - d\underset{1}{x}{}^{\tau} d\underset{2}{x}{}^{\nu} \right)$ is the infinitesimal surface element of the quadrangle.

There exists a similar geometric illustration of the torsion property $S_{\mu\nu}{}^{\lambda} \neq 0$. Consider a crystal with an edge dislocation (see Fig. 12.2). Let us focus attention upon a closed circuit with the form of a parallelogram in the ideal reference crystal (i.e., in the coordinate frame $\mathbf{e}^a$) and suppose its image in the $\mathbf{e}^{\mu}$-frame encloses the dislocation line (see Fig. 12.4).

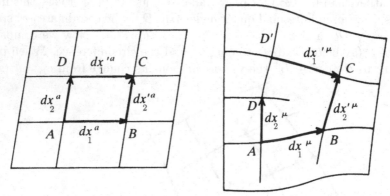

FIGURE 12.4 Illustration of non-closure of a parallelogram after inserting an edge dislocation.

Volterra process of constructing the dislocation, the reference crystal was cut open, and a layer of atoms was inserted. In this process, the original parallelogram is opened such that the dislocation crystal has a gap between the open ends. The gap vector is precisely the Burgers vector. To be specific, let the parallelogram in the ideal reference crystal be spanned by the vectors $AB = d\underset{1}{\overset{a}{x}} = DC, AD = d\underset{a}{\overset{2}{x}} = BC$. In the defected spacetime x^{μ} these become $AB = d\underset{1}{\overset{\mu}{x}}, AD = d\underset{2}{\overset{\mu}{x}}, D'C = d\underset{1}{\overset{\prime\mu}{x}}, BC = d\underset{2}{\overset{\prime\mu}{x}}$. Since $d\underset{1}{\overset{\prime\mu}{x}}, d\underset{2}{\overset{\prime\mu}{x}}$ are parallel in the ideal reference crystal, they are parallel vectors, i.e., the vectors $v^{\mu}(x) = d\underset{2}{\overset{\mu}{x}}, v^{\mu}\left(x^{\mu} + d\underset{1}{\overset{\mu}{x}}\right)$ satisfy (12.70) when going from A to B_j, i.e., $\partial_{\nu} d\underset{2}{\overset{\mu}{x}} = -\Gamma_{\nu\lambda}{}^{\mu} d\underset{2}{\overset{\lambda}{x}}$ and hence

$$dx_2'^{\mu} = d\underset{2}{x}{}^{\mu} - \Gamma_{\nu\lambda}{}^{\mu} d\underset{2}{x}{}^{\nu} d\underset{2}{x}{}^{\lambda}. \tag{12.78}$$

Similarly the vectors $d\underset{1}{\overset{\mu}{x}}$ and $d\underset{1}{\overset{\prime\mu}{x}}$ are parallel and therefore related by

$$d\underset{1}{\overset{\prime\mu}{x}} = d\underset{1}{\overset{\mu}{x}} - \Gamma^{\mu}{}_{\nu\lambda}d\underset{2}{\overset{\nu}{x}}d\underset{1}{\overset{\lambda}{x}}. \tag{12.79}$$

From this it follows that

$$b^{\mu} = \left(d\underset{2}{\overset{\prime}{x}}+d\underset{1}{x}\right)^{\mu} - \left(d\underset{1}{\overset{\prime}{x}}+d\underset{2}{x}\right)^{\mu} = -S_{\nu\lambda}{}^{\lambda}ds^{\nu\lambda}. \tag{12.80}$$

In a Minkowski spacetime, the torsion vanishes and the image is again a closed parallelogram. Einstein assumed the vanishing of torsion in gravitational spacetime.

12.4 Circuit Integrals in Affine Spaces with Curvature and Torsion

In order to establish contact with the circuit definitions of disclinations and dislocations in crystals, let us phrase the differential results (12.77) and (12.80) in terms of contour integrals.

12.4.1 Closed-Contour Integral over Parallel Vector Field

Given a vector field $v^{\mu}(x)$ which is locally *parallel*, i.e., which has $D_{\nu}v^{\mu}(x) = 0$, consider the change of $v^{\mu}(x)$ while going around a closed contour which is

$$\Delta v^{\mu} = \oint dv^{\mu}(x) = \oint dx^{\nu}\partial_{\nu}v^{\mu}(x). \tag{12.81}$$

By decomposing C into a large set of infinitesimal surface elements we can apply (12.77) and find

$$\Delta v^{\mu} = \oint_{C(x^{\mu})} dx^{\nu}\partial_{\nu}v^{\mu} = -\frac{1}{2}\int_{S(x^{\mu})} ds^{\tau\nu} R_{\tau\nu\kappa}{}^{\mu}(x)v^{\kappa}(x). \tag{12.82}$$

Note that the tetrad fields e_a^{μ} are locally parallel by definition such that they satisfy

$$\Delta e_a{}^{\mu} = -\oint_{C(x^{\mu})} dx^{\nu}\partial_{\nu}e_a{}^{\mu} = -\frac{1}{2}\int_{S(x^{\mu})} ds^{\tau\nu} R_{\tau\nu\kappa}{}^{\mu}(x)e_a^{\kappa}(x). \tag{12.83}$$

Actually, this relation follows directly from Stokes' theorem:

$$\Delta e_a^{\mu} = \oint_{C(x^{\mu})} dx^{\nu}\partial_{\nu}e_a{}^{\mu} = \oint_{C(x^{\mu})} ds^{\tau\nu}\partial_{\tau}\partial_{\nu}e_a{}^{\mu} =$$

$$= -\frac{1}{2}\oint_{S(x^{\mu})} ds^{\tau\nu} R_{\tau\nu\kappa}{}^{\mu}e_a{}^{\kappa}. \tag{12.84}$$

For an infinitesimal circuit, we can remove the tetrad from the integral and have

$$\Delta e_a{}^{\mu} \approx \left\{-\frac{1}{2}\oint_{S(x^{\mu})} ds^{\tau\nu} R_{\tau\nu\kappa}{}^{\mu}\right\} e_a{}^{\kappa} \equiv \omega^{\mu}{}_{\kappa}e^{\kappa}{}_a. \tag{12.85}$$

The matrix $\omega^\mu{}_\kappa$ has the property that $\omega_{\mu\kappa} = g_{\mu\lambda}\omega^\lambda{}_\kappa$ is antisymmetric, due to the antisymmetry of $R_{e\nu\kappa\mu}$ in $\kappa\mu$. Hence $\omega^\mu{}_\kappa$ can be interpreted as the parameters of an infinitesimal local Lorentz transformation. In three dimensions, this is a local rotation in agreement with what we observed previously:

Curvature is a signal for disclinations and these are rotational defects.

12.4.2 Closed-Contour Integral over Coordinates

Let us now give an integral characterization of torsion. For this we consider an arbitrary closed contour $C(x^a)$ in the inertial frame (which generalizes the parallelogram used in the previous discussion). In the defected spacetime this contour has an image $C'(x^a)$ which does not necessarily close. In order to find how much is missing we form the integral

$$\Delta x^\mu = \oint_{C(x^a)} dx^\mu = \oint_{C(x^a)} dx^a \frac{\partial x^\mu}{\partial x^a} = \oint_{C(x^a)} dx^a e_a{}^\mu(x^a). \tag{12.86}$$

By Stokes' theorem, this becomes

$$\frac{1}{2}\int_{C(x^a)} ds^{ab}\,(\partial_a e_b{}^\mu - \partial_b e_a{}^\mu) = \frac{1}{2}\oint_{S(x^a)} ds^{ab}\,(e_a{}^\nu\partial_\nu e_b{}^\mu - (a \leftrightarrow b)) = -\oint ds^{ab} S_{ab}{}^\mu. \tag{12.87}$$

The quantity

$$S_{ab}{}^\mu = -\frac{1}{2}e_a{}^\nu\,[\partial_\nu e_b{}^\mu - (a \leftrightarrow b)] \tag{12.88}$$

is related to the torsion $S_{\lambda\kappa}{}^\mu$ conversion of the lower indices from the local to the inertial form

$$S_{ab}{}^\mu = e_a{}^\lambda e_b{}^\kappa S_{\lambda\kappa}{}^\mu = -\frac{1}{2}\left\{e_a{}^\lambda e_b{}^\kappa\,[e^c{}_\kappa\partial_\lambda e_c{}^\mu - (a \leftrightarrow b)]\right\}$$

$$\equiv -\frac{1}{2}\left[e_a{}^\lambda\partial_\lambda e_b{}^\mu - (a \leftrightarrow b)\right]. \tag{12.89}$$

If the tetrad vectors are known as functions of the external coordinates x^a, we can also use $e_a{}^\lambda\partial_\lambda = \partial_a$ and write $S_{ab}{}^\mu$ in the form

$$S_{ab}{}^\mu \equiv -\frac{1}{2}[\partial_a e_b{}^\mu - (a \leftrightarrow b)]. \tag{12.90}$$

Sometimes one also converts the upper Einstein index μ into a local Lorentz index c and works with

$$S_{ab}{}^c = e^c{}_\mu S_{ab}{}^\mu = -\frac{1}{2}\left[e^c{}_\mu\partial_a e_b{}^\mu - (a \leftrightarrow b)\right]. \tag{12.91}$$

If there is no torsion, the integral (12.87) vanishes. Otherwise the image of the closed contour $C(x^a)$ has a gap and thus is defined as the Burgers vector

$$b^\mu = \int_{C'(x^\mu)} dx^\mu = -\oint_{C(x^a)} ds^{ab} S_{ab}{}^\mu. \tag{12.92}$$

12.4.3   Closure Failure and Burgers Vector

It should be mentioned that the circuit integrals measuring curvature and torsion may be executed in the opposite way by forming closed circuits $C(x^\mu)$ around the defect in the spacetime x^μ and studying the properties of the image circuit $C'(x^a)$ in the ideal reference crystal. The torsion measures how much the image $c'(x^a)$ fails to close. The *closure failure* is given by the Burgers vector

$$b^a = \int_{c'(x^a)} dx^a = \int_{c(x'^\mu)} dx^\mu \frac{\partial x^a}{\partial x^\mu} = \int_{c(x^\mu)} dx^\mu e^a{}_\mu \tag{12.93}$$

which can be rewritten, using Stokes' theorem, as

$$b^a \int_{S(x^\mu)} ds^{\nu\mu} \partial_\nu e^a{}_\mu = \int_{S(x^\mu)} ds^{\nu\mu} S_{\nu\mu}{}^a. \tag{12.94}$$

The tensor $S_{\mu\nu}{}^a \equiv S_{\nu\mu}{}^\lambda e^a_\lambda = \frac{1}{2}\left(\partial_\mu e^a_\nu - \partial_\nu e^a_\mu\right)$ is related to (12.90) by the exchange of Einstein and local Lorentz indices.

12.4.4   Alternative Circuit Integral for Curvature

There is an analogous circuit integral characterizing the curvature from the standpoint of the coordinates x^a. For this we introduce the local Lorentz tensor related to $R_{\mu\nu\rho}{}^\kappa$:

$$R_{abc}{}^d \equiv e_a{}^\mu e_b{}^\nu e_c{}^\lambda e^d{}_\kappa R_{\mu\nu\lambda}{}^\kappa. \tag{12.95}$$

Then the circuit integral reads

$$\Delta e_a{}^\mu = -\frac{1}{2}\int_{S(x^\mu)} ds^{ed} R_{eda}{}^b e_b{}^\mu. \tag{12.96}$$

If one wants to calculate $R_{abc}{}^d$ directly in x^a spacetime using differentiations one has to keep in mind that under the multivalued mapping $x^a \to x^\mu$, $R_{\mu\nu\lambda}{}^\kappa$ does not transform like a tensor. In fact, a simple manipulation shows

$$
\begin{aligned}
R_{\mu\nu\lambda}{}^\kappa &= e_a{}^\kappa \left(\partial_\mu\partial_\nu - \partial_\nu\partial_\mu\right) e^d{}_\lambda \\
&= e_d{}^\kappa \left[e^a{}_\mu \partial_a e^b{}_\nu \partial_b - (\mu \leftrightarrow \nu)\right] e^d{}_\lambda \\
&= e^a{}_\mu e^b{}_\nu e_d{}^\kappa \left(\partial_a\partial_b - \partial_b\partial_a\right) e^d{}_\lambda + \left[e^a{}_\mu e_d{}^\kappa \left(\partial_a e^b{}_\nu\right)\left(\partial_b e^d{}_\lambda\right) - (\mu \leftrightarrow \nu)\right] \\
&\quad - e^a{}_\mu e^b{}_\nu \tilde{R}_{ab\lambda}{}^\kappa + \left[e^a{}_\mu e_d{}^\kappa e_a{}^\sigma \Gamma_{\sigma\nu}{}^b e_b{}^\tau \Gamma_{\tau\lambda}{}^d - (\mu \leftrightarrow \nu)\right] \\
&= e^a{}_\mu e^b{}_\nu \tilde{R}_{ab\lambda}{}^\kappa + \left[\Gamma_{\mu\nu}{}^\sigma \Gamma_{\sigma\lambda}{}^\kappa - (\mu \leftrightarrow \nu)\right] \\
&= e^a{}_\mu e^b{}_\nu \tilde{R}_{ab\lambda}{}^\kappa + 2S_{\mu\nu}{}^\sigma \Gamma_{\sigma\lambda}{}^\kappa,
\end{aligned}
\tag{12.97}
$$

where

$$\tilde{R}_{ab\lambda}{}^\kappa \equiv e_d{}^\kappa \left(\partial_a\partial_b - \partial_b\partial_a\right) e^d{}_\lambda \tag{12.98}$$

is evaluated in the same way as $R_{\mu\nu\lambda}{}^{\kappa}$ in Eq. (11.130), but by forming ∂_a derivatives rather than ∂_μ. Expressing also the torsion $S_{\mu\nu}{}^{\sigma}$ in terms of derivatives $\partial/\partial x^a = \partial_a$ as in (12.90) we can write

$$S_{\mu\nu}{}^{\sigma} = e^a{}_\mu e^b{}_\nu e_c{}^\sigma S_{ab}{}^c. \tag{12.99}$$

For the affine connection we may define, similarly,

$$\Gamma_{\mu\nu}{}^{\sigma} \equiv e^a{}_\mu e^b{}_\nu e_c{}^\sigma \Gamma_{ab}{}^c \tag{12.100}$$

with

$$
\begin{aligned}
\Gamma_{ab}{}^c &= e_a{}^\mu e_b{}^\nu e^c{}_\lambda \Gamma_{\mu\nu}{}^\lambda = -e_a{}^\mu e_b{}^\nu e^c{}_\lambda e^d{}_\nu \partial_\mu e_d{}^\lambda = -e_a{}^\mu e^c{}_\lambda \partial_\mu e_b{}^\lambda \\
&= -e^c{}_\lambda \partial_a e_b{}^\lambda \equiv e^c{}_\lambda \Gamma_{ab}{}^\lambda = e_b{}^\lambda \partial_a e^c{}_\lambda.
\end{aligned} \tag{12.101}
$$

Then $R_{abc}{}^d$ of (12.95) can be written as

$$R_{abc}{}^d = \tilde{R}_{abd}{}^d + 2S_{ab}{}^e \Gamma_{ec}{}^d. \tag{12.102}$$

12.4.5 Parallelism in World Crystal

From the standpoint of our world crystal with defects, parallelism has a simple meaning. Consider Fig. 11.1b. We identify the dashed curves $x^a(x^\mu) = $ const. with the crystal planes of an elastically distorted crystal as seen from the local frame with coordinates x^μ. An observer living on the distorted crystal orients himself by the planes $x^a(x^\mu) = $ const. He measures distances and directions by counting atoms along the crystal directions. The above definition of parallelism amounts to vectors being defined as parallel if they are so from his point of view, i.e., if they correspond to parallel vectors in the undistorted crystal. Thus the normal vectors to the dashed coordinate planes $x^a(x^\mu) = $ const. are parallel to each other. Indeed, they form the vector fields $e^\mu_a(x)$, which always satisfy $D_\nu e^\mu_a = 0$ [see (11.93].

If the mapping $x^\mu(x^a)$ contains defects it is, in general, impossible to find a global definition of parallelism. Consider, for example, a wedge disclination which is shown in Fig. 12.2, say the -90^0 one. The crystal has been cut from the left, and new crystalline material has been inserted in the Volterra construction process. The crystalline coordinate planes define parallel lines. Since the right-hand piece remembers the original crystal, there exists a completely consistent definition of parallelism. For example, the almost horizontal planes are all parallel. The lines cutting these vertically are also parallel by definition. On the left-hand side, the vertical lines continue smoothly into the inserted new crystalline material. In the middle, however, they meet and turn suddenly out to be orthogonal. Still, the coordinate planes define parallelism in any small region inside the original as well as the inserted material, except on the disclination line.

12.5   Bianchi Identities for Curvature and Torsion Tensors

Let us derive a few important properties of curvature and torsion tensors. As noted before, the curvature tensor is antisymmetric in $\mu\nu$, by construction, and in $\lambda\kappa$, due to the integrability condition (11.18) for the metric tensor. In addition, it satisfies the so-called *fundamental identity*. This follows directly from the representation (11.129) by adding terms in which $\mu\nu\lambda$ are interchanged cyclically:

$$R_{\nu\mu\lambda}{}^{\kappa} = 2D_{\nu}S_{\mu\lambda}{}^{\kappa} - 4S_{\nu\mu}{}^{\rho}S_{\lambda\rho}{}^{\kappa} \tag{12.103}$$

where the symbol $\llcorner\!\!\lrcorner$ denotes a sum of cyclic permutations of the indicated subscripts. The derivation of the fundamental identity requires commuting derivatives in front of the metric tensor, i.e., it requires that the metric $g_{\mu\nu}$ satisfies the integrability condition

$$(\partial_{\mu}\partial_{\nu} - \partial_{\nu}\partial_{\mu})g_{\lambda\kappa} = 0. \tag{12.104}$$

The fundamental identity is therefore a Bianchi identity [recall the definition given after Eq. (2.89)].

In symmetric spacetimes where $S_{\mu\nu\lambda} = 0$ and $R_{\mu\nu\lambda\kappa} = \bar{R}_{\mu\nu\lambda\kappa}$, the fundamental identity implies the additional symmetry property of the Riemann tensor

$$\bar{R}_{\mu\nu\lambda\kappa} + \bar{R}_{\nu\lambda\mu\kappa} + \bar{R}_{\lambda\mu\nu\kappa} = 0. \tag{12.105}$$

Using the antisymmetry in $\mu\nu$ and $\lambda\kappa$ leads once more to the property (11.147):

$$\bar{R}_{\mu\nu\lambda\kappa} = \bar{R}_{\lambda\kappa\mu\nu}. \tag{12.106}$$

Another important identity is the original *Bianchi identity* which has given the name to all similar identities in this book which are based on the integrability condition of observable fields. The original Bianchi identity follows from the assumption of the single-valuedness of the affine connection which implies that it satisfies the integrability condition

$$\left(\partial_{\mu}\partial_{\nu} - \partial_{\nu}\partial_{\mu}\right)\Gamma_{\lambda\kappa}{}^{\rho} = 0. \tag{12.107}$$

Consider the vector

$$\mathbf{R}_{\sigma\nu\mu} \equiv (\partial_{\sigma}\partial_{\nu} - \partial_{\nu}\partial_{\sigma})\,\mathbf{e}_{\mu}, \tag{12.108}$$

which determines the curvature tensor $R_{\sigma\nu\mu}{}^{\lambda}$ via the scalar product with $\mathbf{e}^{\lambda}$ [recall (11.130)]. Applying the covariant derivative gives

$$D_{\tau}\mathbf{R}_{\sigma\nu\mu} = \partial_{\tau}\mathbf{R}_{\sigma\nu\mu} - \Gamma_{\tau\sigma}{}^{\kappa}\mathbf{R}_{\kappa\nu\mu} - \Gamma_{\tau\nu}{}^{\kappa}\mathbf{R}_{\sigma\nu\kappa}. \tag{12.109}$$

Performing cyclic sums over $\tau\sigma\nu$ and using the antisymmetry of $\mathbf{R}_{\sigma\nu\mu}$ in $\sigma\nu$ leads to

$$D_{\tau}\mathbf{R}_{\sigma\nu\mu} = \partial_{\tau}\mathbf{R}_{\sigma\nu\mu} - \Gamma_{\tau\mu}{}^{\kappa}\mathbf{R}_{\sigma\nu\kappa} + 2S_{\tau\sigma}{}^{\kappa}\mathbf{R}_{\nu\kappa\mu}. \tag{12.110}$$

Now we use

$$\partial_\sigma \partial_\nu \mathbf{e}_\mu = \partial_\sigma \left(\Gamma_{\nu\mu}{}^\alpha \mathbf{e}_\alpha \right) = \Gamma_{\nu\mu}{}^\kappa \mathbf{e}_\kappa, \tag{12.111}$$

to derive

$$\partial_\tau \partial_\sigma \partial_\nu \mathbf{e}_\mu = \partial_\tau \Gamma_{\nu\mu}{}^\kappa \partial_\sigma \mathbf{e}_\kappa + (\tau\sigma) + \partial_\tau \partial_\sigma \Gamma_{\nu\mu}{}^\kappa \mathbf{e}_\alpha + \Gamma_{\nu\mu}{}^\kappa \partial_\tau \partial_\sigma \mathbf{e}_\kappa. \tag{12.112}$$

Antisymmetrizing in $\sigma\tau$ gives

$$\partial_\tau \partial_\sigma \partial_\nu \mathbf{e}_\mu - \partial_\sigma \partial_\tau \partial_\nu \mathbf{e}_\mu = \Gamma_{\nu\mu}{}^\alpha \mathbf{R}_{\tau\sigma\alpha} + \left[(\partial_\tau \partial_\sigma - \partial_\sigma \partial_\tau) \Gamma_{\nu\mu}{}^\alpha \right] \mathbf{e}_\alpha. \tag{12.113}$$

At this point we make use of the integrability condition for the connection (12.107) to drop the last term, resulting in

$$\partial_\tau \mathbf{R}_{\sigma\nu\mu} - \Gamma_{\nu\mu}{}^\alpha \mathbf{R}_{\tau\sigma\alpha} = 0 \tag{12.114}$$

Inserting this into (12.110) and multiplying by $\mathbf{e}^n$ we obtain an expression involving the covariant derivative of the curvature tensor

$$D_\tau R_{\sigma\nu\mu}{}^\kappa - 2S_{\tau\sigma}{}^\lambda R_{\nu\lambda\mu}{}^\kappa = 0. \tag{12.115}$$

This is the *Bianchi identity* which guarantees the integrability of the connection.

Within the defect interpretation of torsion and curvature, we are now prepared to demonstrate that these two identities have a simple physical interpretation. They are the nonlinear versions of the conservation laws for dislocation and disclination densities. These read[2]

$$\partial_i \alpha_{ij} = -\epsilon_{jkl} \theta_{kl}, \tag{12.116}$$

$$\partial_i \theta_{ij} = 0. \tag{12.117}$$

They state that disclination lines never end while dislocation lines can end at most at a disclination line.

Consider now Eq. (12.115). Its linearization gives

$$\partial_\tau R_{\sigma\nu\mu}{}^\lambda + \partial_\sigma R_{\nu\tau\mu}{}^\lambda + \partial_\nu R_{\tau\mu\sigma}{}^\lambda = 0. \tag{12.118}$$

Contracting ν and μ and τ with λ we obtain

$$\partial_\tau R_{\sigma\nu}{}^{\nu\tau} + \partial_\sigma R_{\nu\lambda}{}^{\nu\lambda} + \partial_\nu R_\tau{}^\nu{}_\sigma{}^\tau = 2\partial_\tau R_\sigma{}^\tau + \partial_\sigma R = 2\partial_\tau G_\sigma{}^\tau = 0. \tag{12.119}$$

In three dimensions, the Einstein tensor $G_{\mu\nu}$ corresponds to the disclination density $\theta_{\mu\nu}$ in Eq. (12.34), and (12.119) coincides indeed with the defect conservation law (12.117).

---

[2]See Eqs. (11.90) and (11.91) in Part III of Ref. [1].

The fundamental identity (12.103) has the linearized form

$$2\left(\partial_\nu S_{\mu\lambda}{}^\kappa + \partial_\mu S_{\lambda\nu}{}^\kappa + \partial_\lambda S_{\nu\mu}{}^\kappa\right) = R_{\nu\mu\lambda}{}^\kappa + R_{\mu\lambda\nu}{}^\kappa + R_{\lambda\nu\mu}{}^\kappa. \tag{12.120}$$

Contracting ν and κ gives

$$2\left(\partial_\nu S_{\mu\nu}{}^\nu + \partial_\mu S_{\lambda\nu}{}^\nu - \partial_\lambda S_{\mu\nu}{}^\nu\right) = R_{\nu\mu\lambda}{}^\nu + R_{\mu\lambda\kappa}{}^\kappa + R_{\lambda\nu\mu}{}^\nu = R_{\mu\lambda} - R_{\lambda\mu}, \tag{12.121}$$

where we have used the antisymmetry of $R_{\nu\mu\lambda\kappa}$ in the last two indices which is a consequence of the integrability condition for the metric tensor. The right-hand side is the same as $G_{\mu\lambda} - G_{\lambda\mu}$.

In three dimensions we can contract this equation with the ϵ-tensor and find

$$\epsilon_{jkl}\left(\partial_i S_{kli} + \partial_k S_{lnm} - \partial_l S_{knn}\right) = \epsilon_{ijkl} G_{kl}. \tag{12.122}$$

Inserting here $S_{klj} = (1/2)\epsilon_{kli}\alpha_{ij}$ from (12.34), and Eq. (12.40) for G_{lk}, this becomes the conservation law (12.116) for the dislocation density.

12.6 Special Coordinates in Riemann Spacetime

Since the theory of gravity is independent of the coordinates by which spacetime is parametrized, there are many possible choices of coordinates, depending on the physical problem to be studied. A fw of these will be sketched in this section.

12.6.1 Geodesic Coordinates

To a local observer, curved spacetime looks flat in his immediate neighborhood. After all, this is why men believed for a long time that the earth has the form of a flat disc. In four-dimensional spacetime the equivalent statement is that, in a freely falling elevator one does not experience any gravitational force as long as the elevator cabin is small enough to make higher nonlinear effects negligible. The cabin constitutes an inertial frame of reference for the motion of a mass point. From Eq. (11.21) we see that its coordinates in an arbitrary geometry can be determined from the requirement of a vanishing Christoffel symbol $\{\mu'\lambda', \kappa'\} = 0$, which amounts to

$$\partial_{\lambda'} g_{\mu'\lambda'}(x') = 0, \tag{12.123}$$

$$\partial_{\lambda'} g^{\mu'\lambda'}(x') = -g^{\mu'\sigma'} g^{\lambda'\tau'} \partial_{\lambda'} g_{\sigma'\tau'}(x') = 0. \tag{12.124}$$

Given an arbitrary set of coordinates x, the derivatives are connected by

$$\begin{aligned}
\partial_{\lambda'} g^{\mu'\lambda'}(x') &= \partial_\lambda \left[g^{\mu\nu}(x)\alpha_\mu{}^{\mu'}\alpha_\nu{}^{\nu'}\right]\alpha^\lambda{}_{\lambda'} \\
&= \partial_\lambda g^{\mu\nu}(x)\alpha_\mu{}^{\mu'}\alpha_\nu{}^{\nu'}\alpha^\lambda{}_{\lambda'} \\
&\quad + g^{\mu\nu}\partial_\lambda\alpha_\mu{}^{\mu'}\alpha_\nu{}^{\nu'}\alpha^\lambda{}_{\lambda'} + g^{\mu\nu}\alpha_\mu{}^{\mu'}\partial_\lambda\alpha_\nu{}^{\nu'}\alpha^\lambda{}_{\lambda'}.
\end{aligned} \tag{12.125}$$

Recall that derivative symbols ∂_μ are meant to act only on the first function behind it. Equations (12.123) and (12.124) provide us with $D^2(D+1)/2$ partial differential equations for the D coordinates $x'^{\mu'}(x)$ which do not, in general, have a solution over a finite region. If $\partial_{\lambda'}g^{\mu'\lambda'}$ were to vanish over a finite region, the spacetime would necessarily be flat. So we can, at best, achieve

$$\partial_{\lambda'}g^{\mu'\nu'}(X') = 0 \tag{12.126}$$

at some point $x = X$. This implies, via (12.123), that also $\partial_{\lambda'}g_{\sigma'\tau'}(X') = 0$. and thus the vanishing of the Christoffel symbols at that point. Then a mass point will move force-free at X. In any neighborhood of X there are gravitational forces of order $O(x - X)$.

Let us try and solve (12.126) by an expansion

$$
\begin{aligned}
x'^{\mu'} &= X^\mu + a^\mu{}_\lambda (x-X)^\lambda + \frac{1}{2!}a^\mu_{\lambda\kappa}(x-X)^\lambda(x-X)^\kappa \\
&+ \frac{1}{3!}a^\mu_{\lambda\kappa\delta}(x-X)^\lambda(x-X)^\kappa(x-X)^\delta + \dots .
\end{aligned}
\tag{12.127}
$$

The associated transformation matrix $\alpha_\mu{}^{\mu'} \equiv \partial x'^{\mu'}/\partial x^\mu$ satisfies

$$
\begin{aligned}
\alpha_\mu{}^{\mu'} &= a_\mu{}^{\mu'} + a_{\mu\lambda}{}^{\mu'}(x-X)^\lambda + \frac{1}{2!}a_{\mu\lambda\kappa}{}^{\mu'}(x-X)^\lambda(x-X)^\kappa + \dots , \\
\partial_\lambda\alpha_\mu{}^{\mu'} &= a_{\mu\lambda}{}^{\mu'} + a_{\mu\lambda\kappa}{}^{\mu'}(x-X)^\kappa + \dots , \\
\partial_\kappa\partial_\lambda\alpha_\mu{}^{\mu'} &= a_{\mu\lambda\kappa}{}^{\mu'} + \dots .
\end{aligned}
\tag{12.128}
$$

Inserting this into (12.125) we find

$$
\begin{aligned}
\partial_\lambda g^{\mu'\nu'} &= \partial_\lambda g^{\mu\nu}(X)a_\mu{}^{\mu'}a_\nu{}^{\nu'} + g^{\sigma\tau}(X)\left[a_{\sigma\lambda}{}^{\mu'}a_\tau{}^{\nu'} + \left(\mu' \leftrightarrow \nu'\right)\right] \\
&= 0 + \mathcal{O}(x-X).
\end{aligned}
\tag{12.129}
$$

This is solved by

$$a_\mu{}^{\mu'} = g_\mu{}^{\mu'}(X), \qquad a_{\lambda\kappa}{}^\mu = \frac{1}{2}\bar{\Gamma}_{\lambda\kappa}{}^\mu(X), \tag{12.130}$$

in accordance with (11.97). Hence the coordinates which are locally geodesic at X are given by

$$x'^\mu = X^\mu + (x-X)^\mu + \frac{1}{2}\bar{\Gamma}_{\lambda\kappa}{}^\mu(x-X)^\lambda(x-X)^\kappa + \mathcal{O}\left((x-X)^3\right). \tag{12.131}$$

Note that while the Christoffel symbols vanish in the geodesic frame at X, their derivatives are nonzero if the curvature is nonzero at X.

In order to complete the construction of a freely falling coordinate system we just note that, in the neighborhood of the point X, the geodesic coordinates can always be brought to a Minkowski-form by a further linear transformation

$$\left(x' - X\right)^\mu \rightarrow L^\mu{}_\alpha(x-X)^\alpha \tag{12.132}$$

which transforms $g_{\mu\nu}$ into $g\alpha\beta$, i.e.,

$$g_{\alpha\beta} = L^\mu{}_\alpha L^\nu{}_\beta g_{\mu\nu} = g_{\alpha\beta}. \qquad (12.133)$$

Such a linear transformation does not change the geodesic property of the coordinates such that the coordinates $(x'' - X)^\alpha$ are a local inertial frame, which is what we wanted to find.

As far as the crystalline defects are concerned, the possibility of constructing geodesic coordinates is related to the fact that, in the regions between defects, the crystal can always be distorted elastically to form a regular array of atoms. In the continuum limit, these regions shrink to zero, but so do the Burgers' vectors of the defects. Therefore, even though any small neighborhood does contain some defects, these themselves are infinitesimally small, so that the perfection of the crystal is disturbed only infinitesimally.

12.6.2 Canonical Geodesic Coordinates

The condition of being geodesic determined the coordinates transformation (12.127) up to the coefficients of the quadratic terms.

$$\begin{aligned}
x'^\mu &= X^\mu + (x - X)^\mu + \frac{1}{2}\bar{\Gamma}_{\lambda\kappa}{}^\mu (x - X)^\lambda (x - X)^\kappa \\
&\quad + \frac{1}{3!}a^\mu{}_{\lambda\kappa\delta}(x - X)^\lambda (x - X)^\kappa (x - X)^\delta + \dots \qquad (12.134)
\end{aligned}$$

By construction, the transformation matrix

$$\alpha^\mu{}_\nu = \frac{\partial x'^\mu}{\partial x^\nu} = \delta^\mu{}_\nu + \bar{\Gamma}_{\lambda\nu}{}^\mu (x - X)^\lambda + \frac{1}{2}a^\mu{}_{\lambda\kappa\nu}(x - X)^\lambda (x - X)^\kappa + \dots \quad (12.135)$$

has the property of making the Christoffel symbol of the point X vanish. It is obvious that the higher coefficients $a^\mu_{\lambda\kappa\delta}$ must have an influence upon the derivatives of the Christoffel symbols. In general, these cannot be made zero since the curvature tensor at the point X, where $\bar{\Gamma}_{\mu'\nu'}{}^{\lambda'}$ vanishes, is

$$R_{\mu'\nu'\lambda'}{}^{\kappa'} = \partial_{\mu'}\bar{\Gamma}_{\nu'\lambda'}{}^{\kappa'} - (\mu' \leftrightarrow \nu'). \qquad (12.136)$$

This implies that one can find $a^\mu_{\lambda\kappa\nu}$ to make also $\partial_{\mu'}\bar{\Gamma}_{\nu'\lambda'}{}^{\kappa'} = 0$ only if spacetime is flat.

Even though the derivatives cannot be brought to zero, there is a most convenient coordinate system referred to as *canonical*, in which the derivatives satisfy the following relation

$$\partial_{\mu'}\bar{\Gamma}_{\nu'\lambda'}{}^{\kappa'} + \partial_{\lambda'}\bar{\Gamma}_{\nu'\mu'}{}^{\kappa'} = 0. \qquad (12.137)$$

Before we show how to find such a system, let us first see what its advantages are. The canonical condition allows us to invert the relation (12.136) for $R^{\kappa'}_{\mu'\nu'\lambda'}$ and express (always at a geodesic coordinate point) the derivatives of the Christoffel symbols uniquely in terms of the curvature tensor

$$\partial_{\nu'}\bar{\Gamma}_{\mu'\lambda'}{}^{\kappa'} = -\frac{1}{3}\left(R_{\mu'\nu'\lambda'}{}^{\kappa'} + R_{\lambda'\mu'\nu'}{}^{\kappa'}\right). \tag{12.138}$$

Thus we can expand the Riemann connection $\Gamma_{\mu\nu}{}^{\lambda}(x)$ in the neighborhood of the point X^{μ} uniquely up to first order $x - X$:

$$\bar{\Gamma}_{\mu\lambda}{}^{\kappa} = -\frac{1}{3}\left(R_{\mu\nu\lambda}{}^{\kappa} + R_{\lambda\mu\nu}{}^{\kappa}\right)(x - X)^{\lambda} + \mathcal{O}((x-X)^2). \tag{12.139}$$

For the metric $g_{\mu\nu}(x)$ we find a related expansion up to second order $x - X$. In order to see this we recall Eqs. (11.96). Differentiating this once more we find

$$\partial_{\kappa}\partial_{\lambda}g_{\mu\nu} = \partial_{\kappa}\bar{\Gamma}_{\lambda\mu}{}^{\sigma}g_{\sigma\nu} + \partial_{\kappa}\bar{\Gamma}_{\lambda\nu}{}^{\sigma} + \bar{\Gamma}_{\lambda\mu}{}^{\sigma}\partial_{\kappa}g_{\sigma\nu} + \bar{\Gamma}_{\lambda\nu}{}^{\sigma}\partial_{\kappa}g_{\mu\sigma}. \tag{12.140}$$

At a point where the coordinates are geodesic, this becomes simply

$$\begin{aligned}
\partial_{\kappa'}\partial_{\lambda'}g_{\mu'\nu'} &= -\frac{1}{3}\left(R_{\lambda'\kappa'\mu'}{}^{\sigma'} + R_{\mu'\kappa'\lambda'}{}^{\sigma'}\right)g_{\sigma'\nu'} - \frac{1}{3}\left(R_{\lambda'\kappa'\nu'}{}^{\sigma'} + R_{\nu'\kappa'\lambda'}{}^{\sigma'}\right)g_{\mu'\sigma'} \\
&= \frac{1}{3}\left(R_{\kappa'\mu'\lambda'\nu'} + R_{\kappa'\nu'\lambda'\mu'}\right). \tag{12.141}
\end{aligned}$$

Hence the metric has the expansion

$$\begin{aligned}
g_{\mu'\nu'}(x') &= g_{\mu'\nu'}(X) + \frac{1}{2}\partial_{\kappa'}\partial_{\lambda'}g_{\mu'\nu'}(X)\left(x'-X\right)^{\kappa'}\left(x'-X\right)^{\lambda'} + \dots \\
&= g_{\mu'\nu'}(X) + \frac{1}{3}R_{z'\mu'\lambda'\nu'}\left(x'-X\right)^{\kappa'}\left(x'-X\right)^{\lambda'} + \dots \\
&= g_{\mu'\nu'}(X) + \frac{1}{3}R_{z'\mu'\lambda'\nu'}\left(x'-X\right)^{\kappa'}\left(x'-X\right)^{\lambda'} + \dots. \tag{12.142}
\end{aligned}$$

If we insert the Riemann connection (12.139) into the geodesic equation of motion (11.22), we find

$$g_{\lambda\sigma}\delta\ddot{x}^{\nu}(\sigma) - \frac{1}{3}\left(R_{\mu\nu\lambda}{}^{\kappa} + R_{\lambda\mu\nu}{}^{\kappa}\right)\delta x^{\lambda}(\tau)\dot{\delta x}^{\mu}(\tau)\dot{\delta x}^{\nu}(\tau) = 0, \quad \delta x \equiv x' - X. \tag{12.143}$$

This equation shows that while there is no acceleration at X, since the point with coordinate X falls freely, the neighborhood of X experiences the so-called *tidal forces*. These distort all extended bodies in free fall, such as the planets in orbit around the sun.

Let us now turn to the construction of these canonical coordinates. For this we take the transformation law for the Christoffel symbols (11.101)

$$\bar{\Gamma}_{\mu'\nu'}{}^{\lambda'} = \alpha_{\mu'}{}^{\mu}\alpha_{\nu'}{}^{\nu}\alpha^{\lambda'}{}_{\lambda}\bar{\Gamma}_{\mu\nu}{}^{\lambda} - \alpha_{\mu'}{}^{\mu}\alpha_{\nu'}{}^{\nu}\partial_{\mu}\alpha^{\lambda'}{}_{\nu}, \tag{12.144}$$

and differentiate them once more with

$$\partial_{\kappa'} = \frac{\partial x^{\kappa}}{\partial x^{\kappa'}}\partial_{\kappa} = \alpha_{\kappa'}{}^{\kappa}\partial_{\kappa}. \tag{12.145}$$

This gives

$$
\begin{aligned}
\partial_{\kappa'}\bar{\Gamma}_{\mu'\nu'}{}^{\lambda'} &= \alpha_{\kappa'}{}^{\kappa}\alpha_{\mu'}{}^{\mu}\alpha_{\nu'}{}^{\nu}\alpha^{\lambda'}{}_{\lambda}\partial_{\kappa}\bar{\Gamma}_{\mu\nu}{}^{\lambda} + \alpha_{\kappa'}{}^{\kappa}\partial_{\kappa}\alpha_{\mu'}{}^{\mu}\alpha_{\nu'}{}^{\nu}\alpha^{\lambda'}{}_{\lambda}\bar{\Gamma}_{\mu\nu}{}^{\lambda} \\
&+ \alpha_{\kappa'}{}^{\kappa}\alpha_{\mu'}{}^{\mu}\partial_{\kappa}\alpha_{\nu'}{}^{\nu}\alpha^{\lambda'}{}_{\lambda}\bar{\Gamma}_{\mu\nu}{}^{\lambda} + \alpha_{\kappa'}{}^{\kappa}\alpha_{\mu'}{}^{\mu}\alpha_{\nu'}{}^{\nu}\partial_{\kappa}\alpha^{\lambda'}{}_{\lambda}\bar{\Gamma}_{\mu\nu}{}^{\lambda} \\
&- \alpha_{\kappa'}{}^{\kappa}\partial_{\kappa}\alpha_{\mu'}{}^{\mu}\alpha_{\nu'}{}^{\nu}\partial_{\mu}\alpha^{\lambda'}{}_{\nu} - \alpha_{\kappa'}{}^{\kappa}\alpha_{\mu'}{}^{\mu}\partial_{\kappa}\alpha_{\nu'}{}^{\nu}\partial_{\mu}\alpha^{\lambda'}{}_{\nu} + \alpha_{\kappa'}{}^{\kappa}\alpha_{\mu'}{}^{\mu}\alpha_{\nu'}{}^{\nu}\partial_{\kappa}\partial_{\mu}\alpha^{\lambda'}{}_{\nu}.
\end{aligned} \tag{12.146}
$$

Besides the known transformation coefficient $\alpha^{\mu}{}_{\nu}$, this formula also involves the inverse coefficients $\alpha_{\kappa}{}^{\lambda}$. Since are close to unity for $x \sim X$, $\alpha^{\mu}{}_{\nu}$, the inverse is simply [recall (11.66)]

$$
\alpha_{\mu}{}^{\nu} = \frac{\partial x^{\nu}}{\partial x'^{\mu}} = \delta_{\mu}{}^{\nu} - \bar{\Gamma}_{\lambda\nu}{}^{\mu}(x-X)^{\lambda} + \dots , \tag{12.147}
$$

which shows that, indeed,

$$
\alpha_{\mu}{}^{\nu}\alpha^{\mu'}{}_{\nu} = \delta_{\mu}{}^{\mu'} + \bar{\Gamma}_{\lambda\nu}{}^{\mu}(x-X)^{\lambda} - \bar{\Gamma}_{\lambda\nu}{}^{\mu}(x-X)^{\lambda} + \dots = \delta_{\mu}{}^{\mu'}. \tag{12.148}
$$

Inserting $\alpha_{\mu}{}^{\nu}$ and $\alpha^{\nu}{}_{\nu}$ into the above transformation law gives

$$
\partial_{\kappa'}\bar{\Gamma}_{\mu'\nu'}{}^{\lambda'} = \left[\partial_{\kappa}\bar{\Gamma}_{\mu\nu}{}^{\lambda} + \bar{\Gamma}_{\mu\nu}{}^{\sigma}\bar{\Gamma}_{\sigma\kappa}{}^{\lambda} - a^{\lambda}{}_{\kappa\mu\nu}\right]_{\kappa'=\kappa,\lambda'=\lambda,\mu'=\mu,\nu'=\nu}. \tag{12.149}
$$

Note the appearance of the coefficients $a^{\lambda}{}_{\kappa\mu\nu}$ of the cubic expansion terms. Interchanging on the left-hand side of (12.147) the indices $\kappa'\mu'\nu'$ cyclically, and adding the three expansions, we find

$$
\partial_{\kappa'}\bar{\Gamma}_{\mu'\nu'}{}^{\lambda'} + 2 \text{ cyclic perm. of } \left(\kappa'\mu'\nu'\right) \tag{12.150}
$$
$$
+ \left[\partial_{\kappa}\bar{\Gamma}_{\mu\nu}{}^{\lambda} + \bar{\Gamma}_{\mu\nu}{}^{\sigma}\bar{\Gamma}_{\sigma\kappa}{}^{\lambda} + 2 \text{ cyclic perm. of } (\kappa\mu\nu) - 3a^{\lambda}{}_{\kappa\mu\nu}\right]\Big|_{\kappa'=\kappa,\lambda'=\lambda,\mu'=\mu,\nu'=\nu}.
$$

By setting the left-hand side equal to zero we obtain the desired equation for $a^{b}{}_{\kappa\mu\nu}$.

Thus given an arbitrary coordinate frame x^{μ}, the coefficients $a^{b}{}_{\kappa\mu\nu}$ can indeed be chosen to make the geodesic coordinate frame x'^{μ} canonical in the neighborhood of the point X, thereby determining $g_{\mu\nu}(x')$ in this neighborhood up to the second order in $x - X$ uniquely in terms of the curvature tensor, as stated in Eq. (12.142).

12.6.3 Harmonic Coordinates

While geodesic properties of coordinates can be enforced at most at one point there exists a way of fixing the choice of coordinates in the entire spacetime by choosing what are called *harmonic coordinates*. These were introduced first by T. DeDonder and C. Lanczos and extensively used by V. Fock in his work on gravitation [2]. Given an arbitrary set of coordinates x^{μ}, one asks for d independent scalar functions $f^{a}(x)$ $(a = 0, 1, 2, d)$ which satisfy the Laplace equation in curved spacetime

$$
D^{2}f^{a}(x) = g^{\mu\nu}D_{\mu}D_{\nu}f^{a}(x) = 0. \tag{12.151}
$$

Since $f^a(x)$ are supposed to be scalar functions, we calculate

$$D_\mu D_\nu f = D_\mu \partial_\nu f = \left(\partial_\mu \partial_\nu - \Gamma_{\mu\nu}{}^\lambda \partial_\lambda\right) f, \tag{12.152}$$

so that the Laplace equation reads

$$D^2 f = g^{\mu\nu} D_\mu D_\nu f = \left(g^{\mu\nu}\partial_\mu\partial_\nu - \Gamma^\lambda\partial_\lambda\right) f, \tag{12.153}$$

where we have introduced the contracted affine connection

$$\Gamma^\lambda \equiv \Gamma_\mu{}^{\mu\lambda}. \tag{12.154}$$

In a symmetric spacetime

$$\begin{aligned}
\Gamma^\lambda &= \frac{1}{2} g^{\mu\nu} g^{\lambda\kappa} \left(\partial_\mu g_{\nu\kappa} + \partial_\nu g_{\mu\kappa} - \partial_\kappa g_{\mu\nu}\right)\\
&= g^{\mu\nu} g^{\lambda\kappa} \partial_\mu g_{\nu\kappa} - \frac{1}{2} g^{\lambda\kappa} g^{\mu\nu} \partial_\kappa g_{\mu\nu}\\
&= -\frac{1}{\sqrt{-g}} \partial_\kappa \sqrt{-g} g^{\lambda\kappa},
\end{aligned} \tag{12.155}$$

and

$$D^2 f = \left[g^{\mu\nu}\partial_\mu\partial_\nu + \frac{1}{\sqrt{-g}}\partial_\kappa\sqrt{-g}g^{\lambda\kappa}\right] f = \Delta f, \tag{12.156}$$

where

$$\Delta \equiv \frac{1}{\sqrt{-g}} \partial_\mu g^{\mu\nu} \sqrt{-g} \partial_\nu \tag{12.157}$$

is the *Laplace-Beltrami operator* Δ in curved spacetime. The Laplace operator in a spacetime with torsion is related to the Laplace-Beltrami operator by

$$D^2 f = \Delta f - S_\mu{}^{\mu\nu}\partial_\lambda f. \tag{12.158}$$

Suppose we have found d functions $f^a(x)$ which satisfy (12.151), then we introduce the harmonic coordinates X^a as

$$X^a = f^a(x). \tag{12.159}$$

When transforming the Laplace equation (12.151) from coordinates x^μ to the harmonic coordinates X^a, we obtain

$$\left(g^{bc}\partial_b\partial_c - \Gamma^c\partial_c\right) X^a = -\Gamma^c \delta_c{}^a = 0. \tag{12.160}$$

Thus, harmonic coordinates are characterized by vanishing Γ^a $(a = 1, \ldots, d)$.

12.6.4   Coordinates with $\det(g_{\mu\nu}) = 1$

A further choice of coordinates favored by Einstein is one in which the determinant of the metric g is constant and has the Minkowski value -1 in all spacetime. Since

$$
\begin{aligned}
\bar{\Gamma}_{\mu\nu}{}^{\mu} &= \frac{1}{2}g^{\mu\nu}\left(\partial_{\mu}g_{\nu\lambda} + \partial_{\nu}g_{\mu\lambda} - \partial_{\lambda}g_{\mu\nu}\right) \\
&= \frac{1}{2}g^{\mu\nu}\partial_{\nu}g_{\mu\nu} = \frac{1}{\sqrt{-g}}\partial_{\nu}\sqrt{-g} = \partial_{\nu}\log\sqrt{-g}
\end{aligned}
\tag{12.161}
$$

this condition can be stated in the form

$$
\bar{\Gamma}_{\mu\nu}{}^{\mu} = 0.
\tag{12.162}
$$

Given an arbitrary coordinate system x^{μ}, the favored coordinates $\bar{x}^{\mu}$ are found by a transformation

$$
\bar{x}^{\mu} = \alpha^{\mu}{}_{\nu}x^{\nu}
\tag{12.163}
$$

which fulfills the condition

$$
\sqrt{-\bar{g}} = |\det\left(\alpha^{\mu}{}_{\nu}\right)|\sqrt{-g} = |\det\left(\alpha^{\mu}{}_{\nu}\right)|.
\tag{12.164}
$$

Taking the logarithm and differentiating it gives the d conditions

$$
\bar{\Gamma}_{\mu\nu}{}^{\mu} = \partial_{\nu}\log\det\left(\alpha^{\lambda}{}_{\kappa}\right) = \partial_{\nu}\mathrm{tr}\log\left(\alpha^{\lambda}{}_{\kappa}\right) = \mathrm{tr}\left(\alpha^{-1}\partial_{\nu}\alpha\right) = \alpha_{\lambda}{}^{\kappa}\partial_{\nu}\alpha^{\lambda}{}_{\kappa} = 0.
\tag{12.165}
$$

These are d differential equations determining d new coordinate functions $\bar{x}(x)$.

Note the difference with respect to harmonic coordinates which have $\Gamma^{\mu} = \Gamma_{\lambda}{}^{\lambda\mu} = 0$, i.e., the first two indices of the affine connection are contracted [recall (12.154) and (12.160)]. The present condition, on the other hand, has the first (or the second index) of the Christoffel symbol contracted with the third.

12.6.5   Orthogonal Coordinates

For many calculations, it is useful to employ orthogonal coordinates in which $g_{\mu\nu}$ has only diagonal elements. This makes many entries of the Christoffel symbols equal to zero:

$$
\bar{\Gamma}_{\mu\lambda,\kappa} = 0, \quad \bar{\Gamma}_{\mu\lambda}{}^{\kappa} = 0, \qquad \mu \neq \lambda, \kappa \neq \mu, \kappa \neq \lambda.
\tag{12.166}
$$

The zeros simplifies the calculation of the other components. In a symmetric spacetime, we may use formula (12.1) for the Riemann tensor $\bar{R}_{\mu\nu\lambda}{}^{\kappa}$ and find that it vanishes whenever all its indices are different. The nonvanishing elements can be calculated as follows:

$\nu \neq \kappa, \lambda \neq \mu, \nu \neq \lambda$

$$\bar{R}_{\nu\kappa\lambda\mu} = -\frac{1}{2}\left(\partial_\lambda\partial_\kappa g_{\mu\nu} + \partial_\mu\partial_\nu g_{\kappa\lambda} - \partial_\lambda\partial_\nu g_{\mu\kappa}\right)$$

$$+\frac{1}{2}\partial_\mu\left(\log\sqrt{g_{\nu\nu}}\right)\partial_\nu g_{\lambda\kappa} - \frac{1}{2}\partial_\nu\left(\log\sqrt{g_{\mu\mu}}\right)\left(\partial_\lambda g_{\mu\kappa} - \partial_\mu g_{\lambda\kappa}\right) \qquad (12.167)$$

$$-\frac{1}{2}\partial_\lambda\left(\log\sqrt{g_{\nu\nu}}\right)\left(\partial_\nu g_{\mu\kappa} - \partial_\kappa g_{\mu\nu}\right) - \frac{1}{2}\partial_\nu\left(\log\sqrt{g_{\lambda\lambda}}\right)\left(\partial_\lambda g_{\kappa\mu} - \partial_\mu g_{\kappa\lambda}\right) + \frac{1}{2}\bar{\Gamma}_{\kappa\lambda}{}^\rho\partial_\rho g_{\nu\mu},$$

$\nu \neq \kappa, \lambda \neq \mu, \nu \neq \lambda, \nu \neq \mu$

$$\bar{R}_{\nu\kappa\lambda\mu} = -\frac{1}{2}\left(\partial_\mu\partial_\nu g_{\kappa\lambda} - \partial_\lambda\partial_\nu g_{\mu\kappa}\right)$$

$$+\frac{1}{2}\partial_\mu\left(\log\sqrt{g_{\nu\nu}}\right)\partial_\nu g_{\lambda\kappa} - \frac{1}{2}\partial_\lambda\left(\log\sqrt{g_{\nu\nu}}\right)\partial_\alpha g_{\mu\kappa} \qquad (12.168)$$

$$-\frac{1}{2}\partial_\nu\left(\log\sqrt{g_{\mu\mu}}\right)\left(\partial_\lambda g_{\mu\kappa} - \partial_\mu g_{\lambda\kappa}\right) - \frac{1}{2}\partial_\nu\log\sqrt{g_{\lambda\lambda}}\left(\partial_\lambda g_{\mu\kappa} - \partial_\mu g_{\kappa\lambda}\right),$$

$\nu \neq \kappa, \lambda \neq \mu, \nu \neq \lambda, \nu \neq \mu, \kappa \neq \lambda$

$$\bar{R}_{\nu\kappa\lambda\mu} = \frac{1}{2}\partial_\lambda\partial_\nu g_{\mu\kappa} - \frac{1}{2}\partial_\nu\left(\log\sqrt{g_{\mu\mu}g_{\lambda\lambda}}\right)\partial_\lambda g_{\mu\kappa} - \frac{1}{2}\partial_\lambda\left(\log\sqrt{g_{\nu\nu}}\right)\partial_\nu g_{\mu\kappa}. \quad (12.169)$$

The Ricci tensor becomes

$$\bar{R}_{\mu\nu} = \sum_\lambda \frac{1}{g_{\lambda\lambda}}\bar{R}^{\mu\lambda\lambda\nu} \qquad (12.170)$$

giving the off-diagonal elements

$\mu \neq \nu$

$$\bar{R}_{\mu\nu} = \sum_{\lambda\neq\mu,\lambda\neq\nu}\left[\partial_\mu\partial_\nu\left(\log\sqrt{g_{\lambda\lambda}}\right) - \partial_\mu\left(\log\sqrt{g_{\nu\nu}}\right)\partial_\nu\left(\log\sqrt{g_{\lambda\lambda}}\right)\right.$$

$$\left. - \partial_\mu\left(\log\sqrt{g_{\lambda\lambda}}\right)\partial_\nu\left(\log\sqrt{g_{\mu\mu}}\right) + \partial_\mu\left(\log\sqrt{g_{\lambda\lambda}}\right)\partial_\nu\left(\log\sqrt{g_{\lambda\lambda}}\right)\right]$$

$$= -\partial_\mu\partial_\nu\log\sqrt{-g} + \partial_\mu\left(\log\sqrt{g_{\nu\nu}}\right)\partial_\nu\left(\log\sqrt{-g}\right) + \partial_\mu\log\sqrt{-g}\partial_\nu\log\sqrt{g_{\mu\mu}} \quad (12.171)$$

$$+ \partial_m\partial_\nu\log\sqrt{g_{\mu\mu}g_{\nu\nu}} - 2\partial_\mu\left(\log\sqrt{g_{\nu\nu}}\right)\partial_\nu\log\sqrt{g_{\mu\mu}} - \sum_{\lambda=1}^d\partial_\mu\left(\log\sqrt{g_{\lambda\lambda}}\right)\partial_\nu\left(\log\sqrt{g_{\lambda\lambda}}\right),$$

and the diagonal elements

$$\bar{R}^{\mu\mu} = -\partial^2\log\sqrt{-g} + 2\partial_\mu^2\left(\log\sqrt{g_{\mu\mu}}\right) - 2\left(\partial_\mu\log\sqrt{g_{\mu\mu}}\right)^2$$

$$+2\partial_\mu\left(\log\sqrt{-g}\right)\partial_\mu\left(\log\sqrt{g_{\mu\mu}}\right) - \sum_{\lambda=1}^d\left(\partial_\mu\log\sqrt{g_{\lambda\lambda}}\right)^2$$

$$-g_{\mu\mu}\sum_{\lambda=1}^d\frac{1}{g_{\lambda\lambda}}\left[\partial_\lambda^2\left(\log\sqrt{g_{\mu\mu}}\right) + \partial_\lambda\left(\log\sqrt{-g}\right)\partial_\lambda\left(\log\sqrt{g_{\mu\mu}}\right)\right. \qquad (12.172)$$

$$\left.-2\partial_\lambda\left(\log\sqrt{g_{\lambda\lambda}}\right)\partial_\lambda\left(\log\sqrt{g_{\lambda\lambda}}\right)\right].$$

The curvature scalar reads

$$\bar{R} = \sum_\lambda^d \frac{1}{g_{\lambda\lambda}} \left\{ 2\partial_\lambda{}^2 \left(\log \sqrt{-g} \right) - 2\partial_\lambda{}^2 \log \sqrt{g_{\lambda\lambda}} \right.$$
$$+ 2 \left(\partial_\lambda \log \sqrt{g_{\lambda\lambda}} \right)^2 - 4\partial_\lambda \left(\log \sqrt{g_{\lambda\lambda}} \right) \partial_\lambda \left(\log \sqrt{-g} \right)$$
$$\left. + \left(\partial_\lambda \log \sqrt{-g} \right)^2 + \sum_\kappa^d \left(\partial_\lambda \sqrt{g_{\kappa\kappa}} \right)^2 \right\}. \tag{12.173}$$

12.7 Number of Independent Components of $R_{\mu\nu\lambda}{}^\kappa$ and $S_{\mu\nu}{}^\lambda$

With the antisymmetry in $\mu\nu$ and $\lambda\kappa$, there exist at first $N_d^{\bar{R}} = [d(d-1)/2]^2$ components of $\bar{R}_{\mu\nu\lambda\kappa}$ and $N_d^S = d^2(d-1)/2$ components of $S_{\mu\nu}{}^\lambda$ in d dimensions. Thus $R_{\mu\nu\lambda\kappa}$ may be viewed as a $\frac{1}{2}d(d-1) \times \frac{1}{2}d(d-1)$ matrix $R_{(\mu\nu)(\lambda\kappa)}$ in the index pairs. In symmetric spacetimes, there is in addition symmetry of $\bar{R}_{\mu\nu\lambda\kappa}$ between the index pairs $\mu\nu$ and $\lambda\kappa$, so that it has

$$\frac{1}{2} \left\{ \frac{1}{2}d(d-1) \times \left[\frac{1}{2}d(d-1) - 1 \right] \right\} = \frac{1}{8}d(d-1)(d^2 - d + 2) \tag{12.174}$$

components. Now, the fundamental identity (12.105) not only leads to the symmetry, it contains also the information that the completely antisymmetric part of $\bar{R}_{\mu\nu\lambda\kappa} + \bar{R}_{\nu\lambda\mu\kappa} + \bar{R}_{\lambda\nu\mu\kappa}$ vanishes. This gives $d(d-1)(d-2)(d-3)/4!$ further relations, and one is left with

$$N_d^{\bar{R}} = \frac{1}{8}d(d-1)(d^2 - d + 2) - \frac{1}{24}d(d-1)(d-2)(d-3) = \frac{1}{12}d^2(d^2 - 1) \tag{12.175}$$

independent components of $\bar{R}_{\mu\nu\lambda\kappa}$. In four dimensions, this number is 20, in three dimensions it is 6.

12.7.1 Two Dimensions

In two dimensions, there is only one independent component, for instance $\bar{R}_{1221}$. Indeed, the most general tensor with the above symmetry properties is

$$\bar{R}_{\mu\nu\lambda\kappa} = -\text{const} \times e_{\mu\nu}e_{\lambda\kappa}, \tag{12.176}$$

where $e_{\mu\nu} = \sqrt{-g}\epsilon_{\mu\nu}$ is the covariant version of the Levi-Civita symbol ($\epsilon_{01} = -\epsilon_{10} = 1$ (compare (11A.7)). Contracting this with $g^{\nu\lambda}$ gives the Ricci tensor

$$\bar{R}_{\mu\kappa} = -\text{const} \times g^{\nu\lambda}\epsilon_{\mu\nu}\epsilon_{\lambda\kappa} = -\text{const} \times \left(g_{\mu\kappa} - g_\lambda{}^\lambda g_{\mu\kappa} \right)$$
$$= \text{const} \times g_{\mu\kappa}, \tag{12.177}$$

and the scalar curvature

$$\bar{R} = \text{const} \times g^{\mu\kappa}g_{\mu\kappa} = 2 \times \text{const}. \tag{12.178}$$

Hence const $= \bar{R}/2$, and the single independent element of $\bar{R}_{\mu\nu\lambda\kappa}$ is

$$\bar{R}_{0110} = \bar{R}_{1001} = g\, \frac{\bar{R}}{2}. \tag{12.179}$$

The full curvature tensor is given by

$$\bar{R}_{\mu\nu\lambda\kappa} = -e_{\mu\nu} e_{\lambda\kappa} \frac{\bar{R}}{2}. \tag{12.180}$$

In three dimensions, the number $N_d^{\bar{R}} = 6$ of independent components of $\bar{R}_{\mu\nu\lambda\kappa}$ agrees with the number of independent components of the Ricci tensor $\bar{R}_{\mu\nu}$, whose knowledge is therefore be sufficient to calculate $\bar{R}_{\mu\nu\lambda\kappa}$. Indeed, we can easily see that

$$\bar{R}_{\mu\nu\lambda\kappa} = e_{\mu\nu\delta} e_{\lambda\kappa\tau} \left(\bar{R}^{\tau\delta} - \frac{1}{2} g^{\tau\delta} \bar{R} \right) \tag{12.181}$$

where $e_{\mu\nu\delta} = \sqrt{-g}\,\epsilon_{\mu\nu\delta}$ is the three-dimensional covariant version of the Levi-Civita symbol in Eq. (11A.7). The proof of Eq. (12.181) follows by contraction with $e^{\mu\nu\delta} e^{\lambda\kappa\tau}$, which gives via the identity (1A.17) for products of two Levi-Civita tensors:

$$\bar{R}^{\tau\delta} - \frac{1}{2} g^{\tau\delta} \bar{R} = \frac{1}{4} e^{\mu\nu\delta} e^{\lambda\kappa\tau} \bar{R}_{\mu\nu\lambda\kappa}. \tag{12.182}$$

This equation is trivially valid due to the identity (1A.18) and the definition (11.141) of the Ricci tensor [see also (12.27)].

Since $\bar{R}_{\mu\nu\lambda\kappa}$ is a tensor, its $N_d^{\bar{R}} = d^2(d^2 - 1)/12$ components are different in different coordinates frame. It is useful to find out how many independent invariants one can form which do not depend on the frame. In two dimensions, the scalar curvature $\bar{R}$ determined $\bar{R}_{\mu\nu\lambda\kappa}$ completely. In general, the invariants of $\bar{R}_{\mu\nu\lambda\kappa}$ can all be constructed by suitable contractions with $g^{\mu\nu}$. The tensors $\bar{R}_{\mu\nu\lambda\kappa}$ and $g^{\mu\nu}$ together have $d^2(d^2 - 1)/12 + d(d+1)/2$ matrix elements. There are d^2 arbitrary coordinate transformations matrices $\partial x'^{\mu}/\partial x^{\lambda}$ which can be applied to these tensors. The number of invariants is equal to the number of independent components of both $\bar{R}_{\mu\nu\lambda\kappa}$ and $g_{\mu\nu}$. This number is

$$N_d^{\text{inv}} = \frac{1}{12} d^2(d^2 - 1) + \frac{d(d+1)}{2} - N^2 = \frac{d}{12}(d-1)(d-2)(d+3). \tag{12.183}$$

This formula is valid only for $d \neq 2$. In two dimensions we have seen before that there is only one invariant, the scalar curvature, i.e., invariant, the scalar curvature. The above general counting procedure breaks down since one of the N^2 coordinate transformations subtracted in (12.183) happens to leave R_{1234} and $g_{\mu\nu}$ invariant. For $d = 3, 4$ the numbers are $N_{3,4}^{\text{inv}} = 3,\ 14$, respectively.

12.7.2 Three Dimensions

In three dimensions, the invariants are the eigenvalue of the characteristic equation

$$\det\left(g^{\mu\lambda}\bar{R}_{\lambda\kappa} - \lambda\delta^{\mu}{}_{\kappa}\right) = \det\left(g^{-1}\bar{R} - \lambda\right) = -\lambda^3 + \lambda^2 I_1 - \lambda I_2 + I_3, \quad (12.184)$$

where

$$
\begin{aligned}
I_1 &= \operatorname{tr}\left(g^{-1}\bar{R}\right) = g^{\mu\nu}\bar{R}_{\lambda\mu} = \bar{R}, \\
I_2 &= \frac{1}{2}\left(\bar{R}^{\mu}{}_{\nu}\bar{R}^{\nu}{}_{\lambda} - \bar{R}_{\lambda}{}^{\lambda}\bar{R}_{\kappa}{}^{\kappa}\right), \qquad\qquad (12.185) \\
I_3 &= \det\left(g^{-1}\bar{R}\right) = \det\left(g^{\mu\lambda}\bar{R}_{\lambda\nu}\right) = \frac{\det\left(\bar{R}_{\mu\nu}\right)}{\det\left(g_{\mu\nu}\right)}.
\end{aligned}
$$

12.7.3 Four or More Dimensions

In four or more dimensions, relation (12.181) generalizes to the *Weyl decomposition* of the curvature tensor

$$
\begin{aligned}
\bar{R}_{\mu\nu\lambda\kappa} &= -\frac{1}{d-2}\left(g_{\mu\lambda}\bar{R}_{\nu\kappa} - g_{\nu\lambda}\bar{R}_{\mu\kappa} + g_{\nu\kappa}\bar{R}_{\mu\lambda} - g_{\mu\kappa}\bar{R}_{\nu\lambda}\right) \\
&+ \frac{\bar{R}}{(d-1)(d-2)}\left(g_{\mu\lambda}g_{\nu\kappa} - g_{\nu\lambda}g_{\mu\kappa}\right) + C_{\mu\nu\lambda\kappa}, \qquad (12.186)
\end{aligned}
$$

where $C_{\mu\nu\lambda\kappa}$ is called the *Weyl conformal tensor*, which vanishes for $d = 3$, due to (12.181). Each of the three terms in this decomposition has the same symmetry properties as $\bar{R}_{\mu\nu\lambda\kappa}$. In addition, $C_{\mu\nu\lambda\kappa}$ satisfies the $d(d+1)/2$ conditions

$$C_{\mu}{}^{\kappa} = C_{\mu\nu}{}^{\nu\kappa} = 0, \qquad (12.187)$$

since the Ricci tensor comes entirely from the first two terms. Hence, the Weyl tensor has

$$N_d^C = \frac{1}{12}d^2(d^2 - 1) - \frac{1}{2}d(d+1) = \frac{1}{12}d(d+1)(d+2)(d-3) \qquad (12.188)$$

independent elements which is the same as the number of invariants of $\bar{R}^{\mu\nu\lambda\kappa}$. In many cases, this makes it possible to find all invariants by going to a coordinate frame in which $g_{\mu\nu} = g_{\mu\nu}$ and $\bar{R}_{\mu\nu} = $ diagonal, the first by going to a freely falling frame, the second by performing an appropriate additional Lorentz transformation. This procedure works as long as $\bar{R}_{\mu\nu}$ does not have any degenerate eigenvalues. Otherwise the Lorentz transformations remain independent and the counting does not work [3].

The above results have interesting consequences as far as a possible geometric theory of gravitation in lower-dimensional spacetimes is concerned. It turns out that in a $3 + 1$-dimensional spacetime it is impossible to have a theory which reduces to

Newton's theory in the weak coupling limit. As we shall see in Chapter 15, the crucial geometric quantity in Einstein's theory is the Einstein tensor

$$\bar{G}_{\mu\nu} = \bar{R}_{\mu\nu} - \frac{1}{2}g_{\mu\nu}\bar{R}. \tag{12.189}$$

In the above discussion we have learned that the Ricci tensor in two spacetime dimensions can be expressed in terms of the scalar curvature as

$$\bar{R}_{\mu\nu} = g_{\mu\nu}\frac{\bar{R}}{2}. \tag{12.190}$$

This implies that in two spacetime dimensions, the Einstein tensor $\bar{G}_{\mu\nu}$ *vanishes identically*.

In Eq. (15.63) we shall derive the famous Einstein field equation of gravitation according to which the Einstein tensor in the absence of torsion is proportional to the symmetric energy-momentum tensor of matter Eq. (15.63):

$$\bar{G}_{\mu\nu} = \kappa \stackrel{\mathrm{m}}{T}_{\mu\nu}, \tag{12.191}$$

with a constant κ determines by Newton's gravitational constant. Hence also the energy-momentum tensor vanishes and there is no Einstein theory of gravity. At first sight, there seems to be an escape by allowing for the presence of a so-called *cosmological term*, in which case the Einstein equation reads,

$$\bar{G}_{\mu\nu} = \kappa T_{\mu\nu} + \Lambda g_{\mu\nu}. \tag{12.192}$$

However, even if this is added, the two-dimensional theory has a severe problem: The metric $g_{\mu\nu}$ is determined completely by the local energy-momentum tensor

$$g_{\mu\nu} = -\frac{\kappa}{\Lambda}\stackrel{\mathrm{m}}{T}_{\mu\nu}, \tag{12.193}$$

and hence vanishes in the empty spacetime between mass points. Thus also this version of gravity is unphysical.

How about a geometric theory of gravitation in $2+1$ dimensions? Here the Ricci tensor is independent of scalar such that there does exist a nontrivial Einstein tensor $\bar{G}_{\mu\nu}$. Still, the tensor is almost trivial. In three dimensions there is no Weyl tensor $C_{\mu\nu\lambda\kappa}$ and the full curvature tensor is determined in terms of the Ricci tensor by Eq. (12.186) for $d = 3$

$$\begin{aligned}
\bar{R}_{\mu\nu\lambda\kappa} &= -\left(g_{\mu\lambda}\bar{R}_{\nu\kappa} - g_{\nu\lambda}\bar{R}_{\mu\kappa} + g_{\nu\kappa}\bar{R}_{\mu\lambda} - g_{\mu\kappa}\bar{R}_{\nu\lambda}\right) \\
&\quad + \frac{\bar{R}}{2}\left(g_{\mu\lambda}g_{\nu\kappa} - g_{\nu\lambda}g_{\mu\kappa}\right).
\end{aligned} \tag{12.194}$$

Inserting

$$\bar{R}_{\mu\nu} = \bar{G}_{\mu\nu} - \frac{g_{\mu\nu}}{d-2}\bar{G}, \tag{12.195}$$

this becomes

$$
\begin{aligned}
\bar{R}^{\mu\nu\lambda\kappa} &= -\left(g_{\mu\lambda}\bar{G}_{\nu\kappa} - g_{\nu\lambda}\bar{G}_{\mu\kappa} + g_{\nu\kappa}\bar{G}_{\mu\lambda} - g_{\mu\kappa}\bar{G}_{\nu\lambda}\right) \\
&\quad + \bar{G}\left(g_{\mu\lambda}g_{\nu\kappa} - g_{\nu\lambda}g_{\mu\kappa}\right).
\end{aligned}
\tag{12.196}
$$

Together with Einstein's equation (12.191), this implies that the entire energy-momentum tensor $\bar{R}_{\mu\nu\lambda\kappa}$ is determined by the local energy distribution. Hence there is no curvature in the empty spacetime between masses. As we shall see later, this implies physically that interstellar dust would experience no relative acceleration, i.e., no tidal forces (12.143).

Notes and References

[1] The physics of defects is explained in the textbook
H. Kleinert, *Gauge Fields in Condensed Matter*, Vol. II, *Stresses and Defects*, World Scientific, Singapore, 1989 (kl/b2), where kl is short for the www address http://www.physik.fu-berlin.de/~kleinert.

[2] For the introduction of harmonic coordinates see:
T. DeDonder, *La gravifique Einsteinienne*, Gauthier-Villars, Paris, 1921;
C. Lanczos, Phys. Z. **23**, 537 (1923).
and
V. Fock, *The Theory of Space, Time, and Gravitation*, Elsevier, Amsterdam, 1964.

[3] More on the counting of independent components and invariants can be found in
S. Weinberg, *Gravitation and Cosmology Principles and Applications of the General Theory of Relativity*, John Wiley & Sons, New York, 1972.

Do it big or stay in bed.
LARRY KELLY
(Opera producer in movie "Callas Forever")

13

Curvature and Torsion from Embedding

In the previous chapter we have created spaces with curvature and torsion by performing coordinate transformations $x^a(x^\mu)$ which are not integrable and whose derivative are not integrable. The first nonintegrability introduces torsion, the second curvature [recall Eq. (12.7)]. It is possible to avoid the second type of nonintegrability by starting out from a higher-dimensional spacetime and mapping it into a sub-spacetime of the desired dimension. This procedure is called *embedding*. It is well known how to do this to construct spacetimes with only curvature. This is done by imposing suitable constraints upon the higher-dimensional spacetime. Torsion will arise by allowing these constraints to become nonholonomic [1].

13.1 Spacetimes with Constant Curvature

For a d-dimensional space with constant curvature, the embedding procedure is very simple. We may choose a $D = d + 1$-dimensional Euclidean space as an embedding space, in which the infinitesimal distances are determined by

$$(d\mathbf{x})^2 = (dx^1)^2 + (dx^2)^2 + \ldots + (dx^D)^2. \tag{13.1}$$

A spherical surface of radius r is defined in this space by the constraint

$$(x^1)^2 + (x^2)^2 + \ldots + (x^d)^2 + (x^D)^2 = r^2. \tag{13.2}$$

This can be used to eliminate one of the D coordinates in (13.1), for instance

$$x^D = \sqrt{r^2 - (x^1)^2 - (x^2)^2 - \ldots - (x^d)^2}. \tag{13.3}$$

Then (13.1) becomes

$$(d\mathbf{x})^2 = (dx^1)^2 + (dx^2)^2 + \ldots + (dx^d)^2 + \frac{(x^1 dx^1 + dx^2 + \ldots + x^d dx^d)^2}{r^2 - r_d^2}, \tag{13.4}$$

where

$$r_d^2 \equiv (x^1)^2 + (x^2)^2 + \ldots + (x^d)^2. \tag{13.5}$$

374

Equation (13.4) may be rewritten as

$$(d\mathbf{x})^2 = g_{\mu\nu}(x)^d x^\mu dx^\nu, \tag{13.6}$$

where $g_{\mu\nu}(x)$ is the induced metric on the spherical surface:

$$g_{\mu\nu}(x) = \delta_{\mu\nu} + \frac{x^\mu x^\nu}{r^2 - r_d^2}. \tag{13.7}$$

A spherical surface has a constant scalar curvature $\bar{R}$. This may be found by evaluating Eqs. (11.129). (11.141), and (11.142) on any point on the sphere, for instance in the neighborhood of $x^\mu = 0$ ($\mu = 1, \ldots, d$), where $r_d = \mathcal{O}((x^\mu)^2)$ and $g_{\mu\nu}(x) \approx \delta_{\mu\nu} + x^\mu x^\nu / r^2 + \mathcal{O}((x^\mu)^4)$. The Christoffel symbols (11.21) are $\Gamma_{\mu\nu}{}^\lambda \approx \Gamma_{\mu\nu\lambda} \approx \delta_{\mu\nu} x_\lambda / r^2 + \mathcal{O}((x^\mu)^3)$. These are inserted into Eq. (11.129) to yield the curvature tensor for small x^μ:

$$\bar{R}_{\mu\nu\lambda\kappa} \approx \frac{1}{r^2}\left(\delta_{\mu\kappa}\delta_{\nu\lambda} - \delta_{\mu\lambda}\delta_{\nu\kappa}\right). \tag{13.8}$$

This local expression can be extended covariantly to the full surface of the sphere by replacing $\delta_{\mu\lambda}$ by the metric tensor $g_{\mu\lambda}(x)$, yielding

$$\bar{R}_{\mu\nu\lambda\kappa}(x) = \frac{1}{r^2}\left[g_{\mu\kappa}(x)g_{\nu\lambda}(x) - g_{\mu\lambda}(x)g_{\nu\kappa}(x)\right]. \tag{13.9}$$

This result remains valid in hyperbolic spaces. In particular it holds for spacetime with constant curvature.

Contracting $\bar{R}_{\mu\nu\lambda\kappa}(x)$ with $g^{\mu\kappa}(x)$ yields the Ricci tensor [recall (11.141)]

$$\bar{R}_{\nu\lambda}(x) = \bar{R}_{\mu\nu\lambda}{}^\mu(x) = \frac{d-1}{r^2}g_{\nu\lambda}(x). \tag{13.10}$$

Contracting this with $g^{\nu\kappa}(x)$ gives the curvature scalar [recall (11.142)]:

$$\bar{R} = \frac{(d-1)d}{r^2}. \tag{13.11}$$

13.2   Basis Vectors

The above embedding procedure can be generalized to arbitrary curved spacetimes as follows. We embed the d-dimensional spacetime with coordinates x^μ ($\mu = 1, \ldots, d$) into a higher-dimensional flat spacetime with coordinates x^A ($A = 1, \ldots, D$), whose metric η_{AB} is diagonal with elements equal to ± 1. The embedding is done with D functions $x^A(x^\mu)$. The derivatives $\partial x^A(x^\mu)/\partial x^\kappa$ define $D \times d$ functions

$$\varepsilon^A{}_\lambda(x^\mu) \equiv \frac{\partial x^A}{\partial x^\lambda}, \tag{13.12}$$

from which we form D basis vectors $\mathbf{e}_A$ in the

$$\mathbf{e}_\lambda(x^\mu) = \mathbf{e}_A \varepsilon^A{}_\lambda(x^\mu). \tag{13.13}$$

These are the local tangent vectors to the coordinate lines in the d-dimensional sub-spacetime with coordinates x^μ. There they induce a metric

$$g_{\lambda\kappa}(x^\mu) = \mathbf{e}_\lambda(x^\mu)\,\mathbf{e}_\kappa(x^\mu) \tag{13.14}$$

whose inverse $g^{\lambda\kappa}(x^\mu)$ can be used to raise the index of (13.12) and define the $D \times d$ functions

$$\varepsilon^{A\lambda}(x^\mu) \equiv g^{\lambda\lambda'}(x^\mu)\varepsilon^A{}_{\lambda'}(x^\mu). \tag{13.15}$$

As usual, the metric g_{AB} serves to lower the superscripts $A,\,B,\,\dots$, and the inverse metric, $g^{AB} = (g^{-1})^{AB}$ is equal to g_{AB}, is used to raise subscripts $A,\,B,\,\dots$.

In contrast to the multivalued basis tetrads $\varepsilon^{a\lambda}(x)$ and $e_{b\lambda}(x)$ of Eq. (11.37), there exist now no completeness relation [compare (11.41)]:

$$\varepsilon^{A\lambda}(x^\mu)e_{B\lambda}(x^\mu) \neq \delta^A{}_B. \tag{13.16}$$

This is obvious since λ runs only from 1 to $d < D$, so there are not enough functions $\varepsilon^{A\lambda}(x^\mu)$ to span the D-dimensional embedding spacetime. However, the functions $\varepsilon^{A\lambda}(x^\mu)$ and $e_{A\kappa}(x^\mu)$ do fulfill the orthogonality relation [compare (11.40) and (11.41)]

$$\varepsilon^{A\lambda}(x^\mu)e_{A\kappa}(x^\mu) = e_A{}^\lambda(x^\mu)\varepsilon^A{}_\kappa(x^\mu) = \delta^\lambda{}_\kappa. \tag{13.17}$$

Note that due to the incompleteness relation (13.16), the curvature tensor has to be calculated from (11.129). Formula (11.130) has become meaningless since the derivation of that formula would require an equation

$$\partial_\mu \varepsilon^A{}_\nu = \Gamma_{\mu\nu}{}^\lambda \varepsilon^A{}_\lambda, \tag{13.18}$$

which is no longer true.

Fortunately, the determining equation for the affine connection in the embedded spacetime x^μ has still the same form as before in Eq. (11.91):

$$\Gamma_{\mu\nu}{}^\lambda = \varepsilon_A{}^\lambda \partial_\mu \varepsilon^A{}_\nu = -\varepsilon^A{}_\nu \partial_\mu \varepsilon_A{}^\lambda. \tag{13.19}$$

This can be derived by modifying only slightly the derivation in Subsection 11.3.4. First we observe that it is no longer possible to introduce the covariant derivative of a vector field $v_\mu(x^\sigma)$ from an equation of the form Eq. (11.84). This would be based on an analog of Eq. (11.80). Defining the extension of the vector field into the embedding spacetime by

$$v_A(x^\sigma) \equiv \varepsilon_A{}^\mu(x^\sigma)\,v_\mu(x^\sigma), \tag{13.20}$$

this would read

$$\partial_B v_A = \partial_B \left(\varepsilon_A{}^\mu v_\mu \right) \qquad \text{(undefined)}. \qquad (13.21)$$

But this equation is meaningless since the functions v_A depend only on the subspacetime x^μ of x^A.

The slight modification of Eq. (11.80) which does lead to a meaningful starting equation is obtained by multiplying (11.80) on both sides by $e^b{}_\lambda$, yielding

$$\partial_\lambda v_a = \partial_\lambda \left(e_a{}^\mu v_\mu \right). \qquad (13.22)$$

Evaluation of the derivative leads to the defining equation for the covariant derivative

$$\partial_\lambda v_a = e_a{}^\mu D_\lambda v_\mu. \qquad (13.23)$$

This replaces the defining equation (11.84).

Both equations (13.22) and (13.23) remain meaningful in the embedding scenario, where they read

$$\partial_\lambda v_A = \partial_\lambda \left(\varepsilon_A{}^\mu v_\mu \right), \qquad (13.24)$$

and

$$\partial_\lambda v_A = \varepsilon_A{}^\mu D_\lambda v_\mu. \qquad (13.25)$$

Multiplying the latter by $\varepsilon^A{}_\sigma$ of Eq. (13.12), and using the orthogonality relation (13.17) we obtain

$$D_\lambda v_\mu = \varepsilon^A{}_\mu \partial_\lambda v_A. \qquad (13.26)$$

Expressing v_A in terms of v_μ via Eq. (13.20) leads to

$$D_\lambda v_\mu = \varepsilon^A{}_\mu \partial_\lambda \varepsilon_A{}^\lambda v_\lambda = \partial_\lambda v_\mu + (\varepsilon^A{}_\mu \partial_\lambda \varepsilon_A{}^\lambda) v_\lambda = \partial_\lambda v_\mu - (\varepsilon_A{}^\lambda \partial_\lambda \varepsilon^A{}_\mu) v_\lambda. \qquad (13.27)$$

This shows that the affine connection is indeed given by (13.19), in terms of which the covariant derivative has the same form as in (11.85).

Note that the mappings $x^A(x^\mu)$ may arise from constraints imposed upon the coordinates x^A in the embedding spacetime. This was the case in Section (13.1) where the constraint was the restriction (13.2) to a sphere, from which we derived the mapping function $x^D(x^\mu)$ in Eq. (13.3).

If there is torsion, the constraints leading to the nonintegrable mapping functions $x^D(x^\mu)$ will be nonholonomic in the sense used in classical mechanics. According to the Hertz classification [2], constraints are said to be holonomic if they are integrable (i.e., equivalent to some constraints on the configuration spacetime only). They are called and nonholonomic if they are nonintegrable. Sometimes dynamical systems with nonholonomic constraints are called nonholonomic systems.

Let us illustrate the use of the functions $\varepsilon^A{}_\mu$ by calculating once more the curvature tensor of a sphere. Rather than proceeding as in Section 13.1, we obtain a sphere of radius r in three dimensions from the embedding mapping

$$x^A = \left(x^1, x^2, x^3\right) = r\left(\sin\theta\cos\varphi, \sin\theta, \sin\varphi, \cos\theta\right). \tag{13.28}$$

The tangent vectors of the sphere have the 3×2 components

$$\begin{aligned}
\varepsilon^A{}_1 &= r\left(\cos\theta\cos\varphi, \cos\theta, \sin\varphi, -\sin\theta\right) = \varepsilon_{A1}, \\
\varepsilon^A{}_2 &= r\left(-\sin\theta\sin\varphi, \sin\theta\cos\varphi, 0\right) = \varepsilon_{A2},
\end{aligned} \tag{13.29}$$

where the two coordinates x^μ are chosen to be the spherical angles $\theta \in (0, \pi)$ and $\varphi \in (0, 2\pi)$. The induced metric (13.14) becomes

$$g_{\mu\nu} = r^2\begin{pmatrix} 1 & 0 \\ 0 & \sin^2\theta \end{pmatrix}, \qquad g^{\mu\nu} = \frac{1}{r^2}\begin{pmatrix} 1 & 0 \\ 0 & \sin^{-2}\theta \end{pmatrix} \tag{13.30}$$

such that

$$\varepsilon^1_A = \varepsilon_{A1} = \varepsilon^A{}_1, \qquad \varepsilon^2_A = \frac{1}{r}\left(-\frac{\sin\varphi}{\sin\theta}, \frac{\sin\varphi}{\sin\theta}, 0\right). \tag{13.31}$$

The Riemann connection is symmetric:

$$\begin{aligned}
\bar{\Gamma}_{221} &= \varepsilon_{A1}\partial_2\varepsilon^A{}_2 = r\,\varepsilon_{A1}\left(-\sin\theta\cos\varphi, -\sin\theta\sin\varphi, 0\right) \\
&= -r^2\sin\theta\cos\theta = -\bar{\Gamma}_{212} = -\bar{\Gamma}_{122}.
\end{aligned} \tag{13.32}$$

All other elements vanish. By raising the last index we obtain

$$\bar{\Gamma}_{22}{}^1 = -\sin\theta\cos\theta, \qquad \bar{\Gamma}_{21}{}^2 = \cot\theta. \tag{13.33}$$

The Riemann curvature tensor has the components [recall (11.129)]:

$$\begin{aligned}
\bar{R}_{122}{}^1 &= \partial_1\bar{\Gamma}_{22}{}^1 - \partial_2\bar{\Gamma}_{12}{}^1 - \bar{\Gamma}_{12}{}^1\bar{\Gamma}_{21}{}^1 - \bar{\Gamma}_{12}{}^2\bar{\Gamma}_{22}{}^1 + \bar{\Gamma}_{22}{}^1\bar{\Gamma}_{11}{}^1 + \bar{\Gamma}_{22}{}^2\bar{\Gamma}_{12}{}^1 \\
&= -\cos^2\theta + \sin^2\theta + \cot\theta\sin\theta\cos\theta = \sin^2\theta,
\end{aligned} \tag{13.34}$$

implying that

$$\bar{R}_{12}{}^{21} = \frac{1}{r^2}. \tag{13.35}$$

All other elements can be obtained using the antisymmetry of $\bar{R}_{\mu\nu\lambda\kappa}$ in $\mu \to \nu, \lambda \to \kappa$ and symmetry under $\mu\nu \leftrightarrow \lambda\kappa$, which is a consequence of the symmetry of $\bar{\Gamma}_{\mu\nu}{}^\lambda$ in $\mu\nu$ [recall the derivation of (12.106)]. Thus we can form the Ricci tensor

$$\bar{R}_{\mu\nu}{}^{\lambda\mu} = \bar{R}_\nu{}^\lambda = \frac{1}{a^2}\begin{pmatrix} 1 & 0 \\ 0 & 1 \end{pmatrix}, \tag{13.36}$$

and the curvature scalar

$$\bar{R} = \bar{R}_\mu{}^\nu = \frac{2}{a^2}. \tag{13.37}$$

In general, it is possible to generate any curved spacetime by embedding it in a higher-dimensional flat spacetime. If the curved spacetime has d dimension, the embedding spacetime must have at least $d(d+1)/2$ dimension. This is seen by noticing that, for a given $g_{\mu\nu}(x)$, the equation for the induced metric in (13.14), written in the form

$$g_{\mu\nu}(x) = g_{AB}\frac{\partial x^A}{\partial x^\mu}\frac{\partial x^B}{\partial x^\nu}, \tag{13.38}$$

specifies $d(d+1)/2$ differential equations for the functions $x^A(x^\mu)$.

13.3 Torsion

We may now easily introduce torsion into the embedded space by allowing the embedding mapping $x^A(x^\mu)$ to be multivalued. As a consequence, we obtain a nonzero torsion tensor defined as in (11.107) by the antisymmetric part of the affine connection

$$S_{\mu\nu}{}^\lambda \equiv \frac{1}{2}\left(\Gamma_{\mu\nu}{}^\lambda - \Gamma_{\nu\mu}{}^\lambda\right) = \varepsilon_A{}^\lambda \partial_\mu \varepsilon^A{}_\nu - \varepsilon_A{}^\lambda \partial_\nu \varepsilon^A{}_\mu. \tag{13.39}$$

The metric tensor $g_{\mu\nu}$ has $D(D+1)/2$ independent components. The torsion tensor $S_{\nu\kappa}{}^\mu$ has $D^2(D-1)/2$ independent components. To embed a general metric spacetime with torsion in a larger space, the number $D \times d$ of independent embedding functions $\varepsilon^A{}_\mu$ should be at least equal to $D(D^2+1)/2$.

This leads to the relation between the dimensions of x^μ- and x^A-spaces:

$$2\dim[x^A] \geq (\dim[x^\mu])^2 + 1. \tag{13.40}$$

Notes and References

[1] H. Kleinert and S.V. Shabanov, *Spaces with Torsion from Embedding, and the Special Role of Autoparallel Trajectories*, Phys. Lett B **428** , 315 (1998) (quant-ph/9503004) (www.physik.fu-berlin.de/~kleinert/259).

[2] V.I. Arnold, V.V. Koslov, and A.I. Neishtadt, in: *Encyclopedia of Mathematical Sciences, Dynamical Systems III, Mathematical Aspects of Classical and Celestial Mechanics*, Springer-Verlag, Berlin, 1988;
L.A. Pars, *A Treatise on Analytical Dynamics*, Heinemann, London, 1965;
G. Hamel, *Theoretische Mechanik – Eine einheitliche Einführung in die gesamte Mechanik*, Springer Series in Mathematics, Vol. 57, Springer, Berlin, 1978.

14

Multivalued Mapping Principle

The multivalued, nonholonomic mappings from flat to curved spacetime with torsion enable us to sharpen Einstein's equivalence principle to a more powerful statement. Whereas Einstein postulated that equations written down in flat spacetime with curvilinear coordinates remain valid in curved spacetime, we postulate the *new equivalence principle*:

Fundamental physical laws in curved spacetime are direct images of the laws in flat spacetime under multivalued mappings.

If spacetime has only curvature and no torsion, we re-obtain the well-known laws postulated by Einstein on the basis of coordinate invariance and minimal coupling to gravity. In the presence of torsion, the new equivalence principle makes new predictions, and it will be interesting to investigate these.

It must be noted that the assumption of minimal coupling can only be applied to fundamental particles. This is familiar from electromagnetism, where composite particles such as protons and neutrons do not couple minimally to the vector potential $A^\mu(x)$. Their magnetic moments reflect the nontrivial internal distributions of quark currents. The quarks themselves, however, do couple minimally. So do leptons, whose anomalous magnetic moments can be explained by higher-order electroweak perturbative corrections.

In gravity, only point-like objects couple minimally. Extended objects such as planets do not. Their quadrupole moment couples non-minimally to the tidal forces, i.e., to the curvature tensor of the geometry, leading to an extra precession rate of the spin vector S^μ, in addition to the geodetic or autoparallel precessions to be dicussed in Section 14.40. Also protons and neutrons are subject to tidal forces, although these have so far not been observed experimentally. Quarks and leptons should again couple minimally. Also the photon should do so, and the graviton itself. The Higgs boson, on the other hand, has probably a nonminimal coupling, since it cannot be a fundamental field. This holds also for the W- and Z-bosons of weak interactions, since they are photon-like particles that have become massive by mixing with a Higgs particle.

14.1 Motion of Point Particle

The derivation of the geodesic trajectories of point particles in curved spacetime was performed in Section 11.2. Only minor modifications will be necessary to follow the new equivalence principle. As observed when going from Eq. (11.13) to (11.14), we simplify the discussion by considering the nonrelativistic action $\overset{m}{A}$ of Eq. (11.15) if we use the proper time τ to parameterize the paths.

14.1.1 Classical Action Principle for Spaces with Curvature

Instead of performing an ordinary coordinate transformation in flat spacetime from Minkowski coordinates x^a to curvilinear coordinates x^μ via Eq. (11.6), we perform a multivalued coordinate transformation

$$dx^a = e^a{}_\mu(x)dx^\mu, \tag{14.1}$$

where the basis vectors $e^a{}_\mu$ describe coordinate transformations in which

$$\partial_\mu e^a{}_\nu(x) - \partial_\nu e^a{}_\mu(x) \neq 0. \tag{14.2}$$

This implies that second derivatives in front of the multivalued functions $x^a(x^\mu)$ do not commute as in Eq. (11.30):

$$(\partial_\lambda \partial_\kappa - \partial_\kappa \partial_\lambda)x^a(x) \neq 0, \tag{14.3}$$

thus violating the Schwarz integrability criterion. Such a spacetime has torsion. If the spacetime has also curvature, then the functions $e^a{}_\nu(x) = \partial_\nu x^a(x)$ have also no commuting derivatives [recall (11.31)]:

$$(\partial_\mu \partial_\nu - \partial_\nu \partial_\mu)e^a{}_\lambda(x) = (\partial_\mu \partial_\nu - \partial_\nu \partial_\mu)\partial_\lambda x^a(x) \neq 0. \tag{14.4}$$

In either case, the metric in the image space has the same form as in Eq. (11.8), and the derivation of the extremum of the action seems, at first, to follow the same pattern as in Section 11.2, leading to the equation of motion (11.24) for geodesic trajectories. The nonholonomically transformed action (11.2) is independent of the torsion fields $S_{\mu\nu}{}^\lambda$, and for this reason also the equation of motion (11.24) is indifferent to the presence of torsion.

This result would be perfectly acceptable, were it not for an apparent inconsistency with another result obtained by applying the new variational principle. Instead of transforming the action and varying it in the usual way, we may transform the equation of motion of a free particle (11.1) in flat space nonholonomically into a spacetime with curvature and torsion.

14.1.2 Autoparallel Trajectories in Spaces with Torsion

In the absence of external forces, the equation of motion (11.1) in flat space states that the second derivative of $x^i(\tau)$ vanishes. In spacetime, the free equation of motion reads $\ddot{x}^a(\tau) = 0$, where the dot denotes the derivative with respect to the proper time $\tau = s/c$, where s is the invariant length of the path. The equation of motion $\ddot{x}^a(\tau) = 0$ is transformed by the multivalued mapping (14.1) as follows:

$$\frac{d^2 x^a}{d\tau^2} = \frac{d}{d\tau}(e^a{}_\mu \dot{x}^\mu) = e^a{}_\mu \ddot{x}^\mu + e^a{}_{\mu,\nu}\dot{x}^\mu \dot{x}^\nu = 0, \tag{14.5}$$

or as

$$\ddot{x}^\mu + e_a{}^\mu e^a{}_{\kappa,\lambda}\dot{x}^\kappa \dot{x}^\lambda = 0. \tag{14.6}$$

The subscript λ separated by a comma denotes the partial derivative: $f_{,\lambda}(x) \equiv \partial_\lambda f(x)$. The quantity in front of $\dot{x}^\kappa \dot{x}^\lambda$ is the *affine connection* (11.91). Thus we arrive at the transformed flat spacetime equation of motion

$$\ddot{x}^\mu + \Gamma_{\kappa\lambda}{}^\mu \dot{x}^\kappa \dot{x}^\lambda = 0. \tag{14.7}$$

The solutions of this equation are called *autoparallel* trajectories. They differ from the geodesic trajectories in Eq. (11.24) by an extra torsion term. Inserting the decomposition (11.115) and using the antisymmetry of $S_{\mu\nu}{}^\lambda$ in the first two indices, we may rewrite (14.7) as

$$\ddot{x}^\mu + \bar{\Gamma}^\mu{}_{\kappa\lambda}\dot{x}^\kappa \dot{x}^\lambda - 2S^\mu{}_{\kappa\lambda}\dot{x}^\kappa \dot{x}^\lambda = 0. \tag{14.8}$$

Note the index positions of the torsion tensor, which may be written more explicitly as $S^\mu{}_{\kappa\lambda} \equiv g^{\mu\sigma}g_{\lambda\kappa}S_{\sigma\kappa}{}^\kappa$. This is not antisymmetric in the last two indices so that it possesses a symmetric part which contributes to Eq. (14.7).

How can we reconcile this result with an application of the new equivalence principle applied to the action. Since the transformed action is independent of the torsion and carries only information on the Riemann part of the spacetime geometry, torsion can enter the equations of motion only via some overlooked feature of the variation procedure. Indeed, a moment's thought convinces us that this was applied incorrectly in the previous section. According to the new equivalence principle we must also transform the variational procedure nonholonomically to spacetimes with curvature and torsion. We must find the image of the flat spacetime variations $\delta x^a(\tau)$ under the multivalued mapping

$$\dot{x}^\mu = e_a{}^\mu(x)\dot{x}^a. \tag{14.9}$$

The images are quite different from ordinary variations as illustrated in Fig. 14.1(a). The variations of the Cartesian coordinates $\delta x^a(\tau)$ are done at fixed endpoints of the paths. Thus they form *closed paths* in the x-spacetime. Their images, however, lie in a spacetime with defects and thus possess a closure failure indicating the amount of torsion introduced by the mapping. This property will be emphasized by writing the images $\delta^S x^\mu(\tau)$ and calling them *nonholonomic variations*. The superscript indicates the special feature caused by torsion.

Let us calculate them explicitly. The paths in the two spaces are related by the integral equation

$$x^\mu(\tau) = x^\mu(\tau_a) + \int_{\tau_a}^{\tau} d\tau' e_a{}^\mu(x(\tau'))\dot{x}^a(\tau'). \tag{14.10}$$

For two neighboring paths in x-space differing from each other by a variation $\delta x^a(\tau)$, Eq. (14.10) determines the nonholonomic variation $\delta^S x^\mu(\tau)$:

$$\delta^S x^\mu(\tau) = \int_{\tau_a}^{\tau} d\tau' \delta^S [e_a{}^\mu(x(\tau'))\dot{x}^a(\tau')]. \tag{14.11}$$

A comparison with (14.9) shows that the variation δ^S and the derivatives with respect to the parameter s of $x^\mu(\tau)$ commute with each other:

$$\delta^S \dot{x}^\mu(\tau) = \frac{d}{d\tau}\delta^S x^\mu(\tau), \tag{14.12}$$

just as for ordinary variations δx^a in Eq. (2.7):

$$\delta \dot{x}^a(\tau) = \frac{d}{d\tau}\delta x^a(\tau). \tag{14.13}$$

Let us also introduce *auxiliary nonholonomic variations* of the paths $x^\mu(\tau)$ in x^μ-space:

$$\delta x^\mu \equiv e_a{}^\mu(x)\delta x^a. \tag{14.14}$$

In contrast to $\delta^S x^\mu(\tau)$, these vanish at the endpoints,

$$\delta x(\tau_a) = \delta x(\tau_b) = 0, \tag{14.15}$$

just as the usual variations $\delta x^a(\tau)$, i.e., they form *closed* paths with the unvaried orbits.

Using (14.12), (14.13), and the fact that $\delta^S x^a(\tau) \equiv \delta x^a(\tau)$ by definition, we derive from (14.11) the relation

$$\begin{aligned}
\frac{d}{d\tau}\delta^S x^\mu(\tau) &= \delta^S e_a{}^\mu(x(\tau))\dot{x}^a(\tau) + e_a{}^\mu(x(\tau))\frac{d}{d\tau}\delta x^a(\tau) \\
&= \delta^S e_a{}^\mu(x(\tau))\dot{x}^a(\tau) + e_a{}^\mu(x(\tau))\frac{d}{d\tau}[e^a{}_\nu(\tau)\,\delta x^\nu(\tau)].
\end{aligned} \tag{14.16}$$

After inserting

$$\delta^S e_a{}^\mu(x) = -\Gamma_{\lambda\nu}{}^\mu \delta^S x^\lambda e_a{}^\nu, \qquad \frac{d}{d\tau}e^a{}_\nu(x) = \Gamma_{\lambda\nu}{}^\mu \dot{x}^\lambda e^a{}_\mu, \tag{14.17}$$

this becomes

$$\frac{d}{d\tau}\delta^S x^\mu(\tau) = -\Gamma_{\lambda\nu}{}^\mu \delta^S x^\lambda \dot{x}^\nu + \Gamma_{\lambda\nu}{}^\mu \dot{x}^\lambda \delta x^\nu + \frac{d}{d\tau}\,\delta x^\mu. \tag{14.18}$$

It is useful to introduce the difference between the nonholonomic variation $\delta^S x^\mu$ and an auxiliary closed nonholonomic variation δx^μ:

$$\delta^S b^\mu \equiv \delta^S x^\mu - \delta x^\mu. \tag{14.19}$$

Then we can rewrite (14.18) as a first-order differential equation for $\delta^S b^\mu$:

$$\frac{d}{d\tau}\delta^S b^\mu = -\Gamma_{\lambda\nu}{}^\mu \delta^S b^\lambda \dot{x}^\nu + 2S_{\lambda\nu}{}^\mu \dot{x}^\lambda \,\delta x^\nu. \tag{14.20}$$

After introducing the matrices

$$G^\mu{}_\lambda(\tau) \equiv \Gamma_{\lambda\nu}{}^\mu(x(\tau))\dot{x}^\nu(\tau) \tag{14.21}$$

and

$$\Sigma^\mu{}_\nu(\tau) \equiv 2S_{\lambda\nu}{}^\mu(x(\tau))\dot{x}^\lambda(\tau). \tag{14.22}$$

Equation (14.20) can be written as a vector differential equation:

$$\frac{d}{d\tau}\delta^S b = -G\delta^S b + \Sigma(\tau)\,\delta x^\nu(\tau). \tag{14.23}$$

Although not necessary for further development, we solve this equation by

$$\delta^S b(\tau) = \int_{\tau_a}^{\tau} d\tau'\, U(\tau,\tau')\,\Sigma(\tau')\,\delta x(\tau'), \tag{14.24}$$

with the matrix

$$U(\tau,\tau') = \hat{T}_s \exp\left[-\int_{\tau'}^{\tau} d\tau''G(\tau'')\right], \tag{14.25}$$

where $\hat{T}_s$ denotes the time-ordering operator for the parameter s. In the absence of torsion, $\Sigma(\tau)$ vanishes identically and $\delta^S b(\tau) \equiv 0$, and the variations $\delta^S x^\mu(\tau)$ coincide with the auxiliary closed nonholonomic variations $\delta x^\mu(\tau)$ (see Fig. 14.1b). In a spacetime with torsion, the variations $\delta^S x^\mu(\tau)$ and $\delta x^\mu(\tau)$ are different from each other (see Fig. 14.1c).

We now calculate the variation of the action (11.10) under an arbitrary nonholonomic variation $\delta^S x^\mu(\tau) = \delta x^\mu + \delta^S b^\mu$. Since s is the invariant path length, we may just as well use the auxiliary action (11.15) to calculate this quantity (it differs only by a trivial factor 2):

$$\delta^S \bar{A} = M \int_{\tau_a}^{\tau_b} d\tau \left(g_{\mu\nu}\dot{x}^\nu \delta^S \dot{x}^\mu + \frac{1}{2}\partial_\mu g_{\lambda\kappa}\delta^S x^\mu \dot{x}^\lambda \dot{x}^\kappa \right). \tag{14.26}$$

After a partial integration of the $\delta\dot{x}$-term we use (14.15), (14.12), and the identity $\partial_\mu g_{\nu\lambda} \equiv \Gamma_{\mu\nu\lambda} + \Gamma_{\mu\lambda\nu}$, which follows directly from the definitions $g_{\mu\nu} \equiv e^a{}_\mu e^a{}_\nu$ and $\Gamma_{\mu\nu}{}^\lambda \equiv e_a{}^\lambda \partial_\mu e^a{}_\nu$, to obtain

$$\delta^S \bar{A} = M\int_{\tau_a}^{\tau_b} d\tau \left[-g_{\mu\nu}\left(\ddot{x}^\nu + \bar{\Gamma}_{\lambda\kappa}{}^\nu \dot{x}^\lambda \dot{x}^\kappa\right)\delta x^\mu + \left(g_{\mu\nu}\dot{x}^\nu \frac{d}{d\tau}\delta^S b^\mu + \Gamma_{\mu\lambda\kappa}\delta^S b^\mu \dot{x}^\lambda \dot{x}^\kappa\right)\right]. \tag{14.27}$$

To derive the equation of motion we first vary the action in a spacetime without torsion. Then $\delta^S b^\mu(\tau) \equiv 0$, and (14.27) becomes

$$\delta^S \bar{A} = -M \int_{t_a}^{t_b} d\tau g_{\mu\nu}(\ddot{x}^\nu + \bar{\Gamma}_{\lambda\kappa}{}^\nu \dot{x}^\lambda \dot{x}^\kappa)\,\delta x^\nu. \tag{14.28}$$

Thus, the action principle $\delta^S \bar{A} = 0$ produces the equation for the geodesics (11.24), which are the correct particle trajectories in the absence of torsion.

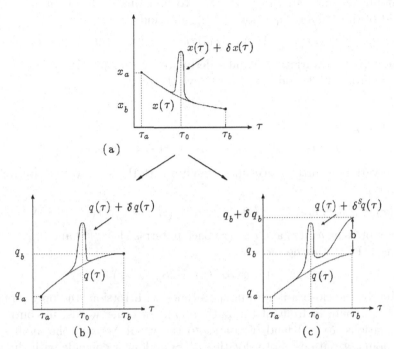

FIGURE 14.1 Images under holonomic and nonholonomic mapping of fundamental δ-function path variation. In the holonomic case, the paths $x(\tau)$ and $x(\tau) + \delta x(\tau)$ in (a) turn into the paths $x(\tau)$ and $x(\tau) + \delta x(\tau)$ in (b). In the nonholonomic case with $S^{\lambda}_{\mu\nu} \neq 0$, they go over into $x(\tau)$ and $x(\tau) + \delta^S x(\tau)$ shown in (c) with a closure failure b^{μ} at t_b analogous to the Burgers vector b^{μ} in a solid with dislocations.

In the presence of torsion, $\delta^S b^{\mu}$ is nonzero, and the equation of motion receives a contribution from the second parentheses in (14.27). After inserting (14.20), the nonlocal terms proportional to $\delta^S b^{\mu}$ cancel and the total nonholonomic variation of the action becomes

$$
\begin{aligned}
\delta^S \bar{\mathcal{A}} &= -M \int_{\tau_a}^{\tau_b} d\tau\, g_{\mu\nu} \left[\ddot{x}^{\nu} + \left(\bar{\Gamma}_{\lambda\kappa}{}^{\nu} + 2S^{\nu}{}_{\lambda\kappa} \right) \dot{x}^{\lambda} \dot{x}^{\kappa} \right] \delta x^{\mu} \\
&= -M \int_{\tau_a}^{\tau_b} d\tau\, g_{\mu\nu} \left(\ddot{x}^{\nu} + \Gamma_{\lambda\kappa}{}^{\nu} \dot{x}^{\lambda} \dot{x}^{\kappa} \right) \delta x^{\mu}.
\end{aligned}
\tag{14.29}
$$

The second line follows from the first after using the identity $\Gamma_{\lambda\kappa}{}^{\nu} = \bar{\Gamma}_{\{\lambda\kappa\}}{}^{\nu} + 2S^{\nu}{}_{\{\lambda\kappa\}}$. The curly brackets indicate the symmetrization of the enclosed indices. Setting $\delta^S \bar{\mathcal{A}} = 0$ and inserting for $\delta x(\tau)$ the image under (14.14) of an arbitrary δ-function variation $\delta x^a(\tau) \propto \epsilon^a \delta(\tau - s_0)$ gives the autoparallel equations of motion (14.7), which is what we wanted to show.

The above variational treatment of the action is still somewhat complicated and calls for a simpler procedure [1, 2]. The extra term arising from the second parenthesis in the variation (14.27) can be traced to a simple property of the auxiliary closed nonholonomic variations (14.14). To find this we form the time derivative $d_t \equiv d/dt$ of the defining equation (14.14) and find

$$d_t \, \delta x^\mu(\tau) = \partial_\nu e_a{}^\mu(x(\tau)) \, \dot{x}^\nu(\tau) \delta x^a(\tau) + e_a{}^\mu(x(\tau)) d_\tau \delta x^a(\tau). \tag{14.30}$$

Let us now perform variation δ and s-derivative in the opposite order and calculate $d_\tau \, \delta x^\mu(\tau)$. From (14.9) and (11.40) we have the relation

$$d_\tau x^\lambda(\tau) = e_i{}^\lambda(x(\tau)) \, d_\tau x^i(\tau). \tag{14.31}$$

Varying this gives

$$\delta d_\tau x^\mu(\tau) = \partial_\nu e_a{}^\mu(x(\tau)) \; \delta x^\nu d_t x^a(\tau) + e_a{}^\mu(x(\tau)) \, \delta d_\tau x^a. \tag{14.32}$$

Since the variations in x^a-spacetime commute with the s-derivatives [recall (14.13)], we obtain

$$\delta d_\tau x^\mu(\tau) - d_\tau \, \delta x^\mu(\tau) = \partial_\nu e_a{}^\mu(x(\tau)) \, \delta x^\nu d_t x^a(\tau) - \partial_\nu e_a{}^\mu(x(\tau)) \, \dot{x}^\nu(\tau) \delta x^a(\tau). \tag{14.33}$$

After re-expressing $\delta x^a(\tau)$ and $d_t x^a(\tau)$ back in terms of $\delta x^\mu(\tau)$ and $d_t x^\mu(\tau) = \dot{x}^\mu(\tau)$, and using (11.91), this becomes

$$\delta d_\tau x^\mu(\tau) - d_\tau \, \delta x^\mu(\tau) = 2 S_{\nu\lambda}{}^\mu \dot{x}^\nu(\tau) \, \delta x^\lambda(\tau). \tag{14.34}$$

Thus, due to the closure failure in spacetimes with torsion, the operations d_τ and δ do not commute in front of the path $x^\mu(\tau)$. In other words, in contrast to the open variations $\delta x^\mu(\tau)$ [and of course to the usual $\delta x^\mu(\tau)$], the auxiliary closed nonholonomic variations δ of velocities $\dot{x}^\mu(\tau)$ no longer coincide with the velocities of variations.

This property is responsible for shifting the trajectory from geodesics to autoparallels. Indeed, let us vary an action

$$\bar{A} = \int_{\tau_a}^{\tau_b} d\tau \, L\left(x^\mu(\tau), \dot{x}^\mu(\tau)\right) \tag{14.35}$$

directly by $\delta x^\mu(\tau)$ and impose (14.34). Then we find

$$\delta \bar{A} = \int_{\tau_a}^{\tau_b} d\tau \left\{ \frac{\partial L}{\partial x^\mu} \delta x^\mu + \frac{\partial L}{\partial \dot{x}^\mu} \frac{d}{d\tau} \, \delta x^\mu + 2 \, S^\mu{}_{\nu\lambda} \frac{\partial L}{\partial \dot{x}^\mu} \dot{x}^\nu \, \delta x^\lambda \right\}. \tag{14.36}$$

After a partial integration of the second term using the vanishing $\delta x^\mu(\tau)$ at the endpoints, we obtain the Euler-Lagrange equation

$$\frac{\delta \bar{A}}{\delta x^\mu} = \frac{\partial L}{\partial x^\mu} - \frac{d}{dt} \frac{\partial L}{\partial \dot{x}^\mu} = \frac{\delta \bar{A}}{\delta x^\mu} - 2 S_{\mu\nu}{}^\lambda \dot{x}^\nu \frac{\partial L}{\partial \dot{x}^\lambda} = -2 S_{\mu\nu}{}^\lambda \dot{x}^\nu \frac{\partial L}{\partial \dot{x}^\lambda}. \tag{14.37}$$

This differs from the standard Euler-Lagrange equation by the additional torsion force. For the action (11.10), we thus obtain the equation of motion

$$\ddot{x}^\mu + \left[g^{\mu\kappa} \left(\partial_\nu g_{\lambda\kappa} - \frac{1}{2} \partial_\kappa g_{\nu\lambda} \right) + 2 S^\mu{}_{\nu\lambda} \right] \dot{x}^\nu \dot{x}^\lambda = 0, \tag{14.38}$$

which is once more the Eq. (14.7) for autoparallels.

Thus a consistent application of the new equivalence principle yields consistently autoparallel trajectories for point particles in spacetime with curvature and torsion.

14.1.3 Equations of Motion For Spin

In Eq. (1.294) we have derived the time derivative of the spin four-vector of a spinning point particle in Minkowski spacetime. The multivalued or nonholonomic mapping principle transforms this to a general affine geometry:

$$\frac{DS_\mu}{d\tau} = S_\kappa \frac{Du^\kappa}{d\tau} u_\mu. \tag{14.39}$$

This equation shows that in the absence of external forces, the spin four-vector of a point particle remains always parallel to its initial orientation along the entire autoparallel trajectory:

$$\frac{DS_\mu}{d\tau} = 0. \tag{14.40}$$

In Einstein's theory, the same equation holds with D_μ replaced by the Riemann-covariant derivative $\bar{D}_\mu$. As a consequence of the curvatur, the spinning top shows a so-called *geodetic precession*. If spacetime has also torsion, Eq. (14.40) predicts an *autoparallel precession*.

14.1.4 Special Properties of Gradient Torsion

Consider a special torsion tensor which consists of an antisymmetric combination of gradients of a scalar field $\theta(x)$:

$$S_{\mu\nu}{}^\lambda(x) = \frac{1}{2} \left[\delta_\nu{}^\lambda \partial_\mu \theta(x) - \delta_\mu{}^\lambda \partial_\nu \theta(x) \right]. \tag{14.41}$$

This expression is called *gradient torsion* [3]. If spacetime possesses only gradient torsion, its effect upon the equations of motion of a point particle can be simulated in a purely Riemannian spacetime, provided that the action is modified by the scalar field $\theta(x)$ in a peculiar way to be specified below in Eq. (14.44). By extremizing the modified action in the usual way, the resulting equation of motion coincides with the autoparallel equation derived in the initial spacetime with torsion from the modified variational principle in Eqs. (14.8):

$$\ddot{x}^\mu + \bar{\Gamma}^\mu{}_{\kappa\lambda} \dot{x}^\kappa \dot{x}^\lambda - 2S^\mu{}_{\kappa\lambda} \dot{x}^\kappa \dot{x}^\lambda = 0. \tag{14.42}$$

For the pure gradient torsion (14.41), this becomes

$$\ddot{x}^\lambda(s) + \bar{\Gamma}_{\mu\nu}{}^\lambda(x(s)) \dot{x}^\mu(s) \dot{x}^\nu(s) = -\dot{\theta}(x(s)) \dot{x}^\lambda(s) + g^{\lambda\kappa}(x(s)) \partial_\kappa \theta(x(s)), \tag{14.43}$$

with the extra terms on the right-hand side reflecting the closure failure of parallel-ograms caused by the torsion.

The same trajectory is found from the following alternative action in a purely Riemannian spacetime

$$\overset{m}{\mathcal{A}} = -mc \int_{\sigma_a}^{\sigma_b} d\sigma \, e^{\theta(x)} \left[g_{\mu\nu}(x(\sigma)) \dot{x}^\mu(\sigma) \dot{x}^\nu(\sigma) \right]^{\frac{1}{2}}. \tag{14.44}$$

The extra factor $e^{\theta(x)}$ has precisely the same effect in a Riemannian spacetime as the gradient torsion (14.41) in a Riemann-Cartan spacetime. Indeed, the extremum of this action can be derived from the geodesic trajectory without the σ-field by introducing, for a moment, an auxiliary metric

$$\tilde{g}_{\mu\nu}(x) \equiv e^{2\theta(x)} \equiv g_{\mu\nu}(x). \tag{14.45}$$

The invariant line element remains, of course,

$$ds = \left[g_{\mu\nu}(x(\sigma)) \dot{x}^\mu(\sigma) \dot{x}^\nu(\sigma) \right]^{\frac{1}{2}} = e^{-\theta(x)} \left[\tilde{g}_{\mu\nu}(x(\sigma)) \dot{x}^\mu(\sigma) \dot{x}^\nu(\sigma) \right]^{\frac{1}{2}} = e^{-\theta(x)} d\tilde{s}. \tag{14.46}$$

By varying the action as in Eqs. (11.13)–(11.18), we obtain the modified equation of motion (11.20):

$$\tilde{g}_{\lambda\nu} \frac{d^2 x^\nu(\sigma)}{d\tilde{\sigma}^2} + \left(\partial_\mu \tilde{g}_{\lambda\nu} - \frac{1}{2} \partial_\lambda \tilde{g}_{\mu\nu} \right) \frac{dx^\mu(\sigma)}{d\tilde{\sigma}} \frac{dx^\nu(\sigma)}{d\tilde{\sigma}} = 0. \tag{14.47}$$

Inserting (14.45) and (14.46), this becomes

$$g_{\lambda\nu} \left(\frac{d^2 x^\nu(\sigma)}{d\sigma^2} - \dot{\theta} \frac{dx^\nu(\sigma)}{d\sigma} \right) + \left(\partial_\mu g_{\lambda\nu} - \frac{1}{2} \partial_\lambda g_{\mu\nu} \right) \frac{dx^\mu(\sigma)}{d\sigma} \frac{dx^\nu(\sigma)}{d\sigma}$$

$$+ 2\dot{\theta}(x) \frac{dx^\nu(\sigma)}{d\sigma} - \partial_\lambda \theta(x) g_{\mu\nu} \frac{dx^\mu(\sigma)}{d\sigma} \frac{dx^\nu(\sigma)}{d\sigma} = 0, \tag{14.48}$$

which coincides with the autoparallel trajectory (14.43).

14.2   Autoparallel Trajectories from Embedding

There exists another way of deriving autoparallel trajectories. Instead of using multivalued mappings to carry physical laws from flat spacetime to spacetimes with curvature and torsion, we may use the embedding procedure of Section 13.2.

14.2.1   Special Role of Autoparallels

Let us first remark that apart from extremizing a length between two fixed endpoints, geodesics in a Riemann spacetime can be obtained by embedding the Riemann spacetime in a flat spacetime of a higher dimension. This is done by imposing certain constraints on the coordinates spanning the flat spacetime. The points on the constraint hypersurface constitute the embedded Riemann spacetime. Straight lines in the flat spacetime, which are geodesic and autoparallel, determine a free motion in that spacetime. They become geodesics when the motion is restricted to the constraint hypersurface. This restriction is done in the conventional way by adding the equations of constraint to the equations of motion. When the constraining force is removed, geodesic trajectories turn into straight lines in the embedding spacetime.

For curved spacetime with torsion the embedding procedure was described in Chapter 13. The consequences for the trajectories were worked out in Ref. [4]. It turns out that, also from this point of view, autoparallel curves are specially favored geometric curves in the embedded spacetime. They are the images of straight lines in the embedding spacetime.

14.2.2　Gauss Principle of Least Constraint

There is also a classical mechanical argument favoring autoparallel over geodesic motion. This is intrinsically linked with the concept of inertia. Inertia favors trajectories whose acceleration deviates *minimally* from the acceleration of the corresponding unconstrained motion. This property can be formulated mathematically by means of Gauss' principle of least constraint [5, 6].

Consider a Lagrangian system in the spacetime x^A with a Lagrangian $L = L(x^A, \dot{x}^A) = L(x^A, u^A)$. At each moment of time, a state of the system can be labeled by a point in phase space $(x^A(\tau), u^A(\tau))$. Let $H_{AB}(x^A(\tau), u^A(\tau)) \equiv \partial^2 L/\partial u^A \partial u^B$ be the *Hessian matrix* of the system. Let $x_1^A(\tau)$ and $x_2^A(\tau)$ be two slightly different paths. Gauss has defined a deviation function for the two paths:

$$G = \frac{1}{2}\left(\dot{v}_1^A - \dot{v}_2^A\right) H_{AB} \left(\dot{v}_1^B - \dot{v}_2^B\right) , \qquad (14.49)$$

where a dot denotes the derivative with respect to τ. It measures the deviation of two possible motions from one another [5, 6].

Now, let the motion in x^A-spacetime be subject to constraints. All paths $x^A(\tau)$ allowed by the constraints are called *conceivable* motions. A path $\bar{x}^A(\tau)$ is called *released* if it satisfies the Euler-Lagrange equations without constraint. Gauss' principle of least constraint says that the deviation of conceivable motion from a released motion has a stationary value at the physical orbit. The released motion in the embedding spacetime x^A is a free motion with zero acceleration $\ddot{x}^A = \dot{u}^A = 0$, i.e., it runs along straight lines. The principle says that the physical orbit wants to be as close as possible to a straight line.

Calculating the accelerations $\dot{u}^A$ of the conceivable motions with the help of (13.20), we find

$$\dot{u}^A = e^A{}_\mu \dot{u}^\mu + \partial_\nu e^A{}_\mu u^\mu u^\nu . \qquad (14.50)$$

Recalling Eqs. (13.19), (11.85), and (11.118), we may write this also as

$$\dot{u}^A = e^A{}_\mu \frac{D}{d\tau} u^\mu. \qquad (14.51)$$

Since H_{AB} has only constant diagonal elements equal to ± 1 for the flat spacetime, Gauss' deviation function (14.49) assumes the form

$$G = \frac{1}{2}\left[\dot{u}^A\right]^2 = \frac{1}{2}\left[\frac{Du^\mu}{d\tau}\right]^2 , \qquad (14.52)$$

where an infinitesimal factor $d\tau^2$ has been removed. This function has a minimum at $G = 0$ at paths satisfying the equation of motion

$$\frac{Du^\mu}{d\tau} = 0 . \qquad (14.53)$$

This is once more the equation for an autoparallel trajectory.

Another derivation of autoparallel trajectories rests on the d'Alembert-Lagrange principle [5, 6]. In theoretical mechanics, one defines a *Lagrange derivative*

$$[L]_A \equiv \frac{d}{d\tau}\frac{\partial L}{\partial u^A} - \frac{\partial L}{\partial x^A}. \tag{14.54}$$

The d'Alembert-Lagrange principle asserts that motion of a system with the Lagrangian L proceeds such that

$$u^A [L]_A = 0 \tag{14.55}$$

for all velocities allowed by the constraints. Taking the free Lagrangian $L = u^A u_A/2$ with $[L]_A = \dot{u}_A$, and recalling (14.51), we find that the autoparallel equation (14.53) satisfies Eq. (14.55).

Finally we point out that the motion of a *holonomic* system is completely determined by the restriction of the Lagrangian to the constraining surface [5]. Thus, holonomic constrained systems are indistinguishable from ordinary unconstrained Lagrangian systems. This is not true for nonholonomic systems, meaning that the Euler-Lagrange equations for the Lagrangian restricted on the constraining surface do not coincide with the original equations for the constrained motion. This difficulty prevents us from applying a conventional Hamiltonian formalism to the autoparallel motion, and subjecting it to a canonical quantization. In other words, Dirac's method of quantizing constrained systems [7] is inapplicable to nonholonomic systems since these do not follow the conventional Lagrange formalism [5].

14.3   Maxwell-Lorentz Orbits as Autoparallel Trajectories

It is rather straightforward to set up Maxwell-Lorentz equations for the motion of a charged particle in curved spacetime. We rewrite the flat spacetime equation of motion (1.170) as

$$\ddot{x}^a(\tau) = \frac{e}{c}F^a{}_b(x(\tau))\dot{x}^b(\tau), \tag{14.56}$$

and subject this to a multivalued mapping. This adds the geometric force of Eq. (14.7), leading to

$$\ddot{x}^\lambda(\tau) + \bar{\Gamma}_{\mu\nu}{}^\lambda(x(\tau)\dot{x}^\mu(\tau)\dot{x}^\nu(\tau) = \frac{e}{mc}F^\lambda{}_\kappa(x(\tau))\,\dot{x}^\kappa(\tau). \tag{14.57}$$

It is now interesting to observe that this equation of motion may be viewed as an autoparallel motion in an affine geometry with torsion. Torsion is created only along the orbit of the particle according to the equation [8]

$$S_{\mu\nu}{}^\lambda(x(\tau)) = \frac{e}{mc}F_{\mu\nu}(x(\tau))\dot{x}^\lambda(\tau). \tag{14.58}$$

Indeed, if we insert this torsion into the autoparallel equation in the form (14.8), we obtain the Maxwell-Lorentz equation (14.57) in curved spacetime.

Note that this type of torsion does not propagate into spacetime.

14.4 Bargmann-Michel-Telegdi Equation from Torsion

Interestingly enough, also the spin precession equation (1.307) can be understood as a purely geometric equation in a spacetime with torsion. If we transform the flat spacetime equation (1.307) to curved spacetime, it becomes

$$\frac{\bar{D}}{d\tau} S^\mu = \frac{e}{2mc} \left[g F^{\mu\nu} S_\nu + \frac{g-2}{m^2 c^2} p^\mu S_\lambda F^{\lambda\kappa} p_\kappa \right] = 0. \tag{14.59}$$

For a classical particle, which has $g = 1$, this equation is the same as for a spin vector undergoing a parallel transport along the trajectory $q^\mu(\tau)$ according to the law (14.40). Decomposing the covariant derivative into a Riemannian part and a contribution of torsion according to Eq. (11.120), we find

$$\frac{DS_\mu}{d\tau} = \frac{\bar{D}S_\mu}{d\tau} + S^\mu{}_{\nu\lambda} S^\nu \dot{x}^\lambda - S_{\nu\lambda}{}^\mu S^\nu \dot{x}^\lambda + S_\lambda{}^\mu{}_\nu S^\nu \dot{x}^\lambda. \tag{14.60}$$

Inserting here the torsion (14.58) yields

$$\frac{DS_\mu}{d\tau} = \frac{\bar{D}S_\mu}{d\tau} + \frac{e}{mc^3} \left(F^\mu{}_\nu \dot{x}_\lambda S^\nu \dot{x}^\lambda - F_{\nu\lambda} \dot{x}^\mu S^\nu \dot{x}^\lambda + F_\lambda{}^\mu \dot{x}_\nu S^\nu \dot{x}^\lambda \right). \tag{14.61}$$

Recalling that $\dot{x}^\lambda \dot{x}_\lambda = c^2$ [see (1.151)], and that the spin vector has the transversality property (1.288), the last term vanishes, we arrive at

$$\frac{DS_\mu}{d\tau} = \frac{\bar{D}S_\mu}{d\tau} + \frac{e}{mc} \left(F^\mu{}_\nu S^\nu - \frac{1}{c^2} \dot{x}^\mu S^\nu F_{\nu\lambda} \dot{x}^\lambda \right), \tag{14.62}$$

which is indeed the same as (14.59) for $g = 1$.

Notes and References

[1] H. Kleinert and A. Pelster, Gen. Rel. Grav. **31**, 1439 (1999) (gr-qc/9605028); H. Kleinert, Mod. Phys. Lett. **A 4**, 2329 (1989) (kl/199), where kl is short for the www address http://www.physik.fu-berlin.de/~kleinert; H. Kleinert, *Quantum Equivalence Principle for Path Integrals in Spaces with Curvature and Torsion*, Lecture at the XXVth International Symposium Ahrenshoop on *Elementary Particles* held in Gosen/Germany, CERN report 1991, ed. H. J. Kaiser (quant-ph/9511020); H. Kleinert, *Quantum Equivalence Principle*, Lecture presented at the Summer School *Functional Integration: Basics and Applications* in Cargèse/France (1996) (kl/199).

[2] See the discussion in Chapter 10 of the textbook H. Kleinert, *Path Integrals in Quantum Mechanics, Statistics, Polymer Physics, and Financial Markets*, 4th ed., World Scientific, Singapore, 2006 (kl/b5/psfiles/pthic10.pdf).

[3] S. Hojman, M. Rosenbaum, and M.P. Ryan, Phys. Rev. D **19**, 430 (1979).

[4] H. Kleinert and S.V. Shabanov, *Spaces with Torsion from Embedding, and the Special Role of Autoparallel Trajectories*, Phys. Lett B **428** , 315 (1998) (quant-ph/9503004) (**kl/259**).

[5] V.I. Arnold, V.V. Koslov and A.I. Neishtadt, in: *Encyclopedia of Mathematical Sciences, Dynamical Systems III, Mathematical Aspects of Classical and Celestial Mechanics*, Springer, Berlin, 1988;
L.A. Pars, *A Treatise on Analytical Dynamics*, Heinemann, London, 1965;
G. Hamel, *Theoretische Mechanik – Eine einheitliche Einführung in die gesamte Mechanik*, Springer Series in Mathematics, Vol. 57, Springer, Berlin, 1978.

[6] L.S. Polak (ed.), *Variational principles of mechanics. Collection of papers*, Fizmatgiz, Moscow, 1959.

[7] P.A.M. Dirac, *Lectures on Quantum Mechanics*, Yeshiva University Press, NY, 1964.

[8] H.I. Ringermacher, Class. Quant. Grav. **11**, 2383 (1994).

15

Field Equations of Gravitation

In the previous chapter, we derived the equations of motion for a particle subject to a gravitational field. These look precisely the same as those of a particle in Minkowski spacetime expressed in curvilinear coordinates. All information on the gravitational field is contained in certain properties of the metric. We may now ask the question how to find the metric caused by a gravitational massive object. For this, the ten components of the metric tensor $g_{\mu\nu}(x)$ have to be considered as dynamical variables and we need an action principle to determine them [1, 2, 3].

15.1 Invariant Action

The equation of motion for $g_{\mu\nu}(x)$ must be independent of the general coordinates employed. This is guaranteed if the action is invariant under Einstein transformations $x^\mu \to x'^{\mu'}(x^\mu)$. The coordinate increments transform like

$$dx^\mu \quad \to \quad dx'^{\mu'} = \alpha^{\mu'}{}_\mu(x)dx^\mu, \quad \alpha^{\mu'}{}_\mu(x) = \partial x'^{\mu'}/\partial x^\nu. \tag{15.1}$$

We want to set up a *local action* for the gravitational field. According to the definition in Subsection 2.3.1, it must be an integral over a Lagrangian density

$$\mathcal{A} = \int d^4x\, \mathcal{L}(x). \tag{15.2}$$

The Lagrangian density $\mathcal{L}(x)$ can only depend on the metric and its first derivatives $\partial_\lambda g_{\mu\nu}(x)$, modulo integrations by parts.

Under the coordinate transformations (15.1), the volume element transforms as

$$d^4x \to d^4x' = d^4x \det\alpha. \tag{15.3}$$

The simplest Lagrangian density $\mathcal{L}(x)$ which leaves the action (15.2) invariant can be formed from the determinant of the metric

$$g = \det(g_{\mu\nu}). \tag{15.4}$$

Since the metric changes under (15.1) to $g'_{\mu'\nu'}(x') = g_{\mu\nu}(\alpha^{-1})^\mu{}_{\mu'}(\alpha^{-1})^\nu{}_{\nu'}$, we see that

$$g \to g' = g \det^{-2}\alpha, \tag{15.5}$$

393

so that the integral

$$\mathcal{A}_\Lambda = \int d^4x\, \mathcal{L}_\Lambda(x) \equiv \frac{\Lambda}{\kappa} \int d^4x \sqrt{-g} \tag{15.6}$$

is invariant under coordinate transformations (15.1). However, this expression cannot yet serve as an action for gravity since it does not depend on the derivatives of the metric $g_{\mu\nu}(x)$ and is therefore unable to yield equations of motion. In order to allow gravity to propagate through spacetime, we must find a scalar Lagrangian density $\mathcal{L}$ containing $g_{\mu\nu}$ and $\partial_\lambda g_{\mu\nu}$ and an action of the form

$$\mathcal{A} = \int d^4x \sqrt{-g}\mathcal{L}(g, \partial g). \tag{15.7}$$

The only fundamental scalar quantity which occurred in the previous geometric analysis and which contains the derivatives $\partial_\lambda g_{\mu\nu}$ is R, the scalar curvature. Therefore, Hilbert and Einstein postulated the following gravitational field action

$$\overset{\text{f}}{\mathcal{A}} = -\frac{1}{2\kappa} \int d^4x \sqrt{-g}\, \bar{R}. \tag{15.8}$$

Here κ is a constant related to Newton's gravitational coupling $G_{\mathrm{N}} \approx 6.673 \cdot 10^{-8}$ cm^3 g^{-1} s^{-2} of Eq. (1.3) by

$$\frac{1}{\kappa} = \frac{c^3}{8\pi G_{\mathrm{N}}}. \tag{15.9}$$

It can be expressed in terms of the Planck length (12.42) as

$$\frac{1}{\kappa} = \frac{\hbar}{8\pi l_P^2}. \tag{15.10}$$

From the fundamental point of view, the Einstein-Hilbert action (15.8) has a problem which has so far not been solved. It does not allow for a quantization of gravity. Attempts to quantize the theory runs into severe difficulties at very short distances of the order of the Planck length l_{P}. There it develops infinities which cannot be absorbed in the coupling constant κ. This property is called *nonrenormalizability* of gravity. Such a theory makes only sense as a classical effective theory.

From the practical point of view the quantization of gravity is irrelevant. At present it is unimaginable that such short length scales can ever be explored experimentally. In addition, it is quite possible that the quantum of gravitational waves, the *graviton*, is undetectable in principle during the lifetime of the universe [4]. Thus one can lead a perfectly comfortable life as a theoretical physicist without ever feeling the need to quantize gravity [4]. It must be said, however, that part of the theoretical physics community feels the need to construct a *Theory of Everything* which explains all physics down to any small distance. They assume that the presently known principles preclude the discovery of completely new properties of matter at the Planck scale. The author does not share this view and believes that nature will, fortunately, keep surprising us forever, and experimentalists will find phenomena which no theorist can dream of at present.

If one insists, for reasons of theoretical satisfaction, on making gravity a quantizable theory one may simply add to the Einstein-Hilbert action (15.8) an additional invariant term quadratic in the curvature tensor of the general form

$$\overset{f}{\mathcal{A}} = -\frac{1}{2} \int d^4x \sqrt{-g} \left(\frac{1}{\kappa_{2,1}} \bar{R}^2 + \frac{1}{\kappa_{2,1}} \bar{R}_{\mu\nu} \bar{R}^{\mu\nu} + \frac{1}{\kappa_{2,3}} \bar{R}_{\mu\nu\lambda\kappa} \bar{R}^{\mu\nu\lambda\kappa} \right), \qquad (15.11)$$

where the coupling constants $\kappa_{2,i}$ $(i = 1, 2, 3)$ are dimensionless. This theory is renormalizable, i.e., all infinities can be absorbed in the coupling constants. And it remains meaningful down to very small distances of the order of the Planck length. There it possesses unphysical properties, such as states with negative norm. These should not bother us, however, since the physics at such short distances will remain unknown for many more years. It must be recalled that also quantum electrodynamics, the most accurate quantum field theory so far, which is perfectly renormalizable, has unphysical properties at very short distances. These are the famous Landau ghosts, which can never be detected since long before they can show up, quantum electrodynamics receives much larger corrections from strong interactions which are not contained in the action of quantum electrodynamics. It is a gratifying feature of renormalizable theories that they make predictions which do not depend on the physics of unexplored short distances.

Thus we may restrict our attention to the classical theory of gravity implied by the Einstein-Hilbert action (15.8). For a system consisting of a set of mass points $m_1, \ldots, m_N$, we add the particle action (11.2) and obtain a total action

$$\mathcal{A} = \overset{f}{\mathcal{A}} - \sum_{n=1}^{N} m_n c \int ds_n \equiv \overset{f}{\mathcal{A}} + \overset{m}{\mathcal{A}}. \qquad (15.12)$$

In the following formulas it will be convenient to set $\kappa = 1$ since κ can always be reintroduced as a relative factor between field and matter parts in all field equations to be derived.

Variation of the particle paths $x_n(s_n)$ at fixed $g_{\mu\nu}(x)$ gives the equations of motion of a point particle in an external gravitational field as discussed in the beginning. In addition, the action (15.12) permits to find out which gravitational field is generated by the presence of these points. They are obtained from the variational equation

$$\frac{\delta \mathcal{A}}{\delta g_{\mu\nu}(x)} = 0. \qquad (15.13)$$

There are 10 independent components of $g_{\mu\nu}$. Four of them are unphysical, representing merely reparametrization degrees of freedom.

Equations (15.13) are not sufficient to determine the geometry of spacetime. The curvature tensor $R_{\mu\nu\lambda}{}^\kappa$ also contains torsion tensors $S_{\mu\nu}{}^\lambda$ combined to a contortion tensor $K_{\mu\nu}{}^{\lambda}$. It has 24 independent components, which are determined by the equation of motion

$$\frac{\delta \mathcal{A}}{\delta K_{\mu\nu}{}^{\lambda}(x)} = 0. \tag{15.14}$$

Einstein avoided this problem by considering only symmetric (Riemannian) space-times from the outset. For spinning matter, however, this may not be sufficient, and a determination of torsion fields from the spin densities may be necessary for a complete dynamical theory.

15.2   Energy-Momentum Tensor and Spin Density

It is useful to study separately the derivatives of the different pieces of the action with respect to $g^{\mu\nu}$ and $K^{\mu\nu\lambda}$. In view of the physical interpretations to be given later we introduce

$$\frac{\delta \overset{m}{\mathcal{A}}}{\delta g_{\mu\nu}} \equiv -\frac{1}{2}\sqrt{-g}\, \overset{m}{T}{}^{\mu\nu}, \tag{15.15}$$

$$\frac{\delta \overset{f}{\mathcal{A}}}{\delta g_{\mu\nu}} \equiv -\frac{1}{2}\sqrt{-g}\, \overset{f}{T}{}^{\mu\nu}, \tag{15.16}$$

respectively, as the *symmetric energy-momentum tensors* of matter and field, and

$$\frac{\delta \overset{m}{\mathcal{A}}}{\delta K_{\mu\nu}{}^{\lambda}} \equiv -\frac{1}{2}\sqrt{-g}\, \overset{m}{\Sigma}{}^{\nu}{}_{\lambda}{}^{,\mu}, \tag{15.17}$$

$$\frac{\delta \overset{f}{\mathcal{A}}}{\delta K_{\mu\nu}{}^{\lambda}} \equiv -\frac{1}{2}\sqrt{-g}\, \overset{f}{\Sigma}{}^{\nu}{}_{\lambda}{}^{,\mu}, \tag{15.18}$$

as the *spin current density* of matter and field, respectively.

We have remarked before that the identity (11A.24) implies a change of sign if we calculate the energy-momentum tensors from a variation $\delta g^{\mu\nu}$ rather than $\delta g^{\mu\nu}$ so that Eqs. (15.15) and (15.16) go over into

$$\frac{\delta \overset{m}{\mathcal{A}}}{\delta g^{\mu\nu}} \equiv \frac{1}{2}\sqrt{-g}\, \overset{m}{T}{}_{\mu\nu}, \tag{15.19}$$

$$\frac{\delta \overset{f}{\mathcal{A}}}{\delta g^{\mu\nu}} \equiv \frac{1}{2}\sqrt{-g}\, \overset{f}{T}{}_{\mu\nu}. \tag{15.20}$$

Let us calculate these quantities for a point particle. For a specific world line $x^{\mu}(\sigma)$ parameterized by an arbitrary timelike variable σ, the action reads [recall (11.10), (11.11)]

$$\begin{aligned} \overset{m}{\mathcal{A}} &= -mc\int ds = -mc^2\int d\sigma\, \sqrt{g_{\mu\nu}(x(\sigma))\dot{x}^{\mu}(\sigma)\dot{x}^{\nu}(\sigma)} \\ &= -mc\sqrt{-g}\int d\sigma\int d^4x\sqrt{-g}\,\sqrt{g_{\mu\nu}(x)\dot{x}^{\mu}(\sigma)\dot{x}^{\nu}(\sigma)}\,\delta^{(4)}(x-x(\sigma)). \end{aligned} \tag{15.21}$$

Variation with respect to $g_{\mu\nu}(x)$ and $K_{\mu\nu}{}^{\lambda}(x)$ gives

$$\frac{\delta \overset{m}{\mathcal{A}}}{\delta g_{\mu\nu}(x)} \equiv -\frac{1}{2}\sqrt{-g}\,mc\int d\sigma\,\frac{1}{\sqrt{g_{\mu\nu}(x(\sigma))\dot{x}^{\mu}(\sigma)\dot{x}^{\nu}(\sigma)}}\dot{x}^{\mu}(\sigma)\dot{x}^{\nu}(\sigma)\delta^{(4)}(x-x(\sigma)) \quad (15.22)$$

$$= -\frac{1}{2}\sqrt{-g}\,m\int d\tau\,\dot{x}^{\mu}(\tau)\dot{x}^{\nu}(\tau)\delta^{(4)}(x-x(\tau)),$$

where $\tau = s/c$ is the proper time (1.141). The functional derivative with respect to $K_{\mu\nu}{}^{\lambda}(x)$ vanishes identically:

$$\frac{\delta \overset{m}{\mathcal{A}}}{\delta K_{\mu\nu}{}^{\lambda}(x)} \equiv 0. \quad (15.23)$$

Thus we identify energy-momentum tensor and spin current densities

$$\overset{m}{T}{}^{\mu\nu}(x) \equiv m\int d\tau\,\dot{x}^{\mu}(\tau)\dot{x}^{\nu}(\tau)\delta^{(4)}(x-x(\tau)), \quad (15.24)$$

$$\overset{f}{\Sigma}{}^{\nu,\mu}{}_{\lambda}(x) \equiv 0. \quad (15.25)$$

We now determine these quantities for the gravitational field in the Einstein-Hilbert action (15.8). First we perform the variation of $\sqrt{-g}$ with respect to $\delta g_{\mu\nu}$. We vary

$$\delta\sqrt{-g} = -\frac{1}{2\sqrt{-g}}\delta g, \quad (15.26)$$

and use Eq. (11A.25) to express this as

$$\delta\sqrt{-g} = \frac{1}{2}\sqrt{-g}\,g^{\mu\nu}\delta g_{\mu\nu} = -\frac{1}{2}\sqrt{-g}\,g_{\mu\nu}\delta g^{\mu\nu}. \quad (15.27)$$

We now vary the action (15.8), rewritten in the form

$$\overset{f}{\mathcal{A}} = -\frac{1}{2}\int d^4x\sqrt{-g}\,g^{\mu\nu}R_{\mu\nu}, \quad (15.28)$$

and find

$$\delta\overset{f}{\mathcal{A}} = -\frac{1}{2}\int d^4x\sqrt{-g}\left\{-\frac{1}{2}g_{\mu\nu}\delta g^{\mu\nu}R + \delta g^{\mu\nu}R_{\mu\nu} + g^{\mu\nu}\delta R_{\mu\nu}\right\}$$

$$= -\frac{1}{2}\int d^4x\sqrt{-g}\left[\delta g^{\mu\nu}(R_{\mu\nu} - \frac{1}{2}g_{\mu\nu}R) + g^{\mu\nu}\delta R_{\mu\nu}\right]. \quad (15.29)$$

The factor accompanying $\delta g^{\mu\nu}$ is known as the *Einstein tensor*

$$G_{\mu\nu} \equiv R_{\mu\nu} - \frac{1}{2}g_{\mu\nu}R. \quad (15.30)$$

Note that this tensor is symmetric only in symmetric spacetimes. The variation in $\delta g^{\mu\nu}$, however, picks out only the symmetrized part of it.

Consider now the variation of the Ricci tensor in (15.29)

$$\delta R_{\mu\nu} = \partial_\kappa \delta \Gamma_{\mu\nu}{}^\kappa - \partial_\mu \delta \Gamma_{\kappa\nu}{}^\kappa - \delta \Gamma_{\kappa\nu}{}^\tau \Gamma_{\mu\tau}{}^\kappa - \Gamma_{\kappa\nu}{}^\tau \delta \Gamma_{\mu\tau}{}^\kappa + \delta \Gamma_{\mu\nu}{}^\tau \Gamma_{\kappa\tau}{}^\kappa + \Gamma_{\mu\nu}{}^\tau \delta \Gamma_{\kappa\tau}{}^\kappa .$$
(15.31)

The left-hand side is a tensor. Let us express also the right-hand side in a covariant way. We know from the transformation law (11.104) that the affine connection $\Gamma_{\mu\nu}{}^\kappa$ is not a tensor. Its variation $\delta \Gamma_{\mu\nu}{}^\kappa$, however, is a tensor. [1] This follows directly from the transformation law (11.104), whose last term $\partial_\mu \partial_\nu \xi^\kappa$ disappears in $\delta \Gamma_{\mu\nu}{}^\kappa$ since it is the same for $\Gamma_{\mu\nu}{}^\kappa$ and $\Gamma_{\mu\nu}{}^\kappa + \delta \Gamma_{\mu\nu}{}^\kappa$. For this reason we may rewrite (15.31) covariantly as

$$\delta R_{\mu\nu} = D_\kappa \delta \Gamma_{\mu\nu}{}^\kappa - D_\mu \delta \Gamma_{\kappa\nu}{}^\kappa + 2 S_{\kappa\mu}{}^\tau \delta \Gamma_{\tau\nu}{}^\kappa .$$
(15.32)

Indeed, by working out the covariant derivatives we find

$$\delta R_{\mu\nu} = -\partial_\kappa \delta \Gamma_{\mu\nu}{}^\kappa - \partial_\mu \delta \Gamma_{\kappa\nu}{}^\kappa - \Gamma_{\kappa\mu}{}^\tau \delta \Gamma_{\tau\nu}{}^\kappa - \Gamma_{\kappa\nu}{}^\tau \delta \Gamma_{\mu\tau}{}^\kappa$$
$$+ \Gamma_{\kappa\tau}{}^\kappa \delta \Gamma_{\mu\nu}{}^\tau + \Gamma_{\mu\kappa}{}^\tau \delta \Gamma_{\tau\nu}{}^\kappa + \Gamma_{\mu\nu}{}^\tau \delta \Gamma_{\kappa\tau}{}^\kappa - \Gamma_{\mu\tau}{}^\kappa \delta \Gamma_{\kappa\nu}{}^\tau + 2 S_{\kappa\mu}{}^\tau \delta \Gamma_{\tau\nu}{}^\kappa , \quad (15.33)$$

thus recovering (15.31). In symmetric spacetimes, the covariant relation (15.31) was first used by Palatini.

We now have to express $\delta R_{\mu\nu}$ in terms of $\delta g^{\mu\nu}$ and $\delta K_{\mu\nu}{}^\lambda$. It is useful to perform all operations underneath the integral in (15.29):

$$-\frac{1}{2} \int d^4x \sqrt{-g} g^{\mu\nu} \delta R_{\mu\nu} .$$
(15.34)

Due to the tensor nature of $\delta \Gamma_{\mu\nu}{}^\kappa$ we can take $g^{\mu\nu}$ through the covariant derivative and write (15.34) as

$$-\frac{1}{2} \int d^4x \sqrt{-g} \left(D_\kappa \delta \Gamma_\mu{}^{\mu\kappa} - D_\mu \delta \Gamma_\kappa{}^{\mu\kappa} + 2 S_\kappa{}^{\mu\tau} \delta \Gamma_{\tau\mu}{}^\kappa \right) .$$
(15.35)

The covariant derivatives can now be removed by a partial integration. In spacetime with torsion, partial integration has some particular features which requires a special discussion.

Take any tensors $U^{\mu...\nu}, V_{...\nu...}$ and consider an invariant volume integral

$$\int d^4x \sqrt{-g} \, U^{\mu...\nu...} D_\mu V_{...\nu...} .$$
(15.36)

A partial integration gives

$$- \int d^4x \left[(\partial_\mu \sqrt{-g} \, U^{\mu...\nu...}) V_{...\nu...} + \sum_i U^{\mu...\nu_i...} \Gamma_{\mu\nu_i}{}^{\lambda_i} V_{...\lambda_i...} \right] + \text{surface terms}, (15.37)$$

---

[1]Note that the tensor character holds only for independent variations of $\Gamma_{\mu\nu\lambda}$ at a fixed metric. This is in contrast to the nontensorial behavior of the difference $\Gamma'_{\mu\nu\kappa} - \Gamma_{\mu\nu\kappa}$, where $\Gamma'_{\mu\nu\lambda}$ is formed from the varied metric $g_{\mu\nu} + \delta g_{\mu\nu}$.

where the sum over i runs over all indices of $V_{\ldots\lambda_i\ldots}$, linking them via the affine connection with the corresponding indices of $U^{\mu\ldots\nu_i\ldots}$. Now we use the relation

$$\partial_\mu\sqrt{-g} = \sqrt{-g}\,\bar{\Gamma}_{\mu\kappa}{}^\kappa = \sqrt{-g}\,\Gamma_{\mu\kappa}{}^\kappa = \sqrt{-g}\left(2S_\mu + \Gamma_{\kappa\mu}{}^\kappa\right) \tag{15.38}$$

and (15.37) becomes

$$-\int d^4x\sqrt{-g}\left[(\partial_\mu U^{\mu\ldots\lambda_i\ldots} - \Gamma_{\kappa\mu}{}^\kappa U^{\mu\ldots\lambda_i\ldots} + \sum_i \Gamma_{\mu\nu_i}{}^{\lambda_i} U^{\mu\ldots\nu_i\ldots})V_{\ldots\lambda_i\ldots}\right.$$
$$\left. +2S_\mu \sum_i U^{\mu\ldots\lambda_i\ldots} V_{\ldots\lambda_i\ldots}\right] + \text{surface terms.} \tag{15.39}$$

The terms in parentheses can be collected to the covariant derivative of $U^{\mu\ldots\nu_i\ldots}$, such that we arrive at the rule of partial integration

$$\int d^4x\sqrt{-g}\,U^{\mu\ldots\nu\ldots}D_\mu V_{\ldots\nu\ldots} = -\int d^4x\sqrt{-g}\,D^*_\mu U^{\mu\ldots\nu\ldots}V_{\ldots\nu\ldots} + \text{surface terms,} \tag{15.40}$$

where D^*_μ is defined as

$$D^*_\mu \equiv D_\mu + 2S_\mu, \tag{15.41}$$

and we have abbreviated:

$$S_\kappa \equiv S_{\kappa\lambda}{}^\lambda, \quad S^\kappa \equiv S^\kappa{}_\lambda{}^\lambda. \tag{15.42}$$

It is easy to show that (15.40) holds also if the operators D_μ and D^*_μ are interchanged, i.e.,

$$\int d^4x\sqrt{-g}\,U^{\mu\ldots\nu\ldots}D^*_\mu V_{\ldots\nu\ldots} = -\int d^4x\sqrt{-g}\,D_\mu U^{\mu\ldots\nu\ldots}V_{\ldots\nu\ldots} + \text{surface terms.} \tag{15.43}$$

For the particular case that $V_{\ldots\nu\ldots}$ is equal to 1, the latter rule tells us that

$$\int d^4x\sqrt{-g}D_\mu U^\mu = -\int d^4x\sqrt{-g}\,2S_\mu U^\mu + \text{surface terms.} \tag{15.44}$$

This allows us to replace the covariant derivatives of the tensors $\delta\Gamma_\mu{}^{\mu\kappa}$ and $\delta\Gamma_\kappa{}^{\mu\kappa}$ in Eq. (15.35) by $-2S_\kappa$ and $-2S_\mu$, respectively, and we obtain

$$-\frac{1}{2}\int d^4x\sqrt{-g}g^{\mu\nu}\delta R_{\mu\nu} = -\frac{1}{2}\int d^4x\sqrt{-g}\left(-2S_\kappa\delta\Gamma^\nu{}_\nu{}^\kappa + 2S_\mu\delta\Gamma_\kappa{}^{\mu\kappa} + 2S_\kappa{}^{\nu\tau}\delta\Gamma_{\tau\nu}{}^\kappa\right). \tag{15.45}$$

The result can also be stated as follows:

$$-\frac{1}{2}\int d^4x\sqrt{-g}g^{\mu\nu}\delta R_{\mu\nu} = -\frac{1}{2}\int d^4x\sqrt{-g}S^\mu{}_\kappa{}^{,\tau}\delta\Gamma_{\tau\mu}{}^\kappa \tag{15.46}$$

where $S_{\mu\kappa}{}^{,\tau}$ is the following combination of torsion tensors:

$$\frac{1}{2}S_{\mu\kappa}{}^{,\tau} \equiv S_{\mu\kappa}{}^{\tau} + \delta_{\mu}{}^{\tau}S_{\kappa} - \delta_{\kappa}{}^{\tau}S_{\mu}. \qquad (15.47)$$

This tensor is referred to as the *Palatini tensor*. The relation can be inverted to

$$S_{\mu\nu\lambda} = \frac{1}{2}\left(S_{\mu\nu,\lambda} + \frac{1}{2}g_{\mu\lambda}S_{\nu\kappa}{}^{,\kappa} - \frac{1}{2}g_{\nu\lambda}S_{\mu\kappa}{}^{,\kappa}\right). \qquad (15.48)$$

We now proceed to express $\delta\Gamma_{\tau\mu}{}^{\kappa}$ in terms of $\delta g_{\mu\nu}$ and $\delta K_{\mu\nu\lambda}$. For this purpose we note that the varied metric $g_{\mu\rho} + \delta g_{\mu\rho}$ certainly satisfies the identity (11.95),

$$D_{\tau}{}^{\Gamma+\delta\Gamma}\left(g_{\mu\rho} + \delta g_{\mu\rho}\right) = 0, \qquad (15.49)$$

where $D^{\Gamma+\delta\Gamma}$ is the covariant derivative formed with the varied connection $\Gamma_{\mu\nu}{}^{\lambda} + \delta\Gamma_{\mu\nu}{}^{\lambda}$. For variations $\delta g_{\mu\rho}$ this implies

$$\overset{\Gamma}{D}_{\tau}\,\delta g_{\mu\rho} = \delta\Gamma_{\tau\mu\rho} + \delta\Gamma_{\tau\rho\mu} \qquad (15.50)$$

where we have introduced

$$\delta\Gamma_{\mu\tau\rho} \equiv g_{\rho\lambda}\,\delta\Gamma_{\mu\tau}{}^{\lambda}. \qquad (15.51)$$

This gives

$$\frac{1}{2}\left(\overset{\Gamma}{D}_{\tau}\,\delta g_{\mu\rho} + \overset{\Gamma}{D}_{\mu}\,\delta g_{\tau\rho} - \overset{\Gamma}{D}_{\rho}\,\delta g_{\tau\mu}\right) = \delta\Gamma_{\tau\mu\rho} - \delta S_{\tau\mu\rho} + \delta S_{\mu\rho\tau} - \delta S_{\rho\tau\mu}$$
$$= \delta\Gamma_{\tau\mu\rho} - \delta K_{\tau\mu\rho}, \qquad (15.52)$$

where

$$\delta S_{\tau\mu\rho} \equiv g_{\rho\lambda}\,\delta S_{\tau\mu}{}^{\lambda} \equiv g_{\rho\lambda}\frac{1}{2}\left(\Gamma_{\tau\mu}{}^{\lambda} - \Gamma_{\mu\tau}{}^{\lambda}\right) \qquad (15.53)$$

and

$$\delta K_{\tau\mu\rho} \equiv \delta S_{\tau\mu\rho} - \delta S_{\mu\mu\rho} - \delta S_{\mu\rho\tau} + \delta S_{\rho\tau\mu} \qquad (15.54)$$

are the results of a variation of $S_{\mu\nu}{}^{\lambda}$ at fixed $g_{\mu\nu}$. Note that even though $\bar{\Gamma}_{\mu\nu}{}^{\lambda} = \Gamma_{\mu\nu}{}^{\lambda} - K_{\mu\nu}{}^{\lambda}$, the left-hand side of (15.52) cannot be identified with $g_{\rho\kappa}\delta\bar{\Gamma}_{\tau\mu}{}^{\lambda}$ since $\delta K_{\mu\nu}{}^{\lambda}$ contains contribution from $\delta S_{\mu\nu}{}^{\lambda}$ at fixed $\delta g_{\mu\nu}$ and from $\delta g_{\mu\nu}$ at fixed $S_{\mu\nu}{}^{\lambda}$. The first term in (15.52) is, in fact, equal to $g_{\rho\kappa}\delta\bar{\Gamma}_{\tau\mu}{}^{\kappa} + \delta K_{\tau\mu}{}^{\kappa}|_{S_{\mu\nu}^{\lambda}=\text{fixed}}$. Using (15.52), we rewrite (15.46) as

$$-\frac{1}{2}\int d^4x\sqrt{-g}g^{\mu\nu}\delta R_{\mu\nu} = \qquad (15.55)$$
$$-\frac{1}{2}\int d^4x\sqrt{-g}S^{\mu\kappa,\tau}\left[\delta K_{\tau\mu,\kappa} + \frac{1}{2}\left(D_{\tau}\delta g_{\mu\kappa} + D_{\mu}\delta g_{\tau\kappa} - D_{\kappa}\delta g_{\tau\mu}\right)\right].$$

The first term shows that the Palatini tensor $S^\mu{}_\kappa{}^{,\tau}$ plays the role of the spin current of the gravitational field [recall the definition (15.18) up to a factor $1/\kappa$]

$$\overset{f}{\Sigma}{}^\mu{}_\kappa{}^{,\tau} = -\frac{1}{\kappa}S^\mu{}_\kappa{}^{,\tau}. \qquad (15.56)$$

The second term can now be partially integrated, leading to

$$\frac{1}{4}\int d^4x\sqrt{-g}\left\{D^*_\mu S^{\mu\rho,\epsilon}\delta g_{\mu\rho} + D^*_\mu S^{\mu\rho,\tau}\delta g_{\tau\rho} - D^*_q S^{\mu\rho,\tau}\delta g_{\mu\tau}\right\} + \text{surface term.} \qquad (15.57)$$

After relabeling the indices in (15.52), we arrive at the following variation of the action with respect to $\delta g_{\mu\nu}$, using the identity $\delta g^{\mu\nu}G_{\mu\nu} = -\delta g_{\mu\nu}G^{\mu\nu}$ following from (11A.24),

$$-\frac{1}{2}\int d^4x\sqrt{-g}\left[G^{\mu\nu} - \frac{1}{2}D^*_\lambda\left(S^{\mu\nu,\lambda} - S^{\nu\lambda,\mu} + S^{\lambda\mu,\nu}\right)\right]\delta g_{\mu\nu}, \qquad (15.58)$$

so that the complete energy-momentum tensor of the field reads

$$\overset{f}{T}{}^{\mu\nu} = -\frac{1}{\kappa}\left[G^{\mu\nu} - \frac{1}{2}D^*_\lambda\left(S^{\mu\nu,\lambda} - S^{\nu\lambda,\mu} + S^{\lambda\mu,\nu}\right)\right]. \qquad (15.59)$$

Actually, the variation $\delta g^{\mu\nu}$ can yield only the symmetrized part of $\overset{f}{T}_{\mu\nu}$. This specification is, however, unnecessary. We shall demonstrate later that the conservation law for the spin current density, to be derived in Eq. (18.60), makes $\overset{f}{T}{}^{\mu\nu}$ symmetric as it stands (even if $G^{\mu\nu}$ is not symmetric).

Thus we arrive at the Einstein-Cartan field equations

$$-\kappa\overset{f}{\Sigma}_{\mu\kappa}{}^{,\tau} = S_{\mu\kappa}{}^{,\tau} = \kappa\overset{m}{\Sigma}_{\mu\kappa}{}^{,\tau}, \qquad (15.60)$$

$$-\kappa\overset{f}{T}{}^{\mu\nu} = G^{\mu\nu} - \frac{1}{2}D^*_\lambda\left(S^{\mu\nu,\lambda} - S^{\nu\lambda,\mu} + S^{\lambda\mu,\nu}\right) = \kappa\overset{m}{T}{}^{\mu\nu}, \qquad (15.61)$$

which for a set of spinless point particles reduce to

$$S_{\mu\kappa}{}^{,\tau} = 0, \qquad (15.62)$$

$$G^{\mu\nu} = \kappa\overset{m}{T}{}^{\mu\nu}. \qquad (15.63)$$

15.3 Symmetric Energy-Momentum Tensor and Defect Density

In defect physics, the symmetric energy-momentum tensor obtained in (15.59) has a direct physical interpretation. In three Euclidean dimensions, the linearized version of (15.59) reads

$$-\kappa\overset{f}{T}{}^{ij} = G_{ij} - \frac{1}{2}\partial_\kappa\left(S_{ij,k} - S_{jk,i} - S_{ki,j}\right), \qquad (15.64)$$

with the spin density (15.47)

$$-\frac{1}{2}\kappa\,\overset{f}{\Sigma}_{ij,k}=\frac{1}{2}S_{ij,k}=S_{ijk}+\delta_{ik}S_j-\delta_{jk}S_i.\tag{15.65}$$

Let us insert the dislocation density according to

$$S^{ijk}=\frac{1}{2}\left(\partial_i\partial_j-\partial_j\partial_i\right)u_k=\frac{1}{2}\epsilon_{ij}\alpha_{lk}.\tag{15.66}$$

Then the spin density reads

$$S_{ij,k}=\epsilon_{ijl}\alpha_{lk}+\delta_{ik}\epsilon_{jpl}\alpha_{lp}-\delta_{jk}\epsilon_{ipl}\alpha_{lp}.\tag{15.67}$$

Since both sides are antisymmetric in ij, we can contract them with ϵ_{ijn},

$$\begin{aligned}\epsilon_{ijn}S_{ij,k}&=2\alpha_{nk}+\epsilon_{kjn}\epsilon_{jpl}\alpha_{lp}-\epsilon_{ikn}\epsilon_{ipl}\alpha_{lp}=2\alpha_{nk}-2\left(\delta_{kp}\delta_{nl}-\delta_{kl}\delta_{np}\right)\alpha_{lp}\\&=2\alpha_{kn},\end{aligned}\tag{15.68}$$

and see that $S_{ij,k}$ becomes simply

$$S_{ij,k}=\epsilon_{ijl}\alpha_{kl}.\tag{15.69}$$

Thus the spin density is equal to the dislocation density.

The spin density has a vanishing divergence

$$\partial_k S_{ij,k}=\epsilon_{ijl}\partial_k\alpha_{kl}=0.\tag{15.70}$$

In terms of the derivatives of the displacement field $u_i(x)$, the spin density reads

$$S_{ij,k}=\epsilon_{ijl}\epsilon_{kmn}\partial_m\partial_n u_l.\tag{15.71}$$

In this expression, the conservation law (15.70) is trivially fulfilled.

Let us now form the three combinations of ij,k appearing in (15.64)

$$\frac{1}{2}\left(S_{ij,k}-S_{jk,i}+S_{ki,j}\right)=\frac{1}{2}\left(\varepsilon_{ijl}\alpha_{kl}-\epsilon_{jkl}\alpha_{ij}+\epsilon_{kil}\alpha_{jl}\right).\tag{15.72}$$

By contracting the identity

$$\epsilon_{ijl}\delta_{km}+\epsilon_{jkl}\delta_{im}+\epsilon_{kil}\delta_{jm}=\epsilon_{ijk}\delta_{lm}\tag{15.73}$$

with α_{ml}, we find

$$\epsilon_{ijl}\alpha_{kl}+\epsilon_{jkkl}\alpha_{il}+\epsilon_{kil}\alpha_{jl}=\epsilon_{ijk}\alpha_{ll}\tag{15.74}$$

so that

$$\frac{1}{2}\left(S_{ij,k}-S_{jk,i}+S_{ki,j}\right)=-\epsilon_{jkl}\alpha_{il}+\frac{1}{2}\epsilon_{ijk}\alpha_{ll}.\tag{15.75}$$

The right-hand side is recognized to be

$$\epsilon_{jkl} K_{li} \tag{15.76}$$

where

$$K_{lj} = -\alpha_{jl} + \frac{1}{2}\delta_{lj} K_{kk}$$

is Nye's contortion tensor. With this notation, equation (15.64) becomes

$$-\kappa \overset{\text{f}}{T}_{ij} = G_{ij} - \epsilon_{jhl}\partial_n K_{li}. \tag{15.77}$$

Now we recall that the Einstein tensor G_{ij} for a metric $g_{ij} = \delta_{ij} + \partial_i u_j + \delta_j u_i$ coincides with the disclination density Θ_{ji}. But then, comparison with Eq. (12.41) shows that the total energy-momentum tensor multiplied by $-\kappa$ is the total defect density η_{ij}:

$$-\kappa \overset{\text{f}}{T}_{ij} = \eta_{ij} \tag{15.78}$$

Notes and References

[1] R. Utiyama, Phys. Rev. **101**, 1597 (1956).

[2] T.W.B. Kibble, J. Math. Phys. **2**, 212 (1961).

[3] H. Kleinert, *Gauge Fields in Condensed Matter*, Vol. II: *Stresses and Defects, Differential Geometry, Crystal Defects*, World Scientific, Singapore, 1989 (kl/b2), where kl is short for the www address (http://www.physik.fu-berlin.de/~kleinert/b2).

[4] Read Freeman Dyson's article *The World on a String*, New York Review of Books, **51**, 8 (2004), where he criticizes a book by Brian Greene entitled *The Fabric of the Cosmos: Space, Time, and the Texture of Reality*, Knopf, New York, 2004, and offers many critical remarks on the attempts to quantize gravity. Readable on the www at kl/papers/dyson2004.txt. See also the related paper
T. Rothman and S. Boughn, *Can Gravitons Be Detected?*, Found. Phys. **36**, 1801 (2006) (gr-qc/0601043).

16

Minimally Coupled Fields of Integer Spin

So far we have discussed the gravitational field interacting with classical relativistic massive point particles. If we want to include quantum effects, we must describe these particles by relativistic fields such as the scalar field in Section 2.3, or the Maxwell field in Section 2.4. These fields are then quantized, so that incoming negative-energy waves describe outgoing antiparticles (recall p. 55).

If we want to couple these fields to gravity, we follow the multivalued mapping principle of Chapter 14, according to which action in flat spacetime must simply be transformed to spacetimes with curvature and torsion by means of a multivalued coordinate transformation. The result is a minimal coupling of the gravitational field.

In this text we shall not discuss the quantum aspect of the relativistic fields, confining our attention to the coupling problem which can be discussed at the level of classical fields.

16.1 Scalar Fields in Riemann-Cartan Space

The action (2.25) of a charged scalar field in flat spacetime is most easily transformed to general metric-affine spacetime. The partial derivative ∂_a is equal, via Eq. (14.1), to

$$\partial_a = e_a{}^\mu(x)\partial_\mu, \tag{16.1}$$

and the volume element d^4x^a in flat spacetime becomes

$$d^4x^a = d^4x^\mu \, |\det e^a{}_\mu(x)|. \tag{16.2}$$

Since $e^a{}_\mu(x)$ is the square root of the metric $g_{\mu\nu}(x)$ [recall Eq. (11.38)], the determinants are also related by a square root, so that (16.2) implies the replacement rule for the flat-spacetime volume:

$$d^4x \to d^4x\sqrt{-g}. \tag{16.3}$$

Hence the action (2.25) is mapped into

$$\mathcal{A} = \int d^4x\sqrt{-g}\left[\hbar^2 e^{a\mu}(x)\partial_\mu\phi^*(x)e_a{}^\nu(x)\partial_\nu\phi(x) - M^2c^2\phi^*(x)\phi(x)\right]. \tag{16.4}$$

404

This expression cannot yet be used for field-theoretic calculations since the fields $e_a{}^\nu(x)$ are multivalued. However, we can use Eq. (11.42) to rewrite the action as

$$\mathcal{A} = \int d^4x \sqrt{-g}\, \left[\hbar^2 g^{\mu\nu}(x)\partial_\mu\phi^*(x)\partial_\nu\phi(x) - M^2 c^2\phi^*(x)\phi(x)\right]. \tag{16.5}$$

This expression contains only the single-valued metric tensor.

The equation of motion is derived most simply by applying an integration by parts to the gradient term. Ignoring the contribution from the boundaries we obtain

$$\mathcal{A} = \int d^4x \sqrt{-g}\, \left[-\hbar^2 \phi^*(x)\Delta\phi(x) - M^2 c^2\phi^*(x)\phi(x)\right], \tag{16.6}$$

where $\Delta = \sqrt{-g}^{-1}\partial_\mu\sqrt{-g}g^{\mu\nu}\partial_\nu$ is the Laplace-Beltrami differential operator (12.157). From the action (16.6) we obtain directly the equation of motion [compare (2.38)]:

$$\begin{aligned}
\frac{\delta\mathcal{A}}{\delta\phi^*(x)} &= \int d^4x'\,\sqrt{-g'}\left[-\hbar^2\delta^{(4)}(x'-x)\Delta'\phi(x') - m^2 c^2\delta^{(4)}(x'-x)\phi(x')\right] \\
&= (-\hbar^2\Delta - M^2 c^2)\phi(x) = 0.
\end{aligned} \tag{16.7}$$

This equation of motion contains an important prediction. There is no extra R−term in the wave equation, which would be allowed by Einstein's covariance principle. In many textbooks [1], the Klein-Gordon equation is therefore written as

$$(-\hbar^2\Delta - \xi\hbar^2 R - M^2 c^2)\phi(x), \tag{16.8}$$

with a parameter ξ for which several numbers have been proposed in the literature: $1/6$, $1/12$, $1/8$. The same R-term would of course appear in the nonrelativistic limit of (16.8). This limit is obtained by setting $\phi(x) \equiv e^{-iMc^2 t/\hbar}\psi(x)$ and letting the light velocity c go to infinity. Assuming that $g_{0i} = 0$ and choosing $g_{00} = 1$, this leads to the Schrödinger equation:

$$\left(-\frac{1}{2M}\hbar^2\Delta - \xi\hbar^2 R_d\right)\psi(x) = i\hbar\partial_t\psi(x), \tag{16.9}$$

where R_d is the curvature scalar of space in the $D = d + 1$-dimensional spacetime. On a sphere of radius r in D dimensions, we know from Eq. (13.10) that R_d is equal to $d(d-1)/r^2$.

The choice $\xi = (D-2)/4(D-1)$ makes the wave equation (16.8) for $M = 0$ conformally invariant in D spacetime dimensions [2], so that $\xi = 1/6$ is a preferred value of field theorists. When Bryce DeWitt set up a time-sliced path integral in curved space [3], he obtained the value $\xi = 1/6$ from his particular slicing assumptions. A slightly different slicing led to $\xi = 1/12$ [4]. In more recent work, DeWitt prefers the value $\xi = 1/8$ [5] which is motivated by a perturbative treatment of the path integral with dimensional regularization [6]. The value $\xi = 0$ in the action (16.4) was predicted on the basis of the multivalued mapping in Ref. [6].

So far there is no direct experimental confirmation of the $\xi = 0$-prediction. It is a challenge to experimentalists to measure ξ. The x-dependence of $R(x)$ stemming from the gravitational fields of celestial bodies is too small to have an effect on atomic spectra. At present, the only realistic possibility to measure ξ seem to require a study of the energy spectrum of electrons confined to a thin ellipsoidal surface. The presence of an R_d-term in the Schrödinger equation (16.9) would lead to an observable ξ-dependent distortion of the spectrum.

At this point we make an important observation. The Lagrangian density (16.5) does not contain the affine connection $\Gamma_{\mu\nu}{}^{\lambda}$. It does not require its presence since for scalar fields, the ordinary derivative $\partial_{\mu}\phi(x)$ is perfectly covariant. As a consequence, the eikonal approximation will lead to classical trajectories which do not couple to torsion. They will be geodesics, not autoparallels, as required by the action principle in Section 14.1. This clash of field theory and classical orbits can so far only be resolved in a dynamical way. We are forced to postulate that matter can create only antisymmetric torsion. Then autoparallels coincide with geodesics and the there is no contradiction between classical trajectories of scalar particles and the orbits coming from the eikonal approximation to scalar fields. This postulate can, in fact, be fulfilled quite naturally by a theory in which torsion coupled only to fundamental spin-$\frac{1}{2}$ fermions such as quarks and leptons, as will be explained in more detail in Section 20.3.1.

16.2   Electromagnetism in Riemann-Cartan Space

Let us go through the same procedure for the electromagnetic action (2.83). The volume element is again mapped according to the rule (16.3). The covariant curl is treated as follows. First we introduce vector fields transforming like the coordinate differentials dx_{μ}

$$A_{\mu}(x) = e^{a}{}_{\mu}(x)A_{a}(x), \qquad j_{\mu}(x) = e^{a}{}_{\mu}(x)j_{a}(x), \tag{16.10}$$

and rewrite the field strengths with the help of (11.84) as

$$
\begin{aligned}
F_{ab}(x) &= \partial_{a}A_{b}(x) - \partial_{b}A_{a}(x) = e_{a}{}^{\mu}(x)\partial_{\mu}e_{b}{}^{\nu}(x)A_{\nu}(x) - e_{b}{}^{\mu}(x)\partial_{\mu}e_{a}{}^{\nu}(x)A_{\nu}(x) \\
&= e_{a}{}^{\mu}(x)e_{b}{}^{\nu}(x)\left[D_{\mu}A_{\nu}(x) - D_{\nu}A_{\mu}(x)\right] \equiv e_{a}{}^{\mu}(x)e_{b}{}^{\nu}(x)F_{\mu\nu}(x).
\end{aligned}
\tag{16.11}
$$

Then the action (2.83) becomes

$$\overset{\text{em}}{A} = \int d^{4}x \sqrt{-g}\, \overset{\text{em}}{\mathcal{L}}(x) \equiv \int d^{4}x \sqrt{-g}\left[-\frac{1}{4c}F^{\mu\nu}(x)F_{\mu\nu}(x) - \frac{1}{c^{2}}j^{\mu}(x)A_{\mu}(x)\right], \tag{16.12}$$

Inserting into (16.11) the explicit expressions for the covariant derivatives from Eq. (11.85), and decomposing $\Gamma_{\mu\nu}{}^{\lambda}$ into Christoffel symbols and torsion according to Eqs. (11.114) and (11.115), we may write

$$F_{\mu\nu}(x) = \bar{F}_{\mu\nu}(x) - 2S_{\mu\nu}{}^{\lambda}A_{\lambda}(x), \tag{16.13}$$

where $\bar{F}_{\mu\nu}(x)$ is the field strength calculates from the Riemannian covariant derivatives (11.116):

$$\bar{F}_{\mu\nu}(x) = \bar{D}_\mu A_\nu(x) - \bar{D}_\nu A_\mu(x). \tag{16.14}$$

The Christoffel symbols contained on the right-hand side cancel each other, so that

$$\bar{F}_{\mu\nu} \equiv \bar{D}_\mu A_\nu - \bar{D}_\nu A_\mu = \partial_\mu A_\nu - \bar{\Gamma}_{\mu\nu}{}^\lambda A_\lambda - \partial_\nu A_\mu + \bar{\Gamma}_{\nu\mu}{}^\lambda A_\lambda = \partial_\mu A_\nu - \partial_\nu A_\mu. \tag{16.15}$$

Thus the Riemann-covariant field strength agrees with the Maxwell field strength in flat space

$$\bar{F}_{\mu\nu} = \partial_\mu A_\nu - \partial_\nu A_\mu. \tag{16.16}$$

The covariant field strength (16.13) reads now

$$F_{\mu\nu}(x) = \partial_\mu A_\nu(x) - \partial_\nu A_\mu(x) - 2S_{\mu\nu}{}^\lambda A_\lambda(x). \tag{16.17}$$

The last term destroys gauge invariance as noted first by Schrödinger [8] (see the remarks in our Preface). It gives the photon a spacetime-dependent tensorial mass term in the action (16.12)

$$\frac{1}{2}m_A^{2\,\lambda\kappa}(x)A_\lambda(x)A_\kappa(x), \quad \text{with} \quad m_A^{2\,\lambda\kappa}(x) = 2S_{\mu\nu}{}^\lambda(x)S^{\mu\nu\kappa}(x). \tag{16.18}$$

This prompted Schrödinger to estimate upper bounds for the photon mass allowed by experimental observations. Present observations imply that the mass is extremely small:

$$m_A < 3 \times 10^{-27}\text{eV}, \tag{16.19}$$

which corresponds to an immense Compton wavelength of the photon

$$l_A = \frac{\hbar}{m_A c} > 6952 \text{ light years.} \tag{16.20}$$

The estimate comes from observations of the range of magnetic fields emanating into spacetime from pulsars. Thus he concluded that the torsion field in spacetime is very small.

This is the reason why most authors [9] advocate that theories of gravity should not contain the torsion field in the field strength. They assume that the Maxwell action is formed with the Riemann-covariant field strength (16.15) rather than the metric-affine-covariant field strength $F_{\mu\nu}$ of Eq. (16.11). Thus the action of the electromagnetic field is taken to be

$$\overset{\text{em}}{\mathcal{A}} = \int d^4x \sqrt{-g}\, \overset{\text{em}}{\mathcal{L}}(x) \equiv \int d^4x \sqrt{-g} \left[-\frac{1}{4c}\bar{F}^{\mu\nu}(x)\bar{F}_{\mu\nu}(x) - \frac{1}{c^2}j^\mu(x)A_\mu(x) \right]. \tag{16.21}$$

This action is invariant under general coordinate transformation and electromagnetic gauge transformations

$$A_\mu(x) \to A_\mu(x) + \partial_\mu \Lambda(x). \tag{16.22}$$

Now the photon does not couple to torsion at all and remains certainly massless. In the philosophy of minimal coupling to ensure electromagnetic gauge invariance and Einstein invariance. this action is of course more minimal than (16.12), which justifies the omission of the torsion.

The same conclusion applies to the bare fields of the vector bosons W and Z which mediate weak interactions. In the standard unified theory of weak and electromagnetic interactions, the bare vector fields $A_\mu(x)$, $W_\mu(x)$, and $Z_\mu(x)$ appear on equal footing in a gauge-invariant way. Thus also their covariant derivatives should be free of torsion.

The physical vector bosons are massive due to the Meissner-Higgs effect. Their mass stems from "eating up" the massless Goldstone bosons of the Higgs field. Since these are scalar particles, they do not couple to torsion either so that the physical vector bosons remain decoupled from torsion [10]. This is in contrast to the massive vector mesons composed of quark-antiquark pairs, such as ρ and ω. This issue will be discussed further in Section 20.3.1.

Later we shall find that torsion is a nonpropagating field. At first sight this suggests that empty space cannot carry any torsion, so that even with the action (16.12), photons would propagate with light velocity though the vacuum. This conclusion, however, would be false. The coupling between $A_\mu(x)$ and $S_{\mu\nu\lambda}(x)$ in $F_{\mu\nu}(x)$ of Eq. (16.13) would have the consequence that the microwave background radiation in the universe would create also a torsion field. Recalling the spin current density (3.238) of the vector potential, the Einstein-Cartan field equation (15.60) determines the Palatini tensor as

$$S^{\mu\kappa,\tau} = \kappa \overset{\text{em}}{\Sigma}{}^{\mu\kappa,\tau} = -\frac{\kappa}{c}\left[F^{\tau\mu}A^\kappa - (\mu \leftrightarrow \kappa)\right]. \tag{16.23}$$

This may be inserted into Eq. (15.48) to find the torsion tensor, and from Eq. (16.18) a nonzero photon mass, thus destroying gauge invariance.

Another source of a local torsion comes from the cosmological constant. As will be discussed in Section 22.2, this has its origin in the nonzero vacuum fluctuations of all field. By Eq. (16.23), these will create a torsion field pervading spacetime, and this would also give the photons a small mass if the action were (16.12).

Although the photon mass which could be generated by a torsion field in this way would be extremely small and experimentally unobservable, the renormalizability of the unified theory of electromagnetic and weak interactions hinges on the gauge invariance of the theory. Thus we must reject the action (16.12) in spite of the fact that torsion does not propagate, and can only accept the action (16.21), which is free of the torsion field. In contrast to the coupling derived from the multivalued mapping principle, which we called minimal, the covariant expression without torsion field may be referred to as *truly minimal*.

Notes and References

[1] N.D. Birell and P.C.W. Davies, *Quantum Fields in Curved Space*, Cambridge University Press, Cambridge, 1982.

[2] See Eq. (13.241) in
H. Kleinert, *Path Integrals in Quantum Mechanics, Statistics, Polymer Physics, and Financial Markets*, World Scientific, Singapore 2004, 4th ed. kl/b5).

[3] B.S. DeWitt, Rev. Mod. Phys., **29**, 337 (1957).

[4] K.S. Cheng, J. Math. Phys. **13**, 1723 (1972).

[5] B.S. DeWitt, *Supermanifolds*, Cambridge Univ. Press, Cambridge, 1984.

[6] See Chapters 10 and 11 in textbook [2] (kl/b5/psfiles/pthic10.pdf).

[7] See Section 13.10 in textbook [2] (kl/b5/psfiles/pthic13.pdf).

[8] E. Schrödinger, Proc. R. Ir. Acad. A **49**, 43, 135 (1943); **49**, 135 (1943); **52**, 1 (1948); **54**, 79 (1951).

[9] See the review article:
F.W. Hehl, P. von der Heyde, G.D. Kerlick, and J.M. Nester, Rev. Mod. Phys. **48**, 393 (1976).

[10] This physical conclusion is in contrast to the assumption in Ref. [9].

We can lick gravity,
but sometimes the paperwork is overwhelming.
WERNHER VON BRAUN (1912–1977)

17

Particles with Half-Integer Spin

Let us now see how electrons and other particles of half-integer spin are coupled to gravity [1].

17.1 Local Lorentz Invariance and Anholonomic Coordinates

Spin is defined in Lorentz-invariant theories as the total angular momentum in the rest frame of the particle. To measure the spin s of a particle moving with velocity $\mathbf{v}$, we should go to a comoving frame by a local Lorentz transformation. Then the particle is at rest and the quantum mechanical description of its spin requires $2s+1$ states $|s, s_3\rangle$ with $s_3 = -s, \ldots, s$. Under rotations, these transform according to an irreducible representation of the rotation group with angular momentum s.

For a particle of spin $s = 1/2$ such as an electron, a muon, or any other massive lepton in Minkowski spacetime, this transformation property is automatically accounted for by the quanta of the Dirac field $\psi_\alpha(x)$, with the action (2.141):

$$\overset{m}{\mathcal{A}} = \int d^4 x^a \, \bar{\psi}(x^a) \left(i\gamma^a \partial_a - m \right) \psi(x^a), \tag{17.1}$$

where the Dirac matrices γ^a satisfy the algebra (1.224):

$$\left\{ \gamma^a, \gamma^b \right\} = 2g^{ab}. \tag{17.2}$$

The Dirac equation is obtained, as in Eq. (2.143), by extremizing this action:

$$\frac{\delta \overset{m}{\mathcal{A}}}{\delta \bar{\psi}(x^a)} = \left(i\gamma^a \partial_a - m \right) \psi(x^a) = 0. \tag{17.3}$$

17.1.1 Nonholonomic Image of Dirac Action

By complete analogy with the treatment of the action of a scalar field in Section 16.1 we can immediately write down the action in spacetime with curvature and torsion:

$$\overset{m}{\mathcal{A}} = \int d^4 x \, \sqrt{-g} \, \bar{\psi}(x) \left[i\gamma^a e_a{}^\mu(x) \partial_\mu - m \right] \psi(x), \tag{17.4}$$

410

where x are the physical coordinates x^μ. In contrast to the scalar case, however, this transformed action contains the multivalued tetrad fields for which the field-theoretic formalism is invalid. We must find a way of transforming away the multivalued content in $e_a{}^\mu(x)$. This is done by the introduction, at each point x^μ, of infinitesimal coordinates dx^α associated with a freely falling Lorentz frame. We may simply imagine infinitesimal freely falling elevators inside of which there is no gravity. The removal of the gravitational force holds only at the center of mass of a body. At any distance away from it there are tidal forces where either the centrifugal force or the gravitational attraction becomes active [recall Eq. (12.143)]. At the center of mass, the coordinates dx^α are Minkowskian, but the affine connections are nonzero and have in general a nonzero curvature which causes the tidal forces.

Intermediate Theory

We proceed as in Section 4.5 and observe that the modified Dirac Lagrangian density

$$\overset{m}{\mathcal{L}} = \bar\psi(x)\left\{i\gamma^\alpha\left[\partial_\alpha - D(\Lambda(x))^{-1}\partial_\alpha D(\Lambda(x))\right] - m\right\}\psi(x) \tag{17.5}$$

describes electrons just as well as the original Lagrangian density in the action (17.4). Here $\Lambda(x)$ is an arbitrary x-dependent Lorentz transformation which connects the flat-spacetime differentials $dx^a = e^a{}_\mu dx^\mu$ in (17.1) with the new coordinate differentials dx^α:

$$dx^a = \Lambda^a{}_\alpha(x)dx^\alpha, \quad dx^\alpha = (\Lambda^{-1})^\alpha{}_a(x)dx^a \equiv \Lambda_a{}^\alpha(x)dx^a, \tag{17.6}$$

and $D(\Lambda)$ is the representation of the local Lorentz transformations defined in Eq. (1.229). The metrics in the two coordinate systems are Minkowskian for any choice of $\Lambda^a{}_\alpha(x)$ [compare (1.28)]:

$$g_{\alpha\beta}(x) = \Lambda^a{}_\alpha(x)\Lambda^b{}_\beta(x)g_{ab} = (\Lambda^T)_\alpha{}^a(x)\,g_{ab}\,\Lambda^b{}_\beta(x) \equiv g_{ab}, \tag{17.7}$$

Let $\psi_\Lambda(x)$ be the solutions of Dirac equation associated with (17.5):

$$\left\{i\gamma^\alpha\left[\partial_\alpha - D(\Lambda(x))^{-1}\partial_\alpha D(\Lambda(x))\right] - m\right\}\psi(x) = 0. \tag{17.8}$$

They are related to the fields $\psi(x)$ which extremize the original action (17.4) by the local spinor transformation

$$\psi_\Lambda(x) = D(\Lambda(x))\psi(x). \tag{17.9}$$

This reflects the freedom of solving Dirac's anticommutation rules (1.224) by the x-dependent γ-matrices [recall (1.235)]:

$$\gamma^\alpha(x) \equiv D(\Lambda(x))^{-1}\gamma^\alpha D(\Lambda(x)) = \Lambda^\alpha{}_\beta\gamma^\beta. \tag{17.10}$$

Indeed, using Eq. (1.28) we verify that

$$\{\gamma^\alpha(x), \gamma^\beta(x)\} = \Lambda^\alpha{}_a(x)\Lambda^\beta{}_b(x)\{\gamma^a, \gamma^b\} = \Lambda^\alpha{}_a(x)\Lambda^\beta{}_b(x)g^{ab} = g^{\alpha\beta}. \tag{17.11}$$

We now recall Eq. (1.338) according to which, in a slightly different notation,

$$D(\Lambda(x))^{-1}\partial_\alpha D(\Lambda(x)) = -i\frac{1}{2}\omega_{\alpha;\delta\sigma}(x)\left(\Sigma^{\delta\sigma}\right)_B{}^C. \tag{17.12}$$

The right-hand side may be defined as the *spin connection for Dirac fields*:

$$\overset{D}{\Gamma}_{\alpha B}{}^C(x) \equiv i\frac{1}{2}\omega_{\alpha;\delta\sigma}(x)\left(\Sigma^{\delta\sigma}\right)_B{}^C. \tag{17.13}$$

Here $\omega_{\alpha;\beta\gamma}$ are the generalized angular velocities obtained by relations of the type (1.333)–(1.335) from the x-dependent tensor parameters $\omega_{\beta\gamma}(x)$ of the local Lorentz transformations $\Lambda(x) = e^{-i\omega_{\beta\gamma}(x)\Sigma^{\beta\gamma}}$.

According to Eq. (1.337), the generalized angular velocities $\omega_{\alpha;\beta}{}^\gamma$ appear also in the derivatives of the local Lorentz matrices $\Lambda^a{}_\alpha(x)$ as

$$\Lambda^{-1\gamma}{}_a(x)\partial_\alpha\Lambda^a{}_\beta(x) = \omega_{\alpha;}{}^\gamma{}_\beta(x) = -\omega_{\alpha;\beta}{}^\gamma(x). \tag{17.14}$$

Thus, if we define

$$\overset{\Lambda}{\Gamma}_{\alpha\beta}{}^\gamma \equiv \Lambda_a{}^\gamma\partial_\alpha\Lambda^a{}_\beta = -\Lambda^a{}_\beta\partial_\alpha\Lambda_a{}^\gamma, \tag{17.15}$$

we can write the Dirac spin connection as

$$\overset{D}{\Gamma}_{\alpha B}{}^C(x) \equiv -\frac{i}{2}\overset{\Lambda}{\Gamma}_{\alpha\delta}{}^\sigma(x)\left(\Sigma^\delta{}_\sigma\right)_B{}^C. \tag{17.16}$$

The transformation has produced a Lagrangian density

$$\overset{m}{\mathcal{L}} = \bar\psi(x)\left(i\gamma^\alpha D_\alpha - m\right)\psi(x), \tag{17.17}$$

with the covariant derivative matrix

$$(D_\alpha)_B{}^C = \delta_B{}^C\partial_\alpha - \overset{D}{\Gamma}_{\alpha B}{}^C(x). \tag{17.18}$$

The Lagrangian density (17.17) is completely equivalent to the original Dirac Lagrangian density in Eq. (17.4), as long as the spin connection is given by (17.15) with single-valued Lorentz transformations $\Lambda_a{}^\gamma(x)$. This is analogous to the situation in the discussion of the Schrödinger Lagrangian (4.81) which was the starting point for the introduction of electromagnetism by multivalued gauge transformations.

We may now proceed in the same way as before by allowing $\omega_{\beta\gamma}(x)$ to be multivalued. Then the components of the spin connection are no longer generalized angular velocities (17.14), but form new fields

$$A_{\alpha\beta}{}^\gamma(x) \equiv \omega_{\alpha;\beta}{}^\gamma(x) = \overset{\Lambda}{\Gamma}_{\alpha\beta}{}^\gamma. \tag{17.19}$$

Since $\omega_\beta{}^\gamma(x)$ does not have commuting derivatives, these cannot be calculated by solving (17.19) as a differential equation for $\omega_\beta{}^\gamma(x)$. In fact, the Lagrangian density (17.17) with the covariant derivative (17.18) describes now a theory in which the field is coupled to torsion. The affine connection (17.15) will be seen in Eq. (17.71) to coincide with the contortion of the local Minkowski differentials dx^α. These have so far no Riemannian curvature, so that the Riemann-Cartan curvature tensor is determined completely by the contortion tensor via Eq. (11.146). The covariant derivative formed with the Christoffel symbols allows for the definition of parallel vector fields over any distance. This theory is a counterpart of the famous *telepar-allel theory* developed by Einstein after 1928 under the influence of a famous letter exchange with Cartan (recall Preface). There the situation is the opposite: the Riemann-Cartan curvature tensor vanishes identically, and the Riemann curvature is given via Eq. (11.146) by

$$-\bar{R}_{\mu\nu\lambda}{}^{\kappa} = \bar{D}_{\mu}K_{\nu\lambda}{}^{\kappa} - \bar{D}_{\nu}K_{\mu\lambda}{}^{\kappa} + \left(K_{\mu\lambda}{}^{\rho}K_{\nu\rho}{}^{\kappa} - K_{\nu\lambda}{}^{\rho}K_{\mu\rho}{}^{\kappa}\right), \qquad (17.20)$$

17.1.2 Vierbein Fields

In order to describe the observable gravitational forces, we must go one step further.

Following the standard procedure of Section 4.5 we first perform the analog of a single-valued gauge transformation which is here an ordinary coordinate transformation from x^α to x^μ:

$$dx^\alpha = dx^\mu h^\alpha{}_\mu(x). \qquad (17.21)$$

The transformation has an inverse

$$dx^\mu = dx^\alpha h_\alpha{}^\mu(x), \qquad (17.22)$$

and the matrix elements $h_\alpha{}^\mu(x)$ and $h^\alpha{}_\mu(x)$ satisfy, at each x, the orthonormality and completeness relations

$$h_\alpha{}^\mu(x)h^\beta{}_\mu(x) = \delta_\alpha{}^\beta, \qquad h^\alpha{}_\mu(x)h_\alpha{}^\nu(x) = \delta_\mu{}^\nu. \qquad (17.23)$$

The 4×4-transformation matrices $h_\alpha{}^\mu(x)$ and $h^\alpha{}_\mu(x)$ are called *vierbein fields* and *reciprocal vierbein fields*, respectively. As in the case of the multivalued basis tetrads $e^a{}_\mu(x), e_a{}^\mu(x)$ we shall freely raise or lower the indices $\alpha, \beta, \gamma, \ldots$ by contraction with the metric $g_{\alpha\beta}$ or its inverse $g^{\alpha\beta}$:

$$h^{\alpha\mu}(x) \equiv g^{\alpha\beta}h_\beta{}^\mu(x), \qquad h_{\alpha\beta}(x) \equiv g_{\alpha\beta}h^\beta{}_\mu(x). \qquad (17.24)$$

Since the transformation functions are, for the moment, single-valued, they satisfy

$$\partial_\mu h^\alpha{}_\nu(x) - \partial_\nu h^\alpha{}_\mu(x) = 0, \qquad \partial_\mu h_\alpha{}^\nu(x) - \partial_\nu h_\alpha{}^\mu(x) = 0. \qquad (17.25)$$

For spinor fields depending on the final physical coordinates x^μ, the covariant derivative (17.18) becomes

$$(D_\alpha)_B{}^C = \delta_B{}^C h_\alpha{}^\mu(x)\partial_\mu - \overset{\mathrm{D}}{\Gamma}_{\alpha B}{}^C(x). \qquad (17.26)$$

The flat spacetime x^a coordinates are now related to the physical coordinates x^μ by the equation

$$dx^a = e^a{}_\mu dx^\mu = \Lambda^a{}_\alpha(x)h^\alpha{}_\mu(x)dx^\mu = \Lambda^a{}_\alpha(x)dx^\alpha, \qquad (17.27)$$

where the matrix elements $\Lambda^a{}_\alpha(x)$ of the local Lorentz transformation are multivalued. Now the action

$$\overset{m}{\mathcal{A}} = \int d^4x \sqrt{-g}\, \bar\psi(x) \left(i\gamma^\alpha h_\alpha{}^\mu(x)D_\mu - m \right) \psi(x) \qquad (17.28)$$

contains only the single-valued geometric fields $h_\alpha{}^\mu(x)$ and $A_{\alpha\delta\sigma}(x)$ inside the spin connection

$$\overset{D}{\Gamma}{}_{\alpha B}{}^C(x) = i\omega_{\alpha;\delta\sigma}(x)\frac{1}{2}\left(\Sigma^{\delta\sigma}\right)_B{}^C = iA_{\alpha\delta\sigma}(x)\frac{1}{2}\left(\Sigma^{\delta\sigma}\right)_B{}^C. \qquad (17.29)$$

17.1.3   Local Inertial Frames

This is now the place where we can introduce curvature by allowing the coordinates x^α to be multivalued functions of the physical coordinates of x^μ. Then the vierbein fields satisfy no longer the relation (17.25). The vierbein fields themselves, however, are single-valued functions, so that the action (17.28) is now perfectly suited to describe electrons and other Dirac particles in spacetimes with curvature and torsion.

Since the differentials dx^α are related to dx^a by a Lorentz transformation (17.27), the square length of the nonholonomic coordinates dx^α

$$ds^2 = g_{\alpha\beta}dx^\alpha dx^\beta \qquad (17.30)$$

is measured everywhere by the Minkowski metric: where

$$g_{\alpha\beta} = \Lambda^a_\alpha(x)\Lambda^b_\beta(x)g_{ab} \equiv \begin{pmatrix} 1 & & & \\ & -1 & & \\ & & -1 & \\ & & & -1 \end{pmatrix}_{\alpha\beta}, \qquad (17.31)$$

due to Eq. (1.28). Combining (17.30) with (17.21), we obtain the relation

$$g_{\alpha\beta} = h_\alpha{}^\mu(x)h_\beta{}^\nu(x)g_{\mu\nu}(x), \qquad (17.32)$$

whose inverse

$$g_{\mu\nu}(x) = h^\alpha{}_\mu(x)h^\beta{}_\nu(x)g_{\alpha\beta} \equiv h^\alpha{}_\mu(x)h_{\beta\nu}(x). \qquad (17.33)$$

Recall the similar relation (11.38) where we expressed the metric as a square of the multivalued basis tetrads $e^a{}_\mu(x)$. Thus both $e^\alpha{}_\mu(x)$ and $h^a{}_\mu(x)$ are "matrix square roots" of the metric $g_{\mu\nu}(x)$.

There is a simple physical relation between the spacetime coordinates x^μ and the infinitesimal coordinates dx^α. The latter are associated with small freely falling small Lorentz frames at each x^μ. Such frames are the inertial frames. They may be imagined as small elevators in which the free fall removes all gravitational forces.

The removal is perfect only at the center of mass of the elevators. At any distance away from it there are tidal forces where either the centrifugal force or the gravitational attraction becomes dominant. These aspects have been discussed before in Subsections 12.6.1 and 12.6.2 when constructing geodesic coordinates.

Let study the tidal forces once more the present context. In a small neighborhood of an arbitrary point X^μ we solve the differential equation (17.22) by the functions

$$x^\alpha(X;x) = a^\alpha + h^\alpha{}_\mu(X)(x^\mu - X^\mu) + \frac{1}{2}h^\alpha{}_\lambda(X)\Gamma_{\mu\nu}{}^\lambda(X)(x^\mu - X^\mu)(x^\nu - X^\nu) + \ldots . \quad (17.34)$$

The derivatives

$$\frac{\partial x^\alpha(X;x)}{\partial x^\mu} = h^\alpha{}_\mu(X) + h^\alpha{}_\lambda(X)\Gamma_{\mu\nu}{}^\lambda(X)(x^\nu - X^\nu) \equiv h^\alpha{}_\mu(X;x) \quad (17.35)$$

fulfill Eq. (17.22) at $x = X$. Consider now a point particle satisfying the equation of motion (14.7). In the coordinates (17.34), the trajectory satisfies the equation

$$\dot{x}^\alpha = h^\alpha{}_\mu(X)\dot{x}^\mu + h^\alpha{}_\lambda(X)\Gamma_{\mu\nu}{}^\lambda(X)\,\dot{x}^\mu(x^\nu - X^\nu) + \cdots , \quad (17.36)$$

and

$$\ddot{x}^\alpha = h^\alpha{}_\mu(X)\ddot{x}^\mu + h^\alpha{}_\lambda(X)\Gamma_{\mu\nu}{}^\lambda(X)\,\ddot{x}^\mu(x^\nu - X^\nu) + h^\alpha{}_\lambda(X)\Gamma_{\mu\nu}{}^\lambda(X)\,\dot{x}^\mu\dot{x}^\nu + \cdots . \quad (17.37)$$

Inserting here (14.7), the first and third terms cancel each other, and the trajectory experiences no acceleration at the point X. In the neighborhood, there are the tidal forces. Thus the increments dx^α constitute an inertial frame in an infinitesimal neighborhood of the point X.

The metric in the coordinates $x^\alpha(X;x)$ is

$$g_{\alpha\beta}(X;x) = \frac{\partial x^\alpha(X;x)}{\partial x^\mu}\frac{\partial x^\beta(X;x)}{\partial x^\mu}g^{\mu\nu}(x) = h^\alpha{}_\mu(X)h^\beta{}_\nu(X)g^{\mu\nu}(x) \quad (17.38)$$

$$+ \left[h^\alpha{}_\lambda(X)h^\beta{}_\nu(X)\Gamma_{\mu\kappa}{}^\lambda(X)(x^\kappa - X^\kappa) + (\alpha \leftrightarrow \beta) \right] g^{\mu\nu}(x).$$

We now expand the metric $g^{\mu\nu}(x)$ in the neighborhood of X as

$$\begin{aligned}
g^{\mu\nu}(x) &= g^{\mu\nu}(X) + \partial_\lambda g^{\mu\nu}(X)(x^\lambda - X^\lambda) + \cdots \\
&= g^{\mu\nu}(X) - \left[g^{\mu\lambda}\Gamma_{\kappa\lambda}{}^\nu(X) + g^{\nu\lambda}\Gamma_{\kappa\lambda}{}^\mu(X) \right](x^\kappa - X^\kappa) + \cdots .
\end{aligned} \quad (17.39)$$

Inserting this into (17.38) and using (17.32), we obtain

$$g_{\alpha\beta}(X;x) = g_{\alpha\beta} + \mathcal{O}(x - X)^2. \quad (17.40)$$

This ensures, that the affine connection formed from $g_{\alpha\beta}(X;x)$ vanished at $x = X$, so that there are no forces at this point. In any neighborhood of X, however, there will be the tidal forces.

In the coordinates dx^a, there are no tidal forces at all. This is possible everywhere only due to defects which make $\Lambda^a{}_\alpha(x)$ multivalued.

The coordinates $x^\alpha(X; x)$ are functions of x which depend on X. There exists no single function $x^\alpha(x)$, so that derivatives in front of $x^\alpha(x)$ do not commute:

$$\left(\partial_\mu\partial_\nu - \partial_\nu\partial_\mu\right)x^\alpha(x) \neq 0, \tag{17.41}$$

implying that

$$\partial_\mu h^\alpha{}_\nu(x) - \partial_\nu h^\alpha{}_\mu(x) \neq 0. \tag{17.42}$$

The functions $h^\alpha{}_\mu(x)$ and $h^\mu_\alpha(x)$, however, which describe the transformation to the freely falling elevators are single-valued. They obey the integrability condition

$$\left(\partial_\mu\partial_\nu - \partial_\nu\partial_\mu\right)h^\alpha{}_\lambda(x) = 0, \qquad \left(\partial_\mu\partial_\nu - \partial_\nu\partial_\mu\right)h_\alpha{}^\lambda(x) = 0. \tag{17.43}$$

This condition has the consequence that if we construct a tensor $\overset{h}{R}_{\mu\nu\lambda}{}^\kappa(x)$ from the transformation matrices $h_\alpha{}^\mu(x)$ in the same way as $R_{\mu\nu\lambda}{}^\kappa(x)$ was made from $e_a{}^\mu(x)$ in Eq. (11.130), we find an identically vanishing result:

$$\overset{h}{R}_{\mu\nu\lambda}{}^\kappa = h_\alpha{}^\kappa(\partial_\mu\partial_\nu - \partial_\nu\partial_\mu)h^\alpha{}_\lambda \equiv 0. \tag{17.44}$$

17.1.4   Difference between Vierbein and Multivalued Tetrad Fields

Note that although multivalued basis tetrads $e^a{}_\mu(x)$ and the vierbein fields $h^\alpha{}_\mu(x)$ are both "square roots" of the metric $g_{\mu\nu}(x)$, they are completely different mathematical objects. They differ from one another by a local Lorentz transformation $\Lambda^a{}_\mu(x)$. The relation follows from Eq. (17.27):

$$e^a{}_\mu(x) = \Lambda^a{}_\alpha(x)h^\alpha{}_\mu(x), \tag{17.45}$$

which implies that

$$\Lambda^a{}_\alpha(x) \equiv e^a{}_\mu(x)h_\alpha{}^\mu(x), \tag{17.46}$$

This is precisely the freedom one has in defining square root of a matrix.

Most importantly, $e^a{}_\mu(x)$ and $h^\alpha{}_\mu(x)$ possess different integrability properties. While $h^\alpha{}_\lambda(x)$ satisfies the Schwarz integrability condition (17.43), the multivalued basis tetrads $e^a{}_\lambda(x)$ do not. The commutator of the derivatives in front of $e^a{}_\lambda(x)$ determines the curvature tensor via Eq. (11.130). Hence the Lorentz transformation matrices $\Lambda^a{}_\mu(x)$ are multivalued. They introduce defects into the mapping $dx^\alpha = dx^a\Lambda_a{}^\alpha(x)$.

Since $h^\alpha{}_\mu(x)$ and $h_\alpha{}^\mu(x)$ are single-valued functions with commuting derivatives, the curvature tensor $R_{\mu\nu\lambda}{}^\kappa(x)$ in (11.130) may be expressed completely in terms of the noncommuting derivatives of the local Lorentz transformations $\Lambda^a{}_\alpha(x)$. To see this we insert Eq. (17.45) into (11.130), and use (17.43) to find for the curvature tensor the alternative expression

$$R_{\mu\nu\lambda}{}^\sigma = h_\gamma{}^\sigma\left[\Lambda_a{}^\gamma\left(\partial_\mu\partial_\nu - \partial_\nu\partial_\mu\right)\Lambda^a{}_\alpha\right]h^\alpha{}_\lambda \equiv h_\gamma{}^\sigma R_{\mu\nu\alpha}{}^\gamma h^\alpha{}_\lambda. \tag{17.47}$$

From the defect point of view, the single-valued matrices $h_\alpha{}^\mu(x)$ create an intermediate coordinate system dx^α which, by the integrability condition (17.43), has the same disclination content as the coordinates x^μ, but is completely free of dislocations. The metric in the new coordinate system x^α is Minkowski-like at each point in spacetime. Still, the coordinates x^α do not form a Minkowski spacetime since they differ from the inertial coordinates dx^a by the presence of disclinations, i.e., there are wedge-like pieces missing with respect to an ideal reference crystal. The coordinates x^α cannot be defined globally from x^μ. Only the differentials dx^α are uniquely related to dx^μ at each spacetime point by Eqs. (17.21) and (17.22). The local Lorentz transformations $\Lambda^a{}_\alpha(x)$ have noncommuting derivatives on account of the disclinations residing in the coordinates dx^α. The coordinate system dx^α can only be used to specify *derivatives* with respect to x^α, and thus the *directions* of vectors (and tensors) with respect to the intermediate local axes

$$\mathbf{e}_\alpha(x) \equiv \mathbf{e}_\mu(x)\frac{\partial x^\mu}{\partial x^\alpha} = \mathbf{e}_\mu(x)h_\alpha{}^\mu(x)$$
$$\equiv \mathbf{e}_a e^a{}_\mu(x)h_\alpha{}^\mu(x) \equiv \mathbf{e}_a \Lambda^a{}_\alpha(x). \tag{17.48}$$

We can go back to the local basis via the reciprocal vierbein fields

$$\mathbf{e}_\mu(x) = \mathbf{e}_\alpha(x)\frac{\partial x^\alpha}{\partial x^\mu} = \mathbf{e}_\alpha(x)h^\alpha{}_\mu(x). \tag{17.49}$$

17.1.5 Covariant Derivatives in Intermediate Basis

It is instructive to derive the covariant derivatives of the components of an arbitrary vector field $v_\alpha(x)$ in the intermediate local basis $\mathbf{e}_\alpha(x)$. The relation with the components $v_a(x)$ is

$$\mathbf{v}(x) \equiv \mathbf{e}_a v^a(x) = \mathbf{e}_a e^a{}_\mu(x)v^\mu(x) = \mathbf{e}_a \Lambda^a{}_\alpha(x)h^\alpha{}_\mu(x)v^\mu(x)$$
$$= \mathbf{e}_a \Lambda^a{}_\alpha(x)h^{\alpha\mu}(x)v_\mu(x) = \mathbf{e}_a \Lambda^{a\alpha}(x)v_\alpha(x) = \mathbf{e}_a \Lambda^a{}_\alpha(x)v^\alpha(x), \tag{17.50}$$

where we have introduced the co- and contravariant components

$$v_\alpha(x) \equiv v_\mu(x)h_\alpha{}^\mu(x), \quad v^\alpha(x) \equiv v^\mu(x)h^\alpha{}_\mu(x). \tag{17.51}$$

The orthogonality relations (17.23) imply the inverse relations

$$v_\mu(x) = v_\alpha(x)h^\alpha{}_\mu(x), \quad v^\mu(x) \equiv v_\mu(x)h_\alpha{}^\mu(x). \tag{17.52}$$

From these relations we derive the covariant derivatives of the vector fields $v_\beta(x), v^\beta(x)$

$$D_\alpha v_\beta = \partial_\alpha v_\beta - \overset{\Lambda}{\Gamma}{}_{\alpha\beta}{}^\gamma v_\gamma, \quad D_\alpha v^\beta = \partial_\alpha v^\beta + \overset{\Lambda}{\Gamma}{}_{\alpha\gamma}{}^\beta v^\gamma, \tag{17.53}$$

where $\overset{\Lambda}{\Gamma}{}_{\alpha\beta}{}^\gamma$ is precisely the spin connection (17.15). In Eq. (17.16) it was introduced to create covariant derivatives of Dirac fields. Here it appears in the covariant derivatives of vector fields $v_\beta(x), v^\beta(x)$. Remembering Eq. (17.19) we may require the covariant derivatives (17.53) also as

$$D_\alpha v_\beta = \partial_\alpha v_\beta - A_{\alpha\beta}{}^\gamma v_\gamma, \quad D_\alpha v^\beta = \partial_\alpha v^\beta + A_{\alpha\gamma}{}^\beta v^\gamma, \qquad (17.54)$$

By analogy with the pure gradient (4.51) of a single-valued gauge function in the abelian gauge theory of magnetism, the spin connection (17.15) reduces to a trivial gauge field for single-valued local Lorentz transformations $\Lambda(x)$. Indeed, it is easily verified that the field strength associated with this gauge field, the covariant curl

$$F_{\mu\nu\alpha}{}^\gamma \equiv \partial_\mu \overset{\Lambda}{\Gamma}_{\nu\alpha}{}^\gamma - \partial_\nu \overset{\Lambda}{\Gamma}_{\mu\alpha}{}^\gamma - [\overset{\Lambda}{\Gamma}_\mu, \overset{\Lambda}{\Gamma}_\nu]_\alpha{}^\gamma \qquad (17.55)$$

vanishes for single-valued $\Lambda(x)$'s. As in Eqs. (11.129), (11.126), the commutator is defined by considering $\overset{\Lambda}{\Gamma}_{\nu\alpha}{}^\gamma$ as matrices $(\overset{\Lambda}{\Gamma}_\nu)_\alpha{}^\gamma$. For multivalued Lorentz transformations $\Lambda(x)$, the covariant curl is nonzero, and $\overset{\Lambda}{\Gamma}_{\nu\alpha}{}^\gamma(x) = A_{\nu\alpha}{}^\gamma(x)$ is a nonabelian gauge field, whose field strength (17.55) is nonzero and transforms like a tensor under single-valued local Lorentz transformations.

From Eq. (17.15) it follows that $\Lambda^a{}_\alpha(x)$ and $\Lambda_a{}^\alpha(x)$ satisfy identities similar to those for $e^a{}_\nu(x)$ and $e_a{}^\nu(x)$ in Eqs. (11.92) and (11.93):

$$D_\alpha \Lambda^a{}_\beta = 0, \quad D_\alpha \Lambda_a{}^\beta = 0. \qquad (17.56)$$

It is instructive to rewrite the spin connection in matrix notation using the notation (17.21) for the local Lorentz transformations. Then the relation (17.50) shows that

$$v^a(x) = \Lambda^a{}_\alpha v^\alpha(x), \quad v_a(x) = \Lambda_a{}^\alpha v_\alpha(x) = (g\Lambda g)_a{}^\alpha v_\alpha(x) = \left(\Lambda^{T-1}\right)_a{}^\alpha v_\alpha(x), \qquad (17.57)$$

and therefore

$$\begin{aligned}
\partial_\alpha v^a(x) &= \Lambda^a{}_\beta D_\alpha v^\beta(x) &=& \Lambda^a{}_\beta \left[\partial_\alpha \delta^\beta{}_\alpha + (\Lambda^{-1}\partial_\alpha\Lambda)^\beta{}_\gamma\right] v^\gamma(x). \qquad (17.58) \\
\partial_\alpha v_a(x) &= \Lambda_a{}^\beta D_\alpha v_\beta(x) &=& \Lambda_a{}^\beta \left[\partial_\alpha \delta_\beta{}^\gamma + (\Lambda^T\partial_\alpha\Lambda^{T-1})_\beta{}^\gamma\right] v_\gamma(x) \\
& &=& \Lambda_a{}^\beta \left[\partial_\alpha \delta_\beta{}^\gamma - (\Lambda^{-1}\partial_\alpha\Lambda)^\gamma{}_\beta\right] v_\gamma(x). \qquad (17.59)
\end{aligned}$$

From this we identify

$$(\overset{\Lambda}{\Gamma}_\alpha)_\beta{}^\gamma = (\Lambda^{-1}\partial_\alpha\Lambda)^\beta{}_\gamma = -(\Lambda^{-1}\partial_\alpha\Lambda)^\gamma{}_\beta, \qquad (17.60)$$

which is the same as (17.15) expressed in matrix form.

If a field has several local Lorentz indices $\alpha, \beta, \gamma, \ldots$, each index receives an own contribution proportional to the gauge field $A_{\alpha\beta}{}^\gamma$. If it has, in addition, Einstein indices $\mu, \nu, \lambda, \ldots$, there are also additional terms proportional to the affine connection

$\Gamma_{\mu\nu}{}^{\lambda}$. As an example, the covariant derivatives of the fields v_{β}^{μ} and v_{μ}^{β} with respect to the nonholonomic coordinates dx^{α} are from (17.53) and (12.67), (12.68):

$$D_{\alpha}v_{\beta}^{\mu} = \partial_{\alpha}v_{\beta}^{\mu} - \overset{\Lambda}{\Gamma}_{\alpha\beta}{}^{\gamma}v_{\gamma} + h_{\alpha}{}^{\kappa}\Gamma_{\kappa\nu}{}^{\mu}v_{\beta}^{\nu}, \tag{17.61}$$

$$D_{\alpha}v_{\mu}^{\beta} = \partial_{\alpha}v^{\beta} + \overset{\Lambda}{\Gamma}_{\alpha\gamma}{}^{\beta}v^{\gamma} - h_{\alpha}{}^{\kappa}\Gamma_{\kappa\mu}{}^{\nu}v_{\nu}^{\beta}. \tag{17.62}$$

The covariant derivatives with respect to the physical coordinates x^{λ} are

$$D_{\lambda}v_{\beta}^{\mu} = \partial_{\lambda}v_{\beta}^{\mu} - h^{\alpha}{}_{\lambda}\overset{\Lambda}{\Gamma}_{\alpha\beta}{}^{\gamma}v_{\gamma} + \Gamma_{\lambda\nu}{}^{\mu}v_{\beta}^{\nu}, \tag{17.63}$$

$$D_{\lambda}v_{\mu}^{\beta} = \partial_{\lambda}v_{\mu}^{\beta} + h^{\alpha}{}_{\lambda}\overset{\Lambda}{\Gamma}_{\alpha\gamma}{}^{\beta}v_{\mu}^{\gamma} - \Gamma_{\lambda\mu}{}^{\nu}v_{\nu}^{\beta}. \tag{17.64}$$

Let us express the spin connection (17.15) for vector fields in terms of $e^{a}{}_{\mu}$ and $h_{a}{}^{\mu}$. With the help of (17.46), we calculate

$$\begin{aligned}
\overset{\Lambda}{\Gamma}_{\alpha\beta}{}^{\gamma} &= e_{a}{}^{\lambda}h^{\gamma}{}_{\lambda}h_{\alpha}{}^{\mu}\partial_{\mu}(e^{a}{}_{\nu}h_{\beta}{}^{\nu}) \\
&= h^{\gamma}{}_{\lambda}h_{\alpha}{}^{\mu}h_{\beta}{}^{\nu}\Gamma_{\mu\nu}{}^{\lambda} + h^{\gamma}{}_{\lambda}h_{\alpha}{}^{\mu}\delta^{\lambda}{}_{\nu}\partial_{\mu}h_{\beta}{}^{\nu} \\
&= h^{\gamma}{}_{\lambda}h_{\alpha}{}^{\mu}h_{\beta}{}^{\nu}(\Gamma_{\mu\nu}{}^{\lambda} + h^{\delta}{}_{\nu}\partial_{\mu}h_{\delta}{}^{\lambda}).
\end{aligned} \tag{17.65}$$

Employing the covariant derivatives (17.61) and (17.62), this equation can be recast as

$$D_{\alpha}h_{\beta}{}^{\mu} = 0, \qquad D_{\alpha}h^{\beta}{}_{\mu} = 0, \tag{17.66}$$

so that $h_{\alpha}{}^{\mu}$ satisfies similar identities as $e_{a}{}^{\mu}$ in (11.93) and as $\Lambda_{a}{}^{\alpha}$ in (17.56).

At this place it is useful to introduce the symbols

$$\overset{h}{\Gamma}_{\mu\nu}{}^{\lambda} \equiv h_{\alpha}{}^{\lambda}\partial_{\mu}h^{\alpha}{}_{\nu} \equiv -h^{\alpha}{}_{\nu}\partial_{\mu}h_{\alpha}{}^{\lambda}. \tag{17.67}$$

They are defined in terms of $h^{\alpha}{}_{\mu}$ in the same way as $\Gamma_{\mu\nu}{}^{\lambda}$ is defined in terms of $e^{a}{}_{\mu}$ in Eq. (11.91). Then we may rewrite the spin connection (17.65) as

$$\begin{aligned}
\overset{\Lambda}{\Gamma}_{\alpha\beta}{}^{\gamma} &= h^{\gamma}{}_{\lambda}h_{\alpha}{}^{\mu}\left(\Gamma_{\mu\nu}{}^{\lambda} - h_{\delta}{}^{\lambda}\partial_{\mu}h^{\delta}{}_{\nu}\right) \\
&= h^{\gamma}{}_{\lambda}h_{\alpha}{}^{\mu}h_{\beta}{}^{\nu}(\Gamma_{\mu\nu}{}^{\lambda} - \overset{h}{\Gamma}_{\mu\nu}{}^{\lambda}).
\end{aligned} \tag{17.68}$$

If we now decompose the two expressions on the right-hand side into Christoffel parts and contortion tensors in the same way as in Eqs. (11.110)–(11.112), we realize that due to the identity

$$g_{\mu\nu}(x) = e^{a}{}_{\mu}(x)e^{b}{}_{\nu}(x)g_{ab} \equiv h^{\alpha}{}_{\mu}(x)h^{\beta}{}_{\nu}(x)g_{\alpha\beta}, \tag{17.69}$$

the two Christoffel parts in $\Gamma_{\mu\nu}{}^{\lambda}$ and $\overset{h}{\Gamma}_{\mu\nu}{}^{\lambda}$ are the same:

$$\bar{\Gamma}_{\mu\nu}{}^{\lambda} \equiv \overset{h}{\bar{\Gamma}}_{\mu\nu}{}^{\lambda}. \tag{17.70}$$

As a consequence, $\overset{\Lambda}{\Gamma}_{\alpha\beta}{}^{\gamma}$ becomes

$$
\begin{aligned}
\overset{\Lambda}{\Gamma}_{\alpha\beta}{}^{\gamma} &= h^{\gamma}{}_{\lambda}h_{\alpha}{}^{\mu}h_{\beta}{}^{\nu}(\overset{\Lambda}{\bar{\Gamma}}_{\mu\nu}{}^{\lambda} + K_{\mu\nu}{}^{\lambda} - \overset{h}{\bar{\Gamma}}_{\mu\nu}{}^{\lambda} - \overset{h}{K}_{\mu\nu}{}^{\lambda}) \\
&= h^{\gamma}{}_{\lambda}h_{\alpha}{}^{\mu}h_{\beta}{}^{\nu}(K_{\mu\nu}{}^{\lambda} - \overset{h}{K}_{\mu\nu}{}^{\lambda}),
\end{aligned}
\tag{17.71}
$$

where $K_{\mu\nu}{}^{\lambda}$ is the contortion tensor (11.114), and $\overset{h}{K}_{\mu\nu}{}^{\lambda}$ denotes the expression (11.112) with $e^{a}{}_{\mu}$, $e_{a}{}^{\mu}$ replaced by $h^{a}{}_{\mu}$, $h_{a}{}^{\mu}$. Explicitly, these tensors are

$$
K_{\mu\nu}{}^{\lambda} = S_{\mu\nu}{}^{\lambda} - S_{\nu}{}^{\lambda}{}_{\mu} + S^{\lambda}{}_{\mu\nu},
\tag{17.72}
$$

$$
\overset{h}{K}_{\mu\nu}{}^{\lambda} = \overset{h}{S}_{\mu\nu}{}^{\lambda} - \overset{h}{S}_{\nu}{}^{\lambda}{}_{\mu} + \overset{h}{S}{}^{\lambda}{}_{\mu\nu},
\tag{17.73}
$$

where

$$
\overset{h}{S}_{\mu\nu}{}^{\lambda} \equiv \frac{1}{2}\left(h_{\alpha}{}^{\lambda}\partial_{\mu}h^{\alpha}{}_{\nu} - h_{\alpha}{}^{\lambda}\partial_{\nu}h^{\alpha}{}_{\mu}\right).
\tag{17.74}
$$

This is the so-called *object of anholonomy*, often denoted by $\Omega_{\mu\nu}{}^{\lambda}$. They are anti-symmetric in the first two indices. As a consequence, the combination (17.73) with lowered last index has the same antisymmetry in the last two indices $\nu\lambda$ as the contortion $K_{\mu\nu\lambda}$. The tensors (17.74) and (17.73) have therefore the same symmetry properties as the contortion and torsion tensors. The spin connection (17.71) in which the last index is lowered by a contraction with the Minkowski metric $g_{\alpha\beta}$ [recall (17.32)] is then antisymmetric in the last two indices.

In standard theories of gravity without torsion, the spin connection contains only the last term in (17.71). For this reason, $\overset{h}{K}_{\mu\nu\lambda}$ itself is often referred to as spin connection. In spacetimes with torsion, it should be referred to as *torsionless spin connection*.

Note that the antisymmetric part of the spin connection (17.71) is

$$
\overset{\Lambda}{S}_{\alpha\beta}{}^{\gamma} \equiv \frac{1}{2}\left(\overset{\Lambda}{\Gamma}_{\alpha\beta}{}^{\gamma} - \overset{\Lambda}{\Gamma}_{\beta\alpha}{}^{\gamma}\right) = h^{\gamma}{}_{\lambda}h_{\alpha}{}^{\mu}h_{\beta}{}^{\nu}(S_{\mu\nu}{}^{\lambda} - \overset{h}{S}_{\mu\nu}{}^{\lambda}).
\tag{17.75}
$$

It will be helpful to freely use $h^{\alpha}{}_{\mu}$, $h_{\alpha}{}^{\mu}$ for changing indices α into μ, for instance,

$$
K_{\alpha\beta}{}^{\gamma} \equiv h^{\gamma}{}_{\lambda}h_{\alpha}{}^{\mu}h_{\beta}{}^{\nu}K_{\mu\nu}{}^{\lambda},
\tag{17.76}
$$

$$
\overset{h}{K}_{\alpha\beta}{}^{\lambda} = h^{\gamma}{}_{\lambda}h_{\alpha}{}^{\mu}h_{\beta}{}^{\nu}\,\overset{h}{K}_{\mu\nu}{}^{\lambda}.
\tag{17.77}
$$

17.2   Dirac Action in Riemann-Cartan Space

Inserting (17.71) into (17.26) we obtain now the single-valued image of the flat-spacetime action (17.1):

$$
\overset{m}{\mathcal{A}} = \int d^{4}x\,\sqrt{-g}\bar{\psi}(x)\left[i\gamma^{\alpha}h_{\alpha}{}^{\mu}(x)D_{\mu} - m\right]\psi(x),
\tag{17.78}
$$

with the covariant derivative of Eqs. (17.26) and (17.29):

$$D_\mu = \partial_\mu - iA_{\mu\beta}{}^\gamma \frac{1}{2}\Sigma^\beta{}_\gamma \equiv \partial_\mu - ih^\gamma{}_\lambda h_\beta{}^\nu (K_{\mu\nu}{}^\lambda - \overset{h}{K}_{\mu\nu}{}^\lambda)\frac{1}{2}\Sigma^\beta{}_\gamma, \qquad (17.79)$$

Note that with the help of the gauge fields (17.15), the curvature tensor $R_{\mu\nu\alpha}{}^\gamma$ defined in Eq. (17.47) can be rewritten as

$$R_{\mu\nu\alpha}{}^\gamma = \Lambda_a{}^\gamma \left(\partial_\mu\partial_\nu - \partial_\nu\partial_\mu\right)\Lambda^a{}_\alpha = \partial_\mu\overset{\Lambda}{\Gamma}_{\nu\alpha}{}^\gamma - \partial_\nu\overset{\Lambda}{\Gamma}_{\mu\alpha}{}^\gamma - \overset{\Lambda}{\Gamma}_{\mu\alpha}{}^\delta\overset{\Lambda}{\Gamma}_{\nu\delta}{}^\gamma + \overset{\Lambda}{\Gamma}_{\nu\alpha}{}^\delta\overset{\Lambda}{\Gamma}_{\mu\delta}{}^\gamma. \quad (17.80)$$

This follows directly by performing the derivatives successively and inserting (17.15), while using the pseudo-orthogonality of the Lorentz matrices $\Lambda(x)$. On the right-hand side we recognize the standard covariant curl formed from the nonabelian gauge field $\overset{\Lambda}{\Gamma}_{\nu\alpha}{}^\gamma$ in the same way as in Eq. (17.55). Thus we shall denote the right-hand side by

$$F_{\mu\nu\alpha}{}^\gamma \equiv \partial_\mu\overset{\Lambda}{\Gamma}_{\nu\alpha}{}^\gamma - \partial_\nu\overset{\Lambda}{\Gamma}_{\mu\alpha}{}^\gamma - \overset{\Lambda}{\Gamma}_{\mu\alpha}{}^\delta\overset{\Lambda}{\Gamma}_{\nu\delta}{}^\gamma + \overset{\Lambda}{\Gamma}_{\nu\alpha}{}^\delta\overset{\Lambda}{\Gamma}_{\mu\delta}{}^\gamma = R_{\mu\nu\alpha}{}^\gamma = h_\alpha{}^\lambda R_{\mu\nu\lambda}{}^\kappa h^\gamma{}_\kappa. \quad (17.81)$$

It is instructive to prove this equality in another way using Eq. (17.68). This leads to the complicated expression

$$F_{\mu\nu\beta}{}^\gamma = \left\{\partial_\mu\left[(\Gamma - \overset{h}{\Gamma})_{\nu\lambda}{}^\kappa h_\beta{}^\lambda h^\alpha{}_\kappa\right] - (\mu \leftrightarrow \nu)\right\}$$
$$- \left\{(\Gamma - \overset{h}{\Gamma})_{\mu\lambda}{}^\tau (\Gamma - \overset{h}{\Gamma})_{\nu\tau}{}^\kappa h_\beta{}^\lambda h^\gamma{}_\kappa - (\mu \leftrightarrow \nu)\right\}, \qquad (17.82)$$

which may be regrouped to

$$\left[\partial_\mu\Gamma_{\nu\lambda}{}^\kappa - (\Gamma_\mu\Gamma_\nu)_\lambda{}^\kappa - (\mu \leftrightarrow \nu)\right]h_\beta{}^\lambda h^\gamma{}_\kappa$$
$$+ \left\{\Gamma_{\nu\lambda}{}^\kappa\partial_\mu\left(h_\beta{}^{\beta\lambda}h^\gamma{}_\kappa\right) - \partial_\mu\left(\overset{h}{\Gamma}_{\nu\lambda}{}^\kappa h_\beta{}^\lambda h^\gamma{}_\kappa\right) - (\mu \leftrightarrow \nu)\right\}$$
$$+ \left\{\left(\Gamma_\mu\overset{h}{\Gamma}_\nu + \overset{h}{\Gamma}_\mu\Gamma_\nu - \overset{h}{\Gamma}_\mu\overset{h}{\Gamma}_\nu\right)_\lambda{}^\kappa h_\beta{}^\lambda h^\gamma{}_\kappa - (\mu \leftrightarrow \nu)\right\}. \qquad (17.83)$$

Recalling (11.130) we see that the equality (17.81) is verified if we demonstrate the vanishing of the terms in curly brackets. The first term inside these brackets is

$$\Gamma_{\nu\lambda}{}^\kappa\partial_\mu h_\beta{}^\lambda h^\gamma{}_\kappa + \Gamma_{\nu\lambda}{}^\kappa h_\beta{}^\lambda\partial_\mu h^\gamma{}_\kappa - (\mu \leftrightarrow \nu) = -\Gamma_{\nu\lambda}{}^\gamma\overset{h}{\Gamma}_{\mu\beta}{}^\lambda + \Gamma_{\nu\beta}{}^\lambda\Gamma_{\nu\beta}{}^\kappa\overset{h}{\Gamma}_{\mu\kappa}{}^\gamma + (\mu \leftrightarrow \nu),$$

to which the second contributes [using (17.07)]

$$-\partial_\mu\left(h_\beta{}^\lambda\partial_\nu h^\gamma{}_\lambda\right) - (\mu \leftrightarrow \nu) = \overset{h}{\Gamma}_{\mu\beta}{}^\lambda\overset{h}{\Gamma}_{\mu\beta}{}^\gamma - (\mu \leftrightarrow \nu) - h_\beta{}^\lambda\left(\partial_\mu\partial_\nu - \partial_\nu\partial_\mu\right)h^\gamma{}_\lambda.$$

Thus we find indeed, recalling (17.43) and (17.44),

$$F_{\mu\nu\beta}{}^\gamma = \left(R_{\mu\nu\lambda}{}^\kappa - \overset{h}{R}_{\mu\nu\lambda}{}^\kappa\right)h_\beta{}^\lambda h^\gamma{}_\kappa = R_{\mu\nu\lambda}{}^\gamma h_\beta{}^\lambda h^\gamma{}_\kappa. \qquad (17.84)$$

17.3   Ricci Identity

The equality of the covariant curls of $F_{\mu\nu\alpha}{}^\gamma$ and $R_{\mu\nu\lambda}{}^\kappa$ up to a coordinate transformation of the last two indices is related to a fundamental algebraic property of covariant derivatives. Consider a vector field v_λ and apply a commutator of covariant derivatives to it, yielding

$$\left[D_\mu,\ D_\nu\right]v_\lambda = \partial_\mu\left(\partial_\nu v_\lambda - \Gamma_{\nu\lambda}{}^\kappa v_\kappa\right) - \Gamma_\mu{}^\tau D_\tau v_\lambda - \Gamma_{\mu\lambda}{}^\tau\left(\partial_\nu v_\tau - \Gamma_{\nu\tau}{}^\kappa v_\kappa\right) - (\mu \leftrightarrow \nu).$$
(17.85)

For a single-valued vector field satisfying $\left(\partial_\mu\partial_\nu - \partial_\nu\partial_\mu\right)v_\lambda = 0$, we obtain the so-called *Ricci identity*

$$\left[D_\mu,\ D_\nu\right]v_\lambda = -R_{\mu\nu\lambda}{}^\kappa v_\kappa - 2S_{\mu\nu}{}^\tau D_\tau v_\lambda.$$
(17.86)

For a general tensor, $R_{\mu\nu\lambda}{}^\kappa$ and $S_{\mu\nu}{}^\tau$ act separately on each index. Now, a similar relation may be calculated for the components of the vector in the nonholonomic basis $e_\alpha{}^\beta$:

$$\begin{aligned}
\left[D_\mu,\ D_\nu\right]v_\beta &= \partial_\mu\left(\partial_\nu v_\beta - \overset{\Lambda}{\Gamma}_{\nu\beta}{}^\gamma v_\gamma\right) - \Gamma_{\mu\nu}{}^\tau D_\tau v_\beta - \overset{\Lambda}{\Gamma}_{\mu\beta}{}^\gamma\left(\partial_\nu v_\gamma - \overset{\Lambda}{\Gamma}_{\nu\gamma}{}^\delta v_\delta\right) - (\mu \leftrightarrow \nu) \\
&= -F_{\mu\nu\beta}{}^\gamma v_\gamma - 2S_{\mu\nu}{}^\tau D_\tau v_\beta.
\end{aligned}$$
(17.87)

For a field of arbitrary spin, this generalizes to

$$\left[D_\mu,\ D_\nu\right]\psi = \frac{i}{2}F_{\mu\nu\beta}{}^\gamma\Sigma^\beta{}_\gamma\psi - 2S_{\mu\nu}{}^\tau D_\tau\psi.$$
(17.88)

Due to the complete covariance of (17.85) and (17.87), we may multiply (17.85) by $h_\beta{}^\lambda$ and pass this factor through the covariant derivatives (which, in this process, change their connection since they are applied to different objects before and after the passage). The R-term in (17.86) and the F-term in (17.88) remain simply related by (17.80).

17.4   Alternative Form of Coupling

Let us compare the above derived minimal coupling of the vierbein field $h_\alpha{}^\mu(x)$ to a spinning particle with the theory of Weyl, Fock, and Iwanenko [2, 3, 4] in Riemann spacetimes. These authors proposed to express the Dirac theory in curved spacetime in terms of x-dependent Dirac matrices similar to those in Eq. (17.10), but defined by

$$\gamma^\mu(x) = \gamma^\alpha h_\alpha{}^\mu(x).$$
(17.89)

These satisfy the local Dirac algebra [compare (17.11)]:

$$\{\gamma^\mu(x), \gamma^\nu(x)\} = g^{\mu\nu}(x).$$
(17.90)

In terms of these $\mu(x)$,, the Dirac action can be written as

$$\overset{m}{\mathcal{A}} = \int d^4x \, \sqrt{-g} \, \bar{\psi}(x) \left\{ i\gamma^\mu(x) D_\mu - m \right\} \psi(x), \tag{17.91}$$

where D_μ is the covariant derivative (omitting the Dirac spin indices)

$$D_\mu = \partial_\mu \delta_\mu - \Gamma_\mu(x), \tag{17.92}$$

with the Dirac spin connection [compare (17.16)]

$$\Gamma_\mu(x) \equiv -\frac{1}{4}\gamma_\lambda(x) D_\mu \gamma^\lambda(x) = -\frac{1}{4}\gamma_\lambda(x) \left[\partial_\mu \gamma^\lambda(x) + \Gamma_{\mu\nu}{}^\lambda(x)\gamma^\nu(x) \right]. \tag{17.93}$$

Let us verify that the action (17.91) is equivalent to the previous one in (17.78) if there is no torsion. Inserting (17.89) into (17.93) we find

$$\Gamma_\mu = -\left(h_{\alpha\lambda}\partial_\mu h_\beta{}^\lambda + h_{\lambda\alpha} h_\beta{}^\nu \Gamma_{\mu\nu}{}^\lambda \right) \frac{1}{4}\gamma^\alpha\gamma^\beta. \tag{17.94}$$

We now use the analog of (11.86) for $h_\alpha{}^\nu(x)$ [which follows from the completeness relation (17.23)]

$$\partial_\mu h_\beta{}^\lambda = -h_\beta{}^\nu \left(h_\gamma{}^\lambda \partial_\mu h^\gamma{}_\nu \right). \tag{17.95}$$

Then we can rewrite (17.94) as

$$\Gamma_\mu = \left(\overset{h}{\Gamma}_{\mu\nu}{}^\lambda - \Gamma_{\mu\nu}{}^\lambda \right) \frac{1}{4}\gamma^\lambda\gamma^\nu, \tag{17.96}$$

where we have used the definition (17.67). Comparison with (17.68) shows that

$$\Gamma_\mu = -\overset{\Lambda}{\Gamma}_{\mu\beta\alpha} \frac{1}{4}\gamma^\alpha\gamma^\beta. \tag{17.97}$$

Since $\overset{\Lambda}{\Gamma}_{\mu\alpha\beta}$ is antisymmetric in $\alpha\beta$, this is the same as [recall (1.228)]

$$\Gamma_\mu = -\frac{i}{2}\overset{\Lambda}{\Gamma}_{\mu\alpha\beta}\Sigma^{\alpha\beta}, \tag{17.98}$$

in agreement with (17.18), if the Dirac indices are added.

17.5   Invariant Action for Vector Fields

Any theory which is invariant under general coordinate transformations can be recast in such a way that its derivatives refer to the nonholonomic coordinates dx^α. Since the metric in these coordinates is $g^{\alpha\beta}$, the action has the same form as those in flat spacetime, except that derivatives of vector and tensor fields are replaced by covariant ones, for example

$$\partial_\alpha v_\beta \to D_\alpha v_\beta = \partial_\alpha v_\beta - \overset{\Lambda}{\Gamma}_{\alpha\beta}{}^\gamma v_\gamma. \tag{17.99}$$

For example,

$$\mathcal{A} = \int d^4x^\alpha D^\alpha v_\beta(x) D_\alpha v^\beta(x) \tag{17.100}$$

is the nonholonomic form of a general invariant action. As stated in Subsection (17.1), the specification of spacetime points must be made with x^μ-coordinates. For this reason the action is preferably written as

$$\mathcal{A} = \int d^4x^\mu \sqrt{-g} D_\alpha v_\beta(x^\mu) D^\alpha v^\beta(x^\mu). \tag{17.101}$$

Under a general coordinate transformation à la Einstein, $dx^\mu \to dx'^{\mu'} = dx^\mu \alpha_\mu{}^{\mu'}$, the indices α are inert. For instance, $h_\alpha{}^\mu$ itself transforms as

$$h_\alpha{}^\mu(x) \xrightarrow[E]{} h_\alpha{}^{\mu'}(x') = h_\alpha{}^\mu(x)\alpha_\mu{}^{\mu'}. \tag{17.102}$$

Vectors and tensors with indices $\alpha, \beta, \ldots$ experience only changes of their arguments $x \to x - \xi$ so that their infinitesimal substantial changes are

$$\delta_E v_\alpha(x) = \xi^\lambda \partial_\lambda v_\alpha(x) \tag{17.103}$$
$$\delta_E D_\alpha v_\beta(x) = \xi^\lambda \partial_\lambda D_\alpha v_\beta(x). \tag{17.104}$$

The freedom in choosing $h_\alpha{}^\mu(x)$ up to a local Lorentz transformation in the "matrix square root" (2.50) of $g_{\mu\nu}(x)$ implies that the theory should be invariant under the infinitesimal Lorentz transformations

$$\delta_L dx^\alpha = \omega^\alpha{}_\beta(x)dx^\beta, \tag{17.105}$$
$$\delta_L h_\alpha{}^\mu(x) = \omega_\alpha{}^\beta(x)h_\beta{}^\mu(x). \tag{17.106}$$

Here $\omega_\alpha{}^{\alpha'}(x)$ are infinitesimal transformation parameters (11.55) and (11.56).

Indeed the action (17.101) is automatically invariant if every index α is transformed accordingly:

$$\delta_L v_\alpha(x) = \omega_\alpha{}^{\alpha'} v_{\alpha'}(x), \qquad \delta_L v^\alpha = \omega^\alpha{}_{\alpha'}(x)v^{\alpha'}. \tag{17.107}$$

and

$$\delta_L D_\alpha v_\beta(x) = \omega_\alpha{}^{\alpha'}(x)D_{\alpha'}v_\beta(x) + \omega_\beta{}^{\beta'}(x)D_\alpha v_{\beta'}(x), \tag{17.108}$$
$$\delta_L D_\alpha v^\beta(x) = \omega_\alpha{}^{\alpha'}(x)D_{\alpha'}v^\beta(x) + \omega^\beta{}_{\beta'}(x)D_\alpha v^{\beta'}(x). \tag{17.109}$$

The variables x^μ are unchanged since the local Lorentz transformations (17.105) affect only the intermediate local directions defined by the differentials dx^α. They leave the physical coordinate x^μ unchanged.

Let us verify explicitly the properties (17.108) and (17.109) of the covariant derivatives under local Lorentz transformations. The substantial variations of the ordinary derivative $\partial_\alpha v_\beta$ is:

$$
\begin{aligned}
\delta_L \partial_\alpha v_\beta &= (\delta_L \partial_\alpha) v_\beta + \partial_\alpha (\delta_L v_\beta) \\
&= \omega_\alpha{}^{\alpha'} \partial_{\alpha'} v_\beta + \partial_\alpha (\omega_\beta{}^{\beta'} v_{\beta'}) \\
&= \omega_\alpha{}^{\alpha'} \partial_{\alpha'} v_\beta + \omega_\beta{}^{\beta'} \partial_\alpha v_{\beta'} + (\partial_\alpha \omega_\beta{}^{\beta'}) v_{\beta'}.
\end{aligned} \tag{17.110}
$$

The spin connection in the covariant derivative $D_\alpha v_\beta$ of Eq. (17.99) contains two terms [recall (17.68)] which we consider separately. Both are multiplied by a factor $h_\lambda{}^\gamma h_\alpha{}^\mu h_\beta{}^\nu$. The first term, call it $\overset{\Lambda}{\Gamma}{}^{(1)}{}_{\alpha\beta}{}^\gamma$, contains the contortion tensor $K_{\mu\nu}{}^\lambda$ multiplied by $h_\lambda{}^\gamma h_\alpha{}^\mu h_\beta{}^\nu$, and transforms therefore like a tensor

$$
\delta_L \overset{\Lambda}{\Gamma}{}^{(1)}{}_{\alpha\beta}{}^\gamma = \omega_\alpha{}^{\alpha'} \overset{\Lambda}{\Gamma}{}^{(1)}{}_{\alpha'\beta}{}^\gamma + \omega_\beta{}^{\beta'} \overset{\Lambda}{\Gamma}{}^{(1)}{}_{\alpha\beta'}{}^\gamma + \omega^\gamma{}_{\gamma'} \overset{\Lambda}{\Gamma}{}^{(1)}{}_{\alpha\beta}{}^{\gamma'}. \tag{17.111}
$$

The second term in (17.68) contains $\overset{h}{\Gamma}_{\mu\nu}{}^\lambda$ multiplied by $h_\lambda{}^\gamma h_\alpha{}^\mu h_\beta{}^\nu$, and is not a tensor. Its substantial variation contains a nontensorial derivative contribution:

$$
\begin{aligned}
\delta_L \overset{h}{\Gamma}_{\mu\nu}{}^\lambda &= (\delta_L h_\delta{}^\lambda) \partial_\mu h^\delta{}_\nu + h_\delta{}^\lambda \partial_\mu (\delta_L h^\delta{}_\nu) \\
&= \omega_\delta{}^{\delta'} h_{\delta'}{}^\lambda \partial_\mu h^\delta{}_\nu + h_\delta{}^\lambda \partial_\mu (\omega^\delta{}_{\delta'} h^{\delta'}{}_\nu) \\
&= \omega_\delta{}^{\delta'} h_{\delta'}{}^\lambda \partial_\mu h^\delta{}_\nu + \omega^\delta{}_{\delta'} h_\delta{}^\lambda \partial_\mu h^{\delta'}{}_\nu + \partial_\mu \omega^\delta{}_{\delta'} (h_\delta{}^\lambda h^{\delta'}{}_\nu) \\
&= \partial_\mu \omega^\delta{}_{\delta'} h^{\delta'}{}_\nu = -\partial_\mu \omega_{\delta'}{}^\delta h_\delta{}^\lambda h^{\delta'}{}_\nu,
\end{aligned} \tag{17.112}
$$

the cancellation in the third line being due to the antisymmetry of $\omega_\delta{}^{\delta'} = -\omega^{\delta'}{}_\delta$. Multiplication with $h_\lambda{}^\gamma h_\alpha{}^\mu h_\beta{}^\nu$ and use of use of Eq. (17.106) yields

$$
\delta_L \overset{h}{\Gamma}_{\mu\nu}{}^\lambda = \delta_{L_0} \overset{h}{\Gamma}_{\mu\nu}{}^\lambda - \partial_\alpha \omega_\beta{}^\gamma \tag{17.113}
$$

where $\delta_{L_0} \overset{h}{\Gamma}_{\mu\nu}{}^\lambda$ indicates the terms of the type (17.111) which would arise if $\overset{h}{\Gamma}_{\mu\nu}{}^\lambda$ were a tensor.

Inserting (17.113) together with (17.111) into Eq. (17.68) we obtain

$$
\delta_L \overset{\Lambda}{\Gamma}_{\alpha\beta}{}^\gamma = \delta_{L_0} \overset{\Lambda}{\Gamma}_{\alpha\beta}{}^\gamma + \partial_\alpha \omega_\beta{}^\gamma. \tag{17.114}
$$

The last term cancels the last nontensorial part in (17.110), so that the covariant derivative $D_\alpha v_\beta$ has indeed the covariant transformation law (17.108). The law (17.109) follows by raising the index β with the help of the inverse Minkowski metric $g^{\alpha\beta}$. In the notation (17.114) for the spin connection, the transformation law (17.114) reads

$$
\delta_L A_{\alpha\beta}{}^\gamma = \omega_\beta{}^{\beta'} A_{\alpha\beta'}{}^\gamma + \omega_\alpha{}^{\alpha'} A_{\alpha'\beta}{}^\gamma + \omega^\gamma{}_{\gamma'} A_{\alpha\beta}{}^{\gamma'} + \partial_\alpha \omega_\beta{}^\gamma. \tag{17.115}
$$

For later convenience, we convert the subscript α to the spacetime index μ and define the $A_{\mu\beta}{}^\gamma \equiv h^\alpha{}_\mu A_{\alpha\beta}{}^\gamma$. It changes under local Lorentz transformations like

$$
\delta_L A_{\mu\beta}{}^\gamma = \omega_\beta{}^{\beta'} A_{\mu\beta'}{}^\gamma + \omega^\gamma{}_{\gamma'} A_{\mu\beta}{}^{\gamma'} + \partial_\mu \omega_\beta{}^\gamma. \tag{17.116}
$$

17.6 Verifying Local Lorentz Invariance

Let us study the invariance under local Lorentz transformations in more detail. These serve to go at an arbitrary point x^μ from one freely falling elevator to another. A spinor field $\psi(x)$ transforms under them as follows:

$$\delta_L \psi(x) = -\frac{i}{2}\omega^{\alpha\beta}(x)\Sigma_{\alpha\beta}\psi(x). \tag{17.117}$$

Here $\Sigma_{\alpha\beta}$ are the spin representation matrices (1.227) of the local Lorentz group. They are antisymmetric in α, β and satisfy the commutation relations (1.226):

$$\left[\Sigma_{\alpha\beta}, \Sigma_{\alpha\gamma}\right] = -ig_{\alpha\alpha}\Sigma_{\beta\gamma}, \quad \text{no sum over } \alpha. \tag{17.118}$$

They may be expressed as commutators of Dirac matrices [recall (1.227) and (1.228)]:

$$\Sigma_{\alpha\beta} = \frac{i}{4}\left[\gamma_\alpha, \gamma_\beta\right]. \tag{17.119}$$

The infinitesimal Lorentz transformation of the derivative of ψ is

$$\begin{aligned}
\delta_L \partial_\alpha \psi &= \omega_\alpha{}^{\alpha'}\partial_{\alpha'}\psi + \partial_\alpha \delta_L \psi \\
&= \omega_\alpha{}^{\alpha'}\partial_{\alpha'}\psi - \frac{i}{2}\partial_\alpha\left(\omega^{\beta\gamma}\Sigma_{\beta\gamma}\right)\psi \\
&= \omega_\alpha{}^{\alpha'}\partial_{\alpha'}\psi - \frac{i}{2}\omega^{\beta\gamma}\Sigma_{\beta\gamma}\partial_\alpha\psi - \frac{i}{2}\left(\partial_\alpha\omega^{\beta\gamma}\right)\Sigma_{\beta\gamma}\psi.
\end{aligned} \tag{17.120}$$

The first two terms exhibit the Lorentz transformation properties of $\partial_\alpha\psi$ for fixed angles $\omega^{\beta\gamma}$. The last term is due to the dependence of $\omega^{\beta\gamma}(x)$ on x. It can be removed by going to the covariant derivative (17.18) formed with the spin connection (17.16):

$$D_\alpha \psi(x) \equiv \partial_\alpha \psi(x) + \frac{i}{2}\overset{\Lambda}{\Gamma}_{\alpha\beta}{}^\gamma\Sigma^\beta{}_\gamma\psi(x). \tag{17.121}$$

Indeed, if we calculate the variation of the second term in $D_\alpha\psi(x) \equiv \partial_\alpha\psi(x)$:

$$\delta_L \frac{i}{2}\overset{\Lambda}{\Gamma}_{\alpha\beta}{}^\gamma\Sigma^\beta{}_\gamma\psi(x), \tag{17.122}$$

we obtain two terms. There is a term with the regular Lorentz transformation property

$$\delta_{L0}\frac{i}{2}\overset{\Lambda}{\Gamma}_{\alpha\beta}{}^\gamma\Sigma^\beta{}_\gamma\psi = -\frac{i}{2}\omega^{\sigma\tau}\Sigma_{\sigma\tau}\left(\frac{i}{2}\overset{\Lambda}{\Gamma}_{\alpha\beta}{}^\gamma\Sigma^\beta{}_\gamma\psi\right). \tag{17.123}$$

This follows from

$$\frac{i}{2}\delta_L\overset{\Lambda}{\Gamma}_{\alpha\beta}{}^\gamma\Sigma^\beta{}_\gamma\psi + \frac{i}{2}\overset{\Lambda}{\Gamma}_{\alpha\beta}{}^\gamma\Sigma^\beta{}_\gamma\delta_L\psi, \tag{17.124}$$

and an application of the commutation rule (17.118). A second term arises from $\partial_\alpha\omega_\beta{}^\gamma$, which is

$$\frac{i}{2}\partial_\alpha\omega_\beta{}^\gamma\Sigma^\beta{}_\gamma\psi \tag{17.125}$$

and cancels against the last term in (17.120). Thus $D_\alpha \psi$ behaves like

$$\delta_L D_\alpha \psi = \omega_\alpha{}^{\alpha'}(x) D_{\alpha'} \psi - \frac{i}{2} \omega^{\beta\gamma}(x) \Sigma_{\beta\gamma} D_\alpha \psi, \tag{17.126}$$

and represents, therefore, a proper covariant derivative which generalizes the standard Lorentz transformation behavior to the case of local transformations $\omega_\alpha{}^\beta(x)$.

The transformation law (17.117) holds for fields of any spin, if $\Sigma_{\alpha\beta}$ is replaced by the appropriate representation of the Lorentz group. As an example, take a vector field, where the spin representation matrices are given by the defining Lorentz generators (1.51):

$$\left(L_{\alpha\beta}\right)_{\alpha'\beta'} = i\left[g_{\alpha\alpha'} g_{\beta\beta'} - (\alpha \leftrightarrow \beta)\right], \quad \left(L_{\alpha\beta}\right)_{\alpha'}{}^{\beta'} = i\left[g_{\alpha\alpha'} \delta_\beta{}^{\beta'} - (\alpha \leftrightarrow \beta)\right]. \tag{17.127}$$

If these matrices are inserted for $\Sigma_{\alpha\beta}$ in the transformation law (17.117), we recover the symmetry transformation law (17.107) for vector fields:

$$\delta_L v_\alpha = -\frac{i}{2} \omega^{\gamma\delta} i \left(g_{\gamma\alpha} \delta_\delta{}^\beta - g_{\delta\alpha} \delta_\gamma{}^\beta\right) v_\beta = \omega_\alpha{}^\beta v_\beta. \tag{17.128}$$

17.7 Field Equations with Spinning Matter

Consider the action of a spin-1/2 field interacting with a gravitational field:

$$\begin{aligned} \mathcal{A}[h, K, \psi] &= -\frac{1}{2K} \int d^4x \sqrt{-g} R + \frac{1}{2} \int d^4x \sqrt{-g} \bar\psi \gamma^\alpha D_\alpha \psi(x) + \text{h.c.} \\ &= \overset{\text{f}}{\mathcal{A}}[h, K] + \overset{\text{m}}{\mathcal{A}}[h, K, \psi], \end{aligned} \tag{17.129}$$

where the covariant derivatives are expressed in terms of $K_{\mu\nu}{}^\lambda - \overset{h}{K}_{\mu\nu}{}^\lambda$ as in Eq. (17.79). The action (17.129) is a functional of the vierbein field $h_\alpha{}^\mu$, the contortion $K_{\mu\nu}{}^\lambda$ and the Dirac field $\psi(x)$. Varying $\mathcal{A}$ with respect to $\bar\psi$ we obtain the equation of motion

$$\frac{\delta \overset{\text{m}}{\mathcal{A}}}{\delta \bar\psi} = \sqrt{-g}(\gamma^\alpha D_\alpha - m)\psi(x) = 0 \tag{17.130}$$

of a Dirac particle in a general affine spacetime.

To obtain the gravitational field equations we again define the spin current density, just as we did in (15.18), by differentiating with respect to $K_{\mu\nu}{}^\lambda$ at fixed $h_\alpha{}^\mu$, thus obtaining for the gravitational field

$$\frac{\delta \overset{\text{f}}{\mathcal{A}}}{\delta K_{\mu\nu}{}^\lambda} = -\frac{1}{2}\sqrt{-g} \overset{\text{f}}{\Sigma}{}^\nu{}_\lambda{}^{,\mu}, \tag{17.131}$$

as in (15.65).

From the matter action (17.18) we obtain

$$\sqrt{-g}\,\overset{m}{\Sigma}{}^{\nu}{}_{\lambda}{}^{,\mu} \equiv 2\frac{\delta\overset{m}{\mathcal{A}}}{\delta K_{\mu\nu}{}^{\lambda}} = \sqrt{-g}\left[-\frac{i}{2}\bar{\psi}(x)\gamma^{\mu}(x)\Sigma^{\nu}{}_{\lambda}(x)\psi(x) + \text{h.c.}\right]$$

$$= h_{\gamma}{}^{\lambda}h_{\alpha}{}^{\mu}h_{\beta}{}^{\nu}\sqrt{-g}\left[-\frac{i}{2}\bar{\psi}(x)\gamma^{\alpha}\Sigma^{\beta}{}_{\gamma}\psi(x) + \text{h.c.}\right]$$

$$= h_{\gamma}{}^{\lambda}h_{\alpha}{}^{\mu}h_{\beta}{}^{\nu}h_{\beta}{}^{\nu}\sqrt{-g}\,\overset{m}{\Sigma}{}^{\beta}{}_{\gamma}{}^{,\alpha}. \qquad (17.132)$$

The expression in brackets is recognized as the canonical spin current density $\overset{m}{\Sigma}{}^{\beta}{}_{\gamma}{}^{,\alpha}$ of a Dirac particle in Minkowski spacetime derived in Eq. (3.228). Equation (17.132) provides us with the generally covariant version of this. Thus, for the spin-1/2 field, the definition (15.17) of the spin current density is consistent with the canonical definition:

$$\overset{m}{\Sigma}{}^{\nu}{}_{\lambda}{}^{,\mu} \equiv -i\sum_{i}\pi_{i}{}^{\mu}\Sigma^{\nu}{}_{\lambda}\varphi_{i} = -i\sum_{i}\frac{\partial\overset{m}{\mathcal{L}}}{\partial D_{\mu}\varphi_{i}}\Sigma^{\nu}{}_{\lambda}\varphi_{i}, \qquad (17.133)$$

where the sum over i runs over all matter fields of the system. This is true for all matter fields due to the fact that the general Einstein-invariant matter action has the functional form [compare (17.18)]

$$\overset{m}{\mathcal{A}} = \overset{m}{\mathcal{A}}[h, K, \varphi_{i}] = \int d^{4}x\sqrt{-g}\,\overset{m}{\mathcal{L}}\left(h_{\alpha}{}^{\mu}, \varphi_{i}, D_{\mu}\varphi_{i}\right). \qquad (17.134)$$

This implies indeed that for fixed $h_{\alpha}{}^{\mu}$:

$$2\frac{\delta\overset{m}{\mathcal{A}}}{\delta K_{\mu\nu}{}^{\lambda}}\bigg|_{h_{\alpha}{}^{\mu}} = 2\sqrt{-g}\sum_{i}\frac{\partial\overset{m}{\mathcal{L}}}{\partial D_{\mu}\varphi_{i}}\frac{i}{2}\Sigma^{\nu}{}_{\lambda}\varphi_{i}$$

$$\equiv i\sqrt{-g}\sum_{i}\pi_{i}{}^{\mu}\Sigma^{\nu}{}_{\lambda}\varphi_{i} = -\overset{m}{\Sigma}{}^{\nu}{}_{\lambda}{}^{,\mu}. \qquad (17.135)$$

By varying the total action of fields and matter with respect to $\delta K_{\mu\nu}{}^{\lambda}$, we therefore obtain the field equation

$$-\kappa\overset{f}{\Sigma}{}^{\nu}{}_{\lambda}{}^{,\mu} = \kappa\overset{m}{\Sigma}{}^{\nu}{}_{\lambda}{}^{,\mu}, \qquad (17.136)$$

thus extending the field equation (15.60) to systems with spinning matter. Together with Eq. (15.56), this determines the Palatini tensor (15.47):

$$S_{\mu\nu,\lambda} = -\kappa\overset{m}{\Sigma}_{\mu\nu,\lambda}, \qquad (17.137)$$

and thus, via Eq. (15.48), the torsion of spacetime by the field equation:

$$S_{\mu\nu\lambda} = \frac{\kappa}{2}\left(\overset{m}{\Sigma}_{\mu\nu,\lambda} + \frac{1}{2}g_{\nu\lambda}\overset{m}{\Sigma}_{\mu\kappa}{}^{,\kappa} - \frac{1}{2}g_{\nu\lambda}\overset{m}{\Sigma}_{\mu\kappa}{}^{,\kappa}\right). \qquad (17.138)$$

Let us now turn to the field equations arising from extremization with respect to $h_{\alpha}{}^{\mu}$. We define the total energy-momentum tensor as

$$\sqrt{-g}T_{\mu}{}^{\alpha}(x) \equiv \frac{\delta\mathcal{A}}{\delta h_{\alpha}{}^{\mu}(x)}\bigg|_{S_{\mu\nu}{}^{\lambda}}, \qquad (17.139)$$

with the derivative formed at fixed $S_{\mu\nu}{}^{\lambda}$. Due to the relation (17.95), we may use the chain rule of differentiation to write alternatively

$$\sqrt{-g}\,T_{\alpha}{}^{\mu}(x) = -\left.\frac{\delta\mathcal{A}}{\delta h^{\alpha}{}_{\mu}(x)}\right|_{S_{\mu\nu}{}^{\lambda}}. \tag{17.140}$$

For the pure gravitational action which depends only on $g^{\mu\nu} = h^{\alpha\mu}h_{\alpha}{}^{\nu}$ and $K_{\mu\nu}{}^{\lambda}$, this definition leads trivially to the same symmetric energy-momentum tensor as the one introduced earlier in (15.16), except that one index refers to the basis $\mathbf{e}_{\alpha}(x)$. This follows from the chain rule of differentiation, and using (15.16):

$$\sqrt{-g}\,\overset{f}{T}_{\mu}{}^{\alpha} \equiv \frac{\delta\overset{f}{\mathcal{A}}}{\delta h_{\alpha}{}^{\mu}} = \frac{\delta\overset{f}{\mathcal{A}}}{\delta g^{\lambda\kappa}}\frac{\partial g^{\lambda\kappa}}{\partial h_{\alpha}{}^{\mu}} = \sqrt{-g}\,\overset{f}{T}_{\mu\kappa}\,h^{\alpha\kappa}. \tag{17.141}$$

There is, of course, a similar rule involving the derivative with respect to $h^{\alpha}{}_{\mu}$ as in (17.140):

$$\sqrt{-g}\,\overset{f}{T}_{\alpha}{}^{\mu} \equiv -\frac{\delta\overset{f}{\mathcal{A}}}{\delta h^{\alpha}{}_{\mu}} = -\frac{\delta\overset{f}{\mathcal{A}}}{\delta g_{\lambda\kappa}}\frac{\partial g_{\lambda\kappa}}{\partial h^{\alpha}{}_{\mu}} = \sqrt{-g}\,\overset{f}{T}{}^{\kappa\mu}h_{\alpha\kappa}. \tag{17.142}$$

For matter fields, the actual calculation of the symmetric energy-momentum tensor is most conveniently performed in two steps. Take, for instance, the Dirac field. As a first step we differentiate $\sqrt{-g}$ and $\gamma^{\alpha}h_{\alpha}{}^{\mu}\partial_{\mu}$ with respect to $h_{\alpha}{}^{\mu}$ while keeping, for the moment, $D_{\mu} = \text{const.}$ The result is the so-called canonical energy-momentum tensor:

$$\sqrt{-g}\,\overset{m}{\Theta}_{\mu}{}^{\alpha} \equiv \sqrt{-g}\frac{1}{2}\left(\bar{\psi}\gamma^{\alpha}iD_{\mu}\psi - h^{\alpha}{}_{\mu}\overset{m}{\mathcal{L}}\right) + \text{h.c.} \tag{17.143}$$

This is a general feature of the formalism: The derivative of (17.134) with respect to the $h_{\alpha}{}^{\mu}$ fields contained in the covariant derivative $D_{\mu}\varphi_i = h_{\mu}^{\alpha}D_{\alpha}\varphi_i$ gives

$$\frac{\delta\overset{m}{\mathcal{A}}}{\delta h_{\alpha}{}^{\mu}} \longrightarrow \sqrt{-g}\sum_{i}\frac{\partial\overset{m}{\mathcal{L}}}{\partial D_{\nu}\varphi_i}D_{\mu}\varphi_i\,h^{\alpha\nu}. \tag{17.144}$$

The derivative of (17.134) with respect to $h_{\alpha}{}^{\mu}$ contained in the $\sqrt{-g}$-term adds to this

$$\frac{\delta\overset{m}{\mathcal{A}}}{\delta h_{\alpha}{}^{\mu}} \longrightarrow -\sqrt{-g}g_{\mu\nu}\overset{m}{\mathcal{L}}\,h^{\alpha\nu}. \tag{17.145}$$

The sum of the two contributions yields

$$\overset{m}{\Theta}_{\mu}{}^{\alpha} = \left(\sum_{i}\frac{\partial L}{\partial D_{\nu}\varphi_i}D_{\mu}\varphi_i - g_{\mu\nu}\overset{m}{\mathcal{L}}\right)h^{\alpha\nu}, \tag{17.146}$$

which is indeed the canonical energy-momentum tensor for an arbitrary Lagrangian containing covariant derivatives.

Applying this formalism to a pure gravitational field we can compare the first step of differentiation at fixed D_μ with the variation (15.29) and find the symmetric part of the equation

$$\overset{f}{\Theta}_\mu{}^\alpha = -\frac{1}{\kappa} G_{\mu\nu} h^\alpha{}_\nu. \tag{17.147}$$

We will see below that this holds, in fact, without symmetrization. Thus the canonical energy-momentum tensor of the gravitational field is equal to minus $1/\kappa$ times the Einstein tensor.

We now turn to the second step, the calculation of the remaining functional derivative with respect to $h_\alpha{}^\mu$. This is somewhat tedious. Let us write the additional contribution to $\overset{m}{\Theta}_\kappa{}^\delta$ as

$$\sqrt{-g}\, \delta \overset{m}{\Theta}_\kappa{}^\delta = \int d^4x \frac{\delta \overset{m}{\mathcal{A}}}{\delta K_{\mu\beta}{}^\gamma} \frac{\delta \overset{\Lambda}{\Gamma}_{\mu\beta}{}^\gamma}{\delta h_\delta{}^\kappa}\bigg|_{S_{\mu\nu\lambda}} = -\frac{1}{2} \int d^4x \sqrt{-g}\, \overset{m}{\Sigma}{}^\beta{}_\gamma{}^{,\mu} \frac{\delta \overset{\Lambda}{\Gamma}_{\mu\beta}{}^\gamma}{\delta h_\delta{}^\kappa}\bigg|_{S_{\mu\nu\lambda}}, \tag{17.148}$$

and use for the spin connection the explicit form

$$\overset{\Lambda}{\Gamma}_{\mu\beta}{}^\gamma = h^\gamma{}_\lambda h_\beta{}^\nu (\Gamma_{\mu\nu}{}^\lambda - \overset{h}{\Gamma}_{\mu\nu}{}^\lambda) = -h_\beta{}^\nu \overset{\Gamma}{D}_\mu h^\gamma{}_\nu = h^\gamma{}_\nu \overset{\Gamma}{D}_\mu h_\beta{}^\nu \tag{17.149}$$

where $\overset{\Gamma}{D}_\mu$ denotes the part of the covariant derivative containing only the ordinary connection $\Gamma_{\mu\beta}{}^\lambda$. If we vary $\delta h_{\mu\beta}{}^\gamma$ and hold $\Gamma_{\mu\nu}{}^\lambda$ fixed, we have

$$\delta \overset{\Lambda}{\Gamma}_{\mu\beta}{}^\gamma\big|_{\Gamma_{\mu\nu}{}^\lambda} = \delta h^\gamma{}_\nu \overset{\Gamma}{D}_\mu h_\beta{}^\nu + h^\gamma{}_\nu \overset{\Gamma}{D} \delta h_\beta{}^\nu. \tag{17.150}$$

Since $D_\mu h^\gamma{}_\nu = 0$ [recall (17.66)], we see that $\overset{\Gamma}{D}_\mu h_\beta{}^\nu = \overset{\Lambda}{\Gamma}_{\mu\beta}{}^\lambda h_\lambda{}^\nu$ and we may write

$$\delta \overset{\Lambda}{\Gamma}_{\mu\beta}{}^\gamma\big|_{\Gamma_{\mu\nu}{}^\lambda} = h^\gamma{}_\nu D_\mu \delta h_\beta{}^\nu. \tag{17.151}$$

Inserting this into (17.148), a partial integration gives the first contribution

$$\Delta_1 \overset{m}{\Theta}_\kappa{}^\delta = -(1/2) D_\mu \overset{m}{\Sigma}_\kappa{}^{\delta,\mu}. \tag{17.152}$$

We now include the contribution from $\delta\Gamma_{\mu\nu}{}^\lambda$. Using the decomposition (15.52) with $\delta S_{\mu\nu\lambda} = 0$, i.e. $\delta K_{\mu\nu\lambda} = 0$, we find

$$\Delta_2 \overset{m}{\Theta}_\kappa{}^\delta = \frac{1}{4} \left[D_\mu \left(\overset{m}{\Sigma}{}^{\nu\sigma,\mu} - \overset{m}{\Sigma}{}^{\sigma\mu,\nu} + \overset{m}{\Sigma}{}^{\mu\nu,\sigma} \right) \right] \frac{\partial g_{\nu\sigma}}{\partial h_\delta{}^\kappa}. \tag{17.153}$$

With

$$\frac{\partial g_{\nu\sigma}}{\partial h_\delta{}^\kappa} = g_{\nu\kappa} h^\delta{}_\sigma + (\nu \leftrightarrow \sigma), \tag{17.154}$$

this gives, altogether,

$$\Delta \overset{m}{\Theta}{}_\kappa{}^\delta(x) = -\frac{1}{2} D^*_\mu \left(\overset{m}{\Sigma}{}_\kappa{}^{\delta,\,\mu} - \overset{m}{\Sigma}{}^{\delta\mu}{}_{,\,\kappa} + \overset{m}{\Sigma}{}^\mu{}_{\kappa,\,\delta} \right). \tag{17.155}$$

This is precisely the same type of correction $\Delta\Theta_\kappa{}^\delta = \Delta\Theta_\kappa{}^\nu h^\delta{}_\nu$ that was added to the canonical energy-momentum tensor $\Theta_{\kappa\delta}$ of the gravitational field in Eq. (15.59), in order to produce the symmetric one $T_{\kappa\delta}$. Here it is obtained for arbitrary spinning matter fields:

$$\overset{m}{T}{}_{\kappa\nu} = \overset{m}{\Theta}{}_{\kappa\nu} + \Delta \overset{m}{\Theta}{}_{\kappa\nu} + \Delta \overset{m}{\Theta}{}_{\kappa\nu} = -\frac{1}{2} D^*_\mu \left(\overset{m}{\Sigma}{}_{\kappa\nu}{}^{,\,\mu} - \overset{m}{\Sigma}{}_\nu{}^\mu{}_{,\,\kappa} + \overset{m}{\Sigma}{}^\mu{}_{\kappa,\,\nu} \right). \tag{17.156}$$

For spin $\frac{1}{2}$, this is the expression (3.231) found by Belinfante in 1939. We have lowered the index ν on both sides which is permissible due to the covariant form of the equation.

In terms of $\overset{m}{T}{}_{\mu\nu}$, the field equations which follows from variations of the action with respect to $\delta h_\alpha{}^\mu$ have once more the simple form (15.63):

$$G^{\mu\nu} = \kappa \overset{m}{T}{}^{\mu\nu}, \tag{17.157}$$

with the energy-momentum tensors (17.156) of spinning matter.

Notes and References

[1] This Chapter follows largely the textbook
H. Kleinert, *Gauge Fields in Condensed Matter*, Vol. II, *Stresses and Defects*, World Scientific, Singapore, 1989 (kl/b2), See in particular pp. 1338–1377 (kl/b1/gifs/v1-1338s.html).

[2] H. Weyl, Z. Phys. **56**, 330 (1929).

[3] V. Fock, Z. Phys. **57**, 261 (1929).

[4] V. Fock and D. Iwanenko, Phys. Z. **30**, 648 (1929).

[5] See S. Weinberg's article in Physics Today, April 2006, p. 10, where he comments on various letters responding to his previous November 2005 article on Einstein's mistakes. In the April article he remarks that he never understood what is physically so important about the torsion tensor. This prompted F.W. Hehl to a further comment in the March 2007 issue where he reminded Weinberg of Sciama and Kibble's way of expressing Einstein-Cartan gravity in terms of gauge fields of Lorentz transformations and translations, in which curvature and torsion are the gauge-invariant field strengths. Hehl's answer did not satisfy Weinberg who could only be convinced unless there is an invariance principle which requires the Christoffel symbol to be accompanied by a torsion tensor in the affine connection. As we have mentioned in Section 20.2, the multivalued mapping principle of Chapter 14 is precisely such an invariance principle.

18

Covariant Conservation Law

According to Noether's theorem derived in Chapter 3, the invariance of the action under general coordinate transformations as well as local Lorentz transformations must be associated with certain conservation laws. For the following considerations, we shall consider the vierbein field $h_\alpha{}^\mu(x)$ and the spin connection $\overset{\Lambda}{\Gamma}{}_{\alpha\beta}{}^\gamma$ as independent variables. The latter appear in this chapter always in the notation $A_{\alpha\beta}{}^\gamma$ of Eq. (17.19). Moreover, we shall convert the subscript α to the spacetime index μ and express all equations in terms of the gauge field $A_{\mu\beta}{}^\gamma \equiv h^\alpha{}_\mu A_{\alpha\beta}{}^\gamma$.

From the derivation of the canonical energy-momentum tensor in (17.141) it follows that the functional derivative of the action with respect to $h_\alpha{}^\mu(x)$ at fixed $A_{\mu\beta}{}^\gamma(x)$ yields the canonical energy-momentum tensor

$$\frac{\delta \mathcal{A}[h_\alpha{}^\mu, A_{\mu\beta}{}^\gamma]}{\delta h_\alpha{}^\mu} = \sqrt{-g}\, \Theta_\mu{}^\alpha. \tag{18.1}$$

The functional derivative with respect to $A_{\mu\beta}{}^\gamma = \overset{\Lambda}{\Gamma}{}_{\mu\beta}{}^\gamma$ at fixed $h_\alpha{}^\mu$, on the other hand, is equivalent to a functional derivative with respect to $K_{\mu\beta}{}^\gamma$, as can be seen from (17.71). It therefore produces, according to Eq. (17.131), the spin current density[1]

$$\frac{\delta \mathcal{A}[h_\alpha{}^\mu, A_{\mu\beta}{}^\gamma]}{\delta A_{\mu\beta}{}^\gamma} = -\frac{1}{2}\sqrt{-g}\, \Sigma^\beta{}_\gamma{}^{,\alpha}\, h_\alpha{}^\mu \equiv -\frac{1}{2}\sqrt{-g}\, \Sigma^\beta{}_\gamma{}^{,\mu}. \tag{18.2}$$

These quantities will now be shown to satisfy covariant conservation laws.

Note that the action in these equations is the sum $\mathcal{A} = \overset{f}{\mathcal{A}} + \overset{m}{\mathcal{A}}$ of gravitational field and matter action.

18.1   Spin Density

Consider first local Lorentz transformations. Under these the vierbein fields $h_\alpha{}^\alpha(x)$ ($\mu = 0, \ldots, 3$) behave like vectors in the index α,

$$\delta_L h_\alpha{}^\alpha(x) = \omega_\alpha{}^{\alpha'}(x) h_{\alpha'}{}^\mu(x). \tag{18.3}$$

---

[1]Recall the decomposition $A_{\mu\beta}{}^\gamma = h^\gamma{}_\lambda h_\beta{}^\nu (K_{\mu\nu}{}^\lambda - \overset{h}{K}{}_{\mu\nu}{}^\lambda)$ following from Eq. (17.71) makes $A_{\mu\beta}{}^\gamma$ antisymmetric in β, γ.

Similarly, the field $A_{\mu\beta}{}^{\gamma}$ ($\mu = 0, \ldots, 3$) behaves like a tensor in the local Lorentz indices β, γ. The subscript μ it is not a tensor index, since it picks up the typical additional derivative of gauge fields [see (17.116)]

$$\delta_L A_{\mu\beta}{}^{\gamma} = \omega_{\beta}{}^{\beta'}(x) A_{\mu\beta'}{}^{\gamma} + \omega^{\gamma}{}_{\gamma'}(x) A_{\mu\beta}{}^{\gamma'} + \partial_{\mu}\omega_{\beta}{}^{\gamma}(x). \tag{18.4}$$

The symmetry variations $\delta_L \mathcal{A}$ of the action have to vanish as a consequence of the Euler-Lagrange equations. Inserting (18.3) and (18.4) into an arbitrary invariant action $\mathcal{A}$, we obtain

$$
\begin{aligned}
\delta_L \mathcal{A} &= \int d^4x \left\{ \frac{\delta \mathcal{A}}{\delta h_{\alpha}{}^{\mu}(x)} \omega_{\alpha}{}^{\alpha'}(x) h_{\alpha'}{}^{\mu}(x) \right. \\
&\quad \left. + \frac{\delta \mathcal{A}}{\delta A_{\mu\beta}{}^{\gamma}(x)} \left(\omega_{\beta}{}^{\beta'}(x) A_{\mu\beta'}{}^{\gamma}(x) + \omega^{\gamma}{}_{\gamma'}(x) A_{\mu\beta}{}^{\gamma'}(x) + \partial_{\mu}\omega_{\beta}{}^{\gamma}(x) \right) \right\} \\
&= \int d^4x \sqrt{-g} \left\{ \Theta_{\mu}{}^{\alpha} \omega_{\alpha}{}^{\alpha'} h_{\alpha'}{}^{\mu} - \frac{1}{2}\Sigma^{\beta}{}_{\gamma}{}^{,\mu} \left(\omega_{\beta}{}^{\beta'} A_{\mu\beta'}{}^{\gamma} + \omega^{\gamma}{}_{\gamma'} A_{\mu\beta}{}^{\gamma'} + \partial_{\mu}\omega_{\beta}{}^{\gamma} \right) \right\}.
\end{aligned}
\tag{18.5}
$$

Partially integrating the last term gives

$$
\begin{aligned}
\int d^4x \left\{ \sqrt{-g}\, \Theta_{\mu}{}^{\alpha} \omega_{\alpha}{}^{\alpha'} h_{\alpha'}{}^{\mu} + \frac{1}{2}\partial_{\mu}\left(\sqrt{-g}\Sigma^{\beta}{}_{\gamma}{}^{,\mu} \right) \omega_{\beta}{}^{\gamma} \right. \\
\left. - \frac{1}{2}\sqrt{-g}\, \Sigma^{\beta}{}_{\gamma}{}^{,\mu} \left(\omega_{\beta}{}^{\beta'} A_{\mu\beta'}{}^{\gamma} + \omega^{\gamma}{}_{\gamma'} A_{\mu\beta}{}^{\gamma'} \right) \right\}.
\end{aligned}
\tag{18.6}
$$

Since $\omega_{\beta}{}^{\gamma}(x')$ is an arbitrary antisymmetric function of x' it can be chosen to be zero everywhere except at some place x and we find

$$
\begin{aligned}
\frac{1}{2}\sqrt{-g}\left(\Theta_{\mu}{}^{\beta} h_{\gamma}{}^{\mu} - \Theta_{\mu\gamma} h^{\beta\mu} \right) + \frac{1}{2}\partial_{\mu}\sqrt{-g}\Sigma^{\beta}{}_{\gamma}{}^{,\mu} \\
- \frac{1}{2}\sqrt{-g}\left(\Sigma^{\beta}{}_{\delta'}{}^{\mu} A_{\mu\gamma}{}^{\delta} + \Sigma^{\delta}{}_{\beta'}{}^{\mu} A_{\mu\delta}{}^{\beta} \right).
\end{aligned}
\tag{18.7}
$$

Defining

$$\Theta_{\gamma}{}^{\beta} \equiv \Theta_{\mu}{}^{\beta} h_{\gamma}{}^{\mu} \tag{18.8}$$

and raising the index γ with the Minkowski metric $\eta^{\gamma\gamma'}$, this reads

$$\frac{1}{2}\left[\Theta^{\gamma\beta} - \Theta^{\beta\gamma} \right] + \frac{1}{2}\Gamma_{\mu\sigma}{}^{\sigma}\Sigma^{\beta\gamma,\mu} + \frac{1}{2}D_{\mu}^{L}\Sigma^{\beta\gamma\mu} = 0 \tag{18.9}$$

where $\overset{L}{D}_{\mu}$ is the covariant derivative for the local Lorentz index γ, i.e., for a vector

$$
\begin{aligned}
\overset{L}{D}_{\mu} v_{\alpha} &= \partial_{\mu} v_{\alpha} - A_{\mu\alpha}{}^{\beta} v_{\beta} = h^{\beta}{}_{\mu} D_{\beta} v_{\alpha}, \tag{18.10} \\
\overset{L}{D}_{\mu} v^{\alpha} &= \partial_{\mu} v^{\alpha} - A_{\mu}{}^{\alpha}{}_{\beta} v^{\beta} = \partial_{\mu} v^{\alpha} + A_{\mu\beta}{}^{\alpha} v^{\beta} = h^{\beta}{}_{\mu} D_{\beta} v^{\alpha}. \tag{18.11}
\end{aligned}
$$

The derivative $\overset{L}{D}_\mu \, \sigma^{\beta\gamma,\nu}$ can be made completely covariant also in the Einstein index μ, by going to

$$D_\mu \Sigma^{\beta\gamma,\nu} \equiv \overset{L}{D}_\mu \, \Sigma^{\beta\gamma,\nu} - \Gamma_{\mu\lambda}{}^\nu \Sigma^{\beta\gamma,\lambda}. \qquad (18.12)$$

If we contract μ with ν, and apply Eq. (18.9), the last term in (18.12) cancels the middle term in (18.9), and we obtain using D_μ^* of Eq. (15.41):

$$\frac{1}{2} D_\mu^* \Sigma^{\beta\gamma,\mu} = \frac{1}{2} \left[\Theta^{\beta\gamma} - \Theta^{\gamma\beta} \right]. \qquad (18.13)$$

Multiplying this by $h_\beta{}^\lambda h_\gamma{}^\kappa$, and moving the vierbeins to the right of the covariant derivative, which is possible due to relation (17.66), we find the local conservation law for the spin current density:

$$\frac{1}{2} h_\beta{}^\lambda h_\gamma{}^\kappa D_\mu \Sigma^{\beta\gamma,\mu} - \Theta^{[\lambda,\kappa]} = \frac{1}{2} D_\mu^* \Sigma^{\lambda\kappa,\mu} - \Theta^{[\lambda,\kappa]} = 0. \qquad (18.14)$$

18.2   Energy-Momentum Density

Let us now deduce the consequence of local Einstein invariance. In this case the spacetime coordinates must be transformed as well, and the action is invariant in the following sense:

$$\mathcal{A} = \int d^4x \sqrt{-g(x)} \mathcal{L}(h(x), A(x)) = \int d^4x' \sqrt{-g'(x')} \mathcal{L}\left(h'(x'), A'(x')\right). \qquad (18.15)$$

If we change the variables x' to x in the second integral we see that the difference

$$\int d^4x \left\{ \sqrt{-g'(x)} \mathcal{L}\left(h'(x), A'(x)\right) - \sqrt{-g(x)} \mathcal{L}\left(h(x), A(x)\right) \right\} \qquad (18.16)$$

must be concentrated on the immediate neighborhood of the surface of the integration volume. This is due to the fact that the integrations $\int d^4x'$ and $\int d^4x$ cover the *same* volume. Thus, after the change of variables $x' \to x$, the first integral runs through a slightly different region. Infinitesimally this amounts to the statement that

$$\delta_E \mathcal{A} = \int d^4x \, \delta_E \left[\sqrt{-g(x)} \mathcal{L}\left(h(x), A(x)\right) \right] \qquad (18.17)$$

is a pure surface term. The symbol δ_E denotes the substantial change under Einstein transformations at a *fixed* argument x [see (3.132),(11.78)], i.e.,

$$\delta_E g_{\mu\nu}(x) = \bar{D}_\mu \xi_\nu(x) + \bar{D}_\nu \xi_\mu(x). \qquad (18.18)$$

Under these Einstein transformations, the metric transforms as

$$\delta_E \sqrt{-g} = -\frac{1}{2} \sqrt{-g} \, g_{\mu\nu} \delta_E g^{\mu\nu} = \frac{1}{2} \sqrt{-g} \, g^{\mu\nu} \delta_E g_{\mu\nu}, \qquad (18.19)$$

which, upon inserting (18.18), yields

$$\frac{1}{2}\sqrt{-g}g^{\mu\nu}\left[\xi^\lambda\partial_\lambda g_{\mu\nu} + \left(\partial_\mu\xi^\lambda\right)g_{\lambda\nu} + \left(\partial_\nu\xi^\lambda\right)g_{\mu\lambda}\right]. \tag{18.20}$$

Therefore

$$\delta_E\sqrt{-g} = \xi^\lambda\partial_\lambda\sqrt{-g} + \sqrt{-g} + \partial_\lambda\xi^\lambda = \partial_\lambda\left(\xi^\lambda\sqrt{-g}\right) \tag{18.21}$$

and

$$\delta_E\int d^4x\sqrt{-g} = \int d^4x\sqrt{g}D_\lambda\xi^\lambda = \int d^4x\partial_\lambda\left(\xi^\lambda\sqrt{-g}\right). \tag{18.22}$$

This shows that the trivial action $\int d^4x\sqrt{-g}$ indeed changes by a pure surface term. There is complete invariance if we require $\xi^\lambda(x)$ to vanish at the surface.

The same result holds for a general action if $\mathcal{L}$ is a scalar Lagrangian density satisfying $\mathcal{L}'(x') = \mathcal{L}(x)$, so that

$$\delta_E\mathcal{L}(x) \equiv \mathcal{L}'(x) - \mathcal{L}(x) = \mathcal{L}'(x') - \mathcal{L}(x') = \mathcal{L}(x) - \mathcal{L}(x') = \xi^\lambda\partial_\lambda\mathcal{L}(x). \tag{18.23}$$

The variation of $\mathcal{A}$ is

$$\begin{aligned}\delta_E\mathcal{A} &= \delta_E\int d^4x\left(\sqrt{-g}\mathcal{L}(x)\right) = \int d^4x\left\{\left[\delta_E\sqrt{-g}\right]\mathcal{L}(x) + \sqrt{-g}\delta_E\mathcal{L}(x)\right\}\\ &= \int d^4x\left\{\partial_\lambda\left[\xi^\lambda\sqrt{-g}\right]\mathcal{L}(x) + \sqrt{-g}\xi^\lambda\partial_\lambda\mathcal{L}(x)\right\}\\ &= \int d^4x\partial_\lambda\left(\xi^\lambda\sqrt{-g}\mathcal{L}(x)\right). \end{aligned} \tag{18.24}$$

We can now derive the covariant conservation law associated with Einstein invariance by performing the substantial variations $\delta_E h_\alpha{}^\mu$ and of $\delta_E A_{\mu\beta}{}^\gamma$ and calculating $\delta_E\mathcal{A}$ once more as follows:

$$\begin{aligned}\delta_E\mathcal{A} &= \int d^4x\left(\frac{\delta\mathcal{A}}{dh_\alpha{}^\mu}\delta_E h_\alpha{}^\mu + \frac{d\mathcal{A}}{\delta A_{\mu\beta}{}^\gamma}\delta_E A_{\mu\beta}{}^\gamma\right)\\ &= \int d^4x\left(\sqrt{-g}\Theta_\mu{}^\alpha\delta_E h_\alpha{}^\mu - \frac{1}{2}\sqrt{-g}\Sigma^\beta{}_\gamma{}^{,\mu}\delta_E A_{\mu\beta}{}^\gamma\right). \end{aligned} \tag{18.25}$$

The substantial variations of the vierbein fields $h_\alpha{}^\mu$ and $A_{\mu\beta}{}^\gamma$ are those of a vector with a super- and subscript μ, respectively [recall (11.73), (11.74)]:

$$\delta_E h_\alpha{}^\mu = \xi^\lambda\partial_\lambda h_\alpha{}^\mu - \partial_\kappa\xi^\mu h_\alpha{}^\kappa, \quad \delta_E A_{\mu\beta}{}^\gamma = \xi^\lambda\partial_\lambda A_{\mu\beta}{}^\gamma + \partial_\mu\xi^\lambda A_{\lambda\beta}{}^\gamma. \tag{18.26}$$

Inserting these into (18.25), we find

$$\delta_E\mathcal{A} = \int d^4x\left\{\sqrt{-g}\Theta_\mu{}^\alpha\left(\xi^\lambda\partial_\lambda h_\alpha{}^\mu - \partial_\kappa\xi^\mu h_\alpha{}^\kappa\right) - \frac{1}{2}\sqrt{-g}\Sigma_\gamma^{\beta,\mu}\left(\xi^\lambda\partial_\lambda A_{\mu\beta}{}^\gamma + \partial_\mu\xi^\lambda A_{\lambda\beta}{}^\gamma\right)\right\}. \tag{18.27}$$

After partial integrations and setting ξ^λ equal to zero everywhere, except for a δ-function at some place x, we find

$$\partial_\kappa \left(\sqrt{-g} \Theta_\lambda{}^\alpha h_\alpha{}^\kappa \right) + \sqrt{-g} \Theta_\mu{}^\alpha \partial_\lambda h_\alpha{}^\mu$$
$$+ \frac{1}{2} \partial_\mu \left(\sqrt{-g} \Sigma^\beta{}_\gamma{}^{,\mu} A_{\lambda\beta}{}^\gamma \right) - \frac{1}{2} \sqrt{-g} \Sigma^\beta{}_\gamma{}^{,\mu} \partial_\lambda A_{\mu\beta}{}^\gamma = 0. \tag{18.28}$$

The second line can be rewritten as

$$\frac{1}{2} \partial_\mu \left(-\sqrt{-g} \Sigma^\beta{}_\gamma{}^{,\mu} \right) A_{\lambda\beta}{}^\gamma + \frac{1}{2} \sqrt{-g} \Sigma^\beta{}_\gamma{}^{,\mu} \left(\partial_\mu A_{\lambda\beta}{}^\gamma - \partial_\lambda A_{\mu\beta}{}^\gamma \right). \tag{18.29}$$

If we introduce the covariant curl of field $A_{\lambda\beta}$,

$$F_{\mu\lambda\beta}{}^\gamma \equiv \partial_\mu A_{\lambda\beta}{}^\gamma - \partial_\lambda A_{\mu\beta}{}^\gamma - \left[A_{\mu\beta}{}^\gamma A_{\lambda\delta}{}^\gamma - (\mu \leftrightarrow \lambda) \right], \tag{18.30}$$

we may rewrite (18.29) as

$$\frac{1}{2} \partial_\mu \left(\sqrt{-g} \Sigma^\beta{}_\gamma{}^{,\mu} \right) A_{\lambda\beta}{}^\gamma + \frac{1}{2} \sqrt{-g} \Sigma^\beta{}_\gamma{}^{,\mu} \left[A_{\mu\beta}{}^\delta A_{\lambda\delta}{}^\gamma - (\mu \leftrightarrow \lambda) \right] + \frac{1}{2} \sqrt{-g} \Sigma^\beta{}_\gamma{}^{,\mu} F_{\mu\lambda\beta}{}^\gamma. \tag{18.31}$$

The first three terms in this can be collected into a covariant derivative D_μ^* defined in (15.41):

$$\frac{1}{2} \sqrt{-g} D_\mu^* \Sigma^\beta{}_\gamma{}^{,\mu} A_{\lambda\beta}{}^\gamma, \tag{18.32}$$

where we have used (15.38) and (17.75) for $A_{\alpha\beta}{}^\gamma = \overset{\Lambda}{\Gamma}_{\alpha\beta}{}^\gamma$. Recalling the conservation law (18.13), the expression (18.31) [which is still equal to the second line in (18.28)] reduces to

$$-\sqrt{-g} \, \Theta_\gamma{}^\beta A_{\lambda\beta}{}^\gamma + \frac{1}{2} \sqrt{-g} \Sigma^\beta{}_\gamma{}^{,\mu} F_{\mu\lambda\beta}{}^\gamma. \tag{18.33}$$

In the first line of (18.28) we write

$$\Theta_\mu{}^\alpha \partial_\lambda h_\alpha{}^\mu = \Theta_\mu{}^\alpha D^L{}_\lambda h_\alpha{}^\mu + \Theta_\mu{}^\alpha A_{\lambda\alpha}{}^\beta h_\beta{}^\mu \tag{18.34}$$

and (18.28) turns into

$$\partial_\kappa \left(\sqrt{-g} \Theta_\lambda{}^\kappa \right) + \sqrt{-g} \Theta_\mu{}^\alpha D^L{}_\lambda h_\alpha{}^\mu - \frac{1}{2} \sqrt{-g} \Sigma^\beta{}_\gamma{}^{,\mu} F_{\lambda\mu\beta}{}^\gamma = 0. \tag{18.35}$$

This equation is covariant under local Lorentz transformations but not yet manifestly so under Einstein transformations. In order to verify the latter we observe that the derivative D^L of h can be rewritten as

$$\begin{aligned} D^L{}_\lambda h_\alpha{}^\mu &= \partial_\lambda h_\alpha{}^\mu - A_{\lambda\alpha}{}^\beta h_\beta{}^\mu \\ &= -\overset{h}{\Gamma}_{\lambda\kappa}{}^\mu h_\alpha{}^\kappa - \left(\Gamma_{\lambda\sigma}{}^\mu - \overset{h\mu}{\Gamma}_{\lambda\sigma} \right) h_\alpha{}^\sigma = -\Gamma_{\lambda\sigma}{}^\mu h_\alpha{}^\sigma, \end{aligned} \tag{18.36}$$

in accordance with the identity $D_\lambda h_\alpha{}^\mu = 0$. Then the second term is

$$-\sqrt{-g}\,\Gamma_{\lambda\sigma}{}^\mu \Theta_\mu{}^\sigma\,. \tag{18.37}$$

We now rewrite the first term as

$$\sqrt{-g}\left(D^*_\kappa \Theta_\lambda{}^\kappa + \Gamma_{\kappa\lambda}{}^\tau \Theta_\tau{}^\kappa\right), \tag{18.38}$$

and obtain the completely covariant conservation law for the energy momentum tensor [1, 2, 3].

$$D^*_\kappa \Theta_\lambda{}^\kappa + 2S_{\kappa\lambda}{}^\tau \Theta_\tau{}^\kappa - \frac{1}{2}\Sigma^\beta{}_{\gamma}{}^{,\mu} F_{\lambda\mu\beta}{}^\gamma = 0. \tag{18.39}$$

18.3 Covariant Derivation of Conservation Laws

The conservation laws of energy, momentum, and angular momentum can be derived somewhat more efficiently, if some initial effort is spent in preparing the Einstein and local Lorentz transformations (18.26), (18.3), (18.4) of $h_\alpha{}^\mu$ and $A_{\mu\alpha}{}^\beta$ in a covariant form. Take $\delta_E h_\alpha{}^\mu$. It can be rewritten as

$$\delta_E h_\alpha{}^\mu = \xi^\lambda \partial_\lambda h_\alpha{}^\mu + \Gamma_{\lambda\kappa}{}^\mu h_\alpha{}^\lambda \xi^\kappa - D_\lambda \xi^\mu h_\alpha{}^\lambda. \tag{18.40}$$

Using the identity

$$\partial_\lambda h_\alpha{}^\mu = -\overset{h}{\Gamma}_{\lambda\nu}{}^\mu h_\alpha{}^\nu = A_{\lambda\alpha}{}^\beta h_\beta{}^\mu - \Gamma_{\lambda\nu}{}^\mu h_\alpha{}^\nu, \tag{18.41}$$

we can rewrite (18.40) in the covariant form

$$\delta_E h_\alpha{}^\mu = -D_\alpha \xi^\mu + \left(A_{\lambda\alpha}{}^\mu + 2S_{\lambda\alpha}{}^\mu\right)\xi^\beta. \tag{18.42}$$

The reciprocal vierbein field $h^\alpha{}_\mu$ transforms as

$$\delta_E h^\alpha{}_\mu = D_\mu \xi^\alpha - \left(A_{\beta\mu}{}^\alpha - 2S_{\beta\mu}{}^\alpha\right)\xi^\beta. \tag{18.43}$$

Similarly, we find

$$\begin{aligned}
\delta_E A_{\mu\alpha}{}^\beta &= \xi^\lambda \partial_\lambda A_{\mu\alpha}{}^\beta + D_\mu \xi^\lambda A_{\lambda\alpha}{}^\beta - \Gamma_{\mu\kappa}{}^\lambda A_{\lambda\alpha}{}^\beta \xi^\kappa \\
&= D_\mu\left(\xi^\lambda A_{\lambda\alpha}{}^\rho\right) - \xi^\lambda\left(D_\mu A_{\lambda\alpha}{}^\beta - \partial_\lambda A_{\mu\alpha}{}^\beta\right) - \Gamma_{\mu\kappa}{}^\lambda A_{\lambda\alpha}{}^\beta \xi^\kappa \\
&= D_\mu\left(\xi^\lambda A_{\lambda\alpha}{}^\beta\right) - \xi^\lambda F_{\mu\lambda\alpha}{}^\beta.
\end{aligned} \tag{18.44}$$

Under local Lorentz transformations, the vierbein field has already its simplest possible form,

$$\delta_L h_\alpha{}^\mu = \omega_\alpha{}^\beta h_\alpha{}^\mu, \tag{18.45}$$

while $A_{\mu\alpha}{}^{\beta}$ acquires the typical additive term of a gauge field

$$\delta_L A_{\mu\alpha}{}^{\beta} = D_\mu \omega_\alpha{}^{\beta}. \tag{18.46}$$

Using these covariant transformation rules, the variations of the action (18.6), (18.27) become

$$\delta_L \mathcal{A} = \int d^4x \sqrt{-g} \left\{ \Theta_\beta{}^\alpha \omega_\alpha{}^\beta h_\beta{}^\mu - \frac{1}{2} \Sigma^\alpha{}_\beta{}^{,\mu} D_\mu \omega_\alpha{}^\beta \right\}, \tag{18.47}$$

$$\delta_E \mathcal{A} = \int d^4x \sqrt{-g} \left\{ \Theta_\mu{}^\alpha \left(-D_\lambda \xi^\mu h_\alpha{}^\lambda + (A_{\lambda\alpha}{}^\mu - 2S_{\lambda\alpha}{}^\mu) \xi^\lambda \right) \right.$$
$$\left. - \frac{1}{2} \Sigma^\alpha{}_\beta{}^{,\mu} \left[D_\mu \left(\xi^\lambda A_{\lambda\alpha}{}^\beta \right) - \xi^\lambda F_{\mu\lambda\alpha}{}^\beta \right] \right\}. \tag{18.48}$$

A partial integration of (18.47) [using (15.36), (15.40)] gives directly the divergence of the spin current (18.13). A partial integration of (18.48) leads to

$$D_\lambda^* \Theta_\mu{}^\lambda + \left(A_{\mu\alpha}{}^\beta - 2S_{\mu\alpha}{}^\beta \right) \Theta_\beta{}^\alpha + \frac{1}{2} D_\nu^* \Sigma^\alpha{}_\beta + \frac{1}{2} \Sigma^\alpha{}_\beta{}^{,\nu} F_{\nu\mu\alpha}{}^\beta = 0, \tag{18.49}$$

which, after inserting (18.13), reduces correctly to the covariant conservation law (18.39) for the canonical energy-momentum tensor.

18.4   Matter with Integer Spin

If matter fields only carried integer spin it would not be necessary to introduce the $h^\alpha{}_\mu$, $A_{\mu\alpha}{}^\beta$ fields. Then the theory could be formulated only with indices μ in an Einstein-invariant way. Let us derive the conservation of angular momentum for this situation. The action may be viewed as a functional of $g_{\mu\nu}$ and $K_{\mu\nu}{}^\lambda$, which enters via the affine connection $\Gamma_{\mu\nu}{}^\lambda = \bar{\Gamma}_{\mu\nu}{}^\lambda + K_{\mu\nu}{}^\lambda$. Einstein invariance implies the vanishing of the symmetry variation

$$\delta_E \mathcal{A} = \int d^4x \left(\left. \frac{\delta \mathcal{A}}{\delta g_{\mu\nu}} \right|_{S_{\mu\nu}{}^\lambda} \delta_E g_{\mu\nu} + \left. \frac{\delta \mathcal{A}}{\delta K_{\mu\nu}{}^\lambda} \right|_{g_{\mu\nu}} \delta_E K_{\mu\nu}{}^\lambda \right)$$
$$= -\frac{1}{2} \int d^4x \sqrt{-g} \left\{ T^{\mu\nu} \left(\xi^\lambda \partial_\lambda g_{\mu\nu} + \partial_\mu \xi^\lambda g_{\lambda\nu} + \partial_\nu \xi^\lambda g_{\mu\lambda} \right) \right.$$
$$\left. + \Sigma^\nu{}_\kappa{}^{,\mu} \left(\xi^\lambda \partial_\lambda K_{\mu\nu}{}^\kappa + \partial_\mu \xi^\lambda K_{\lambda\nu}{}^\kappa + \partial_\nu \xi^\lambda K_{\mu\lambda}{}^\kappa - \partial_\lambda \xi^\kappa K_{\mu\nu}{}^\lambda \right) \right\}, \tag{18.50}$$

if the Euler-Lagrange equations are fulfilled. Here we have used the definitions (15.15) and (15.17) of the energy momentum tensor and the current density, and inserted the infinitesimal transformation laws (11.76) of $g_{\mu\nu}$ and (11.109) of $S_{\mu\nu}{}^\lambda$ (which holds also for $K_{\mu\nu}{}^\lambda$). We have omitted the superscripts m since the equations in this section apply just as well to the gravitational field action $\mathcal{A}$, if we use the definitions (15.16) and (15.18). The further calculations are simplified by defining the symmetrized canonical energy-momentum tensor as follows:

$$\left. \frac{\delta \mathcal{A}}{\delta g_{\mu\nu}} \right|_{\Gamma_{\mu\nu}{}^\lambda=\text{const.}} \equiv -\frac{1}{2} \sqrt{-g} \left(\Theta_{\mu\nu} + \Theta_{\nu\mu} \right). \tag{18.51}$$

It is easy to see that this definition agrees with (17.143) if there are no spin$-1/2$ fields. This is done by differentiating $\mathcal{A}$ with respect to $h_\alpha{}^\mu$ at fixed $A_{\mu\alpha}{}^\beta$ and converting the index α to ν. It may also be verified by forming

$$\frac{\delta\mathcal{A}}{\delta g_{\mu\nu}(x)}\bigg|_{S_{\mu\nu}{}^\lambda=\text{const.}} = \frac{\delta\mathcal{A}}{\delta g_{\mu\nu}(x)}\bigg|_{\Gamma_{\mu\nu}{}^\lambda=\text{const.}} + \int dy\, \frac{\delta\mathcal{A}}{\delta\Gamma_{\sigma\tau}{}^\lambda(x)}\bigg|_{g_{\mu\nu}=\text{const.}} \frac{\delta\Gamma_{\sigma\tau}{}^\lambda(y)}{\delta g_{\mu\nu}(x)}\bigg|_{S_{\mu\nu}{}^\lambda=\text{const.}}, \tag{18.52}$$

so that one obtains the standard Belinfante relation (17.156) between $T_{\mu\nu}$ and $\Theta_{\mu\nu}$, now derived from geometric arguments for spaces with curvature and torsion:

$$T^{\mu\nu} = \Theta^{\mu\nu} - \frac{1}{2}\partial_\lambda(\Sigma^{\mu\nu,\lambda} - \Sigma^{\nu\lambda,\mu} + \Sigma^{\lambda\mu,\nu}). \tag{18.53}$$

For pure gravity, (18.51) is in accord with (17.147) which states that $\Theta_{\mu\nu}$ is the Einstein tensor [recall (17.147)] up to a factor $-\kappa$

$$-\kappa\Theta_{\mu\nu} = G_{\mu\nu} = R_{\mu\nu} - \frac{1}{2}g_{\mu\nu}R.$$

This can be seen from Eq. (15.29). The Belinfante relation (17.156) again coincides with (15.59).

Thus we can evaluate the consequences of Einstein invariance by using Θ and Σ, and considering, instead of (18.50), the variation

$$\begin{aligned}
0 &= \delta_E\mathcal{A} = \int d^4x\left\{\frac{\delta\mathcal{A}}{\delta g_{\mu\nu}}\bigg|_{\Gamma_{\mu\nu}{}^\lambda}\delta_E g_{\mu\nu} + \frac{\delta\mathcal{A}}{\delta\Gamma_{\mu\nu}{}^\lambda}\bigg|_{g_{\mu\nu}}\delta_E\Gamma_{\mu\nu}{}^\lambda\right\} \\
&= -\frac{1}{2}\int d^4x\sqrt{-g}\left\{\Theta^{\mu\nu}\left(\xi^\lambda\partial_\lambda g_{\mu\nu} + \partial_\mu\xi^\lambda g_{\lambda\nu} + \partial_\nu\xi^\lambda g_{\mu\lambda}\right)\right. \\
&\quad \left. -\Sigma^\nu{}_\kappa{}^{,\mu}\left(\xi^\lambda\partial_\lambda\Gamma_{\mu\nu}{}^\kappa + \partial_\mu\xi^\lambda\Gamma_{\lambda\nu}{}^\kappa + \partial_\nu\xi^\lambda\Gamma_{\mu\lambda}{}^\kappa - \partial_\lambda\xi^\kappa\Gamma_{\mu\nu}{}^\lambda + \partial_\mu\partial_\nu\xi^\kappa\right)\right\}. \tag{18.54}
\end{aligned}$$

It is again useful to bring the variations $\delta_E g_{\mu\nu}, \delta_E\Gamma_{\mu\nu}{}^\lambda$ into a covariant form. We rewrite the Einstein variation of the metric as

$$\begin{aligned}
\delta_E g_{\mu\nu} &= \bar{D}_\mu\xi_\nu + \bar{D}_\nu\xi_\mu = D_\mu\xi_\nu + D_\nu\xi_\mu + \left[K_{\mu\nu}{}^\lambda + (\mu\leftrightarrow\nu)\right]\xi_\lambda \\
&= D_\mu\xi_\nu + D_\nu\xi_\mu + 2\left[S_{\lambda\mu\nu} + (\mu\leftrightarrow\nu)\right]\xi^\lambda, \tag{18.55}
\end{aligned}$$

and the variation of the connection

$$\delta_E\Gamma_{\mu\nu}{}^\kappa = D_\mu D_\nu\xi^\kappa - 2D_\mu\left(S_{\nu\lambda}{}^\kappa\xi^\lambda\right) + R_{\lambda\mu\nu}{}^\kappa\xi^\lambda. \tag{18.56}$$

Inserting this into (18.52) gives

$$\begin{aligned}
\delta_E\mathcal{A} &= \int d^4x\sqrt{-g}\left\{(\Theta^{\nu\mu} + \Theta^{\mu\nu})\left(D_\nu\xi_\mu + 2S_{\lambda\mu\nu}\right)\xi^\lambda\right. \\
&\quad \left. +\Sigma^\nu{}_\kappa{}^{,\mu}\left[D_\mu D_\nu\xi^\kappa - 2D_\mu\left(S_{\nu\lambda}{}^\kappa\xi^\lambda\right) + R_{\lambda\mu\nu}{}^\kappa\xi^\lambda\right]\right\}. \tag{18.57}
\end{aligned}$$

By partially integrating the Σ term and using the spin divergence law (18.13), we obtain immediately

$$\delta_E \mathcal{A} = 2 \int d^4x \sqrt{-g} \left\{ -D_\mu \Theta_\lambda{}^\mu - 2 S_{\mu\lambda}{}^\nu \Theta_\nu{}^\mu + \frac{1}{2} \Sigma^\nu{}_{\kappa}{}^{,\mu} R_{\lambda\mu\nu}{}^\kappa \right\} \xi^\lambda, \qquad (18.58)$$

leading directly to the covariant conservation law

$$D_\mu^* \Theta_\lambda{}^\mu + 2 S_{\mu\lambda}{}^\nu \Theta_\nu{}^\mu - \frac{1}{2} \Sigma^\nu{}_{\kappa}{}^{,\mu} R_{\lambda\mu\nu}{}^\kappa = 0. \qquad (18.59)$$

Due to Eq. (17.81), the last term $\Sigma^\nu{}_{\kappa}{}^{,\mu} R_{\lambda\mu\nu}{}^\kappa$ can be rewritten as $\Sigma^\beta{}_{\gamma}{}^{,\mu} F_{\lambda\mu\beta}$, so that (18.59) coincides with (18.39).

The covariant conservation laws (18.13) and (18.59) hold for the total canonical energy momentum tensor $\Theta^{\mu\nu}$ and the total spin current density $\Sigma^{\mu\nu,\lambda}$. Remarkably, the same equations are found for the spin current density and the energy-momentum tensor of the gravitational field by itself. For this we must only apply the above derivation procedure to the gravitational field action $\overset{f}{\mathcal{A}}$ alone. The result can be written down directly by expressing the energy-momentum tensor and the spin current density in Eqs. (18.13) and (18.59) in terms of the Einstein and the Palatini tensor $G^{\mu\nu} = R^{\mu\nu} - g^{\mu\nu} R$ and $(1/2)S^\nu{}_\kappa{}^{,\mu} = S^\nu{}_\kappa{}^\mu + g^{\nu\mu} S_\kappa - \delta_\kappa{}^\mu S^\nu$, respectively. The equations for this are (15.56) and (17.147). This brings the covariant conservations laws (18.13) and (18.59) to the form

$$\frac{1}{2} D_\mu^* S^{\lambda\kappa,\mu} = G^{[\lambda,\kappa]}, \qquad (18.60)$$

$$D_\mu^* G_\lambda{}^\mu + 2 S_{\nu\lambda}{}^\kappa G_\kappa{}^\nu - \frac{1}{2} S^\nu{}_\kappa{}^{,\mu} R_{\lambda\mu\nu}{}^\kappa = 0. \qquad (18.61)$$

18.5   Relations between Conservation Laws and Bianchi Identities

The two covariant conservation laws (18.60) and (18.61) for the gravitational field itself are actually satisfied automatically, irrespective of the presence of matter. They are consequences of the fundamental identity (12.103), and of the Bianchi identity (12.115), respectively. To see this we apply the covariant derivative (15.41) to the Palatini tensor (15.47) and obtain

$$\begin{aligned} \frac{1}{2} D_\lambda^* S_{\nu\mu}{}^{,\lambda} &= D_\lambda^* (S_{\nu\mu}{}^\lambda + S_\nu{}^\lambda S_\mu - S_\mu{}^\lambda S_\nu) \\ &= D_\lambda S_{\nu\mu}{}^\lambda + D_\nu S_\mu - D_\mu S_\nu \\ &= D_\lambda S_{\nu\mu}{}^\lambda + D_\nu S_\mu - D_\mu S_\nu + 2 S_\lambda S_{\nu\mu}{}^\lambda. \end{aligned} \qquad (18.62)$$

Now we take (12.103) and contract the subscript ν with the superscript κ to obtain

$$\begin{aligned} R_{\mu\lambda} - R_{\lambda\mu} &= 2 \left(D_\kappa S_{\mu\lambda}{}^\kappa + D_\mu S_{\lambda\kappa}{}^\kappa + D_\lambda S_{\kappa\mu}{}^\kappa \right) - 4 \left(S_{\kappa\mu}{}^\rho S_{\lambda\rho}{}^\kappa + S_{\mu\lambda}{}^\rho S_{\kappa\rho}{}^\kappa + S_{\lambda\kappa}{}^\rho S_{\mu\rho}^\kappa \right) \\ &= 2 \left(D_\kappa S_{\mu\lambda}{}^\kappa + D_\mu S_\lambda - D_\lambda S_\mu \right) + 4 S_\rho S_{\mu\lambda}{}^\rho = D_\lambda^* S_{\nu\mu}{}^{,\lambda}. \end{aligned} \qquad (18.63)$$

Since $R_{\mu\lambda}$ differs from $G_{\mu\lambda}$ only by the symmetric tensor $g_{\mu\lambda}/2$, the same equation holds for $G_{\mu\lambda} - G_{\lambda\mu}$, so that we find the gravitational field version of the conservation law (18.60):

$$D_\lambda^* S_{\nu\mu}{}^{,\lambda} = G_{\nu\mu} - G_{\mu\nu} \qquad (18.64)$$

in agreement with (18.60).

Similarly, using (12.115) and permuting the indices we have

$$D_\tau R_{\sigma\nu\mu}{}^\tau + D_\sigma R_{\nu\tau\mu}{}^\tau + D_\nu R_{\tau\sigma\mu}{}^\tau = 2S_{\tau\sigma}{}^\lambda R_{\nu\lambda\mu}{}^\tau + 2S_{\sigma\nu}{}^\lambda R_{\tau\lambda\mu}{}^\tau + 2S_{\nu\tau}{}^\lambda R_{\sigma\lambda\mu}{}^\tau . \quad (18.65)$$

Contracting ν and μ, this becomes

$$
\begin{aligned}
2D_\tau R_\sigma{}^\tau - D_\sigma R &= 2D_\tau G_\sigma{}^\tau = -2S_{\tau\sigma}{}^\lambda + 2S_\sigma{}^{\mu\lambda} R_{\lambda\mu} + 2S^\mu{}_\tau{}^\lambda R_{\sigma\lambda\mu}{}^\tau \\
&= -4S_{\tau\sigma}{}^\lambda R_\lambda{}^\tau + 2S^\mu{}_\tau{}^\lambda R_{\sigma\lambda\mu}{}^\tau
\end{aligned}
\qquad (18.66)
$$

or

$$D_\mu^* G_\sigma{}^\mu - 2S_\mu \left(R_\lambda{}^\mu - \frac{1}{2}\delta_\lambda{}^\mu R \right) + S_{\tau\sigma}{}^\lambda \left(G_\lambda{}^\tau + \frac{1}{2}\delta_\lambda{}^\tau R \right) - S^\mu{}_\tau{}^\lambda R_{\sigma\lambda\mu}{}^\tau = 0, \quad (18.67)$$

in agreement with (18.61).

Within the defect interpretation of curvature and torsion, we observed before that the fundamental identities are nonlinear generalizations of the conservation laws of defect densities. From what we have just learned, the same equation can be obtained as conservation laws of energy-momentum and angular momentum from an Einstein action.

The two laws follow from the invariance of the Einstein action under general coordinate transformations, which are local translations, and under local Lorentz transformations, respectively.

These transformations correspond to elastic deformations (translational and rotational) of the world crystal and the invariance of the action expresses the fact that elastic deformations do not change the defect structure.

It is important to realize that due to the intimate relationship between the conservation laws and the fundamental identities for the gravitational fields, they remain valid in the presence of any matter distribution. Then, by the field equations (17.136) and (17.157), the spin density and energy-momentum tensor of the matter fields have to satisfy the same divergence laws by themselves. Indeed, it can easily be seen that this is a direct consequence of the Einstein invariance of the matter action in an arbitrary but fixed affine space, i.e., in a space whose geometry is specified from the outset rather than being determined by the matter fields via the field equations.

18.6 Particle Trajectories from Energy-Momentum Conservation

The classical equations of motion for a point particle have the consequence that its energy-momentum tensor $\overset{m}{T}_{\lambda}{}^{\mu} = 0$ in Eq. (15.24) satisfies the covariant conservation law (18.39) all by itself. Otherwise the Einstein equations (18.51) would not be satisfied. Since the symmetric energy-momentum tensors $\overset{f}{T}_{\lambda}{}^{\mu} = 0$ and $\overset{m}{T}_{\lambda}{}^{\mu} = 0$ of gravitational field and matter are proportional to each other, they must separately satisfy the covariant conservation law.

Consider first a particle without spin in a space without torsion. Starting point is the covariant conservation law (18.39) which reads now

$$\bar{D}_{\kappa}\overset{m}{T}_{\lambda}{}^{\kappa}(x) = 0. \tag{18.68}$$

Expressing the covariant derivative in terms of the Riemann connection, and this further in terms of derivatives of the metric using the identity

$$\frac{1}{\sqrt{-g}}\partial_{\nu}\sqrt{-g} = \frac{1}{2}g^{\lambda\kappa}\partial_{\nu}g_{\lambda\kappa} = \bar{\Gamma}_{\nu\lambda}{}^{\lambda}, \tag{18.69}$$

Eq. (18.68) becomes

$$\partial_{\nu}[\sqrt{-g}\,\overset{m}{T}{}^{\mu\nu}(x)] + \sqrt{-g}\,\bar{\Gamma}_{\nu\lambda}{}^{\mu}(x)\,\overset{m}{T}{}^{\lambda\nu}(x) = 0, \tag{18.70}$$

which must hold for the energy-momentum tensor of the particle trajectories (15.24). Inserting that tensor into (18.70) gives

$$m\int d\tau\,[\dot{x}^{\mu}(\tau)\dot{x}^{\nu}(\tau)\partial_{\nu}\delta^{(4)}(x - x(\tau)) + \bar{\Gamma}_{\nu\lambda}{}^{\mu}(x)\dot{x}^{\nu}(\tau)\dot{x}^{\lambda}(\tau)\,\delta^{(4)}(x - x(\tau))] = 0. \tag{18.71}$$

The first term in the integrand can also be written as $-\dot{x}^{\mu}(\tau)\partial_{\tau}\delta^{(4)}(x - x(\tau))$, so that a partial integration leads to

$$m\int d\tau\,[\ddot{x}^{\mu}(\tau) + \bar{\Gamma}_{\nu\lambda}{}^{\mu}(x)\dot{x}^{\nu}(\tau)\dot{x}^{\lambda}(\tau)]\,\delta^{(4)}(x - x(\tau)) = 0. \tag{18.72}$$

Integrating this over a thin tube around the trajectory $x^{\mu}(\tau)$, we obtain the equation (11.24) for the geodesic trajectory [4].

The same result may be derived from the following consideration. According to Eq. (18.58), the variation of the action under Einstein transformations of the coordinates $\delta_{E}x^{\mu} = -\xi^{\mu}$ is in Riemannian space

$$\delta_{E}\mathcal{A} = -2\int d^{4}x\sqrt{-g}\bar{D}_{\mu}T_{\lambda}{}^{\mu}\xi^{\lambda}. \tag{18.73}$$

Due to Einstein's equation (15.63), this holds separately for field and matter parts. If matter consists of point particles only, we obtain:

$$\delta_{E}\overset{m}{\mathcal{A}} = -\int d\tau\frac{\delta\overset{m}{\mathcal{A}}}{\delta x^{\mu}(\tau)}\xi^{\mu}(x(\tau)). \tag{18.74}$$

This vanishes along the geodesic trajectory implying that the energy-momentum tensor is covariantly conserved.

Let us now allow for torsion in the curved spacetime, where the covariant conservation law (18.68) becomes (18.59):

$$D^*_\mu \overset{m}{\Theta}{}_\lambda{}^\mu + 2S_{\mu\lambda}{}^\nu \overset{m}{\Theta}{}_\nu{}^\mu - \frac{1}{2} \overset{m}{\Sigma}{}^\nu{}_\kappa{}^{,\mu} R_{\lambda\mu\nu}{}^\kappa = 0. \tag{18.75}$$

Recalling the Belinfante relation (18.53) and the definition of D^*_μ in Eq. (15.41), we obtain for scalar particles with $\Sigma^\nu{}_\kappa{}^{,\mu} = 0$:

$$\bar{D}_\mu \overset{m}{T}{}_\lambda{}^\mu = 0. \tag{18.76}$$

This coincides with the conservation law (18.68) in Riemannian space and leads once more to the geodesic trajectories (11.24).

How can we remove the discrepancy with respect to the autoparallel trajectory found by the multivalued mapping procedure in Eq. (14.7)? The answer to this question was indicated before at the end of Section 16.1 and will be given in Section 20.3.1.

Notes and References

[1] R. Utiyama, Phys. Rev. **101**, 1597 (1956).

[2] T.W.B. Kibble, J. Math. Phys. **2**, 212 (1961).

[3] H. Kleinert, *Gauge Fields in Condensed Matter*, Vol. II *Stresses and Defects*, World Scientific, Singapore, 1989, pp. 744-1443 (http://www.phy-sik.fu-berlin.de/~kleinert/b2).

[4] F.W. Hehl, Phys. Lett. A **36**, 225 (1971).

Gravitation cannot be held responsible
for people falling in love.
ALBERT EINSTEIN (1879–1955)

19

Gravitation of Spinning Matter as a Gauge Theory

The reader will have noticed by now that the theory of gravity of spinning matter, when formulated in terms of fields $h_\alpha{}^\mu$, $A_{\mu\alpha}{}^\mu$ introduced in Eqs. (17.78) and (18.30), is really a gauge theory of local Lorentz transformations. Gauge properties have become apparent before when we observed in Eq. (11.104) that the connection $\Gamma_{\mu\nu}{}^\lambda$ transforms like a nonabelian gauge field under general coordinate transformations. But at that early stage, we could not have really spoken about a gauge theory since the connection $\Gamma_{\mu\nu}{}^\lambda$ was not an independent field of the system. When introducing spinning particles, the metric $g_{\mu\nu}(x)$ as a fundamental field was replaced by the vierbein field $h_\alpha{}^\mu(x)$ which transforms like a gauge field under translations. The Dirac theory in curved space, on the other hand, in which the covariant derivative contains the spin connection $A_{\alpha\beta}{}^\gamma$ of Eq. (17.79), is a bona fide gauge theory of local Einstein and Lorentz transformations. As discussed in Section 20.2, this is true even without torsion, i.e., without the contortion tensor $K_{\mu\nu}{}^\lambda$ in the combination (17.71), i.e., if it only contains the objects of anholonomy Eq. (17.73). The latter is sufficient to compensate the nontensorial gradient terms arising from local Lorentz transformations. The only feature brought about by the presence of torsion is that the spin connection $A_{\alpha\beta}{}^\gamma$ becomes an *independent* gauge field in addition to $h_\alpha{}^\mu$. Let us study the properties of this gauge theory in more detail.

19.1   Local Lorentz Transformations

Recall that under infinitesimal Lorentz transformations a vector field behaves like [see Eq. (17.107)]

$$\delta_L v_\alpha(x) = \omega_\alpha{}^\beta v_\beta(x), \quad \delta_L v^\alpha(x) = \omega^\alpha{}_\beta v^\beta(x), \tag{19.1}$$

where the physical coordinates x^μ remain unchanged since only the local directions are transformed. Due to the antisymmetry of the matrix $\omega_{\alpha\beta}$, we may write

$$\delta_L v_\alpha(x) = -v_\beta(x)\omega^\beta{}_\alpha, \quad \delta_L v^\alpha(x) = -v^\beta(x)\omega_\beta{}^\alpha. \tag{19.2}$$

For the spin connection, the transformation law was given in Eq. (17.114). With the notation (17.19), this can be written as

$$\delta_L A_{\alpha\beta}{}^\gamma = \omega_\beta{}^{\beta'} A_{\alpha\beta'}{}^\gamma + \omega_\alpha{}^{\alpha'} A_{\alpha'\beta'}{}^\gamma + \omega^\gamma{}_{\gamma'} A_{\alpha\beta}{}^{\gamma'} + \partial_\alpha \omega_\beta{}^\gamma. \tag{19.3}$$

$$\delta_L A_{\mu\beta}{}^\gamma = \omega_\beta{}^{\beta'} A_{\mu\beta'}{}^\gamma + \omega^\gamma{}_{\gamma'} A_{\mu\beta}{}^{\gamma'} + \partial_\mu \omega_\beta{}^\gamma. \tag{19.4}$$

Observe that the spacetime variables x^μ are not transformed so that the Lorentz group acts like an internal symmetry group. There is, however, a certain similarity with external gauge symmetries discussed in Section 3.5, Part I. This is so because $h_\alpha{}^\mu$ can couple Lorentz and Einstein indices, just as in (I.3.135), thus giving rise to more invariants. For instance, there is no need of forming $(F_{\mu\nu\alpha}{}^\beta)^2$ in order to get an invariant action. There also exists an invariant expression linear in the field strength,

$$\overset{f}{\mathcal{A}} = -\frac{1}{2\kappa} \int d^4x \sqrt{-g}\, h^{\alpha\mu} h_\beta{}^\nu F_{\nu\mu\alpha}{}^\beta. \tag{19.5}$$

In fact, from (17.80) we see that this is just the Einstein-Cartan action (15.8).

For completeness, let us show once more how the spin current and energy-momentum tensor follow from this action for the independent fields $h_\alpha{}^\mu$, $A_{\mu\alpha}{}^\beta$, rather than $h_\alpha{}^\mu$, $K_{\mu\nu}{}^\lambda$ in Section 17.7.

First we calculate the spin current of the field which is defined by the functional derivative with respect to the gauge field $A_\mu{}^{\alpha\beta}(x)$ [compare (17.131), (17.137)]:

$$\frac{1}{2}\sqrt{-g}\, \overset{f}{\Sigma}_{\alpha\beta}{}^{,\mu}(x) = -\frac{\delta \overset{f}{\mathcal{A}}}{\delta A_\mu{}^{\alpha\beta}(x)} \tag{19.6}$$

$$= \frac{1}{2\kappa}\frac{\delta}{\delta A_\mu{}^{\alpha\beta}(x)} \int d^4x \sqrt{-g}\, h^{\alpha'\mu'} h_{\beta'}{}^{\nu'} \left(\partial_{\nu'} A_{\mu'\alpha'}{}^{\beta'} - \partial_{\mu'} A_{\nu'\alpha'}{}^{\beta'} - A_{\nu'\alpha'}{}^\gamma A_{\mu'\gamma}{}^{\beta'} + A_{\mu'\alpha'}{}^\gamma A_{\nu'\gamma}{}^{\beta'} \right)$$

$$= -\frac{1}{2\kappa}\left\{ \partial_\nu \sqrt{-g}\left[h_\alpha{}^\mu h_\beta{}^\nu - (\alpha \leftrightarrow \beta) \right] + \sqrt{-g}\left[(A_{\nu\alpha'\alpha} h^{\alpha'\mu} h_\beta{}^\nu + A_{\nu\beta'\beta} h_\alpha{}^\mu h^{\beta'\nu}) - (\alpha \leftrightarrow \beta) \right] \right\}.$$

We may write this in terms of the partially covariant derivatives (18.10), (18.11) as

$$-\kappa \overset{f}{\Sigma}_{\alpha\beta}{}^{,\mu} = \overset{L}{D}_\nu \left[h_\alpha{}^\mu h_\beta{}^\nu - (\alpha \leftrightarrow \beta) \right] + \Gamma_{\nu\sigma}{}^\sigma \left[h_\alpha{}^\mu h_\beta{}^\nu - (\alpha \leftrightarrow \beta) \right]. \tag{19.7}$$

Applying the chain rule of differentiation this becomes

$$-\kappa \overset{f}{\Sigma}_{\alpha\beta}{}^{,\mu} = \left(\overset{L}{D}_\beta h_\alpha{}^\mu - h_\alpha{}^\mu \overset{L}{D}_\nu h_\beta{}^\nu + h_\alpha{}^\mu \Gamma_{\beta\sigma}{}^\sigma \right) - (\alpha \leftrightarrow \beta). \tag{19.8}$$

We now observe that, due to the identity $D_\mu h_\alpha{}^\nu \equiv 0$, the connection can be rewritten as

$$\Gamma_{\mu\nu}{}^\lambda = h^{\alpha\lambda} \overset{L}{D}_\mu h_{\alpha\nu} = -h^\alpha{}_\nu \overset{L}{D}_\mu h_\alpha{}^\lambda. \tag{19.9}$$

This relation is complementary to the relation $\overset{\Lambda}{\Gamma}_{\mu\beta}{}^\gamma = h^\gamma{}_\nu \overset{\Gamma}{D}_\mu h_\beta{}^\nu$ of Eq. (17.149). Using (19.9), the spin current of the field becomes

$$-\kappa \overset{f}{\Sigma}_{\alpha\beta}{}^{,\mu} = 2\left(S_{\alpha\beta}{}^\mu + h_\alpha{}^\mu S_\beta - h_\beta{}^\mu S_\alpha\right) = S_{\alpha\beta}{}^\mu, \qquad (19.10)$$

in agreement with (15.47), (15.56).

We now calculate the functional derivative of the action with respect to $h_\alpha{}^\mu$ [compare (17.141), (17.147)]. It shows directly that the canonical energy-momentum tensor of the gravitational field coincides with the Einstein tensor,

$$\sqrt{-g}\,\Theta_\mu{}^\alpha = \frac{\delta \overset{f}{\mathcal{A}}}{\delta h_\alpha{}^\mu} = \sqrt{-g}\left(h^{\delta\nu} F_{\mu\nu\delta}{}^\alpha - h^\beta{}_\mu F_{\beta\delta}{}^{\delta\alpha}\right) = h^{\alpha\nu}\sqrt{-g}\left(R_{\mu\nu} - g_{\mu\nu}R\right)$$

$$= \sqrt{-g}\,G_\mu{}^\alpha. \qquad (19.11)$$

The use of the field $h_\alpha{}^\mu$ has made it possible to retrieve the Einstein tensor without projecting out the symmetric part of it, as was the case in the previous formulas (15.29) and (18.51).

19.2   Local Translations

Let us now show that the vierbein field plays the role of a *gauge field of local translations*. In Eq. (11.60) we have written the Einstein transformations $x \to x'(x)$ in the form of a local translation

$$x' = x - \xi(x). \qquad (19.12)$$

The vierbein field ensures that the theory is invariant under these transformations. In fact, the covariant derivative

$$D_\alpha \equiv h_\alpha{}^\mu \partial_\mu + \frac{i}{2} A_{\alpha\beta}{}^\gamma \Sigma^\beta{}_\gamma \qquad (19.13)$$

may be viewed as a combination of $h_\alpha{}^\mu$ times the translational "functional matrix" ∂_μ, and $iA_{\alpha\beta}{}^\gamma$ times the Lorentz matrix $(1/2)\Sigma^\beta{}_\gamma$. This viewpoint becomes most transparent by considering the expression in (17.88), the commutator of two covariant derivatives with respect to the dislocation coordinates,

$$\left[D_\alpha, D_\beta\right]\psi = iF_{\alpha\beta\gamma}{}^\delta \tfrac{1}{2}\Sigma^\gamma{}_\delta \psi + i2S_{\alpha\beta}{}^\gamma iD_\gamma\psi. \qquad (19.14)$$

Since the factor $F_{\alpha\beta\gamma}{}^\delta$ is the curl of the gauge field of Lorentz transformations generated by $\tfrac{1}{2}\Sigma^\gamma{}_\delta$. By analogy, the factor $2S_{\alpha\beta}{}^\gamma$ of the generator of translations $iD_\gamma\psi$ may be considered as the curl of the gauge field of translations. Indeed, if we write $2S_{\alpha\beta}{}^\gamma$ in the form

$$2S_{\alpha\beta}{}^\gamma = -h^\gamma{}_\nu\left[h_\alpha{}^\mu \overset{L}{D}_\mu h_\beta{}^\nu - (\alpha \leftrightarrow \beta)\right]$$

$$= h_\alpha{}^\mu h_\beta{}^\nu\left[\overset{L}{D}_\mu h^\gamma{}_\nu - (\mu \leftrightarrow \nu)\right], \qquad (19.15)$$

we arrive at the standard form of a curl, and the present formulation of gravity of spinning matter is indeed a gauge theory of both local Lorentz transformations and local translations.

If the space has no torsion, then $A_{\mu\alpha}{}^\beta$ is completely composed of derivatives of vierbein fields [recall Eq. (17.68)]

$$A_{\alpha\beta}{}^\gamma = -h^\gamma{}_\lambda h_\alpha{}^\mu h_\beta{}^\nu \overset{h}{\Gamma}_{\mu\nu}{}^\lambda. \tag{19.16}$$

Inserting this into (19.15) we verify that this is equivalent to a vanishing torsion.

In recent years, this aspect of gravitational theory has received increasing attention, due to the shift in emphasis from geometric principles to gauge principles.

Notes and References

The gauge aspects of gravity are discussed in
D.W. Sciama, *The Physical Structure of General Relativity*, in Recent Developments in General Relativity, Festschrift for Infeld, Pergamon Press, Oxford, 1962, p. 463; Rev. Mod. Phys. **36**, 463, 1103 (1964);
T.W.B. Kibble, J. Math. Phys. **2**, 212 (1961);
R. Utiyama, Phys. Rev. **101**, 1597 (1956);
F.W. Hehl, P. von der Heyde, G.D. Kerlick, and J.M. Nester, *Rev. Mod. Phys.* **48**, 393 (1976);
H. Kleinert, *Gauge Fields in Condensed Matter*, Vol. II, *Stresses and Defects*, World Scientific, Singapore, 1989 (`http://www.physik.fu-berlin.de/~kleinert/b1/contents2.html`).

20

Evanescent Properties of Torsion in Gravity

What additional physics is brought about by torsion? If the field action is of the Einstein-Cartan type (19.5), the consequences turn out to be practically unobservable. The situation cannot be improved by adding higher powers of the curvature tensor to the Lagrangian density, such as the terms in the action (15.11), but with the Riemannian $\bar{R}_{\mu\nu\lambda\kappa}$ replaced by the Riemann-Cartan $R_{\mu\nu\lambda\kappa}$. Also additional powers of the torsion tensor $S_{\mu\nu}{}^{\lambda}$ do not help, nor terms containing different squares of covariant derivatives $D_{\kappa}S_{\mu\nu}{}^{\lambda}$, or mixed terms composed of curvature and torsion tensors. The dominant action is always the Einstein-Cartan action (15.8), which leads to the field equation (17.138) and determines the torsion as follows

$$S_{\mu\nu\lambda} = \frac{\kappa}{2}\left(\overset{\mathrm{m}}{\Sigma}_{\mu\nu,\lambda} + \frac{1}{2}g_{\nu\lambda}\overset{\mathrm{m}}{\Sigma}_{\mu\kappa}{}^{,\kappa} - \frac{1}{2}g_{\nu\lambda}\overset{\mathrm{m}}{\Sigma}_{\mu\kappa}{}^{,\kappa}\right). \tag{20.1}$$

This is a local equation. Torsion sits on top of particle spins and never reaches out into empty space by more than a Planck length (12.42). If the particles are quantized, their spin will be smeared out over a Compton wavelength $\lambda_C = \hbar/mc$ corresponding to their mass, extending the volume of nonzero torsion, which however, remains bound to matter.

For massless fields such the photon field, this could in principle be different since magnetic fields are observed to reach out quite far in spacetime. But, as discussed at the end of Section 16.2, photons must be assumed to be coupled truly minimally to gravity, i.e., be decoupled from torsion, to guarantee electromagnetic gauge invariance.

Let us discuss some more physical consequences.

20.1   Local Four-Fermion Interaction due to Torsion

A nontrivial effect of torsion can be derived for Dirac fields, where the spin density of matter is, from (17.133) and (1.227),

$$\overset{\mathrm{m}}{\Sigma}_{\alpha\beta,\gamma} = -i\frac{1}{2}\bar{\psi}[\gamma_{\gamma}, \Sigma_{\alpha\beta}]_{+}\psi, \tag{20.2}$$

with $\Sigma_{\alpha\beta} = (i/4)[\gamma_\alpha, \gamma_\beta]$. More explicitly, the spin density reads

$$\overset{m}{\Sigma}_{\alpha\beta,\gamma} = \frac{1}{2}\bar{\psi}\gamma_{[\alpha}\gamma_\beta\gamma_{\gamma]}\psi = \frac{1}{2}\varepsilon_{\alpha\beta\gamma\lambda}\bar{\psi}\gamma^\lambda\gamma_5\psi \tag{20.3}$$

with $\gamma_5 \equiv (1/4!)\varepsilon_{\alpha\beta\gamma\delta}\gamma^\alpha\gamma^\beta\gamma^\gamma\gamma^\delta$, where the brackets around the subscripts denote their complete antisymmetrization. Due to the antisymmetry, the Palatini tensor divided by 2, the torsion, and the contortion tensor are all equal to $(\kappa/2)\overset{m}{\Sigma}_{\alpha\beta,\gamma}$:

$$\frac{1}{2}S_{\alpha\beta,\gamma} = S_{\alpha\beta\gamma} = K_{\alpha\beta,\gamma} = \frac{\kappa}{2}\overset{m}{\Sigma}_{\alpha\beta,\gamma} . \tag{20.4}$$

In Eq. (11.146) we have expressed the curvature tensor in terms of the Riemann curvature tensor plus the contortion. Two contractions give the corresponding decomposition of the scalar curvature

$$R = \bar{R} + \bar{D}_\mu K_\nu{}^{\nu\mu} - \bar{D}_\nu K_\mu{}^{\nu\mu} + \left(K_\mu{}^{\mu\rho}K_{\nu\rho}{}^\nu - K_\nu{}^{\mu\rho}K_{\mu\rho}{}^\nu\right) . \tag{20.5}$$

In the gravitational Einstein-Cartan action, R is integrated over the total invariant volume of the universe. In this process, the terms $\bar{D}_\mu K_\nu{}^{\nu\mu}$ and $\bar{D}_\nu K_\mu{}^{\nu\mu}$ produce irrelevant surface terms and can be ignored. The action can therefore be separated into a Hilbert-Einstein action

$$\overset{f}{\mathcal{A}} = -\frac{1}{2\kappa}\int d^4x\sqrt{-g}\bar{R}, \tag{20.6}$$

plus a field torsion action

$$\overset{f}{\mathcal{A}}{}^S = \int d^4x\sqrt{-g}\,\overset{f}{\mathcal{L}}{}^S, \tag{20.7}$$

with a Lagrangian density

$$\overset{f}{\mathcal{L}}{}^S = -\frac{1}{2\kappa}\left(K_\mu{}^{\mu\rho}K_\rho{}^\nu - K_\nu{}^{\mu\rho}K_{\mu\rho}{}^\nu\right). \tag{20.8}$$

The latter can be rearranged to

$$\overset{f}{\mathcal{L}}{}^S = \frac{1}{2\kappa}S_{\mu\nu,\lambda}K^{\lambda\nu\mu}, \tag{20.9}$$

where $S_{\mu\nu,\lambda}$ is the Palatini tensor (15.48). As a cross check we differentiate this with respect to $K^{\lambda\nu\mu}$ and obtain

$$\frac{\partial\overset{f}{\mathcal{L}}{}^S}{\partial K^{\lambda\nu\mu}} = \frac{1}{2\kappa}S_{\mu\nu,\lambda}, \tag{20.10}$$

in accordance with Eqs. (15.18) and (15.56).

From the Dirac action (17.18) we extract the interaction density between matter and torsion:

$$\overset{m}{\mathcal{L}}{}^{S} = \frac{1}{2} \overset{m}{\Sigma}_{\mu\nu,\lambda} K^{\lambda\nu}. \tag{20.11}$$

Adding this to (20.9) and extremizing the combined Lagrangian density $\mathcal{L} = \overset{f}{\mathcal{L}}{}^{S} + \overset{m}{\mathcal{L}}{}^{S}$, we recover once more (20.4). Inserting this back into the total Lagrangian density gives the effective torsion Lagrangian at the extremum:

$$\mathcal{L}^{\text{eff}} = \frac{\kappa}{4} \overset{m}{\Sigma}_{\mu\nu\lambda} K^{\lambda\nu\mu} = \frac{\kappa}{8} \overset{m}{\Sigma}_{\mu\nu\lambda} \overset{m\mu\nu\lambda}{\Sigma}. \tag{20.12}$$

Inserting (20.3), this becomes [1]:

$$\mathcal{L}^{\text{eff}} = \frac{3\kappa}{16} \bar{\psi}\gamma_{\mu}\gamma_{5}\psi\bar{\psi}\gamma^{\mu}\gamma_{5}\psi. \tag{20.13}$$

Unfortunately, this interaction is too weak to be detectable by present-day experiments, and probably will be for many generations to come. The interaction (20.13) will hide under the much larger weak interactions. For electrons and their neutrinos ν_{e}, these have a Lagrangian

$$
\begin{aligned}
\mathcal{L}^{\text{weak}} &= -\frac{G}{\sqrt{2}} \Big\{ \big[\bar{e}\gamma_{\mu}(1-\gamma_{5})\nu_{e}\big]\big[\bar{\nu}_{e}\gamma^{\mu}(1-\gamma_{5})\nu\big] \\
&\quad + 2\big[\bar{e}\gamma^{\mu}e + \bar{\nu}_{e}\gamma^{\mu}\nu_{e}\big]\big[2\sin^{2}\theta_{W}\bar{e}_{R}\gamma_{\lambda}e_{R} - \cos2\theta_{W}\bar{e}_{R}\gamma_{\mu}e_{R} - \bar{e}_{L}\gamma_{\mu}e_{L}\big] \Big\}. \tag{20.14}
\end{aligned}
$$

Here $e_{L} \equiv (1-\gamma_{5})e/2$ and $\nu_{eL} \equiv (1-\gamma_{5})\nu_{e}/2$ are the left-handed Dirac spinors of electron and its neutrino, θ_{W} is the Weinberg angle with

$$\sin^{2}\theta_{W} \approx 0.23120(15), \tag{20.15}$$

and G is the coupling constant

$$G = (1.14730 \pm 0.000641) \times 10^{-5}\text{GeV}^{-2}. \tag{20.16}$$

This may be expressed in terms of the electron charge as

$$G \approx \frac{e^{2}}{m_{W}^{2}}\frac{4\pi\alpha}{m_{W}^{2}} \tag{20.17}$$

where

$$m_{W} = 80.403 \pm 0.029\,\text{GeV} \tag{20.18}$$

is the mass of the charged vector mesons $W^{\pm}$. The ratio of this and the heavier vector boson mass

$$m_{Z} = 91.1876 \pm 0.0021\,\text{GeV} \tag{20.19}$$

fixes the Weinberg angle:

$$\cos\theta_W = m_W/m_Z, \tag{20.20}$$

Comparison of (20.14) with the four-fermion interaction due to torsion in Eq. (20.13) shows that the latter is smaller than the weak interaction by the extremely small factor

$$\frac{m_W^2}{m_P^2} \approx 4.34 \times 10^{-35}, \tag{20.21}$$

where m_P is the Planck mass (12.43). Thus any hope for a detection in the foreseeable future is an illusion. Thus it will also be impossible for a long time to measure the coupling constant γ to torsion in the spin connection (20.22) for the composite particles such as protons and neutrons.

An additional problem is that a four-fermion interaction such as (20.13) is not renormalizable, so that it cannot possibly be a fundamental interaction. It can at best be a phenomenological approximation to some more fundamental theory which for high energies will show important deviations from the four-fermion expression (20.13).

20.2 No Need for Torsion in Gravity

An important observation is the following: The covariant derivative (17.18), if expressed as in (17.71) in terms of the difference between the contortion $K_{\mu\nu}{}^{\lambda}$ and the field $\overset{h}{K}_{\mu\nu}{}^{\lambda}$, does not really need the contortion field $K_{\mu\nu}{}^{\lambda}$ to be covariant under local Lorentz transformations. For this, the field $\overset{h}{K}_{\mu\nu}{}^{\lambda}$ is completely sufficient. It supplies, via the transformation property Eq. (17.113), the compensating nontensorial term to make the derivative in front of a Dirac field a vector. Thus, a consistent theory that is invariant under local Lorentz transformations exists also in spacetime with without torsion and only curvature. As far as covariance is concerned, torsion is a pure luxury of the theory.

This has important implications for the coupling of torsion to spin. Whereas the coupling of the torsionless spin connection $\overset{h}{K}_{\mu\nu}{}^{\lambda}$ is uniquely fixed by the local Lorentz transformation properties of the field, the contortion field $K_{\mu\nu}{}^{\lambda}$ in the spin connection (17.71) can be multiplied by an *arbitrary* coupling constant γ:

$$\overset{\Lambda}{\Gamma}_{\alpha\beta}{}^{\gamma} = h^{\gamma}{}_{\lambda}h_{\alpha}{}^{\mu}h_{\beta}{}^{\nu}(\gamma\, K_{\mu\nu}{}^{\lambda} - \overset{h}{K}_{\mu\nu}{}^{\lambda}). \tag{20.22}$$

The unit strength of $\overset{h}{K}_{\mu\nu}{}^{\lambda}$ in this expression is responsible for the fact that ordinary torsionless gravity couples *universally* to spin and orbital angular momentum. Indeed, this seems to be a natural property of gravitational forces [2]. A rotating test particle with angular momentum **j** far away from a rotating celestial body with angular momentum **J** experiences a precession due to the *Lense-Thirring effect*. The speed of precession should not depend on the origin of the angular momenta $\mathbf{j}$ and $\mathbf{J}$. The gravitational phenomena are the same if $\mathbf{j}$ and $\mathbf{J}$ are caused by oriented spins of the constituents or by their orbital rotations. This universality is ensured by the spin part of the symmetric Belinfante energy-momentum tensor (3.231) and by the way this spin part arises from the variational derivation of the energy-momentum tensor in Eq. (17.141). Recall that the functional derivative of the matter action with respect to $h_\alpha{}^\mu(x)$ has two terms, an orbital part which is the canonical energy-momentum tensor $\overset{m}{\Theta}_\mu{}^\alpha(x)$ [recall Eqs. (17.143) and (17.146)], and the spin part (17.148) coming from the derivative of the action with respect to $h_\alpha{}^\mu(x)$ inside the spin connection. The unit strength of $\overset{h}{K}_{\mu\nu}{}^\lambda$ in (20.22) is essential to obtain the symmetry of the Belinfante tensor and the covariant conservation law for the spin current density $\frac{1}{2}D_\mu^*\Sigma^{\lambda\kappa,\mu} = \Theta^{[\lambda,\kappa]}$ in Eq. (18.14).

The torsion field violates this universality principle. Torsion is able to probe directly the spin of the fundamental particles in a body. This is a remarkable property of torsion, which is also a severe handicap for the theoretician: it is an obstacle to constructing a consistent field theory of torsion before it is known which particles are truly fundamental. A spin-$\frac{1}{2}$ particle that looks fundamental at present, so that one includes it into the action with $\gamma = 1$ in (20.22), may eventually turn out to be composite. Then its spin would be partly due to orbital motion of its constituents, so that γ would be different from unity.

Faced with the non-universality of γ, its value will be a measurable extra property of each particle, such as mass, spin, and charge, unless some higher symmetry principle fixes it. Such a principle was invoked at the end of Section 16.2, to fix $\gamma = 0$ for photons and W and Z bosons. This was necessary to guarantee gauge invariance without which the theory could not be renormalized. By analogy, one may well assume that all elementary particles have $\gamma = 0$, so that the universe is completely torsionless, as assumed by Einstein, for simplicity.

The possibility of setting $\gamma = 0$ in (20.22) led Weinberg to reject torsion as a special geometrical field altogether [3]. According to him, torsion is just an ordinary tensor field of rank three with antisymmetry in the first two indices. This field may or may not be present in nature. He would only be convinced of a special role of torsion if there exists some higher symmetry principle which requires that the torsion tensor always accompanies the Christoffel symbol in the combination (11.102), i.e., with $\gamma = 1$ in (20.22).

Such a higher symmetry principle is only provided by the multivalued mapping principle of Chapter 14. The image of classical equations and field theories in Euclidean space under such a mapping are theories in a Riemann-Cartan space. These contain the general metric-affine connection (11.115) that automatically includes the torsion tensor with $\gamma = 1$.

However, it depends on the fundamental nature of the considered particles whether the mapping principle can be applied to them. The situation is the same as in the minimal-coupling principle of electrodynamics. This holds perfectly for bare electron, muon, and quark fields, since they are fundamental. It does not hold for composite particles, such as protons, neutrons, and other hadrons.

20.3   Scalar Fields

Since torsion fields propagate only over Planck scales, it is merely of academic interest to study their couplings to other particles. Still, one would like the theory to be internally consistent. In Eq. (15.62) we have already noted that in minimal coupling. scalar fields $\phi(x)$ do not interact with torsion. For scalar fields, the ordinary field derivatives of scalar fields, $\partial_\mu \phi(x)$, are automatically covariant derivatives. Thus the minimally coupled Lagrangian density contains only the metric tensor $g_{\mu\nu}(x)$, and no affine connection $\Gamma_{\mu\nu}{}^\lambda(x)$. It is automatically coupled in the same truly minimal way as the photon field in the action (16.21).

   This contradicts our finding in Section 14.1.2 that classical particle trajectories are coupled to torsion, making them autoparallel rather than geodesic. At the end of Section 16.1 we have indicated how this conflict can be resolved, and this will now be done.

20.3.1   Only Spin-1/2 Sources

The key to the resolution of the conflict is Eq. (20.3). The spin current density $\overset{m}{\Sigma}_{\alpha\beta,\gamma}$ of a spin-$\frac{1}{2}$ particle is completely antisymmetric. Thus, if all particles in nature are ultimately composed of spin-$\frac{1}{2}$ particles, then the torsion field $S_{\mu\nu\lambda}$ is completely antisymmetric [5]. As a consequence, it decouples from the classical equation of motion of point particles in Eq. (14.8). Autoparallel trajectories become geodesic trajectories, and the conflict disappears.

   This scenario is physically quite plausible. In the standard model of weak and electromagnetic interactions, photons and vector bosons W and Z are contained as massless spin-1 gauge fields which are not coupled to torsion, to guarantee gauge invariance [recall Section 16.2]. The vector bosons become massive by a Meissner-Higgs effect, where they mix with the Goldstone modes of the Higgs fields. The latter were introduced originally by analogy with the Ginzburg-Landau field of superconductivity. This field describes "Cooper pairs" of electrons. It is then natural to conjecture that also the Higgs fields will ultimately be composed of pairs of some fundamental spin-$\frac{1}{2}$ constituents. Indeed, such models have been around in the literature for many years involving so-called *technicolor quarks* [6]. Just as in the Cooper pairs of conventional superconductors, the spins of the technicolor quarks are antiparallel in an s-wave bound state. This may well be the reason why Higgs particles do not couple to torsion.

   There is also another evidence that Higgs particles cannot be fundamental. Their interacting field theory is renormalizable, but the propagators of Higgs fields possess unphysical Landau poles at large momenta. This shows that the theory is incomplete, a conclusion which applies also to any other scalar field.

In this context it is worth remarking that there are also theoretical models [7] trying to explain photons and gravitons as being composed of more fundamental spin-$\frac{1}{2}$ constituents. This scenario is incompatible with gravity with torsion. The spin-$\frac{1}{2}$ fermions in the photon would be in an s-wave bound state with parallel spins. This would cause a nonzero coupling to torsion, which would destroy gauge invariance.

The conclusion of Section (16.2) that the photon and the fundamental vector bosons W and Z do not couple to torsion to guarantee gauge invariance remains valid also under the assumption that torsion fields are completely antisymmetric. This follows from Eq. (16.18) which gives a nonzero mass also for antisymmetric $S_{\mu\nu\lambda}(x)$. This would destroy the necessary electromagnetic gauge invariance. By analogy with electromagnetism, the fields of the vector bosons W and Z mediating the weak interactions in the standard model of these interactions must also remain decoupled from torsion if $S_{\mu\nu\lambda}(x)$ is completely antisymmetric.

Note that this property makes the gauge bosons quite different from massive vector mesons composed of quark-antiquark pairs, such as ρ- and ω-mesons. Their wave functions have a large amplitude in a state of a quark and an antiquark in an s-wave spin triplet channel. Such a state will couple to torsion via its quark content, so that these particles can be described effectively by a spin-1 field with a nonzero γ in the spin connection (20.22). The value of γ will, however, certainly be smaller than unity. The reason is that only part of the wave function consists of a quark-antiquark pair in an s-wave. A sizable part consists of two pions in a p-wave. Pions are pseudoscalar particles in which a quark-antiquark pair is coupled to a spin singlet in an s-wave. This part does not couple to torsion and reduces γ to a value $\gamma < 1$. Another small part of the ρ and ω wave functions consists of a sea of quark-antiquark pairs [8] which further reduces γ. By a similar argument, all strongly interacting particles have $\gamma < 1$, i.e., they are not coupled minimally.

The situation may be compared with that of the magnetic moments in quantum electrodynamics. The bare electrons, muons, and quarks are coupled minimally to the vector currents $\bar{\psi}(x)\gamma^{\mu}\psi(x)$. From these couplings one can easily derive the ratio of the magnetic moments between proton and neutron of $-3/2$, knowing only the three-quark composition of these particles. But the size of the two magnetic moments depends on the wave functions of the quarks inside these particles. It is *not* universal. In the same way, a proton and a neutron cannot be expected to have $\gamma = 1$ in the spin connection (20.22), nor can any other composite particle such as Δ or Ω^-.

All these considerations are, of course, based on the assumption that fundamental spin-$\frac{1}{2}$ particles such as leptons and quarks do really follow the multivalued mapping principle and couple to torsion with the universal value $\gamma = 1$. This assumption may well be false, and nature may have chosen to treat all particles like photons and W, Z mesons, giving them all the truly minimal value $\gamma = 0$, thus making gravity completely torsionless as assumed by Einstein.

20.4   Modified Energy-Momentum Conservation Law

For the sake of academic curiosity one may search for a field theory of scalar particles which does couple to torsion. This is only possible if one modifies, in the presence of torsion, the variational procedure which led to the Einstein equation (15.63). Since spacetime with torsion has a closure failure, a modification could well be justified. If we recall the derivation of the geodesics equation of motion (18.72) from the covariant conservation law (18.68), we realize that autoparallel trajectories would emerge if the covariant conservation law for the energy-momentum tensor of a free spinless point particle $\bar{D}_\kappa \overset{m}{T}_\lambda{}^\kappa(x) = 0$ of Eq. (18.68) would be modified to

$$D_\nu^* \overset{m}{T}{}^{\mu\nu}(x) = 0, \tag{20.23}$$

rather than (18.76).

In order to see how this conservation law could be obtained let us go once more through its derivation. We calculated in Eq. (18.54) the change of the total action under Einstein transformations (11.78). For a point particle, this may be written explicitly as

$$\delta_E \mathcal{A} = \int d^4x \left\{ \frac{\delta \mathcal{A}}{\delta g_{\mu\nu}}\bigg|_{\Gamma_{\mu\nu}{}^\lambda} \delta_E g_{\mu\nu} + \frac{\delta \mathcal{A}}{\delta \Gamma_{\mu\nu}{}^\lambda}\bigg|_{g_{\mu\nu}} \delta_E \Gamma_{\mu\nu}{}^\lambda \right\} + \int d\tau \frac{\delta \overset{m}{\mathcal{A}}}{\delta x^\mu(\tau)} \delta_E x^\mu(\tau). \tag{20.24}$$

Initially, the last term does not vanish on autoparallel trajectories. However, if we account for the closure failure of spacetime by changing $\delta x(\tau)$ to the nonholonomic $\delta x(\tau)$ defined in Eq. (14.14) with the property (14.30), then it does.

The vanishing of the first term in (20.24) for arbitrary infinitesimal $\delta_E g_{\mu\nu}(x)$ of Eq. (18.18) is responsible for the covariant conservation law (18.59), leading to geodesic trajectories.

It is now interesting to realize the following: if we were to replace the Einstein transformation (18.18) in (20.24) by a transformation defined by

$$\delta_E g_{\mu\nu}(x) = D_\mu \xi_\nu(x) + D_\nu \xi_\mu(x) = \bar{D}_\mu \xi_\nu(x) + \bar{D}_\nu \xi_\mu(x) - 4S^\lambda{}_{\mu\nu}\xi_\lambda(x), \tag{20.25}$$

which looks like (18.18), but with the Riemann covariant derivative replaced by the full covariant derivative D_μ, the variation would contain an extra term,

$$\delta_E g_{\mu\nu}(x) = \delta_E g_{\mu\nu}(x) - 4S^\lambda{}_{\mu\nu}\xi_\lambda(x). \tag{20.26}$$

The change of the matter part of the action would become

$$\delta_E \overset{m}{\mathcal{A}} = -\frac{1}{2}\int d^4x \sqrt{-g}\, \overset{m}{T}{}^{\mu\nu}(x)\, \delta_E g_{\mu\nu}(x) = -\int d^4x \sqrt{-g}\, \overset{m}{T}{}^{\mu\nu}(x)\, D_\nu \xi_\mu(x). \tag{20.27}$$

Integrals over invariant expressions containing the covariant derivative D_μ can be integrated by parts according to the rule (15.40). If we neglect the surface terms we find

$$\delta_E \overset{m}{\mathcal{A}} = \int d^4x \sqrt{-g}\, D_\nu^* \overset{m}{T}{}^{\mu\nu}(x)\xi_\mu(x), \tag{20.28}$$

where $D_\nu^* = D_\nu + 2S_{\nu\lambda}{}^\lambda$. From the vanishing of this for all $\xi_\mu(x)$ we would indeed derive the covariant conservation law (20.23) for a spinless point particle in spacetime which corresponds to autoparallel trajectories.

The question arises whether the modified conservation law (20.23) allows for the construction of a suitable modification of the original Einstein field equation

$$\bar{G}^{\mu\nu} = \kappa \overset{m}{T}{}^{\mu\nu} \tag{20.29}$$

which would remain valid in spacetimes with torsion. Which tensor will stand on the left-hand side of the field equation (20.29) if the energy-momentum tensor satisfies the conservation law (20.23) instead of (18.76)?

20.4.1   Solution for Gradient Torsion

At present, we can give an answer to this question only for a pure gradient torsion [9], which has the general form (14.41):

$$S_{\mu\nu}{}^\lambda = \frac{1}{2}(\delta_\mu{}^\lambda \partial_\nu \theta - \delta_\nu{}^\lambda \partial_\mu \theta). \tag{20.30}$$

Then we may simply replace (20.29) by

$$e^\theta G^{\mu\nu} = \kappa \overset{m}{T}{}^{\mu\nu}, \tag{20.31}$$

where $G^{\mu\nu}$ is the full Einstein tensor $G^{\mu\nu} \equiv R^{\mu\nu} - \frac{1}{2}g^{\mu\nu}R$ formed from the curvature tensor (11.129) in Einstein-Cartan space. For gradient torsion, this is symmetric, thus matching the symmetry of the right-hand side. This is easily proved with the help of the fundamental identity (18.60):

$$D^*{}_\lambda S_{\mu\nu}{}^{;\lambda} = G_{\mu\nu} - G_{\nu\mu}. \tag{20.32}$$

Indeed, inserting (20.30) into (15.48), we find the Palatini tensor

$$S_{\lambda\mu}{}^{;\kappa} \equiv -2[\delta_\lambda{}^\kappa \partial_\mu \theta - (\lambda \leftrightarrow \mu)]. \tag{20.33}$$

This has a vanishing covariant derivative

$$D^*{}_\lambda S_{\mu\nu}{}^{;\lambda} = -2[D^*{}_\mu \partial_\nu \theta - D^*{}_\nu \partial_\mu \theta] = 2[S_{\mu\nu}{}^\lambda \partial_\lambda \theta - 2S_{\mu\lambda}{}^\lambda \partial_\nu \theta + 2S_{\nu\lambda}{}^\lambda \partial_\mu \theta] = 0. \tag{20.34}$$

The terms on the right-hand side cancel after using (20.30) and $S_{\mu\lambda}{}^\lambda \equiv S_\mu = -3\partial_\mu \theta/2$.

Now we insert (20.30) into the Bianchi identity (18.61), with the result

$$\bar{D}^*_\nu G_\lambda{}^\nu + \partial_\lambda \theta G_\kappa{}^\kappa - \partial_\nu \theta G_\lambda{}^\nu + 2\partial_\nu \theta R_\lambda{}^\nu = 0. \tag{20.35}$$

Setting here $R_{\lambda\kappa} = G_{\lambda\kappa} - \frac{1}{2}g_{\lambda\kappa}G_\nu{}^\nu$, this becomes

$$D^*_\nu G_\lambda{}^\nu + \partial_\nu \theta G_\lambda{}^\nu = 0. \tag{20.36}$$

Thus we find the Bianchi identity

$$D_\nu^*(e^\theta G_\lambda{}^\nu) = 0. \tag{20.37}$$

This makes the left-hand side of the new field equation (20.31) compatible with the right hand side, which guarantees autoparallel particle trajectories.

If spacetime contains torsion which is not of the gradient type, the only solution of the problem is the one indicated in Section 20.3.1, that all sources of torsion are fundamental spin-$\frac{1}{2}$ particles, such as quarks and leptons. These create only completely antisymmetric torsion fields making autoparallel orbits equal to geodesics.

20.4.2 Gradient Torsion coupled to Scalar Fields

If torsion is of the gradient type (20.30), then there is a way in which torsion can enter the scalar Higgs field action

$$\mathcal{A}[\phi] = \int d^4x \, \sqrt{-g} \left(\frac{1}{2} g^{\mu\nu} |\nabla_\mu \phi \nabla_\nu \phi| - \frac{m^2}{2} |\phi|^2 - \frac{\lambda}{4} |\phi^2|^2 \right). \tag{20.38}$$

To avoid inessential complications, the Higgs field is assumed to be a complex scalar field with a Ginzburg-Landau-type action. As usual, $g = \det g_{\mu\nu}$ is the determinant of the metric $g_{\mu\nu}(x)$. The symbol ∇_μ denotes the covariant electromagnetic derivative $\nabla_\mu = \partial_\mu - ieV_\mu$. For a scalar field, no Christoffel symbol is needed to achieve covariance.

The square mass is negative, so that the Higgs field has a nonzero expectation value with $|\phi|^2 = -m^2/\lambda$. The derivative term provides the vector field with a mass term $e^2|\phi|^2 V^\mu V_\mu / 2$, leading to the free part of the vector boson action

$$\mathcal{A}[V] = \int d^4x \, \sqrt{-g} \left(-\frac{1}{4} \bar{F}_{\mu\nu} \bar{F}^{\mu\nu} - \frac{e^2 m^2}{2\lambda} V_\nu V^\mu \right), \tag{20.39}$$

where $\bar{F}_{\mu\nu}$ is the Riemann-covariant curl which by Eq. (16.14) is equal to the ordinary curl $\bar{F}_{\mu\nu} = \partial_\mu V_\nu - \partial_\nu V_\mu$. Of course, the covariant curl of the nonabelian vector bosons of electromagnetic and weak interactions would also have self-interactions, which may be ignored in the present discussion, being only interested in the free-particle propagation.

The Meissner effect creates the masses of the vector bosons by mixing the uncoupled bare vector boson with the scalar Higgs field. It is then obvious that the massive vector bosons couples to torsion if and only if the scalar Higgs field has such a coupling.

Our goal is to introduce a coupling of the gradient torsion in such a way that the scalar particles move along autoparallel trajectories, as required by the study of particle orbits in spacetime with a closure failure [10, 11]. This is possible by introducing a new $\theta(x)$-dependent scalar product.

20.4.3   New Scalar Product

In textbook [10] it was pointed out that there exists a consistent Schrödinger formulation for a particle in a space with torsion if torsion is completely antisymmetric, which is the case discussed in Subsection 20.3.1, or if it has the restricted gradient form (20.30). The Schrödinger equation is driven by the Laplace operator $g^{\mu\nu} D_\mu D_\nu$, where D_μ is the covariant derivative involving the full affine connection $\Gamma_{\mu\nu}{}^\lambda$, including torsion. It differs from the Laplace-Beltrami operator in torsion-free spaces $\Delta \equiv \sqrt{|g|}^{-1} \partial_\mu \sqrt{|g|} g^{\mu\nu} \partial_\nu$ by a term $-2 S^{\nu\lambda}{}_\lambda \partial_\nu = -3(\partial^\nu \theta) \partial_\nu$. The important observation here is that this operator is hermitian only in a scalar product which contains a factor $e^{-3\theta}$ [15, 16].

The gradient torsion (20.30) has the advantage that it can be coupled to a scalar particle in such a way that the modification of the variational procedure found in [12, 13] becomes superfluous, while still producing autoparallel trajectories. For a massive particle, the coupled action reads [17]

$$\overset{m}{\mathcal{A}} [x] = -mc \int d\sigma\, e^{\theta(x(\sigma))} \sqrt{g_{\mu\nu}(x))\dot{x}^\mu(\sigma)\dot{x}^\nu(\sigma)} = -mc \int ds\, e^{\theta(x(s))}, \qquad (20.40)$$

where σ is an arbitrary parameter, and $s = c\tau$ the invariant length of the orbit. We have shown in Subsection 14.1.4 that extremizing the action (20.40) yields the autoparallel equation of motion (14.43) in the presence of gradient torsion (14.41):

$$\ddot{x}^\lambda + \bar{\Gamma}_{\mu\nu}{}^\lambda \dot{x}^\mu \dot{x}^\nu = -\dot{\theta}(x)\dot{x}^\lambda + g^{\lambda\kappa}(x)\partial_k \theta(x). \qquad (20.41)$$

Along this trajectory, the Lagrangian under the action integral is a constant of motion whose value is, moreover, fixed by the mass shell constraint

$$\overset{m}{L} = e^{\theta(x)} \sqrt{g_{\mu\nu}(x))\dot{x}^\mu \dot{x}^\nu} \equiv 1, \qquad \tau = s. \qquad (20.42)$$

The same classical equation of motion emerges from the eikonal approximation of a scalar field theory which contains the scalar field $\theta(x)$ in a peculiar way:

$$\mathcal{A}[\phi] = \int d^4x\, \sqrt{-g}\, e^{-3\theta} \left(\frac{1}{2} g^{\mu\nu} |\nabla_\mu \phi \nabla_\nu \phi| - \frac{m^2}{2} |\phi|^2 e^{-2\theta} \right). \qquad (20.43)$$

The necessity of a factor $e^{-3\theta(x)}$ in the scalar product was discovered in Ref. [10], and became the basis of a series of studies in general relativity [18]. The action (20.43) is extremized by the Euler-Lagrange equation

$$D_\mu D^\mu \phi + m^2 e^{-2\theta(x)} \phi = 0. \qquad (20.44)$$

Let us find the equation of motion for the classical particles described by the scalar field $\phi(x)$. For this we make the ansatz $\phi(x) \approx e^{iS(x)}$, and find the eikonal equation for the phase $S(x)$ [19]:

$$e^{2\theta(x)} g^{\mu\nu}(x)[\partial_\mu S(x)][\partial_\nu S(x)] = m^2. \qquad (20.45)$$

Since $\partial_\mu S$ is the momentum of the particle, the replacement $\partial_\mu S \to m\dot{x}_\mu$ shows that the eikonal equation (20.45) guarantees the constancy of the Lagrangian (20.42), which is characteristic for autoparallel classical trajectories.

20.4.4 Self-Interacting Higgs Field

Apart from the factor $e^{-3\theta(x)}$ accompanying the volume integral, the θ-field couples to the scalar field like a dilaton, the power of $e^{-\theta}$ being determined by the dimension of the associated term. If we add a quartic self-interaction to the free-field action (20.43), and go to negative m^2 to have a Meissner effect, the self-interaction will not carry an extra factor $e^{-\theta}$, so that the proper Higgs action in the presence of gradient torsion reads

$$\mathcal{A}[\phi] = \int d^4x \sqrt{-g} e^{-3\theta} \left(\frac{1}{2} g^{\mu\nu} |\nabla_\mu \phi \nabla_\nu \phi| - \frac{m^2}{2} |\phi|^2 e^{-2\theta} - \frac{\lambda}{4} |\phi^2|^2 \right). \tag{20.46}$$

If m^2 is negative, and the torsion depends only weakly on spacetime, the Higgs field has a smooth vacuum expectation value

$$|\phi|^2 = -\frac{m^2}{\lambda} e^{-2\theta}. \tag{20.47}$$

The smoothness of the torsion field is required over a length scale of the Compton wavelength of the Higgs particle, i.e., over a distance of the order $1/20\text{GeV} \approx 10^{-15}$ cm. For a torsion field of gravitational origin, this smoothness will certainly be guaranteed. From the gradient term in (20.46) we then extract in the gauge $\phi =$real the mass term of the vector bosons

$$\int d^4x \sqrt{-g} e^{-3\theta} \frac{1}{2} m_V^2 e^{-2\theta(x)} V^\mu V_\mu \tag{20.48}$$

where

$$m_V^2 = -\frac{e^2}{\lambda} m^2, \quad m^2 < 0. \tag{20.49}$$

Taking the physical scalar product in the presence of torsion into account, we obtain for the massive vector bosons the free-field action

$$\mathcal{A}[V] = \int d^4x \sqrt{-g} e^{-3\theta} \left(-\frac{1}{4} F_{\mu\nu} F^{\mu\nu} + m_V^2 e^{-2\theta(x)} V_\nu V^\mu \right). \tag{20.50}$$

The appearance of the factor $e^{-2\theta}$ in the mass term guarantees again the same autoparallel trajectories in the eikonal approximation as for spinless particles in the action (20.43).

Note that the scalar product factor $e^{-3\theta(x)}$ implies a coupling to torsion also for the initial massless vector boson fields W and Z which does not destroy gauge invariance. Due to the similarity between the bare fields of W- and Z-bosons with the photon field, the factor $e^{-3\theta(x)}$ must be present also in the electromagnetic action.

20.5 Summary

What have we learned about torsion in this chapter? Is there a chance of its influencing gravitational physics? Ultimately, experiments will have to decide this question. This is a very difficult task, due to the weakness of the coupling of torsion to the intrinsic spins of the fundamental constituents a celestial body.

To illustrate the difficulty more quantitatively, take for example the satellite experiment *Gravity Probe B* which went into orbit on April 20, 2004. It attempts to measure the small frame-dragging effect predicted in 1918 by Lense and Thirring [7]. The satellite carries four spherical quartz gyroscopes of diameter 3.8 cm around the earth in an orbit 740 km above the poles. According to Einstein's theory, the axes will precess 6.6" per year due to the geodetic precession [recall Section 14.1.3]. In addition there is a precession caused by the rotation of the earth which drags the freely falling frames with it and causes an extra 0.042". This is the famous *Lense-Thirring·effect*. The mass of the gyroscopes is roughly 63 g $\approx$ 1 mol, and their moment of inertia $I \approx 90\,\mathrm{g\,cm^2}$. The rotation frequency is $\omega = 2\pi \times 10\,000\,\mathrm{min^{-1}}$, implying an angular momentum $L \approx 10^6\,\mathrm{g\,cm^2/s}$. The number of molecules in the gyroscopes is close to the Avogadro number 6×10^{23}. From this we derive the angular momentum per molecule $l \approx 40\,000\,\hbar$.

Suppose now that we want to detect the influence of torsion upon this experiment. The only way of doing this is to replace the material of the gyroscopes by a completely polarized magnetic material. At the same rate of rotation, the spin per atom s is of the order of a few $\hbar$. Hence the effect of torsion, even it it were to exist along the orbit with a similar strength as Einstein's gravitational field (which it cannot since torsion does not propagate), would be suppressed by a factor 10^{-4}. This makes it unmeasurable with present-day technology.

Thus, all theories with torsion will remain purely speculative for a long time to come. The probability of its existence can only be guessed by the esthetical appeal of the theory. This is often governed by a minimality principle. Nature makes use of a mathematical structure only if it is needed by some symmetry principle. The only argument pro torsion is that it enters naturally when flat-space theories are transformed into spaces with curvature and torsion by multivalued coordinate transformations, and that the same transformations carry ideal crystals into crystals with dislocations and disclinations. Since nature uses often the same mathematical structures in different phenomena, this could also happen here.

If, on the other hand, the minimality principle is invoked as a dominant criterion, torsion will not be coupled at all, since it is not necessary to guarantee the local symmetries of gravitation.

Notes and References

[1] F.W. Hehl and B.K. Datta, J. Math. Phys. **12**, 1334 (1971); R.F. O'Connell, Phys. Rev. Lett. **37**, 1653 (1976).

[2] H. Kleinert, *Universality Principle for Orbital Angular Momentum and Spin in Gravity with Torsion*, Gen. Rel. Grav. **32**, 1271 (2000) kl/271j.pdf.

[3] See S. Weinberg's article in Physics Today, April 2006, p. 10, where he comments on various letters responding to his previous November 2005 article on Einstein's mistakes. In the April article he remarks that he never understood what is physically so important about the torsion tensor. This prompted F.W. Hehl to a further comment in the March 2007 issue where he reminded Weinberg of Sciama and Kibble's way of expressing Einstein-Cartan gravity in terms of gauge fields of Lorentz transformations and translations, in which curvature and torsion are the gauge-invariant field strengths. Hehl's answer did not satisfy Weinberg who could only be convinced unless there is an invariance principle which requires the Christoffel symbol to be accompanied by a torsion tensor in the affine connection. As we have mentioned in Section 20.2, the multivalued mapping principle of Chapter 14 is precisely such an invariance principle.

[4] F.W. Hehl, Phys. Lett. A **36**, 225 (1971).

[5] This was observed in Section 11.4 of Ref. [10].

[6] For reviews see
S. Eidelmann et al., Phys. Lett. B **592**, 1 (2004) (http://pdg.lbl.gov), and the web pages en.wikipedia.org/wiki/Technicolor_(physics) and pdg.lbl.gov/2004/listings/s057.pdf;
C.T. Hill and E.H. Simmons, Phys. Rep. **381**, 235 (2003);
D.K. Hong, S.D.H. Hsu, F. Sannino, *Opening the Window for Technicolor*, (hep-ph/0410310).

[7] J.D. Bjorken, Ann. Phys. (N.Y.) **24**, 174 (1963); (hep-th/0111196).
For a review see
W.A. Perkins, *Interpreted History Of Neutrino Theory Of Light And Its Future*, in *Lorentz Group, CPT and Neutrinos*, eds. A.E. Chubykalo, V.V. Dvoeglazov, D.J. Ernst, V.G.Kadyshevsky, and Y.S. Kim, World Scientific, Singapore, 2000, p.115 (hep-ph/0107122).
See also the more recent paper by
P. Kraus and E.T. Tomboulis, *Photons and Gravitons as Goldstone Bosons and the Cosmological Constant*, Phys. Rev. D **66**, 045015 (2002), and references therein.

[8] H. Kleinert, *Quark Pairs inside Hadrons*, Phys. Letters B **59**, 163 (1975) http://www.physik.fu-berlin.de/~kleinert/50.

[9] H. Kleinert and A. Pelster, Acta Phys. Polon. B **29** , 1015 (1998).

[10] H. Kleinert, *Path Integrals in Quantum Mechanics, Statistics, Polymer Physics, and Financial Markets*, 4th ed., World Scientific, Singapore 2006 (kl/b5).

[11] H. Kleinert, *Nonholonomic Mapping Principle for Classical and Quantum Mechanics in Spaces with Curvature and Torsion*, Gen. Rel. Grav. **32**, 769 (2000) (`kl/258`).
Short version presented as a lecture at the Workshop on *Gauge Theories of Gravitation*, Jadwisin, Poland, 4-10 September 1997, Acta Phys. Pol. B **29**, 1033 (1998) (gr-qc/9801003).

[12] P. Fiziev and H. Kleinert, *New Action Principle for Classical Particle Trajectories in Spaces with Torsion*, Europhys. Lett. **35**, 241 (1996) (hep-th/9503074).

[13] H. Kleinert and A. Pelster, *Autoparallels From a New Action Principle*, Gen. Rel. Grav. **31**, 1439 (1999) (gr-qc/9605028).

[14] Our notation of field-theoretic and geometric quantities is the same as in the textbook
H. Kleinert, *Gauge Fields in Condensed Matter*, Vol. II, *Stresses and Defects*, World Scientific, Singapore 1989, pp. 744-1443 (`kl/b1/contents1.html`).

[15] See Section 11.4 in Ref. [10]. Note that the normalization of the $\theta(x)$-field is different from the present one by a factor $2/3$. There we introduced θ (under the name of σ which is used here for different purposes) via the relation $S_{\mu\nu}{}^{\nu} = \partial_{\mu}\theta$, whereas here $S_{\mu\nu}{}^{\nu} = (3/2)\partial_{\mu}\theta$.

[16] In the case of a totally antisymmetric torsion advocated in Subsection 20.3.1, a change of the scalar product is superfluous since the two Laplace operators $g^{\mu\nu}D_{\mu}D_{\nu}$ and Δ are equal, and the original scalar product ensures hermiticity and thus unitarity of time evolution.

[17] H. Kleinert and A. Pelster, *Novel Geometric Gauge-Invariance of Autoparallels*, Lectures presented at Workshop *Gauge Theories of Gravitation*, Jadwisin, Poland, 4-10 September 1997, Acta Phys. Pol. B **29**, 1015 (1998) (gr-qc/9801003).

[18] A. Saa, Mod. Phys. Lett. A **8**, 2565 (1993); ibid. 971 (1994); Class. Quant. Grav. **12**, L85 (1995); J. Geom. and Phys. **15**, 102 (1995); Gen. Rel. and Grav. **29**, 205 (1997).

[19] P. Fiziev, *Spinless Matter in Transposed-Equi-Affine Theory of Gravity*, Gen. Rel. Grav. **30**, 1341 (1998) (gr-qc/9712004). See also *Gravitation Theory with Propagating Torsion* (gr-qc/9808006).

21

Teleparallel Theory of Gravitation

At this point it is useful to remind the reader of an alternative theory of gravity proposed by Einstein in the thirties, the so-called *theory of teleparallelism*. As remarked in the Preface, this theory was inspired by Cartan's work in 1922 and a subsequent letter communication between Einstein and Cartan [1]. In this theory, spacetime is generated from flat space by assuming the absence of the multivalued local Lorentz transformations $\Lambda^a{}_\alpha(x)$ in the basis tetrads $e^a{}_\mu(x)$ of Eq. (17.45). Hence $e^a{}_\mu(x)$ becomes single-valued, and coincides with the vierbein field $h^{\alpha=a}{}_\mu(x)$. It follows from Eq. (11.130) that the Riemann-Cartan curvature tensor vanishes identically:

$$R_{\mu\nu\lambda}{}^\kappa \equiv 0 \quad \text{(in teleparallel spacetime)}. \tag{21.1}$$

This property has the pleasant consequence that it allows for the definition of parallel vector fields in *all* spacetime, hence the name teleparallelism. Since the vanishing of the Riemann-Cartan curvature tensor is caused by commuting derivatives in (11.130), the vanishing of $R_{\mu\nu\lambda}{}^\kappa$ may be considered as a *Bianchi identity* of teleparallel spacetime. The single-valuedness of e $\Lambda_\alpha{}^\beta(x)$ permits choosing at each point a Lorentz frame where $\Lambda_\alpha{}^\beta(x) = 1$ so that $\overset{\Lambda}{\Gamma}_{\alpha\beta}{}^\gamma = 0$ by Eq. (17.15). Then we deduce from Eq. (17.68) that in this gauge the affine connection $\Gamma_{\mu\nu}{}^\lambda$ coincides with the quantities $\overset{h}{\Gamma}_{\mu\nu}{}^\lambda$ of Eq. (17.67):

$$\Gamma_{\mu\nu}{}^\lambda = \overset{h}{\Gamma}_{\mu\nu}{}^\lambda \equiv h_\alpha{}^\lambda \partial_\mu h^\alpha{}_\nu \equiv -h^\alpha{}_\nu \partial_\mu h_\alpha{}^\lambda, \tag{21.2}$$

and the torsion tensor reduces to the object of anholonomy (17.74):

$$S_{\mu\nu}{}^\lambda = \frac{1}{2}\left(h_\alpha{}^\lambda \partial_\mu h^\alpha{}_\nu - h_\alpha{}^\lambda \partial_\nu h^\alpha{}_\mu\right) \equiv -\frac{1}{2}\left(h^\alpha{}_\nu \partial_\mu h_\alpha{}^\lambda - h^\alpha{}_\mu \partial_\nu h_\alpha{}^\lambda\right). \tag{21.3}$$

21.1 Torsion Form of Einstein Action

Recalling the decomposition of the Riemann-Cartan curvature tensor (11.146), the vanishing of $R_{\mu\nu\lambda}{}^\kappa$ implies further that the Riemann curvature tensor can be expressed in terms of the contortion tensor as

$$\bar{R}_{\mu\nu\lambda}{}^\kappa = \bar{D}_\mu K_{\nu\lambda}{}^\kappa - \bar{D}_\nu K_{\mu\lambda}{}^\kappa - \left(K_{\mu\lambda}{}^\rho K_{\nu\rho}{}^\kappa - K_{\nu\lambda}{}^\rho K_{\mu\rho}{}^\kappa\right). \tag{21.4}$$

For the scalar curvature this becomes

$$\bar{R} = \bar{R}_{\mu\nu\lambda}{}^{\mu}g^{\nu\lambda} = -\bar{D}_{\mu}K_{\nu}{}^{\nu\mu} + \bar{D}_{\nu}K_{\mu}{}^{\nu\mu} + \left(K_{\mu}{}^{\nu\rho}K_{\nu\rho}{}^{\mu} + K_{\nu}{}^{\nu\rho}K_{\mu}{}^{\mu}{}_{\rho}\right), \qquad (21.5)$$

or, expressing everything in terms of the torsion tensor using (11.114) and $K_{\mu}{}^{\mu\nu} = 2S^{\nu}$ [recall (15.42)]:

$$\bar{R} = -4\bar{D}_{\mu}S^{\mu} + \left(S_{\mu\nu\lambda}S^{\mu\nu\lambda} + 2S_{\mu\nu\lambda}S^{\mu\lambda\nu} - 4S^{\rho}S_{\rho}\right). \qquad (21.6)$$

Thus, in a teleparallel spacetime, the Einstein action (15.8) can be replaced by

$$\overset{f}{\mathcal{A}}_{E,S} = -\frac{1}{2\kappa}\int d^4x\sqrt{-g}\left[-4\bar{D}_{\mu}S^{\mu} + \left(S_{\mu\nu\lambda}S^{\mu\nu\lambda} + 2S_{\mu\nu\lambda}S^{\mu\lambda\nu} - 4S^{\rho}S_{\rho}\right)\right]. \qquad (21.7)$$

Integrating this by parts yields the action

$$\overset{f}{\mathcal{A}}_{E,S} = -\frac{1}{2\kappa}\int d^4x\sqrt{-g}\left(S_{\mu\nu\lambda}S^{\mu\nu\lambda} + 2S_{\mu\nu\lambda}S^{\mu\lambda\nu} - 4S^{\rho}S_{\rho}\right), \qquad (21.8)$$

where we have dropped the pure surface term

$$\overset{f}{\mathcal{A}}_{E,S,\text{surface}} = \frac{1}{2\kappa}\int d^4x\, 4\partial_{\mu}\left(\sqrt{-g}S^{\mu}\right), \qquad (21.9)$$

which does not contribute to the field equations.

It is useful to decompose the Lagrangian density into irreducible parts of $S_{\mu\nu\lambda}$ constructed with the help of Young tableaux [2]. An obviously irreducible part of $S_{\mu\nu\lambda}$ is the vector $S_{\mu} = S_{\mu\nu}{}^{\nu}$ of Eq. (15.42). It is associated with the mixed Young tableau:

$$\boxed{\begin{array}{cc}\mu & \nu\end{array}} \atop \boxed{\lambda} \quad : \quad S_{\mu\nu}{}^{\nu}. \qquad (21.10)$$

A second irreducible part is the totally symmetric combination:

$$\boxed{\mu}\,\boxed{\nu}\,\boxed{\lambda} \quad : \quad t_{\mu\nu\lambda} = \frac{1}{2}\left(S_{\mu\nu\lambda} + S_{\mu\lambda\nu}\right) + \frac{1}{6}\left(g_{\mu\lambda}S_{\nu} + g_{\mu\nu}S_{\lambda} - 2g_{\nu\lambda}S_{\mu}\right). \qquad (21.11)$$

The third is an axial vector arising from the totally antisymmetric combination:

$$\boxed{\begin{array}{c}\mu\\\nu\\\lambda\end{array}} \quad : \quad a^{\mu} = \frac{1}{6}e^{\mu\nu\lambda\kappa}S_{\nu\lambda\kappa} \qquad (21.12)$$

where $e^{\mu\nu\lambda\kappa}$ is the covariant Levi-Civita tensor (11A.1). The torsion can be recovered from these tensors and vectors as follows:

$$S_{\mu\nu\lambda} = \frac{2}{3}\left(t_{\mu\nu\lambda} - t_{\nu\mu\lambda}\right) - \frac{1}{3}\left(g_{\mu\lambda}S_{\nu} - g_{\nu\lambda}S_{\mu}\right) - e_{\mu\nu\lambda\kappa}a^{\kappa}. \qquad (21.13)$$

The three invariants in the action (21.8) can be expressed in terms of the irreducible invariants $t_{\mu\nu\lambda}t^{\mu\nu\lambda}$, $S_\mu S^\mu$, and $a_\mu a^\mu$ as follows:

$$S_{\mu\nu\lambda}S^{\mu\nu\lambda} = \frac{4}{3}t_{\mu\nu\lambda}t^{\mu\nu\lambda} + \frac{2}{3}S_\mu S^\mu - 6a_\mu a^\mu, \quad S_{\mu\nu\lambda}S^{\mu\lambda\nu} = \frac{2}{3}t_{\mu\nu\lambda}t^{\mu\nu\lambda} + \frac{1}{3}S_\mu S^\mu + 6a_\mu a^\mu.$$
(21.14)

The inverse relations are

$$t_{\mu\nu\lambda}t^{\mu\nu\lambda} = \frac{1}{2}S_{\mu\nu\lambda}S^{\mu\nu\lambda} + \frac{1}{2}S_{\mu\nu\lambda}S^{\mu\lambda\nu} - \frac{1}{2}S_\mu S^\mu, \quad a_\mu a^\mu = -\frac{1}{18}(S_{\mu\nu\lambda}S^{\mu\nu\lambda} - 2S_{\mu\nu\lambda}S^{\mu\lambda\nu}).$$
(21.15)

Thus we can rewrite the action (21.8) as

$$\overset{f}{\mathcal{A}}_{E,S} = -\frac{1}{2\kappa}\int d^4x\sqrt{-g}\left(\frac{8}{3}t_{\mu\nu\lambda}t^{\mu\nu\lambda} - \frac{8}{3}S_\mu S^\mu + 6a_\mu a^\mu\right).$$
(21.16)

All results of Einstein's theory can be rederived from this action in Einstein-Cartan spacetime with $R_{\mu\nu\lambda\kappa} \equiv 0$, in which the vierbein fields $h^\alpha{}_\mu$ are four teleparallel vector fields.

This reformulation of Einstein gravity would only become interesting if future experiments were to discover deviations from Einstein's theory. Since $S_{\mu\nu\lambda}$ is a tensor field, the action (21.16) does not necessarily have to contain the three invariants in the specific combination implied by Eq. (21.6). Any other combination is invariant [2]:

$$\overset{f}{\mathcal{A}}_{E,S} = \int d^4x\sqrt{-g}\,\overset{f}{\mathcal{L}}_S = -\frac{1}{2\kappa}\int d^4x\sqrt{-g}\left(\gamma_1 t_{\mu\nu\lambda}t^{\mu\nu\lambda} + \gamma_2 S_\mu S^\mu + \gamma_3 a_\mu a^\mu\right).$$
(21.17)

Of course, the three parameters γ_i are not completely free. There will be one constraint between them fixed by Newton's law, so that the generalized theory has two free parameters. These can be fixed, for example, by the post-Newtonian expansion of the gravitational field around a mass point.

Expressed in terms of the original invariants in (21.17), the generalized action (21.17) reads

$$\overset{f}{\mathcal{A}}_S = \int d^4x\sqrt{-g}\,\overset{f}{\mathcal{L}}_S = -\frac{1}{2\kappa}\int d^4x\sqrt{-g}\left(\sigma_1 S_{\mu\nu\lambda}S^{\mu\nu\lambda} + \sigma_2 S_{\mu\nu\lambda}S^{\mu\lambda\nu} + \sigma_3 S_\mu S^\mu\right),$$
(21.18)

with

$$\sigma_1 = \tfrac{1}{2}\gamma_1 - \tfrac{1}{18}\gamma_3, \quad \sigma_2 = \tfrac{1}{2}\gamma_1 + \tfrac{1}{9}\gamma_3, \quad \sigma_3 = \gamma_2 - \tfrac{1}{2}\gamma_1.$$
(21.19)

For the sake of deriving the equations of motion there exist several convenient forms of writing the action (21.18) to be distinguished by superscripts for better reference. The first is

$$\overset{f}{\mathcal{A}}_S^{(1)} = \int d^4x\sqrt{-g}\,\overset{f}{\mathcal{L}}^{(1)} = -\frac{1}{2\kappa}\int d^4x\sqrt{-g}\,S_{\mu\nu\lambda}\,P^{\mu\nu\lambda,\mu'\nu'\lambda'}\,S_{\mu'\nu'\lambda'},$$
(21.20)

where the tensor $P^{\mu\nu\lambda,\mu'\nu'\lambda'}$ is a combination of contravariant metric tensors:

$$P^{\mu\nu\lambda,\mu'\nu'\lambda'} = \sigma_1 \, g^{\mu\mu'} g^{\nu\nu'} g^{\lambda\lambda'} + \sigma_2 \, g^{\mu\mu'} g^{\nu\lambda'} g^{\lambda\nu'} + \sigma_3 \, g^{\mu\mu'} g^{\nu\lambda} g^{\nu'\lambda'}. \tag{21.21}$$

The second form is

$$\overset{f}{\mathcal{A}}{}_S^{(2)} = -\frac{1}{2\kappa} \int d^4x \sqrt{-g} \, S_{\mu\nu\lambda} F^{\mu\nu\lambda}, \tag{21.22}$$

with

$$
\begin{aligned}
F^{\mu\nu\lambda} &\equiv \frac{\kappa}{2} \frac{\partial \overset{f}{\mathcal{L}}_S}{\partial S_{\mu\nu\lambda}} = P^{[\mu\nu]\lambda,\mu'\nu'\lambda'} S_{\mu'\nu'\lambda'} \\
&= \sigma_1 S^{\mu\nu\lambda} + \frac{\sigma_2}{2}\left(S^{\mu\lambda\nu} - S^{\nu\lambda\mu}\right) + \frac{\sigma_3}{2}\left(g^{\nu\lambda}S^\mu - g^{\mu\lambda}S^\nu\right) = -F^{\nu\mu\lambda}. \tag{21.23}
\end{aligned}
$$

The indices in brackets are antisymmetrized, as usual. This tensor may also be expressed as a modification of (21.13):

$$F^{\mu\nu\lambda} = \gamma_1\left(t^{\mu\nu\lambda} - t^{\nu\mu\lambda}\right) - \gamma_2\left(g^{\mu\lambda}S^\nu - g^{\nu\lambda}S^\mu\right) - \frac{\gamma_3}{3}e^{\mu\nu\lambda\kappa}a_\kappa = -F^{\nu\mu\lambda}. \tag{21.24}$$

Using the asymmetry of $F^{\mu\nu\lambda}$ we can replace the torsion tensor in (21.22) by $h_{\alpha\lambda}\partial_\mu h^\alpha{}_\nu$ via formula (21.3), and rewrite the action (21.22) also as

$$\overset{f}{\mathcal{A}}{}_S^{(3)} = \int d^4x \sqrt{-g} \, \overset{f}{\mathcal{L}}{}_S^{(3)} = -\frac{1}{2\kappa} \int d^4x \sqrt{-g} \, h_{\alpha\lambda}\partial_\mu h^\alpha{}_\nu F^{\mu\nu\lambda}, \tag{21.25}$$

or as

$$\overset{f}{\mathcal{A}}{}_S^{(4)} = \int d^4x \sqrt{-g} \, \overset{f}{\mathcal{L}}{}_S^{(4)} = -\frac{1}{2\kappa} \int d^4x \sqrt{-g} \, h_{\alpha\lambda}\partial_\mu h^\alpha{}_\nu \, P^{[\mu\nu]\lambda,[\mu'\nu']\lambda'} h_{\alpha'\lambda'}\partial_{\mu'} h^{\alpha'}{}_{\nu'}, \tag{21.26}$$

where $P^{[\mu\nu]\lambda,[\mu'\nu']\lambda'}$ is the tensor (21.21) antisymmetrized in $\mu\nu$ and $\mu'\nu'$.

In order to have a better comparison with Einstein's theory, we shall add to the above action also the Einstein action multiplied with a parameter γ_0, thus working with the Lagrangian density of the gravitational field

$$
\begin{aligned}
\overset{f}{\mathcal{A}} &= \int d^4x \sqrt{-g} \, \overset{f}{\mathcal{L}} \equiv \gamma_0 \overset{f}{\mathcal{A}}_E + \overset{f}{\mathcal{A}}_S = \int d^4x \sqrt{-g} \left(\gamma_0 \overset{f}{\mathcal{L}}_E + \overset{f}{\mathcal{L}}_S\right) \\
&= \int d^4x \sqrt{-g} \left[-\frac{\gamma_0}{2\kappa}\bar{R} - \frac{1}{2\kappa}\left(\gamma_1 t_{\mu\nu\lambda}t^{\mu\nu\lambda} + \gamma_2 S_\mu S^\mu + \gamma_3 a_\mu a^\mu\right)\right], \tag{21.27}
\end{aligned}
$$

where the subscripts E and S indicate Einstein's terms and their possible torsion corrections, respectively. The parametrization involves now a redundant parameter due to the equality of $\overset{f}{\mathcal{L}}_E$ and $\overset{f}{\mathcal{L}}_S$ for $\gamma_1 = -8/3, \gamma_2 = 8/3, \gamma_3 = -6$.

According to Eq. (17.142), the energy-momentum tensor of the gravitational field is obtained by varying the field action with respect to $h_\alpha{}^\nu$ and multiplying the result by $h^{\alpha\mu}$:

$$\overset{f}{\Theta}{}^{\mu\nu} = -\frac{1}{\sqrt{-g}} h^{\alpha\mu} \frac{\delta \overset{f}{\mathcal{A}}}{\delta h^\alpha{}_\nu}. \tag{21.28}$$

Due to the absence of the object of anholonomy in the teleparallel formulation, the analog of the spin contribution (17.155) to the energy-momentum tensor is absent, so that the result is here the canonical energy-momentum tensor, not the symmetric one. Applying formula (21.28) to the Einstein term in the action (21.27), which depends only on the metric tensor $g_{\mu\nu}$, we may evaluate the functional derivative $h^{\alpha\mu}\delta\overset{f}{\mathcal{A}}/\delta h^{\alpha}{}_{\nu}$ in (21.28) as $(1/2)\delta\overset{f}{\mathcal{A}}_E/\delta g_{\mu\nu}$, and obtain with the help of Eq. (17.147) $-(\gamma_0/\kappa)$ times the Einstein tensor $\bar{G}^{\mu\nu}$.

It will now be convenient to define the complete right-hand side of (21.28) as a generalized Einstein tensor $\tilde{G}^{\mu\nu}$ in the teleparallel Einstein-Cartan space, i.e., we set

$$\overset{f}{\Theta}{}^{\mu\nu} \equiv -\frac{1}{\kappa}\tilde{G}^{\mu\nu} = -\frac{1}{\kappa}\left(\gamma_0\bar{G}^{\mu\nu} + \tilde{\delta}_S G_S^{\mu\nu}\right), \tag{21.29}$$

where

$$\frac{1}{\kappa}\Delta_S G^{\mu\nu} = \frac{1}{\sqrt{-g}}h^{\alpha\mu}\left[\frac{\partial\overset{f}{\mathcal{L}}_S}{\partial h^{\alpha}{}_{\nu}} - \partial_\sigma\frac{\partial\overset{f}{\mathcal{L}}_S}{\partial\partial_\sigma h^{\alpha}{}_{\nu}}\right]. \tag{21.30}$$

One contribution to this is trivially found from the expression (21.25) of the action. By forming the derivatives with respect to explicit $h^{\alpha}{}_{\nu}$s, we obtain

$$\frac{1}{\kappa}\Delta_S^{(3)}G^{\mu\nu} = -\frac{1}{\kappa}\left(\Gamma_{\sigma\lambda\mu}F^{\sigma\lambda\nu} - h^{\alpha\mu}\frac{1}{\sqrt{-g}}\partial_\sigma\sqrt{-g}h_{\alpha\lambda}F^{\sigma\nu\lambda}\right). \tag{21.31}$$

The prefactor $1/2$ in (21.25) is canceled by the fact that $S_{\mu\nu\lambda}$ is contained once more in $F^{\mu\nu\lambda}$ in a symmetric fashion, as is manifest in the expression (21.26) of the same action. Applying to the second term of (21.31) the derivative rule (12.161) in the form

$$\frac{1}{\sqrt{-g}}\partial_\nu\sqrt{-g} = \bar{\Gamma}_{\mu\nu}{}^{\mu} = \Gamma_{\mu\nu}{}^{\mu} - K_{\mu\nu}{}^{\mu} = \Gamma_{\mu\nu}{}^{\mu} + 2S_\nu, \tag{21.32}$$

and exploiting further the antisymmetry of $F^{\sigma\lambda\nu}$, we may also write

$$\frac{1}{\kappa}\Delta_S^{(3)}G^{\mu\nu} = -\frac{1}{\kappa}\left(S_{\sigma\lambda}{}^{\mu}F^{\sigma\lambda\nu} + S_{\sigma\lambda}{}^{\nu}F^{\sigma\lambda\mu} - D_\sigma F^{\sigma\nu\mu} - 2S_\sigma F^{\sigma\nu\mu}\right). \tag{21.33}$$

A further contribution to $\kappa^{-1}\Delta_S G^{\mu\nu}$ comes from the dependence of the tensor $P^{\mu\nu\lambda,\mu'\nu'\lambda'}$ in the action (21.20) on the inverse metric $g^{\mu\nu} = h^{\alpha\mu}h_\alpha{}^{\nu}$. Here we find from (21.30)

$$\frac{1}{\kappa}\Delta_S^{(1)}G^{\mu\nu} \equiv -\frac{2}{\sqrt{-g}}\frac{\delta\overset{f}{\mathcal{A}}_S^{(1)}}{\delta g^{\mu\nu}}, \tag{21.34}$$

which contains the three terms:

$$\frac{1}{\kappa}\Delta_S^{(1)}G^{\mu\nu} = \frac{1}{\kappa}\left[\sigma_1\left(2S^{\mu}{}_{\lambda\kappa}S^{\nu\lambda\kappa} + S_{\lambda\kappa}{}^{\mu}S^{\lambda\kappa\nu}\right) + \sigma_2\left(S^{\mu}{}_{\lambda\kappa}S^{\nu\kappa\lambda} - S^{\mu\lambda\kappa}S_{\lambda\kappa}{}^{\nu} - S^{\nu\lambda\kappa}S_{\lambda\kappa}{}^{\mu}\right)\right.$$
$$\left. + \sigma_3\left(S^{\mu}S^{\nu} + S^{\lambda\mu\nu}S_\lambda + S^{\lambda\nu\mu}S_\lambda\right)\right]. \tag{21.35}$$

A final contribution is due to the variation of $\sqrt{-g}$ in the action (21.18) [compare (17.145)], which adds to the energy-momentum tensor (21.28) a further term $-g^{\mu\nu}\overset{f}{\mathcal{L}}_S$, changing the Einstein tensor by

$$\frac{1}{\kappa}\Delta^{(4)}G^{\mu\nu} = g^{\mu\nu}\mathcal{L}_S. \tag{21.36}$$

In order to add all terms we define several tensors:

$$\begin{aligned}
H_1^{\mu\nu} &\equiv S^{\mu\lambda\kappa}S^{\nu}{}_{\lambda\kappa} = H_1^{T\,\mu\nu}, \\
H_2^{\mu\nu} &\equiv S^{\mu\lambda\kappa}S^{\nu}{}_{\kappa\lambda} = H_2^{T\,\mu\nu}, \\
H_3^{\mu\nu} &\equiv S^{\lambda\kappa\mu}S_{\kappa\lambda}{}^{\nu} = H_3^{T\,\mu\nu}, \\
H_4^{\mu\nu} &\equiv S^{\mu\lambda\kappa}S_{\kappa\lambda}{}^{\nu} \neq H_4^{T\,\mu\nu}, \\
H_5^{\mu\nu} &\equiv S^{\mu}S^{\nu} = H_5^{T\,\mu\nu}, \\
H_6^{\mu\nu} &\equiv S^{\lambda\mu\nu}S_{\lambda} \neq H_6^{T\,\mu\nu}.
\end{aligned} \tag{21.37}$$

Using these and the decomposition (21.23), we rewrite the first two terms in the parentheses of (21.33) as

$$S_{\sigma\lambda\mu}F^{\sigma\lambda\nu} + (\mu \leftrightarrow \nu) = 2\sigma_1 H_3^{\mu\nu} - \sigma_2(H_4^{\mu\nu} + H_4^{T\,\mu\nu}) + \frac{\sigma_3}{2}(H_6^{\mu\nu} + H_6^{T\,\mu\nu}). \tag{21.38}$$

The brackets on the right-hand side of (21.35) become

$$\left[\sigma_1(2H_1^{\mu\nu} + H_3^{\mu\nu}) + \sigma_2\left(H_2^{\mu\nu} - H_4^{\mu\nu} - H_4^{T\,\mu\nu}\right) + \sigma_3\left(H_5^{\mu\nu} + H_6^{\mu\nu} + H_6^{T\,\mu\nu}\right)\right]. \tag{21.39}$$

Adding all contributions and also the energy-momentum tensor of matter, we finally obtain the field equation

$$\frac{1}{\kappa}\tilde{G}_{\mu\nu} \equiv \frac{\gamma_0}{\kappa}\bar{G}_{\mu\nu} + \frac{1}{\kappa}\left[D_\lambda F^{\lambda}{}_{\nu\mu} + 2S_\lambda F^{\lambda}{}_{\nu\mu} + H_{\mu\nu}\right] + g_{\mu\nu}\mathcal{L}_S = \overset{m}{\Theta}_{\mu\nu}, \tag{21.40}$$

where

$$H_{\mu\nu} = H_{\nu\mu} = 2S_{\mu\sigma}{}^{\rho}F_{\nu}{}^{\sigma}{}_{\rho} - S_{\sigma\rho\nu}F^{\sigma\rho}{}_{\mu}. \tag{21.41}$$

This result can be checked by noting that the two terms in the tensor (21.41) have the decomposition

$$S_{\mu\sigma}{}^{\rho}F_{\nu}{}^{\sigma}{}_{\rho} = \sigma_1 H_1^{\mu\nu} + \frac{\sigma_2}{2}(H_2^{\mu\nu} - H_4^{\mu\nu}) + \frac{\sigma_3}{2}(H_5^{\mu\nu} + H_6^{\mu\nu}), \tag{21.42}$$

$$S_{\sigma\rho\nu}F^{\sigma\rho}{}_{\mu} = \sigma_1 H_3^{\mu\nu} - \sigma_2 H_4^{\mu\nu} + \sigma_3 H_6^{\mu\nu}, \tag{21.43}$$

so that they combine to

$$H^{\mu\nu} = \sigma_1(2H_1^{\mu\nu} - H_3^{\mu\nu}) + \sigma_2 H_2^{\mu\nu} + \sigma_3 H_5^{\mu\nu}, \tag{21.44}$$

which is the same as the difference between (21.38) and (21.39). Note that $H^{\mu\nu}$ is a symmetric tensor.

For a point particle of mass M at the origin, the energy-momentum tensor $\overset{m}{\Theta}_{\mu\nu}$ is symmetric and coincides with the symmetric energy-momentum tensor $\overset{m}{T}_{\mu\nu}$. Recalling the expression (1.243) it reads explicitly

$$\overset{m}{\Theta}_{\mu\nu} = Mc\,\delta_\mu{}^\nu\delta_\mu{}^0\delta^{(3)}(\mathbf{x}). \tag{21.45}$$

21.2 Schwarzschild Solution

A spherically symmetric solution in empty space up to a point source at the origin is obtained setting the right-hand side in Eq. (21.40) equal to zero everywhere in space except at the origin, and inserting the spatially isotropic ansatz for the vierbein field

$$
h^\alpha{}_\mu = \begin{pmatrix} \sqrt{H(r'')}c & 0 & 0 & 0 \\ 0 & \sqrt{J(r'')} & 0 & 0 \\ 0 & 0 & \sqrt{J(r'')} & 0 \\ 0 & 0 & 0 & \sqrt{J(r'')} \end{pmatrix}. \tag{21.46}
$$

The vierbein field $h_\alpha{}^\mu$ is given by the inverse of this matrix. The metric $g_{\mu\nu} = \eta_{\alpha\beta} h^\alpha{}_\mu h^\beta{}_\nu$ yields the invariant length

$$
ds^2 = H(r'')c^2(dt)^2 - J(r'')[(dr'')^2 + r''^2(d\theta)^2 + r''^2 \sin^2\theta(d\phi)^2]. \tag{21.47}
$$

The affine connection (21.2) has only a few nonzero matrix elements. The matrices $\Gamma_{t\mu}{}^\nu$ and $\Gamma_{3\mu}{}^\nu$ are zero, $\Gamma_{\theta\mu}{}^\nu$ has only one nonzero element $\cot\theta$, and $\Gamma_{r\mu}{}^\nu$ has only diagonal elements:

$$
\Gamma_{r\mu}{}^\nu = \begin{pmatrix} \dot{h}(r'')/2 & 0 & 0 & 0 \\ 0 & \dot{j}(r'') & 0 & 0 \\ 0 & 0 & 1/r'' & 0 \\ 0 & 0 & 0 & 1/r'' \end{pmatrix}, \tag{21.48}
$$

where $h \equiv \log H$, $j \equiv \log J$, and a dot denotes derivatives with respect to r''. Multiplying the field equation (21.40) by κ, we obtain the only nonzero diagonal elements

$$
\bar{G}_t{}^t = -\frac{C}{J}\left\{ \epsilon\ddot{h} + (1 - 2\epsilon)\ddot{j} + \frac{2}{r''}\left[\epsilon\dot{h} + (1 - 2\epsilon)\dot{j}\right] + \frac{\epsilon}{4}\dot{h}^2 + \frac{\epsilon}{2}\dot{h}\dot{j} + \frac{1 - 4\epsilon}{4}\dot{j}^2 \right\}, \tag{21.49}
$$

$$
\bar{G}_r{}^r = -\frac{C}{2J}\left\{ \frac{2}{r''}\left[(1 - 2\epsilon)\dot{h} + \dot{j}\right] + (1 - 2\epsilon)\dot{h}\dot{j} + \frac{\epsilon}{2}\dot{h}^2 + \frac{1}{2}\dot{j}^2 \right\}, \tag{21.50}
$$

$$
\bar{G}_\theta{}^\theta = \bar{G}_\phi{}^\phi = \frac{1}{2}\bar{G}_r{}^r + \frac{C}{2J}\left[(1 - 2\epsilon)\ddot{h} + \ddot{j} + \frac{1 - 7\epsilon/2}{2}\dot{h}^2 - \frac{1 - 4\epsilon}{2}\dot{h}\dot{j} - \frac{1}{4}\dot{j}^2 \right], \tag{21.51}
$$

where

$$
C \equiv \gamma_0 - \gamma_2 - \gamma_1/4, \qquad \epsilon \equiv -(\gamma_1 + \gamma_2)/4C. \tag{21.52}
$$

The vanishing of the combination $\bar{G}_r{}^r + \bar{G}_\theta{}^\theta$ (except at the origin where the mass point lies) yields now the differential equation

$$
(1 - 2\epsilon)\ddot{h} + \ddot{j} + \frac{3}{r''}\left[(1 - 2\epsilon)\dot{h} + \dot{j}\right] + \frac{1}{2}\left(\dot{h} + \dot{j}\right)\left[(1 - 2\epsilon)\dot{h} + \dot{j}\right] = 0, \tag{21.53}
$$

which can be rewritten as

$$
\frac{d}{dr''}\left\{ r''^3[(1 - 2\epsilon)\dot{h} + \dot{j}]\right\} + \frac{r''^3}{2}(\dot{h} + \dot{j})[(1 - 2\epsilon)\dot{h} + \dot{j}] = 0, \tag{21.54}
$$

and solved by

$$(1 - 2\epsilon)\dot{h} + \dot{j} = \frac{1}{\sqrt{HJ}} \frac{c_1^2}{r''^3}, \tag{21.55}$$

where c_1^2 is a constant of integration. From the vanishing of $G_t{}^t + 2G_\theta{}^\theta - 3G_r{}^r$ (except at the mass point at the origin) we obtain

$$\frac{d}{dr''}\left\{r''^2[(1 - 3\epsilon)\dot{h} + 2\epsilon\dot{j}]\right\} + \frac{1}{2r''^2}(\dot{h} + \dot{j})[(1 - 3\epsilon)\dot{h} + 2\epsilon\dot{j}] = 0, \tag{21.56}$$

which is solved by

$$(1 - 3\epsilon)\dot{h} + 2\epsilon\dot{j} = \frac{1}{\sqrt{HJ}} \frac{c_2}{r''^2}, \tag{21.57}$$

where c_2 is a second constant of integration.

The combination $(1 - 5\epsilon) \times (21.55) + 2\epsilon \times (21.57)$ implies that

$$(1 - \epsilon)(1 - 4\epsilon)(\dot{h} + \dot{j}) = \frac{1}{\sqrt{HJ}}\left[(1 - 5\epsilon)\frac{c_1^2}{r''^3} + 2\epsilon\frac{c_2}{r''^2}\right], \tag{21.58}$$

or

$$\dot{\sqrt{HJ}} = \frac{1}{2(1 - \epsilon)(1 - 4\epsilon)}\left[(1 - 5\epsilon)\frac{c_1^2}{r''^3} + 2\epsilon\frac{c_2}{r''^2}\right], \tag{21.59}$$

from which we find

$$\sqrt{HJ} = \left[1 - \frac{1}{4}\frac{(1 - 5\epsilon)\bar{c}_1^2}{r''^2} - \epsilon\frac{\bar{c}_2}{r''}\right] = \left(1 + \frac{a_+}{2r''}\right)\left(1 - \frac{a_-}{2r''}\right), \tag{21.60}$$

where $\bar{c}_1^2 \equiv c_1^2/(1 - \epsilon)(1 - 4\epsilon)$, $\bar{c}_2 \equiv c_2/(1 - \epsilon)(1 - 4\epsilon)$, and

$$a_\pm = \sqrt{(1 - 5\epsilon)\bar{c}_1^2 + \epsilon^2\bar{c}_2^2} \mp \epsilon\bar{c}_2. \tag{21.61}$$

Anticipating that the generalization of the previous relation $c_2 = 2c_1$ is now $\bar{c}_2 = 2\bar{c}_1$, we obtain

$$a_\pm = \left[\sqrt{(1 - \epsilon)(1 - 4\epsilon)} \mp 2\epsilon\right]\bar{c}_1. \tag{21.62}$$

Inserting this into (21.57) yields the differential equation

$$(1 - 3\epsilon)\dot{h} + 2\epsilon\dot{j} = \frac{2}{a_+ + a_-}\left(\frac{1}{1 - a_-/2r''} - \frac{1}{1 + a_+/2r''}\right)\frac{c_2}{r''^2}. \tag{21.63}$$

This is integrated to

$$H^{1-3\epsilon}J^{2\epsilon} = \left[\frac{1 - a_-/2r''}{1 + a_+/2r''}\right]^\nu, \quad \nu \equiv \frac{2c_2}{a_+ + a_-} = 2\sqrt{(1 - \epsilon)(1 - 4\epsilon)}. \tag{21.64}$$

The third constant of integration has been set equal to unity to ensure that the metric is Minkowskian at infinity.

Together with (21.60) we obtain

$$H = \frac{\left(1 - a_-/2r''\right)^{(\nu-4\epsilon)/(1-5\epsilon)}}{\left(1 + a_+/2r''\right)^{(\nu+4\epsilon)/(1-5\epsilon)}}, \qquad J = \frac{\left(1 + a_+/2r''\right)^{(2+\nu-6\epsilon)/(1-5\epsilon)}}{\left(1 - a_-/2r''\right)^{(-2+\nu+6\epsilon)/(1-5\epsilon)}}, \qquad (21.65)$$

so that

$$HJ = \left(1 + a_+/2r''\right)^2\left(1 - a_-/2r''\right)^2. \qquad (21.66)$$

Defining the powers

$$p_\pm = \frac{\sqrt{(1-\epsilon)(1-4\epsilon)} \pm 2\epsilon}{1 - 5\epsilon} = \frac{1}{\sqrt{(1-\epsilon)(1-4\epsilon)} \mp 2\epsilon} = \frac{\bar{c}_1}{a_\pm}, \qquad (21.67)$$

we can write

$$H = \frac{\left(1 - a_-/2r''\right)^{2p_-}}{\left(1 + a_+/2r''\right)^{2p_+}}, \qquad J = \frac{\left(1 - a_-/2r''\right)^{2(1-p_-)}}{\left(1 + a_+/2r''\right)^{-2(1+p_+)}}. \qquad (21.68)$$

We now impose the condition that the asymptotic behavior

$$H = 1 - \frac{p_+ a_+ + p_- a_-}{r''} + \dots \qquad (21.69)$$

yields Newton's law. This fixes

$$p_+ a_+ + p_- a_- = 2\bar{c}_1 = \frac{2MG}{c^2}. \qquad (21.70)$$

The parameter ϵ of the generalized theory changes the post-Newtonian approximation of the metric with respect to Einstein's theory. This is usually parametrized expressing the asymptotic part of the metric in terms of the three constants α, β, and γ as

$$\begin{aligned} ds^2 &= \left(1 - 2\alpha\frac{MG}{r''c^2} + 2\beta\frac{M^2G^2}{r''^2c^4} + \dots\right)c^2dt^2 \\ &- \left(1 + 2\gamma\frac{MG}{r''c^2} + 2\delta\frac{M^2G^2}{r''^2c^4} + \dots\right)(dr''^2 + r''^2d\theta^2 + r''^2\sin^2\theta d\phi^2). \end{aligned} \qquad (21.71)$$

Expanding H and J to higher orders in $1/r''$, we obtain

$$\begin{aligned} H &= 1 - \frac{2GM}{c^2r''} + \frac{1}{4}\left(a_+^2 p_+ - a_-^2 p_- + 2a_-^2 p_-^2 + 4a_+ a_- p_+ p_- + 2a_+^2 p_+^2\right)\left(\frac{GM}{c^2r''}\right)^2 + \dots \\ &= 1 - \frac{2GM}{c^2r''} + 2\left(1 - \frac{\epsilon}{2}\right)\left(\frac{GM}{c^2r''}\right)^2 + \dots, \end{aligned} \qquad (21.72)$$

and

$$
\begin{aligned}
J &= 1 + \left(a_+ - a_- + a_+ p_+ + a_- p_-\right)\frac{GM}{c^2 r''} + \frac{1}{4}\left[a_-^2 + a_+^2 - 4a_- a_+(1 + p_+ - p_- - p_+ p_-)\right. \\
&\quad \left. -3(a_-^2 p_- - a_+^2 p_+) + 2(a_-^2 p_-^2 + 2a_+^2 p_+^2)\right]\left(\frac{GM}{c^2 r''}\right)^2 + \dots \\
&= 1 + (1 - 2\epsilon)\frac{2GM}{c^2 r''} + 2\left[\frac{3}{4}(1 - 3\epsilon + 8\epsilon^2/3)\right]\left(\frac{GM}{c^2 r''}\right)^2 + \dots .
\end{aligned} \tag{21.73}
$$

By comparison with (21.71) we identify the post-Newtonian parameters

$$
\beta = 1 - \frac{1}{2}\epsilon, \quad \gamma = 1 - 2\epsilon, \quad \delta = \frac{3}{4}\left(1 - 3\epsilon + 8\epsilon^2/3\right). \tag{21.74}
$$

Let us finally verify that the two constants of integration $\bar{c}_1$ and $\bar{c}_2$ are related by $\bar{c}_2 = 2\bar{c}_1$, as anticipated in Eq. (21.62). Combining (21.55) with (21.57), we find

$$
\dot{h} = \frac{1}{\sqrt{HJ}}\left(\frac{\bar{c}_2}{r''^2} - 2\epsilon\frac{\bar{c}_1^2}{r''^3}\right), \quad \dot{j} = \frac{1}{\sqrt{HJ}}\left[(1 - 3\epsilon)\frac{\bar{c}_1^2}{r''^3} - (1 - 2\epsilon)\frac{\bar{c}_2}{r''^2}\right]. \tag{21.75}
$$

Inserting these equations into $\bar{G}_r{}^r = 0$ of Eq. (21.50), we obtain

$$
\bar{G}_r{}^r = -\frac{C}{2J}\frac{1}{HJ}(1 - \epsilon)(1 - 4\epsilon)\left(2\bar{c}_1^2 - \frac{1}{2}\bar{c}_2^2\right), \tag{21.76}
$$

which vanishes for $\bar{c}_2 = 2\bar{c}_1$, as it should.

It remains to fix the parameter C in Eqs. (21.49)–(21.51). For this we go to large-r'' where, according to Eqs. (21.72) and (21.73),

$$
H = -\frac{2GM}{c^2 r''} + \dots , \quad J = 1 + (1 - 2\epsilon)\frac{2GM}{c^2 r''} + \dots , \tag{21.77}
$$

and consider the field equation Eq. (21.40) for $\mu = \nu = t$

$$
\frac{1}{\kappa}\tilde{G}_t{}^t = Mc\,\delta(\mathbf{x}''). \tag{21.78}
$$

The point source changes the vanishing expression (21.56) to

$$
\frac{C}{J}\frac{d}{dr''}(r''^2\dot{h}) + \frac{1}{2r''^2}\dot{h}(\dot{h} + \dot{j}) = \kappa Mc\frac{\delta(r'')}{4\pi}, \tag{21.79}
$$

where we have replaced $\delta^{(3)}(\mathbf{x}'') \to \delta(r'')/4\pi r''^2$. Here we recall that the Laplace equation for the Coulomb potential reads

$$
-\Delta\frac{1}{r} = 4\pi\delta^{(3)}(\mathbf{x}). \tag{21.80}
$$

In spherical coordinates, this becomes

$$
-\frac{1}{r^2}\partial_r r^2 \partial_r \frac{1}{r} = \frac{1}{r^2}\delta(r), \tag{21.81}
$$

as is easily verified by integrating this equation from zero to a small nonzero radius r_0 and performing, on the left-hand side, an integration by parts. As a consequence, the solution (21.57) of the homogenous Eq. (21.56) for $r'' \neq 0$ acquires at $r'' = 0$ an inhomogeneous part

$$\frac{C}{J}\frac{d}{dr''}(r''^2 \dot{h}) + \frac{1}{2r''^2}h(\dot{h} + \dot{j}) = \frac{1}{\sqrt{HJ}}\frac{Cc_2}{J}\delta(r''). \qquad (21.82)$$

Since H and J have unit values at the origin, comparison with (21.79) fixes $Cc_2 = \kappa Mc/4\pi = 2G_{\mathrm{N}}M/c^2$ [recall (15.9)], or

$$Cc_2 = C\bar{c}_2(1 - \epsilon)(1 - 4\epsilon) = C\,2\bar{c}_1(1 - \epsilon)(1 - 4\epsilon) = 2\frac{G_{\mathrm{N}}M}{c^2}. \qquad (21.83)$$

Inserting $\bar{c}_1 = GM/c^2$ from (21.70), the constant C must satisfy

$$C(1 - \epsilon)(1 - 4\epsilon) = (c_0 - c_2 - c_1/4)(1 - \epsilon)(1 - 4\epsilon) = 1. \qquad (21.84)$$

For recent work on teleparallel geometry see Ref. [3].

Notes and References

[1] A. Einstein, *Auf die Riemann-Metrik und den Fernparallelismus gegrndete einheitliche Feldtheorie*, Math. Ann. **102**, 685 (1930).
Translation can be downloaded from http://www.lrz-muenchen.de/~aunzicker/ae1930.html.

[2] K. Hayashi and T. Shirafuji, Phys. Rev. D **19**, 3524 (1979).

[3] A.A. Sousa, J.S. Moura, and R.B. Pereira, *Energy and Angular Momentum in an Expanding Universe in the Teleparallel Geometry*, (arXiv:gr-qc/0702109);
Yu-Xiao Liu, Zhen-Hua Zhao, Jie Yang, Yi-Shi Duan,*The total energy-momentum of the universe in teleparallel gravity*, (arXiv:0706.3245);
R. Ferraro and F. Fiorini, *Modified teleparallel gravity: inflation without inflaton*, Phys. Rev. D **75**, 084031 (2007) (arXiv:gr-qc/0610067);
E.E. Flanagan and E. Rosenthal, *Can Gravity Probe B Usefully Constrain Torsion Gravity Theories?*, Phys. Rev. D **75**, 124016 (2007);
R. Ferraro and F. Fiorini, *Modified Teleparallel Gravity: Inflation without an Inflaton*, Phys. Rev. D **75**, 084031 (2007);
M. Sharif and M.J. Amir, *Teleparullel Versions of Friedmann and Lewis-Papapetrou Spacetimes*, Gen. Rel. Grav. **38**, 1735 (2006);
M. Leclerc, *One-parameter Teleparallel Limit of Poincaré Gravity*, Phys. Rev. D **72**, 044002 (2005).
See also the webpage
http://prola.aps.org/forward/PRD/v19/i12/p3524_1.

22

Emerging Gravity

In Eq. (12.42) we have introduced the Planck length $l_P \approx 1.616 \times 10^{-33}$cm as the fundamental length scale of gravitational physics. We observed that spacetime with curvature and torsion may be imagined as a world crystal of lattice spacing l_P with defects without contradicting any experiment. In fact, it makes no physical sense to produce theories which predict properties of the universe at smaller length scales. Since the times of Galileo Galilei, such theories fall into the realm of philosophy, or even religion. The history of science shows us that nature has always surprised us with new discoveries as observations invaded into shorter and shorter distances. So far, all theories in the past which claimed for a while to be *theories of everything* have been falsified by completely unexpected discoveries.

The presently most popular examples for the claim to be the theory of everything are string theories. Mathematically, they describe the physics ranging from cosmological length scales down to the trans-Planckian regime of zero length. In the experimentally accessible energy range these theories require a large number of extra spacetime dimensions which have so far unobserved effects. Unfortunately, they also predict many particles which are not found in nature. In particular, the assumption of a string representing fundamental particles makes only sense if there are *overtones*, which any string must have. In string theory, these overtones lie all in the inaccessible Planck regime. Being unable to explain correctly the observed low-energy particle spectra it is unclear how anybody can believe the predictions for this ultra-high energy regime.

One of the most important features of string theories is that they predict the validity of Lorentz invariance at *all* energies in the trans-Planckian regime. In this chapter we would like to point out that at this level of speculations, an entirely different scenario is possible.

22.1   Gravity in the World Crystal

Let us suppose, just for fun, that we live in a world crystal with a lattice constant of the order of the Planck length [1]. Up to now we would have been unable to notice this. And this would remain so for a long time to come. None of the present-day relativistic physical laws would be observably violated. The gravitational forces and their geometric description would arise from variants of the plastic forces in this *world crystal*. The observed curvature of spacetime would be just a signal of the presence of disclinations in the crystal. Matter would be sources of disclinations [2].

For simplicity, we shall present such a construction only for a system without torsion. If the world crystal is distorted by an infinitesimal displacement field

$$x^\mu \to x'^\mu = x^\mu + u^\mu(x), \tag{22.1}$$

it has a strain energy

$$A = \frac{\mu}{4} \int d^4x \, (\partial_\mu u_\nu + \partial_\nu u_\mu)^2, \tag{22.2}$$

where μ is some elastic constant. We assume the second possible elastic constant, the Poisson ratio, to be zero, to shorten calculations. If the distortions are partly plastic, the world crystal contains defects defined by Volterra surfaces, where crystalline sections have been cut out. The displacement field is multivalued. The Euclidean action of the world crystal becomes the spacetime generalization of the crystal energy (10.9) (without the λ-term);

$$A = \mu \int d^4x \, (u_{\mu\nu} - u^p_{\mu\nu})^2, \tag{22.3}$$

where $u_{\mu\nu} \equiv (\partial_\mu u_\nu + \partial_\nu u_\mu)/2$ is the elastic strain tensor, and $u^p_{\mu\nu}$ the plastic strain tensor. As explained in Section 9.11, the plastic strain tensor is a gauge field of plastic deformations. The energy density is invariant under the single-valued *defect gauge transformations* [the continuum limit of (10.19) and (10.20)]:

$$u_{\mu\nu}{}^p \to u_{\mu\nu}{}^p + (\partial_\mu\lambda_\nu + \partial_\nu\lambda_\mu)/2, \qquad u_\mu \to u_\mu + \lambda_\mu. \tag{22.4}$$

Physically, they express the fact that defects are not affected by elastic distortions of the crystal. Only multivalued gauge functions λ_μ change the defect content in the plastic gauge field $u^p_{\mu\nu}$.

We now rewrite the action (22.3) in a canonical form [the analog of (10.15)] by introducing an auxiliary symmetric stress tensor field $\sigma_{\mu\nu}$ as

$$A = \int d^3x \left[\frac{1}{4\mu} \sigma_{\mu\nu}\sigma^{\mu\nu} + i\sigma^{\mu\nu}(u_{\mu\nu} - u^p_{\mu\nu}) \right]. \tag{22.5}$$

After a partial integration and extremization in u_μ, the second term in the action yields the equation

$$\partial_\nu \sigma^{\mu\nu} = 0. \tag{22.6}$$

This may be guaranteed identically, as a Bianchi identity, by an ansatz generalizing (10.17)

$$\sigma_{\mu\nu} = \epsilon_\mu{}^{\kappa\lambda\sigma} \epsilon_\nu{}^{\kappa\lambda'\sigma'} \partial_\lambda \partial_{\lambda'} \chi_{\sigma\sigma'}. \tag{22.7}$$

Inserting (22.7) into (22.5), we obtain the analog of (10.18):

$$A = \int d^4x \left\{ \frac{1}{4\mu} \left[\epsilon^{\mu\kappa\lambda\sigma} \epsilon^{\nu\kappa\lambda'\sigma'} \partial_\lambda \partial_{\lambda'} \chi_{\sigma\sigma'} \right]^2 - i\epsilon^{\nu\kappa\lambda\sigma} \epsilon^{\mu\kappa\lambda'\sigma'} \partial_\lambda \partial_{\lambda'} \chi_{\sigma\sigma'} u^p_{\mu\nu} \right\}. \tag{22.8}$$

A further partial integration brings this to the form

$$\mathcal{A} = \int d^4x \left\{ \frac{1}{4\mu} \left[\epsilon^{\mu\kappa\lambda\sigma} \epsilon^{\nu\kappa\lambda'\sigma'} \partial_\lambda \partial_{\lambda'} \chi_{\sigma\sigma'} \right]^2 - i\chi_{\sigma\sigma'} \left[\epsilon^{\sigma\kappa\lambda\nu} \epsilon^{\sigma'\kappa\lambda'\mu} \partial_\lambda \partial_{\lambda'} u^p_{\mu\nu} \right] \right\}. \quad (22.9)$$

This action is now double-gauge theory invariant under the defect gauge transformation (22.4), and under stress gauge transformations

$$\chi_{\sigma\tau} \to \chi_{\sigma\tau} + \partial_\sigma \Lambda_{\sigma'} + \partial_\tau \Lambda_\sigma. \quad (22.10)$$

The action can further be rewritten as

$$\mathcal{A} = \int d^4x \left\{ \frac{1}{4\mu} \sigma_{\mu\nu} \sigma^{\mu\nu} - i\chi_{\mu\nu} \eta^{\mu\nu} \right\}, \quad (22.11)$$

where $\eta_{\mu\nu}$ is the four-dimensional extension of the defect density η_{ij} in Eq. (9.97):

$$\eta_{\mu\nu} = \epsilon_\mu{}^{\kappa\lambda\sigma} \epsilon_{\nu\kappa}{}^{\lambda'\tau} \partial_\lambda \partial_{\lambda'} u^p_{\sigma\tau}. \quad (22.12)$$

This is invariant under defect gauge transformations (22.4), and satisfies the conservation law

$$\partial_\nu \eta^{\mu\nu} = 0. \quad (22.13)$$

We may now replace $u^p_{\sigma\sigma'}$ by half the metric field $g_{\mu\nu}$ in (12.20) and, recalling Eq. (12.30), we recognize the tensor $\eta_{\mu\nu}$ as the Einstein tensor associated with the metric tensor $g_{\mu\nu}$.

Let us eliminate the stress gauge field from the action (22.11). For this we use the identity (1A.23) for the product of two Levi-Civita tensors, and rewrite the stress field (22.7) as [compare (10.23)]

$$\begin{aligned} \sigma_{\mu\nu} &= \epsilon_\mu{}^{\kappa\lambda\sigma} \epsilon_\nu{}^{\kappa\lambda'\tau} \partial_\lambda \partial_{\lambda'} \chi_{\sigma\tau} \\ &= -(\partial^2 \chi_{\mu\nu} + \partial_\mu \partial_\nu \chi_\lambda{}^\lambda - \partial_\mu \partial_\lambda \chi_\mu{}^\lambda - \partial_\nu \partial_\lambda \chi_\mu{}^\lambda) + \eta_{\mu\nu}(\partial^2 \chi_\lambda{}^\lambda - \partial_\lambda \partial_\kappa \chi^{\lambda\kappa}). \end{aligned} \quad (22.14)$$

Introducing the field $\phi_\mu{}^\nu \equiv \chi_\mu{}^\nu - \frac{1}{2}\delta_\mu{}^\nu \chi_\lambda{}^\lambda$, and going to the Hilbert gauge $\partial^\mu \phi_\mu{}^\nu = 0$, the stress tensor reduces to

$$\sigma_{\mu\nu} = -\partial^2 \phi_{\mu\nu}, \quad (22.15)$$

and the action of an arbitrary distribution of defects becomes

$$\mathcal{A} = \int d^4x \left\{ \frac{1}{4\mu} \partial^2 \phi^{\mu\nu} \partial^2 \phi_{\mu\nu} + i\phi_\mu{}^\nu (\eta^\mu{}_\nu - \frac{1}{2}\delta^\mu{}_\nu \eta^\lambda{}_\lambda) \right\}. \quad (22.16)$$

Extremization with respect to the field $\phi^{\mu\nu}$ yields the interaction of an arbitrary distribution of defects [the analog of (10.28)]:

$$\mathcal{A} = \mu \int d^4x \, (\eta^\mu{}_\nu - \frac{1}{2}\delta^\mu{}_\nu \eta^\lambda{}_\lambda) \frac{1}{(\partial^2)^2} (\eta_\mu{}^\nu - \frac{1}{2}\delta_\mu{}^\nu \eta^\lambda{}_\lambda). \quad (22.17)$$

This is not the Einstein action for a Riemann spacetime. It would be so if the derivatives $(\partial^2)^2$ in the denominator would be replaced by $\sigma_{\mu\nu}(-\partial^2)\sigma^{\mu\nu}$, which would change $(\partial^2)^2$ in (22.17) into $-\partial^2$. An index rearrangement would then lead to the interaction

$$\mathcal{A} = \mu \int d^4x \left(\eta^\mu{}_\nu - \tfrac{1}{2}\delta^\mu{}_\nu \eta^\lambda{}_\lambda\right) \frac{1}{-\partial^2} \eta_\mu{}^\nu. \tag{22.18}$$

The defect tensor $\eta_{\mu\nu}$ is composed of the plastic gauge fields $u^p_{\mu\nu}$ in the same way as the stress tensor is in terms of the stress gauge field in Eq. (22.14), i.e.:

$$\eta_{\mu\nu} = \epsilon_\mu{}^{\kappa\lambda\sigma} \epsilon_\nu{}^{\kappa\lambda'\tau} \partial_\lambda \partial_{\lambda'} u^p_{\sigma\tau}.$$
$$= -(\partial^2 u^p_{\mu\nu} + \partial_\mu \partial_\nu u^{p\lambda}_\lambda - \partial_\mu \partial_\lambda u^{p\lambda}_\nu - \partial_\nu \partial_\lambda u^{p\lambda}_\mu) + \eta_{\mu\nu}(\partial^2 u^{p\lambda}_\lambda - \partial_\lambda \partial_\kappa u^{p\lambda\kappa}). \tag{22.19}$$

If we introduce the auxiliary field $w^{p\nu}_\mu \equiv u^{p\nu}_\mu - \tfrac{1}{2}\delta_\mu{}^\nu u^{p\lambda}_\lambda$, and chose the Hilbert gauge $\partial^\mu w^p_{\mu\nu} = 0$, the defect density reduces to

$$\eta_{\mu\nu} = -\partial^2 w^p_{\mu\nu}, \qquad \eta_\mu{}^\nu - \tfrac{1}{2}\delta_\mu{}^\nu \eta^\lambda{}_\lambda = -\partial^2 u^p_{\mu\nu}, \tag{22.20}$$

and the interaction (22.18) of an arbitrary distribution of defects would become

$$\mathcal{A} = \mu \int d^4x \, u^p{}_{\mu\nu}(x) \eta^{\mu\nu}(x). \tag{22.21}$$

This coincides with the linearized Einstein-Hilbert action

$$\mathcal{A} = -\frac{1}{2\kappa} \int d^4x \sqrt{-g}\,\bar{R}, \tag{22.22}$$

if we identify the elastic constant μ with $1/4\kappa$, where κ is the gravitational constant (15.9). Indeed, in the linear approximation $g_\mu{}^\nu = \delta_\mu{}^\nu + h_\mu{}^\nu$ with $|h_\mu{}^\nu| \ll 1$, where the Christoffel symbols can be approximated by

$$\bar{\Gamma}_{\mu\nu}{}^\lambda \approx \frac{1}{2}\left(\partial_\mu h_{\nu\lambda} + \partial_\nu h_{\mu\lambda} - \partial_\lambda h_{\mu\nu}\right), \tag{22.23}$$

and the Riemann curvature tensor becomes

$$\bar{R}_{\mu\nu\lambda\kappa} \approx \frac{1}{2}\left[\partial_\mu \partial_\lambda h_{\nu\kappa} - \partial_\nu \partial_\kappa h_{\mu\lambda} - (\mu \leftrightarrow \nu)\right], \tag{22.24}$$

as can be seen directly from Eq. (11.151). This gives the Ricci tensor

$$\bar{R}_{\mu\kappa} \approx \frac{1}{2}(\partial_\mu \partial_\lambda h_{\lambda\kappa} + \partial_\kappa \partial_\lambda h_{\lambda\mu} - \partial_\mu \partial_\kappa h - \partial^2 h_{\mu\kappa}), \tag{22.25}$$

where $h \equiv h_\lambda{}^\lambda$ is the trace of the tensor $h_{\mu\nu}$. The associated scalar curvature is

$$\bar{R} \approx -(\partial^2 h - \partial_\mu \partial_\nu h^{\mu\nu}). \tag{22.26}$$

In combination with (22.25) we obtain the Einstein tensor

$$\bar{G}_{\mu\kappa} = \bar{R}_{\mu\kappa} - \frac{1}{2}g_{\mu\kappa}\bar{R} \tag{22.27}$$

$$\approx -\frac{1}{2}(\partial^2 h_{\mu\kappa} + \partial_\mu\partial_\kappa h - \partial_\mu\partial_\lambda h^\lambda{}_\kappa - \partial_\kappa\partial_\lambda h^\lambda{}_\mu) + \frac{1}{2}\eta_{\mu\kappa}(\partial^2 h - \partial_\nu\partial_\lambda h^{\nu\lambda}).$$

This can be written as a four-dimensional version of a double curl

$$\bar{G}_{\mu\kappa} = \frac{1}{2}\epsilon_{\mu\delta}{}^{\nu\lambda}\epsilon_\kappa{}^{\delta\sigma\tau}\partial_\nu\partial_\sigma h_{\lambda\tau}, \tag{22.28}$$

as can be verified using the identity (1A.23).

Thus the Einstein-Hilbert action has the linear approximation

$$\mathcal{A} \approx \frac{1}{4\kappa}\int d^4x\, h_{\mu\nu}G^{\mu\nu}. \tag{22.29}$$

Recalling the previously established identifications of plastic field and defect density with metric and Einstein tensor, respectively, the interaction between defects (22.21) is indeed the linearized version of the Einstein-Hilbert action (22.22) if $\mu = 1/4\kappa$.

The world crystal with the elastic energy (22.3) does not lead to the interaction (22.18). But it is easy to modify it to do so. We may simply introduce two more derivatives and assign to the world crystal the higher-gradient elastic energy

$$\mathcal{A}' = \mu\int d^4x\, [\partial(u_{\mu\nu} - u^p_{\mu\nu})]^2. \tag{22.30}$$

In the canonical form (22.5) of the energy it changes the term $\sigma_{\mu\nu}\sigma^{\mu\nu}$ to $\sigma_{\mu\nu}(-\partial^2)\sigma^{\mu\nu}$. This has the desired effect of removing one power of $-\partial^2$ from the denominator in the interaction (22.17).

A world crystal whose elastic energy is governed by the higher-gradient action (22.30) by be called *floppy world crystal*. Thus we have shown that defects in the floppy world crystal are capable of creating a Riemannian spacetime with an action of the Einstein-Hilbert type. For more work on the world crystal see Ref. [3].

22.2 Gravity Emerging from Fluctuations of Matter and Radiation in Closed Friedmann Universe

In 1967 Sacharov put forward the interesting idea [4, 5] that geometry does not possess a dynamics of its own, but that the stiffness of spacetime could be entirely due to the vacuum fluctuations of all quantum fields in the universe. Each of these gives rise to an Einstein action proportional to R, and to all possible higher powers of $R_{\mu\nu\lambda\kappa}$ contracted to scalars such as R^2, $R_{\mu\nu}R^{\mu\nu}$, $R_{\mu\nu\lambda\kappa}R^{\mu\nu\lambda\kappa}$, R^3, ... In addition, they generate a cosmological term without R. Its coefficient and those of the linear and quadratic terms diverge in the ultraviolet, but if all fluctuating fields stem from a renormalizable quantum field theory, all infinities can be subtracted to leave a finite value to be fixed by experiment. The renormalization procedure was discussed before in the context of Eq. (15.11).

The cosmological term without R changes the gravitational action (15.8) to

$$\overset{f}{A} = -\frac{1}{2\kappa} \int d^4x \sqrt{-g}(R + 2\lambda). \tag{22.31}$$

The constant λ is the so-called *cosmological constant*. It changes the energy-momentum tensor of the gravitational field from $-(1/\kappa)G_{\mu\nu}$ to the combination $-(1/\kappa)\left(G_{\mu\nu} - \lambda g_{\mu\nu}\right)$.

When calculating the effect of fluctuations for any of the fundamental fields one finds a contribution to the cosmological constant corresponding to an action density

$$\Lambda \equiv \frac{\lambda}{\kappa} = \frac{\lambda c^3}{8\pi G_{\rm N}} \tag{22.32}$$

which is of the order of $\pm \hbar/l_{\rm P}^4$, where $l_{\rm P}$ is the Planck length. For bosons, the sign is positive, for fermions negative, reflecting the filling of all negative-energy states in the vacuum.

A constant of this size is much larger than the present experimental estimate. In the literature one usually finds estimates for the dimensionless quantity

$$\Omega_{\lambda 0} \equiv \frac{\lambda c^2}{3H_0^2} \tag{22.33}$$

where H_0 is the *Hubble constant* which parametrizes the expansion velocity of the universe as a function of the distance r from the earth by *Hubble's law*:

$$v = H_0 r. \tag{22.34}$$

The inverse of H_0 is roughly equal to the lifetime of the universe

$$H_0^{-1} \approx 14 \times 10^9 \,\text{years}. \tag{22.35}$$

Present fits to distant supernovae and other cosmological data yield the estimate [6]

$$\Omega_{\lambda 0} \equiv 0.68 \pm 0.10. \tag{22.36}$$

As a result, the experimental number for Λ is

$$\Lambda = \Omega_{\lambda 0} \frac{3H_0^2}{c^2} \frac{l_{\rm P}^2}{8\pi} \approx 10^{-122} \frac{\hbar}{l_{\rm P}^4}. \tag{22.37}$$

Such a small prefactor in front of the "natural" action density $\hbar/l_{\rm P}^4$ can only arise from an almost perfect cancellation of the contributions of boson and fermion fields. This cancellation is the main reason why some people postulate the existence of a broken supersymmetry in the universe, in which every boson has a fermionic counterpart. So far, the known particle spectra show no trace of such a symmetry. Thus there is need to explain it by some other not yet understood mechanism.

A mechanical model for Sacharov's idea of emerging gravity would be an infinitely thin massless plastic bag filled with water. The bag represents the geometry which does not have any dynamics of its own. All its movements are controlled by the dynamics of the water contained in it.

Sacharov's idea is very appealing. Unfortunately, the calculation of the emerging gravitational action along his lines would require the knowledge of all elementary fields in nature and, in addition, their properties at arbitrarily short distances. This will never be available. In addition, also the vacuum fluctuations of all composite particles will contribute. This comprises for instance all elements in the periodic system, such as gold and platinum. The problem of calculating the vacuum energy a la Sacharov seems therefore beset by the same difficulties as the application of set theory to the set of all sets.

The problem can only be avoided by using only renormalizable quantum field theories of matter to describe elementary particles. The renormalizability is necessary to remove the dependence of the theories on the unknown ultra-short distance properties whose influence is impossible to guess. In a renormalizable theory, the vacuum energy will always be a free parameter, as discussed before in the context of Eq. (15.11).

Notes and References

[1] H. Kleinert, *Gravity as Theory of Defects in a Crystal with Only Second-Gradient Elasticity*, Ann. d. Physik, **44**, 117(1987) (kl/172).

[2] For an attempt to explain why no torsion is found in the world crystal see H. Kleinert and J. Zaanen, *World Nematic Crystal Model of Gravity Explaining the Absence of Torsion*, Phys. Lett. A **324**, 361 (2004) (gr-qc/0307033).

[3] M. Danielewski, *The Planck-Kleinert Crystal*, Z. Naturfoschung **62a**, 546 (2007).

[4] A.D. Sacharov, DAN SSSR **177**, 70, (1967). Reprinted in Gen. Rel. Grav. **32**, 365, (2000).

[5] A cosmological model based on Sacharov's idea is discussed in H. Kleinert and H.-J. Schmidt, *Cosmology with Curvature-Saturated Gravitational Lagrangian $R/(1 + l^4)^{1/2}$*, Gen. Rel. Grav. **34**, 1295 (2002) (gr-qc/0006074).

[6] For cosmological data see the internet page http://super.colorado.edu/~michaele/Lambda/links.html.

[7] J. Lense and H. Thirring, Phys. Zeitschr. **19**, 156 (1918).

Index